Statistical Thermodynamics of Nonequilibrium Processes

Joel Keizer

Statistical Thermodynamics of Nonequilibrium Processes

With 59 Illustrations

Springer-Verlag
New York Berlin Heidelberg
London Paris Tokyo

Joel Keizer
Department of Chemistry and
Institute of Theoretical Dynamics
University of California
Davis, CA 95616
U.S.A.

Library of Congress Cataloging-in-Publication Data
Keizer, Joel.
Statistical thermodynamics of nonequilibrium processes.
Bibliography: p.
Includes index.
1. Nonequilibrium thermodynamics. 2. Statistical
thermodynamics. I. Title.
QC318.I7K42 1987 536'.7'015195 87-9469

Typeset by Asco Trade Typesetting Ltd., Hong Kong.
Printed and bound by R. R. Donnelley and Sons, Harrisonburg, Virginia.
Printed in the United States of America.

9 8 7 6 5 4 3 2 1

ISBN 0-387-96501-7 Springer-Verlag New York Berlin Heidelberg
ISBN 3-540-96501-7 Springer-Verlag Berlin Heidelberg New York

To Susan, who has contributed more than I know how to say.

Preface

The structure of the theory of thermodynamics has changed enormously since its inception in the middle of the nineteenth century. Shortly after Thomson and Clausius enunciated their versions of the Second Law, Clausius, Maxwell, and Boltzmann began actively pursuing the molecular basis of thermodynamics, work that culminated in the Boltzmann equation and the theory of transport processes in dilute gases. Much later, Onsager undertook the elucidation of the symmetry of transport coefficients and, thereby, established himself as the father of the theory of nonequilibrium thermodynamics. Combining the statistical ideas of Gibbs and Langevin with the phenomenological transport equations, Onsager and others went on to develop a consistent statistical theory of irreversible processes. The power of that theory is in its ability to relate measurable quantities, such as transport coefficients and thermodynamic derivatives, to the results of experimental measurements. As powerful as that theory is, it is linear and limited in validity to a neighborhood of equilibrium.

In recent years it has been possible to extend the statistical theory of nonequilibrium processes to include nonlinear effects. The modern theory, as expounded in this book, is applicable to a wide variety of systems both close to and far from equilibrium. The theory is based on the notion of elementary molecular processes, which manifest themselves as random changes in the extensive variables characterizing a system. The theory has a hierarchical character and, thus, can be applied at various levels of molecular detail. The simplest is the thermodynamic level, in which the extensive variables are the internal energy, volume, and particle numbers. This description corresponds to the variables in classical thermodynamics, but contains information about both the time rate of change of the variables and their statistical fluctuations. Descriptions at the hydrodynamic and Boltzmann levels involve considerably more molecular detail and provide progressively more information about the thermodynamic state of the system.

This book has been written with two purposes in mind. The first is to present a unified account of the ideas that go into the statistical theory of nonequilibrium thermodynamics. In doing so, I have tried to unify the various approaches that have been used to treat molecular fluctuations in macroscopic systems. Thus the basic *ansatz* introduced in Chapters 3 and 4 deals with the probability of changes in the extensive variables induced by elementary processes. Chapter 4 discusses the thermodynamic implications of the *ansatz*, including the relationship to master equation theories, stochastic diffusion equations, nonlinear Langevin equations, and, in the thermodynamic limit, the fluctuation–dissipation theory.

The second purpose of the book is to bring the theory to the attention of a wider audience of scientists. With this in mind, I have endeavored to make the book accessible to graduate students in the physical sciences. The main prerequisites are good undergraduate courses in thermodynamics and statistical mechanics or an introductory graduate course in statistical thermodynamics of the sort offered by most physics and chemistry departments in the United States. The necessary background material on stochastic processes is provided in Chapter 1, and I have made every effort to provide detailed derivations of the equations in the text. While this will make the text easy going for more advanced researchers, I hope that they will find sufficient new material to hold their attention, too.

In order to illustrate the utility of the theory for analyzing physical measurements, I have included a number of specific applications of the theory in physics, chemistry, and biophysics. The applications include material on Brownian motion in Chapter 1; hydrodynamics in Chapters 2, 6, and 10; the theory of chemical thermodynamics and chemical reaction fluctuations in Chapters 3, 5, 6, 7, 8, and 10; electrochemistry in Chapters 5, 8, and 10; ion channels in biological membranes in Chapters 5 and 10; gas-phase transport coefficients and the fluctuating Boltzmann equation in Chapters 2, 3, and 9; and diffusion effects on chemical reactions in Chapter 9. Chapter 10 deals with transient phenomena, limit cycles, and chaotic trajectories, focusing on the molecular mechanisms that give rise to such behavior in dissipative systems. Following the basic development of the theory in the first four chapters, a judicious choice from these applications should make the book useful as a graduate text in chemistry, physics, or biophysics.

Although the book contains no exercises, there are numerous worked-out examples in the text. It is hoped that these will help instruct the reader in how to use the theory for his or her own problems. I have also eschewed footnoting references to the literature in favor of general references at the end of each chapter. Many of the topics discussed in the text can be pursued in greater detail in the references, and it is hoped that the reader will make use of them for further study.

Over the years a number of people have contributed in significant ways to my thinking about statistical thermodynamics. I owe a great intellectual debt to Lars Onsager, and although I spoke to him only once, his admonition to pursue the molecular, rather than the phenomenological, point of view figures prominently in the theory developed in this book. Similarly, George Uhlenbeck, who continues to uphold the traditions of Boltzmann, insisted that the entropy be a consequence of the theory, rather than an ingredient. In this regard, I hope that he approves of Chapters 4 and 8. Terrell Hill is the culprit who encouraged me to write this book, and Peter Mazur is the one who counseled me not to rush through it. I have learned much from both of them and from their books and hope that they will find something in this book to please them.

Perhaps my greatest intellectual debt is to Ronald Fox. Already as an undergraduate, he introduced me to the joys of Tolman's *Statistical Mechanics*, and later his own doctoral thesis, *Contributions to the Theory of Non-Equilibrium Thermodynamics*, introduced me to the theory of irreversible processes. It is fair to say that everything in the present book has been discussed with him at one time or another, and although we have not always agreed, I am forever in his debt for his intellectual camaraderie.

Besides Ron Fox, Gerhard Magnus, Enrique Peacock-Lopez, and Atilla Szabo have read and commented on various portions of the manuscript. I am indebted to them for their suggestions and hopeful that most of my errors have been expunged.

Special thanks go to Carolyn Mikashus and Louanne Scribner, who transformed thousands of pages of handwritten equations into flawless typed copy and remained cheerful to the end, and to Susan Keizer, who prepared the drawings—although over the years this has been the least of her contributions. Finally, I would like to thank the University of California, the National Science Foundation, the National Institutes of Health, and the John Simon Guggenheim Memorial Foundation for financial support during the writing of this book.

Joel Keizer

Contents

CHAPTER 1

Ensembles and Stochastic Processes

1.1. The Mechanical Description of Matter

The study of matter which is large with respect to molecular size is one of the oldest of the sciences. It originated in the early metallurgy of China and the Middle East and permeates modern research in both the physical and biological sciences. From a theoretical point of view, the reversible or mechanical properties of matter were the first to be understood. These are the properties characterized by Newton's laws of motion which relate forces, velocities, and spatial positions. These properties are termed reversible because if one reverses all velocities, the magnetic field, and the time, the resulting motion is the reverse of what was observed up to that time—like seeing a movie run backwards. This reversibility of the mechanical theory suggested for many years that any observable motion should have a twin reversed motion which can also be observed. The consequences of reversibility, however, seem contrary to our intuition: Although salt spontaneously dissolves in water, no one has ever seen salt precipitate from an unsaturated solution.

This irreversible characteristic of matter was recognized in early Greek science. Indeed, in the physics of Aristotle it is implied that a force must be continuously applied to matter to keep it moving in a straight line. This is in accord with the observation that a ball falling toward the earth reaches a definite terminal velocity. However, it is at variance with the observation that the same ball falling in a vacuum continuously increases its velocity. The difference, of course, is explained in the Newtonian theory in terms of the frictional force exerted by the air on the ball. The frictional force is said to be "dissipative" in that the potential energy released in falling is not transformed into kinetic energy but is released or dissipated as heat. This sort of dissipation is present in one form or another in all our observations and partially explains why it took so long to uncover the intrinsic reversibility in the equations which describe how matter changes.

If all our observations of matter consisted of following velocities and positions, the problem of describing how matter changes in time would have been solved by Newton. This, however, is not the case. The observation that matter is composed of atoms and molecules gave rise to a more complete mechanical description of matter. In quantum mechanics one uses a field-theoretic picture in which the state of matter is represented by a wave function. Nonetheless, most of the Newtonian mechanical concepts, such as position, momentum, and energy, are retained, and the theory is still demonstrably reversible. In both the quantum and classical theories, irreversibility cannot occur except through interactions of one system of matter with another. Apparent irreversibility, or at least very complicated motions, occur in a variety of simple interacting mechanical systems. Yet, if one really solves the equations of motion, these trajectories will have a companion reversed motion which traverses the trajectory backwards in time.

The desire to understand irreversibility in terms of mechanics goes back to Newton, who is responsible for introducing the idea of viscosity into fluid dynamics. The dynamical theory of fluids developed before the atomic make-up of matter was known. It was, thus, a continuum theory and it made sense to assign velocities to each material point in space. The rate of change of these velocities was determined by forces which volume elements exerted on each other. After chemists bestowed reality on atoms and molecules, there were several attempts to develop a mechanical picture of matter based on collisions and the interaction of molecules. Later this kinetic molecular picture was developed in great detail by Boltzmann, as is discussed in Chapter 2.

A complete mechanical description of a large collection of molecules would be very complicated. For simplicity, consider the classical mechanical description. Because Newtonian mechanics is deterministic, the description would require as an ingredient a knowledge of all the initial positions and velocities of the particles. Another ingredient would be some sort of solution of the equations of motion either in analytical or tabulated form. While solving the equations of motion is merely an awesome task, accurately assigning the initial positions and velocities is actually impossible. Even in a sample of matter containing a few hundred molecules it would not be possible, as a practical matter, to measure the initial coordinates and momenta of all the molecules exactly. As a consequence, even with the fastest, most accurate computers this type of mechanical calculation of the properties of a large collection of molecules cannot succeed.

For a mechanical calculation to be successful it must get around the problem of specification of initial conditions. The first attempt to do this was made by Gibbs, who introduced the concept of an *ensemble* into mechanics. An ensemble consists of a large number of conceptual copies of the system of interest, all of which are prepared in the same macroscopic fashion. For example, if the system is a uniform single-component fluid, one might specify the initial volume, temperature, and number of molecules. Such a coarse specification of the initial condition of a system clearly is compatible with many different initial positions and momenta for the molecules. Gibbs called the space of positions and momenta *phase space*, and thus an ensemble corresponds to a distribution of points in phase space. In this way the notion of an ensemble permits the introduction of statistical ideas into mechanics: The distribution of initial phase points is used to assign probabilities to points or regions in phase space based on their frequency in the initial ensemble. As each member of the ensemble changes over time, its phase point is carried into a new phase point. The evolution of this probability cloud in phase space is described by a Newtonian equation of motion called the Liouville equation.

Gibbs' idea was to use the statistical properties of the evolving ensemble to characterize the system of interest. He suggested that an estimate of the value of any measurable property in the system could be obtained by averaging over the same property in the ensemble. Clearly this idea does not work very well for arbitrary ensembles. For example, if only the volume and molecule number are used to specify the initial phase points in the ensemble, the ensemble will be distributed over phase points corresponding to differing temperature or, perhaps, states for which the temperature is not defined. Thus the average dynamics of the ensemble will bear little relevance to the system of interest, which initially may have some particular temperature. Two questions naturally arise in the Gibbs ensemble approach. First, when is the specification of an ensemble complete enough so that the resulting average will give good estimates for any system so prepared? And, second, for such a good specification of an ensemble, what is the corresponding distribution in phase space?

To answer the first of these questions requires a knowledge of which initial macroscopic specifications suffice to reliably predict the future. Otherwise, such a range of conditions exist in the ensemble that predictions from ensemble averages will bear no relationship to measurement on a particular system. Indeed, there is no first-principles way to answer this question short of examining the details of the evolution of the ensemble for various initial macroscopic conditions. Consequently, in the Gibbs approach it is necessary to use empirical information to determine which macroscopic specifications lead to reliable prediction.

The basic empirical guide to good ensemble specifications has been equilibrium thermodynamics. In thermodynamics the state of a large system at equilibrium is determined by its entropy as a function of the extensive variables—the internal energy E, the volume V, and the number of molecules N. Moreover, except near critical points of phase transitions, it is known that measurements of such systems are predictable to a high degree of accuracy. Thus thermodynamics suggests that the specification of E, V, and N for systems that are well-aged should give rise to a reproducible ensemble. This is the *microcanonical ensemble* which forms the cornerstone of the Gibbs picture of equilibrium statistical mechanics.

Given this empirical justification of the reliability of the microcanonical ensemble, the question still remains as to what distribution in phase space corresponds to this specification. This question is surely an extra-mechanical one and has no answer within the confines of the mechanical theory. Its answer, as a result, must be relegated to the status of a postulate. For the microcanonical ensemble Gibbs postulated that a region of phase points corresponding to an energy in the range E to $E + dE$ has an ensemble probability proportional to its volume. On this basis, there is equal probability for regions of the same Euclidean phase space volume. Such a specification corresponds correctly to a stationary ensemble, since the probability density depends only on the energy. Liouville's equation implies that such a probability distribution will be time independent.

Since the form of Gibbs' microcanonical distribution has been elevated to a postulate, there is no way to verify it independently, except by comparison to experimental deductions based on it. That is, its use is to be justified *a posteriori* like the postulates of all physical theories. In fact, the results based on Gibbs' microcanonical distribution are in excellent agreement with the experimental properties of matter at equilibrium. Actually, most calculations of thermodynamic properties use the *canonical distribution* corresponding to an equilibrium ensemble of systems with well-defined values of V, N, and the temperature. Since the canonical distribution can be derived from the microcanonical one, a test of its predictions serves to test the basic postulate. From averages in the canonical ensemble one can calculate pressure–volume equations of state, heat capacities, and other quantities—such as radial distribution functions. These calculations agree both qualitatively and quantitatively with experimental results. In this way Gibb's microcanonical distribution has been abundantly justified *a posteriori*.

One can still ask if Gibbs' microcanonical distribution is the unique distribution which describes the microcanonical ensemble. In other words, is it the only distribution that works? The answer is certainly no. In fact, a canonical distribution with a correctly chosen temperature parameter gives results completely equivalent to the microcanonical distribution in the limit of infinite N and V with N/V fixed. This is the so-called *thermodynamic limit*, which is usually taken to justify the use of the canonical distribution. This result, however, also suggests that the reason Gibbs' postulated microcanonical distribution works so well is not due to its particular form but, perhaps, to something deeper.

In any case, one is left in the awkward position in mechanical ensemble theory of having to introduce postulates about the nature of initial distribution functions which cannot themselves be measured. For a nonequilibrium ensemble, this becomes an especially serious problem since no one knows how to assign a correct distribution to a given nonequilibrium specification. Even if such a prescription existed, it would be necessary to solve the mechanical equations of motion to find out how such an ensemble evolved in time. For a large nonuniform system, such a solution cannot presently be obtained.

1.2. Macroscopic Descriptions and Contractions

The previous section gives an outline of some of the ingredients which are necessary in a mechanical theory of matter. Because of the vast number of molecules involved in any reasonably sized piece of matter, it is not possible to obtain the empirical information, namely, the initial point in phase space which is required to predict material properties. This difficulty is inherent in our system of measurement. Since there is no way to keep track of all the positions and momenta of a large molecular system, measurements are limited to the specification of a small number of characteristic quantities. For example, in a well-stirred chemical reaction one might keep the pressure and temperature fixed and follow quantities like the concentrations of various chemical species in solution. Thus instead of the order of Avogadro's number of variables, as few as four or five may be sufficient to characterize such a reacting system. This example is a typical contracted description of a large molecular system. In fact, because of the limited number of simultaneous measurements that can be made, every description of a macroscopic system must of necessity be contracted.

The fact that measurements on macroscopic systems are restricted to a small number of variables raises a question similar to that raised by Gibbs' ensemble theory. Namely, under what conditions will a given set of variables lead to a self-contained description? In other words, what variables will provide a good contracted description of a macroscopic system, in the sense that future behavior is reasonably well predicted by a knowledge of initial values? From the macroscopic point of view, this is an empirical question which must be answered for each system individually. For an isothermal solution undergoing reasonably slow chemical reactions at constant pressure, it is known empirically that the concentrations of the various reacting species give a good contracted description. Similarly, for fluid flows that are not too rapid, a hydrodynamic-level description involving mass, energy, and momentum densities as a function of spatial position provides a good contracted description. In fact, the probing of matter at the macroscopic level over the last few centuries has yielded a variety of self-contained contracted descriptions of matter.

Preeminent among contracted descriptions of matter is the thermodynamic description of equilibrium. Classical thermodynamics involves a description of matter in terms of *extensive variables*. These include such quantities as the internal energy, volume, and the numbers of molecules which make up the matter. These variables are called extensive because they depend on the extent of the system. Thus when two identical pieces of matter are combined, the value of an extensive variable will be just twice that of either system individually. This is to be contrasted with intensive variables, like the temperature, which stay the same when two identical systems are combined. Because the extensive variables provide a good contracted description of matter at equilibrium, they are said to characterize the macroscopic equilibrium state of a system. In other words, two systems which are at equilibrium and possess the same values of the extensive variables are, for the purposes of most measurements, identical.

It is clear, however, that two systems which have the same contracted description cannot be identical in all respects. This follows from the fact that the molecular state, or location in phase space, will not be the same for two such systems. As a consequence, the contracted description has built into it *molecular fluctuations*. These fluctuations measure the differences which appear in any collection of systems that have the same macroscopic description. For a contracted description to be a good one, these fluctuations must be small, although they cannot be vanishingly small. In this way, contracted descriptions, which are forced upon us by our limited ability to make measurements, are inherently statistical.

To deal with the differences between systems with identical contracted descriptions, it is natural to introduce the idea of a physical ensemble. A physical ensemble consists of a collection of systems, identically prepared with respect to their contracted description. This differs somewhat from the Gibbs ensemble which corresponds to a mathematical distribution of systems in phase space that is used to estimate the properties of a particular system. A physical ensemble, on the other hand, is introduced because of the coarseness of the contracted description, which corresponds not to a single system but to a collection of systems, namely, the ensemble itself. In other words, when adopting a contracted description one has given up the idea of describing a single system and focuses instead on the statistical properties of the ensemble. Measurements on systems within the ensemble, thus, are relevant only insofar as they contribute to an understanding of the statistical behavior of the ensemble.

In adopting the physical ensemble point of view, we take a more avowed operational approach than in the Gibbsian ensemble point of view. For a Gibbsian ensemble one requires knowledge of the statistical distribution in phase space. For a physical ensemble one requires only a knowledge of the statistical distribution for the variables which one has chosen to use in the contracted description. Thus, for a system at fixed volume in contact with a temperature bath it sometimes makes sense to form a contracted description from the usual extensive variables, i.e., the internal energy (E) and numbers of molecules ($N_1, \ldots, N_k$). For such a physical ensemble one needs only to have information about the statistical distribution of E, N_1, ..., N_k. Since these quantities form the basis of a contracted description, this sort of information is accessible to measurement in a way that the distribution in phase space can never be.

One of the major themes of this book is the development of the physical ensemble point of view for simple contracted descriptions of matter. In doing so, we stay far away from the classical and quantum Hamiltonian descriptions of matter. Nonetheless, the theory of physical ensembles described here has associated with it dynamical equations of motion. As we will see, these equations describe the probabilities of various processes and are inherently irreversible.

1.3. Stochastic Processes and Physical Ensembles

Physical ensembles, as discussed in the preceding section, are the embodiment of the contracted picture of macroscopic systems. The natural mathematical setting for treating such ensembles is the theory of stochastic processes. In this section we review some of the basic ideas about stochastic processes which are useful for describing physical ensembles.

Consider first, for illustration, a real-valued stochastic process, $n(t)$. Thus $n(t)$ is a random variable which depends parametrically on the continuous value of t. By *random variable* it is meant that a distribution of possible values of $n(t)$ exists, any one of which might be encountered in an actual measurement on a system in the ensemble. For example, $n(t)$ might be the number of hydrogen ions at time t in a beaker of water at equilibrium at a given temperature and pressure. Even if two beakers contained an identical number of water molecules, careful measurement of the number of hydrogen ions in each beaker would produce two answers which would differ slightly. The causes of this difference are the molecular fluctuations which govern the process of ionization. In addition to this variability of measured values within the ensemble at time t, there is the additional variability of $n(t)$ as a function of time. It is this dependence on the parameter t, which we will always take to be the time, that is characteristic of stochastic processes and that makes them useful for describing physical ensembles.

Since contracted descriptions of macroscopic systems usually involve several variables, we need to consider more complicated stochastic processes. Thus let $n_1(t), n_2(t), \ldots, n_k(t)$ be random variables which depend on the time, t. As these variables make up the contracted description of a macroscopic system, we will often think of them as being continuous variables. We shall also treat the number of variables, k, in the contracted description as being finite. This is not a serious limitation, and later it will be shown how to treat systems whose properties vary from point to point in space by taking the limit $k \to \infty$. It is convenient to collect the k variables n_i together as a column vector which is written as $\mathbf{n}(t)$ a *vector-valued stochastic process*. Since the time is a continuous variable, this means that $\mathbf{n}(t)$ really represents an uncountable infinity of random variables. To characterize these variables for all values of the time, it is necessary to specify the statistical properties of all finite subsets of these variables, e.g., $\mathbf{n}(t_1)$, $\mathbf{n}(t_2)$, ..., $\mathbf{n}(t_m)$. The stochastic process is characterized by the joint distribution functions for all possible finite specifications of times. Thus we need to know

$$\text{Prob}[\mathbf{n}(t_1) \leq \mathbf{n}_1, \mathbf{n}(t_2) \leq \mathbf{n}_2, \ldots, \mathbf{n}(t_m) \leq \mathbf{n}_m], \tag{1.3.1}$$

where $\text{Prob}[\cdot]$ is the joint probability that the variables are less than the indicated values $\mathbf{n}_1$, ..., $\mathbf{n}_m$. In general, we will restrict our attention to stochastic processes for which these joint probabilities are derivable from probability densities. Thus, for example, we will have

$$\text{Prob}[\mathbf{n}(t) \leq \mathbf{n}'] = \int_{-\infty}^{n'_1} dn'_1 \cdots \int_{-\infty}^{n'_k} dn'_k \, W_1(\mathbf{n}, t), \tag{1.3.2}$$

where $W_1(\mathbf{n}, t)$ is the *single-time probability density* for $\mathbf{n}(t)$. The name density comes from the fact that by subtracting Eq. (1.3.2) for $\mathbf{n}' = \mathbf{n}$ from Eq. (1.3.2) for $\mathbf{n}' = \mathbf{n} + d\mathbf{n}$ one gets $\text{Prob}[\mathbf{n} \leq \mathbf{n}(t) \leq \mathbf{n} + d\mathbf{n}] = W_1(\mathbf{n}, t)\, d\mathbf{n}$. Thus, $W_1(\mathbf{n}, t)\, d\mathbf{n}$ is the probability that $\mathbf{n}(t)$ takes on values between $\mathbf{n}$ and $\mathbf{n} + d\mathbf{n}$. Multiple-time densities are also assumed to exist, and they describe the *joint distribution* of $\mathbf{n}(t_1)$, ..., $\mathbf{n}(t_m)$. In a similar fashion one finds that

$$\begin{aligned} &W_m(\mathbf{n}_1, t_1; \mathbf{n}_2, t_2; \ldots; \mathbf{n}_m, t_m)\, d\mathbf{n}_1\, d\mathbf{n}_2 \ldots d\mathbf{n}_m \\ &\quad = \text{Prob}[\mathbf{n}_1 \leq \mathbf{n}(t_1) \leq \mathbf{n}_1 + d\mathbf{n}_1, \\ &\qquad \mathbf{n}_2 \leq \mathbf{n}(t_2) \leq \mathbf{n}_2 + d\mathbf{n}_2, \ldots, \mathbf{n}_m \leq \mathbf{n}(t_m) \leq \mathbf{n}_m + d\mathbf{n}_m]. \end{aligned} \tag{1.3.3}$$

The joint probability densities for m contain all the statistical information that the densities for $m' < m$ contain, as well as information about additional instants of time. This gives rise to the following consistency relationship which is obtained by integrating over $\mathbf{n}_m$ in Eq. (1.3.3):

$$W_{m-1}(\mathbf{n}_1, t_1; \ldots; \mathbf{n}_{m-1}, t_{m-1}) = \int W_m(\mathbf{n}_1, t_1; \ldots; \mathbf{n}_m, t_m)\, d\mathbf{n}_m. \tag{1.3.4}$$

A knowledge of all the joint probability densities provides a complete statistical description for measurements of $\mathbf{n}$ made at a finite number of different times.

The joint probability densities can be used to obtain average values for a function of the random variable $\mathbf{n}(t)$. For example, the single-time average of a scalar function, $f(\mathbf{n})$, is given by

$$\langle f(\mathbf{n}(t)) \rangle_1 \equiv \int f(\mathbf{n}) W_1(\mathbf{n}, t)\, d\mathbf{n}. \tag{1.3.5}$$

Similarly, two-time averages of the scalar functions f and g are defined by

$$\langle f(\mathbf{n}(t)) g(\mathbf{n}(t')) \rangle \equiv \int f(\mathbf{n}) g(\mathbf{n}') W_2(\mathbf{n}, t; \mathbf{n}', t')\, d\mathbf{n}\, d\mathbf{n}' \tag{1.3.6}$$

with three-time and higher-order averages being defined in the obvious fashion. It is usually clear from the values of the time parameter in the angular brackets which multiple-time average is being carried out. Thus we will often leave off the subscript 1, 2, ..., m from the brackets.

There are several special cases of these averages that are commonly used to characterize stochastic processes. The ensemble average of $n_i(t)$, namely,

$$\langle n_i(t)\rangle = \int n_i W_1(\mathbf{n}, t)\, d\mathbf{n}, \tag{1.3.7}$$

is one of these. Another group of important averages are the central moments of $n_i(t)$. These are given by

$$\langle [n_i(t) - \langle n_i(t)\rangle]^l \rangle$$

for $l = 1, 2, \ldots$. The first central moment is obviously zero and the second central moment is called the variance. Central moments are particularly useful quantities and it is customary to introduce new variables,

$$\delta\mathbf{n}(t) \equiv \mathbf{n}(t) - \langle \mathbf{n}(t)\rangle, \tag{1.3.8}$$

which we will call the *fluctuations* about the average.

Besides these moments, there are various measures of correlations among the variables $n_i(t)$ that are in use. The single-time covariance matrix, usually called just the covariance matrix, is written

$$\sigma_{ij}(t) = \langle \delta n_i(t)\delta n_j(t)\rangle \tag{1.3.9}$$

or, in vector notation,

$$\sigma(t) = \langle \delta\mathbf{n}(t)\delta\mathbf{n}^T(t)\rangle, \tag{1.3.10}$$

where the superscript T represents the transpose, i.e., $\delta\mathbf{n}^T$ is a row vector. The covariance gives a measure of the statistical dependence of $n_i(t)$ and $n_j(t)$. These two variables are said to be *uncorrelated* if

$$\langle n_i(t)n_j(t)\rangle = \langle n_i(t)\rangle\langle n_j(t)\rangle. \tag{1.3.11}$$

This would be the case, for example, for the density fluctuations of molecules which were located in two different beakers. Using the definition of $\delta\mathbf{n}$ in Eq. (1.3.8), one easily verifies that

$$\langle \delta n_i(t)\delta n_j(t)\rangle = \langle n_i(t)n_j(t)\rangle - \langle n_i(t)\rangle\langle n_j(t)\rangle. \tag{1.3.12}$$

Thus the covariance of the fluctuations of uncorrelated random variables will vanish. A related measure of the correlation of random variables is the *single-time correlation function*, $g_{ij}(t)$. It is defined by

$$g_{ij}(t) = \langle n_i(t)n_j(t)\rangle / \langle n_i(t)\rangle\langle n_j(t)\rangle \tag{1.3.13}$$

and is identically one when n_i and n_j are uncorrelated. Using this definition of g_{ij} in Eq. (1.3.12) shows that the covariance of n_i and n_j can be written in terms of the correlation function as

$$\langle \delta n_i(t)\delta n_j(t)\rangle = \langle n_i(t)\rangle\langle n_j(t)\rangle(g_{ij}(t) - 1). \tag{1.3.14}$$

The single-time correlation function measures correlations between pairs of variables at the same instant in time.

The two-time correlation function, often just called the *time correlation function*, measures correlations among variables at different instants in time. It is defined by the relationship

$$C_{ij}(t, t') = \langle n_i(t) n_j(t') \rangle, \tag{1.3.15}$$

in which the right-hand side involves a two-time average. A related quantity, based on the central moments, is the two-time covariance

$$\sigma_{ij}(t, t') \equiv \langle \delta n_i(t) \delta n_j(t') \rangle. \tag{1.3.16}$$

These correlation functions are connected by the relationship

$$\sigma_{ij}(t, t') = C_{ij}(t, t') - \langle n_i(t) \rangle \langle n_j(t') \rangle. \tag{1.3.17}$$

Finally a two-time version of $g_{ij}(t)$ has the obvious definition

$$g_{ij}(t, t') = C_{ij}(t, t') / \langle n_i(t) \rangle \langle n_j(t') \rangle.$$

Comparing this equation with Eqs. (1.3.13) and (1.3.15) shows that $g_{ij}(t, t) = g_{ij}(t)$. These three functions contain similar information about the way correlations between variables change in the ensemble during the time interval, $t' - t$, between two measurements. For many situations encountered experimentally, the irreversibility inherent in contracted descriptions causes all three of these correlation functions to diminish with time.

In addition to joint probabilities, other kinds of probabilities, called *conditional probabilities*, are useful for characterizing an ensemble. Remember that an ensemble consists of a collection of systems for each of which we keep track of the variable $\mathbf{n}(t)$. The probability of measuring a value of $\mathbf{n}(t)$ between $\mathbf{n}$ and $\mathbf{n} + d\mathbf{n}$ is interpreted in the ensemble picture as the fraction of systems having a value of $\mathbf{n}$ in this infinitesimal range, or window, at time t. By measuring the fraction of systems that have values of $\mathbf{n}$ within two prescribed windows at two different times we similarly obtain the two-time joint probability density. Conditional probabilities are also obtained by examining the fractions of systems having certain values of $\mathbf{n}$ within the ensemble at several different times. However, it is necessary to process the data differently. Let us examine the ensemble at time t^0 and select from it a subensemble, all of whose members have $\mathbf{n}(t^0) = \mathbf{n}^0$. This is a single conditional ensemble. Since $\mathbf{n}$ forms only a contracted description of the system, this does not mean that the members of the conditional ensemble are identical to one another, but only that they possess at time t^0 identical values of $\mathbf{n}$. Indeed, as time progresses, molecular interactions will cause each system to evolve slightly differently. Thus, at a later time t the conditional ensemble will have a distribution of values of $\mathbf{n}$. If at t we then measure the fraction of members of the conditional ensemble with values of $\mathbf{n}$ in the window $\mathbf{n}$ to $\mathbf{n} + d\mathbf{n}$, we will obtain the probability density *conditional* on our precise knowledge that $\mathbf{n} = \mathbf{n}^0$ at time t^0. A schematic

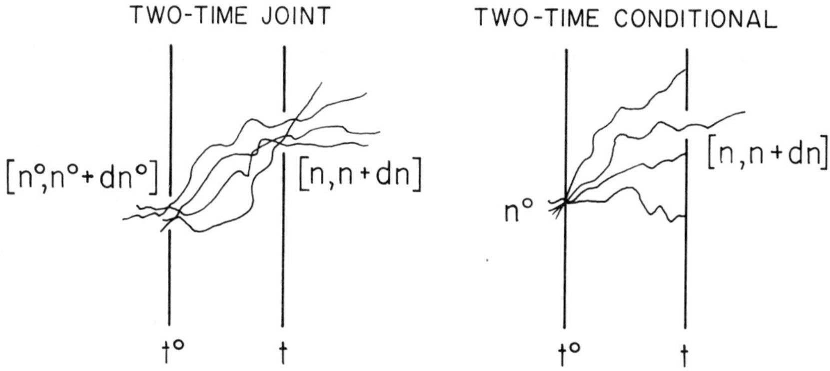

FIGURE 1.1. A schematic representation of trajectories that contribute to the two-time joint and conditional probability densities. For the joint probability, trajectories pass through the observation window $[n^0, n^0 + dn^0]$ at time t^0 and the observation window $[n, n + dn]$ at time t. For the conditional probability, trajectories begin precisely at n^0 at time t^0 and pass through the window $[n, n + dn]$ at time t.

diagram of the relationship between the single-time conditional probability and the two-time conditional probability is given in Fig. 1.1.

Based on this ensemble picture we can formally define the conditional probability for the window $\mathbf{n}$ to $\mathbf{n} + d\mathbf{n}$ as the following probability

$$\begin{aligned}&\text{Prob}[n_1(t^0) = n_1^0, \ldots, n_k(t^0) = n_k^0,\\&\qquad n_1 \le n_1(t) \le n_1 + dn_1, \ldots, n_k \le n_k(t) \le n_k + dn_k]\\&\equiv P_2(\mathbf{n}^0, t^0 | \mathbf{n}, t)\, d\mathbf{n}.\end{aligned} \tag{1.3.18}$$

The function $P_2(\mathbf{n}^0, t^0 | \mathbf{n}, t)$ is the *two-time conditional probability density*. Notice that $\mathbf{n}^0$ is only a parameter and that the probability density refers to the variable $\mathbf{n}$ at time t. The single-time and two-time joint probability densities are closely related to the two-time conditional density. In fact, because the probability density for $\mathbf{n} = \mathbf{n}^0$ at t^0 is given by $W_1(\mathbf{n}^0, t^0)$, it follows that

$$W_2(\mathbf{n}^0, t^0; \mathbf{n}, t) = W_1(\mathbf{n}^0, t^0) P_2(\mathbf{n}^0, t^0 | \mathbf{n}, t). \tag{1.3.19}$$

The two-time conditional density is sometimes called the *transition probability*. The significance of this can be seen by combining Eqs. (1.3.4) and (1.3.19) to give

$$W_1(\mathbf{n}, t) = \int W_1(\mathbf{n}^0, t^0) P_2(\mathbf{n}^0, t^0 | \mathbf{n}, t)\, d\mathbf{n}^0. \tag{1.3.20}$$

Thus with the aid of P_2 we can calculate how the single-time probability density evolves over time.

Actually we can define whole hierarchies of conditional ensembles involving more and more conditions. For example,

$$P_3(\mathbf{n}_1, t_1; \mathbf{n}_2, t_2 | \mathbf{n}_3, t_3)\, d\mathbf{n}_3$$

is the probability of finding $\mathbf{n}(t_3)$ in the range $\mathbf{n}_3$ to $\mathbf{n}_3 + d\mathbf{n}_3$ in the subensemble which was known to have $\mathbf{n}(t_1) \equiv \mathbf{n}_1$ *and* $\mathbf{n}(t_2) \equiv \mathbf{n}_2$ with $t_1 \leq t_2 \leq t_3$. P_m involves $m - 1$ prior conditions and, in general,

$$\begin{aligned} &W_m(\mathbf{n}_1, t_1; \ldots; \mathbf{n}_m, t_m) \\ &\quad = W_{m-1}(\mathbf{n}_1, t_1; \ldots; \mathbf{n}_{m-1}, t_{m-1}) P_m(\mathbf{n}_1, t_1; \ldots; \mathbf{n}_{m-1}, t_{m-1} | \mathbf{n}_m, t_m). \end{aligned} \tag{1.3.21}$$

The average values in a conditional subensemble contain useful information that is related to averages taken from the entire ensemble. The connection is implicit in Eqs. (1.3.19) and (1.3.21). For example, the conditional average of a function $f(\mathbf{n}(t))$ is defined by

$$\langle f(\mathbf{n}(t)) \rangle^0 = \int f(\mathbf{n}) P_2(\mathbf{n}^0, t^0 | \mathbf{n}, t) d\mathbf{n}, \tag{1.3.22}$$

where the superscript naught reminds us of the condition at t^0. If we multiply the left-hand side of Eq. (1.3.22) by another function $g(\mathbf{n}^0)$ and then average over $\mathbf{n}^0$ at t^0, we obtain

$$\begin{aligned} \langle g(\mathbf{n}^0(t^0)) \langle f(\mathbf{n}(t)) \rangle^0 \rangle_1 &= \int d\mathbf{n}^0 \int d\mathbf{n} g(\mathbf{n}^0) f(\mathbf{n}) W_1(\mathbf{n}^0, t^0) P_2(\mathbf{n}^0, t^0 | \mathbf{n}, t) \\ &= \int d\mathbf{n}^0 \int d\mathbf{n} g(\mathbf{n}^0) f(\mathbf{n}) W_2(\mathbf{n}^0, t; n, t) \\ &= \langle g(\mathbf{n}(t^0)) f(\mathbf{n}(t)) \rangle_2. \end{aligned} \tag{1.3.23}$$

If $f(\mathbf{n}) = n_j$ and $g(\mathbf{n}) = n_i$ in Eq. (1.3.23), we discover that the two-time correlation function defined in Eq. (1.3.15) can be written

$$C_{ij}(t, t') = \langle n_i^0(t) \langle n_j(t') \rangle^0 \rangle_1. \tag{1.3.24}$$

This is a fundamental connection between two-time conditional averages and two-time joint averages which is often useful in applications.

An important subclass of stochastic processes, especially for certain physical conditions, is *Markov processes*. A Markov process is defined by the property that conditions prior to the most recent one in a conditional ensemble do not affect the subsequent probability distribution. More precisely, for all $m \geq 2$

$$P_m(\mathbf{n}_1, t_1; \ldots; \mathbf{n}_{m-1}, t_{m-1} | \mathbf{n}_m t_m) = P_2(\mathbf{n}_{m-1}, t_{m-1} | \mathbf{n}_m, t_m). \tag{1.3.25}$$

This condition is often said to characterize a process without memory, since the history of observations prior to the most recent is irrelevant for the future of the stochastic processes. A simple Markov process is the flipping of a coin, for which one imagines that even the most recent outcome of a flip is irrelevant for determining the probability of the next outcome. We will encounter many other physical examples of Markov processes, such as Brownian motion, in the ensuing chapters. It should be clear from Eqs. (1.3.21) and (1.3.25) that a Markov process is completely defined by $W_1(\mathbf{n}, t)$ and $P_2(\mathbf{n}^0, t^0|\mathbf{n}, t)$. For example, combining these equations for $m = 3$ gives

$$W_3(\mathbf{n}_1, t_1; \mathbf{n}_2, t_2; \mathbf{n}_3, t_3) = W_1(\mathbf{n}_1, t_1)P_2(\mathbf{n}_1, t_1|\mathbf{n}_2, t_2)P_2(\mathbf{n}_2, t_2|\mathbf{n}_3, t_3), \quad (1.3.26)$$

with analogous results for larger values of m. For this reason, a Markov stochastic process is one of the simplest to characterize completely.

Another stochastic property which is often relevant for physical systems is stationarity. A *stationary* stochastic processes is one which is invariant to time translations by the same interval τ. Thus the variables $\mathbf{n}(t_1 + \tau), \ldots, \mathbf{n}(t_m + \tau)$ are statistically indistinguishable from the untranslated variables $\mathbf{n}(t_1), \ldots, \mathbf{n}(t_m)$ for all m and τ. This implies that all single-time averages are constants and, thus, the single-time probability density, W_1, is independent of time. An ensemble of systems at thermodynamic equilibrium is, for example, stationary as is an ensemble in which the component systems are maintained at a nonequilibrium steady state—at least after the ensemble has been sufficiently aged. In addition to W_1 being time independent, the joint probability densities in a stationary ensemble depend only on pairwise time differences. Thus

$$W_2(\mathbf{n}_1, t_1; \mathbf{n}_2, t_2) = W_2(\mathbf{n}_1, 0; n_2, t_2 - t_1)$$

and

$$P_2'(\mathbf{n}^0, t^0|\mathbf{n}, t) = P_2(\mathbf{n}^0, 0|\mathbf{n}, t - t^0). \quad (1.3.27)$$

Because of this, for a stationary ensemble we will often suppress the initial time in these expressions and write $P_2(\mathbf{n}^0|\mathbf{n}, t)$, etc.

One of the simplest forms that a probability density can take is the *Gaussian form*. This is the multivariable generalization of Gauss's well-known bell-shaped curve, which is represented by

$$G(\mathbf{n}) = [(2\pi)^m \det \sigma]^{-1/2} \exp[-(\mathbf{n} - \bar{\mathbf{n}})^T \sigma^{-1}(\mathbf{n} - \bar{\mathbf{n}})/2], \quad (1.3.28)$$

where $\det \sigma$ is the determinant of the nonsingular, symmetric, positive definite matrix σ and σ^{-1} is its inverse. The Gaussian density, $G(\mathbf{n})$, is written in its normalized form, that is,

$$\int G(\mathbf{n})\, d\mathbf{n} = 1. \quad (1.3.29)$$

The two values which characterize the Gaussian are its average and covariance, which are given by

$$\int \mathbf{n}G(\mathbf{n})\, d\mathbf{n} = \bar{\mathbf{n}} \quad (1.3.30)$$

$$\int (\mathbf{n} - \bar{\mathbf{n}})(\mathbf{n} - \bar{\mathbf{n}})^T G(\mathbf{n})\, d\mathbf{n} = \sigma. \quad (1.3.31)$$

For a single variable, n, these equations follow from the well-known integral identities

$$\int_{-\infty}^{+\infty} \exp(-x^2/2\sigma)\,dx = (2\pi\sigma)^{1/2}$$
$$\int_{-\infty}^{+\infty} x\exp(-x^2/2\sigma)\,dx = 0 \qquad (1.3.32)$$

and

$$\int_{-\infty}^{+\infty} x^2\exp(-x^2/2\sigma)\,dx = (2\pi\sigma^3)^{1/2}.$$

By changing variables so as to diagonalize σ, using the fact that the determinant of σ is the product of its eigenvalues, and employing the integrals in Eq. (1.3.32), one can also verify that Eqs. (1.3.29)–(1.3.31) are true when $\mathbf{n}$ is a vector process.

A stochastic process is called *Gaussian* if all the joint and conditional probability densities have a Gaussian form. For a stationary Gaussian process, the lowest-order joint density will be time independent. Consequently, the average values and covariance that characterize the Gaussian density will be constants. This sort of density is well known in equilibrium statistical mechanics, where its form was first derived in the thermodynamic limit by Einstein. Conditional probabilities are also often found to be Gaussian for physical processes. For conditional ensembles, the average values and covariance of $\mathbf{n}(t)$ will be time dependent. As we will see in subsequent chapters, the differential equations solved by these time-dependent quantities can be related to macroscopic transport equations.

Gaussian probabilities are characteristic of physical ensembles only in the limit that the systems in the ensemble are, in some sense, large. Although the exact meaning of "large" will need to await a precise description, the sense of the meaning comes from the classical central limit theorem for sums of independent random variables. The original context for this result was coin tossing, which is a discrete random process for which the time can be thought of as taking on the discrete values $t = 1, 2, \ldots$ corresponding to the instant at which a toss is made. Thus for any t we might write $n(t) = 1$ or 0 corresponding to heads (1) or tails (0) appearing. Each toss is independent of all others and the outcomes of any one toss are assumed to have probabilities $P(1) = p$ and $P(0) = 1 - p \equiv q$ with $0 \leq p \leq 1$.

Let us now imagine that we are not observing the outcome of a single toss, but rather the *net* outcome of the tossing of n identical coins at the same instant. In other words, we observe only the total number of heads that come up, $n_+ = 0, 1, 2, \ldots, n$. Notice that $n_+(t) = \sum_{i=1}^{n} n_i(t)$, where n_i gives the result of the toss (either 1 or 0) for the ith coin. It is thus a sum of random variables. In this sense, it provides a contraction of the coin-tossing problem, which would otherwise require that we keep track of each coin. From the independence of tosses, we can calculate the probability of any sequence of heads and tails, labeled by coin number, e.g., $(+1,\ldots,0)$, to be

$$p(+1,\ldots,0) = p^m q^{n-m}, \tag{1.3.33}$$

where m is the number of heads appearing in the sequence of n tosses.

The probability that $n_+ = m$ can thus be obtained by adding up all ways this might occur. The result is

$$p(m) = [n!/m!(n-m)!]p^m q^{n-m}, \tag{1.3.34}$$

since there are $[n!/m!(n-m)!]$ outcomes with exactly m heads. Equation (1.3.34) gives the single-time probability for $n_+(t)$ and is independent of time. Moreover, $n_+(t)$ and $n_+(t')$ for $t \neq t'$ are independent random variables; thus

$$\begin{aligned} W_1(m,t) &= p(m) \\ W_2(m,t;m',t') &= p(m)p(m'), \end{aligned} \tag{1.3.35}$$

etc.

The *central limit theorem* asserts that in the limit of many coins, $n_+(t)$ becomes, in some sense, a Gaussian process. The real result is this: In the limit that $n \to \infty$, the random variable $x(t) \equiv \lim_{n\to\infty} [n_+(t) - np]/\sqrt{(npq)}$ has a Gaussian density given by

$$W_1(x) = (2\pi)^{-1/2}\exp(-x^2/2). \tag{1.3.36}$$

In obtaining this limit from Eq. (1.3.35) it is essential to scale $n^+(t)$ by $1/\sqrt{n}$. This result is called the *DeMoivre–Laplace limit theorem* and is a special case of the central limit theorem for sums of independent random variables. The proof of Eq. (1.3.36) is straightforward. First we use Stirling's formula

$$n! \sim (2\pi)^{1/2} n^{n+(1/2)} e^{-n}, \tag{1.3.37}$$

where the $\sim$ sign means that the ratio of the two sides tends to one in the limit. Applying this to the factorials in Eq. (1.3.34) gives

$$p(m) \sim \left[\frac{n}{2\pi m(n-m)}\right]^{1/2} \left(\frac{np}{m}\right)^m \left(\frac{qn}{n-m}\right)^{n-m} \tag{1.3.38}$$

Now define the random variable $\chi_n - (n_+ - np)/\sqrt{(npq)}$, so that $n_+ = m = x\sqrt{(npq)} + np$ when $\chi_n = x$. If we substitute this into Eq. (1.3.38) and take the logarithm we obtain

$$\begin{aligned} \ln p(x) \sim{} & \ln\left[2\pi\left(\sqrt{\frac{pq}{n}}x + p\right)\left(-\sqrt{\frac{pq}{n}}x + q\right)n\right]^{-/2} \\ & - (np + \sqrt{npq}\,x)\ln\left(1 + x\sqrt{\frac{q}{np}}\right) \\ & - (nq - \sqrt{npq}\,x)\ln\left(1 - x\sqrt{\frac{p}{nq}}\right). \end{aligned} \tag{1.3.39}$$

Finally, using the Taylor series expansion of $\ln(1+\alpha) = \alpha - \alpha^2/2 + \alpha^3/3\ldots$ for the second and third terms leads to

$$\ln p(x) \sim \ln(2\pi npq)^{-1/2} - \frac{x^2}{2} + \mathcal{O}\left(\frac{1}{\sqrt{n}}\right), \tag{1.3.40}$$

where the notation $\mathcal{O}(1/\sqrt{n})$ means terms that involve higher powers of $1/\sqrt{n}$. Equating the exponential of both sides of Eq. (1.3.40) gives

$$\lim_{n\to\infty} p(x)\sqrt{npq} = (2\pi)^{-1/2}\exp(-x^2/2), \tag{1.3.41}$$

and since, from its definition, $dx = 1/\sqrt{(npq)}$, the limit theorem in Eq. (1.3.36) is proven.

The central limit theorem shows that in the limit of large n, the contracted variable $n_+(t)$ is a Gaussian process. Since Eq. (1.3.35) shows that its joint probabilities are independent, it is also a stationary process. Further more, comparing Eq. (1.3.35) and Eq. (1.3.21) shows that the conditional probability is $P_2(m|m't) = p(m')$ and that $n_+(t)$ is a Markov process. Thus, succinctly, the total number of heads that appear in n tosses is a *stationary, Gaussian, Markov* process in the limit that n becomes large. Actually, since all the joint probabilities are independent of time, it is a very simple stationary, Markov process called *time independent*.

We have belabored the proof of the central limit theorem for coin tossing because it typifies certain results that hold for molecular processes. In the thermodynamic limit the contracted description of molecular systems will involve Gaussian conditional probability densities. Furthermore, near stable steady states it is usually possible to find a contracted description that will be stationary, Gaussian, and Markovian.

1.4. Brownian Motion and the Wiener Process

In 1827 the Scottish botanist Robert Brown noticed a peculiar phenomenon. Examining pollen grains in a drop of water with his microscope, Brown observed that these particles did not stay in a fixed position but, instead, moved around in jagged, irregular fashion. His first thoughts were that the motion had something to do with a life-force contained in the pollen. However, this idea was dispelled when inorganic matter of a similar size was shown to behave in a similar fashion. This motion subsequently became known as Brownian motion. A variety of explanations were advanced to explain this phenomenon, and by late in the nineteenth century a number of people had suggested Brownian motion was a reflection of the molecular nature of matter. The idea was that a pollen grain—which measures about 10^{-6} m in diameter—was small enough to be buffeted around by water molecules. Thus the particle could act as a sensor of fluctuations in the velocity of water molecules in its immediate neighborhood. Because of the small size of water molecules, the resulting motion would be rapid and erratic.

The first quantitative explanation of Brownian motion was due to Einstein. In his picture the spatial position of a Brownian particle was a stochastic process. He modeled the stochastic process by suggesting that the probability density for position changed in time as if the particle were undergoing diffusion. Einstein argued that small particles, just like molecules, should diffuse and that the diffusion constant for spheres of radius ρ should be given by what is now called the Stokes–Einstein formula

$$D = k_B T/6\pi\eta R, \tag{1.4.1}$$

where k_B is Boltzmann's constant, T the absolute temperature, and η the viscosity coefficient of the fluid. Thus adopting a statistical ensemble picture, the evolving probability density for a single particle located originally at $\mathbf{r}_0$ should be just the same as the spreading mass density of a group of particles all initially at $\mathbf{r}_0$. The latter solves Fick's diffusion equation; consequently, so should the probability density. Thus in the Einstein picture the conditional probability density solves the equation

$$\partial P_2(\mathbf{r}_0|\mathbf{r},t)/\partial t = D\nabla^2 P_2(\mathbf{r}_0|\mathbf{r},t) \tag{1.4.2}$$

with the initial condition that $P_2(\mathbf{r}_0|\mathbf{r},0) = \delta(\mathbf{r} - \mathbf{r}_0)$, a delta function centered at r_0. The beauty of Einstein's model is that it permits a quantitative comparison to experiment. For example, the solution of Eq. (1.4.2) is well known to be

$$P_2(\mathbf{r}_0|\mathbf{r},t) = (4\pi Dt)^{-3/2}\exp(-|\mathbf{r} - \mathbf{r}_0|^2/4Dt), \tag{1.4.3}$$

as can be verified directly by substitution and the fact that as $t \to 0$ Eq. (1.4.3) approaches a delta function at $\mathbf{r}_0$. From this result the mean square distance that a particle moves in time t can be calculated to be

$$\langle|\mathbf{r}(t) - \mathbf{r}_0|^2\rangle^0 = 3(4\pi Dt)^{-1/2}\int_{-\infty}^{+\infty} x^2 \exp(-x^2/4Dt)\,dx = 6Dt, \tag{1.4.4}$$

where the value of the integral can be gotten from Eq. (1.3.32). This provides an independent was of measuring the diffusion constant, which can be checked against the Stokes–Einstein formula. In fact, if one accepts both Eqs. (1.4.4) and (1.4.2) as valid, it provides a way of measuring Boltzmann's constant.

Einstein's picture of Brownian motion is typical of contracted descriptions. One keeps track only of the position of the Brownian particle and ignores the positions and momenta of all the water molecules, as well as the momentum of the particle itself. As a consequence, the position of the particle becomes a random variable. The Einstein model involves a rather severe contraction, since it ignores the momentum of the particle. This, in fact, leads to an unphysical result for the average velocity of the particle, pointed out by

Einstein. Using Eq. (1.4.4) the average velocity is found to be

$$d(\langle|\mathbf{r}(t) - \mathbf{r}_0|^2\rangle^0)^{1/2}/dt = (3D/2t)^{1/2}. \tag{1.4.5}$$

This means that for short times the velocity of the particle becomes unbounded, which certainly cannot be the case. By including the momentum of the Brownian particle as an additional variable in the contracted description this defect can be removed.

An elegant mathematical description of Einstein's picture of Brownian motion is contained in the work of Wiener. Rather than constructing the stochastic processes by looking at a differential equation for the probability density, Wiener focused on the trajectories of the Brownian particle. The trajectories are simply the position, $\mathbf{r}(t)$, of a particle for the entire time course $-\infty \leq t \leq \infty$. This is the detailed motion of an individual particle, i.e., one member of the ensemble, and provides a single realization of the stochastic process. The collection of all possible trajectories along with the probability of each trajectory defines the entire stochastic process. In Wiener's way of looking at things, $\mathbf{r}(t)$ is a vector-valued stochastic process with components $x(t)$, $y(t)$, and $z(t)$ that are equivalent and independent of one another. Thus it suffices to consider, say, just the x-coordinate. A Wiener, or Brownian motion, process is one with independent increments that depend only on elapsed time and trajectories that are continuous. More precisely, a *Wiener process* is defined by properties (i)–(iii):

(i) For $t_1 < t_2 < \cdots < t_k$, $x(t_2) - x(t_1), x(t_3) - x(t_2), \ldots, x(t_k) - x(t_{k-1})$ are independent random variables
(ii) $x(t + \tau) - x(t)$ does not depend on t
(iii) $\lim_{\tau \to 0} \text{Prob}(|x(t + \tau) - x(t)| \geq \Delta)/\tau = 0$ for all $\Delta > 0$. (1.4.6)

A great mathematical literature has been devoted to the properties of the Wiener process.

It is not difficult to see that the Wiener process is a Markov process. This is contained in property (i) which asserts that the increments are independent of the increments in all prior time intervals. Thus assume that $x(t)$ has the known values $x_m, \ldots, x_0$ at the times $t_m < \cdots < t_0 < t$. Since $x(t) - x(t_0)$ is independent of all the previous increments, it must be independent of the individual values $x_m, \ldots, x_1$. This is just the verbal statement of the Markov property. Another important property is that the Wiener process $x(t) - x(0)$ is Gaussian. This follows from a generalization of the central limit theorem for the independent increments. The result can be derived by considering the increments $z_k = x(kt/n) - x((k-1)t/n)$ for $k = 1, \ldots, n$, which can be used to write $x(t) - x(0) = \sum_{k=1}^{n} z_k$. By property (i) the z_k are independent and by property (ii) they are identically distributed, just like the n coin tosses in the preceding section. Since this decomposition is possible for all n, the example of the $n \to \infty$ limit for coin tossing suggests that $x(t) - x(0)$ might be Gaussian.

In fact, a judicious application of the continuity requirement (iii) (see Breiman in the references) can be used to make this suggestion precise.

Because the initialized Wiener process, $x(t) - x(0)$, is Gaussian, it is fully characterized by its average and variance. These two moments are easily determined. For example,

$$a(t+\tau) = \langle x(t+\tau) - x(0)\rangle = \langle x(t+\tau) - x(\tau)\rangle + \langle x(\tau) - x(0)\rangle$$
$$= \langle x(t) - x(0)\rangle + \langle x(\tau) - x(0)\rangle = a(t) + a(\tau), \tag{1.4.7}$$

where the third equality is based on property (ii) and $a(t)$ is the indicated average. Since $a(t)$ is continuous by property (iii), the equality $a(t+\tau) = a(t) + a(\tau)$ implies that $a(t)$ is linear in t. Since $a(0) = 0$, it follows that

$$\langle x(t+\tau) - x(\tau)\rangle = \mu t, \tag{1.4.8}$$

where μ is a constant, called the drift constant. A similar argument for the variance of the increments in x gives

$$\langle (x(t+\tau) - x(\tau))^2\rangle = \sigma t, \tag{1.4.9}$$

where σ is also a constant, necessarily non-negative since the left-hand side of Eq. (1.4.9) is non-negative. Because both the average and variance depend on time, the Wiener process is not stationary.

Property (iii) for the Wiener process guarantees that the trajectories, or sample paths $x(t)$, are continuous functions of time. Actually, the result is not quite so strong but says only that one can always find a realization of a Wiener process that has continuous paths. Continuity means that the position variable is a continuous function of time and is a requirement for differentiability. However, continuity does not imply differentiability and the Wiener process possesses the perversity that its sample paths are *nowhere* differentiable. In other words, the velocity of a Wiener process is undefined. The proof of this result is technical. However, the basis for it is identical to what bothered Einstein about his result for Brownian motion, namely, the variance in Eq. (1.4.9) is linear in the time. Hence the process $[x(t+\tau) - x(t)]/\tau$ has a variance equal to σ/τ, which become unbounded as τ goes to zero.

The connection between the Wiener process and Einstein's model of Brownian motion is very close. In fact, the conditional probability for a three-dimensional Wiener process is the same as that which Einstein obtained using the diffusion equation (Eq. 1.4.2). To see the equivalence, we treat the three coordinates $x(t)$, $y(t)$, and $z(t)$ as independent Wiener processes. Thus they are Gaussian and uncorrelated with each other. To find $P_2(\mathbf{r}_0|\mathbf{r}, t)$, it suffices to know the averages and variances of the individual coordinates, conditioned on $x(0) = x_0$, etc. These are given by the results in Eqs. (1.4.8) and (1.4.9). Using the expression in Section 1.3 for the Gaussian density and assuming that σ is the same for the three coordinates gives

$$P_2(\mathbf{r}_0|\mathbf{r}, t) = (2\pi t\sigma)^{-3/2} \exp[-|\mathbf{r} - \boldsymbol{\mu}t - \mathbf{r}_0|^2/2\sigma t]. \tag{1.4.10}$$

where $\boldsymbol{\mu} = (\mu_x, \mu_y, \mu_z)$. This is identical to the result of Einstein if the drift vector $\boldsymbol{\mu}$ is set equal to zero and we identify

$$\sigma = 2D.$$

For this reason σ is sometimes referred to as the diffusion constant.

The drift term in Eq. (1.4.10) refers to a systematic velocity, i.e., a convective motion, superimposed on the diffusive character of Brownian motion. Such systematic terms often occur for physical processes, and they contribute separately to the differential equation that the conditional probability density satisfies. The form of this contribution is quite general and can be obtained by differentiating Eq. (1.4.10) with respect to the position variable twice. The simplest way to do this is to notice that $\mathbf{r} - \boldsymbol{\mu}t - \mathbf{r}_0 \equiv \boldsymbol{\delta}(t)$ satisfies Eq. (1.4.2) if we differentiate with respect to $\boldsymbol{\delta}$. Thus

$$(\partial P_2/\partial t)_\delta = D\nabla_{\boldsymbol{\delta}}^2 P_2, \tag{1.4.11}$$

where $\boldsymbol{\delta}$ is held constant on the left-hand side and the subscript $\boldsymbol{\delta}$ represents differentiation with respect to $\boldsymbol{\delta}$. To change variables back to $\mathbf{r}$ requires that we use the identities $\nabla_{\boldsymbol{\delta}} = \nabla_{\mathbf{r}}$ and $(\partial/\partial t)_{\mathbf{r}} = (\partial/\partial t)_{\boldsymbol{\delta}} - \boldsymbol{\mu}\cdot\nabla_{\mathbf{r}}$. Thus as a function of $\mathbf{r}$ the conditional density satisfies

$$\partial P_2/\partial t = -\boldsymbol{\mu}\cdot\nabla P_2 + D\nabla^2 P_2. \tag{1.4.12}$$

This equation is called the *Fokker–Planck*, or forward, equation for the Wiener process, including the drift term $-\boldsymbol{\mu}\cdot\nabla P_2$. The drift term represents a completely systematic velocity as can be seen by setting D equal to zero. In this case the solution in Eq. (1.4.10) reduces to the delta function

$$\lim_{D\to 0} P_2(\mathbf{r}_0|\mathbf{r}, t) = \delta(\mathbf{r} - \boldsymbol{\mu}t - \mathbf{r}_0), \tag{1.4.13}$$

which represents a sharp probability density centered at $\mathbf{r} = \mathbf{r}_0 + \boldsymbol{\mu}t$ that does not spread with time. This is just what one would expect for an ensemble of particles each moving with a fixed velocity $\boldsymbol{\mu}$.

The Fokker–Planck equation for the Wiener process can be written in a slightly different form, which helps reveal its basic structure. Since $\boldsymbol{\mu}$ and D are constant, they commute with the divergence operator. Thus Eq. (1.4.12) can be rewritten as

$$\partial P_2/\partial t = -\nabla\cdot(\boldsymbol{\mu}P_2 - D\nabla P_2) = -\nabla\cdot\mathbf{j}. \tag{1.4.14}$$

In this form the Fokker–Planck equation has the same appearance as the continuity equation for mass density with the flux of probability defined by

$$\mathbf{j} = \boldsymbol{\mu}P_2 - D\nabla P_2. \tag{1.4.15}$$

This flow of probability is entirely analogous to the flow of mass points in a fluid, which is the analogy originally used by Einstein. Smoluchowski extended this analogy by allowing the drift velocity to depend on position. For a mass point subject to a force, **F**, the terminal velocity in a fluid is given by $\mathbf{v} = \mathbf{F}/\zeta$, where ζ is the friction coefficient. If the force is caused by a potential field, like gravity, then it can be represented as $\mathbf{F} = -\nabla\phi$ and so $\mathbf{v} = -\nabla\phi/\zeta$. This expression for velocity can be further rewritten using the *Nernst–Einstein formula*

$$D = k_B T/\zeta, \tag{1.4.16}$$

where k_B is Boltzmann's constant. Using this relationship, the velocity of a mass point becomes

$$\mathbf{v} = -(D/k_B T)\nabla\phi. \tag{1.4.17}$$

Smoluchowski created a new stochastic process by identifying **v** with $\boldsymbol{\mu}$. Thus Smoluchowski wrote

$$\begin{aligned} \partial P_2/\partial t &= -\nabla\cdot\mathbf{j} \\ \mathbf{j} &= -D\left(\frac{P_2\nabla\phi}{k_B T} + \nabla P_2\right). \end{aligned} \tag{1.4.18}$$

In this form Eq. (1.4.18) is called the *Smoluchowski equation.* It differs from the Fokker–Planck equation for a Wiener process in that the drift term depends on position. The Smoluchowski equation provides one sort of contracted description of the position of a Brownian particle in a potential field.

1.5. The Langevin Equation and Stochastic Integrals

The Einstein picture of Brownian motion involves a contracted description in which the position of the particle is the only variable that is considered. This ignores the coupling between the velocity and position variable and leads to the anomalous result, quoted in Section 1.4, that the velocity of the particle is not defined. A simple way to remedy this situation is to use a less-contracted picture of the ensemble that includes the dynamic coupling of the velocity and position variables. This approach was taken by Langevin in 1907, shortly after Einstein's papers on Brownian motion had appeared. Langevin's approach was to consider the mechanical equations of motion for a Brownian particle of mass m. According to Newton they would be

$$d\mathbf{r}/dt = \mathbf{p}/m \tag{1.5.1}$$

$$d\mathbf{p}/dt = \mathbf{F}, \tag{1.5.2}$$

where **F** is the force acting on the particle and $\mathbf{p} = m\mathbf{v}$ is its momentum. Since the particle is contained in a fluid, Langevin argued that the force consisted of three parts. First, a systematic part caused by the viscous forces in the fluid that tends to damp the motion of the particle. Using a hydrodynamic model, this force can be written

$$\mathbf{F}_v = -\zeta \mathbf{v}, \tag{1.5.3}$$

where ζ is the friction constant. A second systematic part of the force is due to outside or external forces, $\mathbf{F}_e$. And finally, there will be forces due to the molecular nature of the surrounding fluid which are rapid and unpredictable. This is termed the random component of the force and is written $\tilde{\mathbf{f}}$, where the tilde reminds us that $\tilde{\mathbf{f}}$ is a random variable. Thus Langevin wrote

$$d\mathbf{r}/dt = \mathbf{p}/m \tag{1.5.4}$$

$$d\mathbf{p}/dt = -(\zeta/m)\mathbf{p} + \mathbf{F}_e + \tilde{\mathbf{f}}. \tag{1.5.5}$$

Notice that $\mathbf{r}$ and $\mathbf{p}$ are coupled and that it is in the momentum equation that the random force actually appears. Because it is anticipated that on the average the ensemble of particles exhibit a damping of their momentum to a terminal value, it is natural to assume that

$$\langle \tilde{\mathbf{f}}(t) \rangle \equiv \mathbf{0}. \tag{1.5.6}$$

This only partially specifies the properties of the random term and unless they are specified in complete detail, the stochastic processes $\mathbf{r}(t)$ and $\mathbf{v}(t)$ are not defined.

To complete the specification of the random force, we write it in terms of the increments that it produces in the momentum of the particle, i.e.,

$$d\mathbf{w} = \mathbf{w}(t + dt) - \mathbf{w}(t) \equiv \tilde{\mathbf{f}}(t)\, dt. \tag{1.5.7}$$

Since this force is caused by molecular impacts, it makes sense that it changes rapidly and randomly. Thus the increments in the momentum, $d\mathbf{w}$, should be uncorrelated, yet continuous, random variables. Further, these impulses originate in the fluid. If the fluid is in equilibrium, then the stochastic properties of the force should be independent of the time t. These properties are just the ones which we used to define a Wiener process in Eq. (1.4.6) and thus it seems natural to take $\mathbf{w}(t)$ to be a Wiener process. Unfortunately, Eq. (1.5.7) shows that $\tilde{\mathbf{f}}(t)$ would be the derivative of a Wiener process, which we know does not exist!

To get around this difficulty the Langevin equation (1.5.5) needs to be written in an equivalent, but mathematically well-defined, fashion. For example, in terms of differentials one has

$$d\mathbf{p} = -(\zeta/m)\mathbf{p}\, dt + \mathbf{F}_e\, dt + d\mathbf{w}, \tag{1.5.8}$$

where we have used Eq. (1.5.7). In this form the incremental changes in momentum $d\mathbf{p}$ are defined in terms of the value of $\mathbf{p}$ and $\mathbf{F}_e$ at t and the well-defined increments of the Wiener processes. To obtain a solution to this differential expression, each side must be integrated. Here, again, there is a difficulty. Since $\mathbf{w}(t)$ is nowhere differentiable, it is not of bounded variation. Thus the Stieltjes integral, $\int d\mathbf{w}$, on the right-hand side, does not make sense.

There is a clever way out of this dilemma. To see what it is, we return to the Langevin equation. If we assume for the time being that $\mathbf{F}_e$ is zero, Eqs. (1.5.4) and (1.5.5) uncouple, giving

$$d\mathbf{p}/dt = -(\zeta/m)\mathbf{p} + \tilde{\mathbf{f}}(t). \tag{1.5.9}$$

Using the integrating factor $\exp(t\zeta/m)$, this equation can be solved, formally, to give

$$\mathbf{p}(t) = \mathbf{p}(0) + \int_0^t \exp[-\alpha(t-\tau)]\tilde{\mathbf{f}}(\tau)\,d\tau, \tag{1.5.10}$$

where $\alpha = \zeta/m$. If we write $\tilde{\mathbf{f}}(\tau)\,d\tau = d\mathbf{w}$, we can get rid of the Stieltjes integral involving $d\mathbf{w}$ if we integrate by parts. This gives

$$\int_0^t \exp[-\alpha(t-\tau)]\,d\mathbf{w} = \mathbf{w}(t) - \alpha\int_0^t \exp[-\alpha(t-\tau)]\mathbf{w}(\tau)\,d\tau, \tag{1.5.11}$$

where we have assumed that $\mathbf{w}(0) \equiv \mathbf{0}$.

The right-hand side of this equation is well defined, since $\mathbf{w}(\tau)$ is a continuous function. Because of this we can use the right-hand side of Eq. (1.5.11) as the *definition* of the integral when $\mathbf{w}$ is a Wiener process. Thus the momentum of the Brownian particle as defined in Langevin's approach is

$$\mathbf{p}(t) = \mathbf{p}(0) + \mathbf{w}(t) - \alpha\int_0^t \exp[-\alpha(t-\tau)]\mathbf{w}(\tau)\,d\tau. \tag{1.5.12}$$

Equation (1.5.12) is the solution to the Langevin equation (1.5.9) when $\tilde{\mathbf{f}}(t)\,dt$ is interpreted as the increment of a Wiener process. Notice that the momentum at time t is given completely in terms of the values of $\mathbf{p}(0)$ and $\mathbf{w}(\tau)$ for $0 \le \tau \le t$. Since $w(\tau)$ is a Gaussian stochastic process, then so is $\mathbf{p}(t) - \mathbf{p}(0)$ since the sum and integral of a Gaussian process are also Gaussian. Moreover, since $\mathbf{p}(t)$ depends only on $\mathbf{p}(0)$ and not the value of $\mathbf{p}$ at earlier times, it is a Markov process.

The integral in Eq. (1.5.12) is an example of a stochastic integral, namely, it is the integral of a random variable and produces a new random variable. For many purposes the Wiener process is sufficiently rich that integrals defined from it will have interesting physical applications. These processes provide a natural mathematical generalization of the Wiener process. Consider, for example, a vector stochastic process with increments defined by

$$d\mathbf{n} = \mathbf{h}(\mathbf{n}, t)\,dt + g(\mathbf{n}, t)\,d\mathbf{w}, \tag{1.5.13}$$

where $\mathbf{w}$ is a vector-valued Wiener process with $\sigma_i = 1$ and $\mathbf{w}(0) = \mathbf{0}$, and $\mathbf{h}$ is a vector function of $\mathbf{n}$ and t and g a matrix function of $\mathbf{n}$ and t. If one can find a reasonable definition of a stochastic integral, Eq. (1.5.13) can be thought of as the integral equation

$$\mathbf{n}(t) = \mathbf{n}(t_0) + \int_{t_0}^t \mathbf{h}(\mathbf{n}, t)\,dt + \int_{t_0}^t g(\mathbf{n}, t)\,d\mathbf{w}. \tag{1.5.14}$$

The trick, of course, is finding an acceptable definition of a Stieltjes-like integral over a function of unbounded variation—just as we had to for the Langevin equation. The difficulty was not as severe in that case, however, since $\mathbf{h}$ was a linear function and $g(\mathbf{n})$ was independent of $\mathbf{n}$. As it turns out, two choices for the meaning of the stochastic integral are in common use, and they produce entirely different, although related, stochastic processes for $\mathbf{n}(t)$. The choice suggested by Itô has been popular with mathematicians because the stochastic integral has the martingale property. This property is a technical, but useful, one which relates conditional averages to the prior history of the stochastic process. For the time being we adopt the Itô definition of the integral, which is described in more detail in Section 4.8.

Using Itô's interpretation of the stochastic integral, it is possible to show that the stochastic process defined by Eqs. (1.5.13) and (1.5.14) is a Markov process. Furthermore, it can be shown under appropriate conditions of differentiability on $\mathbf{h}$ and g that the process $\mathbf{n}$ has a probability density which solves a Fokker–Planck type equation. That equation has a form analogous to Eq. (1.4.14), namely,

$$\begin{aligned}\partial P_2(\mathbf{n}_1, t_1 | \mathbf{n}, t)/\partial t = & -\partial h_i(\mathbf{n}, t) P_2(\mathbf{n}_1, t_1 | \mathbf{n}, t)/\partial n_i \\ & + \frac{1}{2}\partial^2 g_{ik}(\mathbf{n}, t) g_{kj}(\mathbf{n}, t) P_2(\mathbf{n}_1, t_1 | \mathbf{n}, t)/\partial n_i \partial n_j, \end{aligned} \tag{1.5.15}$$

where the summation convention on repeated indices has been used. The Fokker–Planck equation is to be solved with the usual initial condition $P_2(\mathbf{n}_1 | \mathbf{n}, 0) = \delta(\mathbf{n} - \mathbf{n}_1)$ on the conditional density. If one is willing to accept the validity of Eq. (1.5.15), it is not necessary to proceed further into the details of the development of Itô stochastic integral. Instead, it suffices to solve Eq. (1.5.15) for the conditional density, which will then describe a particular Markov process. As we shall see in subsequent chapters, only the special case when $\mathbf{h}$ is linear in $\mathbf{n}$ and g is independent of $\mathbf{n}$ will be of interest for physical ensembles, i.e., in the thermodynamic limit. For such restricted functions, the definition of the stochastic integral using the trick of integrating by parts, as illustrated in Eq. (1.5.11), is equivalent to Itô's definition.

There is another way of generating the same class of stochastic processes that Itô has produced with his stochastic integral. These processes are defined in terms of their transition probabilities, $P_2(\mathbf{n}_1, t_1 | \mathbf{n}, t)$, and are defined by a Fokker–Planck equation rather than a stochastic differential equation. These processes are called *stochastic diffusions*, and they provide an independent, but equivalent, generalization of the Wiener process. A stochastic process is called a diffusion process if it is a Markov process and the following limits exist

$$\lim_{\tau \to t} \frac{1}{\tau - t} \int P_2(\mathbf{n}, t | \mathbf{n}', \tau)(\mathbf{n}' - \mathbf{n})\, d\mathbf{n}' = \mathbf{h}(\mathbf{n}, t) \tag{1.5.16}$$

$$\lim_{\tau \to t} \frac{1}{\tau - t} \int P_2(\mathbf{n}, t | \mathbf{n}', \tau)(\mathbf{n}' - \mathbf{n})(\mathbf{n}' - \mathbf{n})^T\, d\mathbf{n}' = \gamma(\mathbf{n}, t) \tag{1.5.17}$$

with $\mathbf{h}(\mathbf{n}, t)$ a once differentiable vector-valued function of $\mathbf{n}$ and $\gamma(\mathbf{n}, t)$ a twice differentiable matrix-valued function. The moments in Eqs. (1.5.16) and (1.5.17) are called *transition moments* since they measure the average increment of change of $\mathbf{n}$ from its value at t. In addition to the existence of these limits, it is also necessary that the higher-order transition moments vanish in this limit, i.e., for $k > 2$

$$\lim_{\tau \to t} \frac{1}{\tau - t} \int P_2(\mathbf{n}, t | \mathbf{n}', \tau) \prod_{l=1}^{k} (n'_{i_l} - n_{i_l})\, d\mathbf{n}' = 0. \tag{1.5.18}$$

Equations (1.5.16) and (1.5.17) require that for short intervals of time both the average and average square of the incremental change in $\mathbf{n}$ be proportional to the elapsed time. Furthermore, the higher transition moments must increase with some higher power of the lapsed time.

The Wiener process is Markovian and satisfies Eqs. (1.5.16) and (1.5.17) as can be seen by referring to Eqs. (1.4.8) and (1.4.9) which define the moments of the increments $x(t + \tau) - x(\tau)$. Moreover, the Wiener process is a Gaussian process so that higher-order odd moments vanish and even moments are of order $2n$ in the elapsed time. This means that Eq. (1.5.18) is also satisfied. Thus a Wiener process is one type of stochastic diffusion. The generalization of the Wiener process that is provided by stochastic diffusions is that the *drift* term, $\mathbf{h}(\mathbf{n}, t)$, and the *diffusion* term, $\gamma(\mathbf{n}, t)$, no longer need to be constant. Thus stochastic diffusions act like a Wiener process, but only locally.

To derive the form of the Fokker–Planck equation satisfied by stochastic diffusions, we need to use the Chapman–Kolmogorov equation. This identity is satisfied by all Markov processes and can be derived from the integral formula Eq. (1.3.4) connecting two-time and three-time probabilities densities, namely

$$W_2(\mathbf{n}_1, t_1; \mathbf{n}_3, t_3) = \int W_3(\mathbf{n}_1, t_1; \mathbf{n}_2, t_2; \mathbf{n}_3, t_3)\, d\mathbf{n}_2. \tag{1.5.19}$$

For a Markov process, Eqs. (1.3.19) and (1.3.26) can be used to rewrite Eq. (1.5.19) as

$$P_2(\mathbf{n}_1, t_1 | \mathbf{n}_3, t_3) = \int P_2(\mathbf{n}_1, t_1 | \mathbf{n}_2, t_2) P_2(\mathbf{n}_2, t_2 | \mathbf{n}_3, t_3)\, d\mathbf{n}_2, \tag{1.5.20}$$

which is the *Chapman–Kolmogorov equation*, occasionally called the *Smoluchowski equation.* To derive the Fokker–Planck equation we look at the integral

$$\lim_{\delta \to 0} \frac{1}{\delta} \int f(\mathbf{n}) [P_2(\mathbf{n}_1, t_1 | \mathbf{n}, t + \delta) - P_2(\mathbf{n}_1, t_1 | \mathbf{n}, t)]\, d\mathbf{n}$$
$$= \int f(\mathbf{n}) [\partial P_2(\mathbf{n}_1, t_1 | \mathbf{n}, t)/\partial t]\, d\mathbf{n}, \tag{1.5.21}$$

where $f(\mathbf{n})$ is any analytic function which vanishes, along with its derivatives, at plus and minus infinity. The left-hand side of this equation can be rewritten, with the help of the Chapman–Kolmogorov identity, as

$$\lim_{\delta\to 0}\frac{1}{\delta}\left[\int f(\mathbf{n})\int P_2(\mathbf{n}_1,t_1|\mathbf{n}_2,t)P_2(\mathbf{n}_2,t|\mathbf{n},t+\delta)\,d\mathbf{n}_2\,d\mathbf{n}\right.$$
$$\left.-\int f(\mathbf{n})P_2(\mathbf{n}_1,t_1|\mathbf{n},t)\,d\mathbf{n}\right]. \tag{1.5.22}$$

In the first integral we expand $f(\mathbf{n})$ in a Taylor series around $\mathbf{n}_2$ and integrate first over $\mathbf{n}$. Since $\mathbf{n}(t)$ is a stochastic diffusion process, Eqs. (1.5.16)–(1.5.18) show that only the first three terms in the expansion need be retained. The first of these cancels the second integral in Eq. (1.5.22). Using Eqs. (1.5.16) and (1.5.17), the remaining terms can be written

$$\int P_2(\mathbf{n}_1,t|\mathbf{n},t)\left[(\partial f/\partial \mathbf{n}_i)h_i(\mathbf{n},t)+\frac{1}{2}(\partial^2 f/\partial n_i\,\partial n_j)\gamma_{ij}(\mathbf{n},t)\right]d\mathbf{n}, \tag{1.5.23}$$

where summation on repeated indices is understood. Integrating both terms in this expression by parts, the first once and the second twice, and combining with Eq. (1.5.21) then gives

$$\int f(\mathbf{n})\left[(\partial P_2/\partial t)+(\partial h_i P_2/\partial n_i)-\frac{1}{2}\partial^2\gamma_{ij}P_2/\partial n_i\,\partial n_j\right]d\mathbf{n}=0. \tag{1.5.24}$$

Since $f(\mathbf{n})$ is arbitrary, the only way that Eq. (1.5.24) can hold true is if

$$\partial P_2(\mathbf{n}_1,t_1|\mathbf{n},t)/\partial t = -\partial h_i(\mathbf{n},t)P_2(\mathbf{n}_1,t_1|\mathbf{n},t)/\partial n_i$$
$$+\frac{1}{2}\partial^2\gamma_{ij}(\mathbf{n},t)P_2(\mathbf{n}_1,t_1|\mathbf{n},t)/\partial n_i\,\partial n_j. \tag{1.5.25}$$

This is the Fokker–Planck equation satisfied by the diffusion process $\mathbf{n}(t)$. For the Wiener process, $x(t)$, with no drift, we recall that $h_i=\mu_i=0$ and $\gamma_{ij}=\sigma\delta_{ij}=2D\delta_{ij}$. Thus, as expected, the Fokker–Planck equation for the Wiener process is identical to Einstein's diffusion equation (1.4.2).

This Fokker–Planck equation should be compared with the Fokker–Planck equation (1.5.15) satisfied by Itô's stochastic differential process. Identifying γ_{ij} with $g_{ik}g_{kj}$, they are seen, in fact, to be identical. Hence these two generalizations of the Wiener process are equivalent, and they can be used interchangeably.

1.6. White Noise

There is another way of giving a meaning to the Langevin equation which yields results that are equivalent to the interpretation in terms of stochastic integrals. Let us begin again with the Langevin equation in the absence of external forces, i.e.,

$$d\mathbf{p}/dt = -(\zeta/m)\mathbf{p} + \tilde{\mathbf{f}}(t). \tag{1.6.1}$$

This time we want to look for a well-defined random force for which in some limit $\tilde{\mathbf{f}}\,dt$ will mimic the increments of a Wiener process. Since $d\mathbf{w}$ is Gaussian, we need $\tilde{\mathbf{f}}$ to be Gaussian and, as before, to have average value zero. To correspond to the independence of increments, we choose a two-time correlation function of the form

$$\langle \tilde{f}_i(t)\tilde{f}_j(t')\rangle = (1/2\tau_c)\gamma_{ij}\exp(-|t-t'|/\tau_c). \tag{1.6.2}$$

This means that $\tilde{f}_i$ and $\tilde{f}_j$ are uncorrelated for time separations much larger than τ_c. This becomes even more reminiscent of the correlation function for the increments of a Wiener process if we consider τ_c to be small. Using $dw_i = w_i(t+dt) - w_i(t) \equiv \tilde{f}_i(t)\,dt$ and taking $dt' \approx \tau_c$, Eq. (1.6.2) can be written

$$\langle dw_i(t)\,dw_j(t')\rangle \approx (\gamma_{ij}/2)\exp(-|t'-t|/\tau_c)\,dt. \tag{1.6.3}$$

The right-hand side of Eq. (1.6.3) vanishes for $|t'-t| \gg \tau_c$ and has the value $(\gamma_{ij}/2)\,dt$ for $|t'-t| \ll \tau_c$, just as do the increments of a Wiener process. In the limit that the correlation time τ_c approaches zero, Eq. (1.6.3) suggests that the $\tilde{f}_i$ approach the correct formula for increments of a Wiener process. In this limit one has, in fact,

$$\langle \tilde{f}_i(t)\tilde{f}_j(t')\rangle = \gamma_{ij}\delta(t'-t), \tag{1.6.4}$$

where δ is the Dirac delta function. The resulting stochastic process is called *white noise.* White noise is not an ordinary stochastic process but a generalized one since its time correlation function is a distribution or generalized function. The infinite spike at equal times for the white noise correlation function corresponds to the lack of differentiability of the Wiener process. By expanding the definition of stochastic processes to include generalized sample functions, the Langevin equation takes on a well-defined operational meaning.

White noise considered as a function of its time argument is a Gaussian process, just like the Wiener process. This means that all of its higher moments are defined in terms of the covariance matrix, namely, the two-time correlation function given in Eq. (1.6.4). Thus consider any series of time points $t_1, t_2, \ldots, t_n, \ldots$ and the random variables $\tilde{\mathbf{f}}(t_1), \tilde{\mathbf{f}}(t_2), \ldots, \tilde{\mathbf{f}}(t_n), \ldots$, each of which vanishes on the average. Because the covariance of any two depends only on the time difference, white noise is a stationary, Gaussian process. Since it is Gaussian, its higher odd moments vanish, i.e.,

$$\langle \tilde{f}_{i_l}(t_l)\ldots\tilde{f}_{i_k}(t_k)\rangle = 0 \tag{1.6.5}$$

for any $2m+1$ times $t_l, \ldots, t_k$. Similarly, for any $2m$ times $t_l, \ldots, t_k$ the even moments satisfy

$$\langle \tilde{f}_{i_l}(t_l)\ldots\tilde{f}_{i_k}(t_k)\rangle = \sum_{\text{all pairs}} \langle \tilde{f}_{i_l}(t_l)\tilde{f}_{i_{l'}}(t_{l'})\rangle\cdots\langle \tilde{f}_{i_k}(t_k)\tilde{f}_{i_{k'}}(t_{k'})\rangle, \tag{1.6.6}$$

where the sum is over all distinct pairs of the $2m$ times.

The reason that white noise is called "white" has to do with the frequency spectrum which makes up its Fourier transform. For each stochastic process defined on the infinite interval $-\infty < t < +\infty$, there corresponds the Fourier transform, which is a random function of frequency, defined by

$$\hat{\mathbf{n}}(\omega) \equiv \frac{1}{2\pi} \int_{-\infty}^{+\infty} \mathbf{n}(t) \exp(i\omega t)\, dt. \tag{1.6.7}$$

Although the inversion formula is valid only for certain well-behaved functions, we will write it in a formal sense as

$$\mathbf{n}(t) = \int_{-\infty}^{+\infty} \hat{\mathbf{n}}(\omega) \exp(-i\omega t)\, d\omega. \tag{1.6.8}$$

Even for generalized functions these formulas can be made meaningful because they involve the integral of the generalized function. For white noise we have

$$\hat{\tilde{\mathbf{f}}}(\omega) = \frac{1}{2\pi} \int_{-\infty}^{+\infty} \tilde{\mathbf{f}}(t) \exp(i\omega t)\, dt. \tag{1.6.9}$$

Since $\hat{\tilde{\mathbf{f}}}$ is a weighted integral over Gaussian processes, it is a Gaussian process, too. By interchanging the operation of averaging and integration, its average value, like that of $\tilde{\mathbf{f}}(t)$, is zero for all frequencies. Its covariance—or frequency–frequency correlation function—is also easy to calculate. By definition

$$\begin{aligned}
\langle \hat{\tilde{\mathbf{f}}}(\omega)\hat{\tilde{\mathbf{f}}}^T(\omega')\rangle &= \left(\frac{1}{2\pi}\right)^2 \int_{-\infty}^{+\infty} dt \int_{-\infty}^{+\infty} dt'\, e^{i(\omega t + \omega' t')} \langle \tilde{\mathbf{f}}(t)\tilde{\mathbf{f}}^T(t')\rangle \\
&= \left(\frac{1}{2\pi}\right)^2 \int_{-\infty}^{+\infty} dt \int_{-\infty}^{+\infty} dt'\, e^{i(\omega t + \omega' t')} \gamma\delta(t' - t) \\
&= \frac{\gamma}{(2\pi)^2} \int_{-\infty}^{+\infty} dt\, e^{i(\omega + \omega')t} \\
&= \gamma\delta(\omega + \omega')/2\pi,
\end{aligned} \tag{1.6.10}$$

where the final step uses the representation of the delta function

$$\delta(\omega + \omega') = \frac{1}{2\pi} \int_{-\infty}^{+\infty} dt'\, e^{i(\omega + \omega')t}. \tag{1.6.11}$$

Since $\hat{\tilde{\mathbf{f}}}(\omega)$ is Gaussian, it is completely described by its average and covariance.

The attribution of whiteness to white noise comes about by the inverse transformation. For any stationary stochastic process the *power spectrum* is defined as the function $S(\omega)$ in the expression

$$\langle \mathbf{n}(t)\mathbf{n}^T(t)\rangle = \int_{-\infty}^{+\infty} S(\omega)\, d\omega/2\pi, \tag{1.6.12}$$

which forms the equal time variance matrix. Thus it gives a measure of the contribution of various frequencies to the mean square value of n. For noise in electrical currents, in fact, $S(\omega)$ is proportional through the resistance to the electrical power contained in the fluctuations of frequency ω, whence the name "power spectrum." For white noise it is straightforward, using Eq. (1.6.10) and the inversion formula Eq. (1.6.8), to calculate that

$$\langle \tilde{\mathbf{f}}(t)\tilde{\mathbf{f}}^T(t)\rangle = \int_{-\infty}^{+\infty} \gamma \, d\omega/2\pi. \tag{1.6.13}$$

Thus for white noise $\gamma = S(\omega)$ and so the power spectrum is the same for all frequencies. By analogy to the fact that idealized white light consists of an equal mixing of all frequencies of visible light, a power spectrum which is the same for all frequencies is called white—hence the name white noise.

1.7. Solution of the Langevin Equation

Having characterized the random force in the Langevin equation as white noise, we can return to the question of how an ensemble of Brownian particles evolves in time in the Langevin picture. In the absence of external force fields, the momentum satisfies

$$d\mathbf{p}/dt = -(\zeta/m)\mathbf{p} + \tilde{\mathbf{f}}(t). \tag{1.7.1}$$

For any given sample function—i.e., any member of the ensemble—this equation can be formally integrated since it is first order in the time, linear, and has the inhomogeneous term $\tilde{\mathbf{f}}(t)$. Using the integrating factor $\exp(\zeta t/m)$ and integrating between t' and t then gives

$$\mathbf{p}(t) = \exp[-(t - t')\zeta/m]\mathbf{p}(t') + \int_{t'}^{t} \exp[-(t - \tau)\zeta/m]\tilde{\mathbf{f}}(\tau)\, d\tau. \tag{1.7.2}$$

Thus the value of the momentum at time t depends only on the value of $\mathbf{p}(t')$ and the values of $\tilde{\mathbf{f}}(\tau)$ for $t' \leq \tau \leq t$. Clearly, to specify $\mathbf{p}(t)$ unambiguously, the stochastic process $\mathbf{p}(t')$ at the earlier time must be specified as well as the statistical relationship between $\mathbf{p}(t')$ and $\tilde{\mathbf{f}}(\tau)$.

Fortunately, the nature of the physical problem can help us out with this. We have imagined the ensemble of Brownian particles to be immersed in a fluid that is itself at thermal equilibrium. If we wait long enough, that is, we let the ensemble of Brownian particles age, they too will achieve a stationary distribution corresponding to equilibrium. If we consider only this stationary ensemble, then $\mathbf{p}(t')$ must be distributed with its equilibrium Maxwell distribution. Thus

$$W_1(\mathbf{p}) = (2\pi m k_B T)^{-3/2} \exp(-p^2/2mk_B T) \tag{1.7.3}$$

is the single-time probability density. For the complete specification of the stochastic process, the relationship between $\mathbf{p}(t')$ and $\tilde{\mathbf{f}}(\tau)$ must also be given. Since the random force is caused by the molecules of the fluid and varies on a rapid time scale, it is natural to assume for $\tau > t'$ that the force is independent of $\mathbf{p}(t')$. Explicitly, it will be assumed that

$$\langle \mathbf{p}(t')\tilde{\mathbf{f}}^T(\tau)\rangle \equiv 0 \tag{1.7.4}$$

for $t' < \tau$. In fact, this allows us to calculate the two-time correlation function. Combining Eqs. (1.7.2) and (1.7.4) gives

$$\langle \mathbf{p}(t)\mathbf{p}^T(t')\rangle = \exp[-(t - t')\zeta/m]\langle \mathbf{p}(t')\mathbf{p}^T(t')\rangle \tag{1.7.5}$$

where the independence of $\mathbf{p}(t')$ and $\tilde{\mathbf{f}}(\tau)$ eliminates the terms involving the integrals. The equal-time correlation function can be gotten from the covariance of the Maxwell distribution. Thus

$$\langle \mathbf{p}(t)\mathbf{p}^T(t')\rangle = \exp[-|t - t'|\zeta/m]mk_BTI \tag{1.7.6}$$

where I is the identity matrix.

The assumptions that we have made about the independence of the random force and the momentum allow us to prove that $\mathbf{p}(t)$ is also a Markov process. To show this we only need to demonstrate that if $\mathbf{p}(t')$ has a known value, then $\mathbf{p}(t)$ is independent of the values of $\mathbf{p}(s)$ for $s < t'$. This is almost obvious from Eq. (1.7.2) since it involves only $\mathbf{p}(t')$ and $\tilde{\mathbf{f}}(\tau)$ for $\tau > t' > s$. In fact, if $\mathbf{p}(t')$ is known precisely, then it follows from Eq. (1.7.2) that

$$\begin{aligned}\langle \mathbf{p}(t)\mathbf{p}^T(s)\rangle &= \exp[-(t - t')\zeta/m]\mathbf{p}(t')\langle \mathbf{p}^T(s)\rangle \\ &\quad + \int_{t'}^{t} \exp[-(t - \tau)\zeta/m]\langle \mathbf{f}(\tau)\mathbf{p}^T(s)\rangle \, d\tau, \\ &= 0\end{aligned} \tag{1.7.7}$$

since the single-time average of $\mathbf{p}$ is zero and $\mathbf{p}$ and $\tilde{\mathbf{f}}$ are uncorrelated for $s < \tau$. Thus $\mathbf{p}(t)$ is a stationary, Gaussian, Markov process.

Since $\mathbf{p}(t)$ is Gaussian and Markov, we need only specify its average and covariance in a conditional ensemble to finish its characterization. Since the ensemble is stationary, we are free to set the time t_1 equal to zero in Eq. (1.7.2). Then assuming that $\mathbf{p}(0)$ is known precisely to be $\mathbf{p}^0$, the conditional average is

$$\langle \mathbf{p}(t)\rangle^0 = \exp(-t\zeta/m)\mathbf{p}^0. \tag{1.7.8}$$

This is an exponential relaxation with the relaxation time $\tau_R = m/\zeta$. Similarly, we may deduce the conditional covariance of the fluctuation $\delta\mathbf{p}(t) = \mathbf{p}(t) - \langle \mathbf{p}(t)\rangle^0$ using

$$\begin{aligned}\langle \delta\mathbf{p}(t)\delta\mathbf{p}^T(t)\rangle^0 &= \int_0^t d\tau \int_0^t d\tau' \, e^{-(t-\tau)\zeta/m}e^{-(t-\tau')\zeta/m}\langle \tilde{\mathbf{f}}(\tau)\tilde{\mathbf{f}}^T(\tau')\rangle \\ &= \gamma e^{-2t\zeta/m} \int_0^t d\tau \int_0^t d\tau' \, e^{(\tau+\tau')\zeta/m}\delta(\tau - \tau')\end{aligned}$$

$$= \gamma e^{-2t\zeta/m} \int_0^t d\tau\, e^{2\tau\zeta/m}$$

$$= (\gamma m/2\zeta)(1 - e^{-2t\zeta/m}) \equiv \sigma(1 - e^{-2t\zeta/m}) \tag{1.7.9}$$

Putting Eqs. (1.7.8) and (1.7.9) together with the form for a Gaussian density in Eq. (1.3.28) shows that the conditional density of $\mathbf{p}$ is

$$P_2(\mathbf{p}^0 | \mathbf{p}, t) = [(2\pi)^3 \det \sigma(1 - e^{-2t\zeta/m})]^{-1/2}$$
$$\cdot \exp\{-(\mathbf{p} - e^{-t\zeta/m}\mathbf{p}^0)^T [\sigma(1 - e^{-2t\zeta/m})]^{-1} (\mathbf{p} - e^{-t\zeta/m}\mathbf{p}^0)/2\} \tag{1.7.10}$$

Together with Eq. (1.7.3) for the single-time probability density this completely defines the stochastic processes for the momentum.

So far we have not used an explicit expression for the power spectrum, γ_{ij}, of the white noise used to describe Brownian motion. Although it may appear that the matrix is arbitrary except for being positive definite, its value is actually implicit in the results we have already obtained. To see this, we need to use a general result that pertains to stationary processes for which the conditional density becomes independent of its initial condition asymptotically in time. For such processes it follows that

$$\lim_{t\to\infty} P_2(\mathbf{n}^0 | \mathbf{n}, t) = W_1(\mathbf{n}). \tag{1.7.11}$$

To prove this, we write Eq. (1.3.20) in a form appropriate for a stationary ensemble as

$$W_1(\mathbf{n}) = \int W_1(\mathbf{n}^0) P_2(\mathbf{n}^0 | \mathbf{n}, t)\, d\mathbf{n}^0. \tag{1.7.12}$$

Since this is true for all $t \geq 0$, it is true in the limit that $t \to \infty$. Thus

$$W_1(\mathbf{n}) = \int W_1(\mathbf{n}^0) \lim_{t\to\infty} P_2(\mathbf{n}^0 | \mathbf{n}, t)\, d\mathbf{n}^0.$$

But if P_2 is independent of $\mathbf{n}^0$ in this limit, it may be taken outside the integral, and since W_1 is normalized, one is left with Eq. (1.7.11).

To make use of this result we return to Eq. (1.7.10), which shows that as $t \to \infty$ the conditional density for the momentum becomes independent of $\mathbf{p}^0$. Hence, knowing that the stationary distribution is the Maxwell distribution, it follows that

$$\lim_{t\to\infty} P_2(\mathbf{p}^0 | \mathbf{p}, t) = (2\pi m k_B T)^{-3/2} \exp(-p^2/2mk_B T). \tag{1.7.13}$$

This limit is easily carried out by inspection of Eq. (1.7.10) and leads to the identity

$$\sigma - mk_B T I = \langle \mathbf{p}(t)\mathbf{p}^T(t) \rangle \tag{1.7.14}$$

However, σ was defined in Eq. (1.7.9) in terms of γ. This leads to the desired expression for γ, namely,

$$\gamma = 2k_B T\zeta I. \tag{1.7.15}$$

This equation relates the strength of the random force γ to the dissipative friction coefficient and is an example of the *fluctuation–dissipation theorem* for an equilibrium ensemble. Because of this consistency condition, the value of γ is completely fixed in the Langevin theory by the dissipative parameters.

We close this section by returning to the problem that brought us from the Einstein picture of Brownian motion to the more complete Langevin picture, namely, that the mean square distance traveled in the Einstein picture is proportional to the elapsed time. In the Langevin picture the velocity is given by Eq. (1.5.4) as $\mathbf{v}(t) = \mathbf{p}(t)/m$. To compare with the Einstein results, we look at Brownian particles located at $\mathbf{r}_0$ at $t = 0$ and use $\mathbf{r}(t) - \mathbf{r}_0 = \int_0^t [\mathbf{p}(\tau)/m]\, d\tau$ to calculate

$$\langle |\mathbf{r}(t) - \mathbf{r}_0|^2 \rangle = \int_0^t d\tau \int_0^t d\tau' \langle \mathbf{p}(\tau)\cdot\mathbf{p}(\tau')\rangle/m^2. \tag{1.7.16}$$

The integrand can be obtained from the two-time correlation function which was found in Eq. (1.7.6). Thus

$$\langle |\mathbf{r}(t) - \mathbf{r}_0|^2 \rangle = 3\int_0^t d\tau \int_0^t d\tau' \exp[-|\tau - \tau'|\zeta/m](k_B T/m). \tag{1.7.17}$$

This integral can be carried out as a two-dimensional integral by integrating separately over the two triangles shown in Fig. 1.2, each of which yields the same value. This gives

$$\langle |\mathbf{r}(t) - \mathbf{r}_0|^2 \rangle = 6\int_0^t d\tau \int_0^\tau d\tau' \exp[-(\tau - \tau')\zeta/m](k_B T/m). \tag{1.7.18}$$

In this form the two integrals are easily performed, with the result that

$$\langle |\mathbf{r}(t) - \mathbf{r}_0|^2 \rangle = 6(k_B T/\zeta)[t + (e^{-t\zeta/m} - 1)m/\zeta]. \tag{1.7.19}$$

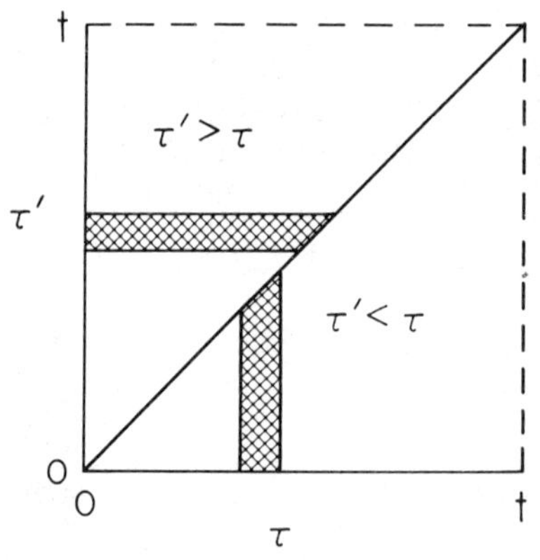

FIGURE 1.2. The domain of integration for the two-dimensional intergral in Eq. (1.7.17). In the upper triangle the intergrand is $\exp[-(\tau' - \tau)\zeta/m]$ and in the lower triangle it is $\exp[-(\tau - \tau')\zeta/m]$. By integrating over the cross-hatched strips in each triangle and then integrating the remaining variable from 0 to t, the contributions from the two triangles are seen to be equal.

Recalling the Nernst–Einstein formula in Eq. (1.4.16), Eq. (1.7.19) can be rewritten as

$$\langle|\mathbf{r}(t) - \mathbf{r}_0|^2\rangle = 6D[t + (e^{-t\zeta/m} - 1)m/\zeta] \qquad (1.7.20)$$

where D is the diffusion constant of the particles.

At short times the mean square distance traveled in the Langevin picture can be obtained from Eq. (1.7.20) by expanding the exponential. The lowest-order nonvanishing term is

$$\langle|\mathbf{r}(t) - \mathbf{r}_0|^2\rangle = \frac{3k_B T}{2m} t^2,$$

a result proportional not to t but to t^2. This means that at short times the root mean square velocity is simply

$$d\langle|\mathbf{r}(t) - \mathbf{r}_0|^2\rangle^{1/2}/dt = (3k_B T/2m)^{1/2}, \qquad (1.7.21)$$

which is the value one would also predict using kinetic theory arguments. Thus the Einstein divergence has been eliminated. Moreover, for times much longer than the relaxation time for the average momentum, $\tau_R = m/\zeta$, Eq. (1.7.20) becomes

$$\langle|\mathbf{r}(t) - \mathbf{r}_0|^2\rangle = 6Dt,$$

just as obtained by Einstein. Thus the Langevin picture of Brownian motion subsumes the Einstein picture. It gives a more complete description of the random motion of small particles because it is less contracted and, therefore, includes more of the relevant molecular processes which affect them. In this sense, Langevin's description lies at a more fundamental level than Einstein's in a hierarchy of descriptions of Brownian motion. This hierarchical point of view is implicit in the use of ensemble to describe observable processes. The implications of these hierarchies will be developed in subsequent chapters.

1.8. Ornstein–Uhlenbeck Processes

Langevin's initial treatment of Brownian motion was not nearly so detailed as the one outlined in the previous section. In fact, it was many years later that Ornstein and Uhlenbeck elaborated the Langevin picture more or less as presented in Section 1.7. There is a nice mathematical generalization of the Ornstein–Uhlenbeck treatment of Langevin's equation, which yields a simple class of stochastic processes that are stationary, Gaussian, and Markovian. In the mathematical literature they are called Ornstein–Uhlenbeck processes.

An *Ornstein–Uhlenbeck process* is generated by a linear stochastic differential equation which includes a white noise term. Consider a stationary ensemble characterized by k random variables $\mathbf{a}(t)$. It is general enough to assume that the single-time average of $\mathbf{a}$ vanishes, since otherwise we can define new variables $\mathbf{a}(t) - \langle \mathbf{a}(t) \rangle$ which have this property. The variables $\mathbf{a}(t)$ satisfy a Langevin-like equation

$$d\mathbf{a}/dt = H\mathbf{a} + \tilde{\mathbf{f}}, \tag{1.8.1}$$

where H is a matrix. To ensure that the stochastic process is bounded, we explicitly assume that the eigenvalues of H have negative real parts. The term $\tilde{\mathbf{f}}(t)$ is a white noise so that

$$\langle \tilde{\mathbf{f}}(t) \rangle = \mathbf{0} \tag{1.8.2}$$

and

$$\langle \tilde{\mathbf{f}}(t')\tilde{\mathbf{f}}^T(t) \rangle = \gamma\delta(t - t'), \tag{1.8.3}$$

where γ is a positive semi-definite matrix. To complete the characterization of the Ornstein–Uhlenbeck process it is assumed that the single-time probability density is Gaussian and that $\mathbf{a}(t)$ is independent of $\tilde{\mathbf{f}}(t')$ for $t < t'$. Thus

$$\langle \tilde{\mathbf{f}}(t')\mathbf{a}^T(t) \rangle = 0 \quad \text{for } t < t' \tag{1.8.4}$$

and

$$W_1(\mathbf{a}) = [(2\pi)^k \det \sigma]^{-1/2} \exp[-\mathbf{a}^T \sigma^{-1} \mathbf{a}/2]. \tag{1.8.5}$$

The Ornstein–Uhlenbeck equation (1.8.1) can be solved just like the Langevin equation (1.7.1). The integrating factor for the equation is now the matrix $\exp(-Ht)$, where the exponential is defined by its convergent power series. This leads directly to the solution

$$\mathbf{a}(t) = \exp(Ht)\mathbf{a}(0) + \int_0^t \exp[H(t - \tau)]\tilde{\mathbf{f}}(\tau)\,d\tau. \tag{1.8.6}$$

From this equation and the independence of $\tilde{\mathbf{f}}$ and $\mathbf{a}$ [property (1.8.4)], it follows that $\mathbf{a}(t)$ is also a Markov process, just as we showed for the Langevin equation. Using Eq. (1.8.6) it is verified, again as in the last section, that the conditional average of $\mathbf{a}(t)$, when $\mathbf{a}(0)$ is known to be $\mathbf{a}^0$, is

$$\langle \mathbf{a}(t) \rangle^0 = \exp(Ht)\mathbf{a}^0. \tag{1.8.7}$$

Furthermore, the covariance of the conditional fluctuation, i.e., $\delta\mathbf{a} = \mathbf{a}(t) - \langle \mathbf{a}(t) \rangle^0$, is easily calculated to be

$$\sigma(t) \equiv \langle \delta\mathbf{a}(t)\delta\mathbf{a}^T(t) \rangle^0 = \exp(Ht) \int_0^t \exp(-H\tau)\gamma \exp(-H^T\tau)\,d\tau \exp(H^T t)\,d\tau. \tag{1.8.8}$$

By differentiating this expression with respect to time, $\sigma(t)$ is easily seen to satisfy the differential equation

$$d\sigma(t)/dt = H\sigma(t) + \sigma(t)H^T + \gamma \tag{1.8.9}$$

with $\sigma(0) \equiv 0$. There are several other useful expressions for $\sigma(t)$. By changing variables to $s = t - \tau$, Eq. (1.8.8) can be rewritten

$$\sigma(t) = \int_0^t \exp(Hs)\gamma \exp(H^T s)\, ds. \tag{1.8.10}$$

From this it follows that

$$\sigma \equiv \lim_{t\to\infty} \sigma(t) = \int_0^\infty \exp(Hs)\gamma \exp(H^T s)\, ds, \tag{1.8.11}$$

which exists because the eigenvalues of H have negative real parts. Finally, anticipating Eq. (1.8.14), we can subtract Eq. (1.8.10) from Eq. (1.8.11) and use the change of variable $\tau = s - t$ to show that

$$\sigma(t) = \sigma - \exp(Ht)\sigma \exp(H^T t). \tag{1.8.12}$$

Because $\mathbf{a}(t)$ is a stationary Gaussian process, the results in Eqs. (1.8.7) and (1.8.8) are enough to completely determine the conditional probability density. It takes the form

$$P_2(\mathbf{a}^0 \,|\, \mathbf{a}, t) = [(2\pi)^k \det \sigma(t)]^{-1/2} \exp[-(\mathbf{a} - e^{Ht}\mathbf{a}^0)^T \sigma^{-1}(t)(\mathbf{a} - e^{Ht}\mathbf{a}^0)/2]. \tag{1.8.13}$$

Because it is a Markov process, the knowledge of W_1 and P_2 suffices to define $\mathbf{a}(t)$.

It is obvious from Eqs. (1.8.11) and (1.8.13) that $P_2(\mathbf{a}^0 \,|\, \mathbf{a}, t)$ becomes independent of $\mathbf{a}^0$ as $t \to \infty$. Thus we can make use of the result in Eq. (1.7.11) on the asymptotic equality of P_2 and W_1. For an Ornstein–Uhlenbeck process characterized by Eqs. (1.8.5) and (1.8.12) this implies that

$$\lim_{t\to\infty} \sigma(t) = \sigma = \langle \mathbf{a}(0)\mathbf{a}^T(0) \rangle,$$

the single-time covariance matrix. This limit can be written in a more transparent way using Eq. (1.8.9). We have already seen from Eq. (1.8.11) that the limit as $t \to \infty$ of $\sigma(t)$ exists. Thus it follows that $\lim d\sigma/dt$ is zero. Taking this limit on both sides of Eq. (1.8.9) and using the preceding equation,

$$H\sigma + \sigma H^T = -\gamma. \tag{1.8.14}$$

This is called the *fluctuation–dissipation theorem* and generalizes the relationship between the random force and the friction constant for Brownian motion. Notice that if γ and H are known, Eq. (1.8.14) can be thought of as a linear equation to be solved for the stationary covariance matrix, σ. Alternatively, as in the case of Brownian motion, σ and H may be known, and the equation then provides an unambiguous expression for γ. We shall have occasion to use this result in both of these ways later.

There is one other property that is easy to calculate from the integrated Ornstein–Uhlenbeck equation, namely, the two-time correlation function defined by Eq. (1.3.15). Since $\mathbf{a}$ is a stationary process, we have for $t' < t$

$$C(t,t') \equiv \langle \mathbf{a}(t)\mathbf{a}^T(t')\rangle = c(t'-t) = \langle \mathbf{a}(t-t')\mathbf{a}^T(0)\rangle.$$

Multiplying Eq. (1.8.6) from the right by $\mathbf{a}^T(0)$ then gives

$$C(t,t') = \exp(H|t'-t|)\sigma. \tag{1.8.15}$$

Thus the correlation function for an Ornstein–Uhlenbeck process relaxes exponentially at a rate determined by the linear regression equations.

Ornstein–Uhlenbeck processes are stationary, Gaussian, and Markovian. In fact, they form the complete class of stationary, Gaussian, Markov processes. This is relatively easy to demonstrate since any stationary, Gaussian, Markov process must have a Gaussian W_1 density of the form of Eq. (1.8.5) and a Gaussian conditional density, P_2. Thus the only trick is to show that the conditional average and covariance of a stationary, Gaussian, Markov process have the same form as those given in Eqs. (1.8.7) and (1.8.12) for an Ornstein–Uhlenbeck process. The first of these follows from the Chapman–Kolmogorov equation, Eq. (1.5.20), as applied to a stationary Markov process, namely,

$$P_2(\mathbf{n}^0|\mathbf{n},t) = \int P_2(\mathbf{n}^0|\mathbf{n}_2,t-\tau)P_2(\mathbf{n}_2|\mathbf{n},\tau)\,d\mathbf{n}_2. \tag{1.8.16}$$

Remember that for a Gaussian process $W_2(\mathbf{n}^0,\mathbf{n},t) = W_1(\mathbf{n}^0)P_2(\mathbf{n}^0|\mathbf{n},t)$ must be Gaussian in *both* the variables $\mathbf{n}^0$ and $\mathbf{n}$. It is shown in Fox's thesis, included in the references, that the only way this is possible is if the conditional average has the form $\langle \mathbf{n}(t)\rangle^0 = E(t)\mathbf{n}^0$ and $\langle \delta\mathbf{n}(t)\delta\mathbf{n}^T(t)\rangle$ is independent of $\mathbf{n}^0$. The conditional average, however, can be calculated from Eq. (1.8.16), which yields

$$\begin{aligned} E(t)\mathbf{n}^0 &= \int P_2(\mathbf{n}^0|\mathbf{n}_2,t-\tau)E(\tau)\mathbf{n}_2\,d\mathbf{n}_2 \\ &= E(\tau)E(t-\tau)\mathbf{n}^0. \end{aligned} \tag{1.8.17}$$

Thus

$$E(t) = E(\tau)E(t-\tau). \tag{1.8.18}$$

Rearranging this formula we find that

$$\lim_{\tau\to 0}\frac{E(t)-E(t-\tau)}{\tau} = \lim_{\tau\to 0}\frac{[E(\tau)-I]E(t-\tau)}{\tau},$$

where I is the identity matrix. Since from its definition $E(0) = I$, it follows that

$$dE/dt = HE(t)$$

where H is the matrix $dE/dt(0)$. Integrating this then gives

$$E(t) = e^{Ht}, \tag{1.8.19}$$

a result which is called *Doob's theorem*. As a consequence,

$$\langle \mathbf{n}(t) \rangle^0 = e^{Ht}\mathbf{n}^0$$

which is the same form as for an Ornstein–Uhlenbeck process.

The remainder of the proof is to verify that $\langle \delta\mathbf{n}(t)\delta\mathbf{n}^T(t) \rangle^0 \equiv \sigma(t)$ satisfies Eq. (1.8.12). Here the Gaussian character of the process is essential. The argument is based on the fact that

$$W_1(\mathbf{n}) = \int W_1(\mathbf{n}^0)P_2(\mathbf{n}^0 | \mathbf{n}, t)\,d\mathbf{n}^0$$

for a stationary process. By writing out both W_1 and P_2 in their Gaussian forms, this equation provides an identity that must be solved by the three matrices $\sigma^{-1}(t)$, $E(t)$, and σ^{-1}. A somewhat lengthy calculation, also found in Fox's thesis, gives the identity

$$\sigma^{-1}(t) = \sigma^{-1} + ME^T(t)\sigma^{-1}(t), \tag{1.8.20}$$

where M has been defined by

$$\sigma^{-1}(t)E(t) = M[\sigma^{-1} + E^T(t)\sigma^{-1}(t)E(t)] \tag{1.8.21}$$

Multiplying Eq. (1.8.20) on the right by $E(t)$ and comparing to Eq. (1.8.21) then gives that

$$M = \sigma^{-1}E(t)\sigma.$$

Substituting this expression into Eq. (1.8.20) shows that $\sigma^{-1}(t)$ must solve

$$\sigma^{-1}(t) = \sigma^{-1} + \sigma^{-1}E(t)\sigma E^T(t)\sigma^{-1}(t). \tag{1.8.22}$$

Finally, multiplying this on the left by σ and on the right by $\sigma(t)$ leads to

$$\sigma(t) = \sigma - E(t)\sigma E^T(t). \tag{1.8.23}$$

Remembering Doob's theorem that $E(t) = e^{Ht}$, Eq. (1.8.23) is seen to be identical to Eq. (1.8.12) for an Ornstein–Uhlenbeck process. Thus stationary, Gaussian, Markov processes and Ornstein–Uhlenbeck processes are one and the same.

An Ornstein–Uhlenbeck process is one of the simplest processes that can be generated by a stochastic differential equation. It is probably not surprising, then, that such processes are also stochastic diffusions. To check this fact, we need to verify that the transition moments of an Ornstein–Uhlenbeck process satisfy Eqs. (1.5.16)–(1.5.18). This is most easily done by rewriting these equations in our notation for the conditional averages. They can be reexpressed as the following conditions in the limit that $t \to 0$:

$$\langle \mathbf{a}(t) - \mathbf{a}^0 \rangle^0 = \mathbf{h}(\mathbf{a}^0)t + \mathcal{O}(t) \tag{1.8.24}$$

$$\langle [\mathbf{a}(t) - \mathbf{a}^0][\mathbf{a}(t) - \mathbf{a}^0]^T \rangle^0 = \gamma(\mathbf{a}^0)t + \mathcal{O}(t) \tag{1.8.25}$$

with all higher-order transitions moments being $\mathcal{O}(t)$. To verify Eq. (1.8.24), we recall the expression for the conditional average of $\mathbf{a}(t)$ in Eq. (1.8.7). For small t this reduces to

$$\langle \mathbf{a}(t) \rangle^0 = \mathbf{a}^0 + H\mathbf{a}^0 t + \mathcal{O}(t).$$

Thus Eq. (1.8.24) is satisfied and the drift term is $\mathbf{h}(\mathbf{a}^0) = H\mathbf{a}^0$. The condition on the average square transition moments can be verified in the following way. Recall that $\sigma(t)$ is the covariance of conditional fluctuation. Thus for small t

$$\begin{aligned}\sigma(t) &= \langle [\mathbf{a}(t) - (\mathbf{a}^0 + H\mathbf{a}^0 t)][\mathbf{a}(t) - (\mathbf{a}^0 + H\mathbf{a}^0 t)]^T \rangle^0 + \mathcal{O}(t) \\ &= \langle [\mathbf{a}(t) - \mathbf{a}^0][\mathbf{a}(t) - \mathbf{a}^0] \rangle^0 + \mathcal{O}(t). \end{aligned} \tag{1.8.26}$$

But from Eq. (1.8.10) we have

$$\sigma(t) = \gamma t + \mathcal{O}(t). \tag{1.8.27}$$

These two equations give

$$\langle [\mathbf{a}(t) - \mathbf{a}^0][\mathbf{a}(t) - \mathbf{a}^0] \rangle^0 = \gamma t + \mathcal{O}(t), \tag{1.8.28}$$

which is the second condition that a Markov process must satisfy to be a diffusion. Moreover, the strength of the white noise, γ, turns out to be the diffusion term. That the higher transition moments all vanish faster than t as $t \to 0$ follows from the Gaussian character of the process. This means that odd moments vanish and that even ones are of order $2n$ in the time.

Because an Ornstein–Uhlenbeck process is a stochastic diffusion, we can use the results of Section 1.5 to write down the Fokker–Planck equation (1.5.25) satisfied by the transition probability, P_2. The drift term is $\mathbf{h} = H\mathbf{a}$ and the diffusion term is γ; thus the Fokker–Planck equation is

$$\partial P_2(\mathbf{a}^0 | \mathbf{a}, t)/\partial t = -\partial H_{ij} a_j P_2(\mathbf{a}^0 | \mathbf{a}, t)/\partial a_i + \frac{1}{2}\gamma_{ij} \partial^2 P_2(\mathbf{a}^0 | \mathbf{a}, t)/\partial a_i \, \partial a_j. \tag{1.8.29}$$

Its solution, of course, is given by the Gaussian expression in Eq. (1.8.13), as can be verified by substituion. In fact, Eq. (1.8.29) is one of the few Fokker–Planck equations for which analytic solutions have been obtained.

There is a simple procedure to obtain the power spectrum for an Ornstein–Uhlenbeck process. To do so we Fourier transform both sides of Eq. (1.8.1). Integrating its left-hand side by parts gives

$$\frac{1}{2\pi} \int_{-\infty}^{+\infty} (d\mathbf{a}/dt) e^{i\omega t} \, dt = -i\omega \hat{\mathbf{a}}(\omega), \tag{1.8.30}$$

where $\hat{\mathbf{a}}(\omega)$ is the Fourier transform of $\mathbf{a}(t)$ defined in Eq. (1.6.7). Since the right-hand side of Eq. (1.8.1) is linear, the complete transform of Eq. (1.8.1) is

$$-i\omega \hat{\mathbf{a}}(\omega) = H\hat{\mathbf{a}}(\omega) + \hat{\hat{\mathbf{f}}}(\omega), \tag{1.8.31}$$

where $\hat{\hat{\mathbf{f}}}(\omega)$ is the Fourier transform of the white noise term discussed in Section 1.6. As noticed there, $\hat{\mathbf{f}}(\omega)$ vanishes on the average and is a Gaussian process. Its covariance is given in Eq. (1.6.10). Equation (1.8.31) is an algebraic equation for $\hat{\mathbf{a}}(\omega)$ and is solved formally by

$$\hat{\mathbf{a}}(\omega) = -(i\omega + H)^{-1}\hat{\mathbf{f}}(\omega). \tag{1.8.32}$$

Since we have assumed that the eigenvalues of H have negative real parts, the inverse of the matrix $(i\omega + H)$ must exist. Thus the spectral resolution of $\mathbf{a}(t)$ is a linear function of the spectral resolution of the white noise term. The *frequency correlation function* of $\hat{\mathbf{a}}(\omega)$ is given by

$$\begin{aligned}\langle \hat{\mathbf{a}}(\omega)\hat{\mathbf{a}}^T(\omega')\rangle &= (i\omega + H)^{-1}\langle \hat{\mathbf{f}}(\omega)\hat{\mathbf{f}}^T(\omega')\rangle(i\omega' + H^T)^{-1} \\ &= (i\omega + H)^{-1}\gamma(-i\omega + H^T)^{-1}\delta(\omega + \omega')/2\pi. \end{aligned}\tag{1.8.33}$$

Now the power spectrum of $\mathbf{a}(t)$ is a matrix $S(\omega)$ and is defined by the obvious generalization of Eq. (1.6.12), i.e.,

$$\langle \mathbf{a}(t)\mathbf{a}^T(t)\rangle = \int_{-\infty}^{+\infty} S(\omega)\, d\omega/2\pi. \tag{1.8.34}$$

$S(\omega)$ can be obtained by taking the inverse Fourier transformation of Eq. (1.8.33) for both ω and ω' using Eq. (1.6.8) and then setting $t = t'$. This gives

$$\langle \mathbf{a}(t)\mathbf{a}^T(t)\rangle = \int_{-\infty}^{+\infty} (i\omega + H)^{-1}\gamma(-i\omega + H^T)^{-1}\, d\omega/2\pi, \tag{1.8.35}$$

$$S(\omega) = (i\omega + H)^{-1}\gamma(-i\omega + H^T)^{-1}. \tag{1.8.36}$$

Unlike white noise the spectrum of $\mathbf{a}(t)$ is frequency dependent, and such a process is said to have a *colored* noise spectrum.

For a scalar process with a real relaxation rate, $H = -\lambda$ and Eq. (1.8.36) reduces to

$$S(\omega) = \gamma/(\omega^2 + \lambda^2). \tag{1.8.37}$$

This power spectrum is peaked at the origin and falls off to zero at high frequency. It is called a *Lorentzian spectrum* and is characterized by a half-width at half-maximum $\omega_{1/2} = \lambda$. Thus a determination of the shape of a power spectrum can lead to the relaxation rate. An Ornstein–Uhlenbeck process with a complex relaxation rate, $H = -\lambda - i\omega_0$, also has a Lorentzian power spectrum. In this case the transpose is replaced by the Hermitian conjugate. Making this change in Eq. (1.8.36), the spectrum can be seen to be

$$S(\omega) = \gamma/[(\omega - \omega_0)^2 + \lambda^2], \tag{1.8.38}$$

which is peaked at ω_0. For a vector-valued Ornstein–Uhlenbeck process, Eq. (1.8.36) gives the generalization of these simple Lorentzian spectra.

For stationary stochastic processes there is a relationship between the power spectrum and the two-time correlation function. Noticed first by Wiener and later rediscovered by Khintchine, it is usually referred to as the *Wiener–Khintchine theorem*. For an Ornstein–Uhlenbeck process the proof of the theorem follows directly from the fluctuation–dissipation theorem and the definition of the correlation function. Equation (1.8.15) gives the two-time correlation function for $t' < t$. In a similar way it can also be shown that for $t' > t$

$$C(t, t') = \sigma \exp[H^T(t' - t)]. \tag{1.8.39}$$

Thus the Fourier transform of the two-time correlation function is proportional to

$$\begin{aligned}\int_{-\infty}^{+\infty} e^{i\omega\tau} C(t+\tau, t)\, d\tau &= \int_0^{\infty} e^{i\omega\tau} e^{H\tau} \sigma\, d\tau + \int_0^{\infty} e^{-i\omega t} \sigma e^{H^T\tau}\, d\tau \\ &= -[(i\omega + H)^{-1}\sigma + \sigma(-i\omega + H^T)^{-1}]. \end{aligned} \tag{1.8.40}$$

The right-hand side of Eq. (1.8.40) can be rearranged to give

$$\begin{aligned}&-(i\omega + H)^{-1}[\sigma(-i\omega + H^T) + (i\omega + H)\sigma](-i\omega + H^T)^{-1} \\ &= -(i\omega + H)^{-1}[H\sigma + \sigma H^T](-i\omega + H^T)^{-1} \\ &= (i\omega + H)^{-1}\gamma(-i\omega + H^T)^{-1} = S(\omega), \end{aligned} \tag{1.8.41}$$

where the final two equalities follow from the fluctuation–dissipation theorem Eq. (1.8.14) and Eq. (1.8.36). Combining Eqs. (1.8.40) and (1.8.41) gives the Wiener–Khintchine theorem

$$\int_{-\infty}^{+\infty} e^{i\omega\tau} C(t+\tau, t)\, d\tau = S(\omega). \tag{1.8.42}$$

This theorem is of practical importance since it allows one to obtain the spectral density function by Fourier transforming the time correlation function. As the time correlation function is often easy to measure, its frequency spectrum can be used to obtain relaxation rates. An example of this is given in Section 5.2.

Finally we note that another form of the Wiener–Khintchine theorem is obtained by combining Eqs. (1.8.33) and (1.8.41), i.e.,

$$\langle \hat{\mathbf{a}}(\omega)\hat{\mathbf{a}}^T(\omega')\rangle = S(\omega)\delta(\omega + \omega')/2\pi. \tag{1.8.43}$$

Experimentally one often examines the correlation of fluctuations of frequency ω in a small band width of frequencies, $[\omega - \Delta\omega/2, \omega + \Delta\omega/2]$, and obtains an average result for the frequency band, $\Delta\omega$. For this situation Eq. (1.8.43) implies that

$$\langle \mathbf{a}\mathbf{a}^T(\omega)\rangle \equiv \frac{1}{\Delta\omega}\int_{\omega-\Delta\omega/2}^{\omega+\Delta\omega/2} \langle \mathbf{a}(\omega)\mathbf{a}^T(\omega')\rangle\, d\omega' = S(\omega)/2\pi\Delta\omega \tag{1.8.44}$$

or

$$S(\omega) = 2\pi\Delta\omega\langle \mathbf{a}\mathbf{a}^T(\omega)\rangle. \tag{1.8.45}$$

Thus the power spectrum can be obtained by measuring the frequency correlation function, averaged over a small band, $\Delta\omega$, of neighboring frequencies.

References

Statistical Mechanics

J.W. Gibbs, *Elementary Principles of Statistical Mechanics* (Yale University Press, New Haven, 1902; reprinted by Dover Press, 1960).
R.C. Tolman, *The Principles of Statistical Mechanics* (Oxford University Press, Oxford, 1938).
T.L. Hill, *Statistical Mechanics* (McGraw Hill, New York, 1956).
D.A. McQuarrie, *Statistical Mechanics* (Harper and Row, New York, 1976).
G.E. Uhlenbeck and G.W. Ford, *Lectures in Statistical Mechanics* (American Mathematical Society, Providence, RI, 1963).

Thermodynamics

A. Sommerfeld, *Thermodynamics and Statistical Mechanics: Lectures in Theoretical Physics*, Vol. V (Academic Press, New York, 1955).
H.B. Callen, *Thermodynamics* (John Wiley, New York, 1960).

Stochastic Processes: Mathematics

W. Feller, *An Introduction to Probability Theory*, Vol. 1, 2nd ed. (Wiley, New York, 1957).
S. Karlin and H.M. Taylor, *A First Course in Stochastic Processes*, 2nd ed. (Academic Press, New York, 1975).
L. Arnold, *Stochastic Differential Equations: Theory and Applications* (Wiley-Interscience, New York, 1974).
N. Wiener, *Collected Works*, Vol. I, P. Masani, ed. (MIT Press, Cambridge, 1976).
L. Breiman, *Probability* (Addison-Wesley, Reading, MA, 1968).

Stochastic Processes: Physical Theory

A. Einstein, *Investigations of the Theory of the Brownian Movement* (Dover, New York, 1956).
N. Wax, *Selected Papers on Noise and Stochastic Processes* (Dover, New York, 1954).
R.F. Fox, *Contributions to the Theory of Nonequilibrium Thermodynamics*, Doctoral dissertation, Rockefeller University, 1969.

Chapter 2

Irreversible Processes: The Onsager and Boltzmann Pictures

2.1. Introduction

The primary objective of this book is to develop a mathematical picture of measurable quantities that can be used to understand macroscopic observations of matter. As we have discussed in Chapter 1, that picture is necessarily stochastic and involves ensembles of systems that are prepared in similar ways. In Chapter 1 we outlined some of the techniques of the theory of stochastic processes that are necessary for understanding physical ensembles. Although we used Brownian motion to illustrate the physical relevance of stochastic processes, the stochastic point of view is essential for understanding all kinds of macroscopic observations. Fluctuations are inherent in all matter because of its molecular constitution. Indeed, one of the lessons of Brownian motion is that these fluctuations are observable and that they are closely related to the irreversible processes caused by molecular motion.

Evidence for molecular fluctuations is well known for many physical and chemical processes. When the sun's rays pass through the atmosphere, or any fluid for that matter, the oscillating electromagnetic field of the light excites electron transitions in molecules. Even when this light is re-radiated at the same frequency, it is radiated as a spherical wave rather than as the incoming plane wave. If the atmosphere were free of fluctuations due to molecular motion, these spherical wave fronts would add up coherently with the net result that the plane wave would propagate with its intensity undiminshed. However, because the local density of molecules is a statistical quantity, the electromagnetic field of the re-radiated light is not coherent, and some of the light is scattered by the fluctuations. As blue light is scattered more strongly than red, the portion of the sky away from the sun has its characteristic blue color. At dusk, however, the blue light is scattered away, leaving the red and orange sunset. In fact, with the advent of laser technology, light scattering has become an accurate method of obtaining information about molecular fluctuations in both pure substances and mixtures.

Another place in which fluctuations can be easily observed is electrical circuits. Indeed, matter which is carrying an electric current exhibits Johnson noise, that is, current fluctuations which are rapid and unpredictable. These are caused by the current carriers and the molecular processes which they undergo in the process of carrying current. Fluctuations have also been observed in the transport of ions across biological membranes. In some cases the fluctuations represent the closing and opening of channels in the membrane, which can be observed electronically. Neutrons are another excellent probe of molecular fluctuations. Since they act as particle waves, neutrons provide information similar to that provided by scattered light. However, neutrons interact most strongly with the nucleus, so they provide data on fluctuations that are very short-range.

Under certain conditions fluctuations can become greatly amplified, making them much easier to observe. For example, when carbon dioxide at equilibrium is held at its gas–liquid critical point, that is, at 304 K and a pressure of 72.8 atmospheres, the fluid takes on a milky, turbid appearance that rapidly vanishes when either the pressure or temperature is changed. This phenomenon is called critical opalescence and is caused by the scattering of light from incipient droplets spread randomly throughout the fluid. A similar phenomenon occurs in systems that are not at equilibrium but are close to critical points of stability. This sort of enhancement of fluctuations, like Brownian motion seen under the microscope, provides one of our clearest windows to the molecular makeup of matter.

The first complete theory of fluctuations for equilibrium ensembles was conceived by Onsager. Although that theory is strictly limited to equilibrium, its basic structure contains many of the features that are needed for more general situations. In its original form, Onsager used it to describe molecular fluctuations for extensive variables, like internal energy, which occur in thermodynamics. This connection with thermodynamics means that the theory is strongly connected to Gibbs' statistical mechanical theory of matter in equilibrium. It also means that the theory, like thermodynamics, is independent of the molecular nature of matter. The Onsager theory is identical to the special case of equilibrium for the general theory described in the remainder of this book. For this reason we develop its content in some detail in this chapter.

A rather different explanation of molecular motion, strongly rooted in collision theory, was given by Boltzmann some 60 years before Onsager. The purpose of Boltzmann's work was to explain in mechanical terms the Second Law of thermodynamics. In doing so he was lead to a kinetic-molecular theory that involved collisional cross sections and a description of mass and momentum transfer for molecules in the gas phase. The theory appears to be much more complicated than the Onsager theory. It also appears to be much more useful since it can be employed to evaluate transport coefficients like the heat conductivity or rate constants for chemical reactions. The Boltzmann theory involves an explicit kinetic picture of molecular events, and as such does not have the thermodynamic character of the Onsager theory. However, the Boltzmann theory can be applied to both equilibrium and nonequilibrium ensembles and, unlike the Onsager theory, which involves linear dynamics, is nonlinear.

Despite these obvious differences, there is a close connection between the Onsager and Boltzmann pictures of irreversible processes. In this chapter we review some of the main features of both theories with an eye to exposing their similarities. In the final section we bring together the central physical ideas of the two theories and provide the motivation for the mechanistic statistical theory of molecular processes which forms the theme of the remaining chapters.

2.2. The Linear Laws

The Onsager theory is concerned with the properties of equilibrium ensembles. To define a stochastic process in such an ensemble it is necessary to specify the physical quantities that will be measured. A whole host of variables suggest themselves—pressure, temperature, mass densities, and so forth—and it may not be obvious which choice is the most sensible. Taking our lead, however, from thermodynamics, it makes sense to consider some subset of the *extensive* variables that characterize the systems. Thus, for example, for a single-component system we could consider the internal energy, the volume, and the mass of the system. This is the simplest contracted description in terms of extensive variables, which we will call the *thermodynamic* level of description. If we are set up to measure the spatial distribution of these quantities, we can make the thermodynamic description more elaborate by splitting up the volume into a large number of cells and use as variables the internal energy and mass in each of the smaller volumes.

Of course, the process of adding extensive variables to produce a more detailed description does not end there. Next, we might add the momentum of the cells. Momentum is an extensive variable since it doubles when the system size is doubled if the temperature, pressure, and local velocity are all held fixed. The thermodynamic level with the local momentum added will be called the *hydrodynamic* level of description. We can continue this process by expanding the hydrodynamic description of the mass in each cell. Thus we might keep track of the internal states of the molecules and instead of the total mass use the masses in the various internal states as extensive variables. Alternatively, it would be possible to consider only mass densities, if we keep track of the distribution of molecules in both position and momentum space. This is the μ- (or molecule) space description, and in honor of its chief originator we will call it the *Boltzmann* level of description.

Let us consider, for the time being, descriptions in which all the variables are extensive variables. For simplicity, we will denote the whole collection by $\mathbf{n}$. We are interested in characterizing the dynamics of extensive variables in the equilibrium ensemble. This means that we will need to examine conditional subensembles so that we can understand the transitions between a well-defined initial value $\mathbf{n}^0$ and possible values $\mathbf{n}$ at a later time t. Since an equilibrium ensemble is stationary, we know from Section 1.3 that such transitions depend only on the elapsed time t. The simplest information that we can obtain about the conditional ensemble is the conditional average, $\langle \mathbf{n}(t) \rangle^0$. To set the stage for the general case, let us return to the well-worn example of Brownian motion. In the present context the most contracted version of Brownian motion involves only its momentum, since its internal energy and mass are to a good approximation constant. This is the Langevin level of description which was treated in detail in Section 1.7. Using Eq. (1.7.1) and recalling that the random force vanishes when we take the conditional average, we find that

$$d\langle \mathbf{p} \rangle^0/dt = -(\zeta/m)\langle \mathbf{p} \rangle^0. \tag{2.2.1}$$

In other words, the rate of change of the conditionally averaged momentum is proportional to the momentum. This is an example of a linear law and is based on the observation that the Stokes' friction of a particle in a fluid is proportional to its momentum. The linear law can be written in an equivalent form as

$$d\langle \mathbf{p}(t) \rangle^0/dt = -\zeta\langle \mathbf{v}(t) \rangle^0, \tag{2.2.2}$$

which shows a similar proportionality to the deviation of the velocity from its equilibrium average value of zero.

A key fact for generalizing the form of the linear laws is that the velocity is closely related to the variable which is *thermodynamically conjugate* to the momentum. That is, if S is the entropy of the particle, then

$$\partial S/\partial p_j = -v_j/T \tag{2.2.3}$$

is called the intensive variable conjugate to the momentum. To verify Eq. (2.2.3), the total differential of the entropy of the Brownian particle is written as

$$dS = dE/T + (p/T)\,dV - (\mu/T)\,dN, \tag{2.2.4}$$

with E internal energy, V volume, N the number of molecules, p the pressure, and μ the chemical potential. We introduce the momentum as a thermodynamic variable by writing $\bar{E} = E + p^2/2m$, where $\bar{E}$ is the total energy. Equation (2.2.4) then becomes

$$dS = d\bar{E}/T - (\mathbf{v}/T)\cdot d\mathbf{p} + (p/T)\,dV - (\mu/T)\,dN, \tag{2.2.5}$$

in which the second term is equivalent to Eq. (2.2.3). Using this fact, the linear law in Eq. (2.2.2) can be rewritten in component form as

$$d\langle p_i(t) \rangle^0/dt = \sum_j L_{ij}[\langle \partial S/\partial p_j(t) \rangle^0 - \langle \partial S/\partial p_j \rangle^e] \tag{2.2.6}$$

with

$$L_{ij} = T\zeta\delta_{ij}. \tag{2.2.7}$$

Written as in Eq. (2.2.6), the linear law for Brownian motion takes on a thermodynamic flavor. In words, the average time derivative of an extensive variable is linearly related to the average deviations of the conjugate intensive variables from their equilibrium values. Moreover, we see from Eq. (2.2.7) that the coupling matrix L is symmetric and non-negative definite. This general relationship is called the *Onsager principle* and defines the linear laws for irreversible processes in an equilibrium ensemble.

As an application of the linear laws, consider the transport of heat by conduction between two solids, one at temperature T_1 and the other at temperature T_2. Here the extensive variables that are changing are the internal energies, E_1 and E_2. According to Eq. (2.2.4), the variables thermodynamically conjugate to these energies are $1/T_1$ and $1/T_2$. Furthermore, the equilibrium value of the temperature in such an ensemble will be the same for both solids, call it T^e. According to the Onsager principle, then, it follows that

$$d\langle E_1(t)\rangle^0/dt = L_{11}\left(\left\langle \frac{1}{T_1}(t)\right\rangle^0 - \frac{1}{T^e}\right) + L_{12}\left(\left\langle \frac{1}{T_2}(t)\right\rangle^0 - \frac{1}{T^e}\right) \tag{2.2.8}$$

$$d\langle E_2(t)\rangle^0/dt = L_{21}\left(\left\langle \frac{1}{T_1}(t)\right\rangle^0 - \frac{1}{T^e}\right) + L_{22}\left(\left\langle \frac{1}{T_2}(t)\right\rangle^0 - \frac{1}{T^e}\right) \tag{2.2.9}$$

where L_{ij} is symmetric and non-negative definite. Thus $L_{12} = L_{21}$. However, by conservation of energy it must also be true that any energy loss from the solid 1 is exactly compensated by a gain in solid 2. Thus the sum of Eqs. (2.2.8) and (2.2.9) must always be zero. This can only be true if $L_{11} = -L_{21}$ and $L_{22} = -L_{12}$. Combining these identities for the coefficients gives $L_{11} = -L_{21} = -L_{12} = L_{22} \equiv L$, where L is non-negative. This allows us to rewrite Eq. (2.2.8) in the more familiar form

$$\begin{aligned} d\langle E_1(t)\rangle^0/dt &= L\left(\left\langle \frac{1}{T_1}(t)\right\rangle^0 - \left\langle \frac{1}{T_2}(t)\right\rangle^0\right) \\ &= (L/T^{e2})(\langle T_2(t)\rangle^0 - \langle T_1(t)\rangle^0), \end{aligned} \tag{2.2.10}$$

where the final equality is valid for small deviations from equilibrium. Equation (2.2.10) states that heat will be transported from solid 1 to solid 2, on the average, if the average temperature of 1 exceeds that of 2; moreover, the rate is proportional to the average temperature difference. This is known as Newton's law of cooling, which in the thermodynamic level of description is the analogue of Fourier's law of heat conduction.

In the two examples of the linear laws considered so far, the variables are the momentum and the internal energy. Although they are both extensive variables, from a mechanical point of view they are quite different. Consider a particular member of the ensemble and the effect of time reversal, i.e., instantaneously reversing all the velocities of the molecules, the magnetic field, and the time. The mechanical equations of motion are invariant under this transformation, and as time progresses, the system will retrace its previous trajectory with all the velocities reversed. The internal energy does not change in this reversal and is called an *even* variable, while the momentum changes its sign and is called an *odd* variable. Symbolically we will write

$$\mathscr{T} n_i(t) = \varepsilon_i n_i(-t) \tag{2.2.11}$$

where $\mathscr{T}$ is the time reversal operator and $\varepsilon_i = \pm 1$ depending on whether the variable is even or odd. In the examples of Brownian motion and heat transport treated in this section, the coupling that occurs between the momentum and the internal energy has been ignored. Thus only variables possessing the same symmetry under time reversal were coupled. For situations in which coupling between variables of different time reversal symmetry is important, the Onsager principle must be extended. Although the linear laws still remain true, the symmetry principle is changed to

$$L_{ij}(\mathbf{B}) = \varepsilon_i \varepsilon_j L_{ji}(-\mathbf{B}), \tag{2.2.12}$$

where the dependence of L on the external magnetic field, $\mathbf{B}$, has been made explicit. In the absence of a magnetic field, Eq. (2.2.12) states that odd and even variables are coupled by an antisymmetric matrix whereas odd variables or even variables are coupled by a symmetric matrix. These relationships are known as the *Onsager–Casimir reciprocal relations.* In Section 2.5 we will provide a derivation of the reciprocal relations based on the reversibility of the mechanical equation of motion.

2.3. Entropy, Dissipation, Fluxes, and Forces

The Onsager principle describes the linear relaxation of the averages of extensive variables to their equilibrium values. For ease in notation we will write the deviations of $\mathbf{n}$ from their equilibrium values as $\mathbf{a} = \mathbf{n} - \mathbf{n}^e$ and indicate the conditional average values of $\mathbf{a}$ by an overbar, i.e., $\bar{\mathbf{a}} = \langle \mathbf{a} \rangle^0$. It has become conventional to call the time derivative of a_i the *flux* of n_i. We also need to consider conditional averages of the conjugate intensive variables corresponding to a_i. If we indicate the variable thermodynamically conjugate to n_i by

$$F_i = \partial S / \partial n_i, \tag{2.3.1}$$

then the variable conjugate to a_i is

$$X_i = \partial S/\partial n_i - (\partial S/\partial n_i)^e, \tag{2.3.2}$$

where the superscript indicates the equilibrium value. The conditional average of X_i will also be indicated by an overbar, so that the verbal statement of the Onsager principle takes on the mathematical form

$$\bar{J}_i \equiv d\bar{a}_i/dt = \sum_j L_{ij}\bar{X}_j. \tag{2.3.3}$$

$\bar{J}_i$ is the average thermodynamic flux and L_{ij} satisfies the Onsager–Casimir reciprocal relationship Eq. (2.2.12). Equation (2.3.3) states that, on the average, the rate of change of deviations of the extensive variables is a linear function of the $\bar{X}_j$. Since the $\bar{X}_j$ "force" relaxation back to equilibrium, they have become known as the *thermodynamic forces*. Expressed in these terms, the linear laws state that the fluxes are a linear function of the conjugate thermodynamic forces.

The linear laws are governed by two additional constraints. First, there are the symmetry relations, Eq. (2.2.12). And, second, the matrix L must be positive semi-definite. This second condition is closely related to the Second Law of thermodynamics. For an isolated system the Second Law says that the entropy cannot decrease in any spontaneous process. Thus the entropy must be maximized at the equilibrium state. Consider the entropy of an isolated system that is displaced from equilibrium by changing the values of the extensive variables. Thus **a** is no longer zero. If **a** is small, the entropy in this state is approximately given by the truncated Taylor series

$$S(\mathbf{a}) = S(\mathbf{0}) + \sum_j F_j^e a_j + \frac{1}{2}\sum_i\sum_j S_{ij}a_i a_j, \tag{2.3.4}$$

where the matrix of second derivatives is

$$S_{ij} = (\partial^2 S/\partial n_i \partial n_j)^e. \tag{2.3.5}$$

Because the entropy is maximized at equilibrium, the first differential term in Eq. (2.3.4) vanishes and the second differential must be negative. As this must be true for arbitrary **a**, it follows that the matrix S_{ij} is negative semi-definite. Thus

$$S(\mathbf{a}) = S(\mathbf{0}) + \frac{1}{2}\sum_i\sum_j S_{ij}a_i a_j. \tag{2.3.6}$$

We see, further, from our definitions that for small **a**

$$X_i = \sum_j S_{ij}a_j. \tag{2.3.7}$$

This allows us to rewrite Eq. (2.3.6) as

$$S(\mathbf{a}) = S(\mathbf{0}) + \frac{1}{2}\sum_i X_i a_i. \tag{2.3.8}$$

According to the temporal version of the Second Law, the entropy of an isolated system must increase in a spontaneous process. This, of course, cannot be literally true since, as Maxwell remarked, "The Second Law has the same degree of truth as the statement that 'If you throw a tumblerful of water into the sea, you cannot get the same tumblerful out again'." In other words, the Second Law holds only on the average. Thus the rate of change of the entropy in a spontaneous process will increase on the average trajectory. The rate of change for a small average deviation can be obtained by combining Eq. (2.3.6) or (2.3.8) with the linear law (2.3.3). This gives

$$dS(\bar{\mathbf{a}})/dt = \sum_i \bar{X}_i\, d\bar{a}_i/dt = \sum_i \sum_j L_{ij}\bar{X}_i\bar{X}_j \geq 0, \tag{2.3.9}$$

where the final inequality is the statement of the Second Law. Since Eq. (2.3.9) is true for any $\bar{\mathbf{a}}$, it must also be true for any $\bar{\mathbf{X}}$. This implies that L is a positive semi-definite matrix.

A little more can be learned about the rate of change of the entropy by considering its behavior under time reversal. According to the ideas of Planck and Boltzmann, which we shall examine in Section 2.5, the form of the entropy function is determined by the statistical distribution of the extensive variables at equilibrium. Since the statistical distribution is stationary, it does not change with time and so is time reversal invariant. This means that the entropy is even under time reversal and so must satisfy

$$\mathscr{T}S(\mathbf{a}, \mathbf{B}) = S(\mathscr{T}\mathbf{a}, -\mathbf{B}) = S(\mathbf{a}, \mathbf{B}), \tag{2.3.10}$$

where we have explicitly included the magnetic field dependence of the entropy. Combining the evenness of S with Eq. (2.3.6) implies that in the absence of a magnetic field

$$\sum_i \sum_j \varepsilon_i \varepsilon_j S_{ij} a_i a_j = \sum_i \sum_j S_{ij} a_i a_j, \tag{2.3.11}$$

where $\varepsilon_i = \pm 1$ depending on whether a_i is even or odd. Indeed, by separately setting all the even variables equal to zero and then all the odd variables equal to zero, Eq. (2.3.11) implies that

$$\sum_{i,j}{}' S_{ij} a_i a_j = -\sum_{i,j}{}' S_{ij} a_i a_j, \tag{2.3.12}$$

where the prime restricts the sums to terms for which a_i and a_j have different symmetries. This can only be true if the elements of S_{ij} that couple even and odd variables vanish. Thus in the absence of external magnetic fields

$$S(\mathbf{a}) - S(\mathbf{0}) = \frac{1}{2} \sum_{\substack{i,j \\ \text{even}}} S_{ij} a_i a_j + \frac{1}{2} \sum_{\substack{i,j \\ \text{odd}}} S_{ij} a_i a_j. \tag{2.3.13}$$

Stated in words, Eq. (2.3.13) says that there is no *thermodynamic coupling* between odd and even variables in the absence of an external magnetic field.

The Onsager–Casimir relations show that there may be a *dynamic coupling* between odd and even variables in the absence of a magnetic field. According to Eq. (2.2.12), that coupling will involve an antisymmetric matrix. The dynamic coupling among the even or among the odd variables separately involves a symmetric matrix. Thus when **B** vanishes the matrix L has the form

$$L = L^e + L^o + L^a = L^s + L^a, \tag{2.3.14}$$

where L^e and L^o are symmetric matrices with zero entries except possibly among the even or among the odd variables, respectively, and L^a is an antisymmetric matrix coupling the odd and even variables. In the presence of magnetic fields there are thermodynamic as well as dynamic couplings between odd and even variables. Thus, in general, odd and even variables can increase the entropy in concert as implied by Eq. (2.3.9).

The quadratic function on the right-hand side of Eq. (2.3.9) was first identified by Lord Rayleigh and is called the Rayleigh–Onsager *dissipation function*. We will write it as

$$\Phi = \sum_i \sum_j L_{ij} \bar{X}_i \bar{X}_j \geq 0. \tag{2.3.15}$$

"Dissipation" is said to occur whenever Φ is positive. Because this is a quadratic form, only the symmetric part of L contributes to the dissipation function. This can be seen by writing

$$L = L^s + L^a \tag{2.3.16}$$

where

$$\begin{aligned} L^s &= \frac{1}{2}(L + L^T) \equiv \operatorname{sym} L \\ L^a &= \frac{1}{2}(L - L^T) \equiv \operatorname{asym} L. \end{aligned} \tag{2.3.17}$$

These are the symmetric and antisymmetric parts of L, and satisfy

$$L^s = (L^s)^T, \quad L^a = -(L^a)^T. \tag{2.3.18}$$

Indeed, substituting Eq. (2.3.18) into Eq. (2.3.16), it follows that

$$\Phi = \sum_i \sum_j L^s_{ij} \bar{X}_i \bar{X}_j \geq 0 \tag{2.3.19}$$

since, according to Eq. (2.3.18), $(L^a_{ij} + L^a_{ji}) = 0$. Thus it is only the symmetric part of the dynamical coupling matrix, L^s, that causes dissipation. As we have seen, the antisymmetric part of L—at least in the absence of a magnetic field—arises from coupling between even and odd variables. According to Eq. (2.3.19), it will not contribute to an increase in the entropy of an isolated system. In fact, in the special case that **B** vanishes we can combine Eqs. (2.3.14) and (2.3.19) to see that

$$\Phi = \sum_{\substack{i,j \\ \text{even}}} L^e_{ij} \bar{X}_i \bar{X}_j + \sum_{\substack{i,j \\ \text{odd}}} L^o_{ij} \bar{X}_i \bar{X}_j. \tag{2.3.20}$$

Thus when there is no external magnetic field, the odd and even variables contribute to the dissipation function independently.

The dissipation function can be written in several equivalent ways. Referring to Eqs. (2.3.9) and (2.3.15), we also have

$$\Phi = \sum_i \bar{X}_i \, d\bar{a}_i/dt = \sum_i \bar{X}_i \bar{J}_i, \tag{2.3.21}$$

where $\bar{J}_i$ is the thermodynamic flux. Recalling the definition of the force, X_i, in Eq. (2.3.7) we can write the dissipation function in terms of the deviation from equilibrium as

$$\Phi = \sum_{i,j} (SLS)_{ij} \bar{a}_i \bar{a}_j. \tag{2.3.22}$$

The matrix SLS has an interesting interpretation in terms of the dynamics of the forces, $\mathbf{X}$. Multiplying the linear law in Eq. (2.3.3) on the left by S and using Eq. (2.3.7) yields

$$d\bar{X}_i/dt = \sum_j \bar{L}_{ij} \bar{a}_j. \tag{2.3.23}$$

where

$$\bar{L} \equiv SLS. \tag{2.3.24}$$

$\bar{L}$ is the matrix congruent to L in the basis set of the forces and shares all the symmetry properties of L. With this notation, Eq. (2.3.22) becomes

$$\Phi(\bar{\mathbf{a}}) = \sum_i \sum_j \bar{L}^s_{ij} \bar{a}_i \bar{a}_j. \tag{2.3.25}$$

Another name often applied to the dissipation function is the *entropy production*. The "entropy" part of this name seems appropriate since Eqs. (2.3.9) and (2.3.15) show that

$$dS(\bar{\mathbf{a}})/dt = \Phi(\bar{\mathbf{a}}). \tag{2.3.26}$$

This identification is only valid, however, for an isolated system. What one has in mind with the terminology "production" is applying this identity to the processes occurring within the system, that is, not including changes in $\bar{\mathbf{a}}$ caused by interaction with the outside. For example, return to the heat flow problem considered in Section 2.2. If the solid labeled by the index 1 is considered the system, then none of the terms in Eq. (2.2.8) can be attributed to processes occurring only within the system. It follows that heat conduction does not involve entropy production for system 1. On the other hand, if we consider the entire system 1 and 2 as isolated, the heat flow within the system leads to a production of entropy given by the dissipation function

$$dS/dt = \Phi = L[(1/\bar{T}_1) - (1/\bar{T}_2)]^2 \geq 0. \tag{2.3.27}$$

Because the concept of entropy depends on our choice of system, the notion of "entropy production" is system dependent. The overall dissipation caused by molecular processes, however, is always given by Φ, and for this reason we will usually refer to Φ as the dissipation function.

2.4. The Hydrodynamic Level of Description

In Section 2.2 the hydrodynamic level for describing a physical system was introduced. At this level the variables in the ensemble are the densities of mass, momentum, and energy as a function of position within the system. These variables are all densities of extensive quantities and so it is possible to identify the linear laws that govern their rate of change near equilibrium. As the overall mass, momentum, and energy are conserved in an isolated molecular system, these variables are sometimes called *conserved* quantities. In fact, the basic equations which govern the rate of change of these quantities can be looked at as conservation equations.

Consider the density, $g(\mathbf{r}, t)$, of any extensive variable. If we fix attention on a stationary volume in space, V, then the total amount of the extensive variable in V is

$$G(V, t) = \int_V g(\mathbf{r}, t)\, d\mathbf{r}. \tag{2.4.1}$$

Its time rate of change is given by

$$dG(V, t)/dt = \int_V (\partial g/\partial t)\, d\mathbf{r} \tag{2.4.2}$$

since V is fixed in space. The quantity of G in the volume can change for only two reasons. It may flow through the surface of V or it may be created, or destroyed, within the volume. Thus we have

$$\int_V (\partial g/\partial t)\, d\mathbf{r} = -\int_A \mathbf{j}_G \cdot d\mathbf{a} + \int_V s_G\, d\mathbf{r}, \tag{2.4.3}$$

where A is the surface area of V and $d\mathbf{a}$ indicates an increment of surface area. The vector $\mathbf{j}_G$ is called the *flux density* and gives the flow rate of G per unit area. The minus sign occurs in front of the surface integral because the direction of $d\mathbf{a}$ is taken as the outward normal of V. Thus when $\mathbf{j}_G$ and $d\mathbf{a}$ have the same direction, there is a loss of G. The term $s_G(\mathbf{r}, t)$ is the *source density* and gives the rate at which G is created per unit volume. The surface integral in Eq. (2.4.3) can be rewritten using Gauss's theorem, which leads to the net balance equation

$$\int_V (\partial g/\partial t)\, d\mathbf{r} = -\int_V \nabla \cdot \mathbf{j}_G\, d\mathbf{r} + \int_V s_G\, d\mathbf{r}. \tag{2.4.4}$$

Equation (2.4.4), however, is true for any volume. Thus taking the limit that V is infinitesimal, we have

$$\partial g/\partial t = -\nabla \cdot \mathbf{j}_G + s_G, \tag{2.4.5}$$

which is the local version of the balance equation. For conserved quantities the source density s_G vanishes in the absence of external interactions. In this special case the local conservation law reduces to

$$\partial g/\partial t = -\nabla \cdot \mathbf{j}_G. \tag{2.4.6}$$

From a phenomenological point of view, the conservation equations of hydrodynamics are the Navier–Stokes equations. The first of these, involving the mass density, $\rho(\mathbf{r}, t)$, is exact since the mass flux density can be expressed in terms of the momentum density. Specifically, consider an infinitesimal area of magnitude da. The flow of mass through da will be in the direction of the local velocity field, $\mathbf{v}(\mathbf{r}, t)$. If the outward normal to da is written as $\mathbf{n}$, then, as is shown in Fig. 2.1, all the mass in the volume element $(\mathbf{v}\,dt)\cdot\mathbf{n}\,da$ will flow through da in a time dt. The mass in this volume element is $\rho\mathbf{v}\cdot\mathbf{n}\,da\,dt$, and, remembering that $d\mathbf{a} = \mathbf{n}\,da$, it follows that the mass flux density is given by

$$\mathbf{j}_m = \rho\mathbf{v}. \tag{2.4.7}$$

However, the momentum of an infinitesimal volume element dV is simply given by $\rho\,dV\,\mathbf{v}$, so that the local momentum density is

$$\mathbf{p}(\mathbf{r}, t) = \rho\mathbf{v}. \tag{2.4.8}$$

Substituting Eqs. (2.4.7) and (2.4.8) into the conservation equation (2.4.6) gives the so-called *continuity equation* for the mass density

$$\partial\rho/\partial t = -\nabla \cdot \mathbf{p}. \tag{2.4.9}$$

To obtain the conservation equation for the momentum density, it is necessary to write an explicit equation for the momentum flux. Just as for the mass, there is a convective part of the flux which has the form

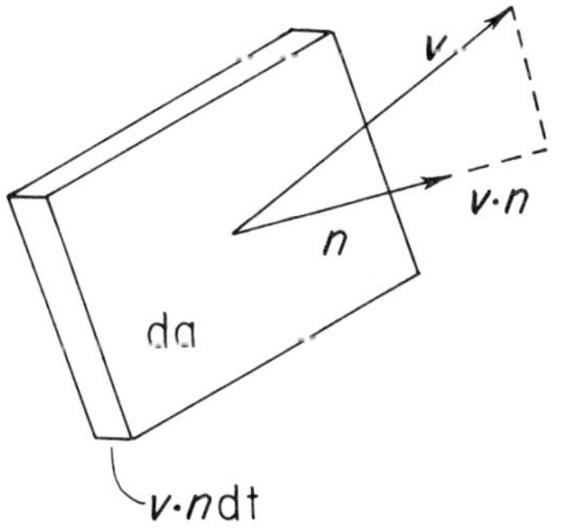

FIGURE 2.1. An infinitesimal element of surface with normal $\mathbf{n}$ and area da. The local velocity of the center of mass is $\mathbf{v}$. All mass points inside the volume $\mathbf{v}\cdot\mathbf{n}\,dt\,da$ will cross the surface in the direction of the normal in the time interval dt.

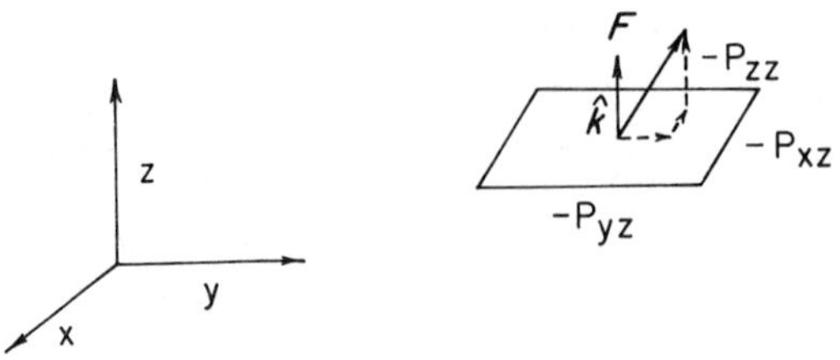

FIGURE 2.2. Decomposition of the force **F** that acts on a surface with normal along the z-axis. The normal component of the pressure tensor is P_{zz} and the tangential components are P_{xz} and P_{yz}.

$$\mathbf{j}_i = p_i \mathbf{v}, \tag{2.4.10}$$

where p_i is the ith coordinate of the momentum density. The other contribution to the momentum flux comes from forces that the fluid volume exerts on its bounding surface. These are conveniently summarized in a position- and time-dependent pressure tensor, $P_{ij}(\mathbf{r}, t)$. The pressure is a second-order tensor because it expresses the force acting on an oriented area. Both the force and the area are vectors. This is illustrated in Fig. 2.2 where a unit area with a normal in the z-direction is drawn. The force, $\mathbf{F}_z$, acting on it can be decomposed into two shearing forces, P_{xz} and P_{yz}, and a normal force, P_{zz}. The normal force on this unit area is usually called the pressure and corresponds to either a compression or an expansion. The shearing forces are given by the off-diagonal elements of the pressure tensor and are related to the two possible sliding motions within the plane of the surface. Using the pressure tensor the net force on an infinitesimal area da_k oriented in the direction $k = x$, y, or z can be written

$$\mathbf{F}_k = (P_{xk}\hat{\mathbf{i}} + P_{yk}\hat{\mathbf{j}} + P_{zk}\hat{\mathbf{k}})\, da_k. \tag{2.4.11}$$

Consider, now, the various ways in which the ith component of the momentum density fluxes through an arbitrary small surface with orientation $\mathbf{n}$. Clearly it can flux through any of the three components of the area da_x, da_y, and da_z. Moreover, by Newton's law, $dt\,\mathbf{F}_k$ is the change of momentum at the component da_k of this area. Thus the total change of the ith component of momentum at the surface $\mathbf{n}\,da$ is given by

$$dt(F_{ix} + F_{iy} + F_{iz}) = dt(P_{ix}\, da_x + P_{iy}\, da_y + P_{iz}\, da_z) = \mathbf{j}_i \cdot d\mathbf{a}\, dt. \tag{2.4.12}$$

Adding Eqs. (2.4.10) and (2.4.12) gives the total flux density of p_i as

$$\mathbf{j}_i = p_i \mathbf{v} + (P_{ix}\hat{\mathbf{i}} + P_{iy}\hat{\mathbf{j}} + P_{iz}\hat{\mathbf{k}}). \tag{2.4.13}$$

In addition to the convective and nonconvective losses of momentum in a volume, external forces create momentum. These forces act as momentum sources, and if $\mathbf{F}(\mathbf{r}, t)$ is the specific force field, i.e., the force per unit mass, then the momentum source density is $\mathbf{F}\rho$. Putting this expression for s_i along with the expression in Eq. (2.4.13) for the flux density into the conservation equation (2.4.5) leads to

$$\partial \mathbf{p}/\partial t = -\nabla \cdot (\mathbf{p}\mathbf{v}^T) - \nabla \cdot P + \mathbf{F}\rho. \tag{2.4.14}$$

This is the equation which expresses conservation of momentum in the fluid.

The conservation equation for the total energy density, $\varepsilon(\mathbf{r}, t)$, can be derived in a similar manner. The flux density depends, however, on three terms. First there is the convective flux density

$$\mathbf{j} = \varepsilon\mathbf{v}$$

resulting from the bulk flow. Second, there is a flux of internal energy caused by forces that do work on the surface of the volume element. The total work on an area, da_k, in the direction k in time dt is given by

$$(P_{xk}\,dx + P_{yk}\,dy + P_{zk}\,dz)\,da_k, \tag{2.4.15}$$

where, for example, dx is the distance through which the surface moves in the x direction during dt. Thus $dx = v_x\,dt$, etc., so that the work in Eq. (2.4.15) can be rewritten

$$(P_{xk}v_x + P_{yk}v_y + P_{zk}v_z)\,da_k\,dt. \tag{2.4.16}$$

Adding the three contributions like Eq. (2.4.16) for an infinitesimal surface with normal $\mathbf{n}$ then gives

$$\mathbf{v}\cdot P\cdot d\mathbf{a}\,dt \tag{2.4.17}$$

for the work done in dt. The resulting contribution to the flux density is

$$\mathbf{j} = \mathbf{v}\cdot P. \tag{2.4.18}$$

The final contribution to the internal energy flux density is caused by heat conduction due to molecular processes, $\mathbf{j}_q$. Just as the external force $\mathbf{F}$ acts as a source of momentum density, it does work on the fluid at the rate $\mathbf{v}\cdot\mathbf{F}$ per unit mass. This term leads to a source density of internal energy equal to $\rho\mathbf{v}\cdot\mathbf{F}$. Combining all these terms leads to the equation of energy conservation

$$\partial\varepsilon/\partial t = -\nabla\cdot(\varepsilon\mathbf{v}) - \nabla\cdot(\mathbf{v}\cdot P) - \nabla\cdot\mathbf{j}_q + \rho\mathbf{v}\cdot\mathbf{F}. \tag{2.4.19}$$

With the exception of the continuity equation (2.4.9), the conservation equations do not define the hydrodynamic level of description. This is due to the presence of the heat flux vector $\mathbf{j}_q$ and the pressure tensor in the momentum and energy equations. Neither of these quantities are expressed as explicit functions of ρ, $\mathbf{p}$, and ε. In fact, in the most general case of fluid flow neither $\mathbf{j}_q$ nor P can be so simply expressed. Empirically it is known that a range of conditions exist for which the heat flux and pressure tensor can be expressed in terms of the velocity, density, and temperature fields. Clearly a necessary condition is that the molecules which compose the fluid must not possess either significant angular momentum or internal degrees of freedom, such as may be encountered in polymer melts. Otherwise the densities of these variables could not be properly neglected. Moreover, the gradients and magnitudes of the various fields must be smaller than typical molecular values since otherwise a more detailed description of molecular collision dynamics would be needed to determine the values of P and $\mathbf{j}_q$. Under conditions such as these, the pressure tensor and heat flux vector can be expressed in terms of the temperarture, density, and velocity gradients. This empirical form of the heat flux is called *Fourier's law* and is

$$\begin{aligned}\mathbf{j}_q &= K(T,\rho)\nabla(1/T)\\ &= -\kappa(T,\rho)\nabla T,\end{aligned} \tag{2.4.20}$$

where $K(T,\rho)$ depends on the temperature, T, and density, ρ, of the fluid at the position $\mathbf{r}$. The function κ is a transport coefficient called the *thermal conductivity*. Fourier's law is usually stated as the second of the equalities in Eq. (2.4.20), although the first form is more fundamental. The simplest empirical form for the pressure tensor is

$$P_{ij} = p(T,\rho)\delta_{ij} - \sigma'_{ij}, \tag{2.4.21}$$

where $p(T,\rho)$ is the pressure of the fluid as determined by the equilibrium equation of state. The term σ'_{ij} vanishes at equilibrium and is the nonequilibrium portion of the *stress tensor*. It has the form

$$\begin{aligned}\sigma'_{ij} &= 2\eta(T,\rho)\left[e_{ij} - \frac{1}{3}e_{ii}\delta_{ij}\right] + \zeta(T,\rho)e_{ii}\delta_{ij}\\ &= 2\eta\mathring{e}_{ij} + \zeta e_{ii}\delta_{ij}\end{aligned} \tag{2.4.22}$$

where e_{ij} is the *rate of strain tensor* defined by

$$e_{ij} = \frac{1}{2}\left(\frac{\partial v_i}{\partial x_j} + \frac{\partial v_j}{\partial x_i}\right), \tag{2.4.23}$$

and $\mathring{e}$ is the traceless part of e. The quantities η and ζ are called the *shear* and *bulk viscosity coefficients*, respectively. Along with the thermal conductivity, they comprise the transport coefficients for the hydrodynamic level of description. As we shall see later on, the first term in Eq. (2.4.22) gives rise to a loss of internal energy caused by pure shearing motions whereas the second term gives rise to energy dissipation due to expansion or dilation. This form of the stress tensor is sometimes referred to as *Newtonian friction*.

Combining the conservation equations (2.4.9), (2.4.14), and (2.4.19) with Fourier's law (Eq. 2.4.20) and the Newtonian form of the stress tensor (Eqs. 2.4.21 and 2.4.22) yields the basic dynamical equations of the hydrodynamic level:

$$\partial\rho/\partial t = -\nabla\cdot\mathbf{p} \tag{2.4.24}$$

$$\begin{aligned}\partial\varepsilon/\partial t = &-\nabla\cdot(\varepsilon\mathbf{v}) - \nabla\cdot(p\mathbf{v} - \zeta\mathbf{v}\nabla\cdot\mathbf{v}) + 2\nabla\cdot(\eta\mathbf{v}\cdot\mathring{e})\\ &- \nabla\cdot[K\nabla(1/T)] + \rho\mathbf{v}\cdot\mathbf{F}\end{aligned} \tag{2.4.25}$$

$$\partial\mathbf{p}/\partial t = -\nabla\cdot(\mathbf{v}\mathbf{p}^T) - \nabla p + 2\nabla\cdot\eta\mathring{e} + \nabla(\zeta\nabla\cdot\mathbf{v}) + \rho\mathbf{F}. \tag{2.4.26}$$

These equations become a closed set of partial differential equations for ρ, ε and $\mathbf{p}$ when the mechanical and thermal equilibrium equations of state are given. Thus one needs to know p and T as a function of ρ and ε.

The hydrodynamic level of description is not limited to conditions that are close to equilibrium, but rather by the molecular size of velocities and gradients. The basic dynamical equations, however, take on a simple form if they are restricted to near a state of equilibrium. At the hydrodynamic level a state of equilibrium is a steady state for which the stress tensor and the heat flux vector vanish identically. According to Fourier's and Newton's laws (Eqs. 2.4.20 and 2.4.22), this implies that the temperature and velocity fields are independent of position. In fact, if the velocity is a constant $\mathbf{v}$ different from zero, we can take it to be zero by changing to a Galilean reference frame of velocity $-\mathbf{v}$. Thus at the hydrodynamic level, equilibrium is characterized by a constant temperature, zero velocity, and mass density and internal energy densities that are time independent. At equilibrium Eq. (2.4.26) implies the further condition that

$$\nabla p^e = \mathbf{F}\rho^e, \tag{2.4.27}$$

where the superscript e implies the equilibrium pressure and density fields. This is sometimes called the *mechanical condition of equilibrium*. It can be used to provide the final thermodynamic criterion for equilibrium, if we recall that p depends on ρ and T as dictated by the equilibrium equation of state. Thus we can use the *Gibbs–Duhem relationship* which follows from Eq. (2.2.4), namely,

$$Md(-\mu/T) + Vd(p/T) + Ed(1/T) = 0, \tag{2.4.28}$$

with M the mass and μ the mass-based chemical potential. By rearranging Eq. (2.4.28) it follows that

$$dp = -T\rho d(-\mu/T) - Thd(1/T), \tag{2.4.29}$$

where ρ is the mass density, $h = e + p$ is the enthalpy density, and $e = E/V$. Applying the fact that $\nabla p \cdot d\mathbf{s} = dp$, we conclude from Eq. (2.4.29) that

$$\nabla p = -T\rho\nabla(-\mu/T) - Th\nabla(1/T). \tag{2.4.30}$$

As a consequence of Eq. (2.4.30) and the fact that the temperature is constant at equilibrium, the mechanical condition of equilibrium, Eq. (2.4.27), takes the form

$$\nabla\mu^e = \mathbf{F}. \tag{2.4.31}$$

Thus if the external force is conservative, so that $\mathbf{F} = -\nabla\phi$, the mechanical condition of equilibrium is

$$\nabla(\mu^e + \phi) = \mathbf{0}. \tag{2.4.32}$$

The combination $\mu + \phi$ is called the electrochemical potential, and according to Eq. (2.4.32) its value at equilibrium is constant throughout a pure substance.

To obtain the linear regression equations for the hydrodynamic level of description, we need only to linearize the Navier–Stokes equations and rearrange into the flux–force form given in Eq. (2.3.3). This means that we must also introduce the variables conjugate to the deviations of the extensive densities from their equilibrium values. Because the entropy of an extended substance will differ from point to point in a nonuniform system, we must also introduce the entropy density, $s(\mathbf{r}, t)$. The entropy density inherits its dependence on position from the densities of the extensive variables ρ and e, where e is the *internal energy density*. Thus we write $s(\rho, e)$. However, we can obtain a dependence of the entropy density on the local value of the momentum by recognizing that the total energy density can be written as $\varepsilon = e + p^2/2\rho + \rho\phi$, where we consider only the special case that $\mathbf{F} = -\nabla\phi$. Thus $s(\rho, \varepsilon, \mathbf{p}) \equiv s(\rho, \varepsilon - p^2/2\rho - \rho\phi)$. Just as for Brownian motion in Section 2.2, we deduce that the variables conjugate to ρ, ε, and $\mathbf{p}$ are

$$\begin{aligned}(\partial s/\partial\rho)_{\varepsilon,\mathbf{p}} &= -(\mu + \phi)/T + v^2/T \\ (\partial s/\partial\varepsilon)_{\rho,\mathbf{p}} &= 1/T \\ (\partial s/\partial p_i)_{\rho,\varepsilon,p_j} &= -v_i/T.\end{aligned} \tag{2.4.33}$$

Conveniently, the gradients of all the conjugate intensive variables vanish at equilibrium.

Using the conjugate variables we can linearize the hydrodynamic equations, (2.4.24)–(2.4.26), around equilibrium. One finds that

$$\partial\Delta\rho/\partial t = \nabla\cdot[(\rho^e T^e)(-\mathbf{v}/T)] \tag{2.4.34}$$

$$\partial\Delta\varepsilon/\partial t = \nabla\cdot[T^e(h^e + \rho^e\phi)(-\mathbf{v}/T)] - \nabla\cdot[K^e\nabla(1/T)] \tag{2.4.35}$$

$$\begin{aligned}\partial\Delta\mathbf{p}/\partial t = {}& \rho^e T^e\nabla\cdot[-(\mu + \phi)/T] + T^e(h^e + \rho^e\phi)\nabla(1/T) \\ & - 2T^e\nabla\cdot\eta^e\mathring{X} - T^e\nabla\zeta^e[\nabla\cdot(-\mathbf{v}/T)].\end{aligned} \tag{2.4.36}$$

In obtaining Eq. (2.4.36) from Eq. (2.4.26) we used the Gibbs–Duhem identity (Eq. 2.4.30) and the fact that $\rho\mathbf{F} = -\rho\nabla\phi$ to write

$$-\nabla p + \rho\mathbf{F} = T\rho\nabla[-(\mu + \phi)/T] + T(h + \rho\phi)\nabla(1/T), \tag{2.4.37}$$

and the shorthand notation

$$\mathring{X}_{ij} = \frac{1}{2}\left[\frac{\partial(-v_i/T)}{\partial x_j} + \frac{\partial(-v_j/T)}{\partial x_i}\right] - \frac{1}{3}\frac{\partial(v_k/T)}{\partial x_k}\delta_{ij}. \tag{2.4.38}$$

These equations are in the form of the linear flux–force regression equations. More explicitly, define $\Delta\rho = a_1$, $\Delta\varepsilon = a_2$, and $\Delta\mathbf{p} = \mathbf{a}_3$ and their corresponding thermodynamic forces $X_1 = -[(\mu + \phi)/T - (\mu^e + \phi)/T^e]$, $X_2 = (1/T - 1/T^e)$, and $\mathbf{X}_3 = -\mathbf{v}/T$. Using this notation Eqs. (2.4.34)–(2.4.36) become

$$\begin{aligned}
\partial a_1/\partial t &= \mathbf{\nabla}\cdot[\rho^e T^e \mathbf{X}_3] \\
\partial a_2/\partial t &= -\mathbf{\nabla}\cdot(K^e \mathbf{\nabla} X_2) + \mathbf{\nabla}\cdot[T^e(h^e + \rho^e\phi)\mathbf{X}_3] \\
\partial \mathbf{a}_3/\partial t &= \rho^e T^e \mathbf{\nabla} X_1 + T^e(h^e + \rho^e\phi)\mathbf{\nabla} X_2 - T^e\mathbf{\nabla}\cdot(\eta^e \mathbf{\nabla}\mathbf{X}_3^T) \\
&\quad - T^e\mathbf{\nabla}([\zeta^e + (\eta^e/3)]\mathbf{\nabla}\cdot\mathbf{X}_3)
\end{aligned} \tag{2.4.39}$$

Equations (2.4.39) are the linear laws for nonequilibrium thermodynamics at the hydrodynamic level of description. In fact, just as for the heat transfer process described in Section 2.3, we could have derived the form of these equations using the Onsager principle instead of basing their derivation on the phenomenological laws of Newton and Fourier. The present derivation, however, provides us with an independent verification of the symmetry properties of the relaxation matrix, L. Reexamining Eqs. (2.4.39), L is seen to have components which are differential operators. Thus, for example, $L_{11} = L_{12} = 0$ and

$$L_{13}\cdot = \frac{\partial}{\partial x}[\rho^e T^e \cdot], \tag{2.4.40}$$

where the dot reminds us that the operator acts on a function in the place where the dot appears. Similarly the matrix element L_{31} has the form

$$L_{31}\cdot = \rho^e T^e \frac{\partial\,\cdot}{\partial x}. \tag{2.4.41}$$

These elements connect an even variable (the mass density) with an odd variable (the x-component of the momentum density). Thus according to the Onsager–Casimir principle, these elements must show antisymmetry. To verify this, it is not sufficient simply to interchange the indices 1 and 3, because the L_{ij} are differential operators. In fact, a more complete representation of these operators is given by their integral form

$$(L_{ij}f)(\mathbf{r}) = \int L_{ij}(\mathbf{r},\mathbf{r}')f(\mathbf{r}')\,d\mathbf{r}',$$

where, for example,

$$L_{13}(\mathbf{r},\mathbf{r}') = \frac{\partial}{\partial x}[T^e\rho^e(\mathbf{r}')\delta(\mathbf{r}-\mathbf{r}')\cdot] \tag{2.4.42}$$

and

$$L_{31}(\mathbf{r},\mathbf{r}') = T^e\rho^e(\mathbf{r})\frac{\partial}{\partial x}[\delta(\mathbf{r}-\mathbf{r}')\cdot]. \tag{2.4.43}$$

The integral representation reflects the fact that the densities of the extensive variables are a continuum representation of an entire field of open thermodynamic systems. Consequently, the matrix L has both the discrete indices i and j and continuous indices $\mathbf{r}$ and $\mathbf{r}'$. Thus to see the antisymmetry of the terms of Eqs. (2.4.40) and (2.4.41) we need to show that

$$L_{13}(\mathbf{r}, \mathbf{r}') = -L_{31}(\mathbf{r}', \mathbf{r}). \tag{2.4.44}$$

This can be proven rather directly by writing

$$\begin{aligned} L_{31}(\mathbf{r}', \mathbf{r}) &= T^e \rho^e(\mathbf{r}') \frac{\partial}{\partial x'} [\delta(\mathbf{r} - \mathbf{r}') \cdot] \\ &= -T^e \rho^e(\mathbf{r}') \frac{\partial}{\partial x} [\delta(\mathbf{r} - \mathbf{r}') \cdot] \\ &= -\frac{\partial}{\partial x} [T^e \rho^e(\mathbf{r}') \delta(\mathbf{r} - \mathbf{r}') \cdot] = -L_{13}(\mathbf{r}, \mathbf{r}'), \end{aligned}$$

where the second equality follows from the definition of the derivative of a delta function. In the same way it is easy to see that the other matrix elements connecting the density and momentum in Eq. (2.4.39) are antisymmetric. Furthermore,

$$L_{23}(\mathbf{r}, \mathbf{r}') = \frac{\partial}{\partial x} \{T^e [h^e(\mathbf{r}') + \rho^e(\mathbf{r}') \phi(\mathbf{r}')] \delta(\mathbf{r} - \mathbf{r}') \cdot\} = -L_{32}(\mathbf{r}', \mathbf{r})$$

is part of an antisymmetric coupling between the total energy and the momentum, as predicted by the Onsager–Casimir principle. The remaining matrix elements couple energy to energy, i.e.,

$$\begin{aligned} L_{22}(\mathbf{r}, \mathbf{r}') = &-\frac{\partial}{\partial x}\left[K^e \frac{\partial}{\partial x'} \delta(\mathbf{r} - \mathbf{r}') \cdot\right] - \frac{\partial}{\partial y}\left[K^e \frac{\partial}{\partial y'} \delta(\mathbf{r} - \mathbf{r}') \cdot\right] \\ &- \frac{\partial}{\partial z}\left[K^e \frac{\partial}{\partial z'} \delta(\mathbf{r} - \mathbf{r}') \cdot\right] = L_{22}(\mathbf{r}', \mathbf{r}) \end{aligned} \tag{2.4.45}$$

and momentum to momentum, i.e.,

$$\begin{aligned} L_{ij}(\mathbf{r}, \mathbf{r}') = &-T^e \delta_{ij} \left(\frac{\partial}{\partial x}\left[\eta^e \frac{\partial}{\partial x'} \delta(\mathbf{r} - \mathbf{r}') \cdot\right] + \frac{\partial}{\partial y}\left[\eta^e \frac{\partial}{\partial y'} \delta(\mathbf{r} - \mathbf{r}') \cdot\right] \right. \\ &\left. + \frac{\partial}{\partial z}\left[\eta^e \frac{\partial}{\partial z'} \delta(\mathbf{r} - \mathbf{r}') \cdot\right]\right) - T^e \frac{\partial}{\partial x_i}\left([\zeta^e + (\eta^e/3)] \frac{\partial}{\partial x_j'} \delta(\mathbf{r} - \mathbf{r}') \cdot\right) \\ = &\ L_{ji}(\mathbf{r}', \mathbf{r}). \end{aligned} \tag{2.4.46}$$

These terms all involve second partial derivatives whose i and j indices are symmetric, and they constitute the symmetric part of L. Since these elements connect variables of the same time reversal symmetry, the fact that they are symmetric is compatible with the Onsager principle.

The symmetric part of L leads to a positive contribution to the dissipation function in Eq. (2.3.19). Thus, Eqs. (2.4.45) and (2.4.46) demonstrate that it is the molecular processes of heat and momentum transport that give rise to dissipation. Although the antisymmetric part of the coupling matrix effects the time rate of change of the extensive densities, it does not lead to dissipation or entropy production. Indeed, in the absence of the viscosity and heat conductivity, the hydrodynamic equations reduce to the *Euler equations*. These equations are completely conservative and give rise to the kinematic terms which comprise the antisymmetric part of L.

One may also verify the second general prediction of the Onsager theory, namely, that the symmetric part of L is positive semi-definite. Consider, as an example, the contribution to L^s due to the transport of heat. To check that its contribution is positive definite, we need to verify that

$$X_2 \cdot L_{22} X_2 \equiv \int d\mathbf{r} \int d\mathbf{r}' \frac{1}{T}(\mathbf{r}) L_{22}(\mathbf{r}, \mathbf{r}') \frac{1}{T}(\mathbf{r}') \geq 0. \tag{2.4.47}$$

Using Eq. (2.4.45) the double integral can be written

$$X_2 \cdot L_{22} X_2 = -\int (1/T) \nabla \cdot K^e \nabla (1/T) \, d\mathbf{r}. \tag{2.4.48}$$

For an equilibrium ensemble the gradient of the temperature vanishes at the boundaries of the system. Thus the integral can be integrated by parts to obtain

$$X_2 \cdot L_{22} X_2 = \int K^e |\nabla(1/T)|^2 \, d\mathbf{r} \tag{2.4.49}$$

which is certainly non-negative. This confirms the general prediction of the Second Law of thermodynamics that the heat conduction—like all molecular processes—can create entropy, but never destroy it.

2.5. Symmetry of the Two-Time Correlation Function and the Reciprocal Relations

In the preceding section the Onsager–Casimir reciprocal relations were derived at the hydrodynamic level of description from the phenomenological laws of Newton and Fourier. Although we will pursue a generalization of the phenomenological approach in subsequent chapters, it is possible to argue for the reciprocal relations in a more general fashion. The idea, due to Onsager and extended by Casimir, is based on the two-time correlation function of the extensive variables in an equilibrium ensemble. According to Eqs. (1.3.15) and (1.3.24) the two-time correlation function can be written in terms of the single-time average

$$C_{ij}(t, t') = \langle a_i^0(t) \langle a_j(t') \rangle^0 \rangle_1,$$

where the superscript indicates the conditional average. Since the equilibrium ensemble is stationary, this simplifies to

$$\begin{aligned} C_{ij}(\tau) &= \langle a_i^0 \langle a_j(\tau) \rangle^0 \rangle_1 \\ &= \langle a_i^0 \overline{a_j(\tau)} \rangle_1 \end{aligned} \tag{2.5.1}$$

where $\tau = t' - t$ and we have returned to the overbar notation for the conditional average introduced in Section 2.3.

To derive the reciprocal relations, we calculate the correlation function using the Onsager theory. According to the linear regression law in Eqs. (2.3.3), $\bar{a}_j$ satisfies the equation

$$d\bar{a}_j/dt = \sum_k L_{jk} \bar{X}_k. \tag{2.5.2}$$

The force $\bar{X}_k$, however, is defined in terms of the conjugate intensive thermodynamic variables by Eq. (2.3.2). Thus to linear order in deviations from equilibrium, it follows that

$$\bar{X}_k = \sum_l S_{kl} \bar{a}_l \tag{2.5.3}$$

where S_{kl} is the matrix of second partial derivatives of the entropy given in Eq. (2.3.5). Combining Eqs. (2.5.2) and (2.5.3) then yields a differential equation solved by the extensive variables. We write it as

$$d\bar{a}_j/dt = \sum_l H_{jl} \bar{a}_l \tag{2.5.4}$$

where

$$H_{jl} = \sum_k L_{jk} S_{kl}. \tag{2.5.5}$$

H is the relaxation matrix governing the return of the average values of the extensive variables to equilibrium. The solution of Eq. (2.5.4), in vector notation, is

$$\bar{a}(t) = \exp(Ht)\mathbf{a}^0, \tag{2.5.6}$$

which satisfies the initial condition that $\bar{\mathbf{a}}(0)$ is identically $\mathbf{a}^0$. Using Eq. (2.5.1) we can write the correlation matrix, again in vector notation, as

$$C(\tau) = \langle \mathbf{a}^0 \bar{\mathbf{a}}^T(\tau) \rangle_1. \tag{2.5.7}$$

Thus Eq. (2.5.6) implies that

$$C(\tau) = \langle \mathbf{a}^0 \mathbf{a}^{0T} \rangle \exp(H^T \tau)$$

or

$$C^T(\tau) = \exp(H\tau) \langle \mathbf{a}^0 \mathbf{a}^{0T} \rangle. \tag{2.5.8}$$

These expressions for the correlation function are still formal since we do not yet have an explicit expression for the single-time correlation function, $\langle \mathbf{a}^0 \mathbf{a}^{0T} \rangle$. To accomplish this in the Onsager theory, it is necessary to introduce an additional assumption about the stationary probability density. According to ideas developed by Boltzmann and enlarged upon by Planck, the probability that an intensive veriable **a** lies in a unit interval of volume Δ in the equilibrium ensemble can be written

$$W_1(\mathbf{a})\Delta = \exp\{[S(\mathbf{a}) - S^e]/k_B\}. \tag{2.5.9}$$

This assumption is called the *Boltzmann–Planck postulate* and gives the single-time probability density, $W_1(\mathbf{a})$, in terms of the local equilibrium entropy, $S(\mathbf{a})$, and the equilibrium value of the entropy, $S(\mathbf{0}) = S^e$. Although this formula can be derived from the microcanonical distribution of Gibbs, since neither has any justification in mechanics, it must be taken as an additional assumption.

Away from critical points of phase transitions, the Boltzmann–Planck formula simplifies considerably. As we saw in Section 2.3, the Second Law of thermodynamics implies the Taylor series expansion of S around equilibrium:

$$\begin{aligned} S(\mathbf{a}) &= S^e + \frac{1}{2}\sum_i \sum_j S_{ij} a_i a_j + \cdots \\ &= S^e + \frac{1}{2}\delta^2 S + \cdots, \end{aligned} \tag{2.5.10}$$

where in the second equality the quadratic form has been indicated by $\delta^2 S$, the *second differential of the entropy*. Critical points occur whenever $\delta^2 S$ vanishes for small nonzero values of **a**. On the other hand, near an equilibrium state that is not a critical point, the entropy maximum principle guarantees that $\delta^2 S$ must be strictly negative, i.e.,

$$\delta^2 S < 0. \tag{2.5.11}$$

This means that the matrix S_{ij} of second partial derivatives is a negative definite matrix, and so to dominant order in the deviations from equilibrium, Eq. (2.5.10) can be written

$$S(\mathbf{a}) - S^e = \frac{1}{2}\delta^2 S. \tag{2.5.12}$$

Thus for equilibrium states that are not too close to critical points, the Boltzmann–planck postulate implies that

$$W_1(\mathbf{a}) = N \exp(\delta^2 S / 2k_B). \tag{2.5.13}$$

The fact that $\delta^2 S$ is a quadratic form means that W_1 is a Gaussian distribution. The value of the normalization constant can be obtained from Eq. (1.3.28). This gives

$$W_1(\mathbf{a}) = [(2\pi)^m \det(E^{-1})]^{-1/2} \exp(-\mathbf{a}^T E \mathbf{a}/2) \tag{2.5.14}$$

where we have introduced the matrix

$$E_{ij} = -S_{ij}/k_B, \tag{2.5.15}$$

and m denotes the number of extensive variables. Equation (2.5.13) was first derived by Einstein and is usually called the *Einstein formula.*

The Einstein formula permits us to calculate the single-time correlation function. Indeed, comparing Eq. (2.5.14) with the Gaussian results in Eqs. (1.3.28) and (1.3.31) gives immediately that

$$\langle \mathbf{a}^0 \mathbf{a}^{0T} \rangle = E^{-1} = -k_B S^{-1}. \tag{2.5.16}$$

This is the result we were looking for. It permits us to write the two-time correlation function in Eq. (2.5.8) as

$$C(\tau) = E^{-1} \exp(H^T \tau)$$

or

$$C^T(\tau) = \exp(H\tau) E^{-1}. \tag{2.5.17}$$

Using Eq. (2.5.5) we find that

$$H = -k_B L E. \tag{2.5.18}$$

Hence in the Onsager theory the two-time correlation function $C(\tau)$ is determined completely by the dynamic coupling matrix, L, and the thermodynamic coupling matrix, E.

To derive the reciprocal relations from Eq. (2.5.17), it is necessary to take advantage of two general symmetry properties of the two-time correlation function. Recall from its definition in Eq. (1.3.15) and the property of stationarity that $C(\tau)$ can be calculated from the two-time average

$$C_{ij}(\tau) \equiv \langle a_i(t) a_j(t') \rangle, \tag{2.5.19}$$

where $t' - t = \tau$. By changing the order of the product inside the average it follows that

$$C_{ij}(\tau) = \langle a_j(t') a_i(t) \rangle \equiv C_{ji}(-\tau).$$

Thus for any stationary process the two-time correlation function satisfies the symmetry property

$$C_{ij}(\tau) = C_{ji}(-\tau). \tag{2.5.20}$$

A second symmetry property is conferred by time reversal symmetry. Recall that the extensive variables encountered so far are either even or odd under time reversal. As described in Section 2.2, the operation of time reversal changes the sign of all the momenta, external magnetic fields, and the time. For an even or odd variable, a_j, this means that for $\tau > 0$ the reversed value is

$$\mathscr{T} a_j(\tau) = \varepsilon_j a_j(-\tau), \tag{2.5.21}$$

where the reversed value of any magnetic field is left implicit, and $\varepsilon_j = 1$ for even variables and -1 for odd variables. To see what this implies for the correlation function, we need to write out Eq. (2.5.19) explicitly in terms of the phase space average

$$C_{ij}(\tau) = \int a_i(x, t) a_j(x, t') \rho^e(x)\, dx, \tag{2.5.22}$$

where x represents all the canonical coordinates and momenta which specify the system. The function $\rho^e(x)$ is the equilibrium probability density in phase space, for example, the canonical or microcanonical distribution. Since ρ^e depends only on the total energy, or other constants of the motion, it is time independent and invariant under time reversal. The time reversal operator $\mathscr{T}$ changes a phase space point to the corresponding point with the same coordinates, reversed momenta, and reversed magnetic field. Thus in Eq. (2.5.22) we change integration variables from x to $x' = \mathscr{T}x$, its reversed value, and find that

$$C_{ij}(\tau) = \int a_i(\mathscr{T}x', t) a_j(\mathscr{T}x', t') \rho^e(x')\, dx' \tag{2.5.23}$$

since ρ^e is invariant under $\mathscr{T}$ and the magnitude of the Jacobian of the transformation $x' = \mathscr{T}x$ is unity. The values of a_i and a_j in Eq. (2.5.23) are time reversed, $a_i(\mathscr{T}x', t) = \mathscr{T}a_i(t)$. Thus they can be replaced in Eq. (2.5.23) using Eq. (2.5.21), which yields

$$\begin{aligned} C_{ij}(\tau) &= \varepsilon_i \varepsilon_j \int a_i(x', -t) a_j(x', -t') \rho^e(x')\, dx' \\ &\equiv \varepsilon_i \varepsilon_j C_{ij}(-\tau), \end{aligned} \tag{2.5.24}$$

where the second equality follows from the definition of C in Eq. (2.5.22). In Eqs. (2.5.21)–(2.5.24) we have suppressed the dependence of C_{ij} on the magnetic field. Because the right-hand side of Eq. (2.5.24) arises from time reversal, it must involve a magnetic field that is reversed with respect to the left-hand side. Thus, more explicitly, the equality in Eq. (2.5.24) becomes

$$C_{ij}(\mathbf{B}, \tau) = \varepsilon_i \varepsilon_j C_{ij}(-\mathbf{B}, -\tau). \tag{2.5.25}$$

Combining this result with Eq. (2.5.20) then gives the second important symmetry property for C_{ij}, namely,

$$C_{ij}(\mathbf{B}, \tau) = \varepsilon_i \varepsilon_j C_{ji}(-\mathbf{B}, \tau). \tag{2.5.26}$$

Because this property arises from the underlying reversibility of the mechanical equations of motion, it is often referred to as *microscopic reversibility*.

The reciprocal relations follow directly from the property of microscopic reversibility. Indeed, taking the time derivative of this equation at $\tau = 0$ implies that

$$dC_{ij}(-\mathbf{B}, 0)/d\tau = \varepsilon_i \varepsilon_j \, dC_{ji}(\mathbf{B}, 0)/d\tau. \tag{2.5.27}$$

As this equality is valid for any two-time correlation function, it must hold for the correlation functions in the Onsager theory. Differentiating the explicit expression for C_{ij} given in Eq. (2.5.17) as indicated in Eq. (2.5.27) implies that

$$[E^{-1}(-\mathbf{B})H^T(-\mathbf{B})]_{ij} = \varepsilon_i \varepsilon_j [H(\mathbf{B})E^{-1}(\mathbf{B})]_{ij}. \tag{2.5.28}$$

If Eq. (2.5.18) is used to eliminate $H = -k_B LE$ in favor of the matrix L in this equality, the result is

$$-k_B[E^{-1}(-\mathbf{B})E(-\mathbf{B})L^T(-\mathbf{B})]_{ij} = -k_B \varepsilon_i \varepsilon_j [L(\mathbf{B})E(\mathbf{B})E^{-1}(\mathbf{B})]_{ij},$$

from which it follows that

$$L_{ij}(\mathbf{B}) = \varepsilon_i \varepsilon_j L_{ji}(-\mathbf{B}). \tag{2.5.29}$$

This is the Onsager–Casimir reciprocal relationship given earlier in Eq. (2.2.12).

2.6. Fluctuations in the Onsager Theory

The Onsager theory as described in the preceding sections provides a partial picture of the dynamics of systems that are close to equilibrium. The picture is partial because the theory indicates only the average time dependence of the extensive variables. In this section we show how the theory can be extended to describe fluctuations around the average. Recall the argument in Section 1.2 that the complete description of a large system is necessarily stochastic and that the best one can do is to understand the properties of an ensemble of similar systems. In Section 2.5 we touched briefly on this aspect of the Onsager theory by adding the Boltzmann–Planck postulate to the linear regression equations. This allowed us to specify the lowest-order joint distribution function, $W_1(\mathbf{a})$, which away from critical points turned out to be Gaussian. Thus we already know that the single-time fluctuations are stationary and Gaussian. To complete the Onsager picture, it is necessary to understand the behavior of fluctuations at several times, that is, to describe the dynamics of the fluctuations.

A partial picture of the dynamics is provided by Eq. (2.5.4). That equation written in matrix notation is

$$d\bar{\mathbf{a}}/dt = H\bar{\mathbf{a}}. \tag{2.6.1}$$

By assumption Eq. (2.6.1) describes the relaxation of the conditional average. Its solution is the exponential

$$\bar{\mathbf{a}}(\mathbf{a}^0, t) = \exp(Ht)\mathbf{a}^0, \tag{2.6.2}$$

where $\mathbf{a}^0$ is the vector of initial deviations from equilibrium in the conditional ensemble. A given system in the ensemble will, of course, not exhibit this exponential relaxation. Thus the fluctuation, $\delta\mathbf{a}(t) = \mathbf{a}(t) - \bar{\mathbf{a}}(\mathbf{a}^0, t)$, for a given system will be nonzero for $t > 0$. However, on the basis of experience we expect that the fluctuation will be small, at least away from a critical point. According to Onsager, this small fluctuation is similar to an impressed macroscopic deviation, only it has appeared spontaneously rather than being deliberately prepared. Thus it should satisfy a dynamical equation similar to that for the average, but with an additional term to describe the random influences of molecular motion. In other words,

$$d\delta\mathbf{a}/dt = H\delta\mathbf{a} + \tilde{\mathbf{f}}, \tag{2.6.3}$$

where $\tilde{\mathbf{f}}$ is a random term. Equation (2.6.3) is *Onsager's regression hypothesis* for fluctuations.

The regression hypothesis is strikingly reminiscent of the Langevin equation for Brownian motion discussed in Chapter 1. Indeed, adding Eq. (2.6.1) to Eq. (2.6.3) shows that

$$d\mathbf{a}/dt = H\mathbf{a} + \tilde{\mathbf{f}}. \tag{2.6.4}$$

If the system of interest is a Brownian particle and its momentum, $\mathbf{p}$, is the only extensive variable being considered, then $\mathbf{a} = \mathbf{p}$ and Eq. (2.6.4) would have the form of the Langevin equation (Eq. 1.6.1). As we saw in Section 1.8, the natural generalization of the stochastic process that satisfies the Langevin equation is a stationary, Gaussian, Markov process. We already know that the equilibrium ensemble is stationary, and the Einstein formula in Eq. (2.5.14) shows that the single-time distribution is Gaussian. Furthermore, the regression hypothesis leads to a linear stochastic differential equation, Eq. (2.6.4), just like that for a stationary, Gaussian, Markov process. Thus it is natural and consistent to complete the Onsager picture with the assumption that fluctuations in the equilibrium ensemble are Gaussian and Markovian. In other words, the random term in Eq. (2.6.4) is a multivariate white noise and the stochastic process is an *Ornstein–Uhlenbeck process.*

The assumption of a stationary, Gaussian, Markov process completely defines the Onsager theory. Because the random component of the time derivative in Eq. (2.6.4) is white noise, its average is zero as are all its correlation functions involving an odd number of times. Its even time correlation functions satisfy the Gaussian equation (1.6.6). Moreover, we can use the fact that the two-time conditional density, P_2, of an Ornstein–Uhlenbeck process approaches the Gaussian equilibrium density, W_1, to obtain an explicit expression for the two-time correlation function of $\tilde{\mathbf{f}}$. This was done in Section 1.8 and yields

$$\langle \tilde{\mathbf{f}}(t)\tilde{\mathbf{f}}^T(t') \rangle = \gamma\delta(t' - t), \tag{2.6.5}$$

where γ satisfies the fluctuation–dissipation theorem in Eq. (1.8.14), i.e.,

$$H\sigma + \sigma H^T = -\gamma, \tag{2.6.6}$$

with σ the equilibrium covariance matrix, $\langle \mathbf{a}\mathbf{a}^T \rangle$.

The fluctuation–dissipation theorem shows that the strength, γ, of the random terms in the time derivatives of the extensive variables is completely determined by the relaxation matrix, H, and the covariance matrix, σ. The fluctuation–dissipation theorem is actually a more remarkable result than that. Returning to the definition of σ in Eq. (1.3.31) we see from the Einstein formula (Eq. 2.5.14) and Eq. (2.5.16) that

$$\sigma = E^{-1} = -k_B S^{-1}. \tag{2.6.7}$$

Moreover, Eq. (2.5.18) shows that $H = -k_B LE$. Putting these two expressions into the left-hand side of Eq. (2.6.6) then gives that

$$\gamma = k_B(L + L^T) = 2k_B L^s, \tag{2.6.8}$$

where L^s is the symmetric part of the coupling matrix L[see Eq. (2.3.17)]. In other words, the strength of the random terms is just proportional to the symmetric part of L. Thus not only does L describe the average dynamics of the extensive variables as given by Eq. (2.5.2) and the dissipation function in Eq. (2.3.19), but it describes the strength of fluctuations in the extensive variables, too! Aside from the dynamic coupling matrix L, the only other quantity required in the Onsager theory is the thermodynamic coupling matrix, $S_{ij} = \partial^2 S/\partial n_i \, \partial n_j$, which is determined by the local equilibrium entropy.

The formal structure of the Onsager theory is easily summarized: In the equilibrium ensemble, the deviations of the extensive variables around their equilibrium values, $\mathbf{a} = \mathbf{n} - \mathbf{n}^e$, are a stationary, Gaussian, Markov process. The single-time probability density of the deviations is a Gaussian centered at zero with the covariance

$$\sigma = \langle \mathbf{a}\mathbf{a}^T \rangle = -k_B S^{-1}. \tag{2.6.9}$$

As a function of time the deviations satisfy the Langevin-type equation

$$d\mathbf{a}/dt = L\mathbf{X} + \tilde{\mathbf{f}} \tag{2.6.10}$$

with

$$\begin{aligned} \langle \tilde{\mathbf{f}}(t) \rangle &= \mathbf{0} \\ \langle \tilde{\mathbf{f}}(t)\tilde{\mathbf{f}}^T(t') \rangle &= 2k_B L^s \delta(t' - t) \end{aligned} \tag{2.6.11}$$

and $\mathbf{X}$ is the thermodynamic force defined by

$$\mathbf{X} = S\mathbf{a}. \tag{2.6.12}$$

Moreover, L is positive semi-definite, determines the Rayleigh–Onsager dissipation fucntion

$$\Phi = \mathbf{X}^T L \mathbf{X} \geq 0, \tag{2.6.13}$$

and satisfies the Onsager–Casimir reciprocal relations

$$L_{ij}(\mathbf{B}) = \varepsilon_i \varepsilon_j L_{ji}(-\mathbf{B}), \tag{2.6.14}$$

where $\varepsilon_i = +1$ for an even variable and -1 for an odd variable. Equations (2.6.9)–(2.6.14) summarize the basic structure of Onsager's statistical theory of equilibrium ensembles.

2.7. The Boltzmann Equation

The Onsager theory provides a statistical description of extensive variables in the equilibrium ensemble. Because fluctuations are small when the equilibrium state is not near a critical point, the dynamic fluctuations solve linear equations. Moreover, the Onsager theory is an avowedly macroscopic theory since it depends only upon the entropy and the transport coefficients which form the coupling matrix, L. Nowhere is there reference to molecules or the underlying mechanical equations of motion. The theory is, nonetheless, quite powerful and can be used, in conjunction with the conservation equations in Section 2.4, to derive the form of the linearized equations of hydrodynamics and to make predictions about light scattering and other experiments that measure fluctuations.

Although Onsager's work provided the first statistical theory of the equilibrium ensemble, questions about the molecular origin of irreversible thermodynamics were being asked by Clausius, Maxwell, and others soon after the Second Law of thermodynamics was formulated. This work, which culminated in Boltzmann's investigations of the molecular dynamics of dilute gases, was primarily concerned with the average behavior of molecules in the gas phase. Fluctuations were not a part of the theory and were held to be small and unimportant. What Boltzmann provided was a consistent picture of how the nonlinear dynamics of molecular collisions affect the average behavior in an ensemble. Nonlinear effects are completely absent in the Onsager theory, although we show in Section 2.11 that the linearized Boltzmann equation is, in fact, just a special case of the Onsager theory.

In Boltzmann's approach the macroscopic variables are a generalization of the local energy, mass, and momenta which enter the hydrodynamic level of description. Consider a single-component substance. A six-dimensional continuum is required to specify simultaneously the position, $\mathbf{r}$, and velocity, $\mathbf{v}$, of the center of mass of a single molecule. For macroscopic purposes we can divide the six-dimensional space—called the *μ-space* or molecule phase space —into small cellular volume elements. If each volume element is assigned an index $i = 1, 2, 3, \ldots$ to distinguish it from other volume elements, then the variables in the Boltzmann level of description are the number of molecules, $N_i(t)$, that occupy the volume elements. These occupancy numbers are extensive variables since they are proportional to the size of the volume element. They are also stochastic variables since specification of the number of molecules in these volume elements does not provide enough information to specify deterministically which collisions will occur. It is the incomplete specification of the mechanical coordinates at the Boltzmann level of description—just as at the thermodynamic or hydrodynamic levels—that leads to the stochastic nature of the Boltzmann picture.

One of Boltzmann's important contributions to statistical mechanics is the nonlinear Boltzmann equation, which describes the effect of collisions on the occupancy numbers of volume elements in μ-space. For short-range forces, collisions take place between molecules in volume elements with the same coordinate but, in general, different velocities. In a dilute gas, binary collisions dominate so that only two volume elements, located at $\mathbf{r}$, $\mathbf{v}$ and $\mathbf{r}$, $\mathbf{v}_1$, are involved. At the end of the collision, each of these two volume elements will contain one less molecule, while a new molecule will appear in volume elements located at $\mathbf{r}$, $\mathbf{v}'$ and $\mathbf{r}$, $\mathbf{v}_1'$. The primes are used to represent the center of mass velocities *after* the collision. If we consider these volume elements to be infinitesimal, the extensive property of the occupancy numbers can be used to introduce the μ-space number density, $\rho(\mathbf{r}, \mathbf{v}, t)$. Thus

$$\rho(\mathbf{r}, \mathbf{v}, t)\,d\mathbf{r}\,d\mathbf{v} = \text{the number of molecules with center of mass position and velocity in the ranges } [\mathbf{r}, \mathbf{r} + d\mathbf{r}] \text{ and } [\mathbf{v}, \mathbf{v} + d\mathbf{v}]. \tag{2.7.1}$$

Since collisions are a binary process in a dilute gas, to describe their effect it is necessary to know the density of pairs of molecules in μ-space, $\rho^{(2)}(\mathbf{r}, \mathbf{v}, \mathbf{r}_1, \mathbf{v}_1, t)$, the so-called *pair distribution function.* In a dilute gas it is plausible that the large separation of molecules combined with previous collisions will conspire to make almost all molecules in the gas statistically independent of one another. This assumption was adopted by Boltzmann and called by him the *stosszahlansatz* or assumption of *molecular chaos.* Statistical independence implies that the pair distribution function is simply the product

$$\rho^{(2)}(\mathbf{r}, \mathbf{v}, \mathbf{r}_1, \mathbf{v}_1, t) = \rho(\mathbf{r}, \mathbf{v}, t)\rho(\mathbf{r}_1, \mathbf{v}_1, t). \tag{2.7.2}$$

The great advantage of this assumption is that it permits a description of the average effect of collisions in terms of ρ without independently introducing $\rho^{(2)}$. There have been several investigations of the validity of the molecular chaos assumption both for Boltzmann-equation-like models and for the dynamics of dilute gases using a Hamiltonian approach. The model calculations have shown that in the thermodynamic limit molecular chaos propagates, that is, if molecular chaos exists at an initial time, it will persist into the future. The Hamiltonian approach, on the other hand, has been used to derive the Boltzmann equation for dilute gases and, thus, argues for molecular chaos *a posteriori.*

Assuming molecular chaos, the dynamics of a collision can be treated in the

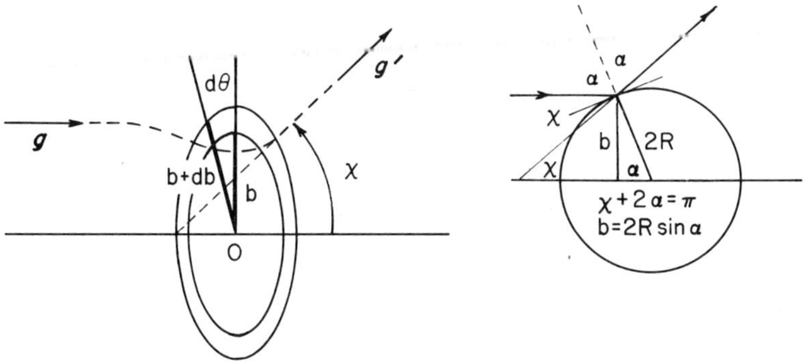

FIGURE 2.3. Geometry of a two-body central force collision. The incoming beam has velocity **g** and impact parameter in the range $[b, b + db]$. After collision the incoming beam is scattered through an angle χ and has a velocity $\mathbf{g}'$. The inset shows specular reflection by an incoming beam of hard spheres of radius R colliding with a stationary sphere of the same radius. The geometry of the collision shows that the scattering angle $\chi = \pi - 2\alpha$ and $b = 2R \sin \alpha$.

same fashion as the dynamics of a molecular beam experiment. A diagram of that kind of experiment is shown in Fig. 2.3. There a beam of molecules of velocity **g** impinges on a stationary scattering center located at the origin and is scattered at an angle χ. In the Boltzmann picture both molecules are moving, so to adopt this analogy it is necessary to change variables to the center of mass and relative coordinates, i.e.,

$$\begin{aligned} \mathbf{R} &= (\mathbf{r} + \mathbf{r}_1)/2, \quad \mathbf{V} = (\mathbf{v} + \mathbf{v}_1)/2 \\ \boldsymbol{\rho} &= \mathbf{r}_1 - \mathbf{r}, \quad \mathbf{g} = \mathbf{v}_1 - \mathbf{v}. \end{aligned} \tag{2.7.3}$$

For simplicity, we assume that the interaction potential, u, between the two molecules produces a central force so that it depends only on $|\mathbf{r} - \mathbf{r}_1|$, i.e., $u(|\boldsymbol{\rho}|)$. Because only a central force acts, the center of mass of the two molecules moves uniformly with a fixed velocity **V**. The relative coordinate $\boldsymbol{\rho}$, however, describes the motion of a fictitious particle with the reduced mass $\mu = m/2$, where m is the molecular mass. The fictitious particle moves under the influence of the central force and appears to scatter at the angle χ with respect to an axis with origin at **r**.

The assumption of a central force implies that the collision has three invariants: the energy, the linear momentum, and the angular momentum. Conservation of energy means that

$$v^2 + v_1^2 = v'^2 + v_1'^2, \tag{2.7.4}$$

while conservation of linear momentum gives that

$$\mathbf{v} + \mathbf{v}_1 = \mathbf{v}' + \mathbf{v}_1'. \tag{2.7.5}$$

From these two equations and the fact that the center of mass velocity, $\mathbf{V}$, is constant one can show that the magnitude of the relative velocity is constant,

$$|\mathbf{g}| = |\mathbf{g}'|. \tag{2.7.6}$$

Conservation of angular momentum implies that the relative motion of the two molecules remains in a plane with a fixed value of θ as shown in Fig. 2.3. The magnitude, l, of the relative angular momentum around the origin at $\mathbf{r}$ is defined by $l = |\mu\boldsymbol{\rho} \times \mathbf{g}|$. Its value is also constant and can be obtained using Fig. 2.3, which gives

$$l = \mu g b, \tag{2.7.7}$$

where b is the impact parameter, that is, the initial off-axis distance of the incoming molecule, and g is the magnitude of the relative velocity.

To describe the scattering process we use the differential scattering cross section, $\sigma(\Omega, g)$. The differential cross section is a proportionality factor between the magnitude of the flux density of molecules in the incoming beam, I, and the number of molecules scattered into the solid angle $d\Omega = \sin\chi\, d\chi\, d\theta$. Thus

$$dN = I\sigma(\Omega, g)\, d\Omega\, dt. \tag{2.7.8}$$

The cross section can be determined directly from the functional relationship between the impact parameter, b, and the scattering angle, χ. Indeed, referring to Fig. 2.3 we see that all the molecules in the beam which pass through the spatial window $[b, b + db] \times [\theta, \theta + d\theta]$ of area $b\, db\, d\theta$ will scatter into the solid angle $d\Omega$. Hence $Ib\, db\, d\theta = dN/dt = I\sigma \sin\chi\, d\chi\, d\theta$, so that

$$\sigma(\Omega, g) = [b(\chi)/\sin\chi]\,|db(\chi)/d\chi|. \tag{2.7.9}$$

In general, evaluating the cross section requires the solution of the mechanical equations of motion. For hard spheres, however, the effect of collisions is easy to describe. The inset in Fig. 2.3 shows the specular reflection that occurs when two spheres of radius R collide. The geometry of the collision dictates that the angles in the diagram satisfy $\chi + 2\alpha = \pi$, so that $\alpha = (\pi/2) - (\chi/2)$. Thus $\sin[(\pi/2) - (\chi/2)] = b/2R$, and so

$$b = 2R\cos(\chi/2), \tag{2.7.10}$$

for $b \leq R$. This is the expression for the impact parameter as a function of scattering angle. If Eq. (2.7.10) is used on the right-hand side of Eq. (2.7.9), the indicated differentiation gives rise to the cross section

$$\begin{aligned} \sigma(\Omega, g) &= \frac{2R^2 \sin(\chi/2)\cos(\chi/2)}{\sin\chi} \\ &= R^2 \end{aligned} \tag{2.7.11}$$

where the second equality uses the half-angle trigonometric identity. Equation (2.7.11) shows that the differential cross section for hard spheres is proportional to their cross-sectional area. The *total cross section* is defined by the integral of the differential cross section over all scattering angles. It is the proportionality factor between the flux of incoming molecules and the total number of molecules scattered, irrespective of angle. Thus

$$\sigma_T(g) = \int \sigma(\Omega, g)\, d\Omega = 2\pi \int b\, db, \tag{2.7.12}$$

where the second equality follows from Eq. (2.7.9). For hard spheres Eq. (2.7.11) shows that σ is independent of angle and velocity. Thus the total cross section for a hard-sphere collision is

$$\sigma_T = 4\pi R^2,$$

which is just the cross-sectional area excluded to the centers of mass in a binary collision.

The Boltzmann equation describes how the density of molecules located at position $\mathbf{r}$ with velocity $\mathbf{v}$ changes with time. A collision changes $\mathbf{v}$ to $\mathbf{v}'$ and the velocity of the collision partner from $\mathbf{v}_1$ to $\mathbf{v}_1'$. Symbolically we write this as $(\mathbf{v}, \mathbf{v}_1) \to (\mathbf{v}', \mathbf{v}_1')$. With each collision of this sort, which is stylized as the *direct* collision, there is associated an *inverse* or *restoring* collision, $(\mathbf{v}'', \mathbf{v}_1'') \to (\mathbf{v}, \mathbf{v}_1)$, in which the final velocities are identical to the initial velocities in the direct collision. It is not difficult to see that the initial velocities in the restoring collision are just $\mathbf{v}'$ and $\mathbf{v}_1'$, i.e., the final velocities in the direct collision. This follows from a consideration of the isotropy of space and the fact that the mechanical equations of motion are time reversal invariant. This is illustrated in Fig. 2.4. There the velocities and time are reversed, to the collision $(-\mathbf{v}', -\mathbf{v}_1') \to (-\mathbf{v}, -\mathbf{v}_1)$. Next the velocities are rotated 180 degrees around an axis in the collision plane which passes through the origin *0* and is perpendicular to the incoming relative velocity. This is followed by a reflection in a plane through the origin perpendicular to the rotation axis, which results in an inversion of the spatial geometry of the time-reversed collision. The net result of time reversal and inversion is the inverse collision $(\mathbf{v}', \mathbf{v}_1') \to (\mathbf{v}, \mathbf{v}_1)$.

The isotropy of space and time reversibility also imply that the inverse collision has the same differential cross section as the direct collision. In the inverse collision the trajectory is first time reversed, which certainly does not change the cross section. Spatial inversion has no effect on σ, either, since the transformation merely changes the absolute orientation of the collision without affecting its dynamics. This means that we can write

$$\sigma \equiv \sigma(\Omega, g) = \sigma(\Omega, g') \equiv \sigma', \tag{2.7.13}$$

where the prime now represents the inverse collision.

To derive the Boltzmann equation we need to account for the effect of both direct and restoring collisions on $\rho(\mathbf{r}, \mathbf{v}, \mathbf{t})$. Consider, first, direct collisions at the position $\mathbf{r}$ involving a molecule of velocity $\mathbf{v}$ and collision partners with velocities in the range $[\mathbf{v}_1, \mathbf{v}_1 + d\mathbf{v}_1]$. Since $\rho(\mathbf{r}, \mathbf{v}_1, t)\, d\mathbf{v}_1$ is the number density

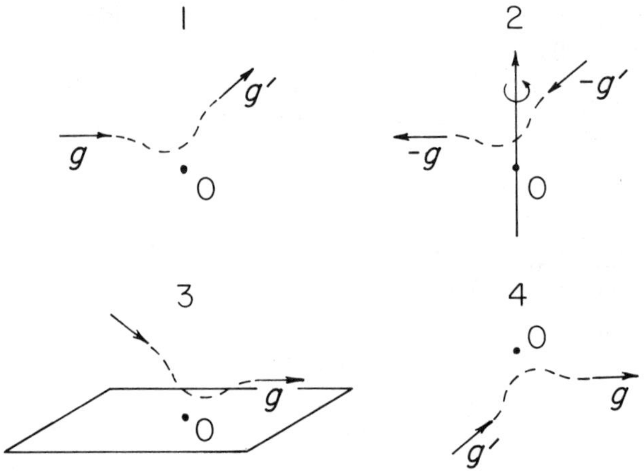

FIGURE 2.4. The sequence of transformations that take the direct collision into the restoring collision as seen in the relative coordinate system. The collision in *1* is time reversed to get the collision in *2*, which is transformed into *3* by rotating 180° about the axis shown in *2*. The restoring collision is obtained from 3 by reflection in the plane perpendicular to the axis of rotation.

in space, the magnitude of the flux density of these molecules impinging on the molecule at $(\mathbf{r}, \mathbf{v})$ is

$$I = g\rho(\mathbf{r}, \mathbf{v}_1, t)\, d\mathbf{v}_1 \tag{2.7.14}$$

Relying on the molecular chaos assumption, we can use the picture of independent scattering by a molecular beam of these molecules given in Fig. 2.3. Thus the number of molecules in the beam scattered in unit time to velocity $\mathbf{v}_1'$ can be obtained by combining Eqs. (2.7.14) and Eq. (2.7.8):

$$(dN/dt)_{\text{beam}} = \rho(\mathbf{r}, \mathbf{v}_1, t) g\sigma(\Omega, g)\, d\Omega\, d\mathbf{v}_1. \tag{2.7.15}$$

This is the result for a single scattering center of velocity $\mathbf{v}$ and a fixed scattering angle. The total number of such centers at $\mathbf{r}$ is $\rho(\mathbf{r}, \mathbf{v}, t)\, d\mathbf{v}\, d\mathbf{r}$. Thus the total number of molecules scattered per unit time is this number multiplied by Eq. (2.7.15) and integrated over all scattering angles, or

$$(dN/dt) = \int d\Omega\, \sigma(\Omega, g) g\rho(\mathbf{r}, \mathbf{v}, t)\rho(\mathbf{r}, \mathbf{v}_1, t)\, d\mathbf{v}\, d\mathbf{v}_1\, d\mathbf{r} \tag{2.7.16}$$

Now for each molecule of velocity $\mathbf{v}_1$ that is scattered from the beam a molecule of velocity $\mathbf{v}$ is also scattered. Thus the total number of molecules lost from the volume element $d\mathbf{r}\, d\mathbf{v}$ per unit time can be obtained by integrating Eq. (2.7.16) over the velocity of the collision partner, $\mathbf{v}_1$. Expressed as a change in the density, this gives the time rate of change of $\rho(\mathbf{r}, \mathbf{v}, t)$ due to direct collisions as

$$[\partial\rho(r, \mathbf{v}, t)/\partial t]_{\text{direct}} = -\iint d\Omega\, \sigma(\Omega, g) g \rho(\mathbf{r}, \mathbf{v}, t) \rho(\mathbf{r}, \mathbf{v}_1, t)\, d\mathbf{v}_1. \quad (2.7.17)$$

The restoring or inverse collisions increase the number of molecules with velocity $\mathbf{v}$ at $\mathbf{r}$, and a similar argument for inverse collisions gives

$$[\partial\rho(\mathbf{r}, \mathbf{v}, t)/\partial t]_{\text{inverse}} = \iint d\Omega\, \sigma(\Omega, g) g \rho(\mathbf{r}, \mathbf{v}', t) \rho(\mathbf{r}, \mathbf{v}_1', t)\, d\mathbf{v}_1. \quad (2.7.18)$$

In obtaining Eq. (2.7.18) it is necessary to make use of the mechanical result that the phase space volume element in a collision, i.e., $d\mathbf{r}\, d\mathbf{v}\, d\mathbf{r}_1\, d\mathbf{v}_1$, is unchanged by the collision. This is a special case of Liouville's theorem. Since the collision occurs at $\mathbf{r}$, we have that $d\mathbf{r} = d\mathbf{r}'$. Next, examining Fig. 2.3 we see that the differential of the beam coordinate can be written as $d\mathbf{r}_1 = 2\pi b\, db g\, dt$ or, using Eq. (2.7.8), $d\mathbf{r}_1 = g\sigma\, d\Omega\, dt$. Thus Liouville's theorem implies that

$$d\mathbf{r}\, d\mathbf{v}\, d\mathbf{v}_1 g\sigma\, d\Omega = d\mathbf{r}'\, d\mathbf{v}'\, d\mathbf{v}_1' g'\sigma'\, d\Omega'. \quad (2.7.19)$$

For the inverse collisions all the variables on the right-hand side of the equation analogous to Eq. (2.7.16) have primes. These can be partially eliminated using Eq. (2.7.19), which gives rise to Eq. (2.7.18).

The direct and inverse collisions are the dissipative processes responsible for change of number density in μ-space. There are other changes in ρ, which are nondissipative, that also must be taken into account. Even if no collisions occurred in the time dt, all molecules at the point $(\mathbf{r}, \mathbf{v})$ would move to a new spatial position, $\bar{\mathbf{r}} = \mathbf{r} + \mathbf{v}\, dt$ since they are moving with a velocity $\mathbf{v}$. Similarly, if an external force, $\mathbf{F}(\mathbf{r})$ is acting, the velocity will be changed to $\bar{\mathbf{v}} = \mathbf{v} + (\mathbf{F}/m)\, dt$. Thus in the absence of collisions all the molecules at the point $(\mathbf{r}, \mathbf{v})$ move to the point $(\bar{\mathbf{r}}, \bar{\mathbf{v}})$, which produces a nondissipative flux in phase space called *streaming*.

To obtain the total local change in the μ-space number density, we must combine the streaming motion with the effect of collisions. This can be done using the conservation equation derived in Section 2.4. Conservation of molecule number in μ-space is governed by an equation of the form of Eq. (2.4.5), which we write as

$$\partial\rho/\partial t = -\nabla_\mu \cdot \mathbf{j}_\mu + s. \quad (2.7.20)$$

The mass flux term, $\mathbf{j}_\mu$, corresponds to the streaming motion and is a six-component vector field on μ-space, so that the gradient operator, ∇_μ, involves derivatives with respect to both position and velocity. Generalizing the expression for the mass flux in position space obtained in Eq. (2.4.7), $\mathbf{j}_\mu$ can be written as

$$\mathbf{j}_\mu = \rho(\mathbf{r}, \mathbf{v}, t)\mathbf{v}_\mu, \quad (2.7.21)$$

where $\mathbf{v}_\mu$ is the six-dimensional μ-space velocity $\mathbf{v}_\mu = d(\mathbf{r}, \mathbf{v})/dt = (\mathbf{v}, \mathbf{F})$, with $\mathbf{F}$ the force per unit mass. The source term, s, on the other hand, represents the dissipative effect of collisions as given in Eqs. (2.7.17) and (2.7.18), so that

$$s = \iint d\Omega\, \sigma(\Omega, g) g [\rho' \rho_1' - \rho \rho_1] \, d\mathbf{v}_1, \tag{2.7.22}$$

where we have introduced the notation $\rho(\mathbf{r}, \mathbf{v}', t) = \rho'$, $\rho(\mathbf{r}, \mathbf{v}_1', t) = \rho_1'$, etc. Combining Eqs. (21.7.20)–(21.7.22), finally, yields the *Boltzmann equation*

$$\partial \rho / \partial t = -\mathbf{v} \cdot \nabla_{\mathbf{r}} \rho - \mathbf{F} \cdot \nabla_{\mathbf{v}} \rho + \int \hat{\sigma}_T g [\rho' \rho_1' - \rho \rho_1] \, d\mathbf{v}_1, \tag{2.7.23}$$

where the gradient operators, $\nabla_{\mathbf{r}}$ and $\nabla_{\mathbf{v}}$, involve derivatives with respect to position and velocity, respectively, and to simplify notation we have introduced the linear operator $\hat{\sigma}_T[h(\Omega)] \equiv \int d\Omega\, \sigma(\Omega, g) h(\Omega)$.

There are many contrasts between the derivation of the Boltzmann equation and the derivation of the Onsager linear regression equations. The most important is that the derivation of the Boltzmann equation requires a clear picture of molecular collisions. The Onsager equations, on the other hand, carry a strong flavor of thermodynamics in which the notion of molecules is completely absent. Indeed, the derivation of the Boltzmann equation appears to be purely mechanical and to rely exclusively on mechanical concepts like center of mass, relative velocity, and collision cross section. Actually, these mechanical features are misleading, and—as we shall see in Chapter 3—the Boltzmann equation is actually a manifestation of the thermodynamic characterization of molecular events in μ-space. Another contrast between the Onsager and Boltzmann pictures is that the Onsager equations are linear, reflecting their limited use in the neighborhood of equilibrium. The Boltzmann equation is nonlinear and because it accounts fully for binary collisions, it is useful for describing processes both near and far from equilibrium. The Onsager picture, however, is rather general and can be applied to all states of matter, not just dilute gases. Moreover, the Onsager theory includes the effect of fluctuations. It is not at once clear to what extent fluctuations have been correctly included in the Boltzmann equation. In the remainder of this chapter we develop some of the properties of the Boltzmann equation to clarify further its relationship to the Onsager theory.

2.8. The *H*-Theorem

One of Boltzmann's motivations in looking at the dynamics of dilute gases was to gain an understanding of the Second Law of thermodynamics. On the surface Boltzmann's approach seems to have little to do with thermodynamics, except for the fact that the Boltzmann equation governs the time dependence of the density of an extensive variable. This is quite different from the Onsager theory in which derivatives of the entropy are used to define thermodynamic forces and appeal is made to the Second Law to show that the coupling matrix L is positive semi-definite. Nonetheless, an entropy-like function—which Boltzmann called the H-function—arises naturally in studying the Boltzmann equation. As we shall see in Chapter 4, the H-function and Boltzmann's H-theorem are a special case of a result which holds for more general nonlinear systems.

Boltzmann's *H-function* is defined by

$$H = \iint \rho \ln \rho \, d\mathbf{r} \, d\mathbf{v}. \tag{2.8.1}$$

What Boltzmann discovered is that for a closed system

$$dH/dt \le 0, \tag{2.8.2}$$

namely, H is a nonincreasing function of time. This is called the *H-theorem*, and its validity follows from the Boltzmann equation. The proof of the H-theorem starts by taking the time derivative of H:

$$\begin{aligned} dH/dt &= \iint (\partial\rho/\partial t) \ln \rho \, d\mathbf{r} \, d\mathbf{v} + \iint (\partial\rho/\partial t) \, d\mathbf{r} \, d\mathbf{v} \\ &= \iint (\partial\rho/\partial t) \ln \rho \, d\mathbf{r} \, d\mathbf{v}. \end{aligned} \tag{2.8.3}$$

The second term in Eq. (2.8.3) vanishes since it equals the time rate of change of $\int \rho \, d\mathbf{r} \, d\mathbf{v} = N$, the total number of molecules, which is constant in a closed system. The first term can be evaluated using the Boltzmann equation (Eq. 2.7.23), which gives

$$\begin{aligned} dH/dt = &-\iint \mathbf{v} \cdot \nabla_{\mathbf{r}} \rho (\ln \rho) \, d\mathbf{r} \, d\mathbf{v} - \iint \mathbf{F} \cdot \nabla_{\mathbf{v}} \rho (\ln \rho) \, d\mathbf{r} \, d\mathbf{v} \\ &+ \iiint \hat{\sigma}_T g(\rho'\rho_1' - \rho\rho_1) \ln \rho \, d\mathbf{r} \, d\mathbf{v} \, d\mathbf{v}_1. \end{aligned} \tag{2.8.4}$$

The first two terms can be integrated by parts and shown to vanish if ρ vanishes as $\mathbf{v} \to \pm\infty$ and if $\nabla_{\mathbf{r}}\rho$ vanishes on the boundary. These conditions are satisfied in a closed system that has a finite number of molecules and no flux of molecules at the boundary. Thus

$$dH/dt = \iiint \hat{\sigma}_T g(\rho'\rho_1' - \rho\rho_1) \ln \rho \, d\mathbf{r} \, d\mathbf{v} \, d\mathbf{v}_1. \tag{2.8.5}$$

Next the integrand on the right-hand side of Eq. (2.8.5) is symmetrized by making three changes of variables. First, change variables from $\mathbf{r}$, $\mathbf{v}$, $\mathbf{v}_1$ to $\mathbf{r}'$, $\mathbf{v}'$, $\mathbf{v}_1'$. This gives

$$\begin{aligned} dH/dt &= \iiint \hat{\sigma}'_T g'(\rho\rho_1 - \rho'\rho'_1) \ln \rho' \, d\mathbf{r}' \, d\mathbf{v}' \, d\mathbf{v}'_1 \\ &= -\iiint \hat{\sigma}_T g(\rho'\rho'_1 - \rho\rho_1) \ln \rho' \, d\mathbf{r} \, d\mathbf{v} \, d\mathbf{v}_1, \end{aligned} \tag{2.8.6}$$

where in the second equality the integrand has been rearranged, and we have used the invariance of the volume element under collision expressed in Eq. (2.7.19). Combining Eqs. (2.8.5) and (2.8.6) gives

$$dH/dt = \frac{1}{2} \iiint \hat{\sigma}_T g(\rho'\rho'_1 - \rho\rho_1) \ln(\rho/\rho') \, d\mathbf{r} \, d\mathbf{v} \, d\mathbf{v}_1. \tag{2.8.7}$$

Next interchange the two particles in Eq. (2.8.7), so that $\mathbf{v}$, $\mathbf{v}_1$ becomes $\mathbf{v}_1$, $\mathbf{v}$, to get

$$dH/dt = \frac{1}{2} \iiint \hat{\sigma}_T g(\rho'\rho'_1 - \rho\rho_1) \ln(\rho_1/\rho'_1) d\mathbf{r} \, d\mathbf{v} \, d\mathbf{v}_1. \tag{2.8.8}$$

Finally adding Eqs. (2.8.7) and (2.8.8) together and dividing by 2 gives

$$dH/dt = \frac{1}{4} \iiint \hat{\sigma}_T g \rho'\rho'_1 \left[\left(1 - \frac{\rho\rho_1}{\rho'\rho'_1}\right) \ln\left(\frac{\rho\rho_1}{\rho'\rho'_1}\right) \right] d\mathbf{r} \, d\mathbf{v} \, d\mathbf{v}_1 \tag{2.8.9}$$

The term in square brackets in the integrand of Eq. (2.8.9) has the form $(1 - x) \ln x$ where x is positive. As $(1 - x) \ln x$ is negative except for $x = 1$, where it vanishes, the term in square brackets is negative semi-definite. The other factors in the integrand are non-negative, so it follows that

$$dH/dt \le 0 \tag{2.8.10}$$

with the equality holding for (and only for) $\rho = \rho^e$ which satisfies

$$\rho^e \rho_1^e = \rho^{e'} \rho_1^{e'}. \tag{2.8.11}$$

The function H acts like an entropy in that it provides a preferential direction for time, that is, the H-theorem shows that time increases in the direction for which H decreases. Moreover, we shall show in Chapter 4 that $-k_B \, dH/dt$ is a special case of the nonlinear dissipation function which generalizes the Onsager–Rayleigh dissipation function in Eq. (2.3.19).

2.9. μ-Space Averages and the Maxwell Distribution

An interesting consequence of the proof of the H-theorem is the result in Eq. (2.8.11). Taking the logarithm it follows that only distributions which satisfy

$$\ln \rho^e + \ln \rho_1^e = \ln \rho^{e'} + \ln \rho_1^{e'} \tag{2.9.1}$$

correspond to a constant value of H. As the H-theorem proves that H is a monotonically decreasing function of time, we infer that the μ-space density which satisfies Eq. (2.9.1) is the asymptotic equilibrium density function. In the absence of an external field, the equilibrium phase space density will be independent of position. To find out how the equilibrium density depends on the velocity we write out Eq. (2.9.1) symbolically as

$$\psi(\mathbf{v}) + \psi(\mathbf{v}_1) = \psi(\mathbf{v}') + \psi(\mathbf{v}'_1), \tag{2.9.2}$$

where $\psi = \ln \rho^e$. A function which satisfies the equality (2.9.2) is called a *collisional invariant*. There are five linearly independent collisional invariants for binary collisions, the mass (or any other function that is independent of velocity), the three components of momentum, and the kinetic energy. The latter four are expressed by the conservation equations (2.7.4) and (2.7.5). As a consequence, it follows that in the absence of an external field the equilibrium μ-space density, $\rho^e(\mathbf{v})$, satisfies

$$\ln \rho^e(\mathbf{v}) = a + \mathbf{b}\cdot\mathbf{v} + cmv^2/2. \tag{2.9.3}$$

If we complete the square in the quadratic function of $\mathbf{v}$ on the right-hand side of Eq. (2.9.3), the equation can be rewritten more conveniently as

$$\ln \rho^e(\mathbf{v}) = \alpha - \beta m|\mathbf{v} - \mathbf{v}_0|^2/2,$$

or

$$\rho^e(\mathbf{v}) = e^\alpha \exp[-\beta m|\mathbf{v} - \mathbf{v}_0|^2/2]. \tag{2.9.4}$$

This Gaussian form is called the *Maxwell distribution.*

To evaluate the constants α and β in Eq. (2.9.4) it is necessary to introduce the idea of a μ-space average. The stochastic ensemble point of view adopted in Chapter 1 replaces an investigation of properties of a single system with those of an ensemble of systems, prepared in a macroscopically identical fashion. At the Boltzmann level of description, that is, using the μ-space number density as the macroscopic variable, the ensemble consists of many N-molecule systems identically prepared with respect to their μ-space density. The Boltzmann equation describes the time evolution of ρ only on the *average.* Thus, for example, $\rho^e(\mathbf{v})$ is really the μ-space density *averaged over the equilibrium ensemble.* In other words, $\rho^e(\mathbf{v})\,d\mathbf{r}\,d\mathbf{v}$ gives the expected number of molecules in the volume element $d\mathbf{r}\,d\mathbf{v}$ at equilibrium.

The μ-space density can be used to carry out an additional averaging, not in the ensemble—which consists of N-molecule systems—but over the N-molecules within a system. Thus if we want to know the average mass density of molecules at $\mathbf{r}$, irrespective of their velocity, we have

$$\langle \rho(\mathbf{r}, t)\rangle \equiv m \int \rho(\mathbf{r}, \mathbf{v}, t)\,d\mathbf{v}. \tag{2.9.5}$$

Similarly the function

$$P(\mathbf{r}, \mathbf{v}, t) \equiv m\rho(\mathbf{r}, \mathbf{v}, t)/\langle\rho(\mathbf{r}, t)\rangle, \tag{2.9.6}$$

which is normalized to unity, can be interpreted as the *probability* that a molecule selected from the average N-molecule system has a velocity in the range $[\mathbf{v}, \mathbf{v} + d\mathbf{v}]$. This was the original interpretation of ρ by Maxwell and Boltzmann which led, in the latter part of the nineteenth century, to confusion and controversy about the meaning of the Boltzmann equation. The completely mechanical picture of the dynamics of an N-molecule gas shows that the system will recurrently visit almost all initial states in its N-molecule phase space, sometimes called the Γ-*phase space* to differentiate it from μ-space. These recurrences seem at odds with the H-theorem which, like the Second Law, predicts a monotonic approach to equilibrium. The ensemble picture, however, makes it clear that the μ-space level of description is inherently stochastic. Thus the ρ in Boltzmann's equation must be an average quantity; otherwise it would satisfy a stochastic equation. What is missing in the Boltzmann picture is a description of fluctuations in the μ-space density. These fluctuations are usually small, which implies that the recurrences that appear in the mechanical theory are extremely unlikely.

With this caveat, we can make use of the μ-space density to average over the properties of the N molecules within a system. For example, using the probability density defined by Eq. (2.9.6) we can write the average velocity of a molecule at $\mathbf{r}$ as

$$\langle\mathbf{v}(\mathbf{r}, t)\rangle = \int \mathbf{v}P(\mathbf{r}, \mathbf{v}, t)\, d\mathbf{v} \tag{2.9.7}$$

and the average energy per molecule (which is all kinetic energy in a dilute gas) as

$$\langle E(\mathbf{r}, t)\rangle = \frac{1}{2}m \int v^2 P(\mathbf{r}, \mathbf{v}, t)\, d\mathbf{v}. \tag{2.9.8}$$

More explicitly, the equilibrium probability density for the Maxwell distribution, which is independent of $\mathbf{r}$, can be written

$$P^e(\mathbf{v}) = (2\pi/\beta m)^{-3/2} \exp[-\beta m|\mathbf{v} - \mathbf{v}_0|^2/2], \tag{2.9.9}$$

where the form of the Gaussian normalization constant was taken from Eq. (1.3.28). The average velocity at equilibrium can be obtained using Eq. (2.9.9) in the integral expression of Eq. (2.9.7). Thus

$$\langle\mathbf{v}\rangle^e = (2\pi/\beta m)^{-3/2} \int \mathbf{v} \exp[-\beta m|\mathbf{v} - \mathbf{v}_0|^2/2]\, d\mathbf{v}. \tag{2.9.10}$$

This is a standard Gaussian integral whose value can be obtained by changing variables to $\mathbf{v}' = \mathbf{v} - \mathbf{v}_0$. The result is that

$$\langle\mathbf{v}\rangle^e = \mathbf{v}_0. \tag{2.9.11}$$

Thus the average equilibrium velocity is a constant independent of position. This corresponds to a uniform translation of the whole system, which can be eliminated by changing variables to a frame of reference moving with velocity $\mathbf{v}_0$. Doing this, we can set $\langle \mathbf{v} \rangle^e = \mathbf{v}_0 = \mathbf{0}$.

To evaluate the constant β in the Maxwell distribution is somewhat more complicated. The first step is to calculate the average value of the kinetic energy per molecule, i.e.,

$$\langle E \rangle^e = (2\pi/\beta m)^{-3/2} \int (mv^2/2) \exp[-\beta m v^2/2] \, d\mathbf{v}. \tag{2.9.12}$$

Since $v^2 = v_x^2 + v_y^2 + v_z^2$, this integral can be written as the sum of three identical Gaussian integrals like the third one in Eq. (1.3.32). This gives

$$\begin{aligned} \langle E \rangle^e &= 3(2\pi/\beta m)^{-1/2}(m/2)[2\pi/(\beta m)^3]^{1/2} \\ &= 3/2\beta. \end{aligned} \tag{2.9.13}$$

To obtain the total energy of all the molecules in the gas, E_T, we must multiply the energy density per molecule by the number of molecules, N. Thus

$$\langle E_T \rangle^e = 3N/2\beta. \tag{2.9.14}$$

An independent expression for E_T can be obtained from a purely mechanical result called the *virial theorem*, originally due to Clausius. For molecules which interact weakly, as in a dilute gas, the virial theorem states that the average of the total energy is proportional to the product of the average pressure, p, and the volume. More specifically

$$\langle E_T \rangle = 3\langle p \rangle V/2. \tag{2.9.15}$$

Now for a dilute gas at equilibrium the equation of state is given by the ideal gas law

$$\langle p \rangle^e V = Nk_B T, \tag{2.9.16}$$

where T is the ideal gas or Kelvin temperature. Substituting this into the virial theorem (Eq. 2.9.15) and comparing with Eq. (2.9.14) yields the identification

$$\beta = 1/k_B T. \tag{2.9.17}$$

This remarkable result shows that the equilibrium Maxwell distribution in μ-space is completely determined by the mass of a molecule and the reading on an ideal gas thermometer. Substituting this result into Eq. (2.9.14) gives a version of the *equipartition theorem*

$$\langle E_T \rangle^e = 3Nk_B T/2, \tag{2.9.18}$$

which states that each quadratic contribution to the mechanical Hamiltonian gives rise to a contribution of $k_B T/2$ to the average energy at equilibrium.

In the presence of an external potential field, $\theta(\mathbf{r})$, the equilibrium distribution is generalized somewhat. It depends on position and becomes the Maxwell–Boltzmann distribution

$$\rho^e(\mathbf{r}, \mathbf{v}) = \left\{ N \bigg/ \int \exp[-\theta(\mathbf{r})/k_B T]\, d\mathbf{r} \right\} (2\pi k_B T/m)^{-3/2} \exp[-mv^2/2k_B T] \\ \times \exp[-\theta(\mathbf{r})/k_B T]. \tag{2.9.19}$$

2.10. Conservation Equations

In the preceding section we discovered how the average μ-space distribution, $\rho(\mathbf{r}, \mathbf{v}, t)$, could be used to obtain the average mass density, velocity, and energy fields in position space. Since the Boltzmann equation describes the average time rate of change of ρ, it can be used to obtain equations which describe how these hydrodynamic densities change over time. This is another example of a contraction, in this case from the Boltzmann level of description to the hydrodynamic level. It is typical of other contractions in that the Boltzmann level involves many more degrees of freedom (a scalar field on six-dimensional μ-space) than the hydrodynamic level (two scalar fields and one vector field on three-dimensional position space). Moreover, the contracted hydrodynamic level of description turns out to be self-contained only on a time scale which is much longer than the collisional time scale of the Boltzmann level. These basic features of contracted descriptions are examined in detail in Chapter 9.

We content ourselves in this section with showing that the conservation equations at the hydrodynamic level, obtained in Section 2.4, can be derived from the Boltzmann equation. To do this we need to investigate the collision term in the Boltzmann equation a bit more thoroughly. The collision term is given in Eq. (2.7.22). It can be thought of as a *bilinear functional* of the μ-space density, i.e.,

$$s(\rho, \rho) = \int \hat{\sigma}_T(g) g [\rho' \rho_1' - \rho \rho_1]\, d\mathbf{v}_1, \tag{2.10.1}$$

which associates a new function of $\mathbf{v}$ and $\mathbf{r}$, namely s, with the function ρ. This nonlinear functional possesses a useful property that relates to the collisional invariants described in Eq. (2.9.2). The property is

$$\int s(\rho, \rho) \psi(\mathbf{v})\, d\mathbf{v} = 0, \tag{2.10.2}$$

where ψ is any one of the five collisional invariants. To prove Eq. (2.10.2) we use Eq. (2.10.1) to write the left-hand side as

$$\iint \hat{\sigma}_T(g) g [\rho' \rho_1' - \rho \rho_1] \psi\, d\mathbf{v}_1\, d\mathbf{v}. \tag{2.10.3}$$

Taking advantage of Eq. (2.7.19), we can change integration variables from $\mathbf{v}_1$ and $\mathbf{v}$ to the final velocities $\mathbf{v}_1'$ and $\mathbf{v}'$ in the integral and so express the left-hand side of Eq. (2.10.3) also as

$$-\iint \hat{\sigma}_T(g)g[\rho'\rho_1' - \rho\rho_1]\psi'\, d\mathbf{v}_1\, d\mathbf{v}, \tag{2.10.4}$$

where we have used the facts that $\sigma_T = \sigma_T'$ and $g = g'$ as noted in Eqs. (2.7.6) and (2.7.13). Combining Eqs. (2.10.2)–(2.10.4) yields

$$\int s(\rho, \rho)\psi(\mathbf{v})\, d\mathbf{v} = \frac{1}{2}\iint \hat{\sigma}_T(g)g[\rho'\rho_1' - \rho\rho_1][\psi - \psi']\, d\mathbf{v}_1\, d\mathbf{v}. \tag{2.10.5}$$

Finally exchanging the variables $\mathbf{v}_1$ and $\mathbf{v}$ in the integral in Eq. (2.10.5) leads to

$$\int s(\rho, \rho)\psi(\mathbf{v})\, d\mathbf{v} = \frac{1}{4}\iint \hat{\sigma}_T(g)g[\rho'\rho_1' - \rho\rho_1][\psi + \psi_1 - \psi' - \psi_1']\, d\mathbf{v}\, d\mathbf{v}'.$$

Since ψ is a collisional invariant, Eq. (2.9.2) shows that the second term in square brackets vanishes identically, which proves Eq. (2.10.2).

To obtain the hydrodynamic-level conservation equations we return to the Boltzmann equation (2.7.23). Multiplying both sides by $\psi(\mathbf{v})$ and integrating over the velocity we find that

$$\partial \int \rho\psi\, d\mathbf{v}/\partial t = -\int \mathbf{v}\psi \cdot \nabla_{\mathbf{r}}\rho\, d\mathbf{v} - \mathbf{F}\cdot \int \psi\nabla_{\mathbf{v}}\rho\, d\mathbf{v}, \tag{2.10.6}$$

where the integral over the collision term vanishes because of the conservation property in Eq. (2.10.2). Equation (2.10.6) is the fundamental conservation equation in position space and is simply another way of writing the conservation of mass, momentum, and energy. For example, let $\psi(\mathbf{v}) = m$. Then according to the definition of average mass density in Eq. (2.9.5) and the average velocity in Eq. (2.9.6) and (2.9.7), Eq. (2.10.6) becomes

$$\partial\langle\rho(\mathbf{r}, t)\rangle/\partial t = -\nabla\cdot\langle\mathbf{v}(\mathbf{r}, t)\rangle\langle\rho(\mathbf{r}, t)\rangle, \tag{2.10.7}$$

where the second term from Eq. (2.10.6) vanishes since $\rho \to 0$ as $\mathbf{v} \to \pm\infty$. Equation (2.10.7) is recognized as the equation of continuity (Eq. 2.4.9) obtained by purely kinematic arguments in Section 2.4. Conservation of momentum is expressed by the three equations which result from Eq. (2.10.6) when ψ is set equal to mv_x, mv_y, and mv_z. Again recalling Eqs. (2.9.5)–(2.9.7) these become

$$\frac{\partial\langle\rho\rangle\langle v_i\rangle}{\partial t} = -\frac{\partial}{\partial x}\langle\rho\rangle\langle v_x v_i\rangle - \frac{\partial}{\partial y}\langle\rho\rangle\langle v_y v_i\rangle - \frac{\partial}{\partial z}\langle\rho\rangle\langle v_z v_i\rangle + F_i\langle\rho\rangle$$

or, in vector notation,

$$\partial\langle\rho\rangle\langle\mathbf{v}\rangle/\partial t = -\nabla\cdot(\langle\rho\rangle\langle\mathbf{v}\mathbf{v}^T\rangle) + \mathbf{F}\langle\rho\rangle. \tag{2.10.8}$$

To put this into a more standard form, we split the velocity $\mathbf{v}$ into its μ-space average and the deviation from the average, i.e.,

$$\mathbf{v} = \langle \mathbf{v} \rangle + \delta \mathbf{v}. \tag{2.10.9}$$

Since the average value of $\delta\mathbf{v}$ is zero, substituting Eq. (2.10.9) into Eq. (2.10.8) leads to

$$\partial \langle \rho \rangle \langle \mathbf{v} \rangle / \partial t = -\nabla \cdot (\langle \rho \rangle \langle \mathbf{v} \rangle \langle \mathbf{v} \rangle^T) - \nabla \cdot P + \mathbf{F} \langle \rho \rangle, \tag{2.10.10}$$

where the pressure tensor, P, is defined by

$$P = \langle \rho \rangle \langle \delta \mathbf{v} \delta \mathbf{v}^T \rangle = m \int \delta \mathbf{v} \delta \mathbf{v}^T \rho \, d\mathbf{v}. \tag{2.10.11}$$

Finally, we can use Eq. (2.10.6) to obtain the hydrodynamic-level conservation equation for the total energy. In defining the average energy, E, in Eq. (2.9.8), both the convective contribution, $\rho\langle \mathbf{v} \rangle^2/2$ and the contribution relative to the convective motion, i.e., $\langle \delta \mathbf{v} \rangle^2/2m$, are lumped together. The μ-space average of $m\delta v^2/2$ is the internal energy (all kinetic in a dilute gas) per molecule. Thus the quantity

$$\langle e(\mathbf{r}, t) \rangle = \langle \rho \rangle \langle \delta v^2/2 \rangle \tag{2.10.12}$$

is the *average internal energy* per unit volume at $\mathbf{r}$. To obtain the conservation equation for the average total density, $\langle \varepsilon \rangle = \langle e \rangle + \langle \rho \rangle \langle \mathbf{v} \rangle^2/2$, we substitute $\psi = mv^2/2$ into Eq. (2.10.6) and integrate the final term by parts to find that

$$\partial \langle \rho \rangle \langle v^2/2 \rangle / \partial t = -\nabla \cdot \langle \rho \rangle \langle \mathbf{v} v^2/2 \rangle + \mathbf{F} \cdot \langle \mathbf{v} \rangle \langle \rho \rangle. \tag{2.10.13}$$

The right- and left-hand sides of Eq. (2.10.13) can be expressed in terms of $\langle \varepsilon \rangle$ and $\langle \mathbf{v} \rangle$ by using Eq. (2.10.9). After a little algebra one obtains

$$\partial \langle \varepsilon \rangle / \partial t = -\nabla \cdot [\langle \mathbf{v} \rangle \langle \varepsilon \rangle + P \cdot \langle \mathbf{v} \rangle + \mathbf{j}_q] + \langle \rho \rangle \langle \mathbf{v} \rangle \cdot \mathbf{F}, \tag{2.10.14}$$

where the average heat flux density is defined by

$$\mathbf{j}_q = \langle \rho \rangle \langle \delta \mathbf{v} \delta v^2/2 \rangle = m \int \delta \mathbf{v} (\delta v^2/2) \rho \, d\mathbf{v}. \tag{2.10.15}$$

Equation (2.10.14) is precisely the kinematic conservation equation for energy, Eq. (2.4.19).

The derivation of the hydrodynamic conservation equations from the Boltzmann equation provides two insights into the equations of hydrodynamics. First, it makes clear that the variables that appear in the conservation equations are average quantities. Although it is intuitively clear that the hydrodynamic-level description must be stochastic, the derivation of the form of the conservation equations demonstrates that the usual phenomenological equations are true only on the average. The derivation also provides explicit expressions for the pressure tensor and heat flux vector in terms of average molecular quantities. Equations (2.10.11) and (2.10.15) express P and $\mathbf{j}_q$ in terms of averages of molecular fluctuations in the velocity, $\delta\mathbf{v}$. This turns out to be a recurring theme in statistical thermodynamics, which we return to in Chapter 9.

Equations (2.10.11) and (2.10.15) can also be used to derive the phenomenological equations of Newton (Eq. 2.4.22) and Fourier (Eq. 2.4.20) for momentum and heat transport and to calculate directly the shear viscosity, η, and thermal conductivity, κ. This is another characteristic of the contraction of a more complete stochastic description to a less complete one, namely, that the transport coefficients in the contracted description (η and κ) can be expressed in terms of more fundamental transport coefficients (the scattering cross section). This program was originally begun by Maxwell and Boltzmann and requires an evaluation of the μ-space density. Although we defer our discussion of the solution to the Boltzmann equation, the gist of the procedure can be seen already by examining the Maxwell distribution, Eq. (2.9.9). If the Maxwell distribution is used in Eq. (2.10.11) we can calculate the equilibrium value of the pressure tensor, P^e. At equilibrium taking $\langle \mathbf{v} \rangle^e = \mathbf{0} = \mathbf{v}_0$, it follows that $\delta \mathbf{v} = \mathbf{v}$. Thus

$$P_{ij}^e = (N/V)m(2\pi/\beta m)^{-3/2} \int v_i v_j \exp(-\beta m \mathbf{v}^2/2)\, d\mathbf{v}. \qquad (2.10.16)$$

The off-diagonal elements of P_{ij}^e vanish because the integrand is an odd function if $i \neq j$. The diagonal elements are all the same, and, recalling from Eq. (2.4.21) that these elements are the pressure, we can write

$$p^e(\rho, T) = (N/V)m(2\pi/\beta m)^{-1/2} \int_{-\infty}^{+\infty} v_x^2 \exp(-\beta m v_x^2/2)\, dv_x. \qquad (2.10.17)$$

The integral in Eq. (2.10.17) has been encountered before in Eq. (2.9.12) and leads to

$$p^e(\rho, T) = (N/V)k_B T, \qquad (2.10.18)$$

where we have used the fact that $\beta = 1/k_B T$. Equation (2.10.18) is the ideal gas equation of state which, it should be emphasized, has been deduced from a knowledge of the dynamical properties of the gas. Since the off-diagonal elements of P_{ij}^e vanish, we conclude from Eq. (2.4.21) that there are no stresses in a dilute gas at equilibrium. Indeed, to obtain information about the dissipative properties of the gas it is necessary to look at *nonequilibrium* distributions in μ-space. This is also evident from looking at the equilibrium expression for the heat flux vector in Eq. (2.10.15), i.e.,

$$\mathbf{j}_q^e = (N/V)m(2\pi/\beta m)^{-3/2} \int \mathbf{v}(v^2/2) \exp(-\beta m v^2/2)\, d\mathbf{v}, \qquad (2.10.19)$$

which also vanishes. Systematic procedures for calculating nonequilibrium corrections to the Maxwell distribution were developed independently by Hilbert, Chapman, and Enskog. The so-called Chapman–Enskog or normal solutions to the Boltzmann equation yield expressions for η and κ which are in good agreement with experiment. We show how to do this in Chapter 9.

2.11. Uniting the Onsager and Boltzmann Pictures

The overtly thermodynamic character of the Onsager picture of irreversible processes is in strong contrast to the mechanical character of the Boltzmann picture. Nonetheless, both pictures share a number of common features. First and foremost, each involves a contracted description of nature and, thus, eschews the deterministic description that is inherent in the mechanical picture of matter. Indeed reviewing the derivation of the Boltzmann equation in Section 2.7, it is evident that Boltzmann's ideas are not really mechanical at all. The total cross section, which is the only variable parameter in the Boltzmann equation, can be looked upon as a transport coefficient in μ-space whose value can be obtained from beam scattering experiments. Thought of in this way, the cross section is an empirical parameter just like the viscosities and thermal conductivity, which are the empirical parameters in the hydrodynamic-level Onsager description. The distance of the Boltzmann picture from mechanics is also made clear by the H-theorem, which predicts an irreversible approach to equilibrium. The equations of mechanics are time reversal invariant and for a bounded system even predict that given enough time, almost all systems will return arbitrarily close to their initial location in the Γ-phase space.

The existence of the H-function provides another point of contact between Boltzmann and Onsager. A knowledge of the form of the entropy is required in the Onsager theory, and the Second Law of thermodynamics is a necessary ingredient in showing that the coupling matrix, L_{ij}, is positive semi-definite. Although the Boltzmann picture does not rely on the existence of an entropy function, the H-theorem does suggest that the negative of H is related to the entropy. In fact, using the form of the Maxwell distribution, It is easy to show that

$$\begin{aligned} H^e &= \int \rho^e \ln \rho^e \, d\mathbf{v} \, d\mathbf{r} \\ &= -N\left[\frac{3}{2}\ln T - \ln(N/V) + 3/2 + \frac{3}{2}\ln(2\pi k_B/m)\right]. \end{aligned} \tag{2.11.1}$$

The first two terms in the brackets in Eq. (2.11.1) give the temperature and density dependence of the entropy of an ideal gas. Using this, Eq. (2.11.1) can be rewritten as

$$S^e(T, V, N) = -k_B H^e + \text{constant}, \tag{2.11.2}$$

where the constant cannot be determined from the Boltzmann equation. Thus even though no entropy is presumed *a priori* in the Boltzmann picture, it is somehow imbedded in it.

Another point of contact between the Boltzmann and the Onsager theories is the recognition of the need for a stochastic description. This is made clearest in the Onsager theory, which treats the equilibrium ensemble as a stationary, Gaussian, Markov process. The stochastic ideas are most apparent in the Boltzmann picture through the probability interpretation of the μ-space number density. Nonetheless, the irreversible approach to equilibrium can be true only on the average. As we have previously emphasized, this means that the Boltzmann equation is true only in some average sense.

A final relationship between the Onsager and Boltzmann theories can be seen if we restrict attention to the Boltzmann equation in a neighborhood of equilibrium. Although the Boltzmann equation is nonlinear, if we look only at small deviations around equilibrium (in the absence of an external field) we can write

$$\rho(\mathbf{r}, \mathbf{v}, t) = \rho^e(\mathbf{v}) + \Delta\rho(\mathbf{r}, \mathbf{v}, t) \tag{2.11.3}$$

where $\Delta\rho(\mathbf{r}, \mathbf{v}, t)$ is assumed to be a small change in μ-space density. If Eq. (2.11.3) is substituted into the Boltzmann equation (Eq. 2.7.23) and only terms linear in $\Delta\rho$ are retained, the linearized Boltzmann equation is obtained:

$$\partial\Delta\rho/\partial t = -\mathbf{v}\cdot\nabla_{\mathbf{r}}\Delta\rho + \int \hat{\sigma}_T g \rho^e \rho_1^e \left[\frac{\Delta\rho'}{\rho^{e'}} + \frac{\Delta\rho_1'}{\rho_1^{e'}} - \frac{\Delta\rho}{\rho^e} - \frac{\Delta\rho_1}{\rho_1^e}\right] d\mathbf{v}_1. \tag{2.11.4}$$

Equation (2.11.4) is a linear integrodifferential equation in μ-space and shares many of the features of linear differential equations. In fact, the collision term on the right-hand side of Eq. (2.11.4) associates a new function of $\mathbf{r}$ and $\mathbf{v}$ with the function $\Delta\rho(\mathbf{r}, \mathbf{v}, t)$. Thus it is a *linear functional* of $\Delta\rho$ which we will write symbolically as

$$C[\Delta\rho] = \int \hat{\sigma}_T g \rho^e \rho_1^e \left(\frac{\Delta\rho'}{\rho^{e'}} + \frac{\Delta\rho_1'}{\rho_1^{e'}} - \frac{\Delta\rho}{\rho^e} - \frac{\Delta\rho_1}{\rho_1^e}\right] d\mathbf{v}_1. \tag{2.11.5}$$

The portion of the integrand in square brackets can be written in terms of velocity translation operators for the collision $(\mathbf{v}, \mathbf{v}_1) \to (\mathbf{v}', \mathbf{v}_1')$. For example, define the linear operator $T(\mathbf{v}, \mathbf{v}_1)$ by

$$T(\mathbf{v}, \mathbf{v}_1)\rho(\mathbf{v}_1) = \rho(\mathbf{v}). \tag{2.11.6}$$

Then T has the formal representation

$$T(\mathbf{v}, \mathbf{v}_1) = \exp[\boldsymbol{\delta}\cdot\nabla_{\mathbf{v}_1}],$$

where $\boldsymbol{\delta}$ is evaluated at $\mathbf{v} - \mathbf{v}_1$ after operating on a function of $\mathbf{v}_1$. This can be verified by expanding the exponential in a power series and noting that this series acting on $\rho(\mathbf{v}_1)$ is just the Taylor series expansion of $\rho(\mathbf{v})$ around $\mathbf{v}_1$. Using this operator we can write the third term in the square bracket of Eq. (2.11.5) as

$$\frac{\Delta\rho(\mathbf{r}, \mathbf{v}, t)}{\rho^e(\mathbf{v})} = T(\mathbf{v}, \mathbf{v}_1)\left(\frac{\Delta\rho(\mathbf{r}, \mathbf{v}_1, t)}{\rho^e(\mathbf{v}_1)}\right).$$

The other translation operators

$$
\begin{aligned}
T'(\mathbf{v}, \mathbf{v}_1)\rho(\mathbf{v}_1) &= \rho(\mathbf{v}') \\
T_1'(\mathbf{v}, \mathbf{v}_1)\rho(\mathbf{v}_1) &= \rho(\mathbf{v}_1') \\
T_1(\mathbf{v}, \mathbf{v}_1)\rho(\mathbf{v}_1) &= \rho(\mathbf{v}_1)
\end{aligned} \tag{2.11.7}
$$

have a comparable representation and can be used to rewrite the other terms in Eq. (2.11.5). Using this notation the linearized collision functional can be written

$$
\begin{aligned}
C[\Delta\rho] &= \int \hat{\sigma}_T |\mathbf{v} - \mathbf{v}_1| \rho^e(\mathbf{v})\rho^e(\mathbf{v}_1)[T'(\mathbf{v}, \mathbf{v}_1) + T_1'(\mathbf{v}, \mathbf{v}_1) - T(\mathbf{v}, \mathbf{v}_1) - T_1(\mathbf{v}, \mathbf{v}_1)] \\
&\quad \times \frac{\Delta\rho(\mathbf{v}_1, \mathbf{r}, t)}{\rho^e(\mathbf{v}_1)} d\mathbf{v}_1 \\
&\equiv \int K(\mathbf{v}, \mathbf{v}_1)[\Delta\rho(\mathbf{v}_1, \mathbf{r}, t)/\rho^e(\mathbf{v}_1)]\, d\mathbf{v}_1,
\end{aligned} \tag{2.11.8}
$$

where K is the indicated linear operator which depends on $\mathbf{v}$ and $\mathbf{v}_1$. In this representation the linearized Boltzmann equation (Eq. 2.11.4) becomes

$$\partial\Delta\rho/\partial t = -\mathbf{v} \cdot \nabla_{\mathbf{r}} \Delta\rho + \int K(\mathbf{v}, \mathbf{v}_1)(\Delta\rho_1/\rho_1^e)\, d\mathbf{v}_1. \tag{2.11.9}$$

To compare this linear regression equation for $\Delta\rho$ to the Onsager regression equations in Section 2.3, we must have a definition of the entropy density at the point $(\mathbf{r}, \mathbf{v}_1)$ in μ-space. This can be obtained from Eqs. (2.11.1) and (2.11.2), if we extend the definition of entropy using the same form of the entropy that works at equilibrium. This extension is called the *local equilibrium* entropy, and allows us to write

$$S = -k_B \int \rho \ln \rho \, d\mathbf{v}\, d\mathbf{r} + \text{constant}. \tag{2.11.10}$$

Thus the *entropy density* in μ-space is

$$s = -k_B \rho \ln \rho. \tag{2.11.11}$$

The intensive variable conjugate to $N(\mathbf{v}, \mathbf{r}) = \rho\, d\mathbf{v}\, d\mathbf{r}$ is obtained by taking the functional derivative of S with respect to ρ, i.e., the derivative of s with respect to ρ. Thus

$$F(\rho) \equiv \delta S/\delta\rho = \partial s/\partial\rho = -k_B(\ln \rho + 1). \tag{2.11.12}$$

This means that in the Onsager theory, the local thermodynamic force in μ-space should be

$$X = F(\rho) - F(\rho^e) = -k_B \ln(\rho/\rho^e). \tag{2.11.13}$$

Finally keeping only the linear term around equilibrium gives

$$X = -k_B \Delta\rho/\rho^e. \tag{2.11.14}$$

We can use this force to rewrite the linearized Boltzmann equation in the form

$$\partial\Delta\rho/\partial t = (\mathbf{v}\rho^e/k_B)\cdot\nabla_\mathbf{r} X + \int L^s(\mathbf{v},\mathbf{v}_1)X_1\,d\mathbf{v}_1,$$

where $L^s(\mathbf{v},\mathbf{v}_1)$ is the linear operator

$$L^s(\mathbf{v},\mathbf{v}_1) = -K(\mathbf{v},\mathbf{v}_1)/k_B. \tag{2.11.15}$$

In an even more compact notation,

$$\partial\Delta\rho/\partial t = L[X] \tag{2.11.16}$$

with

$$L[X] \equiv (\mathbf{v}\rho^e/k_B)\cdot\nabla_\mathbf{r} X + \int L^s(\mathbf{v},\mathbf{v}_1)X_1\,d\mathbf{v}_1. \tag{2.11.17}$$

Equation (2.11.16) is the Onsager regression equation at the Boltzmann level of description.

It is not difficult to show that the relaxation operator L satisfies the requirements of the Onsager theory. The functional L is linear and is composed of an antisymmetric piece, which comes from the streaming term, and a symmetric piece, which comes from the collision term. The antisymmetry of the first term in Eq. (2.11.17) can be seen by writing it as the integral operator

$$L^a[X] = (\mathbf{v}\rho^e/k_B)\cdot\int \nabla_\mathbf{r}\delta(\mathbf{r}-\mathbf{r}_1)\delta(\mathbf{v}-\mathbf{v}_1)X(\mathbf{r}_1,\mathbf{v}_1,t)\,d\mathbf{r}_1\,d\mathbf{v}_1. \tag{2.11.18}$$

Thus the operator is antisymmetric if the kernel is antisymmetric, i.e.,

$$(\mathbf{v}_1\rho_1^e/k_B)\cdot\nabla_{\mathbf{r}_1}\delta(\mathbf{r}_1-\mathbf{r})\delta(\mathbf{v}_1-\mathbf{v}) = -(\mathbf{v}\rho^e/k_B)\cdot\nabla_\mathbf{r}\delta(\mathbf{r}-\mathbf{r}_1)\delta(\mathbf{v}-\mathbf{v}_1). \tag{2.11.19}$$

This, however, is clear since $\nabla_{\mathbf{r}_1}\delta(\mathbf{r}_1-\mathbf{r}) = -\nabla_\mathbf{r}\delta(\mathbf{r}-\mathbf{r}_1)$ by the chain rule and since the velocity delta function commutes with the differential operator. To prove the symmetry of the collision term we need to show that

$$\iint h(\mathbf{v})L^s(\mathbf{v},\mathbf{v}_1)g(\mathbf{v}_1)\,d\mathbf{v}\,d\mathbf{v}_1 = \iint g(\mathbf{v})L^s(\mathbf{v},\mathbf{v}_1)h(\mathbf{v}_1)\,d\mathbf{v}\,d\mathbf{v}_1. \tag{2.11.20}$$

This can be done by referring to the definition of L^s in Eqs. (2.11.15) and (2.11.8). For example, consider the term on the left-hand side of Eq. (2.11.20) that comes from the translation operator T_1', i.e.,

$$I_1' = \iint h(\mathbf{v})[-\hat{\sigma}_T|\mathbf{v}-\mathbf{v}_1|\rho^e\rho_1^e/k_B]g(\mathbf{v}_1')\,d\mathbf{v}_1\,d\mathbf{v}. \tag{2.11.21}$$

Changing integration variables from $(\mathbf{v}_1,\mathbf{v})$ to $(\mathbf{v}',\mathbf{v}_1')$ gives

$$I_1' = \iint h(\mathbf{v}_1')[-\hat{\sigma}_T'|\mathbf{v}' - \mathbf{v}_1'|\rho^{e'}\rho_1^{e'}/k_B]g(\mathbf{v})\,d\mathbf{v}'\,d\mathbf{v}_1'$$

$$= \iint g(\mathbf{v})[-\hat{\sigma}_T|\mathbf{v} - \mathbf{v}_1|\rho^e\rho_1^e/k_B]h(\mathbf{v}_1')\,d\mathbf{v}_1\,d\mathbf{v}, \qquad (2.11.22)$$

where in the second equality we have used Eqs. (2.7.19) and (2.8.11). The second integral in Eq. (2.11.22) is just the term on the right-hand side of Eq. (2.11.20) involving T_1'. In a similar fashion the other terms can be shown to be symmetric, which proves that their sum, L^s, is symmetric. The functional L^s can also be shown to be non-negative definite. In fact, by changing variables as in the proof of the H-theorem in Section 2.8 and using Eq. (2.7.19), it is not hard to see that

$$\iint h(\mathbf{v})L^s(\mathbf{v}, \mathbf{v}_1)h(\mathbf{v}_1)\,d\mathbf{v}_1\,d\mathbf{v}$$

$$= \frac{1}{4k_B}\iint \hat{\sigma}_T[h(\mathbf{v}) + h(\mathbf{v}_1) - h(\mathbf{v}') - h(\mathbf{v}_1')]^2 g\rho^e\rho_1^e\,d\mathbf{v}\,d\mathbf{v}_1, \qquad (2.11.23)$$

which is clearly non-negative. Equation (2.11.23) shows, in addition, that the functional L^s has five zero eigenfunctions, namely, the five collision invariants of mass, momentum, and energy for which the square-bracketed expression in the integrand of Eq. (2.11.23) vanishes identically. Because the other factors in Eq. (2.11.23) are positive, all other eigenvalues of L^s must be positive. Thus the linearized Boltzmann equation has precisely the formal structure predicted by the Onsager theory.

The fact that near equilibrium the Boltzmann equation fits into the formalism of irreversible thermodynamics suggests a stochastic interpretation for the equation, namely, that it governs the *conditional average value* of the μ-space density. With this interpretation the linearized equation can be written

$$\partial\Delta\bar{\rho}/\partial t = L[\bar{X}], \qquad (2.11.24)$$

where, again, the overbars signify the two-time conditional average. Following Fox and Uhlenbeck, the stochastic description of equilibrium at the Boltzmann level can be completed by adopting the Onsager hypothesis for the regression of fluctuations discussed in Section 2.6. Thus $\Delta\rho$ becomes a stationary, Gaussian, Markov process, and Eq. (2.6.10) can be used to write

$$\partial\Delta\rho/\partial t = L[X] + \tilde{f} \qquad (2.11.25)$$

with $\langle\tilde{f}(\mathbf{r}, \mathbf{v}, t)\rangle = 0$ and

$$\langle\tilde{f}(\mathbf{r}, \mathbf{v}, t)\tilde{f}(\mathbf{r}', \mathbf{v}', t')\rangle = 2k_B L^s(\mathbf{v}, \mathbf{v}')\delta(\mathbf{r} - \mathbf{r}')\delta(t - t'), \qquad (2.11.26)$$

which follows from Eq. (2.6.11). In this way the Onsager picture subsumes the Boltzmann picture, at least close to equilibrium. In Chapter 9 this relationship will be shown to imply that fluctuating hydrodynamics is a contraction of the fluctuating Boltzmann equation.

Equations (2.11.25) unite the Onsager and Boltzmann pictures close to equilibrium. Seen in this way, the Boltzmann approach is still the more fundamental one since it provides explicit expressions for the entropy functional, S, and the relaxation functional, L, which are the two ingredients in the Onsager picture. In order to unite the two approaches completely, it is necessary to eliminate the entropy from the Onsager picture and to add fluctuations to the Boltzmann picture. In addition, nonlinear effects need to be included in the Onsager picture, while a clear stochastic interpretation of the nonlinear Boltzmann equation needs to be provided. To do this requires the introduction of a new picture of macroscopic systems which combines the kinetic approach of Boltzmann with the stochastic and thermodynamic approach of Onsager. This new picture provides the content for the remainder of this volume.

References

Nonequilibrium Thermodynamics

S.R. de Groot and P. Mazur, *Non-equilibrium Thermodynamics* (North Holland, Amsterdam, 1962; reprinted by Dover, 1984).
H.B. Callen, *Thermodynamics* (John Wiley, New York, 1960).
D.D. Fitts, *Nonequilibrium Thermodynamics* (McGraw-Hill, New York 1962).
L. Onsager, Reciprocal relations in irreversible processes. I. *Phys. Rev.* **37**, 405 (1931).
L. Onsager, Reciprocal relations in irreversible processes. II. *Phys. Rev.* **38**, 2265 (1931).
H.B.G. Casimir, On Onsager's principle of microscopic reversibility, *Rev. Mod. Phys.* **17**, 343 (1945).

Linear Statistical Theory of Nonequilibrium Thermodynamics

L.D. Landau and E.M. Lifshitz, *Statistical Physics*, 2nd ed. (Pergamon, London 1969).
L. Onsager and S. Machlup, Fluctuations and irreversible processes, *Phys. Rev.* **91**, 1505 (1953).
S. Machlup and L. Onsager, Fluctuations and irreversible processes. II. Systems with kinetic energy, *Phys. Rev.* **91**, 1512 (1953).
R.F. Fox and G.E. Uhlenbeck, Contributions to non-equilibrium thermodynamics. I. Theory of hydrodynamical fluctuations, *Phys. Fluids* **13**, 1893 (1970).
R.F. Fox and G.E. Uhlenbeck, Contributions to nonequilibrium thermodynamics. II. Fluctuation theory for the Boltzmann equation, *Phys. Fluids* **13**, 2881 (1970).

Hydrodynamics

G.K. Batchelor, *An Introduction to Fluid Dynamics* (Cambridge University Press, Cambridge, 1970).

Boltzmann Equation

L. Boltzmann, *Lectures on Gas Theory* (University of California Press, Berkeley, 1964).
S. Chapman and T.G. Cowling, *The Mathematical Theory of Non-uniform Gases* (Cambridge University Press, Cambridge, 1970).
P. Résibois and M. De Leener, *Classical Kinetic Theory of Fluids* (John Wiley, New York, 1977).
G.E. Uhlenbeck and G.W. Ford, *Lectures in Statistical Mechanics* (American Mathematical Society, Providence, RI, 1963).

CHAPTER 3

Elementary Processes and Fluctuations

3.1. Introduction

To understand matter on a macroscopic scale, it is necessary to pay attention to its molecular makeup. That is one of the lessons to be learned from the kinetic theory of Maxwell and Boltzmann. Another lesson from kinetic theory is that the dynamical equations, which describe the change of macroscopic systems, have thermodynamics built into them. Indeed, this is the basic content of Boltzmann's H-theorem. This is also apparent in the Onsager theory which is based on the observation that thermodynamic forces are responsible for the relaxation to equilibrium. In the Onsager theory molecular fluctuations are also related to thermodynamic quantities: The stationary equilibrium fluctuations are determined by the second differential of the entropy, $\delta^2 S$, and the random noise in the thermodynamic fluxes is determined by the dynamic coupling matrix, L_{ij}. All this suggests that both thermodynamics and fluctuations are imbedded in a deeper formalism—a formalism which, like the Boltzmann equation, is based on a picture of molecular events.

In this chapter we develop the mechanistic statistical theory of nonequilibrium thermodynamics for two special cases. First we examine the Boltzmann equation and show how to interpret the effect of molecular collisions stochastically. This leads to a stochastic description of the μ-space number density and to coupled nonlinear equations which we refer to as the fluctuating Boltzmann equation. The stochastic *ansatz* has strong thermodynamic character, and we show that near equilibrium the fluctuating Boltzmann equation reduces to the μ-space stochastic description of Onsager.

The stochastic formulation of the Boltzmann equation depends in a fundamental way on Boltzmann's picture of molecular collisions. Another clear case in which a conception of elementary molecular events is important is chemical reactions at the thermodynamic level of description. The same *ansatz* that generalizes the Onsager theory in μ-space also yields a nonlinear generalization of the linear stochastic theory of chemical reaction fluctuations. Chemical reactions provide a useful, simple archetype of the general theory, and the consequences of the theory for the description of chemical processes are examined in some detail. In doing so, we recover the classical thermodynamic description of chemical equilibrium.

3.2. The Stochastic Description of the Boltzmann Equation

In Chapter 2 the coarseness of the Boltzmann level of description was emphasized by breaking the μ-space into small cellular volume elements. In the stochastic picture this coarseness translates into the fact that collisions cause unpredictable transitions from one cell to another. The reason these transitions are unpredictable is that in μ-space we keep track only of the *number* of molecules in a given μ-space volume element. This does not provide sufficient information to determine the number of binary clusters of molecules, which is required to describe the effect of collisions. Thus the μ-space density provides only a contracted picture of a molecular system. According to the ideas in Section 1.2, this implies that the variables are random and that we are reduced to considering an ensemble of related systems.

It is the collisions, not the streaming motion in μ-space, that make the μ-space description unpredictable. In his original work, Boltzmann ignored these fluctuations and used, as in Section 2.7, a conception of molecular scattering processes to describe the effect of collisions. Boltzmann's conception of a molecular collision event has a strong macroscopic character to it. Consider a direct collision between two molecules with velocities in the range $\mathbf{v}$ to $\mathbf{v} + d\mathbf{v}$ and $\mathbf{v}_1$ to $\mathbf{v}_1 + d\mathbf{v}_1$ in the spatial volume element $d\mathbf{r}$ located at position $\mathbf{r}$. This is a *conditional* event as far as changing the μ-space number density. To clarify this, consider $N(\mathbf{r}, \mathbf{v}, t) = \rho(\mathbf{r}, \mathbf{v}, t) d\mathbf{r}\, d\mathbf{v}$, which is the number of molecules in the volume element $d\mathbf{r}\, d\mathbf{v}$. The scattering event describes what occurs *if* there is a partner molecule which is *poised* for collision in the volume $d\mathbf{r}\, d\mathbf{v}_1$. If a collision occurs, both $N(\mathbf{r}, \mathbf{v}, t)$ and $N(\mathbf{r}, \mathbf{v}_1, t)$ decrease by one whereas $N(\mathbf{r}, \mathbf{v}', t)$ and $N(\mathbf{r}, \mathbf{v}'_1, t)$ both increase by one. Thus the effect on the macroscopic observable $N(\mathbf{r}, \mathbf{v}, t)$ is perfectly deterministic *if* such a collision occurs.

If each volume element in μ-space is labeled by its occupancy number $N(\mathbf{r}, \mathbf{v}, t)$, then a single collision of the type $\mathbf{v}, \mathbf{v}_1 \to \mathbf{v}', \mathbf{v}'_1$ is a jump process in which the occupancy number of the two volume elements at $\mathbf{r}, \mathbf{v}$ and $\mathbf{r}, \mathbf{v}_1$ decreases by one and the occupancy of the two volume elements at $\mathbf{r}, \mathbf{v}'$ and $\mathbf{r}, \mathbf{v}'_1$ increases by one. This is very much like the effect of the outcomes in a coin toss, although in coin tossing only two outcomes are possible, heads or tails, rather than the infinite constellation of outcomes for molecular collisions. To elaborate the analogy, consider n coins which are either heads or tails and for which we keep track of only the number of heads, n_+. If we toss a single coin which was originally tails and heads comes up, symbolically $\mathrm{T} \to \mathrm{H}$, then n_+ changes by $+1$. The event $\mathrm{T} \to \mathrm{H}$ is an allowed process in tossing a coin and if it occurs, there will be a predictable change in the number of heads and tails.

We will refer to any molecular process which causes a deterministic change in an extensive variable as an *elementary process*. The elementary processes that cause change in the occupancy numbers in μ-space are, for a dilute gas, all the possible binary collisions. As we have noted, elementary collision processes come in pairs. Thus for every direct collision, $\mathbf{v}, \mathbf{v}_1 \rightarrow \mathbf{v}', \mathbf{v}'_1$, there is an inverse collision, $\mathbf{v}, \mathbf{v}_1 \leftarrow \mathbf{v}', \mathbf{v}'_1$, which restores the initial velocities. This is typical of all elementary processes. Consequently, it is natural to think of an elementary process and its inverse (or reverse) as a single elementary process, e.g., $\mathbf{v}, \mathbf{v}_1 \rightleftharpoons \mathbf{v}', \mathbf{v}'_1$.

A description of which elementary processes are allowed gives only a catalogue of the events that might occur in any short interval of time, dt. To determine how these processes affect the extensive variables, it is necessary to know the probability that each elementary process occurs in dt. This probability is not given in the Boltzmann picture which, as we have seen in Chapter 2, describes only the average effect of collisions. What Boltzmann provides, instead, is an expression for the rate of change of the occupancy number due to an elementary collision process. Referring back to the Boltzmann equation (Eq. 2.7.23), the rate of change of $N(\mathbf{r}, \mathbf{v}, t)$ due to the elementary process $\mathbf{v}$, $\mathbf{v}_1 \rightleftharpoons \mathbf{v}', \mathbf{v}'_1$ is seen to be

$$R_\kappa = -\hat{\sigma}_T g[\rho\rho_1 - \rho'\rho'_1]\, d\mathbf{v}_1\, d\mathbf{v}\, d\mathbf{r}, \tag{3.2.1}$$

where the first term represents the direct collision and the second term is the restoring collision, and we have labeled this specific collision by the index $\kappa = (\mathbf{v}, \mathbf{v}_1; \mathbf{v}', \mathbf{v}'_1)$. The factor -1 represents the fact that exactly one molecule is lost from $d\mathbf{r}\, d\mathbf{v}$ every time the elementary collision process occurs. The other factor, namely,

$$V_\kappa = \hat{\sigma}_T g[\rho\rho_1 - \rho'\rho'_1]\, d\mathbf{v}_1\, d\mathbf{v}\, d\mathbf{r} \tag{3.2.2}$$

is called the *rate* of the elementary processes, $\mathbf{v}, \mathbf{v}_1 \rightleftharpoons \mathbf{v}', \mathbf{v}'_1$. The contributions to the overall rate of the elementary process arise from the direct and restoring collisions. We will call them the forward and reverse *transition rates*. Symbolically the forward transition rate is

$$V_\kappa^+ = \hat{\sigma}_T g\rho\rho_1\, d\mathbf{v}_1\, d\mathbf{v}\, d\mathbf{r} \tag{3.2.3}$$

and the reverse transition rate is

$$V_\kappa^- = \hat{\sigma}_T g\rho'\rho'_1\, d\mathbf{v}_1\, d\mathbf{v}\, d\mathbf{r}. \tag{3.2.4}$$

The adjective "transition" is appropriate since the events which correspond to these rates are the jumps or transitions that occur in the μ-space occupancy numbers.

How should the expressions for the transition rates in Eqs. (3.2.3) and (3.2.4) be interpreted? According to Boltzmann, they are simply the number of direct or restoring collisions that occur per unit time. As we have seen, this interpretation can be true only in some average sense. There is an alternative, stochastic, interpretation of these transition rates, which generalizes Boltzmann's interpretation. Using this interpretation the Boltzmann equation is recovered *on the average* and one obtains, in addition, a description of fluctuations in the μ-space number density.

To explain this interpretation of the transition rates, we return to our symbolic notation for the occupancy numbers of μ-space volume elements. Labeling each cell by an index i, we keep track of the occupancy numbers by means of a column vector $\mathbf{n}(t) = \text{col}(N_1(t), N_2(t), \ldots, N_k(t))$ where k is the total number of cells and is a very large number. The size of cells should be small, but large enough that $N_i(t)$ is a macroscopic variable. If this condition is not met, the gas will be so dilute that collisions with the walls become important and one is in the so-called Knudsen limit. The variable $\mathbf{n}(t)$ is the vector stochastic process whose time evolution we want to determine. Since it is a random variable, the best we can do is specify the probability that its value will change by any specified amount in the interval dt. Now the specific amounts by which $\mathbf{n}$ can change are completely determined by the elementary processes, labeled by κ. For the forward part of the elementary process only four occupancy numbers in the vector $\mathbf{n}$ change (by plus or minus one) and all other occupancy numbers stay fixed. This vector of changes will be written $\boldsymbol{\omega}_\kappa^+$. For the reverse process there is a corresponding vector of changes, $\boldsymbol{\omega}_\kappa^-$, but

$$-\boldsymbol{\omega}_\kappa^- = \boldsymbol{\omega}_\kappa^+ \equiv \boldsymbol{\omega}_\kappa \tag{3.2.5}$$

because the reverse process restores the effect of the direct collision. Thus the set $\{\boldsymbol{\omega}_\kappa\}$ for all κ summarizes the possible changes in $\mathbf{n}$ which can occur due to collision. Of course, in addition to these collisional changes, which may or may not occur, a change in $\mathbf{n}(t)$ is caused by the inevitable streaming motion in μ-space. If the change in $\mathbf{n}$ in the interval dt due to streaming is $d\mathbf{n}_s$, then the probability of the total change of $\mathbf{n}$, $d\mathbf{n} + d\mathbf{n}_s$, is calculated according to the *ansatz*

$$P_2(\mathbf{n}, t|\mathbf{n}' = \mathbf{n} + d\mathbf{n}_s + d\mathbf{n}, t + dt) = \begin{cases} V_\kappa^\pm\, dt + \mathcal{O}(dt), & \text{if } d\mathbf{n} = \pm\boldsymbol{\omega}_\kappa \\ 1 - \sum_\kappa (V_\kappa^+ + V_\kappa^-)\, dt + \mathcal{O}(dt), & \text{if } d\mathbf{n} = \mathbf{0} \\ 0 \quad \text{otherwise.} \end{cases} \tag{3.2.6}$$

The left-hand side of Eq. (3.2.6) indicates the conditional probability for $\mathbf{n}$ changing in the time interval dt, and κ can be any elementary process. In words Eq. (3.2.6) states that in addition to streaming, only the elementary collision processes cause the occupancy numbers to change. Further, the probability of a collisional change occurring is given by the number of collisions that occur in dt. In the absence of collisions, i.e., if all the rates $V_\kappa^\pm = 0$, Eq. (3.2.6) states that streaming occurs with probability one.

The *ansatz* in Eq. (3.2.6) produces a stochastic process which is, essentially, a diffusion process. Stochastic diffusions are described in Section 1.5 and satisfy conditions (1.5.16)–(1.5.18) on their transition moments. The first transition moment is

$$\int P_2(\mathbf{n}, t|\mathbf{n}', t + dt)(\mathbf{n}' - \mathbf{n})\, d\mathbf{n}' \equiv \mathbf{M}^1(\mathbf{n}). \tag{3.2.7}$$

This can be calculated directly from Eq. (3.2.6), which gives

$$\begin{aligned}\mathbf{M}^1(\mathbf{n}) &= \langle d\mathbf{n}_s + d\mathbf{n}\rangle^{\mathbf{n}} \\ &= \sum_\kappa \boldsymbol{\omega}_\kappa (V_\kappa^+ - V_\kappa^-)\, dt + d\mathbf{n}_s + \mathcal{O}(dt)\end{aligned} \tag{3.2.8}$$

Thus the limit

$$\lim_{dt\to 0} \mathbf{M}^1(\mathbf{n})/dt = \sum_\kappa \boldsymbol{\omega}_\kappa (V_\kappa^+ - V_\kappa^-) + d\mathbf{n}_s/dt \equiv \mathbf{h}(\mathbf{n}, t) \tag{3.2.9}$$

exists as required by Eq. (1.5.16) since $d\mathbf{n}_s$, the streaming change, is proportional to dt. Similarly the second transition moment is

$$\begin{aligned}M^2(\mathbf{n}) &= \langle (d\mathbf{n}_s + d\mathbf{n})(d\mathbf{n}_s + d\mathbf{n})^T\rangle^{\mathbf{n}} \\ &= d\mathbf{n}_s\, d\mathbf{n}_s^T + d\mathbf{n}_s \left[\sum_\kappa \boldsymbol{\omega}_\kappa^T (V_\kappa^+ - V_\kappa^-)\right] dt \\ &\quad + \left[\sum_\kappa \boldsymbol{\omega}_\kappa (V_\kappa^+ - V_\kappa^-)\right] d\mathbf{n}_s^T\, dt \\ &\quad + \sum_\kappa \boldsymbol{\omega}_\kappa \boldsymbol{\omega}_\kappa^T V_\kappa^+\, dt + \sum_\kappa \boldsymbol{\omega}_\kappa \boldsymbol{\omega}_\kappa^T V_\kappa^-\, dt \\ &\quad + \mathcal{O}(dt)\end{aligned} \tag{3.2.10}$$

Since all the terms which involve streaming contributions are of order $(dt)^2$, the limit required in Eq. (1.15.17) exists and

$$\lim_{dt\to 0} M^2(\mathbf{n})/dt = \sum_\kappa \boldsymbol{\omega}_\kappa (V_\kappa^+ + V_\kappa^-)\boldsymbol{\omega}_\kappa^T \equiv \gamma(\mathbf{n}, t). \tag{3.2.11}$$

To verify that the higher-order transition moments vanish, as in Eq. (1.5.18), we need to examine quantities such as

$$M_{ijk}^3(\mathbf{n}) = \langle dn_i\, dn_j\, dn_k\rangle = \sum_\kappa \omega_{\kappa i}\omega_{\kappa j}\omega_{\kappa k}(V_\kappa^+ - V_\kappa^-)\, dt + \mathcal{O}(dt) \tag{3.2.12}$$

since each term $(dn_i)_s$ contributes a factor of dt. Using Eq. (3.2.12), it is easy to see that Eq. (1.5.18) is *not* valid for the extensive variables n_i, since the first term in Eq. (3.2.12) is proportional to dt. In fact, the stochastic process $\mathbf{n}$ is not strictly a stochastic diffusion process except in the limit that a large number of molecules make up the system.

The situation here is quite similar to that encountered in proving the central limit theorem in Section 1.3. There, to achieve a simple description of the outcome of tossing n coins we saw that the number of heads, n_+, had to be scaled by the factor $1/\sqrt{n}$. Furthermore, the central limit theorem is true only in the limit that $n \to \infty$. Clearly, we don't expect such a simple *ansatz* as Eq. (3.2.6) to be valid except for a macroscopic system. Thus all the occupancy numbers n_i must be large, and consequently, the total number of molecules $n = \sum_{i=1}^{k} n_i$ must be, effectively, infinite. Taking our lead from the central limit theorem, it makes sense to introduce the scaled extensive variables $\hat{n}_i = n_i/\sqrt{m}$ where $m = n/k$. The number of subvolumes, k, should be chosen such that the n_i are the order of m and such that $m \gg 1$. The changes in $\hat{n}_i$ during an elementary process κ are $\hat{\omega}_{\kappa i} = \omega_{\kappa i}/\sqrt{m}$. For the scaled variables $\hat{n}_i$ the moment conditions in Eqs. (3.2.8), (3.2.10), and (3.2.12) become

$$\begin{aligned} \mathbf{M}^1(\hat{\mathbf{n}}) &= \sum_\kappa \hat{\boldsymbol{\omega}}_\kappa (V_\kappa^+ - V_\kappa^-)\,dt + d\hat{\mathbf{n}}_s + \mathcal{O}(dt) \\ M^2(\hat{\mathbf{n}}) &= \sum_\kappa \hat{\boldsymbol{\omega}}_\kappa (V_\kappa^+ - V_\kappa^-)\hat{\boldsymbol{\omega}}_\kappa^T\,dt + \mathcal{O}(dt) \\ M_{ijk}^3(\hat{\mathbf{n}}) &= \sum_\kappa \hat{\omega}_{\kappa i}\hat{\omega}_{\kappa j}\hat{\omega}_{\kappa k}(V_\kappa^+ - V_\kappa^-)\,dt + \mathcal{O}(dt). \end{aligned} \tag{3.2.13}$$

The limiting values of these expressions for large m and n depend on the scaling of the rate of the elementary processes, $V_\kappa^\pm$. These rates are extensive quantities given by the collision rates in Eqs. (3.2.3) and (3.2.4). Thus $V_\kappa^\pm$ are the order of n_i, i.e., the order of m. Referring to Eq. (3.2.13) and remembering that $\hat{\omega}_{\kappa i} = \omega_{\kappa i}/\sqrt{m}$ where $\omega_{\kappa i}$ is either $+1$, 0, or -1, the expressions for the first two moments remain unchanged as $dt \to 0$. The first moment, $\mathbf{M}^1$, however, is of order $\sqrt{m}$ while the second moment, M^2, is of order unity. The third moment is of order $1/\sqrt{m}$, and in general the lth moments are of order $m^{(2-l)/2}$. Since $m = n/k$ is much larger than one, this means that

$$M_{ijk}^3(\hat{\mathbf{n}}, t) = \mathcal{O}(dt) \tag{3.2.14}$$

with similar results for higher-order moments. These considerations show that in the thermodynamic limit at least the scaled extensive variables, $\hat{n}_i$, become a stochastic diffusion process.

Let's explore the properties of this stochastic diffusion further. In Section 1.5 it was shown that a stochastic diffusion process satisfies a Fokker–Planck equation, Eq. (1.5.25). The first two moments in Eq. (3.2.13) determine the drift and diffusion terms, h_i and γ_{ij}, for the Fokker–Planck equation which is,

$$\begin{aligned} \partial P_2(\hat{\mathbf{n}}_1, t_1|\hat{\mathbf{n}}, t)/\partial t = &-\partial\left[\sum_\kappa \hat{\omega}_{\kappa i}(V_\kappa^+ - V_\kappa^-) + \hat{s}_i\right] P_2(\hat{\mathbf{n}}_1, t_1|\hat{\mathbf{n}}, t)/\partial \hat{n}_i \\ &+ \frac{1}{2}\partial^2\left[\sum_\kappa \hat{\omega}_{\kappa i}(V_\kappa^+ + V_\kappa^-)\hat{\omega}_{\kappa j}\right] P_2(\hat{\mathbf{n}}_1, t_1|\hat{\mathbf{n}}, t)/\partial \hat{n}_i \partial \hat{n}_j \end{aligned} \tag{3.2.15}$$

where the streaming term $d\hat{\mathbf{n}}_s/dt = \hat{\mathbf{s}}$ and the summation on repeated indices is used for i and j. The first term on the right-hand side of Eq. (3.2.15) is the drift term and the second is the diffusion term. As we have discovered, the drift term is proportional to $\sqrt{m}$, whereas the diffusion is independent of m. Thus for the scaled occupancy numbers, $\hat{\mathbf{n}}$, the drift term dominates. Indeed, if the diffusion term is neglected in Eq. (3.2.15), the equation can be written

$$\partial P_2(\hat{\mathbf{n}}_1, t_1 | \hat{\mathbf{n}}, t)/\partial t = -\partial h_i(\hat{\mathbf{n}}) P_2(\hat{\mathbf{n}}_1, t_1 | \hat{\mathbf{n}}, t)/\partial \hat{n}_i \tag{3.2.16}$$

where for simplicity $h_i(\hat{\mathbf{n}})$ has been written in place of $\sum_\kappa \hat{\omega}_{\kappa i}(V_\kappa^+ - V_\kappa^-) + \hat{s}_i$. It is not difficult to find the solution to Eq. (3.2.16). Recall that the conditional probability density, P_2, satisfies the initial condition $P_2(\hat{\mathbf{n}}_1, t_1 | \hat{\mathbf{n}}, t_1) = \delta(\hat{\mathbf{n}}_1 - \hat{\mathbf{n}})$. Then if $\bar{\hat{\mathbf{n}}}(\mathbf{n}_1, t)$ is the solution to the scaled Boltzmann equation

$$d\bar{\hat{n}}_i/dt = \sum_\kappa \hat{\omega}_{\kappa i}(\bar{V}_\kappa^+ - \bar{V}_\kappa^-) + \bar{\hat{s}}_i = h_i(\bar{\hat{\mathbf{n}}}), \tag{3.2.17}$$

the solution to Eq. (3.2.16) is just

$$P_2(\hat{\mathbf{n}}_1, t_1 | \hat{\mathbf{n}}, t) = \delta(\hat{\mathbf{n}} - \bar{\hat{\mathbf{n}}}(\hat{\mathbf{n}}_1, t)). \tag{3.2.18}$$

This follows from the chain rule and the properties of the delta function:

$$\begin{aligned} \partial\delta[\hat{\mathbf{n}} - \bar{\hat{\mathbf{n}}}(t)]/\partial t &= \{\partial\delta[\hat{\mathbf{n}} - \bar{\hat{\mathbf{n}}}(t)]/\partial\hat{n}_i\}(-d\bar{\hat{n}}_i/dt) \\ &= -\{\partial\delta[\hat{\mathbf{n}} - \bar{\hat{\mathbf{n}}}(t)]/\partial\hat{n}_i\} h_i(\bar{\hat{\mathbf{n}}}) \\ &= -\partial\{h_i(\hat{\mathbf{n}})\delta[\hat{\mathbf{n}} - \bar{\hat{\mathbf{n}}}(t)]\}/\partial\hat{n}_i. \end{aligned} \tag{3.2.19}$$

In other words, neglecting the diffusion term in the Fokker–Planck equation implies that the conditional probability remains a delta function, peaked at the solution to the Boltzmann equation. This means that the occupancy numbers in μ-space satisfy the Boltzmann equation on the conditional average and that, to a first approximation, fluctuations are negligible. Seen in this way, the statistical interpretation of the collision terms in Eq. (3.2.6) implies that for a large system the Boltzmann equation is true on the average.

3.3. The Fluctuating Boltzmann Equation

The stochastic interpretation of the Boltzmann collision terms is restricted to macroscopic systems. For a macroscopic system fluctuations in μ-space density are small, and the Boltzmann equation gives the average behavior in the ensemble, as assumed by Boltzmann. In this section we investigate the nature of fluctuations around the average, which leads to the fluctuating Boltzmann equation. For a macroscopic system the Fokker–Planck equation (3.2.15) describes the stochastic diffusion process for the scaled occupancy numbers. If we remember that this equation is valid only when n and n/k are large, we can rewrite it in terms of the unscaled variables

$$\partial P_2/\partial t = -\partial\left[\sum_\kappa \omega_{\kappa i}(V_\kappa^+ - V_\kappa^-) + S_i\right]P_2/\partial n_i$$
$$+\frac{1}{2}\partial^2\left[\sum_\kappa \omega_{\kappa i}(V_\kappa^+ + V_\kappa^-)\omega_{\kappa j}\right]P_2/\partial n_i\,\partial n_j. \qquad (3.3.1)$$

Equation (3.3.1) is our starting point for examining the fluctuations. Neglecting the diffusion term, we saw in Section 3.2 that the solution to the Fokker–Planck equation is a delta function centered at the solution to the Boltzmann equation. Thus it is reasonable that when the diffusion term, which is of smaller order in $m = n/k$, is included, the conditional probability will exhibit small fluctuations about the average predicted by the Boltzmann equation. To see what the fluctuations are, it is necessary to rescale Eq. (3.3.1) correctly. Because the variables n_i are occupancy numbers, we begin by rescaling by a factor of m^{-1} to make them bounded. Writing $x_i = n_i/m$, Eq. (3.3.1) becomes

$$\partial P_2/\partial t = -\partial\left[\sum_\kappa \omega_{\kappa i}(v_\kappa^+ - v_\kappa^-) + s_i\right]P_2/\partial x_i$$
$$+\frac{m^{-1}}{2}\partial^2\left[\sum_\kappa \omega_{\kappa i}(v_\kappa^+ + v_\kappa^-)\omega_{\kappa j}\right]P_2/\partial x_i\,\partial x_j, \qquad (3.3.2)$$

where $v_\kappa^\pm = V_\kappa^\pm/m$. Since the $V_\kappa^\pm$ are extensive quantities, $v_\kappa^\pm$ is independent of m. For the x_i the solution to the Boltzmann equation

$$d\bar{x}_i/dt = \sum_\kappa \omega_{\kappa i}(\bar{v}_\kappa^+ - \bar{v}_\kappa^-) + \bar{s}_i \qquad (3.3.3)$$

is $\bar{x}_i(\mathbf{x}_1, t)$ with $\bar{x}_i(\mathbf{x}_1, 0) = x_{i1}$, the initial condition. Since we expect that the solution to the Boltzmann equation provides the most probable behavior, it is natural to change variables in Eq. (3.3.3) to the fluctuation about $\bar{x}_i$. Thus consider the new variables q_i, defined by $q_i m^{-1/2} \equiv x_i - \bar{x}_i(\mathbf{x}_1, t)$. The scaling of q_i by the factor $m^{-1/2}$ is chosen because the diffusion term, which is second order in q_i, is smaller than the drift term by a factor of m. The probability density for $\mathbf{q}$ is

$$P(\mathbf{q}, t) \equiv m^{-1/2}P_2(\mathbf{x}_1, t_1|\bar{\mathbf{x}}(\mathbf{x}_1, t) + \mathbf{q}m^{-1/2}, t) \qquad (3.3.4)$$

and $P(\mathbf{q}, 0) = \delta(\mathbf{q})$.

To determine the statistics of fluctuations around the solution to the Boltzmann equation, it is necessary to find the differential equation solved by $P(\mathbf{q}, t)$. This can be done by applying the chain rule to Eq. (3.3.4), which gives,

$$\partial P/\partial t = m^{-1/2}[(\partial P_2/\partial \bar{x}_i)(d\bar{x}_i/dt) + \partial P_2/\partial t]$$
$$= m^{+1/2}(\partial P/\partial q_i)d\bar{x}_i/dt + m^{-1/2}\partial P_2/\partial t, \qquad (3.3.5)$$

since $\partial/\partial x_i = m^{+1/2}\partial/\partial q_i$. The two terms on the right-hand side of Eq. (3.3.5) can be evaluated using Eqs. (3.3.2) and (3.3.3). This leads to

$$\partial P/\partial t = -m^{+1/2}\partial[h_i - \bar{h}_i]P/\partial q_i + \frac{1}{2}\partial^2 \gamma_{ij} P/\partial q_i \partial q_j, \tag{3.3.6}$$

where h_i and γ_{ij} are defined by Eqs. (3.2.9) and (3.2.11) but with $V_\kappa^\pm$ replaced by $v_\kappa^\pm$ and the overbar on $\bar{h}_i$ indicates that this function is evaluated at $\bar{x}_i$. To expose the dependence of the drift and diffusion terms on q_i, we expand them in a Taylor series around $\bar{x}_i$:

$$\begin{aligned} h_i - \bar{h}_i &= (\partial \bar{h}_i/\partial \bar{x}_j) q_j m^{-1/2} + \frac{1}{2}(\partial^2 \bar{h}_i/\partial \bar{x}_j \partial \bar{x}_k) q_j q_k m^{-1} \\ \gamma_{ij} &= \bar{\gamma}_{ij} + (\partial \bar{\gamma}_{ij}/\partial \bar{x}_k) q_k m^{-1/2} + \frac{1}{2}(\partial^2 \bar{\gamma}_{ij}/\partial \bar{x}_k \partial \bar{x}_l) q_k q_l m^{-1}. \end{aligned} \tag{3.3.7}$$

These two series truncate at the quadratic terms because the Boltzmann collision terms are quadratic. Substituting these expressions into Eqs. (3.3.6) then gives

$$\partial P/\partial t + \partial[(\partial \bar{h}_i/\partial \bar{x}_j) q_j P]/\partial q_i - \frac{1}{2}\partial^2 \bar{\gamma}_{ij} P/\partial q_i \partial q_j = \mathcal{O}(m^{-1/2}). \tag{3.3.8}$$

where $\mathcal{O}(m^{-1/2})$ means terms at least of order $m^{-1/2}$. As all the terms on the right-hand side of Eq. (3.3.8) are independent of m, it follows that for $m \gg 1$

$$\partial P/\partial t = -\partial[(\partial \bar{h}_i/\partial \bar{x}_j) q_j P]/\partial q_j + \frac{1}{2}\partial^2 \bar{\gamma}_{ij} P/\partial q_i \partial q_j. \tag{3.3.9}$$

Finally changing variables back to occupancy numbers n_i yields an equation for $P_2(\mathbf{n}_1, t_1 | \bar{\mathbf{n}} + \delta\mathbf{n}, t)$, where $\delta n_i \equiv n_i - \bar{n}_i$ is the fluctuation. The resulting Fokker–Planck type equation is

$$\partial P_2/\partial t = -\partial H_{ij}(\bar{\mathbf{n}}) \delta n_j P_2/\partial \delta n_i + \frac{1}{2}\partial^2 \gamma_{ij}(\bar{\mathbf{n}}) P_2/\partial \delta n_i \partial \delta n_j \tag{3.3.10}$$

with

$$H_{ij}(\bar{\mathbf{n}}) \equiv \partial \left[\sum_\kappa \omega_{\kappa i}(\bar{V}_\kappa^+ - \bar{V}_\kappa^-) + \bar{S}_i \right] \Big/ \partial \bar{n}_j \tag{3.3.11}$$

and

$$\gamma_{ij}(\bar{\mathbf{n}}) \equiv \sum_\kappa \omega_{\kappa i}(\bar{V}_\kappa^+ - \bar{V}_\kappa^-)\omega_{\kappa j}, \tag{3.3.12}$$

where $\bar{n}_i(\mathbf{n}_i, t)$ solves the Boltzmann equation

$$d\bar{n}_i/dt = \sum_\kappa \omega_{\kappa i}(\bar{V}_\kappa^+ - \bar{V}_\kappa^-) + \bar{S}_i. \tag{3.3.13}$$

For a large system, that is, a system for which the occupancy numbers in μ-space, n_i, are all large, Eqs. (3.3.10)–(3.3.13) define the statistical ensemble description.

The Fokker–Planck equation (3.3.10) has a simple structure. Since $\bar{\mathbf{n}}$ is the solution to the Boltzmann equation, it is time dependent, but independent of $\delta\mathbf{n}$. Thus the drift term has a complicated time dependence but is linear in $\delta\mathbf{n}$ Similarly, the diffusion matrix $\gamma_{ij}(\mathbf{n})$ is time dependent, but independent of $\delta\mathbf{n}$. In fact, Eq. (3.3.10) has a structure very similar to that of the Fokker–Planck equation (1.8.29) for an Ornstein–Uhlenbeck process. The only difference is that in Eq. (3.3.10) H_{ij} and γ_{ij} depend explicitly on time. In fact, the stochastic process described by Eq. (3.3.10) is a slight generalization of an Ornstein–Uhlenbeck process. To understand this relationship we use Eq. (3.3.10) to find the conditional average value of $\mathbf{n}$, i.e.,

$$\langle \mathbf{n}(t) \rangle^{\mathbf{n}_1} \equiv \int \mathbf{n} P_2 \, d\mathbf{n}.$$

Actually it is simpler to calculate the average value of the fluctuation, $\delta\mathbf{n} = \int \delta\mathbf{n} P_2 \, d\mathbf{n}$. This can be obtained by multiplying Eq. (3.3.10) by δn_k and integrating over $\delta\mathbf{n}$, which gives

$$\partial \langle \delta n_k(t) \rangle^{\mathbf{n}_1} / \partial t = - \int \delta n_k \partial H_{ij} \delta n_j P_2 / \partial \delta n_j \, d\mathbf{n} + \frac{\gamma_{ij}}{2} \int \delta n_k (\partial^2 P_2 / \partial \delta n_i \delta n_j) \, d\mathbf{n} = 0 \tag{3.3.14}$$

because P_2 vanishes when $\delta\mathbf{n}$ is large, so that integration by parts on the right-hand side of Eq. (3.3.14) gives zero. Thus the average value of the fluctuation is a constant, in fact zero, since at $t = t'$, $\mathbf{n} = \mathbf{n}_1$ with probability one. Thus

$$\langle \delta\mathbf{n}(t) \rangle^{\mathbf{n}_1} = \langle \mathbf{n}(t) \rangle^{\mathbf{n}_1} - \bar{\mathbf{n}}(\mathbf{n}_1, t) = 0. \tag{3.3.15}$$

Equation (3.3.15) states that for a large system *the solution to the Boltzmann equation is identical to the conditional average*. This is the result anticipated based on neglecting the diffusion term.

To describe the fluctuations around the average we need to obtain the complete solution to Eq. (3.3.10). Recall that, according to Itô's generalization of the Langevin equation described in Section 1.5, a diffusion process has an equivalent description given by a stochastic differential equation. The diffusion process for the fluctuations, $\delta\mathbf{n}$, is described by Eq. (3.3.10). Using the equivalence of the Fokker–Planck description in Eq. (1.5.15) and the stochastic differential equation (1.5.13), an equivalent description of the random variable $\delta\mathbf{n}$ is given by

$$d\delta\mathbf{n} = H(\bar{\mathbf{n}}) \delta\mathbf{n} \, dt + \gamma^{1/2}(\bar{\mathbf{n}}) \, d\mathbf{w}, \tag{3.3.16}$$

where $\mathbf{w}$ is a vector-valued Wiener process and $\gamma^{1/2}$ is the square root of the matrix γ, defined $(\gamma^{1/2})^2 = \gamma$. Since Eq. (3.3.12) shows that γ_{ij} is symmetric and positive semi-definite, it has a well-defined square root. The matrices H and $\gamma^{1/2}$ are independent of $\delta\mathbf{n}$. Thus the Itô stochastic differential equation is linear. This means that no ambiguity is introduced by formally dividing Eq. (3.3.16) by dt and writing

$$\gamma^{1/2}\, d\mathbf{w}/dt = \tilde{\mathbf{f}}(t) \tag{3.3.17}$$

as a purely random Gaussian component to the time derivative. Thus

$$d\delta\mathbf{n}/dt = H(\bar{\mathbf{n}}(\mathbf{n}_1, t))\delta\mathbf{n} + \tilde{\mathbf{f}}(t). \tag{3.3.18}$$

The random component, $\tilde{\mathbf{f}}(t)$, is not white noise, but it is closely related to it. Indeed, $d\mathbf{w}/dt$ is a normalized white noise so it follows from Eq. (3.3.17) and Section 1.6 that

$$\langle \tilde{f}_i(t) \rangle \equiv \mathbf{0} \tag{3.3.19}$$

and that

$$\langle \tilde{f}_i(t)\tilde{f}_j(t') \rangle = \gamma_{ij}(\bar{\mathbf{n}}(\mathbf{n}_1, t))\delta(t - t'). \tag{3.3.20}$$

Because the two-time correlation function depends on both $t - t'$ and t, it is called a *colored noise*. In terms of the discussion in Section 1.6, this means that its power spectrum is not a constant.

The derivation of the fluctuating Boltzmann equation from the *ansatz* of Eq. (3.2.6) has been couched, rather abstractly, in terms of elementary processes. We have done so for two reasons. First, many nonlinear transport equations besides the Boltzmann equation can be written in terms of elementary processes. Since the deductions in this and the previous section depend only on the use of elementary processes, the resulting differential equations which describe the fluctuations will have the same form as Eqs. (3.3.18)–(3.3.20). And second, the functional form of the nonlinear operators for the Boltzmann equation is cumbersome. Thus the equations are more easily written down using the shorthand notation of elementary transition rates. Since Eqs. (3.3.18)–(3.3.20) contain the final results we were seeking, we write them out explicitly.

For the Boltzmann equation the transition rates $V_\kappa^{\pm}$ are given in Eqs. (3.2.3) and (3.2.4). To write down the average Boltzmann equation for the μ-space density using Eq. (3.3.13), we need to use changes in the density, $\omega_{\kappa i}^d$, which occur in the elementary process κ. Since we have lumped direct and restoring collisions together in a single elementary process, the sum over κ in Eq. (3.3.13) can be taken over just the direct collisions, say, $\mathbf{v}, \mathbf{v}_1 \to \mathbf{v}', \mathbf{v}_1'$. Such a collision occurs at $\mathbf{r}$ and increases $\mathbf{v}'$ and $\mathbf{v}_1'$ by one and decreases $\mathbf{v}$ and $\mathbf{v}_1$ by one. Thus using $\delta(x, y)$ to represent a Kronecker delta

$$\omega_{\kappa i}^d = [\delta(\mathbf{r}, \mathbf{r}_i)/d\mathbf{r}\, d\mathbf{v}] \times \begin{cases} +1 & \text{if } \mathbf{v}_i = \mathbf{v}' \text{ or } \mathbf{v}_1' \\ -1 & \text{if } \mathbf{v}_i = \mathbf{v} \text{ or } \mathbf{v}_1 \\ 0 & \text{otherwise} \end{cases}$$

$$\equiv [\delta(\mathbf{r}, \mathbf{r}_i)/d\mathbf{r}\, d\mathbf{v}]\chi(\mathbf{v}_i; \mathbf{v}, \mathbf{v}_1) \tag{3.3.21}$$

where we have divided by the size of the volume element in μ-space since we are considering changes in the density. Using this notation and Eqs. (3.2.3) and (3.2.4), Eq. (3.3.13) can be written

$$\partial \bar{\rho}_i/\partial t = \sum_{\mathbf{r}} \sum_{\mathbf{v},\mathbf{v}_1}^{*} \delta(\mathbf{r}, \mathbf{r}_i)\chi(\mathbf{v}_i; \mathbf{v}, \mathbf{v}_1)[\hat{\sigma}_T g(\bar{\rho}\bar{\rho}_1 - \bar{\rho}'\bar{\rho}_1')]\, d\mathbf{v}_1$$
$$- \mathbf{v}_i \cdot \nabla_{\mathbf{r}} \bar{\rho}_i - \mathbf{F}_i \cdot \nabla_{\mathbf{v}} \rho_i, \tag{3.3.22}$$

where the star on the second sum restricts it to direct collisions. The sum over $\mathbf{r}$ is trivial and picks out the value of the densities at $\mathbf{r}_i$. The sum over the velocities is not so straightforward. This sum must be done so as not to double count collisions since direct collisions beginning with $\mathbf{v}$, $\mathbf{v}_1$ and $\mathbf{v}_1$, $\mathbf{v}$ are exactly the same. Double counting can be avoided by summing over only those pairs of velocities $\mathbf{v}$, $\mathbf{v}_1$ with $v_x \le v_{x1}$, $v_y \le v_{y1}$, $v_z \le v_{z1}$. Alternatively we can sum over both $\mathbf{v}$ and $\mathbf{v}_1$ and take 1/2 of the result, if we add to this terms for $\mathbf{v} = \mathbf{v}_1$. Thus the velocity sum can be written

$$\frac{1}{2}\sum_{\mathbf{v}}\sum_{\mathbf{v}_1} \chi(\mathbf{v}_i; \mathbf{v}, \mathbf{v}_1)\hat{\sigma}_T g(\bar{\rho}\bar{\rho}_1 - \bar{\rho}'\bar{\rho}_1')\, d\mathbf{v}_1 + \frac{1}{2}\sum_{\mathbf{v}_1} \chi(\mathbf{v}_i; \mathbf{v}_1, \mathbf{v}_1)\hat{\sigma}_T g(\bar{\rho}_1\bar{\rho}_1 - \bar{\rho}_1'\bar{\rho}_1')\, d\mathbf{v}_1. \tag{3.3.23}$$

For direct collisions, $\chi = -1$ when either $\mathbf{v} = \mathbf{v}_i$ or $\mathbf{v}_1 = \mathbf{v}_i$. Thus the second term in Eq. (3.3.23) becomes

$$\frac{1}{2}\hat{\sigma}_T g(\bar{\rho}_i'\bar{\rho}_i' - \bar{\rho}_i\bar{\rho}_i)\, d\mathbf{v}_1. \tag{3.3.24}$$

The first term in Eq. (3.3.23) can be evaluated by considering terms for which $\mathbf{v}_1 = \mathbf{v}_i$ and $\mathbf{v}_1 \neq \mathbf{v}_i$ separately. If $\mathbf{v}_1 = \mathbf{v}_i$, then according to Eq. (3.3.21) $\chi = -1$ and the sum over $\mathbf{v}$ gives

$$-\frac{1}{2}\sum_{\mathbf{v}} \hat{\sigma}_T g(\bar{\rho}\bar{\rho}_i - \bar{\rho}'\bar{\rho}_i')d\mathbf{v}_1 = -\frac{1}{2}\sum_{\mathbf{v}_1} \hat{\sigma}_T g(\bar{\rho}_i\bar{\rho}_1 - \bar{\rho}_i'\bar{\rho}_1')\, d\mathbf{v}_1, \tag{3.3.25}$$

where in the second writing the dummy summation index was changed from $\mathbf{v}$ to $\mathbf{v}_1$. If $\mathbf{v}_1 \neq \mathbf{v}_i$, then only the term in the sum over $\mathbf{v}$ when $\mathbf{v} = \mathbf{v}_i$ contributes. This gives

$$-\frac{1}{2}\sum_{\mathbf{v}_1 \neq \mathbf{v}_i} \hat{\sigma}_T g(\bar{\rho}_i\bar{\rho}_1 - \bar{\rho}_i'\bar{\rho}_1')\, d\mathbf{v}_1. \tag{3.3.26}$$

Adding the three contributions in Eqs. (3.3.24)–(3.3.26) to the velocity sum gives

$$\sum_{\mathbf{v},\mathbf{v}_1}^{*} \chi(\mathbf{v}_i; \mathbf{v}, \mathbf{v}_1)\hat{\sigma}_T g(\bar{\rho}\bar{\rho}_1 - \bar{\rho}'\bar{\rho}_1')\, d\mathbf{v}_1 = \sum_{\mathbf{v}_1} \hat{\sigma}_T g(\bar{\rho}_i'\bar{\rho}_1' - \bar{\rho}_i\bar{\rho}_1)\, d\mathbf{v}_1. \tag{3.3.27}$$

The left-hand side of (3.3.27) is the discrete version of the velocity integral. Thus using Eq. (3.3.27) in Eq. (3.3.22) and taking the continuum limit yields

$$\partial \bar{\rho}/\partial t = \int \hat{\sigma}_T g(\bar{\rho}'\bar{\rho}_1' - \bar{\rho}\bar{\rho}_1)\, d\mathbf{v}_1 - \mathbf{v} \cdot \nabla_{\mathbf{r}} \bar{\rho} - \mathbf{F} \cdot \nabla_{\mathbf{v}} \bar{\rho}, \tag{3.3.28}$$

where we have dropped the subscript on $\mathbf{r}_i$.

Equation (3.3.28) is the Boltzmann equation, now interpreted as being true on the conditional average. According to Eqs. (3.3.11) and (3.3.18), the conditional fluctuations, $\delta\rho = \rho - \bar{\rho}$, satisfy a linearized version of Eq. (3.3.28). This gives

$$\partial\delta\rho/\partial t = \int \hat{\sigma}_T g[\bar{\rho}_1'\delta\rho' + \bar{\rho}'\delta\rho_1' - \bar{\rho}_1\delta\rho - \bar{\rho}\delta\rho_1]\,d\mathbf{v}_1$$
$$- \mathbf{v}\cdot\nabla_{\mathbf{r}}\delta\rho - \mathbf{F}\cdot\nabla_{\mathbf{v}}\delta\rho + \tilde{f}. \tag{3.3.29}$$

The kernel of the integral operator in Eq. (3.3.29) can be written more succinctly in terms of the translation operators introduced in Eqs. (2.11.6) and (2.11.7). Thus

$$\partial\delta\rho/\partial t$$
$$= \int \hat{\sigma}_T g[\bar{\rho}_1' T' + \bar{\rho}' T_1' - \bar{\rho}_1 T - \bar{\rho} T_1]\delta\rho\,d\mathbf{v}_1 - \mathbf{v}\cdot\nabla_{\mathbf{r}}\delta\rho - \mathbf{F}\cdot\nabla_{\mathbf{v}}\delta\rho + \tilde{f}.$$
$$\equiv H[\bar{\rho}]\delta\rho + \tilde{f}. \tag{3.3.30}$$

where $H[\bar{\rho}]$ is the indicated linear functional which explicitly depends on the conditional average $\bar{\rho}$. The random term, $\tilde{f}$, vanishes on the average, and its two-time correlation function can be obtained from Eq. (3.3.12). Since we are returning from a discrete representation involving occupancy numbers in μ-space to the continuous representation using the μ-space density, the correlation matrix γ_{ij} in Eq. (3.3.12) becomes a functional. Thus we need to find

$$\langle \tilde{f}(\mathbf{r}, \mathbf{v}, t)\tilde{f}(\mathbf{r}_1, \mathbf{v}_1, t')\rangle \equiv \gamma(\mathbf{r}, \mathbf{v}; \mathbf{r}_1, \mathbf{v}_1)\delta(t - t'). \tag{3.3.31}$$

To evaluate the form of γ it is simplest to examine the effect of γ_{ij} on a vector f_j and take the continuum limit. According to Eq. (3.3.12)

$$\sum_j \gamma_{ij} f_j = \sum_\kappa \sum_j \omega_{\kappa i}^d (\bar{V}_\kappa^+ + \bar{V}_\kappa^-)\omega_{\kappa j}^d f_j. \tag{3.3.32}$$

Recall that i that j label volumes in μ-space which include, say, the points $(\mathbf{r}_i, \mathbf{v}_i)$ and $(\mathbf{r}_j, \mathbf{v}_j)$, respectively. The factors $\omega_{\kappa i}^d$ and $\omega_{\kappa j}^d$ are determined by the elementary process κ, i.e., the direct collisions $\mathbf{v}, \mathbf{v}_1 \to \mathbf{v}', \mathbf{v}_1'$ which occur at position $\mathbf{r}$. Evaluation of Eq. (3.3.32) requires that we use the explicit expressions for $\omega_{\kappa i}^d$ in Eq. (3.3.21) and for $\bar{V}_\kappa^\pm$ in Eqs. (3.2.3) and (3.2.4). This gives

$$\sum_j \gamma_{ij} f_j = \sum_{\mathbf{r}} \sum_{\mathbf{v},\mathbf{v}_1}^* \frac{\delta(\mathbf{r}_i, \mathbf{r})\chi(\mathbf{v}_i; \mathbf{v}, \mathbf{v}_1)}{d\mathbf{r}\,d\mathbf{v}} \tilde{\sigma}_T g(\bar{\rho}'\bar{\rho}_1' + \bar{\rho}\bar{\rho}_1)\,d\mathbf{v}_1$$
$$\times \sum_{\mathbf{r}_j} \sum_{\mathbf{v}_j} \delta(\mathbf{r}, \mathbf{r}_j)\chi(\mathbf{v}_j; \mathbf{v}, \mathbf{v}_1) f(\mathbf{r}_j, \mathbf{v}_j). \tag{3.3.33}$$

The sums over $\mathbf{r}$, $\mathbf{r}_j$, and $\mathbf{v}_j$ are straightforward. Using Eq. (3.3.21) the sum over $\mathbf{v}_j$ is seen to have only four nonzero terms. Thus

$$\sum_j \gamma_{ij} f_j d\mathbf{r}\,d\mathbf{v} = \sum_{\mathbf{v},\mathbf{v}_1}^* \chi(\mathbf{v}_i; \mathbf{v}, \mathbf{v}_1)\hat{\sigma}_T g(\bar{\rho}'\bar{\rho}_1' + \bar{\rho}\bar{\rho}_1)[f' + f_1' - f - f_1]\,d\mathbf{v}_1. \tag{3.3.34}$$

The sum over the elementary processes (the direct collisions) can be evaluated just as was done in obtaining Eq. (3.3.27), and this gives

$$\sum_j \gamma_{ij} f_j d\mathbf{r}\, d\mathbf{v} = -\sum_{\mathbf{v}_1} \sigma_T \hat{g}(\bar{\rho}_i' \bar{\rho}_1' + \bar{\rho}_i \bar{\rho}_1)[f_i' + f_1' - f_i - f_1]\, d\mathbf{v}_1. \quad (3.3.35)$$

In the continuum limit both sides of Eq. (3.3.35) can be expressed as the integrals

$$\gamma[f] \equiv \int \gamma(\mathbf{r}, \mathbf{v}; \mathbf{r}_1, \mathbf{v}_1) f(\mathbf{r}_1, \mathbf{v}_1)\, d\mathbf{r}_1\, d\mathbf{v}_1$$

$$= -\int \hat{\sigma}_T g(\bar{\rho}' \bar{\rho}_1' + \bar{\rho}\bar{\rho}_1)[f' + f_1' - f - f_1]\, d\mathbf{v}_1. \quad (3.3.36)$$

The functional γ can be represented by the kernel of the integral operator, $\gamma(\mathbf{r}, \mathbf{v}; \mathbf{r}', \mathbf{v}')$. Finally, we can obtain the form of γ using the right-hand side of Eq. (3.3.36). This gives

$$\gamma(\mathbf{r}, \mathbf{v}; \mathbf{r}_1, \mathbf{v}_1) = -\hat{\sigma}_T g(\bar{\rho}' \bar{\rho}_1' + \bar{\rho}\bar{\rho}_1)[T' + T_1' - T - T_1]\delta(\mathbf{r} - \mathbf{r}_1), \quad (3.3.37)$$

where the T's are the translation operators defined in Eqs. (2.11.6) and (2.11.7). Equation (3.3.37) gives the form of the correlation function for the random term $\tilde{f}$ in the fluctuating Boltzmann equation (3.3.30).

It is an easy exercise to show that near equilibrium the correlation of the random terms in Eq. (3.3.37) reduces to the form based on the Onsager theory in Eq. (2.11.26). Using Eqs. (2.11.15) and (2.11.8), Eq. (2.11.26) can be written

$$2k_B L^s(\mathbf{v}, \mathbf{v}_1)\delta(\mathbf{r} - \mathbf{r}_1) = -2\hat{\sigma}_T g \rho^e \rho_1^e [T' + T_1' - T - T_1]\delta(\mathbf{r} - \mathbf{r}_1). \quad (3.3.38)$$

On the other hand, substituting $\bar{\rho} = \bar{\rho}^e$ into Eq. (3.3.37) and remembering that $\rho^{e'} \rho_1^{e'} = \rho^e \rho_1^e$, Eq. (3.3.37) is seen to reduce at equilibrium to the Onsager result in Eq. (3.3.38). Equations (3.3.28), (3.3.30), (3.3.31), and (3.3.37) provide an explicit theory for fluctuations in the μ-space molecule density which is also valid away from equilibrium.

3.4. Elementary Chemical Reactions

The idea of an elementary process is useful for describing the effect of a great variety of molecular processes. In Section 3.2 we used this idea to derive the Boltzmann equation, for which the molecular events are the collisions that change occupancy numbers in μ-space. An even simpler example of elementary processes are the molecular events that lead to chemical reactions. Although chemists recognized the existence of molecules long before physicists, it was not until after Boltzmann that chemists acknowledged the importance of molecular collisions in causing chemical reactions. One of the early reactions to be studied quantitatively was the oxidation of hydrogen iodide by hydrogen peroxide in aqueous solution,

$$H_2O_2 + 2\,HI = 2\,H_2O + I_2. \tag{3.4.1}$$

The progress of this reaction can be followed using starch as an indicator of the appearance of the iodine. At a temperature of 17°C, the reaction takes about 3.7 min to convert half of the hydrogen peroxide to water.

The chemical equation in Eq. (3.4.1) does not represent an elementary chemical reaction. All it represents is the net change which occurs when H_2O_2 and HI are mixed together in aqueous solution, namely, to produce H_2O and I_2 in relative amounts that preserve the number of atoms in the reactants. An equation like (3.4.1), which expresses a net change, is called a *stoichiometric* equation. The overall process, however, does proceed by a sequence of elementary reactions. These elementary reactions correspond to the actual molecular events which change the chemical identity of the participating molecules. Experimentally it has been found that reactions (3.4.2) and (3.4.3) are sufficient to explain the rate of the overall process:

$$H_2O_2 + I^- \rightleftarrows H_2O + IO^- \tag{3.4.2}$$

$$H_2O_2 + I^- + H^+ \rightleftarrows H_2O + HIO. \tag{3.4.3}$$

The iodine in Eq. (3.4.1) is produced by subsequent reaction of the intermediate species IO^- and HIO.

Reaction (3.4.2) is an elementary reaction and occurs when an iodide ion and an H_2O_2 molecule are adjacent to each other in solution. Like the elementary collisions which cause the μ-space occupancy numbers to change, the elementary reaction in Eq. (3.4.2) is composed of a forward and a reverse reaction. Every time the forward reaction occurs, there is a decrease in the number of I^- and H_2O_2 molecules by one and an increase of the number of IO^- and H_2O molecules, also by one. The elementary reaction does not represent a single mechanical process occurring in solution but rather summarizes the effect on the molecule numbers of a class of related mechanical events. Collisions which cause the forward reaction can occur with a variety of energies and angular orientations and may be aided by the solvent. Nonetheless, their effect on the molecule numbers will be just that represented by Eq. (3.4.2).

In 1864 Guldberg and Waage proposed the law of mass action to describe their observations about the rate of chemical reactions. As applied to the elementary reaction in Eq. (3.4.2), the contemporary version of the mass action law states that the rates of the forward and reverse processes have the form

$$\begin{aligned} V^+ &= Vk^+ \rho_{H_2O_2}\rho_{I^-} \\ V^- &= Vk^- \rho_{H_2O}\rho_{IO^-}, \end{aligned} \tag{3.4.4}$$

where the ρ's are either number or mass densities and V is the volume. These rates give the number of elementary reactions that occur per unit time. The expressions in Eq. (3.4.4) are reminiscent of the rates of the forward and reverse collision processes in Eqs. (3.2.3) and (3.2.4) which give rise to the Boltzmann equation. The quantities k^+ and k^- are called reaction rate constants. Experimentally, at higher concentrations these constants depend on the number densities. A more accurate expression of the rates can be given in terms of activities, a_i. The activity of a substance in solution is a measure of chemical potential, μ_i, with respect to a standard state, and is defined by

$$\mu_i = \mu_i^0 + k_B T \ln a_i, \tag{3.4.5}$$

where μ_i^0 is chemical potential in the standard state. The generalized mass action law for reaction (3.4.2) is

$$\begin{aligned} V^+ &= Vk^+ a_{H_2O_2} a_{I^-} \\ V^- &= Vk^- a_{H_2O} a_{IO^-}. \end{aligned} \tag{3.4.6}$$

In Eq. (3.4.6) the k's are activity-based rate constants. For reactions that are not too rapid, these rate constants are independent of the density.

The effect of chemical reactions on the extensive variables is to change molecule numbers, n_i. The rate of change can be obtained by multiplying the change of n_i that occurs in the elementary reaction times the rate of the reaction. Thus for the rate of change of the number of H_2O_2 molecules, one has for reaction (3.4.2)

$$(dn_{H_2O_2}/dt) = (-1)(V^+ - V^-). \tag{3.4.7}$$

In general, if $\omega_{\kappa i}$ represents the change in n_i due to an elementary chemical reaction labeled by the index κ, then

$$(dn_i/dt)_\kappa = \omega_{\kappa i}(V_\kappa^+ - V_\kappa^-), \tag{3.4.8}$$

where the subscript on the time derivative implies the contribution due to the reaction κ. Summing over all possible reactions gives the total time rate of change of n_i,

$$dn_i/dt = \sum_\kappa \omega_{\kappa i}(V_\kappa^+ - V_\kappa^-). \tag{3.4.9}$$

Equation (3.4.9) has the same formal structure as the collisional terms in the Boltzmann equation, Eq. (3.3.13).

The mass action law reflects the fact that chemical changes are caused by elementary chemical reactions. It also makes clear the dynamical nature of chemical equilibria. A chemical reaction, κ, is said to have achieved *equilibrium* when the forward and reverse rates balance each other, i.e.,

$$V_\kappa^+ = V_\kappa^-. \tag{3.4.10}$$

According to Eq. (3.4.9), overall chemical equilibrium requires that Eq. (3.4.10) be true for all chemical reactions. For the chemical reaction in Eq. (3.4.2), the condition of equilibrium is given by

$$k^+ a_{H_2O_2}^e a_{I^-}^e = k^- a_{H_2O}^e a_{IO^-}^e, \tag{3.4.11}$$

which provides a relationship among the activities of the participating molecules. It should be noted that Eq. (3.4.11) is similar in character to the condition of equilibrium (Eq. 2.8.11) for the Boltzmann equation. It can be derived in a similar way using the generalization of the H-theorem given in Chapter 4.

The second reaction, Eq. (3.4.3), which contributes to the oxidation of HI by H_2O_2 is probabily not an elementary reaction. If it were, it would require the simultaneous coming together of the three chemical species H_2O_2, I^-, and H^+. Even in solution this is an unlikely event. Thus this reaction is probably a complex process involving the elementary reactions

$$H_2O_2 + H^+ \rightleftarrows H_3O_2^+ \tag{3.4.12}$$

$$H_3O_2^+ + I^- \rightleftarrows H_2O + HIO. \tag{3.4.13}$$

The first of these reactions is very rapid and achieves equilibrium before the slower second reaction. Applying the equilibrium condition (Eq. 3.4.10) to Eq. (3.4.12), it follows that

$$a_{H_3O_2^+} = (k_1^+/k_1^-)a_{H_2O_2}a_{H^+}, \tag{3.4.14}$$

where the k's refer to the elementary process in Eq. (3.4.12). The forward rate of the elementary reaction in Eq. (3.4.13) depends on the species $H_3O_2{}^+$,

$$V_2^+ = Vk_2^+ a_{H_3O_2^+} a_{I^-}. \tag{3.4.15}$$

Using the expression for $a_{H_3O_2^+}$ in Eq. (3.4.14), this can be written

$$V_2^+ = Vk^+ a_{H_2O_2} a_{H^+} a_{I^-}, \tag{3.4.16}$$

where $k^+ \equiv (k_2^+ k_1^+/k_1^-)$. Recalling Eq. (3.4.3), Eq. (3.4.16) is seen to be the rate expression predicted by the mass action law. There are many reactions which, like Eq. (3.4.3), proceed via several coupled elementary reactions and have rates which are independent of the concentrations of intermediate chemical species. These are termed *elementary complex* reactions and are often encountered in the gas phase and in solution.

The elementary reaction in Eq. (3.4.2) involves two distinct species and is called bimolecular. The forward process in Eq. (3.4.3) is trimolecular, although we have seen that it is really a complex reaction composed of two bimolecular steps. Unimolecular reactions also exist, e.g.,

$$A \rightleftharpoons B, \tag{3.4.17}$$

where we have used the nonchemical notation A and B to represent reactant and products. Unimolecular reactions often involve collision partners that are unchanged in the reaction. Thus a more complete representation of the mechanism might be

$$X + A \rightleftharpoons X + B, \tag{3.4.18}$$

where X is an inert bath gas or solvent molecule. Since A and B must have the same atomic composition, they are isomers of one another. Collisions with X can give the boost necessary to change one isomeric structure into the other.

3.5. The Canonical Form

An elementary process is a molecular event which produces a well-defined change in the extensive variables. In Section 3.2 we saw that collisions give rise to changes in the number of molecules with particular momenta and in Section 3.4 that chemical reactions give rise to changes in the number of molecules with a given identity. These are both examples of elementary processes. The idea of an elementary process allows us to write down a formal expression for the rate of change of the extensive variables, [cf. Eqs. (3.3.13) and (3.4.9)]. This expression, however, is not very useful unless we know how the transition rates of the elementary processes, $V_\kappa^\pm$, depend on the extensive variables. For the Boltzmann equation we used our knowledge of bimolecular scattering processes to obtain the form of the elementary transition rates. Similarly for chemical reactions we can rely on the mass action law. In this section we explore the thermodynamic connection between these expressions for the transition rates. Each provides an example of the canonical thermodynamic form, which is an ingredient in the mechanistic statistical theory of nonequilibrium thermodynamics.

For the sake of simplicity, we consider first elementary chemical reactions. We can write, symbolically, the general bimolecular chemical reaction as

$$\mathrm{A} + \mathrm{B} \rightleftarrows \mathrm{C} + \mathrm{D}. \tag{3.5.1}$$

For the time being we will neglect other species and assume that our entire system is made up of the chemical species A, B, C, and D. Thus at the thermodynamic level of description the extensive variables are the internal energy, the volume, and the molecule numbers n_A, n_B, n_C, and n_D. To characterize the elementary reaction in Eq. (3.5.1) we notice that one molecule each of A and B is involved in the forward process, whereas neither is involved in the reverse process. We represent this by the notation

$$n_\mathrm{A}^+ = n_\mathrm{B}^+ = 1 \quad \text{and} \quad n_\mathrm{A}^- = n_\mathrm{B}^- = 0. \tag{3.5.2}$$

Similarly for the molecules C and D we write

$$n_\mathrm{C}^+ = n_\mathrm{D}^+ = 0 \quad n_\mathrm{C}^- = n_\mathrm{D}^- = 1. \tag{3.5.3}$$

The n_i^+'s describe the numbers of each molecule involved in the forward rate process and the n_i^-'s describe the fate of those molecules after reaction. Because the back reaction is the reverse of the forward process, this can be looked at in an equivalent way. Namely, the n_i^-'s give the number of molecules involved in the back reaction and the n_i^+'s describe the final result of the back reaction. This is typical of an elementary process, since the choice of "forward" and "reverse" is purely conventional. Using the symbols in Eqs. (3.5.2) and (3.5.3) the elementary reaction in Eq. (3.5.1) can be written

$$(n_\mathrm{A}^+, n_\mathrm{B}^+, n_\mathrm{C}^+, n_\mathrm{D}^+) \rightleftarrows (n_\mathrm{A}^-, n_\mathrm{B}^-, n_\mathrm{C}^-, n_\mathrm{D}^-)$$

or

$$(1,1,0,0) \rightleftarrows (0,0,1,1). \tag{3.5.4}$$

The change, ω_i, of any of the n_i due to the reaction can be obtained from the $n_i^{\pm}$. In the forward direction the change is $n_i^- - n_i^+ \equiv \omega_i$ and in the reverse direction $n_i^+ - n_i^- = -\omega_i$. For example, for the bimolecular reaction in Eq. (3.5.1),

$$\omega_A = \omega_B = -1, \quad \omega_C = \omega_D = +1, \tag{3.5.5}$$

since A and B are decreased by one and C and D are increased by one.

The expressions for the rates of the elementary reaction in Eq. (3.5.1) are given by the mass action law,

$$\begin{aligned} V^+ &= Vk^+ a_A a_B \\ V^- &= Vk^- a_C a_D. \end{aligned} \tag{3.5.6}$$

To write this in the canonical form we need to introduce the chemical potential, μ_i,

$$\frac{-\mu_i}{T} = (\partial S/\partial n_i)_{E,V,\mathbf{n}'} \equiv F_i, \tag{3.5.7}$$

where the notation $\mathbf{n}'$ implies that the n_j other than n_i are held fixed. As in Eq. (2.3.1), F_i is the intensive variable thermodynamically conjugate to the molecule number, n_i. Using the relationship between the activity and the chemical potential in Eq. (3.4.5), we can write

$$a_i = \exp[(\mu_i - \mu_i^0)/k_B T] = \exp(-F_i/k_B)\exp(-\mu_i^0/k_B T). \tag{3.5.8}$$

Putting Eq. (3.5.8) into Eq. (3.5.6) the expressions for the rates take the form

$$\begin{aligned} V^+ &= Vk^+ \exp[-(\mu_A^0 + \mu_B^0)/k_B T]\exp[-(F_A + F_B)/k_B] \\ V^- &= Vk^- \exp[-(\mu_C^0 + \mu_B^0)/k_B T]\exp[-(F_C + F_D)/k_B] \end{aligned} \tag{3.5.9}$$

Since both the rate constants and the standard-state chemical potentials are independent of composition, we combine them into a single term that has the units of inverse time,

$$\begin{aligned} \Omega^+ &= Vk^+ \exp[-(\mu_A^0 + \mu_B^0)/k_B T] \\ \Omega^- &= Vk^- \exp[-(\mu_C^0 + \mu_D^0)/k_B T] \end{aligned} \tag{3.5.10}$$

This yields the *canonical form* of the transition rates for this reaction

$$\begin{aligned} V^+ &= \Omega^+ \exp[-(F_A + F_B)/k_B] \\ V^- &= \Omega^- \exp[-(F_C + F_D)/k_B] \end{aligned} \tag{3.5.11}$$

Written in the canonical form, the expressions for the transition rates have a strong thermodynamic character. Indeed, their dependence on the extensive variables comes exclusively through the intensive variables, F_i, which are thermodynamically conjugate to the extensive variables.

The constants $\Omega^{\pm}$ are the fundamental transport coefficients for the canonical form. They give the intrinsic rate at which the elementary reaction in Eq. (3.5.1) occurs. Moreover, these two constants turn out to be the same, i.e.,

$$\Omega^+ = \Omega^- \equiv \Omega. \tag{3.5.12}$$

This property is called *microscopic reversibility*. For chemical reactions we can derive this property by combining the kinetic condition of equilibrium, $V^+ = V^-$, with the thermodynamic condition of equilibrium. For the reaction in Eq. (3.5.1) the thermodynamic condition for equilibrium is

$$\Delta G_{\mathrm{rx}}^e \equiv \mu_{\mathrm{C}}^e + \mu_{\mathrm{D}}^e - \mu_{\mathrm{A}}^e - \mu_{\mathrm{B}}^e = 0, \tag{3.5.13}$$

where the superscribe e represents evaluation at equilibrium. Putting the canonical form in Eq. (3.5.11) into the kinetic condition of equilibrium, on the other hand, gives

$$\Omega^+ \exp[-(\mu_{\mathrm{A}}^e + \mu_{\mathrm{B}}^e)/k_B T] = \Omega^- \exp[-(\mu_{\mathrm{C}}^e + \mu_{\mathrm{D}}^e)/k_B T]. \tag{3.5.14}$$

Equation (3.5.13) shows that the exponential factors in Eq. (3.5.14) are equal, which gives Eq. (3.5.12). Microscopic reversibility is an important dynamic property which holds for all elementary processes that occur entirely within a system.

The canonical form for the transition rates for the reaction in Eq. (3.5.1) can be written in an equivalent way which turns out to be completely general. The extensive variables that change due to the reaction are n_{A}, n_{B}, n_{C}, and n_{D}. If we represent these by the column vector $\mathbf{n}$, then the conjugate intensive variables are $F_i = \partial S/\partial n_i$. The sums in the exponentials of Eq. (3.5.11) can be written in terms of the F_i and the numbers of molecules involved in the forward and reverse processes, n_i^+ and n_i^-, given in Eq. (3.5.4). Thus

$$\begin{aligned} F_{\mathrm{A}} + F_{\mathrm{B}} &= \sum_j n_j^+ F_j \\ F_{\mathrm{C}} + F_{\mathrm{D}} &= \sum_j n_j^- F_j. \end{aligned} \tag{3.5.15}$$

Putting these expressions into Eqs. (3.5.11) gives

$$V^{\pm} = \Omega^{\pm} \exp\left[-\sum_j n_j^{\pm} F_j/k_B \right]. \tag{3.5.16}$$

Equation (3.5.15) is the general expression for the canonical form of the transition rates.

It is possible to derive the canonical form using an extension of the Boltzmann–Planck postulate in Eq. (2.5.9). In addition to determining the single-time probability density in the equilibrium ensemble, the entropy also determines the number of molecular quantum states compatible with a given assignment of extensive variables. According to Boltzmann and Planck we can write

$$S(\mathbf{n}) = k_B \ln W(\mathbf{n}), \tag{3.5.17}$$

where W is the degeneracy of the quantum states described by the extensive variables $\mathbf{n}$. Equation (3.5.17) is the most general expression of the Boltzmann–Planck postulate. Although the quantity $W(\mathbf{n})$ has a mechanical significance, Eq. (3.5.17) shows that it can be written in terms of the entropy, i.e.,

$$W(\mathbf{n}) = \exp[S(\mathbf{n})/k_B]. \tag{3.5.18}$$

Equation (3.5.18) can be used to obtain an expression for the rate of an elementary process. Using our earlier symbolism, any elementary process can be written as

$$(n_1^+, n_2^+, \ldots) \rightleftarrows (n_1^-, n_2^-, \ldots), \tag{3.5.19}$$

where $\mathbf{n}^+$ describes some molecular amount of the extensive variable $\mathbf{n}$ which is required in the forward portion of the elementary process. The forward rate can be expressed as a product of two factors

$$V^+ = \Omega^+ P^+(\mathbf{n}). \tag{3.5.20}$$

The first factor gives the intrinsic rate at which the process occurs in the forward direction, i.e., the number of times per second that the forward process occurs if molecular-sized clusters of the extensive variables $\mathbf{n}^+$ are preassembled. For example, for the bimolecular reaction $A + B \rightleftarrows C + D$, Ω^+ is the rate at which all adjacent pairs of A and B molecules would react if they were properly arranged for collision. The factor $P^+(\mathbf{n})$, then, gives the probability that such a reactive cluster $\mathbf{n}^+$ actually exists. The number of states compatible with the existence of a reactive cluster $\mathbf{n}^+$ in the system is the same as the number of states in the system with the cluster removed. According to Eq. (3.5.18)

$$W(\mathbf{n} - \mathbf{n}^+) = \exp[S(\mathbf{n} - \mathbf{n}^+)/k_B]. \tag{3.5.21}$$

Dividing this by $W(\mathbf{n})$ gives the probability of the existence of a cluster, or

$$P^+(\mathbf{n}) = \exp\{-[S(\mathbf{n}) - S(\mathbf{n} - \mathbf{n}^+)]/k_B\}. \tag{3.5.22}$$

Now the cluster $\mathbf{n}^+$ is of molecular size, for example, two molecules as in the case of a bimolecular chemical reaction. Thus, for a large system, we can use the Taylor series expansion in the exponential

$$S(\mathbf{n} - \mathbf{n}^+) = S(\mathbf{n}) - \sum_j (\partial S/\partial n_j) n_j^+ + \frac{1}{2} \sum (\partial^2 S/\partial n_i\, \partial n_j) n_i^+ n_j^+ + \ldots \tag{3.5.23}$$

For a large system, we need only keep the first two terms in this expansion since the second and higher-order derivatives of S with respect to extensive variables are proportional at least to the inverse of the size of the system. Thus the higher-order terms are negligible for a large system, and Eq. (3.5.23) can be written

$$S(\mathbf{n}) = S(\mathbf{n} - \mathbf{n}^+) + \sum_j F_j n_j^+. \tag{3.5.24}$$

Combining Eqs. (3.5.22) and (3.5.24) with (3.5.20) yields

$$V^+ = \Omega^+ \exp\left(-\sum_j F_j n_j^+ / k_B \right), \tag{3.5.25}$$

which is the canonical form for the forward rate. Since there is no *a priori* distinction between the forward and the reverse process, the reverse rate has the form

$$V^- = \Omega^- \exp\left[-\sum_j F_j n_j^- / k_B \right]. \tag{3.5.26}$$

Applying the condition of microscopic reversibility, i.e., $\Omega^+ = \Omega^-$, we end up with the general expression of the canonical form in Eq. (3.5.16).

The rates of the elementary collision processes with give rise to the Boltzmann equation can be written in the canonical form of Eq. (3.5.16). This fulfills our quest for incorporating thermodynamics into the μ-space description of molecular collisions. It also provides a clear example of the unity that exists among macroscopic descriptions of molecular processes. According to Eqs. (3.2.3) and (3.2.4), the rates of elementary collision processes in μ-space have the form

$$V_\kappa^+ = \hat{\sigma}_T g \rho \rho_1 \, d\mathbf{v}_1 \, d\mathbf{v} \, d\mathbf{r} \tag{3.5.27}$$

$$V_\kappa^- = \hat{\sigma}_T g \rho' \rho_1' \, d\mathbf{v}_1 \, d\mathbf{v} \, d\mathbf{r}, \tag{3.5.28}$$

where the subscript κ labels the collision at $\mathbf{r}$ indicated by $\mathbf{v}, \mathbf{v}_1 \rightleftarrows \mathbf{v}', \mathbf{v}_1'$. To put these equations into the canonical form, we need to recall the local equilibrium entropy density in μ-space given in Eq. (2.11.11). The intensive variable conjagate to ρ is given in Eq. (2.11.12), i.e.,

$$F(\rho) = -k_B(\ln \rho + 1), \tag{3.5.29}$$

so that

$$\rho = e^{-F/k_B} e^{-1}. \tag{3.5.30}$$

Using Eq. (3.5.30), the transition rates take the form

$$V_\kappa^+ = (e^{-2} \hat{\sigma}_T g \, d\mathbf{v}_1 \, d\mathbf{v} \, d\mathbf{r}) \exp[-(F + F_1)/k_B] \tag{3.5.31}$$

$$V_\kappa^- = (e^{-2} \hat{\sigma}_T g \, d\mathbf{v}_1 \, d\mathbf{v} \, d\mathbf{r}) \exp[-(F' + F_1')/k_B]. \tag{3.5.32}$$

Since precisely one molecule of velocity $\mathbf{v}$ and one of $\mathbf{v}_1$ at position $\mathbf{r}$ are involved in the forward process and one molecule of velocity $\mathbf{v}'$ and one of $\mathbf{v}_1'$ also at $\mathbf{r}$ are involved in the reverse process, the exponentials have the form indicated in Eq. (3.5.16). Thus we can make the identifications

$$\Omega_\kappa^+ = \Omega_\kappa^- \equiv \Omega_\kappa = e^{-2} \hat{\sigma}_T g \, d\mathbf{v}_1 \, d\mathbf{v} \, d\mathbf{r}. \tag{3.5.33}$$

This is microscopic reversibility expressed for the elementary collision processes in μ-space. The equality in Eq. (3.5.33) actually comes from earlier manipulations in our derivation of the Boltzmann equation in Section 2.7. There we used Liouville's theorem to deduce Eq. (2.7.19), i.e., that

$$\sigma\, d\Omega g\, d\mathbf{v}_1\, d\mathbf{v}\, d\mathbf{r} = \sigma'\, d\Omega' g'\, d\mathbf{v}_1'\, d\mathbf{v}'\, d\mathbf{r}'. \tag{3.5.34}$$

Integrating this equation over the scattering angles gives

$$\hat{\sigma}_T g\, d\mathbf{v}_1\, d\mathbf{v}\, d\mathbf{r} = \hat{\sigma}_T' g'\, d\mathbf{v}_1'\, d\mathbf{v}'\, d\mathbf{r}'. \tag{3.5.35}$$

Had we not used this equality to rewrite the rate of the restoring collisions in Eq. (2.7.18) we would have found in Eqs. (3.5.31) and (3.5.32) that

$$\begin{aligned} \Omega_\kappa^+ &= e^{-2} \hat{\sigma}_T g\, d\mathbf{v}_1\, d\mathbf{v}\, d\mathbf{r} \\ \Omega_\kappa^- &= e^{-2} \hat{\sigma}_T' g'\, d\mathbf{v}_1'\, d\mathbf{v}'\, d\mathbf{r}'. \end{aligned} \tag{3.5.36}$$

Thus for collisions in μ-space the principle of microscopic reversibility can be deduced from Liouville's theorem.

The canonical form provides a remarkable unification of the description of dissipative molecular processes. It describes not only the rate of chemical reactions and the effect of molecular collisions in μ-space but is applicable to electrochemical reactions, viscous flow in Newtonian fluids, heat transport, diffusion, and a range of other macroscopic phenomena. The canonical form shows that thermodynamics sits in a natural way in the theory of molecular rate processes. Indeed, it is the conjugate intensive thermodynamic variables that determine the rates of change of the extensive variables. Near equilibrium the canonical form allows us to derive the linear theory of irreversible thermodynamics in precisely the same way that the Boltzmann equation was shown to reduce to the Onsager theory in μ-space.

There still remains a question of the interpretation of the canonical form. If we were to follow Boltzmann's lead and forget about fluctuations, then the canonical form would provide us with a description of the average behavior in an ensemble. A more complete interpretation of the canonical form is that it gives the transition rate for stochastic changes in extensive variables which accompany molecular processes. This is the interpretation given in Section 3.2 to the elementary collision rates in μ-space. This led us to the Boltzmann equation on the conditional average and to a description of the molecular fluctuations that accompany collisions. In the next section we apply this idea to elementary chemical reactions and arrive at a thermodynamic-level description of fluctuations caused by chemical reactions.

3.6. Stochastic Theory of Chemical Reactions at the Thermodynamic Level of Description

The canonical form for the rate of chemical reactions depends only on the exponential of the chemical potentials of the reactant and product molecules. This description is a highly contracted one and, thus, must be accompanied by stochastic fluctuations. The indeterminacy arises because chemical potentials do not provide sufficient information for a precise molecular-level description. To describe the actual sequence of chemical reactions in a macroscopic system would require a mechanical knowledge of the positions and momenta of all atoms on all molecules. Although this would allow us to calculate the timing of all reactive events, it is too complete to be implemented. Thus, as in the Boltzmann-level description of dilute gases, we are forced to rely on an ensemble picture to understand the effect of chemical reactions at the thermodynamic level.

The basic assumption of the stochastic ensemble theory is that the rates of elementary processes give the transition rates for changes in the extensive variables. To apply this to chemical reactions we consider a homogeneous macroscopic system composed of $j = 1, 2, \ldots, k$ chemical species. We suppose further that these species can react with one another through any number of elementary chemical reactions, which we distinguish by the index κ. The molecule numbers of each species will be represented by n_j. Symbolically the chemical reactions can be represented as

$$(n_{\kappa 1}^{+}, n_{\kappa 2}^{+}, \ldots, n_{\kappa k}^{+}) \rightleftarrows (n_{\kappa 1}^{-}, n_{\kappa 2}^{-}, \ldots, n_{\kappa k}^{-}). \tag{3.6.1}$$

The changes in $\mathbf{n}$ due to the reaction κ are $\boldsymbol{\omega}_\kappa = \mathbf{n}_\kappa^- - \mathbf{n}_\kappa^+$. According to Section 3.5, these elementary reactions have a rate given by the canonical form, Eq. (3.5.16). Recalling that for the molecule numbers, $F_j = -\mu_j/T$ (the number-based chemical potential), the rate expressions can be written

$$V_\kappa^{\pm} = \Omega_\kappa \exp\left[\sum_j n_{\kappa j}^{\pm} \mu_j / k_B T\right]. \tag{3.6.2}$$

For the sake of generality, let us also suppose that there are systematic sources or sinks for each of the chemical species. In an interval of time dt these systematic effects will cause a change in $\mathbf{n}$ of the form

$$d\mathbf{n}_s = \mathbf{S}(\mathbf{n}, t)\, dt, \tag{3.6.3}$$

where $\mathbf{s}(\mathbf{n}, t)$ is a column vector of the rates of the sources or sinks. According to our basic assumption, the probability of a transition from a given set of molecule numbers $\mathbf{n}$ at time t to another set $\mathbf{n}'$ at $t + dt$ is

$$P_2(\mathbf{n}, t | \mathbf{n}' = \mathbf{n} + d\mathbf{n}_s + d\mathbf{n}, t + dt) = \begin{cases} V_\kappa^{\pm}\, dt + \mathcal{O}(dt), & \text{if } d\mathbf{n} = \pm \boldsymbol{\omega}_\kappa \\ 1 - \sum_\kappa (V_\kappa^+ + V_\kappa^-) + \mathcal{O}(dt), & \text{if } d\mathbf{n} = \mathbf{0} \\ 0, \quad \text{otherwise} \end{cases} \tag{3.6.4}$$

This *ansatz*, now expressed in the context of chemical reactions, is precisely the *ansatz* used in Eq. (3.2.6) to describe the effect of collisions in dilute gases at the Boltzmann level. As we discussed in Section 3.2, the *ansatz* is applicable only to a contracted description, that is, only for an ensemble in which individual systems contain a large number of molecules. Under this restriction it is shown in Section 3.2 that this assumption gives rise to a simple stochastic diffusion process. Although that result is proven there for elementary collisions in μ-space, the formal structure of the theory is exactly the same for chemical reactions. Thus for a large system we can deduce in an identical fashion that the conditionally averaged molecule numbers, $\bar{n}_i$, satisfy an equation like (3.3.13), i.e.,

$$\begin{aligned} d\bar{n}_i/dt &= \sum_\kappa \omega_{\kappa i}\Omega_\kappa \left\{\exp\left[\sum_j n^+_{\kappa j}\mu_j(\bar{\mathbf{n}})/k_B T\right]\right. \\ &\qquad \left. - \exp\left[\sum_j n^-_{\kappa j}\mu_j(\bar{\mathbf{n}})/k_B T\right]\right\} + S_i(\bar{\mathbf{n}}, t) \\ &\equiv R_i(\bar{\mathbf{n}}, t), \end{aligned} \tag{3.6.5}$$

with $\bar{n}_i(\mathbf{n}^0, 0) = n_i^0$. Similarly we find that the conditional fluctuation, $\delta n_j(t) \equiv n_j - \bar{n}_j(\mathbf{n}^0, t)$, satisfies the Fokker–Planck equation (3.3.10). The Fokker–Planck equation implies that the conditional fluctuation is a time-dependent Ornstein–Uhlenbeck process and, thus, satisfies the stochastic differential equation

$$d\delta n_i/dt = H_{ij}(\bar{\mathbf{n}}, t)\delta n_j + \tilde{f}_i \tag{3.6.6}$$

where

$$H_{ij}(\bar{\mathbf{n}}, t) \equiv \partial R_i(\bar{\mathbf{n}}, t)/\partial \bar{n}_j \tag{3.6.7}$$

and $\tilde{f}_j(t)$ is a nonstationary Gaussian term with

$$\langle \tilde{f}_i(t)\rangle = 0 \tag{3.6.8}$$

and

$$\langle \tilde{f}_i(t)\tilde{f}_j(t')\rangle = \gamma_{ij}(\bar{\mathbf{n}})\delta(t - t'), \tag{3.6.9}$$

with

$$\gamma_{ij}(\bar{\mathbf{n}}) = \sum_\kappa \omega_{\kappa i}\Omega_\kappa\omega_{\kappa j}\left\{\exp\left[\sum_j n^+_{\kappa j}\mu_j(\bar{\mathbf{n}})/k_B T\right] + \exp\left[\sum_j n^-_{\kappa j}\mu_j(\bar{\mathbf{n}})/k_B T\right]\right\}. \tag{3.6.10}$$

Equations (3.6.5)–(3.6.10) provide a mathematical description at the thermodynamic level of molecule number fluctuations due to chemical reactions.

The simplest application of the theory is to the elementary unimolecular reaction

$$\mathrm{A} \rightleftarrows \mathrm{B}. \tag{3.6.11}$$

This reaction keeps the total amount of the two isomers, $n = n_\mathrm{A} + n_\mathrm{B}$, fixed and is characterized by $n^+_\mathrm{A} = n^-_\mathrm{B} = 1$, $n^-_\mathrm{A} = n^+_\mathrm{B} = 0$, and $\omega_\mathrm{A} = -\omega_\mathrm{B} = -1$. Thus Eq. (3.6.5) for the conditional averages takes the explicit form

$$-d\bar{n}_\mathrm{A}/dt = d\bar{n}_\mathrm{B}/dt = \Omega\{\exp[\mu_\mathrm{A}(\bar{\mathbf{n}})/k_B T] - \exp[\mu_\mathrm{B}(\bar{\mathbf{n}})/k_B T]\}. \tag{3.6.12}$$

To solve Eq. (3.6.12) it is necessary to know the functional form of $\mu_i(\bar{\mathbf{n}})$. For simplicity consider an ideal solution for which the chemical potential can be writeen

$$\mu_i(\bar{\mathbf{n}}) = \mu_i^0(T) + k_B T \ln \bar{\rho}_i, \tag{3.6.13}$$

where $\bar{\rho}_i = \bar{n}_i/V$ is the number density. Using Eq. (3.6.13) and introducing the rate constants

$$\begin{aligned} k^+ &= \Omega \exp(\mu_{\mathrm{A}}^0/k_B T)/V \\ k^- &= \Omega \exp(\mu_{\mathrm{B}}^0/k_B T)/V, \end{aligned} \tag{3.6.14}$$

Eq. (3.6.12) becomes

$$-d\bar{n}_{\mathrm{A}}/dt = d\bar{n}_{\mathrm{B}}/dt = k^+ \bar{n}_{\mathrm{A}} - k^- \bar{n}_{\mathrm{B}}. \tag{3.6.15}$$

By using the fact that $n = n_{\mathrm{A}} + n_{\mathrm{B}}$, these linear equations can be written in terms of either n_{A} or n_{B} alone. Thus for molecule B

$$d\bar{n}_{\mathrm{B}}/dt = -\lambda(\bar{n}_{\mathrm{B}} - n_{\mathrm{B}}^e), \tag{3.6.16}$$

where $\lambda = k^+ + k^-$ and $n_{\mathrm{B}}^e = k^+ n/(k^+ + k^-)$. Equation (3.6.16) describes an exponential relaxation to the equilibrium value n_{B}^e and has the solution

$$\bar{n}_{\mathrm{B}}(t) = n_{\mathrm{B}}^e + \exp[-\lambda t](n_{\mathrm{B}}^0 - n_{\mathrm{B}}^e). \tag{3.6.17}$$

According to Eq. (3.6.6) the fluctuations associated with this isomerization reaction solve a linearized version of Eq. (3.6.16), including a random term. Explicitly,

$$d\delta n_{\mathrm{B}}/dt = -\lambda \delta n_{\mathrm{B}} + \tilde{f}_{\mathrm{B}}, \tag{3.6.18}$$

where from Eq. (3.6.9)

$$\langle \tilde{f}_{\mathrm{B}}(t)\tilde{f}_{\mathrm{B}}(t')\rangle = [(k^- - k^+)\bar{n}_{\mathrm{B}}(t) + k^+ n]\delta(t - t') \equiv \gamma(\bar{n}_{\mathrm{B}}(t))\delta(t - t'). \tag{3.6.19}$$

Equation (3.6.18) has the same form as the Langevin equation for Brownian motion in Eq. (1.7.1) and can be solved in a similar fashion [cf. Eq. (1.7.2)]. Thus

$$\delta n_{\mathrm{B}}(t) = \int_0^t \exp[-\lambda(t - \tau)]\tilde{f}_{\mathrm{B}}(\tau)\, d\tau, \tag{3.6.20}$$

where we have used the fact that at $t = 0$ the conditional fluctuation $\delta n_{\mathrm{B}}(0) = \bar{n}_{\mathrm{B}}(0) - n_{\mathrm{B}}^0$ is identically zero. Now $\tilde{f}_{\mathrm{B}}(\tau)$ is a Gaussian stochastic process and since sums or integrals of Gaussian processes are also Gaussian, $\delta n_{\mathrm{B}}(t)$ is a Gaussian process. By taking averages on both sides of Eq. (3.6.20) and using the fact that $\langle \tilde{f}_{\mathrm{B}}(\tau)\rangle \equiv 0$, we find that

$$\langle \delta n_{\mathrm{B}}(t)\rangle^{n_{\mathrm{B}}^0} \equiv 0.$$

The only other quantity needed to characterize a Gaussian process is the variance

$$\sigma(n_{\rm B}^0, t) \equiv \langle \delta n_{\rm B}(t) \delta n_{\rm B}(t) \rangle^{n_{\rm B}^0}. \tag{3.6.21}$$

The variance can be found by squaring both sides of Eq. (3.6.20) and taking the average. This gives

$$\sigma(n_{\rm B}^0, t) = \int_0^t \exp[-\lambda(t-\tau)]\,d\tau \int_0^t \exp[-\lambda(t-\tau')]\gamma(\bar{n}_{\rm B}(\tau))\delta(\tau-\tau')\,d\tau', \tag{3.6.22}$$

where Eq. (3.6.19) was used. The delta function simplifies the integral over τ', and using the explicit expressions in Eqs. (3.6.17) and (3.6.19), Eq. (3.6.22) becomes

$$\sigma(n_{\rm B}^0, t) = \exp(-2\lambda t) \int_0^t \exp(2\lambda\tau)[2k^- n_{\rm B}^e + (k^- - k^+)\exp(-\lambda\tau)(n_{\rm B}^e - n_{\rm B}^e)]\,d\tau,$$

where we have used detailed balance, i.e., $k^+ n_{\rm A}^e = k^- n_{\rm B}^e$, to write the term in square brackets. The integral is straightforward and gives

$$\sigma(n_{\rm B}^0, t) = \left[\frac{k^- n_{\rm B}^e}{\lambda}(1 - e^{-2\lambda t}) + \frac{(k^- - k^+)(n_{\rm B}^0 - n_{\rm B}^e)}{\lambda}(e^{-\lambda t} - e^{-2\lambda t})\right]. \tag{3.6.23}$$

Because the conditional fluctuations are Gaussian, it follows that

$$P_2(0|\delta n_{\rm B}, t) = [2\pi\sigma(n_{\rm B}^0, t)]^{-1/2} \exp[-\delta n_{\rm B}^2/2\sigma(n_{\rm B}^0, t)],$$

where the zero in the argument of P_2 reminds us that $\delta n_{\rm B} \equiv 0$ at $t = 0$. We convert this equation into an equation for the conditional fluctuations in $n_{\rm B}(t)$ using the fact that $\delta n_{\rm B} = n_{\rm B} - \bar{n}_{\rm B}(t)$. Thus

$$P_2(n_{\rm B}^0|n_{\rm B}, t)$$
$$= [2\pi\sigma(n_{\rm B}^0, t)]^{-1/2} \exp\{-[(n_{\rm B} - n_{\rm B}^e) - e^{-\lambda t}(n_{\rm B}^0 - n_{\rm B}^e)]^2/2\sigma(n_{\rm B}^0, t)\}. \tag{3.6.24}$$

Notice that as $t \to \infty$, P_2 approaches the Gaussian distribution

$$W_1(n_{\rm B}) = (2\pi\sigma^e)^{-1/2} \exp[-(n_{\rm B} - n_{\rm B}^e)^2/2\sigma^e], \tag{3.6.25}$$

where from Eq. (3.6.23) and detailed balance

$$\sigma^e = n_{\rm A}^e n_{\rm B}^e / n. \tag{3.6.26}$$

In Section 3.7 it is shown that W_1 is the equilibrium distribution given by the Einstein formula.

The stochastic process caused by the elementary reaction A $\rightleftarrows$ B is different from the Ornstein–Uhlenbeck process predicted by the Onsager theory. Comparing Eq. (2.6.3), which describes $\delta n_{\rm B}$ in the Onsager theory, to Eq. (3.6.18), we see that both equations are linear. However, in the present theory the random term $\tilde{f}_{\rm B}$ in Eq. (3.6.19) depends explicitly on the initial condition $n_{\rm B}^0$ through $\bar{n}_{\rm B}(t)$. In the Onsager theory, on the other hand, the matrix $\gamma = 2k_B L$, which is a constant. The difference comes about because the Onsager theory is restricted to a neighborhood of equilibrium whereas the theory based on the *ansatz* in Eq. (3.6.4) is not restricted in this way. Nonetheless, Eq. (3.6.19) reduces to the result predicted by the Onsager theory for small deviations around equilibrium. In that case $\bar{n}_B(t) \approx n_B^e$ and, using detailed balance, the formula for γ in Eq. (3.6.19) reduces to

$$\gamma = 2k^- k^+ n/\lambda, \tag{3.6.27}$$

which is a constant. Similarly, close to equilibrium the expression for the conditional variance in Eq. (3.6.23) becomes

$$\sigma(n_B^0, t) = \sigma^e(1 - e^{-2\lambda t}), \tag{3.6.28}$$

which is independent of n_B^0 and has the correct form for an Ornstein–Uhlenbeck process [cf. Eq. (1.8.12)]. In Chapter 4 we prove that close to equilibrium the *ansatz* in Eq. (3.6.15) reduces to the Onsager theory for any set of elementary processes.

3.7. Conservation Conditions and the Progress Variables

Chemical reactions provide an archetype for elementary molecular processes. In this section we discuss an aspect of all elementary processes which is especially important for describing chemical reactions, namely, conserved quantities. In Section 2.4 we derived conservation equations at the hydrodynamic level for the conserved extensive variables energy, mass, and momentum. At the Boltzmann-level of description in Section 2.10 we showed, further, that these conservation relations could be traced to the collisional invariants for the Boltzmann equation. Indeed, the collisional invariants satisfy Eq. (2.10.2), an integral condition on the collision term. Using our present language of elementary processes, that condition turns out to be the fundamental conservation condition built into the canonical equations of motion. For the Boltzmann equation that condition implies that the collision operator has five zero eigenvalues. For chemical reactions, the conservation principle is similar to conservation of overall mass for the Boltzmann equation, although it is easy to show that it contains conservation of energy for chemical reactions, too.

The mass of individual chemical species is not preserved in a chemical reaction. Indeed, for the isomerization reaction A $\rightleftarrows$ B treated in the previous section, a molecule of A disappears and a molecule of B appears when the reaction proceeds in the forward direction. However, the overall amount of A and B, $n = n_A + n_B$, is conserved. For chemical reactions the basic conserved quantities are the overall number of atoms of each kind. For example, in the gas-phase reaction of methane and molecular fluorine

$$CH_4 + F_2 \rightleftarrows CH_3F + HF \tag{3.7.1}$$

there are two fluorine atoms both before and after collision, as well as four hydrogens and one carbon. For the symbolic reaction indicated by $n_{\kappa i}^{\pm}$ in Eq. (3.6.1), the conservation condition can be written

$$\sum_i \psi_{Ai}\omega_{\kappa i} = 0, \tag{3.7.2}$$

where ψ_{Ai} gives the number of atoms of kind A in the species i. We can convert Eq. (3.7.2) into a conservation equation by multiplying it by $(\bar{V}_\kappa^+ - \bar{V}_\kappa^-)$ and summing over κ. This gives

$$\sum_i \psi_{Ai} \sum_\kappa \omega_{\kappa i}(\bar{V}_\kappa^+ - \bar{V}_\kappa^-) = 0. \tag{3.7.3}$$

However, on the conditional average trajectory we know from Eq. (3.6.5) that the second summation in Eq. (3.7.3) is the contribution of the elementary processes to the time rate of change of $\bar{n}_i$. Hence we can write Eq. (3.7.3) as

$$\sum_i \psi_{Ai}(d\bar{n}_i/dt)_{\text{rx}} = 0 \tag{3.7.4}$$

where the subscript rx signifies the contribution due to all chemical reactions. In the more general context of elementary processes, Eq. (3.7.3) is valid for any conserved quantity, ψ_A, where the conservation condition is given in Eq. (3.7.2). That equation generalizes the definition of the collisional invariant at the Boltzmann level of description [cf. Eq. (2.9.2)], and Eq. (3.7.3) generalizes the conservation property of the collision integral in Eq. (2.10.2).

For chemical reactions there will be as many conserved quantities, i.e., atom number functions ψ_A, as there are atoms. Thus for the reaction in Eq. (3.7.1) there are three such functions. Such conservation conditions in the context of the Boltzmann equation—or, later, at the hydrodynamic level—are conveniently thought of as helping to characterize the zero eigenvalues for integral or differential operators. For chemical reactions, however, where there are a small discrete number of variables, n_i, it is conventional to use the conservation conditions simply to eliminate dependent extensive variables. For the fluorination of methane in Eq. (3.7.1) the three conservation equations for carbon, fluorine, and hydrogen are

$$\begin{aligned} -d\bar{n}_{\text{MeH}}/dt + d\bar{n}_{\text{MeF}}/dt &= 0 \\ -2d\bar{n}_{\text{F}_2}/dt + d\bar{n}_{\text{MeF}}/dt + d\bar{n}_{\text{HF}}/dt &= 0 \\ -4d\bar{n}_{\text{MeH}}/dt + 3d\bar{n}_{\text{MeF}}/dt + d\bar{n}_{\text{HF}}/dt &= 0, \end{aligned} \tag{3.7.5}$$

where the symbol Me is used to represent the methyl group. These are three linearly independent conditions among the three time derivatives caused by the reaction. If the input terms, S_i, in Eq. (3.6.5) vanish, then Eq. (3.7.5) can be integrated to give a relationship of the general form

$$M[\bar{\mathbf{n}}(t) - \mathbf{n}^0] = \mathbf{0}, \tag{3.7.6}$$

where M is a 3×4 matrix and $\mathbf{n}^0 = \bar{\mathbf{n}}(0)$. Because of linear independence, Eq. (3.7.6) can be solved for three of the molecule numbers in terms of the others. Thus the conservation conditions in Eq. (3.7.5) imply that there is only one independent quantity changing in this reaction.

A more convenient way to handle this situation for chemical reactions is to introduce *progress variables*. The progress variable is the extensive variable, ξ_κ, which measures the net number of times per second that a reaction κ has proceeded, with the forward direction taken as positive. Thus the average time rate of change of ξ_κ is simply given by

$$d\bar{\xi}_\kappa/dt = \bar{V}_\kappa^+ - \bar{V}_\kappa^- . \tag{3.7.7}$$

In the absence of external sources or sinks, Eq. (3.7.7) can be related to the average changes in $\bar{n}_i$ by

$$d\bar{n}_i/dt = \sum_\kappa \omega_{\kappa i}\, d\bar{\xi}_\kappa/dt. \tag{3.7.8}$$

Integrating this equation between zero and t gives

$$\bar{n}_i(t) = \sum_\kappa \omega_{\kappa i}\bar{\xi}_\kappa(t) + n_i^0, \tag{3.7.9}$$

since $\bar{\xi}_\kappa$ vanishes at time zero. Equation (3.7.9), in turn, can be used to eliminate all the $\bar{n}_i(t)$ from the right-hand side of Eq. (3.7.7), leading to a set of coupled partial differential equations involving a progress variable for each reaction. For the reaction of methane and fluorine in Eq. (3.7.1), the progress variable satisfies

$$d\bar{\xi}/dt = \Omega\{\exp[(\bar{\mu}_{\text{MeH}} + \bar{\mu}_{\text{F}_2})/k_B T] - \exp[(\bar{\mu}_{\text{MeF}} + \bar{\mu}_{\text{HF}})/k_B T]\} \tag{3.7.10}$$

and Eq. (3.7.9) gives

$$\begin{aligned} \bar{n}_{\text{MeH}} &= -\bar{\xi} + n^0_{\text{MeH}}, \quad \bar{n}_{\text{F}_2} = -\bar{\xi} + n^0_{\text{F}_2}, \\ \bar{n}_{\text{MeF}} &= \bar{\xi} + n^0_{\text{MeF}}, \quad \bar{n}_{\text{HF}} = \bar{\xi} + n^0_{\text{HF}}. \end{aligned} \tag{3.7.11}$$

Substituting these equations for $\bar{n}_i$ into the chemical potentials in Eq. (3.7.10) gives a differential equation involving only the progress variable $\bar{\xi}$.

The conservation condition in Eq. (3.7.2) implies the summation constraints on the average equations in Eq. (3.7.4). The conservation condition, however, is stronger than this and applies also to the fluctuations. To see this, we examine the differential equations for the conditional fluctuations, Eq. (3.6.6). Multiplying by $\psi_{\text{A}i}$ and summing gives

$$\partial\left(\sum_i \psi_{\text{A}i}\delta n_i\right)\Big/\partial t = \sum_i\sum_j \psi_{\text{A}i}H_{ij}(\bar{\mathbf{n}})\delta n_j + \sum_i \psi_{\text{A}i}\tilde{f}_i. \tag{3.7.12}$$

It is easy to see that the first term on the right-hand side of Eq. (3.7.12) vanishes if the sources or sinks are independent of $\bar{\mathbf{n}}$. Using Eqs. (3.6.5) and (3.6.7) this term can be written

$$\sum_i \sum_\kappa \sum_j \psi_{Ai}\omega_{\kappa i} \frac{\partial(\bar{V}_\kappa^+ - \bar{V}_\kappa^-)}{\partial \bar{n}_j} \delta n_j = \sum_\kappa \left(\sum_i \psi_{Ai}\omega_{\kappa i} \right) \left(\sum_j \frac{\partial(\bar{V}_\kappa^+ - \bar{V}_\kappa^-)}{\partial \bar{n}_j} \delta n_j \right) \tag{3.7.13}$$

Since the first factor in the sum over κ vanishes for all κ by Eq. (3.7.2), the first term in Eq. (3.7.12) also vanishes. To see that the second term in Eq. (3.7.12) vanishes, we examine its average and variance. Since it is a Gaussian stochastic process, if both its average and variance are zero then it is zero. Because it is the sum of terms which vanish on the average, its average value is certainly zero. Its variance can be obtained from Eqs. (3.6.9) and (3.6.10). Using the notation $\bar{V}_\kappa^\pm$ for the transition rates, the variance becomes

$$\begin{aligned} \left\langle \left(\sum_i \psi_{Ai}\tilde{f}_i(t) \right) \left(\sum_j \psi_{Aj}\tilde{f}_j(t') \right) \right\rangle &= \sum_i \sum_j \psi_{Ai} \langle \tilde{f}_i(t)\tilde{f}_j(t') \rangle \psi_{Aj} \\ &= \sum_i \sum_j \sum_\kappa \psi_{Ai}\omega_{\kappa i}(\bar{V}_\kappa^+ + \bar{V}_\kappa^-)\psi_{Aj}\omega_{\kappa j}\delta(t - t') \\ &= \sum_\kappa (\bar{V}_\kappa^+ + \bar{V}_\kappa^-) \left(\sum_i \psi_{Ai}\omega_{\kappa i} \right)^2 \delta(t - t'). \end{aligned} \tag{3.7.14}$$

Again the final factor in the sum over κ vanishes because of Eq. (3.7.2). Thus the right-hand side of Eq. (3.7.12) vanishes identically. Integrating the left-hand side from zero to t, we are left with the identity

$$\sum_i \psi_{Ai}\delta n_i = 0, \tag{3.7.15}$$

since $\delta n_i(0) \equiv 0$. This identity holds in the absence of sinks or sources that depend explicitly on $\mathbf{n}$. For example, for the isomerization reaction $A \rightleftarrows B$ treated in the previous section, both A and B consist of the same atoms. Thus for all atoms ψ_{Ai} is the same for A and B and Eq. (3.7.15) reduces to

$$\delta n_A + \delta n_B = 0. \tag{3.7.16}$$

This shows that using the explicit formula obtained for δn_B in Eq. (3.6.20), we can immediately obtain δn_A using Eq. (3.7.16).

The linear dependence of fluctuations in the molecule numbers expressed by Eq. (3.7.15) implies that the stochastic differential equations governing the fluctuations are not independent. There are several ways to deal with this situation. By systematically applying Eq. (3.7.15) for all atoms one can eliminate certain fluctuations δn_i in favor of others, ending up with an independent set of stochastic differential equations. Alternatively, progress variables can be used. This is a more systematic procedure which recognizes that, fundamentally, fluctuations arise from the random occurrence of the elementary reactions. Equation (3.7.9) relates average molecule number to average progress variables. It also can be used to define fluctuations in the progress variables through

$$\delta n_i(t) = \sum_{\kappa} \omega_{\kappa i} \delta \xi_\kappa(t). \tag{3.7.17}$$

Since we know how to calculate the fluctuations δn_i, we need to solve Eq. (3.7.17) for $\delta\xi_\kappa$ in terms of δn_i. Then we can use Eqs. (3.6.6)–(3.6.10) to find the stochastic differential equations that the $\delta\xi_\kappa$ satisfy. Unfortunately, inversion of Eq. (3.7.17) for $\delta\xi_\kappa$ is not always possible. It is possible, in fact, if and only if the $\delta\xi_\kappa$ are uniquely defined by that equation, i.e., if and only if $\sum_\kappa \omega_{\kappa i} b_\kappa = 0$ for all i implies that $b_\kappa \equiv 0$. Otherwise, more than one set of $\delta\xi_\kappa$ are defined by Eq. (3.7.17). If this condition is satisfied, the elementary reactions are called *linearly independent*. An example of two linearly independent reaction stoichiometrics are given by

$$\begin{aligned} \mathrm{Y} &\rightleftarrows \mathrm{X} \\ \mathrm{X} + \mathrm{Y} &\rightleftarrows \mathrm{Z}. \end{aligned} \tag{3.7.18}$$

To check that these reactions are linearly independent, we use the criterion above. Thus we check that the equations $b_1 - b_2 = 0$, $-b_1 - b_2 = 0$, and $b_2 = 0$ for X, Y, and Z, respectively, have only the solution $b_1 = b_2 = 0$, which is so. The reactions

$$\begin{aligned} \mathrm{Y} &\rightleftarrows \mathrm{X} \\ \mathrm{X} + \mathrm{Y} &\rightleftarrows 2\mathrm{X}, \end{aligned} \tag{3.7.19}$$

on the other hand, are linearly dependent, since the equations $b_1 + b_2 = 0$ and $-b_1 - b_2 = 0$ have solutions other than $b_1 = b_2 = 0$.

To treat fluctuations in the progress variables we first treat linearly independent reactions. In this case, Eq. (3.7.17) can be solved uniquely for $\delta\xi_\kappa(t)$. For example, for the reactions in Eq. (3.7.18),

$$\begin{aligned} \delta\xi_1 &= (\delta n_\mathrm{X} - \delta n_\mathrm{Y})/2 \\ \delta\xi_2 &= \delta n_\mathrm{Z}. \end{aligned} \tag{3.7.20}$$

More generally, then, for linearly independent reactions we can write the unique expressions

$$\begin{aligned} \delta\xi_\kappa &= \sum_i B_{\kappa i} \delta n_i \\ \delta n_i &- \sum_{\kappa'} \omega_{\kappa' i} \delta\xi_{\kappa'}. \end{aligned} \tag{3.7.21}$$

Substituting the second of these equations into the first gives

$$\sum_i B_{\kappa i} \omega_{\kappa' i} = \delta_{\kappa\kappa'}. \tag{3.7.22}$$

From Eq. (3.7.9) we also find that

$$\partial/\partial\bar{\xi}_\kappa = \sum_j (\partial/\partial\bar{n}_j)\omega_{\kappa j}. \tag{3.7.23}$$

Equations (3.7.21) can be used to change variables from δn_i to $\delta\xi_\kappa$ in Eqs. (3.6.6)–(3.6.10). Multiplying Eq. (3.6.6) by $B_{\kappa i}$ and summing over i yields

$$
\begin{aligned}
d\delta\xi_\kappa/dt &= \sum_i B_{\kappa i} \sum_{\kappa'} \omega_{\kappa' i} \frac{\partial}{\partial \bar{n}_j}(\bar{V}_{\kappa'}^{+} - \bar{V}_{\kappa'}^{-})\delta n_j + \sum_i B_{\kappa i}\tilde{f}_i \\
&= \frac{\partial}{\partial \bar{n}_j}(\bar{V}_\kappa^{+} - \bar{V}_\kappa^{-}) \sum_{\kappa'} \omega_{\kappa' j}\delta\xi_{\kappa'} + \tilde{f}_\kappa \\
&= \sum_{\kappa'} \frac{\partial}{\partial \bar{\xi}_{\kappa'}}(\bar{V}_\kappa^{+} - \bar{V}_\kappa^{-})\delta\xi_{\kappa'} + \tilde{f}_\kappa,
\end{aligned}
\tag{3.7.24}
$$

where in the second equality Eqs. (3.7.21) and (3.7.22) were used and the final equality relies on Eq. (3.7.23). The Gaussian random term $\tilde{f}_\kappa$ vanishes on the average, since the $\tilde{f}_i$ do, and has the covariance

$$
\begin{aligned}
\langle \tilde{f}_\kappa(t)\tilde{f}_{\kappa'}(t')\rangle &= \sum_i \sum_j \sum_{\kappa''} B_{\kappa i}\omega_{\kappa'' i}(\bar{V}_{\kappa''}^{+} + \bar{V}_{\kappa''}^{-})B_{\kappa' j}\omega_{\kappa'' j}\delta(t - t') \\
&= \delta_{\kappa\kappa'}(\bar{V}_\kappa^{+} + \bar{V}_\kappa^{-})\delta(t - t'),
\end{aligned}
\tag{3.7.25}
$$

where Eq. (3.7.22) was used to simplify the summations. For linearly independent reactions the covariance matrix of the random terms is diagonal. Since Eq. (3.7.7) is a closed set of differential equations for the progress variables, all the terms which depend on the average in the fluctuation formula can be written solely as functions of $\bar{\xi}_\kappa$. Equations (3.7.7), (3.7.24), and (3.7.25) give a stochastic theory for the progress variables that is independent of the molecule numbers.

For linearly dependent chemical reactions the use of progress variables is complicated by the fact that Eq. (3.7.17) cannot be inverted. The strategy for this case is simply to use the largest subset of elementary reactions which gives rise to a unique definition of the progress variables in Eq. (3.7.17). For example, we can use just the first reaction in Eq. (3.7.19) to define the single progress variable

$$
\begin{aligned}
\delta n_{\mathrm{Y}} &= -\delta\xi \\
\delta n_{\mathrm{X}} &= \delta\xi.
\end{aligned}
\tag{3.7.26}
$$

This single reaction is certainly linearly independent and the stochastic equation satisfied by either δn_{Y} or δn_{X} can be used to obtain the stochastic equation satisfied by $\delta\xi$. More generally we can write

$$
\delta n_i = \sum_\kappa \omega_{\kappa i}^{*}\delta\xi_\kappa \tag{3.7.27}
$$

where the asterisk signifies the truncated set of linearly independent reactions, and

$$
\delta\xi_\kappa = \sum B_{\kappa i}^{*}\delta n_i. \tag{3.7.28}
$$

Since these new progress variables are linearly independent, they satisfy the relationships in Eqs. (3.7.22) and (3.7.23), except that asterisks now appear. Similarly, for the average values, we use Eq. (3.7.28) to write

$$n_i(t) = \sum_{\kappa'} \omega^*_{\kappa' i} \xi_{\kappa'}(t) + n_i^0.$$

Thus multiplying the differential equation for $\bar{n}_i$ by $B^*_{\kappa i}$, and summing, we obtain

$$d\bar{\xi}_\kappa/dt = \sum_{\kappa} \alpha_{\kappa\kappa'}(\bar{V}^+_{\kappa'} - \bar{V}^-_{\kappa'}) \tag{3.7.29}$$

$$\alpha_{\kappa\kappa'} \equiv \sum_i B^*_{\kappa i}\omega_{\kappa' i}. \tag{3.7.30}$$

The matrix $\alpha_{\kappa\kappa'}$ is not a Kronecker delta because to achieve linear independence some of the $\omega_{\kappa i}$ are set equal to zero in the definition of $\omega^*_{\kappa i}$. By using manipulations similar to those that gave Eqs. (3.7.24) and (3.7.25), we can show for linearly dependent reactions that fluctuations in the progress variables satisfy

$$d\delta\xi_\kappa/dt = \sum_{\kappa'}\sum_{\mu} \alpha_{\kappa\kappa'}\left(\frac{\partial \bar{V}^+_{\kappa'} - \bar{V}^-_{\kappa'}}{\partial \bar{\xi}_\mu}\right)\delta\xi_\mu + \tilde{f}_\kappa \tag{3.7.31a}$$

and

$$\langle \tilde{f}_\kappa(t)\tilde{f}_{\kappa'}(t')\rangle = \sum_{\kappa''} \alpha_{\kappa\kappa''}(\bar{V}^+_{\kappa''} + \bar{V}^-_{\kappa''})\alpha_{\kappa'\kappa''}\delta(t - t'). \tag{3.7.31b}$$

Progress variables are extensive quantities and can be connected to thermodynamics rather simply. In fact, for systems in which molecule numbers change only through chemical reactions, they are the natural extensive variables to choose. Consider the general situation characterized by m independent progress variables ξ_κ and k molecule numbers n_i. Because of the conservation of atoms expressed by Eq. (3.7.2), m is less than k. Thus the state of the system can be characterized by the m extensive variables, ξ_κ, which change by reaction and $k - m$ other extensive variables, $\hat{n}_i$, determined by Eqs. (3.7.4), which are left constant by the reaction. Making these changes of variables, the local equilibrium entropy can be written

$$S(\mathbf{n}) = S(\boldsymbol{\xi}, \hat{\mathbf{n}}), \tag{3.7.32}$$

and a differential change in S can be written

$$dS = \sum_{\kappa} (\partial S/\partial \xi_\kappa) d\xi_\kappa + \sum_i (\partial S/\partial \hat{n}_i)\, d\hat{n}_i. \tag{3.7.33}$$

De Donder called the variables thermodynamically conjugate to ξ_κ the *chemical affinities*, $\mathscr{A}_\kappa$. The extensive variables $(\partial S/\partial \hat{n}_i)$, on the other hand, have no special designation. Combining the definition of the progress variables in Eq. (3.7.27) with the definition of the affinity, we can use the chain rule to see that

$$\mathscr{A}_\kappa = (\partial S/\partial \xi_\kappa) = \sum_i (\partial S/\partial n_i)\omega^*_{\kappa i}$$

$$= -\sum_i \frac{\mu_i \omega^*_{\kappa i}}{T} \equiv -\Delta G_\kappa/T, \tag{3.7.34}$$

where we have used the definition of the chemical potential in Eq. (3.5.7). The quantity ΔG_κ was introduced by G.N. Lewis and is called the *Gibbs free energy change for the reaction.*

For a spontaneous chemical reaction, $d\hat{n}_i = 0$, and we can write

$$dS = \sum_\kappa \mathscr{A}_\kappa \, d\xi_\kappa \tag{3.7.35}$$

and

$$\delta^2 S = \frac{1}{2} \sum_\kappa \sum_{\kappa'} (\partial \mathscr{A}_\kappa/\partial \xi_{\kappa'})_{\xi',\hat{\mathbf{n}}} \, d\xi_\kappa \, d\xi_{\kappa'}, \tag{3.7.36}$$

where on the partial derivatives in Eq. (3.7.36) we have indicated that $\xi_\kappa \neq \xi_{\kappa'}$ and all the $\hat{n}_i$ are held fixed in taking the derivative. Since chemical reactions can occur in isolated systems and the reactions are independent, the entropy maximum principle applied to Eq. (3.7.35) gives the equilibrium condition

$$\mathscr{A}^e_\kappa \equiv 0. \tag{3.7.37}$$

Moreover, $\delta^2 S$ must be negative at the maximum, and thus the matrix

$$S_{\kappa\kappa'} = (\partial \mathscr{A}_\kappa/\partial \xi_{\kappa'})^e \tag{3.7.38}$$

is non-negative definite.

The linear theory of irreversible processes is easily expressed in terms of independent progress variables. Referring to Eq. (2.3.2), the conjugate forces are simply equal to the $\mathscr{A}_\kappa$, since $\mathscr{A}^e_\kappa$ vanishes, and according to Eq. (2.3.3), the fluxes are the derivatives of the progress variables. Using the linear stochastic theory of Onsager summarized in Section 2.6., it follows that

$$d\xi_\kappa/dt = \sum_{\kappa'} L_{\kappa\kappa'} \mathscr{A}_{\kappa'} + \tilde{f}_\kappa \tag{3.7.39a}$$

with

$$\langle \tilde{f}_\kappa(t)\tilde{f}_{\kappa'}(t')\rangle = 2k_B L_{\kappa\kappa'}\delta(t-t'). \tag{3.7.39b}$$

Furthermore, the Einstein formula for the covariance of the progress variables at equilibrium in Eq. (2.6.9) takes the form

$$\langle \xi_\kappa \xi_{\kappa'} \rangle^e = -k_B(\partial \mathscr{A}/\partial \xi)^{e-1}_{\kappa\kappa'}, \tag{3.7.40}$$

where both sides are evaluated at equilibrium.

It is instructive to work out a special case of these results in order to make a comparison will the fluctuation theory developed in Section 3.6. Let us return to the isomerization reaction, A $\rightleftarrows$ B. According to the definition of the progress variable for linearly independent reactions in Eqs. (3.7.9) and (3.7.17),

$$\xi(t) = n_B(t) - n_B^0. \tag{3.7.41}$$

Thus, except for a constant, n_B is the progress variable for this reaction. The affinity can be obtained from Eq. (3.7.34), which gives

$$\mathscr{A} = (\mu_A - \mu_B)/T. \tag{3.7.42}$$

The variable, $\hat{n}$, held fixed by the isomerization reaction is obtained from the conservation condition for this reaction given in Eqs. (3.7.9) and (3.7.16), i.e.,

$$\hat{n} = n_A + n_B = n, \tag{3.7.43}$$

the total amount of the isomer. To obtain the connection to the Onsager equations, we need to write the time rate of change of ξ in terms of the affinity. To do this we can rely on the details already worked out in Section 3.6. Thus from Eq. (3.6.16) and the fact that $\xi = n_B - n_B^0$ it follows that

$$d\bar{\xi}/dt = -\lambda(\bar{\xi} - \xi^e). \tag{3.7.44}$$

From the definition of $\mathscr{A}$ in Eq. (3.7.42) it follows that

$$T\mathscr{A} = (\mu_A^0 - \mu_B^0) + k_B T \ln[(n - n_B^0 - \xi)/(\xi + n_B^0)], \tag{3.7.45}$$

where we have used Eq. (3.6.13) for the chemical potentials as well as the conservation condition in Eq. (3.7.43). For conditions which are initially near equilibrium (i.e., $n_B^0 \approx n_B^e$), $\xi - \xi^e$ will be small, and Eq. (3.7.45) can be linearized to give

$$\mathscr{A} = \left(\frac{\mu_A^0 - \mu_B^0}{T}\right) + k_B \ln(n_A^e/n_B^e) - k_B n(\xi - \xi^e)/n_A^e n_B^e \ldots \tag{3.7.46}$$

The first two terms in Eq. (3.7.46) cancel since $\mathscr{A}$ vanishes at equilibrium, so that

$$\mathscr{A} \approx -k_B(\xi - \xi^e)n/n_A^e n_B^e. \tag{3.7.47}$$

Using Eq. (3.7.47), Eq. (3.7.44) can be rewritten

$$d\bar{\xi}/dt = (k^+ k^- n/k_B\lambda)\bar{\mathscr{A}}. \tag{3.7.48}$$

where we have simplified the coefficient of $\mathscr{A}$ using the definition of n_B^e below Eq. (3.6.16) and the fact that $n_A^e + n_B^e = n$. Comparing Eq. (3.7.48) with Eq. (3.7.39a) shows that the coupling coefficient for this reaction is

$$L = k^+ k^- n/k_B\lambda, \tag{3.7.49}$$

and from Eq. (3.7.46) we find that

$$(\partial\mathscr{A}/\partial\xi)^e = -k_B n/n_A^e n_B^e. \tag{3.7.50}$$

Equations (3.7.49) and (3.7.50) provide the two quantities that are required in the Onsager theory. According to the general expressions in Section 2.6., we can write

$$d\xi/dt = (k^+k^-n/k_B\lambda)\mathscr{A} + \tilde{f} \tag{3.7.51a}$$

$$\langle \tilde{f}(t)\tilde{f}(t')\rangle = (2k^+k^-n/\lambda)\delta(t-t'), \tag{3.7.51b}$$

and combining Eq. (3.7.50) with (3.7.40)

$$\langle \xi^2 \rangle^e = n_A^e n_B^e/n. \tag{3.7.52}$$

Equations (3.7.51) and (3.7.52) are the equations which describe the linear theory near equilibrium.

These equations are easily compared to the stochastic theory developed in the previous section. Adding together Eqs. (3.6.16) and (3.6.18) and using the fact that $n_B = \xi + n_B^0$ gives

$$d\xi/dt = -\lambda(\xi - \xi^e) + \tilde{f} \tag{3.7.53a}$$

with

$$\langle \tilde{f}(t)\tilde{f}(t')\rangle = [(k^- - k^+)\bar{n}_B(t) + k^+n]\delta(t-t'). \tag{3.7.53b}$$

Using the facts that $n_B + n_A = n$ and that $n_B^e = k^+n/\lambda$, the term in square brackets can be shown to equal $2k_BL$ near equilibrium, just as in the Onsager theory. Furthermore, changing variables on the right-hand side of Eq. (3.7.53a) from $(\xi - \xi^e)$ to $\mathscr{A}$ using Eq. (3.7.47), Eq. (3.7.53a) becomes identical to the Onsager equation (3.7.51a). Finally, the equilibrium probability obtained for n_B in Eq. (3.6.25) is Gaussian with a variance identical to that for the progress variable given in Eq. (3.7.52). Since n_B is linearly related to ξ, it follows that the theory developed in Section 3.6 agrees with the Einstein formula at equilibrium. For this reaction these results show that the stochastic theory outlined in this chapter reduces to the Onsager theory near equilibrium. The proof of this equivalence for all processes that have the canonical form is given in Chapter 4.

3.8. Thermodynamics of Chemical Equilibria

Equilibrium does not refer to a state of a single system, even if the system is macroscopic, but rather is a characteristic of an entire ensemble of systems. Even when well aged in the absence of inputs a macroscopic system fluctuates *in perpetuity*. It is only the average over an ensemble of identically prepared systems that ultimately becomes time independent and achieves equilibrium. Although in later chapters we concern ourselves with how equilibrium is achieved, in this section our goal is more limited. Using the canonical form we characterize the state of equilibrium for an ensemble of chemically reacting systems. Combined with the condition of detailed balance we show that the canonical form gives rise to the customary theory of chemical thermodynamics.

The *ansatz* in Section 3.6 gives an expression for the rate of change of the conditional average of the molecule numbers in a chemically reacting system. For an equilibrium ensemble there are no source or sink terms, so

$$d\bar{n}_i/dt = \sum_{\kappa} \omega_{\kappa i}(\bar{V}_\kappa^+ - \bar{V}_\kappa^-), \tag{3.8.1}$$

where the canonical form for $\bar{V}_\kappa^\pm$ is given in Eq. (3.6.2). Intuitively we know that after a sufficiently long wait, no matter what the initial condition n_i^0, the molecule numbers will settle down to some constant values n_i^e. This steady state is called chemical equilibrium and is characterized by the condition of *detailed balance*. Detailed balance occurs when the forward and reverse rates of an elementary process just balance, i.e.,

$$V_\kappa^+ = V_\kappa^- . \tag{3.8.2}$$

At chemical equilibrium detailed balance holds for all chemical reactions, so that

$$V_\kappa^+(\mathbf{n}^e) = V_\kappa^-(\mathbf{n}^e), \tag{3.8.3}$$

for all κ. This is a stronger condition than simply the vanishing of the left-hand side of Eq. (3.8.1) and implies that each term in the sum over κ in Eq. (3.8.1) vanishes separately. From the Boltzmann equation we were able to derive the condition of detailed balance in μ-space, Eq. (2.8.11), using Boltzmann's H-theorem. Given that the Boltzmann equation has the canonical form, it should not be surprising that an H-like function exists for chemical reactions and that it can be used to prove the existence of detailed balance at equilibrium. That proof is given in Chapter 4 for any elementary processes having the canonical form.

If the canonical form in Eq. (3.6.2) is substituted in the condition of detailed balance, the equilibrium ensemble is seen to satisfy

$$\exp\left[\sum_j n_{\kappa j}^+ \mu_j^e / k_B T\right] = \exp\left[\sum_j n_{\kappa j}^- \mu_j^e / k_B T\right]. \tag{3.8.4}$$

This equality implies that the arguments of the exponentials must be equal, or

$$\sum_j \omega_{\kappa j} \mu_j^e / k_B T = 0. \tag{3.8.5}$$

Equation (3.8.5) can be reexpressed either in the notation of De Donder or of Lewis using Eq. (3.7.34) as

$$\mathscr{A}_\kappa^e = 0 \text{ or } \Delta G_\kappa^e = 0, \tag{3.8.6}$$

that is, at equilibrium the affinity and the Gibbs free energy change for the reaction vanish identically. Equation (3.8.6) provides the criteria for chemical equilibria. It can be reexpressed using Lewis's definition of the activity in Eq. (3.4.5) as

$$\Delta G_\kappa^0 = -k_B T \sum_j \omega_{\kappa j} \ln a_i^e$$
$$= -k_B T \ln\left(\prod_j (a_j^e)^{\omega_{\kappa j}}\right), \tag{3.8.7}$$

where the symbol $\prod_j$ represents the product over all j and

$$\Delta G_\kappa^0 \equiv \sum_j \omega_{\kappa j} \mu_j^0, \tag{3.8.8}$$

is the standard Gibbs free energy change for the reaction. Equation (3.8.7) is called the *Lewis equation* and expresses a necessary relationship among the activities of reactants and products at equilibrium. Since the standard-state chemical potentials, μ_j^0, are functions only of the temperature, Eq. (3.8.7) can be rearranged to give

$$\prod_j (a_j^e)^{\omega_{\kappa j}} = \exp(-\Delta G_\kappa^0/k_B T) \equiv K_\kappa(T) \tag{3.8.9}$$

with the right-hand side depending only on the temperature. $K_\kappa(T)$ is the activity-based or thermodynamic equilibrium constant. In the case of chemical equilibria, the species for which $\omega_{\kappa j}$ are negative are called *reactants*, since they show up as a net loss upon reaction, whereas for *products* the $\omega_{\kappa j}$ are positive. Notice that species for which $\omega_{\kappa j} = n_{\kappa j}^- - n_{\kappa j}^+$ vanishes do not appear in the condition of equilibrium. Thus the left-hand side of Eq. (3.8.9) has the customary form of a ratio of activities of products to activities of reactants, both taken to the appropriate stoichiometric powers.

We can obtain an equation for the temperature dependence of the equilibrium constant using Eq. (3.8.9). Taking the natural logarithm we have

$$\ln K_\kappa(T) = -\Delta G_\kappa^0/k_B T. \tag{3.8.10}$$

Next differentiating both sides with respect to the temperature gives

$$d \ln K_\kappa/dT = (\Delta G_\kappa^0/k_B T^2) - (d\Delta G_\kappa^0/dT)(1/k_B T). \tag{3.8.11}$$

Equation (3.8.11) can be simplified using the thermodynamic identity

$$(\partial G/\partial T)_p = -S, \tag{3.8.12}$$

where p is the pressure and S the entropy. Since standard states are by convention taken at one atmosphere pressure, Eq. (3.8.12) implies that

$$d\Delta G_\kappa^0/dT = -\Delta S_\kappa^0 \tag{3.8.13}$$

with ΔS_κ^0 the standard entropy change for the reaction. Using this equation the right-hand side of Eq. (3.8.11) can be written

$$(\Delta G_\kappa^0/k_B T^2) + \Delta S_\kappa^0/k_B T = (\Delta G_\kappa^0 + T\Delta S_\kappa^0)/k_B T^2. \tag{3.8.14}$$

Finally recalling that the *standard enthalpy of the reaction* κ is

$$\Delta H_\kappa^0 = \Delta G_\kappa^0 + T\Delta S_\kappa^0, \tag{3.8.15}$$

Eq. (3.8.11) becomes

$$d \ln K_\kappa / dT = \Delta H_\kappa^0 / k_B T^2. \tag{3.8.16}$$

This result is usually called the *van't Hoff equation.*

The equations that we have obtained for chemical equilibria using the condition of detailed balance apply to elementary chemical reactions. However, it is not difficult to show that they also apply to net chemical processes, like the overall reaction of hydrogen peroxide and hydrogen iodide given in Eq. (3.4.1). The balanced chemical equation for that reaction does not represent an elementary process but merely the net result of the elementary reactions in Eqs. (3.4.2), (3.4.12), (3.4.13), and subsequent reactions involving IO^- and HIO. Nonetheless, the condition for chemical equilibrium in Eq. (3.8.6) still applies to net chemical reactions. Indeed, any net chemical reaction occurs through a sequence of elementary chemical reactions. Multiplying these reactions by appropriate whole numbers and adding or subtracting the reaction equations, a balanced equation for the net reaction can be obtained. Thus the stoichiometric coefficients for a net reaction have the form

$$\omega_j = \sum_\kappa \alpha_\kappa \omega_{\kappa j}. \tag{3.8.17}$$

The elementary complex reaction in Eq. (3.4.3), for example, is simply the sum of the two elementary reactions in Eqs. (3.4.12) and (3.4.13). Multiplying Eq. (3.8.17) by μ_j^e and summing over j gives

$$\sum_j \omega_j \mu_j^e = \sum_\kappa \alpha_\kappa \left(\sum_j \omega_{\kappa j} \mu_j^e \right) = 0, \tag{3.8.18}$$

since Eq. (3.8.5) holds for each of the elementary reactions κ. Equation (3.8.18) implies that the criterion of equilibrium for a net reaction is also

$$\Delta G^e = 0. \tag{3.8.19}$$

Just as for elementary reactions it follows from Eq. (3.8.19) that the thermodynamic equilibrium constant is given by

$$K(T) = \exp(-\Delta G^0 / k_B T) \tag{3.8.20}$$

and that its temperature dependence is governed by the van't Hoff equation

$$d \ln K / dT = \Delta H^0 / k_B T^2. \tag{3.8.21}$$

We can derive another useful expression for the thermodynamic equilibrium constant using the canonical form. Introducing activities the canonical form in Eq. (3.6.2) can be written

$$V_\kappa^\pm = \Omega_\kappa \exp\left[\sum_j n_{\kappa j}^\pm \mu_j^0 / k_B T \right] \prod_j a_j^{n_{\kappa j}^\pm} \tag{3.8.22}$$

or

$$V_\kappa^\pm = V k_\kappa^\pm \prod_j a_j^{n_{\kappa j}^\pm}, \tag{3.8.23}$$

where

$$k_\kappa^\pm \equiv \Omega_\kappa \exp\left[\sum_j n_{\kappa j}^\pm \mu_j^0/k_B T\right]\Big/ V \tag{3.8.24}$$

are the activity-based rate constants. Taking the ratio of the forward to reverse rate constants, it follows from Eq. (3.8.24) that

$$\begin{aligned} k_\kappa^+/k_\kappa^- &= \exp\left[-\sum_j \omega_{\kappa j}\mu_j^0/k_B T\right] \\ &= \exp(-\Delta G_\kappa^0/k_B T). \end{aligned} \tag{3.8.25}$$

Combining Eq. (3.8.25) with Eq. (3.8.9) then gives

$$k_\kappa^+/k_\kappa^- = K_\kappa(T). \tag{3.8.26}$$

Thus the ratio of the activity-based rate constants for an elementary reaction is equal to its equilibrium constant. A net reaction does not have a well-defined rate nor rate constants, and Eq. (3.8.26) does not hold for net reactions.

In this section we have given a glimpse of the thermodynamic implications of the canonical form. In fact, it can be shown that the entire theory of equilibrium thermodynamics is contained in the canonical form and the conservation conditions. Although the proof of this is delayed until Chapter 4, there is a clear message to be learned from this section: classical thermodynamics is a special case contained within the general dynamical description of macroscopic systems. Moreover, the fundamental significance of intensive variables, such as chemical potentials, is in determining the transition rates of elementary processes. Their appearance in the thermodynamic theory of equilibrium is merely a consequence of detailed balance.

References

Elementary Processes and Fluctuations

J. Keizer, A theory of spontaneous fluctuations in macroscopic systems, *J. Chem. Phys.* **63**, 398–403 (1975).

J. Keizer, Dissipation and fluctuations in nonequilibrium thermodynamics, *J. Chem. Phys.* **64**, 1679–1687 (1976).

M. Mangel, Fluctuations at chemical instabilities, *J. Chem. Phys.* **69**, 3697–3708 (1978).

J. Keizer, On the macroscopic equivalence of descriptions of fluctuations for chemical reactions, *J. Math. Phys.* **18**, 1316–1321 (1977).

L.D. Landau and E.M. Lifshitz, *Statistical Physics*, 2nd ed. (Pergamon, London, 1969).

Mass Action Law

A.V. Harcourt and W. Esson, On the laws of connexion between the conditions of a chemical change and its amount, *Phil. Trans.* **156**, 193 (1866) in *Selected Readings in Chemical Kinetics*, M.H. Back and K.J. Laidler, eds. (Pergamon, London, 1967), pp. 3–27.

F. Bell, R. Gill, D. Holden, and W.F.K. Wynne-Jones, The primary salt effect in the reaction between hydrogen peroxide and iodide ions, *J. Phys. Chem.* **55**, 874–881 (1951).

C.M. Guldberg and P. Waage, *Forhandlinger i Videnskabs-Selskabet i Christiana*, **1864**: 35–40, 111–120.

A.A. Frost and R.G. Pearson, *Kinetics and Mechanism*, 2nd ed. (Wiley, New York, 1965).

Progress Variables and Chemical Thermodynamics

S.R. de Groot and P. Mazur, *Non-equilibrium Thermodynamics* (North Holland, Amsterdam, 1962), Chapter X.

J. Keizer, Thermodynamic coupling in chemical reactions, *J. Theor. Biol.* **49**, 323–335 (1975).

J. Keizer, Concentration fluctuations in chemical reactions, *J. Chem. Phys.* **63**, 5037–5043 (1975).

P.A. Rock, *Chemical Thermodynamics* (University Science Books, Mill Valley, CA, 1983).

CHAPTER 4

Mechanistic Statistical Theory of Nonequilibrium Thermodynamics

4.1. Introduction

In the preceding chapter we introduced the idea of elementary molecular processes and showed that they provide a natural description of bimolecular collision dynamics and chemical kinetics. By adopting a statistical interpretation of the transition rate for elementary processes, it was possible to develop a statistical description of the Boltzmann theory and to generalize the linear Onsager theory of concentration fluctuations for chemical reactions. The idea of elementary processes seems to be tailor-made for describing transport processes like chemical reactions which are caused directly by molecular collisions. In fact, all dissipative processes which transport extensive variables have their origin in molecular events, and the language of elementary processes turns out to be a natural one to describe their effects. In this chapter we develop the general theory of elementary processes in a form that is applicable to a wide range of transport processes.

For simplicity we base the statistical interpretation of the theory on the same *ansatz* used in Eqs. (3.2.6) and (3.6.4) to describe the Boltzmann equation and chemical reactions. Such an *ansatz* can be correct only for macroscopic systems, and we apply it only in the thermodynamic limit. In this limit the macroscopic transport equations are recovered on the conditional average. The fluctuations satisfy a linearized version of the average equations, which depend on the temporal behavior of the average and involve a random driving term. There is an intimate connection between the elementary molecular processes that cause dissipation in the transport equation and the strength of the random terms. This connection has lead to the theory being dubbed the *fluctuation–dissipation theory*.

The ansatz which we use to describe the stochastic aspects of elementary processes is not unique. Several other statistical interpretations of the transition rates are possible, and for completeness we describe these other possible interpretations. One of these is the *master equation* interpretation, which leads to differential-difference equations describing a Markov process. There are also other interpretations which lead to stochastic differential equations of the Itô or Stratonovich type. These interpretations yield Markov processes of different kinds. The nonlinear dynamics of macroscopic systems is often non-Markovian on observable time scales, so these formulations would seem to be ill-founded. Nonetheless, when the macroscopic limits of these theories are examined, their limiting forms are all equivalent to the basic equations of the fluctuation–dissipation theory.

The canonical form of the transition rates for elementary processes is determined by the intensive thermodynamic variables. This provides the statistical theory with a strong thermodynamic character and, thus, provides a means to examine thermodynamic aspects of time-dependent phenomena. In this chapter we develope a number of important thermodynamic aspects of the theory, including the definition of the entropy for nonequilibrium states, the generalized H-theorem and approach to equilibrium, and the Caratheodory formulation of the Second Law of thermodynamics. We also provide a general proof that the theory reduces to the linear stochastic theory of Onsager near equilibrium and give a proof of the Onsager reciprocity theorem using the canonical form. The theory described in this chapter generalizes the Onsager theory of linear irreversible processes to the nonlinear domain. Because it is based on qualitative molecular mechanisms inherent in the idea of elementary processes, it provides a mechanistic statistical theory of nonequilibrium thermodynamics.

4.2. The Canonical Theory

Canonical means "established by canon" and describes a general rule with wide applicability. The equations which we use to characterize the dissipative aspect of molecular processes have this character. In the previous chapter we demonstrated their applicability at the Boltzmann level of description and for chemical reactions at the thermodynamic level. In this section we write down the theory in general, i.e., canonically. In subsequent chapters application of the theory is made to a variety of other dissipative processes.

The canonical equations involve extensive thermodynamic variables represented by the column vector **n**. These variables will be treated as discrete, although it is often convenient—as in the case of the Boltzmann level—to use densities of the extensive variables and the continuum limit. The choice of which extensive variables are used is a choice of the level of description—thermodynamic, hydrodynamic, Boltzmann, etc. This choice does not concern us here since, being canonical, the equations have the same formal structure for all hierarchical levels of description. The molecular events that cause **n** to change are characterized as elementary processes, labeled by an index, κ. Symbolically

$$(n_{\kappa 1}^{+}, n_{\kappa 2}^{+}, \ldots) \rightleftarrows (n_{\kappa 1}^{-}, n_{\kappa 2}^{-}, \ldots) \tag{4.2.1}$$

where $n_{\kappa i}^{+}$ is the molecular-size amount of the extensive variable n_i involved in the forward step of the elementary process κ and $n_{\kappa i}^{-}$ is the amount of n_i left when the process is over. The choice of which is the "forward" direction of the elementary process is arbitrary, although each forward process has a "reverse" process as indicated by the backward arrow in Eq. (4.2.1). The extensive variable $\mathbf{n}$ changes by an amount

$$\boldsymbol{\omega}_\kappa \equiv \mathbf{n}_\kappa^- - \mathbf{n}_\kappa^+ \tag{4.2.2}$$

where $\mathbf{n}_\kappa^\pm$ are the column vectors of $(n_{1\kappa}^\pm, n_{2\kappa}^\pm, \ldots)$. The transition rate of an elementary process, that is, the number of times per second that it occurs in the forward or reverse direction, is given by the canonical form

$$V_\kappa^\pm = \Omega_\kappa^\pm \exp\left(-\sum_l F_l n_{\kappa l}^\pm / k_B\right). \tag{4.2.3}$$

The constants $\Omega_\kappa^\pm$ are the intrinsic rates of the forward and reverse steps of the elementary process κ and are the fundamental transport coefficients in the canonical theory. The quantities F_j, on the other hand, are intensive variables which are functions of the extensive variables. In Section 4.5 it is shown that the F_j are partial derivatives of a local equilibrium entropy, $S(\mathbf{n})$, and can be expressed as

$$\partial S/\partial n_j = F_j, \tag{4.2.4}$$

where the other $n_i \neq n_j$ are held fixed.

The set of all elementary processes describes the changes that occur in the extensive variables due to molecular events. Other things can cause the extensive variables to change, for example, changes associated with kinematics such as streaming motions or changes caused by inputs which are under separate experimental control. These are determinstic processes, which affect the change

$$d\mathbf{n}_s = \mathbf{S}(\mathbf{n}, t)\, dt \tag{4.2.5}$$

in the interval of time dt. The fact that these changes are deterministic means that they occur in addition to the random changes caused by the elementary molecular processes. The *ansatz* which governs the overall change in $\mathbf{n}$ in the time dt is

$$P_2(\mathbf{n}, t | \mathbf{n}' = \mathbf{n} + d\mathbf{n}_s + d\mathbf{n}, t + dt) = \begin{cases} V_\kappa^\pm\, dt + \mathcal{O}(dt), & \text{if } d\mathbf{n} = \pm\boldsymbol{\omega}_\kappa \\ 1 - \sum_\kappa (V_\kappa^+ + V_\kappa^-)\, dt + \mathcal{O}(dt), & \text{if } d\mathbf{n} = 0 \\ 0 & \text{otherwise.} \end{cases} \tag{4.2.6}$$

This *ansatz* was used in the previous chapter for the Boltzmann equation and chemical reactions and describes the possible transitions of $\mathbf{n}$ in dt and the probability of their outcomes.

As the notions of extensive and intensive variables, entropy, and so forth make sense only for macroscopic systems, it is necessary to restrict the interpretation of Eq. (4.2.6) to the macroscopic domain. This means that the system must be large according to some appropriate system-size, m. For the Boltzmann description in Section 3.2, m is the average number of molecules per subvolume in μ-space, whereas for chemical reactions in Section 3.6 there is no subvolume and m can be taken as either the total number of atoms or the volume of the system. In any case, the same sort of scaling arguments used in Sections 3.2 and 3.3 lead to a Fokker–Planck type equation for the conditional fluctuation, $\delta\mathbf{n} \equiv \bar{\mathbf{n}} - \mathbf{n}$, i.e.,

$$\partial P_2/\partial t = -\partial H_{ij}(\bar{\mathbf{n}}, t)\delta n_j P_2/\partial\delta n_i + \frac{1}{2}\partial^2\gamma_{ij}(\bar{\mathbf{n}}, t)P_2/\partial\delta n_i\partial n_j \tag{4.2.7}$$

where

$$H_{ij}(\bar{\mathbf{n}}, t) \equiv \partial\left\{\sum_\kappa \omega_{\kappa i}[V_\kappa^+(\bar{\mathbf{n}}) - V_k^-(\bar{\mathbf{n}})] + S_i(\bar{\mathbf{n}}, t)\right\}\Big/\partial\bar{n}_j \tag{4.2.8}$$

$$\gamma_{ij}(\bar{\mathbf{n}}, t) \equiv \sum_\kappa \omega_{\kappa i}[V_\kappa^+(\bar{\mathbf{n}}) + V_\kappa^-(\bar{\mathbf{n}})]\omega_{\kappa j} \tag{4.2.9}$$

and $\bar{\mathbf{n}}$ solves the equation

$$d\bar{\mathbf{n}}/dt = \sum_\kappa \boldsymbol{\omega}_\kappa[V_\kappa^+(\bar{\mathbf{n}}) - V_\kappa^-(\bar{\mathbf{n}})] + \mathbf{S}(\bar{\mathbf{n}}, t). \tag{4.2.10}$$

Since $\bar{\mathbf{n}}$ is the conditional average of $\mathbf{n}$, Eq. (4.2.10) must be solved with the condition $\bar{\mathbf{n}}(t^0) = \bar{\mathbf{n}}^0$ in order to obtain the conditional probability density

$$P_2 \equiv P_2(\mathbf{n}^0, t^0|\mathbf{n}, t). \tag{4.2.11}$$

One of the important properties of the canonical equations is microscopic reversibility. When the molecular processes causing $\mathbf{n}$ to change can be ascribed to molecular events occurring solely within the system, then

$$\Omega_\kappa^+ = \Omega_\kappa^- \equiv \Omega_\kappa. \tag{4.2.12}$$

Equation (4.2.12) is called *microscopic reversibility* and reflects the mechanical reversibility of the underlying equations of motion, as we saw for the Boltzmann equation in Eq. (3.5.33). Microscopic reversibility implies that there is only a single transport coefficient for each elementary process, Ω_κ. Microscopic reversibility also implies that the transition rates for elementary processes can be written in the form

$$V_\kappa^\pm = \Omega_\kappa \exp\left(-\sum_j n_{\kappa j}^\pm F_j/k_B\right). \tag{4.2.13}$$

From Eqs. (4.2.2) and (4.2.13) it follows that

$$V_\kappa^+/V_\kappa^- = \exp\left(\sum_j \omega_{\kappa j}F_j/k_B\right), \tag{4.2.14}$$

so that the ratio of the forward and reverse transition rates depends only on the net changes $\omega_{\kappa j}$ and the intensive variables. Equation (4.2.14) has important implications for the relationship of the canonical theory to classical equilibrium thermodynamics.

4.3. Solution of the Fokker–Planck Equation

The statistical interpretation of the transition rates for elementary molecular processes is given by the *ansatz* in Eq. (4.2.6). As applied to macroscopic systems, this *ansatz* yields the Fokker–Planck equation (4.2.7) for the conditional fluctuation, $\delta\mathbf{n} = \bar{\mathbf{n}} - \mathbf{n}$. Because Eq. (4.2.7) is linear, it can be solved explicitly in terms of the solution to the nonlinear equation (4.2.10) that describes the conditional average. If the number of extensive variables is k, the solution is the Gaussian

$$P_2(\mathbf{n}^0, t^0 | \mathbf{n}, t) = [(2\pi)^k \det \sigma(\mathbf{n}^0, t^0, t)]^{-1/2} \times \exp\left\{ -\frac{1}{2}[\mathbf{n} - \bar{\mathbf{n}}(\mathbf{n}^0, t)]^T \sigma^{-1}(\mathbf{n}^0, t^0, t)[\mathbf{n} - \bar{\mathbf{n}}(\mathbf{n}^0, t)] \right\} \tag{4.3.1}$$

where $\bar{\mathbf{n}}(\mathbf{n}^0, t)$ solves Eq. (4.2.10) and the matrix σ is given by the expression

$$\sigma(\mathbf{n}^0, t, t^0) = \int_{t^0}^{t} d\tau' \operatorname{Pexp}\left[\int_{\tau'}^{t} H(\bar{\mathbf{n}}(\mathbf{n}^0, \tau), \tau)\, d\tau \right] \gamma(\bar{\mathbf{n}}(\mathbf{n}^0, \tau')) \times \left\{ \operatorname{Pexp}\left[\int_{\tau'}^{t} H(\bar{\mathbf{n}}(\mathbf{n}^0, \tau), \tau)\, d\tau \right] \right\}^T. \tag{4.3.2}$$

Here H and γ are the matrices in Eqs. (4.2.8) and (4.2.9). Pexp is the time-ordered exponential, which is a generalization of the exponential function of a matrix. It is defined to be the time-dependent matrix that allows the solution of the equation

$$d\mathbf{X}/dt = H(t)\mathbf{X} \tag{4.3.3}$$

to be written as

$$\mathbf{X}(t) = \operatorname{Pexp}\left[\int_{t^0}^{t} H(s)\, ds \right] \mathbf{X}(t^0). \tag{4.3.4}$$

The explicit representation for Pexp is obtained by integrating Eq. (4.3.3) from t^0 to t to get

$$\mathbf{X}(t) = \mathbf{X}(t^0) + \int_{t^0}^{t} H(s)\mathbf{X}(s)\, ds. \tag{4.3.5}$$

Repeatedly iterating Eq. (4.3.5) by inserting the right-hand side under the integral sign then gives

$$\mathbf{X}(t) = \left[1 + \int_{t^0}^{t} H(s)\,ds + \int_{t^0}^{t} H(s) \int_{t^0}^{s} H(s')\,ds\,ds' + \cdots\right]\mathbf{X}(t^0)$$
$$\equiv \mathrm{Pexp}\left[\int_{t^0}^{t} H(s)\,ds\right]\mathbf{X}(t^0), \tag{4.3.6}$$

which is the explicit expression for the time-ordered exponential. Differentiating Eq. (4.3.6) term by term, it is easily verified that

$$\frac{d}{dt}\mathrm{Pexp}\left[\int_{t^0}^{t} H(s)\,ds\right] = H(t)\,\mathrm{Pexp}\left[\int_{t^0}^{t} H(s)\,ds\right]. \tag{4.3.7}$$

It is also easy to see using Eq. (4.3.6) that if H is independent of s, then

$$\mathrm{Pexp}\left[\int_{t^0}^{t} H(s)\,ds\right] = \exp[(t - t^0)H], \tag{4.3.8}$$

which is the usual matrix-valued exponential function.

The fact that the Gaussian conditional density in Eq. (4.3.1) solves Eq. (4.2.7) can be deduced by direct substitution. In doing this it is necessary to use several identities. The first one comes by differentiating Eq. (4.3.2) with respect to t. Using (4.3.7) we find that

$$d\sigma/dt = H\sigma + \sigma H^T + \gamma. \tag{4.3.9}$$

Next, we notice that γ is a non-negative definite, symmetric matrix. The symmetry is obvious from Eq. (4.2.9) and its non-negative definiteness follows by looking at its quadratic form, i.e.,

$$\sum_i \sum_j \alpha_i \gamma_{ij} \alpha_j = \sum_\kappa \left(\sum_i \omega_{\kappa i}\alpha_i\right)^2 (\bar{V}_\kappa^+ + \bar{V}_\kappa^-), \tag{4.3.10}$$

which is non-negative since none of the terms on the right-hand side of Eq. (4.3.11) is negative. From this fact and the definition of σ in Eq. (4.3.2), it also follows that σ is symmetric. Indeed, σ is related to γ by Eq. (4.3.2) through an integrand of the form $A\gamma A^T = (A\gamma A^T)^T$ and is, therefore, symmetric. Furthermore, since A (the time-ordered exponential) is an invertible matrix, it follows that σ is also non-negative definite.

For Eq. (4.3.1) to be valid, the inverse of σ must exist. This requires that σ be not simply non-negative definite, but actually positive definite. In practice, this means that the extensive variables included in **n** must be independent, as described for chemical reactions in Section 3.7, and that the system is not at a critical point of stability. With these stipulations both γ and σ are positive definite and σ^{-1} exists.

Before proving Eq. (4.3.1), several more identities are required. First, we need the fact that for any positive definite, symmetric matrix σ

$$\ln \det \sigma = \operatorname{tr} \ln \sigma = \sum_i \ln \lambda_i \tag{4.3.11}$$

where tr is the trace, $\ln \sigma$ is the matrix-valued logarithm of σ, and λ_i are the eigenvalues of σ. Equation (4.3.11) can be proven by noting that σ, being positive definite and symmetric, can be diagonalized with positive eigenvalues. Since $\det \sigma = \Pi_i \lambda_i$, the left-hand side of Eq. (4.3.11) equals $\sum_i \ln \lambda_i$. Moreover, in the representation in which σ is diagonal, $\ln \sigma$ is also diagonal with eigenvalues $\ln \lambda_i$. Thus $\operatorname{tr} \ln \sigma$ also equals $\sum_i \ln \lambda_i$, which proves Eq. (4.3.11). The second identity follows from Eq. (4.3.11) and a similar argument, namely,

$$\frac{d \ln \det \sigma}{dt} = \sum_i \frac{1}{\lambda_i} \frac{d\lambda_i}{dt} = \operatorname{tr}\left(\sigma^{-1} \frac{d\sigma}{dt}\right). \tag{4.3.12}$$

Finally since $\sigma\sigma^{-1} = I$ with I the identity matrix, it follows that

$$\frac{d\sigma^{-1}}{dt} = -\sigma^{-1} \frac{d\sigma}{dt} \sigma^{-1}. \tag{4.3.13}$$

We are now in a position to verify that Eqs. (4.3.1) and (4.3.2) give the solution to the Fokker–Planck equation (4.2.7). We proceed by differentiating P_2 in Eq. (4.3.1) with respect to time, keeping $\mathbf{n} - \bar{\mathbf{n}} \equiv \delta\mathbf{n}$ constant, as indicated on the left-hand side of Eq. (4.2.7). This gives two terms, each coming from σ:

$$\partial P_2/\partial t = -\frac{1}{2}\left[\frac{d \ln \det \sigma}{dt} + \delta\mathbf{n}^T \frac{d\sigma^{-1}}{dt} \delta\mathbf{n}\right] P_2. \tag{4.3.14}$$

Using Eqs. (4.3.9), (4.3.12), and (4.3.13) the right-hand side of Eq. (4.3.14) can be rewritten as

$$-\frac{1}{2}\{\operatorname{tr}[\sigma^{-1}(H\sigma + \sigma H^T + \gamma)] - \delta\mathbf{n}^T[\sigma^{-1}H + H^T\sigma^{-1} + \sigma^{-1}\gamma\sigma^{-1}]\delta\mathbf{n}\} P_2$$

$$= \left[-\operatorname{tr} H - \frac{1}{2}\operatorname{tr}(\sigma^{-1}\gamma) + \delta\mathbf{n}^T\sigma^{-1}H\delta\mathbf{n} + \frac{1}{2}\delta\mathbf{n}^T\sigma^{-1}\gamma\sigma^{-1}\delta\mathbf{n}\right] P_2 \tag{4.3.15}$$

where we have used the facts that $\operatorname{tr}(\sigma^{-1}H\sigma) = \operatorname{tr} H$ and that the quadratic form of the matrix $\sigma^{-1}H$ and its transpose $H^T\sigma^{-1}$ are the same. Next we differentiate Eq. (4.3.1) with respect to δn_i as indicated on the right-hand side of Eq. (4.2.7). The differentiations are straightforward and give

$$\left[-\operatorname{tr} H + \delta\mathbf{n}^T\sigma^{-1}H\delta\mathbf{n} - \frac{1}{2}\operatorname{tr}(\sigma^{-1}\gamma) + \frac{1}{2}\delta\mathbf{n}^T\sigma^{-1}\gamma\sigma^{-1}\delta\mathbf{n}\right] P_2. \tag{4.3.16}$$

Since the two expressions in Eqs. (4.3.15) and (4.3.16) are identical, it follows that Eqs. (4.3.1) and (4.3.2) are the solution to the Fokker–Planck equation.

The fact that the solution to the Fokker–Planck equation for P_2 is Gaussian reduces the problem of calculating the time dependence of fluctuations to finding the average and covariance. According to Eq. (4.2.10) the conditional average solves a purely deterministic equation. Indeed, as we have already seen in Chapter 3 for chemical reactions and the Boltzmann equation, Eq. (4.2.10) is just the usual phenomenological transport equation derived from the canonical form. Away from equilibrium or stationary states Eq. (4.2.10) usually involves a number of nonlinear terms which make its solution difficult to obtain analytically. Nonetheless, analytical solutions are known for various problems and it is possible, furthermore, to obtain numerical solutions with digital or analog computers. Examples of this sort are given in Section 5.4 and Chapter 10.

To obtain the covariance of the Gaussian P_2 density one needs to peform the integrals indicated in Eq. (4.3.2). There are two impediments to doing this. First, one must have the solution $\bar{\mathbf{n}}(\mathbf{n}^0, t)$ for the average, as both H and γ must be evaluated along the deterministic trajectory beginning at $\mathbf{n}^0$ at time t^0. Second, one needs to calculate the time-ordered exponentials in Eq. (4.3.2) and perform the indicated integrals. Neither of these are done simply and, consequently, the problem of explicitly calculating P_2 is difficult. Another method of calculating the covariance is suggested by Eq. (4.3.9). This is a linear ordinary differential equation for σ which one can attempt to solve numerically. Since H and γ depend on time explicitly through the conditional average $\bar{\mathbf{n}}(\mathbf{n}^0, t)$, this still does not decouple the covariance from the average. Indeed, the conditional average acts as if it carries the conditional fluctuations, $\delta\mathbf{n}$, on its back—exposing them to different portions of the state space as it winds along its own trajectory.

4.4. Fluctuations and Dissipation

The intimate coupling of the conditional average and the covariance in the canonical theory occurs because the same events that cause dissipation on the average also cause molecular fluctuations. This is represented clearly in the *ansatz* of Eq. (4.2.6) where the elementary molecular processes determine the transition probability to nearby states. Certain features of the transitions will be observable on the average and others will appear only as fluctuations, both having their origin in the same molecular transitions.

The connection between dissipation and fluctuations can be seen in several other equivalent ways. The simplest is based on the experimental observation that the deterministic equations provide a good approximation for many measurements. Thus as long as fluctuations are small, we make *postulate I* that the conditional average, $\bar{\mathbf{n}}(\mathbf{n}^0, t)$, solves the usual deterministic equation, i.e.,

$$d\bar{\mathbf{n}}/dt = \sum_{\kappa} \boldsymbol{\omega}_{\kappa}(\bar{V}_{\kappa}^{+} - \bar{V}_{\kappa}^{-}) + \bar{\mathbf{S}} \equiv \bar{\mathbf{R}}. \tag{4.4.1}$$

It is consistent to interpret Eq. (4.4.1) as describing the conditional average since $\bar{\mathbf{n}}(\mathbf{n}^0, t)$ depends parametrically on a single initial condition, $\mathbf{n}^0$, as does the solution to Eq. (4.4.1). The conditional fluctuations $\delta\mathbf{n} \equiv \mathbf{n} - \bar{\mathbf{n}}$ have their origin in the same molecular processes. Thus as long as $\delta\mathbf{n}$ is small, we make *postulate II* that it satisfies a linearized version of Eq. (4.4.1), i.e.,

$$d\delta\mathbf{n}/dt = H(\bar{\mathbf{n}}(\mathbf{n}^0, t), t)\delta\mathbf{n} + \tilde{\mathbf{f}}(t), \tag{4.4.2}$$

where

$$H_{ij}(\bar{\mathbf{n}}(\mathbf{n}^0, t), t) \equiv (\partial \bar{R}_i/\partial \bar{n}_j). \tag{4.4.3}$$

A third postulate is required to determine the random terms, $\tilde{\mathbf{f}}(t)$. Since the conditional fluctuations vanish on the average, it follows that $\tilde{\mathbf{f}}(t)$ must vanish on the average, i.e.,

$$\langle \tilde{\mathbf{f}}(t) \rangle = \mathbf{0}. \tag{4.4.4}$$

It also is consistent physically to assume that the fluctuations are Markovian since the deterministic equations depend only on the present instant of time, not the past. This will be the case if we take $\tilde{\mathbf{f}}(t)$ to be a purely random Gaussian so that

$$\langle \tilde{\mathbf{f}}(t)\tilde{\mathbf{f}}^T(t') \rangle = \gamma(\bar{\mathbf{n}}(\mathbf{n}^0, t))\delta(t - t'), \tag{4.4.5}$$

with γ defined by

$$\gamma_{ij}(\bar{\mathbf{n}}(\mathbf{n}^0, t)) \equiv \sum_{\kappa} \omega_{\kappa i}(\bar{V}_\kappa^+ + \bar{V}_\kappa^-)\omega_{\kappa j}, \tag{4.4.6}$$

which is *postulate III*. The choice of the form of γ is the only one that is non-negative definite and depends linearly on the rates of the elementary processes. The fact that γ is linear in $V_\kappa^\pm$ implies that the fluctuations will scale like $m^{1/2}$, where m is some measure of the system size. As this is known to be the correct scaling near equilibrium, it is a sensible choice. As Eq. (4.4.6) agrees with the Onsager theory at equilibrium, it is the unique choice based on the use of elementary processes.

These three postulates generate the stochastic differential equation (Eq. 4.4.2) which characterizes the conditional fluctuations, $\delta\mathbf{n}$. This equation is linear and has the solution

$$\delta\mathbf{n}(\mathbf{n}^0, t) = \int_{t^0}^{t} \mathrm{Pexp}\left[\int_{\tau'}^{t} H(\bar{\mathbf{n}}(\mathbf{n}^0, \tau))\, d\tau\right]\tilde{\mathbf{f}}(\tau')\, d\tau', \tag{4.4.7}$$

since initially the conditional fluctuation, $\delta\mathbf{n}(\mathbf{n}^0, t^0)$, vanishes. Equation (4.4.7) is similar to the expression for a fluctuation in a stationary, Gaussian, Markov process given in Eq. (1.8.6). However, because of the explicit time dependence in $\gamma(\bar{\mathbf{n}}(\mathbf{n}^0,t))$ and $H(\bar{\mathbf{n}}(\mathbf{n}^0, t))$, $\delta\mathbf{n}$ will not be stationary. Nonetheless it is Gaussian because $\tilde{\mathbf{f}}(t)$ is Gaussian, and it can be shown to be Markovian since $\tilde{\mathbf{f}}(s)$ is uncorrelated with $\delta\mathbf{n}$ for $s \le t^0$. Thus according to these postulates the conditional fluctuations are a nonstatnionary, Gaussian, Markov process—sometimes called a time-dependent Ornstein–Unlenbeck process.

It is easy to see from Eq. (4.4.7) that postulates I–III generate exactly the same conditional fluctuations as the Fokker–Planck equation (4.2.7). Indeed, the Fokker–Planck equation is solved by the Gaussian P_2 density function in Eq. (4.3.1). That function, like the process produced by postulate I, solves the deterministic equation (4.4.1) on the conditional average. Moreover, postulate III guarantees that the conditional fluctuations in Eq. (4.4.7) are Gaussian. Thus to show that the two theories are equivalent, it suffices to show that they have the same covariance. This can be seen using Eq. (4.4.7). Multiplying each side by its own transpose and taking the average using Eq. (4.4.5), one obtains

$$\langle \delta \mathbf{n}(t) \delta \mathbf{n}^T(t) \rangle = \int_{t^0}^{t} d\tau' \int_{t^0}^{t} d\tau'' \, \mathrm{Pexp}\left[\int_{\tau'}^{t} H(\tau)\, d\tau\right] \gamma(\tau')\delta(\tau' - \tau'') \left\{ \mathrm{Pexp}\left[\int_{\tau''}^{t} H(\tau)\, d\tau\right] \right\}^T. \tag{4.4.8}$$

Once the integral over the delta function is performed, this becomes identical to Eq. (4.3.2) for the covariance σ.

The theory based on postulates I, II, and III has been referred to as the *fluctuation–dissipation theory* because of the way in which the fluctuations are associated with the dissipative deterministic equations. Postulate II, in fact, is strongly reminiscent of Onsager's regression hypothesis in Eq. (2.6.3). An important difference between the two hypotheses is that the fluctuation–dissipation theory is based on a conception of elementary molecular events, whereas the Onsager theory is phenomenological and requires no knowledge of molecular processes.

As we have seen, the theory based on the fluctuation–dissipation postulates and the theory based on the thermodynamic limit of the statistical *ansatz* in Eq. (4.2.6) are identical. Both lead to the Gaussian conditional probability density, P_2, in Eq. (4.3.1). This density can be used to describe how the single-time probability density for an arbitrary ensemble changes as a function of time. According to Eq. (1.3.20) the evolution of the single-time probability density, $W_1(\mathbf{n}^0, t^0)$, is determined by

$$W_1(\mathbf{n}, t) = \int W_1(\mathbf{n}^0, t^0) P_2(\mathbf{n}^0, t^0 | \mathbf{n}, t)\, d\mathbf{n}^0. \tag{4.4.9}$$

Although P_2 is Gaussian, there is no reason to believe that the single-time density W_1 at time t^0 will be Gaussian. In fact, for ensembles away from equilibrium there is a great deal of experimental flexibility in the preparation of the ensemble at time t_0. Nonetheless, even if its probability density has several peaks, Eq. (4.4.9) can be used to describe how the ensemble evolves over time. Thus the conditional probability density contains the information needed to decribe the change in a nonequilibrium ensemble over time.

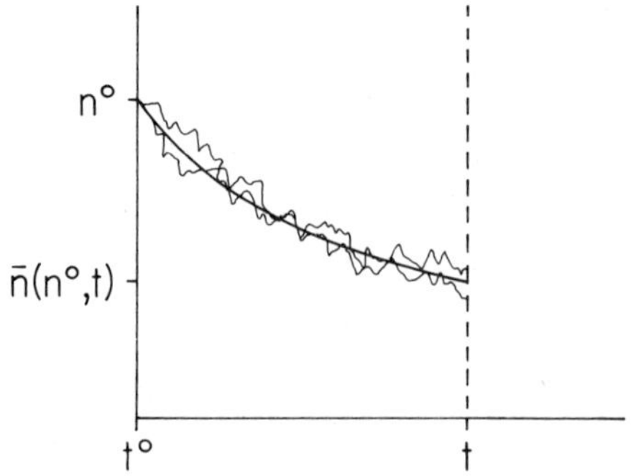

FIGURE 4.1. The conditional average trajectory $\bar{n}(n^0, t)$ shown as a smooth curve between the points n^0 and $\bar{n}(n^0, t)$. The two jagged trajectories are representative of deviations from the average. All three trajectories begin at n^0 since the ensemble is conditional on $n = n^0$ at $t = t^0$.

There is an interesting characterization of the conditional probability density in Eq. (4.3.1) which is based on the actual trajectories that members of the ensemble take between the point $\mathbf{n}^0$ at time t^0 and the point $\mathbf{n}$ at time t. One such trajectory is the average, i.e., the deterministic, trajectory $\bar{\mathbf{n}}(\mathbf{n}^0, t)$. Indeed, since fluctuations around the average trajectory are Gaussian, it follows that the deterministic trajectory is always the most probable one. Other trajectories, which involve fluctuations from the deterministic trajectory, also occur as has been drawn schematically in Fig. 4.1. The probability that such a fluctuating trajectory occurs is determined by a scalar-valued function of the fluctuations, $\mathscr{L}$. This function, which is analogous to the Lagrangian in classical mechanics, is defined by

$$\mathscr{L}(\mathbf{q}, \dot{\mathbf{q}}, t) = [\dot{\mathbf{q}} - H(\mathbf{n}^0, t)\mathbf{q}]^T \gamma^{-1}(\mathbf{n}^0, t)[\dot{\mathbf{q}} - H(\mathbf{n}^0, t)\mathbf{q}], \tag{4.4.10}$$

where the dot represents the time derivative and the matrices H and γ are given in Eqs. (4.2.8) and (4.2.9). Just as with the Lagrangian in classical mechanics, $\mathbf{q}$, and $\dot{\mathbf{q}}$ are independent variables since there are an infinity of trajectories that have the same value $\mathbf{q} = \delta\mathbf{n}$ at any time s between t^0 and t.

Using the statistical "Lagrangian" in Eq. (4.4.10), the conditional probability density can be characterized in the following way:

$$P_2(\mathbf{n}^0, t^0 | \mathbf{n}, t) \propto \exp\left[-\frac{1}{2}\text{extrem} \int_{t^0}^{t} \mathscr{L}(\mathbf{q}, \dot{\mathbf{q}}, s)\, ds\right], \tag{4.4.11}$$

where "extrem" means the extremum with respect to all differentiable paths $\mathbf{q}(s)$ beginning at $\mathbf{q}(t_0) = \mathbf{0}$ and crossing through $\mathbf{q} = \mathbf{n} - \bar{\mathbf{n}}(\mathbf{n}^0, t)$ at time t. To verify Eq. (4.4.11) we need to use Lagrange's equations, which determine the extremal path, $q^*(s)$:

$$\frac{d}{ds}(\partial\mathscr{L}/\partial\dot{q}_i^*) = \partial\mathscr{L}/\partial q_i^*. \tag{4.4.12}$$

Substituting the expression for $\mathscr{L}$ given in Eq. (4.4.10) into Eq. (4.4.12), we find that

$$d\mathbf{X}/ds = -H^T(s)\mathbf{X}, \tag{4.4.13}$$

where

$$\mathbf{X}(s) \equiv \gamma^{-1}(s)[dq^*/ds - H(s)q^*] \tag{4.4.14}$$

(the $\mathbf{n}^0$ dependence of H and γ being suppressed). According to Eqs. (4.3.3)–(4.3.6), Eq. (4.4.13) can be solved using a time-ordered exponential. Thus

$$\mathbf{X}(s) = \phi_-^{(T)}(s)\mathbf{X}(0) \equiv \text{Pexp}\left[-\int_{t^0}^{s} H^T(\tau)\,d\tau\right]\mathbf{X}(0). \tag{4.4.15}$$

Substituting this into the definition of $\mathbf{X}(s)$ yields another first-order equation which can also be solved using a time-ordered exponential. Introducing the notation

$$\phi_\pm(t) \equiv \text{Pexp}\left[\pm\int_{t^0}^{t} H(\tau)\,d\tau\right], \quad \phi_\pm^{(T)}(t) \equiv \text{Pexp}\left[\pm\int_{t^0}^{t} H^T(\tau)\,d\tau\right] \tag{4.4.16}$$

the solution to Eqs. (4.4.13) and (4.4.14) is found to be

$$\mathbf{q}^*(s) = \phi_+(s)\int_{t^0}^{s} d\tau\phi_+^{-1}(\tau)\gamma(\tau)\phi_-^T(\tau)\left[\int_{t^0}^{t} d\tau\phi_+^{-1}(\tau)\gamma(\tau)\phi_-^T(\tau)\right]^{-1}$$
$$\times\ \phi_+^{-1}(t)[\mathbf{n} - \bar{\mathbf{n}}(\mathbf{n}^0, t)]. \tag{4.4.17}$$

It is easy to verify that $q^*(s)$ in Eq. (4.4.17) satisfies the required conditions at $s = t^0$ and $s = t$, i.e., $\mathbf{q}^*(t^0) = \mathbf{0}$ and $\mathbf{q}^*(t) = \mathbf{n} - \bar{\mathbf{n}}(\mathbf{n}^0, t)$. With a bit more effort it can also be verified that the extremal solution satisfies

$$\exp\left[-\frac{1}{2}\int_{t^0}^{t} \mathscr{L}(\mathbf{q}^*, \dot{\mathbf{q}}^*, s)\,ds\right]$$
$$= \exp\left\{-\frac{1}{2}[\mathbf{n} - \bar{\mathbf{n}}(\mathbf{n}^0, t)]^T\sigma^{-1}(\mathbf{n}^0, t)[\mathbf{n} - \bar{\mathbf{n}}(\mathbf{n}^0, t)]\right\}, \tag{4.4.18}$$

where $\sigma(\mathbf{n}^0, t)$ is given by Eq. (4.3.2). Comparing Eqs. (4.3.1) and (4.4.18) shows that the representation of P_2 in terms of the extremal of the Lagrangian is, indeed, valid.

To see how the statistical Lagrangian determines the probability of an entire path we need to use the many-time joint probability densities defined in Section 1.3. The trajectory of a conditional fluctuation is a continuous infinity of random variables, i.e., the set of $\delta\mathbf{n}(s)$ for $0 \le s \le t$ with $\delta\mathbf{n}(0) = \mathbf{0}$. A discrete approximation to a given conditional trajectory can be obtained by specifying $\delta\mathbf{n}(s)$ at a finite number of times $t^0, t_1, \ldots, t_{m-2}, t$. An illustration of this for a scalar stochastic process is given in Fig. 4.2. Although the nature of the trajectory between any two times $t_j < s < t_{j+1}$ is unspecified in this approximation, by spacing the times evenly such that $t_{j+1} - t_j = (t - t^0)/(m - 1)$ and letting m go to infinity we obtain a progressively more accurate specification of the entire trajectory. If m time points are used to specify the trajectory, the

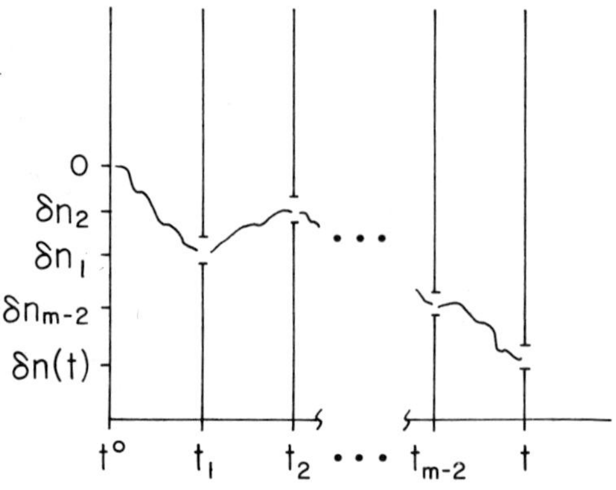

FIGURE 4.2. A discrete approximation of a conditional trajectory, $n(t)$, by specifying the conditional fluctuation $\delta n(t_j) = n(t_j) - \bar{n}(n^0, t_j)$ at m evenly spaced times $t^0, t_1, \ldots, t$.

probability density associated with that trajectory is given by the m-point joint probability density,

$$W_m(\mathbf{0}, t^0; \delta\mathbf{n}_1, t_1; \ldots; \delta\mathbf{n}, t) = W_1(\mathbf{0}, t^0) P_{1,m-1}(\mathbf{0}, t^0 | \delta\mathbf{n}_1, t; \ldots; \delta\mathbf{n}, t)$$

where $P_{1,m-1}$ is the $m-1$-point probability density, conditioned on the absence of a fluctuation at t^0. It is the mth approximation to the probability density which we associate with the trajectory. It can be reexpressed using the Markov property of the conditional fluctuation, $\delta\mathbf{n}(s)$, which allows us to use Eq. (1.3.26) to write

$$P_{1,m-1}(\mathbf{0}, t^0 | \delta\mathbf{n}_1, t_1; \ldots; \delta\mathbf{n}, t) = P_2(\mathbf{0}, t^0 | \delta\mathbf{n}_1, t_1) P_2(\delta\mathbf{n}_1, t_1 | \delta\mathbf{n}_2, t_2) \times \cdots \times P_2(\delta\mathbf{n}_{m-2}, t_{m-2} | \delta\mathbf{n}, t). \tag{4.4.19}$$

Thus in the mth approximation the probability of the trajectory can be written

$$\text{Prob}[\delta\mathbf{n}(s), 0 \le s \le t]_m = P_2(\mathbf{0}, t^0 | \delta\mathbf{n}_1, t_1) \times \cdots \times P_2(\delta\mathbf{n}_{m-2}, t_{m-2} | \delta\mathbf{n}, t)\, d\delta\mathbf{n}_1 \times \cdots \times d\delta\mathbf{n}. \tag{4.4.20}$$

To evaluate the right-hand side of Eq. (4.4.20), we need the conditional probability function for the conditional fluctuation, $\delta\mathbf{n}$. Since $\delta\mathbf{n}$ solves Eq. (4.4.2) it follows that

$$\delta\mathbf{n}(t_{i+1}) = \text{Pexp}\left[\int_{t_i}^{t_{i+1}} H(\tau)\, d\tau\right] \delta\mathbf{n}_i + \int_{t_i}^{t_{i+1}} \text{Pexp}\left[\int_s^{t_{i+1}} H(\tau)\, d\tau\right] \tilde{\mathbf{f}}(s)\, ds, \tag{4.4.21}$$

where $\delta\mathbf{n}_i$ is now arbitrary an fixed and the dependence on $\mathbf{n}^0$ has been left implicit in H and γ. Because the average of $\tilde{\mathbf{f}}$ is zero, the first term gives the conditional average of $\delta\mathbf{n}(t_{i+1})$, and because $\tilde{\mathbf{f}}$ is Gaussian, so is $\delta\mathbf{n}(t_{i+1})$. Following the procedure that led to Eq. (4.4.8), the covariance of $\delta\mathbf{n}(t_{i+1})$ is easily shown to be

$$\begin{aligned}&\langle(\delta\mathbf{n}_{i+1}-\langle\delta\mathbf{n}_{i+1}\rangle)^T(\delta\mathbf{n}_{i+1}-\langle\delta\mathbf{n}_{i+1}\rangle)\rangle\\&\quad=\int_{t_i}^{t_{i+1}}\operatorname{Pexp}\left[\int_s^{t_{i+1}}H(\tau)\,d\tau\right]\gamma(s)\left\{\operatorname{Pexp}\left[\int_s^{t_{i+1}}H(\tau)\,d\tau\right]\right\}^T ds\\&\quad\equiv\sigma(t_i,t_{i+1})\end{aligned}\tag{4.4.22}$$

where

$$\langle\delta\mathbf{n}_{i+1}\rangle=\operatorname{Pexp}\left[\int_{t_i}^{t_{i+1}}H(\tau)\,d\tau\right]\delta\mathbf{n}_i.\tag{4.4.23}$$

Now for large m, $t_{i+1}-t_i=(t-t^0)/(m-1)\equiv d\tau$ is an infinitesimal, and we find using Eqs. (4.3.6), (4.4.22), and (4.4.23) that

$$\operatorname{Pexp}\left[\int_{t_i}^{t_{i+1}}H(\tau)\,d\tau\right]=I+H(\bar{\mathbf{n}}(\mathbf{n}^0,t_i))\,d\tau+\mathcal{O}(d\tau)\tag{4.4.24a}$$

and

$$\sigma(t_i,t_{i+1})=\gamma(\bar{\mathbf{n}}(\mathbf{n}^0,t_i))\,d\tau+\mathcal{O}(d\tau).\tag{4.4.24b}$$

Thus for large m we can use the Gaussian character of $\delta\mathbf{n}$ to write

$$\begin{aligned}&P_2(\delta\mathbf{n}_i,t_i|\delta\mathbf{n}_{i+1},t_{i+1})\\&\propto\exp\left(-\frac{1}{2}\{\delta\mathbf{n}_{i+1}-[I+H(t_i)\,d\tau]\delta\mathbf{n}_i\}^T\frac{\gamma^{-1}(t_i)}{d\tau}\{\delta\mathbf{n}_{i+1}-[I+H(t_i)\,d\tau]\delta\mathbf{n}_i\}\right)\\&=\exp\left\{-\frac{1}{2}\left[\frac{\delta\mathbf{n}_{i+1}-\delta\mathbf{n}_i}{d\tau}-H(t_i)\delta\mathbf{n}_i\right]^T\gamma^{-1}(t_i)\left[\frac{\delta\mathbf{n}_{i+1}-\delta\mathbf{n}_i}{d\tau}-H(t_i)\delta\mathbf{n}_i\right]d\tau\right\}.\end{aligned}\tag{4.4.25}$$

Substituting Eq. (4.4.25) into Eq. (4.4.20), which gives the probability of the trajectory $\delta\mathbf{n}(s)$, we find that the arguments of the exponentials add together to give

$$\begin{aligned}&\operatorname{Prob}[\delta\mathbf{n}(s),0\le s\le t]_m\\&\quad\propto\exp\left\{-\frac{1}{2}\sum_{i=1}^{m}\left[\frac{d\delta\mathbf{n}_i}{d\tau}-H(t_i)\delta\mathbf{n}_i\right]^T\gamma^{-1}(t_i)\left[\frac{d\delta\mathbf{n}_i}{d\tau}-H(t_i)\delta\mathbf{n}_i\right]d\tau\right\},\end{aligned}\tag{4.4.26}$$

or in the limit that $m\to\infty$ and $d\tau\to 0$,

$$\operatorname{Prob}[\delta\mathbf{n}(s),0\le s\le t]\propto\exp\left\{-\frac{1}{2}\int_{t_0}^{t}\mathscr{L}(\delta\dot{\mathbf{n}},\delta\mathbf{n},\tau)\,d\tau\right\},\tag{4.4.27}$$

where $\mathscr{L}$ is the statistical Lagrangian in Eq. (4.4.10). Thus the probability of any conditional trajectory that begins at $\delta\mathbf{n}=\mathbf{n}-\mathbf{n}^0\equiv\mathbf{0}$ can be calculated from the time integral of the Lagrangian of that trajectory, i.e., from the analogue of the classical mechanical action. $\mathscr{L}$ is by definition a positive definite quadratic form, so that the statistical action is positive except on the trajectory $\delta\mathbf{n}(s) \equiv \mathbf{0}$. According to Eq. (4.4.27), this means that the most probable conditional trajectory is the one on which fluctuations about the deterministic average vanish.

The formulation of the canonical theory in terms of the statistical Lagrangian and the action for conditional trajectories is not really useful for practical calculations. Nevertheless, it reveals in a striking way the importance of the dissipative, deterministic trajectory as the most probable route in state space. It also emphasizes that any other realizable trajectories involve fluctuations that are generated by the deterministic trajectory. All that one needs to know is contained in the linearized rate matrix, H, and the covariance matrix, γ, which uniquely define the statistical Lagrangian.

4.5. Thermodynamic Properties of the Canonical Theory

The canonical theory outlined in Section 4.2 provides a mechanistic foundation for thermodynamics. In this section we explore this aspect of the theory for the deterministic trajectories, i.e., we obtain results that are true on the conditional average. These include conservation theorems, a generalization of Boltzmann's H-theorem, a discussion of the approach to equilibrium in closed systems, the entropy maximum principle, and a derivation of the Caratheodory statement of the Second Law. These results all have a dynamical character. This should be contrasted with the classical approach to thermodynamics in which both time and molecular processes are ignored. In large measure our approach is in the spirit of Boltzmann. Indeed, when applied at the Boltzmann level of description the foundation for thermodynamics provided by the canonical theory is identical to that given by Boltzmann a century ago.

According to the canonical theory, the portion of the average transport equations due to molecular processes is given by

$$(d\bar{n}_i/dt)_{\mathrm{mol}} = \sum_{\kappa} \omega_{\kappa i}[V_\kappa^+(\bar{\mathbf{n}}(\mathbf{n}^0, t) - V_\kappa^-(\bar{\mathbf{n}}(\mathbf{n}^0, t)))]. \tag{4.5.1}$$

Since we will concern ourselves here with the approach to equilibrium, the complete average transport equations are

$$d\bar{n}_i/dt = \sum_{\kappa} \omega_{\kappa i}(\bar{V}_\kappa^+ - \bar{V}_\kappa^-) + \bar{S}_i, \tag{4.5.2}$$

where $\bar{S}_i$ is a function only of the average and the overbars imply that the functions are to be evaluated at the average value, $\bar{\mathbf{n}}$. Recall that an equilibrium state, $\mathbf{n}^e$, is one for which detailed balance, i.e.,

$$V_\kappa^+(\mathbf{n}^e) = V_\kappa^-(\mathbf{n}^e) \tag{4.5.3}$$

holds. Thus the terms S_i must satisfy

$$S_i(\mathbf{n}^e) \equiv 0. \tag{4.5.4}$$

As we saw in Section 2.4, this is the case for streaming terms, such as $-\nabla \cdot (\mathbf{v}\mathbf{p}^T)$, or forcing terms, such as $-\nabla p + \mathbf{F}\rho$, that appear at the hydrodynamic level of description. It is also true for the streaming terms in the Boltzmann equation (Eq. 2.7.23). Since these terms correspond to kinematic or mechanical conditions, they further satisfy the fundamental conservation requirements at the given level of description. For example, at the hydrodynamic level they conserve mass, momentum, and energy whereas at the Boltzmann level they conserve particle number. At what we have called the thermodynamic level of description, the variation of matter in space is ignored. Thus there are no streaming-like terms at the thermodynamic level.

The key ingredient in the dynamical connection to thermodynamics is the generalized Onsager–Rayleigh dissipation function

$$\Phi(t) = \sum_j (\bar{F}_j - F_j^e)(d\bar{n}_j/dt)_{\text{mol}}, \tag{4.5.5}$$

where the F_j are intensive variables and the superscript e means evaluated at equilibrium. Like the quadratic Rayleigh–Onsager dissipation function in Eq. (2.3.15), Φ is strictly non-negative. Actually an even more fundamental inequality can be proven, namely,

$$\Phi_\kappa \equiv \sum_j (\bar{F}_j - F_j^e)\omega_{\kappa j}(\bar{V}_\kappa^+ - \bar{V}_\kappa^-) \geq 0. \tag{4.5.6}$$

Thus each elementary process makes a strictly non-negative contribution to the dissipation function. To prove this result we need to invoke the canonical form in Eq. (4.2.3). Thus

$$\Phi_\kappa = \sum_j (\bar{F}_j - \mathrm{F}_j^e)\omega_{\kappa j}\left\{\Omega_\kappa^+ \exp\left[-\sum_l \bar{F}_l n_{\kappa l}^+/k_B\right] - \Omega_\kappa^- \exp\left[-\sum_l \bar{F}_l n_{\kappa l}^-/k_B\right]\right\}. \tag{4.5.7}$$

Introducing the canonical form into the condition of detailed balance in Eq. (4.5.3), we also find that

$$\Omega_\kappa^+ \exp\left[-\sum_l F_l^e n_{\kappa l}^+/k_B\right] = \Omega_\kappa^- \exp\left[-\sum_l F_l^e n_{\kappa l}^-/k_B\right] \tag{4.5.8}$$

or

$$\Omega_\kappa^+ = \Omega_\kappa^- \exp\left[-\sum_l F_l^e \omega_{\kappa l}/k_B\right].$$

Substituting this expression for Ω_κ^+ into Eq. (4.5.7) gives

$$\Phi_\kappa = \Omega_\kappa^- \exp\left[-\sum_l \bar{F}_l n_{\kappa l}^-/k_B\right] \sum_j (\bar{F}_j - F_j^e)\omega_{\kappa j}\left\{\exp\left[\sum_l (\bar{F}_l - F_l^e)\omega_{\kappa l}/k_B\right] - 1\right\} \tag{4.5.9}$$

or

$$\Phi_\kappa = k_B \Omega_\kappa^- \exp\left[-\sum_l \bar{F}_l n_{\kappa l}^-/k_B\right] A_\kappa[\exp(A_\kappa) - 1], \tag{4.5.10}$$

where

$$A_\kappa \equiv \sum_j \omega_{\kappa j}(\bar{F}_j - F_j^e)/k_B. \tag{4.5.11}$$

Since the first three factors on the right-hand side of Eq. (4.5.10) are positive, the sign of Φ_κ is determined by the sign of $A_\kappa[\exp(A_\kappa) - 1]$. This term, however, is positive for $A_\kappa > 0$ as well as $A_\kappa < 0$ and vanishes only if $A_\kappa = 0$. Thus, as stated in Eq. (4.5.6), Φ_κ is strictly non-negative and, furthermore, vanishes if and only if $A_\kappa \equiv 0$. Φ is also strictly non-negative since from Eqs. (4.5.5) and (4.5.6)

$$\Phi(t) = \sum_\kappa \Phi_\kappa \geq 0, \tag{4.5.12}$$

with equality holding if and only if

$$\sum_j \omega_{\kappa j}(\bar{F}_j - F_j^e) \equiv 0 \tag{4.5.13}$$

for all κ. Since Eq. (4.5.13) holds trivially for all κ if $\bar{\mathbf{n}} = \mathbf{n}^e$, it follows that the dissipation function vanishes at an equilibrium state.

The non-negativity of Φ provides in a single expression the generalization of the Rayleigh–Onsager dissipation function and the Boltzmann H-theorem. To see this, let $\bar{\mathbf{n}}$ be close to equilibrium so that we can use Eq. (4.5.11) to write

$$A_\kappa = \sum_j \omega_{\kappa j} X_j/k_B, \tag{4.5.14}$$

where $X_j = \bar{F}_j - F_j^e$ is the thermodynamic force. Using this expression, close to equilibrium Eq. (4.5.10) reduces to

$$\Phi_\kappa = \sum_j \sum_l \omega_{\kappa j}(V_\kappa^-/k_B)\omega_{\kappa l} X_j X_l. \tag{4.5.15}$$

Summing this over κ as in Eq. (4.5.12) gives

$$\Phi = \sum_j \sum_l L_{jl}^s X_j X_l \geq 0, \tag{4.5.16}$$

where L^s is the symmetric matrix defined by

$$L_{jl}^s = \sum_\kappa \omega_{\kappa j}(V_\kappa^-/k_B)\omega_{\kappa l}. \tag{4.5.17}$$

Because of the inequality in Eq. (4.5.16) L^s is also non-negative. This expression for Φ is identical in form to that for the Rayleigh–Onsager dissipation function in Eq. (2.3.19) but provides the bonus in Eq. (4.5.17) of an explicit expression for the symmetric part of L. The antisymmetric part, L^a, comes from the kinematic terms $\bar{S}_i$ as we saw for hydrodynamics in Eqs. (2.4.40)–(2.4.44). The form of L^a is explored further in Section 4.6.

In its nonlinear form the inequality $\Phi \geq 0$ generalizes Boltzmann's H-theorem. Indeed, at the Boltzmann level of description the rates of the individual elementary processes, κ, are written in the canonical form in Eqs. (3.5.27)–(3.5.32). If these are substituted into the expression in Eq. (2.8.5), which gives the time derivative of H, we find that

$$dH/dt = -\iiint (\bar{F}/k_B)\hat{\sigma}_T g(\bar{\rho}'\bar{\rho}_1' - \bar{\rho}\bar{\rho}_1)\,d\mathbf{r}\,d\mathbf{v}\,d\mathbf{v}_1$$

$$= -\frac{1}{k_B}\sum_j \bar{F}_j \sum_\kappa \omega_{\kappa j}(\bar{V}_\kappa^+ - \bar{V}_\kappa^-), \tag{4.5.18}$$

where in the second equality we have used the correspondence between the discrete Boltzmann equation (Eq. 3.3.22) and the continuous Boltzmann equation (Eq. 3.3.28). Thus

$$dH/dt = -\frac{1}{k_B}\sum_j \bar{F}_j(d\bar{n}_j/dt)_{\text{mol}}. \tag{4.5.19}$$

Aside from the factor $-1/k_B$, the left-hand side of Eq. (4.5.19) is nearly the same as in the definition of the dissipation function in Eq. (4.5.5). The difference is that the term F_j^e is missing. However, the elementary processes in the Boltzmann equation involve microscopic reversibility, and for such processes it can be shown that

$$\sum_j F_j^e(d\bar{n}_j/dt)_{\text{mol}} \equiv 0. \tag{4.5.20}$$

Hence

$$dH/dt = -\Phi/k_B \leq 0,$$

which is the H-theorem.

The fact that Eq. (4.5.20) holds for processes satisfying microscopic reversibility can be deduced from the condition of detailed balance in Eq. (4.5.8). Combining this with the equality of Ω_κ^+ and Ω_κ^- in Eq. (4.2.12) gives

$$\exp\left[-\sum_j F_j^e \omega_{\kappa j}/k_B\right] = 1 \tag{4.5.21}$$

or

$$\sum_j F_j^e \omega_{\kappa j} \equiv 0 \tag{4.5.22}$$

for all κ satisfying microscopic reversibility. Multiplying both sides of Eq. (4.5.22) by $(\bar{V}_\kappa^+ - \bar{V}_\kappa^-)$ it follows that

$$\sum_j F_j^e \omega_{\kappa j}(\bar{V}_\kappa^+ - \bar{V}_\kappa^-) \equiv 0 \tag{4.5.23}$$

for all such processes. Finally summing Eq. (4.5.23) over κ gives

$$\sum_j F_j^e\left[\sum_\kappa \omega_{\kappa j}(\bar{V}_\kappa^+ - \bar{V}_\kappa^-)\right] \equiv 0, \tag{4.5.24}$$

which is identical to Eq. (4.5.20). Equation (4.5.20) has a form identical to the condition of conservation of atoms discussed for chemical reaction in Section 3.7. That conservation condition is given in Eq. (3.7.4) and follows from Eq. (3.7.2) just as Eq. (4.5.20) follows from Eq. (4.5.22). The two conservation theorems, however, have a different character since in Eqs. (4.5.20) and (4.5.22) the F_i^e are intensive thermodynamic functions while the quantities ψ_{Ai} in Eqs. (3.7.2) and (3.7.4) depend only on the atomic constitution of individual molecules.

Actually both expressions reflect the same sort of conservation of molecular quantities. Several examples suffice to illustrate this point. Returning to the Boltzmann equation, the extensive variables as molecule numbers and for a given κ the expression in Eq. (4.5.22) involves two particles colliding at point $\mathbf{r}$. Thus Eq. (4.5.22) becomes

$$-F^e(\mathbf{v},\mathbf{r}) - F^e(\mathbf{v}_1,\mathbf{r}) + F^e(\mathbf{v}',\mathbf{r}) + F^e(\mathbf{v}_1',\mathbf{r}) = 0 \tag{4.5.25}$$

since one particle disappears at each velocity $\mathbf{v}$ and $\mathbf{v}_1$ and one reappears at $\mathbf{v}'$ and $\mathbf{v}_1'$. Comparing Eq. (4.5.25) with Eq. (2.9.2), we see that $-F^e$ must be a collisional invariant. Indeed, using the fact that $F = -k_B[\ln \rho + 1]$ and the expression for the Maxwell probability density in Eq. (2.9.3), Eq. (4.5.25) can be written as

$$2a + \mathbf{b}\cdot(\mathbf{v} + \mathbf{v}_1) + \frac{cm}{2}(v^2 + v_1^2) = 2a + \mathbf{b}\cdot(\mathbf{v}' + \mathbf{v}_1') + \frac{cm}{2}(v'^2 + v_1'^2). \tag{4.5.26}$$

Thus, as applied at the Boltzmann level, Eq. (4.5.22) reflects the underlying conservation of mass, momentum, and kinetic energy in a bimolecular collision.

The same pattern reveals itself for chemical reactions. For example, for the isomerization reaction A $\rightleftarrows$ B examined at the end of Section 3.6, Eqs. (4.5.20) and (4.5.22) become

$$-\mu_A^e + \mu_B^e = 0 \tag{4.5.27}$$

$$d[\mu_A^e n_A + \mu_B^e n_B]/dt = 0. \tag{4.5.28}$$

Substituting the first of these into the second gives

$$d(n_A + n_B)/dt = 0. \tag{4.5.29}$$

Equation (4.5.29) states that the reaction conserves the overall amount of the isomer, which is equivalent to the atom conservation condition (Eq. 3.7.4) since isomers have the same number of each kind of atom. This circumstance is quite general and shows that Eq. (4.5.20) has its origin in the conservation of molecular properties during elementary events.

The exchange of energy of a system with a reservoir is not a conservative process. Although the overall molecular process is conservative, energy losses or gains occur that destroy conservation of the energy of the system. Consider, as an example, heat exchange between a thermal reservoir at temperature T_R and a system at temperature T_S. This elementary process requires the assembly of a certain molecular-sized amount of energy ε_S^+ from the system and energy ε_R^+ from the reservoir and converts these energies into $\varepsilon_S^+ + \varepsilon_R^+ = \varepsilon_S^- + \varepsilon_R^-$, which conserves the total energy, but not the energy of the system. The rate of the elementary process, thought of as involving both system and reservoir, is given by the canonical form

$$R = \Omega \exp\left[-\left(\frac{\varepsilon_S^+}{k_B T_S} + \frac{\varepsilon_R^+}{k_B T_R}\right)\right] - \Omega \exp\left[-\left(\frac{\varepsilon_S^-}{k_B T_S} + \frac{\varepsilon_R^-}{k_B T_R}\right)\right], \tag{4.5.30}$$

where the condition $\Omega^+ = \Omega^- = \Omega$, i.e., microscopic reversibility, is consistent with the equality of T_S and T_R at equilibrium. Thus when both system and reservoir are taken into account, the process satisfies microscopic reversibility. If only system variables are considered, as is the point of view adopted in thermodynamics, then the rate expression in Eq. (4.5.30) can be written

$$R = \bar{\Omega}^+ \exp(-\varepsilon_S^+/k_B T_S) - \bar{\Omega}^- \exp(-\varepsilon_S^-/k_B T_S), \tag{4.5.31}$$

where

$$\begin{aligned} \bar{\Omega}^+ &= \Omega \exp(-\varepsilon_R^+/k_B T_R) \\ \bar{\Omega}^- &= \Omega \exp(-\varepsilon_R^-/k_B T_R) \end{aligned} \tag{4.5.32}$$

are constants since the temperature of the reservoir is fixed. Although Eq. (4.5.31) has the canonical form, the intrinsic rates $\bar{\Omega}^+$ and $\bar{\Omega}^-$ are in general no longer equal. This is typical of the exchange of extensive variables with a reservoir. Hence the exchange of an extensive variable with a reservoir violates microscopic reversibility and so does not satisfy the conservation condition in Eq. (4.5.22).

The breakdown of conservation for molecular processes involving reservoirs is closely related to the *Caratheodory statement* of the Second Law of thermodynamics. According to Caratheodory there exists in any neighborhood of any value of the extensive variables, $\mathbf{n}$, a collection of points of nonzero volume which are inaccessible from $\mathbf{n}$ by adiabatic transformations. Although, technically, instead of "volume" one should say "measure," "volume" conveys the correct picture. An *adiabatic transformation* is one in which no heat flows into or out of the system. Thus the Caratheodory statement restricts the states that can be reached when no heat flow is allowed to a set so small that it never occupies a neighboring volume in a significant way. For example, the reversible adiabat for a monotonic ideal gas is a line in the pressure–volume

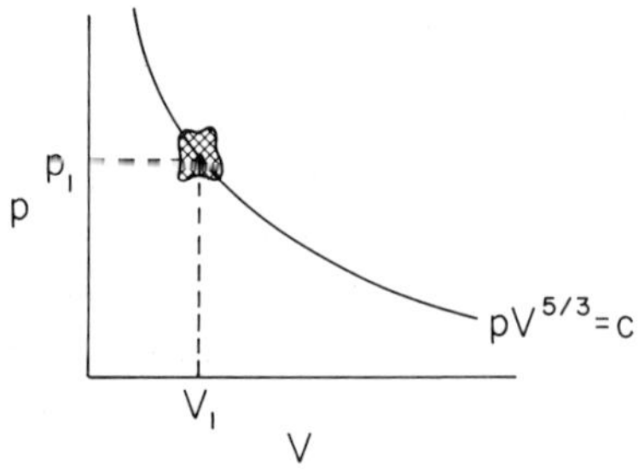

FIGURE 4.3. Schematic representation of the Caratheodory statement of the Second Law for an ideal gas. The cross-hatched area represents a set of states of non-zero measure in the neighborhood of p_1, V_1 that are inaccessible from the point p_1, V_1 by reversible adiabatic transformations falling on the curve $pV^{5/3} = c$.

plane determined by $pV^{5/3} = c$, a constant, as shown in Fig. 4.3. This is compatible with the Caratheodory statement since a line occupies zero volume in two-dimensional state space. Thus a monatomic ideal gas undergoing a reversible, adiabatic transformation from a point (p_1, V_1) cannot reach sets of volume greater than zero, like the cross-hatched one shown in Fig. 4.3.

The Caratheodory statement of the Second Law can be derived from the canonical form of the dissipative part of the transport equations. As we have just seen, microscopic reversibility does not hold for nonadiabatic transformations since heat is exchanged with a reservoir. In other words, if an elementary process satisfies microscopic reversibility, then the process must be adiabatic. However, it is easy to use the conservation principle in Eq. (4.5.20) to show for a point, $\mathbf{n}$, that there exist neighboring sets of points of nonzero volume that are not accessible by processes satisfying microscopic reversibility. The argument goes as follows. Microscopic reversibility implies that Eq. (4.5.20) holds, i.e.,

$$\left(d \sum_j F_j^e \bar{n}_j/dt\right)_{\text{mol}} = 0. \tag{4.5.33}$$

This conservation condition is also satisfied by any mechanical transformations which the system may undergo in performing work. Thus, in general, for a system without external inputs we have

$$d \sum_j F_j^e \bar{n}_j/dt = 0, \tag{4.5.34}$$

where the total time derivative can be written. Integrating Eq. (4.5.34) between the initial state, $\mathbf{n}$, at $t = 0$ and the final equilibrium state, $\mathbf{n}^e$, at $t = \infty$ gives

$$\sum_j F_j^e n_j = \sum_j F_j^e n_j^e. \tag{4.5.35}$$

Equation (4.5.35) provides a functional relationship between the variables n_i and the variables n_i^e. Thus the set of all states, $\mathbf{n}^e$, accessible from $\mathbf{n}$ must lie on a *surface* in the space of extensive variables and so comprise a set of zero volume. Hence, in any neighborhood of $\mathbf{n}$ the set of points that are inaccessible by processes satisfying microscopic reversibility must have a nonzero volume.

Finally, because processes that satisfy microscopic reversibility are necessarily adiabatic, it follows that there are always sets of nonzero measure which are inaccessible by adiabatic transformations. This is the Caratheodory statement of the Second Law.

Three ingredients are required for this proof of the Second Law: the canonical form, the fact that microscopic reversibility implies adiabaticity, and mechanical conservation of extensive variables. The latter is a stronger version of the conservation statement contained in the First Law and implies the overall conservation of mass and momentum for nondissipative processes.

Having seen how the Second Law follows from the canonical form, it is possible to develop the differential form associated with the entropy and relate it to the reversible heat. Recall that the basic postulates that define the canonical form do not assume the existence of the entropy. Indeed, we promised above Eq. (4.2.4) to verify that the intensive variables F_j can be identified as partial derivatives of an entropy function. The existence of the entropy, itself, can be proven using the Caratheodory principle by standard arguments. Using this chain of reasoning we deduce that the canonical form implies the existence of an integrating factor for the reversible heat. Thus

$$\frac{đQ_{\text{rev}}}{T_R} = dS = \sum_j (\partial S/\partial n_j)\, dn_j, \tag{4.5.36}$$

where T_R is the Kelvin temperature of the thermal reservoir to or from which heat flows, and the notation $đQ$ emphasizes that the increment of heat is an inexact differential. To identify $\partial S/\partial n_j$ with intensive variables F_j in the canonical form (Eq. 4.2.3), it is necessary to define a reversible process.

A *reversible* or *quasistatic* process is a hypothetical transformation in which the dissipation function defined in Eq. (4.5.5) vanishes. Thus

$$\sum_j (\bar{F}_j - F_j^e)(d\bar{n}_j/dt)_{\text{rev, mol}} = 0. \tag{4.5.37}$$

Equation (4.5.37) can be extended to the total time derivative of the extensive variables. This can be done since, in general, mechanical changes that the system undergoes, $S_j = (d\bar{n}_j/dt)_{\text{mech}}$, do not change the dissipation function under conditions that lead to equilibrium. Thus for conditions that lead to the equilibrium state, $\mathbf{n}^e$, a reversible transformation is defined by

$$\sum_j (\bar{F}_j - F_j^e)(d\bar{n}_j/dt)_{\text{rev}} = 0. \tag{4.5.38}$$

Consider, now, the special case that the only elementary process that does not satisfy microscopic reversibility is heat transfer to a reservoir. Thus, for all elementary processes except this one, the second term in Eq. (4.5.38) vanishes. Hence

$$\left(\frac{1}{T} - \frac{1}{T_R}\right)(d\bar{E}/dt)_{\text{heat, rev}} + \frac{1}{T}(d\bar{E}/dt)_{\text{other, rev}} + \sum_j{}' F_j(d\bar{n}_j/dt)_{\text{rev}} = 0, \tag{4.5.39}$$

where the prime means that the energy term is missing. Equation (4.5.39) can be rearranged to give

$$\frac{1}{T_R}(dQ/dt)_{\text{rev}} = \sum_j F_j(d\bar{n}_j/dt)_{\text{rev}}, \tag{4.5.40}$$

where $(dE/dt)_{\text{heat, rev}} \equiv (dQ/dt)_{\text{rev}}$. Since a reversible process delineates a transformation in which there is no dissipation, the system must always be in equilibrium. Thus the time becomes simply a mathematical parameter that delineates an infinitely slow path in the state space. Recognizing this, we rewrite Eq. (4.5.40) in the differential form

$$\frac{đQ_{\text{rev}}}{T_R} = \sum_j \bar{F}_j d\bar{n}_j. \tag{4.5.41}$$

Since the intensive variables are independent, we can compare the right-hand side of this equation with the expression in Eq. (4.5.36) for the entropy obtained by following Caratheodory's arguments. This gives

$$dS = \sum_j \bar{F}_j d\bar{n}_j \tag{4.5.42}$$

and the promised identification

$$\partial S/\partial n_j = F_j. \tag{4.5.43}$$

Using the dissipation function we can also establish the *Clausius inequalities* for the time rate of change of the entropy. Using the fact that mechanical processes do not contribute to Φ under conditions that lead to equilibrium, Eqs. (4.5.5) and (4.5.12) can be written

$$\Phi(t) = \sum_j (\bar{F}_j - F_j^e)(d\bar{n}_j/dt) \geq 0. \tag{4.5.44}$$

For an *isolated system* all the elementary molecular processes and any mechanical transformations satisfy the conservation condition in Eq. (4.5.20). Thus the second term on the right-hand side of Eq. (4.5.44) vanishes. Invoking the identification in Eq. (4.5.43) then implies that

$$dS/dt = \sum_j \bar{F}_j(dn_j/dt) \geq 0. \tag{4.5.45}$$

Finally for a system in contact with a *thermal reservoir* reasoning like that which leads from Eq. (4.5.31) to (4.5.40) can be used to deduce that

$$T_R\, dS/dt \geq đQ/dt. \tag{4.5.46}$$

Equality holds in both these equations only when the dissipation function vanishes, i.e., only for a reversible process. Notice that both these results have been deduced from the same principles that give Boltzmann's H-theorem.

4.6. Equivalence to the Onsager Theory at Equilibrium

For an ensemble of systems at equilibrium the fluctuation–dissipation theory is identical with the Onsager theory. We have seen already in the preceding section that the generalized Rayleigh–Onsager dissipation function, Φ, reduces to the quadratic form in Eq. (4.5.16) near equilibrium. The matrix L_{ij}^s which determines the quadratic form is symmetric. It turns out also to be the portion of the flux–force coupling matrix that is due to dissipative processes. As we shall see, the antisymmetric portion of the coupling matrix comes from conservative mechanical processes.

According to our postulate I in Section 4.4 the conditional average, $\bar{\mathbf{n}}(\mathbf{n}^0, t)$, satisfies the nonlinear deterministic Eq. (4.4.1). This equation simplifies considerably if the ensemble is a well-aged equilibrium ensemble that is not close to an equilibrium critical point. This last restriction, as discussed in Section 2.5, implies that the single-time equilibrium probability density, $W_1(\mathbf{n}^0)$, is narrowly distributed around an average value, $\mathbf{n}^e$. The narrowness of this distribution implies that only values of $\mathbf{n}^0$ which are close to $\mathbf{n}^e$ need be considered in examining the conditional average. Hence the right-hand side of Eq. (4.4.1) can be expanded around $\bar{\mathbf{n}} = \mathbf{n}^e$ and only the lowest-order terms need be retained. More explicitly,

$$d\bar{n}_i/dt = \sum_\kappa \omega_{\kappa i}\left(\Omega_\kappa^+ \exp\left[-\sum_l F_l(\bar{\mathbf{n}})n_{\kappa l}^+/k_B\right]\right.$$
$$\left. - \Omega_\kappa^- \exp\left[-\sum_l F_l(\bar{\mathbf{n}})n_{\kappa l}^-/k_B\right]\right) + S_i(\bar{\mathbf{n}}), \tag{4.6.1}$$

where the canonical form of $V_\kappa^\pm$ in Eq. (4.2.3) has been used to rewrite the right-hand side of Eq. (4.4.1). Since $\bar{\mathbf{n}}$ is close to $\mathbf{n}^0$, we write $\bar{\mathbf{n}} = \mathbf{n}^e + \bar{\mathbf{a}}$ and expand the right-hand side of (4.6.1) to lowest order in $\bar{\mathbf{a}}$. If the right-hand side is denoted by $R_i(\bar{\mathbf{n}})$, its Taylor expansion can be written

$$R_i(\bar{\mathbf{n}}) = R_i(\mathbf{n}^e + \bar{\mathbf{a}}) = R_i(\mathbf{n}^e) + \sum_j (\partial \bar{R}_i/\partial \bar{n}_j)^e \bar{a}_j + \cdots \tag{4.6.2}$$

The first term in the expansion, $R_i(\mathbf{n}^e)$, vanishes because the average equilibrium state $\mathbf{n}^e$ is time independent. To calculate the next term we need to calculate the equilibrium value of the derivatives of $\bar{V}_\kappa^\pm$, i.e.,

$$(\partial \bar{V}_\kappa^\pm/\partial \bar{n}_j)^e = -(V_\kappa^{\pm e}/k_B)\sum_l n_{\kappa l}^\pm(\partial \bar{F}_l/\partial \bar{n}_j)^e = -\sum_l (V_\kappa^e/k_B)n_{\kappa l}^\pm(\partial \bar{F}_l/\partial \bar{n}_j)^e, \tag{4.6.3}$$

where we have used detailed balance at equilibrium to write $V_\kappa^{\pm e} = V_\kappa^e$. Using Eq. (4.6.3) it follows that

$$(\partial \bar{R}_i/\partial \bar{n}_j)^e = \sum_\kappa \omega_{\kappa i} \sum_l (n_{\kappa l}^- - n_{\kappa l}^+)(V_\kappa^e/k_B)(\partial \bar{F}_l/\partial \bar{n}_j)^e + (\partial \bar{S}_i/\partial \bar{n}_j)^e$$
$$= \sum_l \sum_\kappa \omega_{\kappa i}(V_\kappa^e/k_B)\omega_{\kappa l} S_{lj} + (\partial \bar{S}_i/\partial \bar{n}_j)^e, \tag{4.6.4}$$

where we have used the fact that $n_{\kappa l}^- - n_{\kappa l}^+ = \omega_{\kappa l}$ and Eq. (4.5.43) to introduce the second differential matrix of the entropy, S_{ij}. Thus, to dominant order in $\bar{\mathbf{a}}$, Eq. (4.6.1) becomes

$$d\bar{a}_i/dt = \sum_l \sum_j L_{il}^s S_{lj} \bar{a}_j + \sum_j (\partial \bar{S}_i/\partial \bar{n}_j)^e \bar{a}_j, \tag{4.6.5}$$

with

$$L_{il}^s \equiv \sum_\kappa \omega_{\kappa i}(V_\kappa^e/k_B)\omega_{\kappa l}. \tag{4.6.6}$$

Recalling that the thermodynamic forces were shown in Eq. (2.3.7) to be

$$X_l = \sum_j S_{lj} a_j, \tag{4.6.7}$$

Eq. (4.6.5) can be written in vector notation as

$$d\bar{\mathbf{a}}/dt = L^s \bar{\mathbf{X}} + L^a \bar{\mathbf{X}} \equiv L\bar{\mathbf{X}} \tag{4.6.8}$$

with

$$L_{ik}^a \equiv \sum_j (\partial \bar{S}_i/\partial \bar{n}_j)^e (S^{-1})_{jk}. \tag{4.6.9}$$

Equation (4.6.8) gives the linear relationship between the fluxes and forces as based on the canonical form.

From its definition the matrix L^s is clearly symmetric. The matrix L^a, on the other hand, arises solely from streaming or other mechanical terms that appear in the transport equations. As demonstrated explicitly for the hydrodynamic equations in Section 2.4, the matrix L^a is antisymmetric. This matrix, in fact, must be antisymmetric in general or otherwise the mechanical terms would contribute to the dissipation function, Φ. We have seen in the previous sections that this does not occur. For example, in the proof of the H-theorem, the streaming terms in Eq. (2.8.4) do not contribute to changes in the H-function if the system is finite and has no imposed fluxes at the boundary. These are just the conditions that lead to equilibrium. They were used in Section 2.11 to show that close to equilibrium the streaming terms lead to an antisymmetric coupling matrix for the Boltzmann equation.

The symmetric part of the coupling matrix, L^s, results from the dissipative elementary processes. The structure of L^s, as revealed by Eq. (4.6.6), depends only on the equilibrium rate of the elementary processes and the net changes, $\boldsymbol{\omega}_\kappa$, in the elementary processes. The net changes are given by the difference of $\mathbf{n}_\kappa^-$ and $\mathbf{n}_\kappa^+$, and two different elementary processes may lead to the same net changes. For example, the two isomerization reactions $Y \rightleftarrows X$ and $X + Y \rightleftarrows 2X$ in Eq. (3.7.19) have the same values of $\boldsymbol{\omega}_\kappa$, i.e., $\omega_Y = -1, \omega_X = 1$. Thus the only feature of the matrix L^s which depends uniquely on mechanism is the term V_κ^e. This term is proportional to the intrinsic rates of the elementary processes, Ω_κ^+. In fact, the matrix elements L_{ij}^s are sometimes referred to as the *transport coefficients*. We can see why this is so by referring to the expression for L obtained for the isomerization reaction $A \rightleftarrows B$ in Eq. (3.7.49). Using Eqs. (3.6.14) and the definition of λ below Eq. (3.6.16), L can be written

$$L = \frac{k^+ \rho \exp(\mu_B^0/k_B T)}{k_B[\exp(\mu_A^0/k_B T) + \exp(\mu_B^0/k_B T)]}, \tag{4.6.10}$$

where ρ is the total number density of the isomer. Thus L is proportional to the rate constant, k^+, through terms that depend only on the choice of the standard state and the overall density of the isomer. A similar kind of result was found for the Onsager representation of the linearized Navier–Stokes equations in Eqs. (2.4.45) and (2.4.46) where, aside from derivatives, L^s is proportional to the thermal conductivity and the shear and bulk viscosity coefficients.

To complete the canonical description of the equilibrium ensemble and show its equivalence to the Onsager theory, we look next at the fluctuations. According to postulate II these satisfy the linear stochastic differential equation (4.4.2). Since we have assumed an equilibrium ensemble that is not close to a critical point, the linearization of the average equation around $\bar{\mathbf{n}}$ to obtain the matrix H_{ij} can again rely on the fact that the probability density, $W_1(\mathbf{n}^0)$, is sharply peaked at $\mathbf{n}^e$. Thus $\bar{\mathbf{n}} \approx \mathbf{n}^0 \approx \mathbf{n}^e$ and the definition of H in Eq. (4.4.3) can be replaced by

$$H_{ij} = (\partial \bar{R}_i/\partial \bar{n}_j)^e. \tag{4.6.11}$$

This is the same matrix which governs the relaxation of the deviation variable $\bar{\mathbf{a}} = \bar{\mathbf{n}} - \mathbf{n}^e$. Thus, noticing that the conditional fluctuation is $\delta \mathbf{a} \equiv \delta \mathbf{n}$, Eq. (4.4.2) becomes

$$d\delta \mathbf{a}/dt = H\delta \mathbf{a} + \tilde{\mathbf{f}}(t) \tag{4.6.12}$$

or

$$d\delta \mathbf{a}/dt = L\delta \mathbf{X} + \tilde{\mathbf{f}}(t), \tag{4.6.13}$$

where we have used Eq. (4.6.7) to define $\delta \mathbf{X} = S\delta \mathbf{a}$ and the fact that

$$H = LS, \tag{4.6.14}$$

which follows from Eqs. (4.6.11), (4.6.4), and (4.6.9).

To obtain the stochastic differential equation satisfied by $\mathbf{a} = \mathbf{n} - \mathbf{n}^e$, we add together Eqs. (4.6.13) and (4.6.8). Since they are linear, this gives

$$d\mathbf{a}/dt = L\mathbf{X} + \tilde{\mathbf{f}}(t), \tag{4.6.15}$$

with

$$\langle \tilde{\mathbf{f}}(t) \rangle = \mathbf{0}. \tag{4.6.16}$$

Equations (4.6.15) and (4.6.16) are identical to the Onsager equations in Section 2.6. Thus all that is needed to complete the demonstration of the equivalence of the two descriptions is to verify that

$$\langle \tilde{\mathbf{f}}(t)\tilde{\mathbf{f}}^T(t')\rangle = 2k_B L^s \delta(t - t'), \tag{4.6.17}$$

as required by Eq. (2.6.11). This is easily done using postulate III of the canonical theory, given in Eqs. (4.4.5) and (4.4.6), i.e.,

$$\langle \tilde{\mathbf{f}}(t)\tilde{\mathbf{f}}^T(t')\rangle = \sum_\kappa \boldsymbol{\omega}_\kappa(\bar{V}_\kappa^+ + \bar{V}_\kappa^-)\boldsymbol{\omega}_\kappa^T \delta(t - t').$$

Again because $\bar{\mathbf{n}} \approx \mathbf{n}^0 \approx \mathbf{n}^e$, we need keep only the lowest-order term in the Taylor expansion of the right-hand side in $\bar{\mathbf{a}} = \bar{\mathbf{n}} - \mathbf{n}^e$. In this case the lowest-order term is the first one and because of detailed balance we find

$$\langle \tilde{\mathbf{f}}(t)\tilde{\mathbf{f}}^T(t')\rangle = 2\sum_\kappa \boldsymbol{\omega}_\kappa V_\kappa^e \boldsymbol{\omega}_\kappa^T \delta(t - t'). \tag{4.6.18}$$

Comparing the right-hand side of Eq. (4.6.18) and the definition of L^s in Eq. (4.6.6), Eq. (4.6.17) is seen to hold. This proves that the canonical theory of an equilibrium ensemble is identical to the Onsager description.

It is worth emphasizing the differences between the reasoning that led Onsager to Eqs. (4.6.15)–(4.6.17) and the present reasoning. Onsager's theory rests on two assumptions: 1) the regression hypothesis, which states that fluctuations change in time like small, random deviations, and 2) the existence of the entropy and the Einstein formula for the single-time probability density. The present reasoning, on the other hand, is based on the assumption that in the limit of a large system the canonical form governs transitions in the intensive variables as described by Eq. (4.2.6). It is not necessary to assume the existence of an entropy function, but rather its existence is deduced from Eq. (4.2.6). Moreover, its relationship to the intensive variables, F_j, which appear in the canonical form is deduced from the basic assumptions, as we saw in the previous section.

Although the canonical theory requires experimental input for the initial probability, $W_1(\mathbf{n}^0)$, unlike the Onsager theory it can be used to deduce the Einstein formula for an equilibrium ensemble. To do so requires that the equilibrium state not be a critical point, so that all the eigenvalues of the relaxation matrix H in Eq. (4.6.14) have negative real parts. This insures that, on the average, an initial deviation from equilibrium will relax back to zero. Under this condition we can rewrite Eq. (4.6.13) as

$$d\mathbf{a}/dt = H\mathbf{a} + \tilde{\mathbf{f}}(t). \tag{4.6.19}$$

Since H is a constant matrix and, according to Eq. (4.6.17), $\tilde{\mathbf{f}}(t)$ is a white noise, it follows that $\mathbf{a}$ is a stationary, Gaussian, Markov process. Thus we can use the results obtained in Section 1.8 to deduce that W_1 has the form

$$W_1(\mathbf{a}) = [(2\pi)^k \det \sigma]^{-1/2} \exp[-\mathbf{a}^T\sigma^{-1}\mathbf{a}/2] \tag{4.6.20}$$

and that the covariance matrix, σ, satisfies Eq. (1.8.14), i.e., the *fluctuation–dissipation theorem*:

$$H\sigma + \sigma H^T = -\gamma = -2k_B L^s. \tag{4.6.21}$$

Under the assumption that the eigenvalues of H have negative real parts, this equation can be solved uniquely for σ. To do so we rewrite it, using the expression for H in Eq. (4.6.14), as

$$L(S\sigma) + (\sigma S)L^T = -2k_B L^s$$

or

$$\mathrm{sym}[L(-S/k_B)\sigma] = L^s, \tag{4.6.22}$$

where sym means the symmetric part of the matrix in brackets. From Eq. (4.6.22) it is obvious that

$$\sigma = -k_B S^{-1}, \tag{4.6.23}$$

which is just the Einstein result derived by other means in Eq. (2.5.16).

The use of the fluctuation–dissipation theorem to obtain the covariance matrix σ is quite different from the use of this theorem in the Onsager theory. There, as described in Eqs. (2.6.5) and (2.6.6), it is used to obtain an explicit expression for the covariance matrix of the random terms, γ, in terms of the matrices H and σ. In the present calculation, on the other hand, γ and H are known, and the fluctuation–dissipation theorem is thought of as an equation to be solved for σ. This is characteristic of the mechanistic statistical theory in which the dynamics of fluctuations are uniquely described by a knowledge of elementary molecular processes.

4.7. The Master Equation Formulation

There are several other ways of translating the physical ideas that govern the canonical theory of Section 4.2 into mathematical terms. One of these, based on the so-called master equation, has been used for many years to describe particular processes in physics and chemistry. Its intellectual father is the discrete time urn model of the Ehrenfests. The basic idea is simple and can be couched, as it often was by the Ehrenfests, in terms of "dogs and fleas."

Imagine two dogs which are populated by fleas. For simplicity, call one dog "A" and the other "B" and let the number of fleas on each dog at any instant be $n_A(t)$ and $n_B(t)$. It is assumed that n_A and n_B change only when fleas jump from one dog to the other, so that the total number of fleas, n, is conserved. The master equation for this model gives the change of the single-time probability, $W_1(n_A, t)$, in a time interval dt due to the fleas jumping. It is expressed as

$$\begin{aligned} W_1(n_A, t + dt) = W_1(n_A, t) &+ \text{Prob}\{(n_A - 1) \to n_A, t, dt\} \\ &+ \text{Prob}\{(n_A + 1) \to n_A, t, dt\} - \text{Prob}\{n_A \to (n_A - 1), t, dt\} \\ &- \text{Prob}\{(n_A \to (n_A + 1), t, dt\}, \end{aligned} \tag{4.7.1}$$

where $\text{Prob}\{x \to y, t, dt\}$ is the joint probability of there being x fleas on A at time t and of a jump occuring in the interval $[t, t + dt]$ which changes n_A from x to y. The size of dt is taken small enough that only single jumps can occur. The basic elementary process is A ⇄ B, that is, an "A" flea becomes a "B" flea or *vice versa*.

Consider the second term on the right-hand side of Eq. (4.7.1), which involves $(n_A - 1) \to n_A$. This is the reverse step of the elementary process. Since fleas are required for this transition, the probability that it occurs in dt will be $V^-(n_A - 1)\,dt \equiv (n_B + 1)k^-\,dt = (n - n_A + 1)k^-\,dt$. However, the probability that there are $n_A - 1$ fleas at t is $W_1(n_A - 1, t)$. Thus the joint probability is

$$\text{Prob}\{(n_A - 1) \to n_A, t, dt\} = (n - n_A + 1)k^- W_1(n_A - 1, t)\,dt + \mathcal{O}(dt). \tag{4.7.2}$$

In a similar fashion, the remaining expressions in Eq. (4.7.1) can be written in terms of jumping rates. The fifth term involves the reverse elementary process, while the third and fourth terms correspond to the forward step. Adding all these together and dividing by dt leads to the differential difference equation

$$\begin{aligned} dW_1(n_A, t)/dt = (n_A + 1)k^+ W_1(n_A + 1, t) &- [n_A k^+ + (n - n_A)k^-] W_1(n_A, t) \\ &+ (n - n_A + 1)k^- W_1(n_A - 1, t), \end{aligned} \tag{4.7.3}$$

which is the so-called *master equation* for this process.

Equation (4.7.4) represents a set of coupled ordinary differential equations for the $n + 1$ variables $W_i(t) \equiv W_1(i, t)$. Thus it can be written symbolically as

$$dW_i/dt = \sum_{j=0}^{n} T_{ij} W_j. \tag{4.7.4}$$

From Eq. (4.7.3) we see that

$$\begin{aligned} T_{i,i+1} &= (i + 1)k^+. \\ T_{ii} &= -[ik^+ + (n - i)k^-] \\ T_{i,i-1} &= (n - i + 1)k^-, \end{aligned} \tag{4.7.5}$$

and that jumps of only one unit are possible. From Eq. (4.7.5) it is easily verified that

$$T_{ii} = -(T_{i-1,i} + T_{i+1,i}), \tag{4.7.6}$$

which allows us to write the master equation in its *gain–loss form* as

$$dW_i/dt = \sum_{j \neq i}^{n} (T_{ij} W_j - T_{ji} W_i). \tag{4.7.7}$$

In this form, summing over i on both sides of the equation, it is easy to see that $d\sum_i W_i/dt = 0$, i.e., that the master equation preserves the total probability. The matrix element T_{ij} is positive for $i \neq j$ and is called the *transition probability* from j to i.

Besides being used to represent the jumping of fleas between two dogs, the master equation in Eq. (4.7.3) can be used to describe the elementary chemical reaction $A \rightleftarrows B$. Then one interprets n_A and n_B as the number of A and B isomers, respectively, and the transition probabilities in Eq. (4.7.5) reflect the canonical form of the rates of the elementary process. It is a short step to writing down a master equation for the general elementary process, $\mathbf{n}_\kappa^+ \rightleftarrows \mathbf{n}_\kappa^-$, which forms the basis of the canonical theory. The net change in an elementary process is $\boldsymbol{\omega}_\kappa = \mathbf{n}_\kappa^- - \mathbf{n}_\kappa^+$. If we consider states of a macroscopic system in which the extensive variable has a value $\mathbf{n}$, these states can be reached by a forward transition in the elementary process which begins at $\mathbf{n} - \boldsymbol{\omega}_\kappa$ and by a reverse transition which begins at $\mathbf{n} + \boldsymbol{\omega}_\kappa$. Following the line of reasoning used in obtaining Eq. (4.7.2), we deduce that

$$\text{Prob}\{(\mathbf{n} \pm \boldsymbol{\omega}_\kappa) \to \mathbf{n}, t, dt\} = V_\kappa^{\mp}(\mathbf{n} \pm \boldsymbol{\omega}_\kappa) W_1(\mathbf{n} \pm \boldsymbol{\omega}_\kappa, t)\, dt + \mathcal{O}(dt) \tag{4.7.8}$$

or, equivalently, that

$$\text{Prob}\{\mathbf{n} \to \mathbf{n} \pm \omega_\kappa, t, dt\} = V_\kappa^{\pm}(\mathbf{n}) W_1(\mathbf{n}, t)\, dt + \mathcal{O}(dt). \tag{4.7.9}$$

Equation (4.7.9) is reminiscent of the *ansatz* in Eq. (4.2.6) which underlies our interpretation of the canonical form. In fact, the two interpretations of the canonical form are actually identical. Indeed, the left-hand side of Eq. (4.7.9) is the joint probability of observing $\mathbf{n}$ at time t and $\mathbf{n}' = \mathbf{n} \pm \boldsymbol{\omega}_\kappa$ at time $t + dt$. Thus, according to the definition of the conditional probability in Eq. (1.3.19), if we divide Eq. (4.7.9) by $W_1(\mathbf{n}, t)$, we obtain

$$P_2(\mathbf{n}, t | \mathbf{n}', t + dt) = V_\kappa^{\pm}(\mathbf{n})\, dt + \mathcal{O}(dt) \tag{4.7.10}$$

if $\mathbf{n}' = \mathbf{n} \pm \boldsymbol{\omega}_\kappa$ and

$$P_2(\mathbf{n}, t | \mathbf{n}, t + dt) = 1 - \sum_\kappa [V_\kappa^+(\mathbf{n}) + V_\kappa^-(\mathbf{n})]\, dt + \mathcal{O}(dt). \tag{4.7.11}$$

This is identical to the *ansatz* of the canonical theory given in Eq. (4.2.6).

To obtain the master equation description of the stochastic process defined by Eq. (4.7.9), we need to calculate the change in $W_1(\mathbf{n}, t)$ caused by the elementary jumps. As in the case of the dog–flea model this is

$$\begin{aligned} W_1(\mathbf{n}, t + dt) = W_1(\mathbf{n}, t) &+ \sum_\kappa V_\kappa^+(\mathbf{n} - \boldsymbol{\omega}_\kappa) W_1(\mathbf{n} - \boldsymbol{\omega}_\kappa, t)\, dt \\ &+ \sum_\kappa V_\kappa^-(\mathbf{n} + \boldsymbol{\omega}_\kappa) W_1(\mathbf{n} + \boldsymbol{\omega}_\kappa, t)\, dt \\ &- \sum_\kappa [V_\kappa^+(\mathbf{n}) + V_\kappa^-(\mathbf{n})] W_1(\mathbf{n}, t)\, dt + \mathcal{O}(dt). \end{aligned} \tag{4.7.12}$$

Dividing Eq. (4.7.12) by dt and rearranging gives

$$dW_1(\mathbf{n}, t)/dt = \sum_{\kappa} V_{\kappa}^{+}(\mathbf{n} - \boldsymbol{\omega}_{\kappa}) W_1(\mathbf{n} - \boldsymbol{\omega}_{\kappa}, t) + V_{\kappa}^{-}(\mathbf{n} + \boldsymbol{\omega}_{\kappa}) W_1(\mathbf{n} + \boldsymbol{\omega}_{\kappa}, t)$$
$$- \sum_{\kappa} [V_{\kappa}^{+}(\mathbf{n}) + V_{\kappa}^{-}(\mathbf{n})] W_1(\mathbf{n}, t), \tag{4.7.13}$$

which is the master equation for this process.

A weakness of the master equation formulation of the *ansatz* in Eq. (4.2.6) is that if used uncritically, it can lead to physically meaningless results. An example is furnished by the coupled chemical reactions

$$\begin{aligned} \mathrm{A} + \mathrm{X} &\rightarrow 2\mathrm{X} \\ 2\mathrm{X} &\rightarrow \mathrm{E}, \end{aligned} \tag{4.7.14}$$

where the absence of the reverse arrows implies that the rates of the reverse reactions are being neglected. Assuming that A and E are held in fixed concentrations, the master equation for these reactions involves only the number of X molecules, n_x. The transition probabilities for X governed by these two reactions are

$$\begin{aligned} P_2(n_x, n_x + 1, dt) &= \bar{k}_1 n_x \, dt \\ P_2(n_x, n_x - 2, dt) &= \bar{k}_2 n_x(n_x - 1) \, dt, \end{aligned} \tag{4.7.15}$$

where the constants $\bar{k}_1$ and $\bar{k}_2$ are related to the bimolecular rate constants k_1 and k_2 by $\bar{k}_1 = k_1(n_{\mathrm{A}}/V)$ and $\bar{k}_2 = (k_2/V)$. Notice that in the second of these equations we have written $n_x(n_x - 1)$ instead of n_x^2 as expected from the canonical form and Eq. (4.7.10). This reflects a slightly different choice for the transition probabilities, which is more customary than Eq. (4.7.10). This choice assumes that the transition probabilities are proportional to the number of ways of forming a particular molecular cluster. Thus for two molecules of the same kind one has $P_2 \propto n_x!/2!(n_x - 2)! = n_x(n_x - 1)/2$. Actually this nuance disappears in the thermodynamic limit where $n_x(n_x - 1) \approx n_x^2$.

From Eqs. (4.7.15) and (4.7.13) we obtain the master equation

$$dW_1(n_x, t)/dt = \bar{k}_1(n_x - 1) W_1(n_x - 1, t) + \bar{k}_2(n_x + 2)(n_x + 1) W_1(n_x + 2, t)$$
$$- [\bar{k}_1 n_x + \bar{k}_2 n_x(n_x - 1)] W_1(n_x, t) \tag{4.7.16}$$

for $n_x \geq 3$, and for $n_x = 0$, 1, and 2

$$\begin{aligned} dW_1(0, t)/dt &= 2\bar{k}_2 W_1(2, t) \\ dW_1(1, t)/dt &= 6\bar{k}_2 W_1(3, t) - \bar{k}_1 W_1(1, t) \\ dW_2(2, t)/dt &= \bar{k}_1 W_1(1, t) + 12\bar{k}_2 W_1(4, t) - [2\bar{k}_1 + 2\bar{k}_2] W_1(2, t). \end{aligned} \tag{4.7.17}$$

The transitions, even in this oversimplified model, couple each state indirectly to each other state. Thus the formal solution to this equation in terms of the transition probability matrix, T_{ij}, namely,

$$\mathbf{W}(t) = \exp(Tt)\mathbf{W}(0) \tag{4.7.18}$$

with $W_j(t) \equiv W_1(j, t)$, is not very useful. Nonetheless, it is easy to find the asymptotic probabilities $\lim_{t\to\infty} W_j(t) \equiv W_j^\infty$. They are time independent and satisfy Eqs. (4.7.17) if the left hand side is set equal to zero. Doing this gives

$$\begin{aligned} 0 &= 2\bar{k}_2 W_2^\infty \\ 0 &= 6\bar{k}_2 W_3^\infty - \bar{k}_1 W_1^\infty \\ 0 &= \bar{k}_1 W_1^\infty + 12\bar{k}_2 W_4^\infty - [2k_1 + 2k_2] W_2^\infty \\ 0 &= \bar{k}_1 W_2^\infty + 20\bar{k}_2 W_5^\infty - [\bar{k}_1 + 6\bar{k}_2] W_3^\infty \end{aligned} \tag{4.7.19}$$

Unlike Eqs. (4.7.16), these equations are easy to solve: The first equation clearly implies that $W_2^\infty = 0$. Using this result in the third equation along with the fact that the W_j^∞'s are positive then implies that $W_1^\infty = W_4^\infty = 0$. Since $W_1^\infty = 0$, the second equation gives that $W_3^\infty = 0$, the fourth that $W_5^\infty = 0$, and so on. Hence $W_j^\infty \equiv 0$ for $j \geq 1$. Since the W_j^∞'s are probabilities, they must add up to unity. We conclude therefore, that $W_0^\infty \equiv 1$, and that the asymptotic probability is

$$W_j^\infty = \delta_{j0}. \tag{4.7.20}$$

This rather singular result corresponds to the fact that the state $n_x = 0$ is an absorbing state, i.e., while the second elementary process in Eq. (4.7.14) leads to that state from $n_x = 2$, neither process leads out of $n_x = 0$. If one waits long enough, all the X disappears with probability one.

All this seems plausible. However, if one carries out the analysis of the chemical reaction scheme in Eq. (4.7.14) using the canonical theory, one reaches quite a different conclusion. According to the canonical theory, the conditional average of n_x, $\bar{n}_x$, satisfies

$$d\bar{\rho}_x/dt = \bar{k}_1 \bar{\rho}_x - 2k_2 \bar{\rho}_x^2, \tag{4.7.21}$$

where $\bar{\rho}_x = \bar{n}_x/V$, with V the volume. There are two time-independent solutions of this equation, that is, two solutions with $d\rho_x^\infty/dt = 0$, namely,

$$\rho_x^\infty = \begin{cases} 0 \\ \bar{k}_1/2k_2. \end{cases} \tag{4.7.22}$$

The state $\rho_x^\infty = 0$ corresponds to the average of the singular probability distribution in Eq. (4.7.20). However, according to Eq. (4.7.21) this state is *unstable*. In fact, if the initial value $\rho_x^0 = \rho_x(0)$ is just slightly larger than zero, Eq. (4.7.21) shows that its value grows in time. Indeed,

$$d\bar{\rho}_x/dt \approx \bar{k}_1 \bar{\rho}_x \tag{4.7.23}$$

since the quadratic term is negligible. Thus, for short times, ρ_x^0 increases exponentially,

$$\bar{\rho}_x(t) = e^{\bar{k}_1 t} \rho_x^0. \tag{4.7.24}$$

The state $\rho_x^\infty = \bar{k}_1/2k_2$, on the other hand, is *stable*. Indeed, if ρ_x^0 is close to $\bar{k}_1/2k_2$, we can write $\bar{\rho}_x = \bar{k}_1/2k_2 + \Delta$ with Δ a small quantity. Equation (4.7.21) then implies that

$$d\bar{\rho}_x/dt \approx -\bar{k}_1\Delta = -\bar{k}_1(\bar{\rho}_x - \rho_x^\infty) \tag{4.7.25}$$

since the term quadratic in Δ is negligible. This means that such a deviation relaxes back to $\rho_x^\infty = \bar{k}_1/2k_2$. Thus the canonical interpretation of the transition probabilities leads to the deduction of a Gaussian probability density which, asymptotically, is centered not at $n_x^\infty = 0$ but at $n_x^\infty = V\bar{k}_1/2k_2$.

The difference between these two results, which stem from the same *ansatz* about the transition probabilities, is that in the canonical theory one has taken the limit of a large system. As we discussed in Section 4.2, the canonical form and its interpretation in terms of transition probabilities make sense only for macroscopic systems. For systems containing only a few molecules the stochastic description does not have a thermodynamic character and a more detailed molecular picture needs to be invoked.

To help clarify what the thermodynamic limit means for the master equation, imagine a numerical simulation of the stochastic process. Starting initially with a large number of X molecules, say 10^3, we let a short interval of time dt transpire. Then we choose at random two numbers between zero and one. If the first number falls in the interval $[0, \bar{k}_1 n_x\, dt]$, we change n_x to $n_x + 1$. If the second random number falls in $[0, \bar{k}_2 n_x(n_x - 1)\, dt]$ we change n_x to $n_x - 2$. Then we let another interval of time dt pass and repeat the process again and *ad infinitum*. The results of the simulation are easy to imagine. Since n_x^0 is large, the probability of n_x taking on other values will grow, and in a time $\tau_1 = 1/\bar{k}_1$ it will become a sharp Gaussian centered at $n_x^\infty = V\bar{k}_1/2k_2$. On a much longer time scale, the order of $\tau_2 = \tau_1 e^{V\bar{k}_1/2k_2}$, the systems of the ensemble in the low-n_x tail of the Gaussian will fall into the absorbing state at $n_x = 0$. This creates a new peak near $n_x = 0$ and, if one waits for the period of time τ_2, almost all members of the ensemble will have no X molecules. The important point here is the difference in the two time scales τ_1 and τ_2. For a large system V becomes infinite. This has no effect on $\bar{k}_1$, which is independent of the volume, whereas τ_2 depends strongly on the volume and becomes infinite like e^V. Thus to witness the probability in Eq. (4.7.20) one would have to wait a time of the order of $10^{10^{23}}$ times longer than it takes to achieve the distribution centered at $n_x^\infty = V\bar{k}_1/2k_2$. This is tantamount to the probability in Eq. (4.7.20) being unobservable.

In Sections 3.2 we showed for a large system that the basic *ansatz* of the canonical theory gives rise to a stochastic diffusion process. In Section 3.3 we went on to verify that this stochastic diffusion process reduces to the generalized Ornstein–Uhlenbeck process postulated in the fluctuation–dissipation theory. There is another route which leads to the same result, but which begins with the master equation implementation of the *ansatz*. We illustrate the general argument for the special case of the reactions in Eq. (4.7.20). As a bonus we will obtain a direct verification that the large-system limit of the master equation for these reactions agrees with the analysis based on the fluctuation–dissipation theory.

Since we are contemplating the limit of a large system, we begin by rewriting the master equation (Eq. 4.7.16) in terms of the density, $\rho_x = n_x/V$, using the fact that $W_1(\rho_x, t) = W_1(n_x, t)$. This gives

$$\begin{aligned} V^{-1}\, dW_1(\rho_x, t)/dt &= \bar{k}_1\left(\rho_x - \frac{1}{V}\right) W_1\left(\rho_x - \frac{1}{V}, t\right) \\ &\quad + k_2\left(\rho_x + \frac{2}{V}\right)\left(\rho_x + \frac{1}{V}\right) W_1\left(\rho_x + \frac{2}{V}, t\right) \\ &\quad - \left[\bar{k}_1 \rho_x + k_2 \rho_x\left(\rho_x - \frac{1}{V}\right)\right] W_1(\rho_x, t). \end{aligned} \tag{4.7.26}$$

(We ignore the equations for $n_x = 0$, 1, and 2 which become irrelevant in the large-system limit.) Notice that Eq. (4.7.26) is valid for any single-time probability. Thus, it is valid for the conditional probability $P_2(\rho_x^0 | \rho_x, t)$, which satisfies the special initial condition $P_2(\rho_x^0 | \rho_x, 0) = \delta_{\rho_x^0, \rho_x}$. We expect that this conditional probability is always peaked near the deterministic solution of Eq. (4.7.21), $\bar{\rho}_x$, which begins at ρ_x^0. To extract the probability of deviations from the deterministic solution, we use a systematic expansion in powers of $V^{-1/2}$, called the *Kramers–Moyal expansion.* Anticipating the thermodynamic limit, we introduce first the probability density for the fluctuation $\delta\rho_x = \rho_x - \bar{\rho}_x$ which is

$$P(\delta\rho_x, t) \equiv P_2(\rho_x^0 | \delta\rho_x + \bar{\rho}_x(\rho_x^0, t), t) V.$$

Using the explicit and implicit dependence of P on t, its partial derivative can be written

$$\begin{aligned} \partial P/\partial t &= V(\partial P_2/\partial \rho_x)(\partial \rho_x/\partial t) + V(\partial P_2/\partial t) \\ &= (\partial P/\partial \delta\rho_x)(d\bar{\rho}_x/dt) + V(\partial P_2/\partial t). \end{aligned} \tag{4.7.27}$$

The term $\partial P_2/\partial t$ is given by the right-hand side of Eq. (4.7.26). The Kramers–Moyal expansion consists in expanding the terms in the expression for $\partial P_2/\partial t$ which involve $\rho_x + 2/V$ and $\rho_x - 1/V$ in a Taylor series around ρ_x. Using Eq. (4.7.26) gives

$$V\partial P_2/\partial t = \sum_{j=1}^{\infty} \frac{V^{-(j-1)}}{j!} \frac{\partial^j}{\partial \delta\rho_x^j} \{[(-1)^j \bar{k}_1 \rho_x + 2^j k_2 \rho_x^2 - 2^j V^{-1} k_2 \rho_x] P\}. \tag{4.7.28}$$

For purposes of scaling the fluctuation, we take our lead from the central limit theorem [cf. Eq. (1.3.36)] and define

$$\delta\rho_x = q V^{-1/2}. \tag{4.7.29}$$

Thus $P(q,t) = P(\delta\rho_x, t)V^{-1/2}$, so that combining Eqs. (4.7.28) and (4.7.27) leads to

$$\partial P(q,t)/\partial t = (\partial P/\partial q)(d\bar{\rho}_x/dt)V^{1/2} + \sum_{j=1}^{\infty} \frac{V^{-(j-2)/2}}{j!}\frac{\partial^j}{\partial q^j}\{[(-1)^j \bar{k}_1(\bar{\rho}_x + qV^{-1/2}) + 2^j k_2(\bar{\rho}_x + qV^{-1/2})^2 - 2^j V^{-1} k_2(\bar{\rho}_x + qV^{-1/2})]P\}. \qquad (4.7.30)$$

Separating out the first and second terms in the summation gives

$$\partial P(q,t)/\partial t = -\frac{\partial(qP)}{\partial q}\left\{\frac{[\bar{k}_1(\bar{\rho}_x + qV^{-1/2}) - 2k_2(\bar{\rho}_x + qV^{-1/2})^2] - [\bar{k}_1\bar{\rho}_x - 2k_2\bar{\rho}_x^2]}{qV^{-1/2}}\right\} + \frac{1}{2}\frac{\partial^2 P}{\partial q^2}\{\bar{k}_1\bar{\rho}_x + 4k_2\bar{\rho}_x^2\} + \mathcal{O}(V^{-1/2}). \qquad (4.7.31)$$

This reorganization of the right-hand side of Eq. (4.7.30) into powers of $V^{-1/2}$ is possible because the phenomenological rate constants, $\bar{k}_1$ and k_2, are independent of the volume. Thus taking the limit $V \to \infty$ with $\bar{\rho}_x$ fixed gives

$$\partial P(q,t)/\partial t = -\frac{\partial}{\partial q}H(\rho_x^0, t)qP + \frac{1}{2}\frac{\partial^2}{\partial q^2}(\bar{k}_1\bar{\rho}_x + 4k_2\bar{\rho}_x^2)P, \qquad (4.7.32)$$

where

$$H \equiv \partial[\bar{k}_1\bar{\rho}_x - 2k_2\bar{\rho}_x^2]/\partial\bar{\rho}_x \equiv \bar{k}_1 - 4k_2\bar{\rho}_x. \qquad (4.7.33)$$

Equation (4.7.32) is a Fokker–Planck equation for q which can be easily transformed into an equation for the fluctuation in the extensive variable, $\delta n_x = qV^{1/2}$. The transformation gives

$$\partial P(\delta n_x, t)/\partial t = -\frac{\partial H(\bar{\rho}_x)\delta n_x P}{\partial \delta n_x} + \frac{1}{2}\frac{\partial^2 \gamma(\bar{\rho}_x)P}{\partial \delta n_x^2} \qquad (4.7.34)$$

with

$$\gamma(\bar{\rho}_x) \equiv (\bar{k}_1\bar{\rho}_x + 4k_2\bar{\rho}_x^2)V. \qquad (4.7.35)$$

Equation (4.7.34) is identical in form to the Fokker–Planck equation (Eq. 4.2.7) satisfied by the fluctuation–dissipation theory. We leave it as an exercise to verify that the expressions for H and γ in Eqs. (4.7.33) and (4.7.35) are a particular case of the general expressions in Eqs. (4.2.8) and (4.2.9).

The *ansatz* in Eq. (4.7.10) involves the canonical rate expression, $V_\kappa^{\pm}$, for an elementary process, which is proportional to the transport coefficient, Ω_κ, for that process. These transport coefficients can be expressed as averages over variables—such as molecular velocities—not included in the extensive variables, **n**, that define the physical ensemble. Indeed, as we demonstrate in Chapter 9, the shear viscosity coefficient and the thermal conductivity can be expressed as averages over the collision cross sections which appear in the canonical equations at the Boltzmann level of description. A similar result holds for chemical reaction rate constants. Since the Boltzmann level of description has a thermodynamic character analogous to that of the thermodynamic level, we see, again, that the *ansatz* in Eq. (4.7.10) on which the master equation is based is valid only for large systems. Since the large-system limit of this *ansatz* is the fluctuation–dissipation theory outlined in Section 4.4, we rely on that formulation as the starting point of our investigation of nonequilibrium ensembles.

4.8. Stochastic Diffusion Processes

Another kind of Markovian process used to describe physical ensembles is the stochastic diffusion process discussed in Chapter 1. These processes satisfy Eqs. (1.5.16)–(1.5.18) and generalize the Wiener process is a natural way. As is shown in Section 1.5, the conditional probability density of a stochastic diffusion process satisfies a Fokker–Planck equation of the form

$$\partial P_2(\mathbf{n}^0, t^0 | \mathbf{n}, t)/t = -\partial h_i(\mathbf{n}, t) P_2(\mathbf{n}^0, t^0 | \mathbf{n}, t)/\partial n_i + \frac{1}{2}\partial^2 \gamma_{ij}(\mathbf{n}, t) P_2(\mathbf{n}^0, t^0 | \mathbf{n}, t)/\partial n_i \partial n_j, \tag{4.8.1}$$

where n_i is an extensive variable and, as usual, the summation convention on repreated indices is used. An equation of this form was encountered in our treatment of the thermodynamic limit of the Boltzmann equation in Sections 3.2 and 3.3. We discovered that the basic *ansatz* governing transitions caused by elementary processes did, in fact, lead to a stochastic diffusion process, but only in the thermodynamic limit. This result is quite general and depends only on the fact that the variables n_i are extensive and so are proportional to the size of the system. Because of this, it seems natural to consider formulations of physical ensembles which start from the Fokker–Planck equation (4.8.1).

An equivalent starting point to the Fokker–Planck equation (Eq. 4.8.1) is the Itô stochastic differential equation (Eq. 1.5.13). Defining the matrix $g_{ij} \equiv \gamma_{ij}^{1/2}$ to be the square root of γ_{ij}, i.e., $\gamma_{ij} = g_{ik} g_{kj}$, (which exists because γ_{ij} must be non-negative definite) the equivalent Itô equation becomes

$$d\mathbf{n} = \mathbf{h}(\mathbf{n}, t)\, dt + \gamma^{1/2}\, d\mathbf{w} \qquad \text{(Itô)} \tag{4.8.2}$$

with $d\mathbf{w}$ a vector Wiener process. With the caveat that Eq. (4.8.2) is to be solved using the manipulations of the Itô calculus described below, it can be written suggestively as

$$d\mathbf{n}/dt = \mathbf{h}(\mathbf{n}, t) + \mathbf{f}(t) \tag{4.8.3}$$

with

$$\langle \mathbf{f}(t) \rangle \equiv \mathbf{0}. \tag{4.8.4}$$

Written in the form of Eq. (4.8.3), the Itô equation is often referred to as a *nonlinear Langevin equation.* Actually it is much more complicated than the Langevin equation because of the difficulties in consistently defining integrals over sample paths of the Wiener process. The definition given by Itô, which underlies Eq. (4.8.2), leads to the result that for the scalar Wiener process

$$\int_{t_0}^{t} w\,dw = \frac{w^2(t) - w^2(t_0)}{2} - \frac{(t - t_0)}{2} \qquad \text{(Itô)} \tag{4.8.5}$$

as opposed to the usual calculus which would give

$$\int_{t_0}^{t} w\,dw = \frac{w^2(t) - w^2(t_0)}{2}. \tag{4.8.6}$$

The "extra" term in the Itô calculus comes from the way in which Itô defined integrals over random variables, namely, to evaluate the function at the lower end point of the infinitesimal intervals. Thus the integral of $w\,dw$ can be defined to be the limit as $t_{i+1} - t_i$ goes to zero of

$$\sum_i w(\tau_i)[w(t_{i+1}) - w(t_i)], \tag{4.8.7}$$

where $t_i \le \tau_i \le t_{i+1}$. Itô's stochastic integral results when $\tau_i = t_i$. Another choice is $\tau_i = (t_{i+1} + t_i)/2$. It can be shown that this choice leads to the usual integration formulas, such as Eq. (4.8.6), and was the choice made by Stratonovich.

These multiple definitions of the stochastic integral lead to very different interpretations of an equation like (4.8.3). A helpful relationship in solving Itô equations is provided by *Itô's theorem.* For a scalar stochastic process, n, satisfying the Itô differential expression

$$dn = h(n)\,dt + \gamma^{1/2}\,dw, \qquad \text{(Itô)} \tag{4.8.8}$$

the theorem states that the function $z(n)$ has the differential

$$dz = z'(n)\,dn + \frac{1}{2}z''(n)\gamma\,dt \qquad \text{(Itô)} \tag{4.8.9}$$

with z' and z'' the usual first and second derivatives of $z(n)$. Again an extra term, of order dt, distinguishes the Itô calculus from normal differentiation. For example, let $n = w - t/2$ so that

$$dn = -dt/2 + dw. \qquad \text{(Itô)}$$

Using Itô's theorem we can calculate the differential of $z \equiv \exp(n)$ to be

$$dz = z\,dn + \frac{1}{2}z\,dt \qquad \text{(Itô)}$$

$$= z\left(\frac{-dt}{2} + dw\right) + \frac{1}{2}z\,dt \tag{4.8.10}$$

or

$$dz = z\,dw. \qquad \text{(Itô)} \tag{4.8.11}$$

Thus the solution of the differential equation (4.8.11) with the initial condition $z(0) = 1$ is

$$z = \exp(n) = \exp(w - t/2). \tag{4.8.12}$$

since $w(0) = 0$. If we attempt to solve the Itô differential equation (4.8.11) with the usual calculus, we do not get the right answer, but instead obtain $z = \exp(w)$. There is, however, a differential equation that can be manipulated by the usual calculus—called a Stratonovich stochastic differential equation—whose solution is identical to the Itô solution of Eq. (4.8.11). It is

$$dz = \frac{-z}{2}dt + z\,dw. \qquad \text{(Stratonovich)} \tag{4.8.13}$$

Since the Stratonovich differential satisfies the usual calculus we can write

$$dz = z\left(dw - \frac{dt}{2}\right) \qquad \text{(Stratonovich)} \tag{4.8.14}$$

or

$$d\ln z = d(w - t/2), \tag{4.8.15}$$

which has Eq. (4.8.12) as its solution.

There is a general correspondence between the two types of stochastic differentials. Thus if the Itô differential of the stochastic process, n, is given by Eq. (4.8.8), the same process has the Stratonovich differential

$$dn = \left[h - \frac{\gamma^{1/2}}{2}(\gamma^{1/2})'\right]dt + \gamma^{1/2}\,dw. \qquad \text{(Stratonovich)} \tag{4.8.16}$$

For example, in the Itô equation (4.8.11) $h = 0$ and $\gamma^{1/2} = z \equiv n$. Carrying out the differentiation indicated in Eq. (4.8.16) we find, as expected, that Eq. (4.8.13) is the corresponding Stratonovich equation. Notice that if $\gamma^{1/2}$ is independent of n, then $(\gamma^{1/2})' = 0$ and the two differential expressions are formally identical. Similarly, the Stratonovich differential equation

$$dn = h\,dt + \gamma^{1/2}\,dw \qquad \text{(Stratonovich)} \tag{4.8.17}$$

and the Itô equation

$$dn = \left[h + \frac{\gamma^{1/2}}{2}(\gamma^{1/2})'\right]dt + \gamma^{1/2}\,dw \qquad \text{(Itô)} \tag{4.8.18}$$

describe the same stochastic process. Because of this correspondence, it is immaterial which definition of the stochastic differential one chooses. In this book we have chosen the Itô definition because of the simple correspondence between the Fokker–Planck equation (4.8.1) and the stochastic differential equation (4.8.2). Nonetheless, in attempting to solve an Itô stochastic differential equation, e.g., Eq. (4.8.11), it is often convenient to translate it into the corresponding Stratonovich equation, e.g., Eq. (4.8.13), for which one can apply the usual calculus.

The correct correspondence between the properties of a physical ensemble and a stochastic differential equation requires the specification of only two functions, $\mathbf{h}(\mathbf{n})$ and $\gamma(\mathbf{n})$. The vector-valued function $\mathbf{h}$ is the drift term in the Fokker–Planck equation (4.8.1) and the matrix γ is the diffusion term. Explicit expressions in terms of the transition rates for elementary processes are given in Eq. (3.3.1). There are, however, other possible choices for $\mathbf{h}$ and γ which yield the same thermodynamic limit. To see how these might arise, let us rescale the extensive variables, n_i, by another extensive variable, m, which is not changed by the elementary processes, e.g., m might be the volume. Thus writing $x_i = n_i/m$, Eq. (4.8.1) becomes

$$\partial P_2/\partial t = -\partial \overline{h}_1(\mathbf{x})P_2/\partial x_i + \frac{1}{2}\partial^2\overline{\gamma}_{ij}(\mathbf{x})P_2/\partial x_i \partial x_j, \tag{4.8.19}$$

where $\overline{h}_i = (h_i/m)$ and $\overline{\gamma}_{ij} = (\gamma_{ij}/m^2)$. Making the choices for $\mathbf{h}$ and γ in Eq. (3.3.1), which we will call the *canonical choice*, gives

$$\begin{aligned} \overline{h}_i &= \sum_\kappa \omega_{\kappa i}(v_\kappa^+ - v_\kappa^-) + s_i \\ \overline{\gamma}_{ij} &= m^{-1} \sum_\kappa \omega_{\kappa i}(v_\kappa^+ + v_\kappa^-)\omega_{\kappa j}, \end{aligned} \tag{4.8.20}$$

where the $v_\kappa^\pm$ are intensive, since the intrinsic rates $\Omega_\kappa^\pm$ are extensive. Thus $\overline{h}_i$ does not change as the system size m is increased, if all the ratios n_i/m are held constant. This is the thermodynamic limit which was detailed for the Boltzmann level of description in Section 3.3. From the form of Eq. (4.8.9) it is clear than any other choices of $\overline{\mathbf{h}}$ and $\overline{\gamma}$ which differ by terms of higher order in m^{-1} will lead to the same thermodynamic limit. Thus as long as

$$\begin{aligned} \overline{h}_i' &= \sum_\kappa \omega_{\kappa i}(v_\kappa^+ - v_\kappa^-) + s_i + \mathcal{O}(1/m^0) \\ \gamma_{ij}' &= m^{-1} \sum_\kappa \omega_{\kappa i}(v_\kappa^+ + v_\kappa^-)\omega_{\kappa j} + \mathcal{O}(1/m), \end{aligned} \tag{4.8.21}$$

the thermodynamic limit of the stochastic diffusion process will be identical to the fluctuation–dissipation theory in Section 4.4.

The use of the functions $\overline{h}_i$ and $\overline{\gamma}_{ij}$ in a Stratonovich differential equation has precisely the effect of adding higher-order terms in m^{-1} to the drift and diffusion terms. This is easy to illustrate with a scalar stochastic process, for example, the coupled chemical reactions in Eq. (4.7.14). For this model $\overline{h}$ is given in Eq. (4.7.21), i.e.,

$$\overline{h}(\rho_x) = \overline{k}_1 \rho_x - 2k_2\rho_x^2 \tag{4.8.22}$$

and $\overline{\gamma}$ can be gotten from Eq. (4.7.35),

$$\bar{\gamma}(\rho_x) = V^{-1}(\bar{k}_1 \rho_x + 4k_2 \rho_x^2). \tag{4.8.23}$$

If we describe these reactions by a Stratonovich stochastic diffusion process using $\bar{h}$ and γ, then we have

$$d\rho_x = (\bar{k}_1 \rho_x - 2k_2 \rho_x^2)\, dt + V^{-1/2}(\bar{k}_1 \rho_x + 4k_2 \rho_x^2)^{1/2}\, dw. \qquad \text{(Stratonovich)} \tag{4.8.24}$$

To see what the drift and diffusion terms in the Fokker–Planck equation are for this process, we need to write down the corresponding Itô equation. Using the relationship in Eqs. (4.8.17) and (4.8.18), we obtain

$$d\rho_x = \left[(\bar{k}_1 \rho_x - 2k_2 \rho_x^2) + \frac{V^{-1}}{4}(k_1 + 8k_2 \rho_x) \right] dt \qquad \text{(Itô)}$$

$$+ V^{-1/2}(\bar{k}_1 \rho_x + 4k_2 \rho_x^2)^{1/2}\, dw \tag{4.8.25}$$

Recalling the association between an Itô equation and the Fokker–Planck equation in Eqs. (4.8.1) and (4.8.2), we see that the drift and diffusion terms are

$$\bar{h}'(\rho_x) = \bar{h}(\rho_x) + \frac{V^{-1}}{4}(\bar{k}_1 + 8\bar{k}_2 \rho_x) \tag{4.8.26}$$

$$\bar{\gamma}'(\rho_x) = \bar{\gamma}(\rho_x). \tag{4.8.27}$$

Since the only difference between these expressions and the canonical choice is a term of order V^{-1}, it follows that the Stratonovich method for generating stochastic diffusions agrees with the fluctuation–dissipation theory in the thermodynamic limit.

To summarize: If the canonical choice is made for the drift and diffusion terms in the Fokker–Planck equation, the thermodynamic limit leads to the fluctuation–dissipation theory. Even other choices that are higher order in the system size than the canonical choice give this result. Nonetheless, it is reasonable to ask whether or not other choices for the drift and diffusion terms are possible. Since the macrosocpic transport laws are well confirmed experimentally, it seems sensible to retain the canonical form for the drift term, $\mathbf{h}$. The correct choice for the form of the diffusion term, γ, seems less clear. All that really appears to be necessary is that the choice agree with the linear Onsager theory near equilibrium. There, as shown in Section 4.6, one has

$$\gamma = 2k_B L^s, \tag{4.8.28}$$

where L is the symmetric part of the relaxation matrix defined by

$$d\mathbf{a}/dt = L\mathbf{X}. \tag{4.8.29}$$

Equations (4.8.28) and (4.8.29) suggest another choice for γ. If we write the nonlinear transport equations using the canonical drift term, then we have

$$d\mathbf{n}/dt = \mathbf{h}(\mathbf{n}). \tag{4.8.30}$$

From this we can write a nonlinear analogue of Eq. (4.8.29). Invoking the fact that $X_i \equiv F_i - F_i^e = X_i(\mathbf{n})$, we can invert this functional relationship and recall that $\mathbf{a} = \mathbf{n} - \mathbf{n}^e$ to write Eq. (4.8.30) in the form

$$d\mathbf{a}/dt = \mathbf{h}[\mathbf{n}(\mathbf{X})] \equiv \mathbf{G}(\mathbf{X}) \equiv L(\mathbf{X})\mathbf{X}. \tag{4.8.31}$$

The matrix $L(\mathbf{X})$ is a function of $\mathbf{X}$ (and, possibly, also of $\mathbf{a}$) and can be thought of as a generalization of the Onsager matrix. Taking our lead from Eq. (4.8.28), we can achieve agreement with the equilibrium theory if we define the diffusion matrix to be

$$\gamma' \equiv 2k_B L^s(\mathbf{X}). \tag{4.8.32}$$

This choice seems attractive because the matrix $L(\mathbf{X})$, as defined by Eq. (4.8.31), is completely phenomenological. Hence this choice for the diffusion matrix means that the Fokker–Planck equation is completely determined by the phenomenological quantities $\mathbf{h}$ and L and, thus, is not dependent upon information about molecular processes. Furthermore, for an equilibrium ensemble the matrix L is a constant and the resulting Fokker–Planck equation is the same as that in the Onsager theory.

Although appealing, the definition of the diffusion matrix in Eq. (4.8.32) disagrees with the diffusion matrix which is obtained from the basic *ansatz* in Section 4.3. The isomerization reaction, $A \rightleftarrows B$, examined at the end of Sections 3.6 and 3.7, clearly illustrates the differences. For that reaction the natural extensive variable is the progress variable $\xi = n_B - n_B^e$, and the thermodynamic force is the affinity, $\mathscr{A} = (\mu_A - \mu_B)/T$. The reaction rate is given by Eq. (3.6.12), which leads to the differential equation

$$\begin{aligned} d\xi/dt &= \Omega[\exp(\mu_A/k_B T) - \exp(\mu_B/k_B T)] \\ &= \Omega \exp(\mu_B/k_B T)[\exp(\mathscr{A}/k_B) - 1] \equiv G(\mathscr{A}). \end{aligned} \tag{4.8.33}$$

This reaction involves only a single progress variable, and so the Onsager matrix L becomes a scalar. Applying the definition in Eq. (4.8.31) gives

$$d\xi/dt = L(\mathscr{A})\mathscr{A}, \tag{4.8.34}$$

where

$$L(\mathscr{A}) = \Omega \exp(\mu_B/k_B T)[\exp(\mathscr{A}/k_B) - 1]/\mathscr{A} \tag{4.8.35}$$

Thus using Eq. (4.8.32)

$$\gamma' = \Omega \exp(\mu_B/k_B T)[\exp(\mathscr{A}/k_B) - 1]/(\mathscr{A}/2k_B). \tag{4.8.36}$$

The expression for the diffusion term based on the canonical choice involves the sum of the rates of the elementary processes and is

$$\begin{aligned} \gamma &= \Omega[\exp(\mu_A/k_B T) + \exp(\mu_B/k_B T)] \\ &= \Omega \exp(\mu_B/k_B T)[\exp(\mathscr{A}/k_B T) + 1]. \end{aligned} \tag{4.8.37}$$

These two expressions are clearly different. Their ratio can be expressed as

$$\gamma'/\gamma = \tanh(\mathscr{A}/2k_B)/(\mathscr{A}/2k_B). \tag{4.8.38}$$

At equilibrium $\mathscr{A}/2k_B$ vanishes and the ratio of γ'/γ is unity. Away from equilibrium, however, the ratio begins to deviate significantly from one when $|\mathscr{A}/k_B| > 0.8$. For an ideal system, we can use Eq. (3.6.13) to write the affinity explicitly as

$$\mathscr{A}/k_B = \ln(n_{\mathrm{A}} n_{\mathrm{B}}^e / n_{\mathrm{A}}^e n_{\mathrm{B}}). \tag{4.8.39}$$

Thus already when $\ln(n_{\mathrm{A}}/n_{\mathrm{A}}^e)$ is of the order of 0.5, there will be measurable differences between the canonical choice for γ and the choice based on Eq. (4.8.28).

Although the theoretical description of nonequilibrium ensembles would be greatly simplified if the phenomenological choice of γ' in Eq. (4.8.32) were correct, this appears not to be the case. Indeed, examination of a Hamiltonian model of the isomerization reaction A $\rightleftarrows$ B in a low-density gas leads directly to the master equation for this elementary process. The master equation turns out to be valid on a time scale which is characteristic of the slowest relaxation time of the internal states of A and B and which is long compared to the collision time. The message of this exercise is that a detailed knowledge of molecular mechanism is required to understand the statistical nature of nonequilibrium processes. A simple extension of the ideas of Onsager, based purely on the phenomenological equations, does not provide a general foundation for the statistical description of matter. What is needed is the union of the Onsager picture with the molecular picture of Boltzmann, as described in the first sections of this chapter. Having these tools in hand, we proceed in the remainder of this volume to explore some of the physical and chemical phenomena which they describe.

References

Fluctuation–Dissipation Theory

J. Keizer, A theory of spontaneous fluctuations in macroscopic systems, *J. Chem. Phys.* **63**, 398–403 (1975).

J. Keizer, Dissipation and fluctuations in nonequilibrium thermodynamics, *J. Chem. Phys.* **64**, 1679–1687 (1976).

D. McQuarrie and J.E. Keizer, Fluctuations in chemically reacting systems, in *Theoretical Chemistry: Advances and Perspectives*, Vol. 6A (Academic Press, New York, 1981), pp. 165–213.

Thermodynamic Limit of Stochastic Theories

N.G. van Kampen, A power series expansion on the master equation, *Can. J. Phys.* **39**, 551–565 (1961).

T. Kurtz, Limit theorems for sequences of jump Markov processes approximating ordinary differential equations, *J. Appl. Prob.* **8**, 344–356 (1971).

J. Keizer, Examination of the stochastic process underlying a simple isomerization reaction, *J. Chem. Phys.* **56**, 5775–4783 (1972).

J. Keizer, On the macroscopic equivalence of descriptions of fluctuations for chemical reactions, *J. Chem. Phys.* **18**, 1316–1321 (1977).

J. Keizer and F.J. Conlan, A fine-grained master equation theory of chemical reaction fluctuations, *Physica* **117A**, 405–426 (1983).

Master Equations and Stochastic Differential Equations

N.G. van Kampen, *Stochastic Processes in Physics and Chemistry* (North Holland, Amsterdam, 1981).

L. Arnold, *Stochastic Differential Equations* (Wiley-Interscience, New York, 1974).

R. Kubo, K. Matsuo, and K. Kitahara, Fluctuation and relaxation of macrovariables, *J. Stat. Phys.* **9**, 51–96 (1973).

Dissipation and the Second Law

S. Chandrasekhar, *An Introduction to the Study of Stellar Structure* (University of Chicago Press, Chicago, 1939), Chapter 1.

J. Keizer, On the kinetic meaning of the second law of thermodynamics, *J. Chem. Phys.* **64**, 4466–4474 (1976).

CHAPTER 5

Thermodynamic-Level Description of Chemical, Electrochemical, and Ion Transport Mechanisms

5.1. Ionic Conduction Noise in Solution

One of the few chemical systems for which concentration fluctuations have been measured is the association–disassociation reaction of beryllium and sulfate ions in aqueous solution. Earlier, using conventional fast-reaction techniques, the mechanism of association was deduced to consist of the two elementary reactions

$$Be^{2+}_{aq} + SO^{2-}_{4aq} \rightleftarrows Be^{2+}_{aq}SO^{2-}_{4aq} \tag{5.1.1}$$

$$Be^{2+}_{aq}SO^{2-}_{4aq} \rightleftarrows BeSO_{4aq}. \tag{5.1.2}$$

The first of these reactions is rapid, and the second, in which a water molecule is expelled from an inner coordination shell, is rate limiting for the overall reaction. Thus adding the two reactions we are led to consider the elementary-complex reaction

$$Be^{2+}_{aq} + SO^{2-}_{4aq} \rightleftarrows BeSO_{4aq}. \tag{5.1.3}$$

This reaction has a nice experimental handle, since association of the two ions decreases the number of current carriers in solution. Consequently, a portion of the fluctuations that occur when current is passed through solution will be attributable to the chemical reaction. In 1973 Feher and Weissman used this idea to measure the density–density correlation function for this reaction.

One of the chief obstacles to measuring density fluctuations by means of the fluctuating current is that other molecular processes contribute to current fluctuations. For ions the chief contribution comes from the Brownian motion of the mobile charges as they conduct current through the solution. This leads to voltage or current fluctuations which are called *Johnson noise*. Johnson noise is observed in systems at or near equilibrium and can be understood in terms of the canonical theory. The extensive variables which change in Brownian motion are the momenta of the charge carriers, $\mathbf{p}_i = m_i\mathbf{v}_i$. In an equilibrium ensemble, the linear equation connecting $\mathbf{p}_i$ and its conjugate variable is

$$d\mathbf{p}_i/dt = -L_i\mathbf{v}_i/T + q_i\mathbf{E} + \tilde{\mathbf{f}}_i, \tag{5.1.4}$$

where from Eq. (2.2.3) $\partial S/\partial \mathbf{p}_i = -\mathbf{v}_i/T$, $\mathbf{E}$ is the electric field, and q_i is the charge of the ith ion. The random force differs for each ion and for dilute solution is uncorrelated between different ions. Thus using Eq. (2.6.11)

$$\langle \tilde{\mathbf{f}}_i(t) \rangle = \mathbf{0} \tag{5.1.5}$$

$$\langle \tilde{\mathbf{f}}_i(t)\tilde{\mathbf{f}}_j^T(t') \rangle = 2k_B T\zeta_i I \delta_{ij}\delta(t - t'), \tag{5.1.6}$$

where I is the 3×3 identity matrix and we have introduced the friction coefficient, $\zeta_i \equiv L_i/T$.

To obtain the electric current density associated with the positive and negative ions, we need to calculate the flux density of ions and multiply the result by their charge. If there is one kind of positive ion of charge q_+ and one kind of negative ion of charge q_-, then the current densities are

$$\mathbf{j}_\pm = q_\pm \sum_{j=1}^{n_\pm} \mathbf{v}_j^{(\pm)}/V, \tag{5.1.7}$$

where the sum is over all the charges of the indicated kind in some small volume, V. If the volume has a cross-sectional area A perpendicular to the current and extends a length l along the direction of average flow, then the electric currents are defined by $\mathbf{i}_\pm = \mathbf{j}_\pm A$; or

$$\mathbf{i}_\pm = (q_\pm/l) \sum_{j=1}^{n_\pm} \mathbf{v}_j^{(\pm)}. \tag{5.1.8}$$

Using the fact that $\mathbf{p}_i^{(\pm)} = m_\pm \mathbf{v}_i^{(\pm)}$, the Langevin equation (5.1.4) gives the following differential equation for the total current $\mathbf{i} = \mathbf{i}_+ + \mathbf{i}_-$,

$$d\mathbf{i}/dt = -\lambda \mathbf{i} + \left[\frac{(q_+^2 n_+/m_+) + (q_-^2 n_-/m_-)}{l}\right]\mathbf{E} + \frac{q_+}{m_+ l}\sum_j \tilde{\mathbf{f}}_j^{(+)}$$
$$+ \frac{q_-}{m_- l}\sum_j \tilde{\mathbf{f}}_j^{(-)}. \tag{5.1.9}$$

In this equation $n_\pm$ are the number of positive and negative ions in the volume $V = Al$ and

$$\lambda = \frac{t_+\zeta_+}{m^+} + \frac{t_-\zeta_-}{m_-}. \tag{5.1.10}$$

The numbers $t_\pm$ give the fraction of the total current carried by the two kinds of ions and are defined by

$$\mathbf{i}_\pm = t_\pm \mathbf{i}. \tag{5.1.11}$$

Introducing the *conductivity*,

$$\sigma = [(q_+^2 n_+/m_+) + (q_-^2 n_-)/m_-]/l\lambda \equiv \sigma_+ + \sigma_-, \tag{5.1.12}$$

Eq. (5.1.9) can be written

$$d\mathbf{i}/dt = -\lambda(\mathbf{i} - \sigma\mathbf{E}) + \tilde{\mathbf{f}}, \tag{5.1.13}$$

where the random term is

$$\tilde{\mathbf{f}} = (q_+/m_+ l)\sum_j \tilde{\mathbf{f}}_j^{(+)} + (q_-/m_- l)\sum_j \tilde{\mathbf{f}}_j^{(-)}. \tag{5.1.14}$$

Since each of the terms $\tilde{\mathbf{f}}_j^{(\pm)}$ is a white noise and satisfies Eqs. (5.1.5) and (5.1.6), $\tilde{\mathbf{f}}$ is also a white noise with the covariance

$$\langle\tilde{\mathbf{f}}(t)\tilde{\mathbf{f}}^T(t')\rangle = 2k_B T[(q_+^2/m_+^2 l^2)n_+\zeta_+ + (q_-^2/m_-^2 l^2)n_-\zeta_-]I\delta(t - t'). \tag{5.1.15}$$

Equations (5.1.13)–(5.1.15) imply that, on the average, the electric current changes from an initial value $\mathbf{i}(0)$ to a value at time t exponentially, i.e.,

$$\bar{\mathbf{i}}(t) = \sigma\mathbf{E} + \exp(-\lambda t)(\mathbf{i}(0) - \sigma\mathbf{E}). \tag{5.1.16}$$

Since λ is positive, $\bar{\mathbf{i}}$ asymptotically approaches the steady current $\mathbf{i}^{ss} = \sigma\mathbf{E}$. The time scale on which this occurs is $\tau = 1/\lambda$. The size of τ can be estimated using Eq. (5.1.10) and the Nernst–Einstein formula for the friction constant in Eq. (1.4.16). For small ions in solution, the diffusion constant D is about 10^{-5} cm^2 s^{-1} and masses m are of the order of 10^{-22} g so that

$$\tau \approx m/\zeta = mD/k_B T \approx 2 \times 10^{-14}\text{ s} \tag{5.1.17}$$

since at room temperature $k_B T = 4 \times 10^{-14}$ erg. This time scale is the order of the time between collisions in solution and means that the steady-state current is achieved almost instantaneously.

To examine the fluctuations around this steady current we substitute

$$\Delta\mathbf{i}(t) = \mathbf{i}(t) - \mathbf{i}^{ss} \tag{5.1.18}$$

into Eq. (5.1.13), which gives

$$d\Delta\mathbf{i}/dt = -\lambda\Delta\mathbf{i} + \tilde{\mathbf{f}}, \tag{5.1.19}$$

with $\tilde{\mathbf{f}}$ a white noise. Because Eq. (5.1.19) is linear, it follows that $\Delta\mathbf{i}$ is a stationary, Gaussian, Markov process. Using this fact we can rely on the results in Section 1.8 to analyze the current fluctuations.

Experimentally it is possible to examine the continuous record of current fluctuations after the average steady current has been attained. This record can be filtered and correlated by electronic devices to obtain the spectral density and the time correlation function or its Fourier transform. For current fluctuations, the spectral density is related to the power that is dissipated by the current at a given frequency. To see this, recall the definition of the spectral density in Eq. (1.8.34), i.e.,

$$\langle\Delta\mathbf{i}(t)\Delta\mathbf{i}^T(t)\rangle = \int_{-\infty}^{+\infty} S(\omega)\,d\omega/2\pi. \tag{5.1.20}$$

If we concentrate on the scalar current, $i(t)$, along the direction of the applied electric field, then we can use the resistance, R, to write

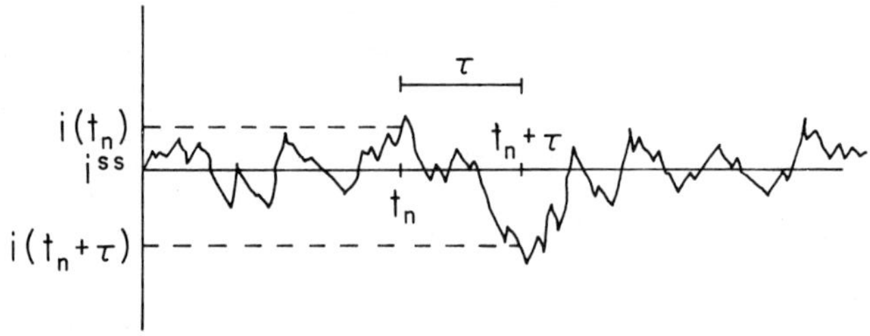

FIGURE 5.1. Schematic representation of a current record as a function of time in a steady-state ensemble. The value of the time correlation function of the current fluctuations at time τ can be obtained from this record by measuring deviations $i(t_n) - i^{ss}$ and $i(t_n + \tau) - i^{ss}$ for a series of times, t_n. The time correlation function at τ is obtained by averaging the product of the deviations as indicated in Eq. (5.1.22).

$$\langle R\Delta i^2(t)\rangle = \int_{-\infty}^{+\infty} [RS(\omega)/2\pi]\, d\omega. \tag{5.1.21}$$

The left-hand side of Eq. (5.1.21) is the average power dissipated by the fluctuating current, while the right-hand side is the spectral resolution of the dissipated power into contributions from various frequencies. Thus $RS(\omega)/2\pi$ is literally the *power spectrum* of the current.

For a stationary current the process of obtaining the power spectrum and the two-time correlation function is straightforward. Consider the time record shown in Fig. 5.1. One chooses a sequence of times, t_n, which will be used as initial current observations, $i(t_n)$. Then setting down a ruler of length τ, the currents at $t = t_n + \tau$ are measured. Taking the product $\Delta i(t_n)\Delta i(t_n + \tau)$ and averaging over a sufficient number of observations gives

$$\lim_{N\to\infty} \sum_{n=1}^{N} \Delta i(t_n)\Delta i(t_n + \tau)/N = \langle \Delta i(t)\Delta i(t + \tau)\rangle. \tag{5.1.22}$$

The spectral density can be obtained using the same data. In fact, the Wiener–Khintchine theorem in Eq. (1.8.42) shows that the spectral density and the time correlation function are Fourier transforms of one another. Thus by performing the spectral analysis of the data in Eq. (5.1.22) with respect to τ, one obtains the power spectrum of the current. Both the correlation of the current and the Fourier transform can be performed automatically using electronic circuitry.

According to the analysis in Section 1.8, the power spectrum for an Ornstein–Uhlenbeck process is given by Eq. (1.8.36). If we consider only current fluctuations along the direction of the applied electric field, then Eq. (5.1.19) implies that

$$d\Delta i/dt = -\lambda\Delta i + \tilde{f}. \tag{5.1.23}$$

The spectral density of $\Delta i(t)$ is the Lorentzian given by Eq. (1.8.37), i.e.,

$$S(\omega) = \gamma/(\omega^2 + \lambda^2), \tag{5.1.24}$$

where γ can be read off from Eq. (5.1.15),

$$\gamma = 2k_B T[(q_+^2/m_+^2 l^2)n_+\zeta_+ + (q_-^2/m_-^2 l^2)n_-\zeta_-]. \tag{5.1.25}$$

Another expression of this result is the relationship in Eq. (1.8.45) between the power spectrum and the frequency correlation function averaged over a small band width of frequencies,

$$\begin{aligned}\langle \Delta i^2(\omega) \rangle \Delta\omega &= S(\omega)/2\pi \\ &= \gamma/(\omega^2 + \lambda^2)2\pi. \end{aligned} \tag{5.1.26}$$

As we have seen, λ is of the order of 10^{14} s^{-1}. For typical experimental frequencies, which are well below 10^{12} Hz, Eq. (5.1.26) reduces to

$$\langle \Delta i^2(\omega) \rangle \Delta\omega = \gamma/\lambda^2 2\pi. \tag{5.1.27}$$

Thus the power spectrum of the current noise appears to be independent of frequency, that is, the noise is white. Equation (5.1.27) is called the *Nyquist formula* and historically was the progenitor of the fluctuation–dissipation theorem.

The Nyquist formula can be written in a somewhat more transparent form using the expressions for λ and γ given in Eqs. (5.1.10) and (5.1.25). To simplify the right-hand side of Eq. (5.1.27), it is necessary to introduce explicit expressions for the current fractions, $t_\pm$, defined by Eq. (5.1.11). At steady state the values of $t_\pm$ can be obtained from the condition $\mathbf{i}^{ss} = \sigma\mathbf{E}$, since the definition of σ in Eq. (5.1.12) shows that the conductivity due to the two charge carriers is additive. Thus

$$t_\pm = \sigma_\pm/\sigma = (q_\pm^2 n_\pm/m_\pm)/[(q_+^2 n_+/m_+) + (q_-^2 n_-/m_-)] = r_\pm/(r_+ + r_-), \tag{5.1.28}$$

where $r_\pm = |q_\pm/m_\pm|$, the charge to mass ratio, and the second equality follows from overall charge neutrality which implies that $q_+n_+ + q_-n_- = 0$. Substituting these expressions for $t_\pm$ into the definition of λ in Eq. (5.1.10) yields

$$\lambda = [(r_+\zeta_+/m_+) + (r_-\zeta_-/m_-)]/(r_+ + r_-). \tag{5.1.29}$$

Using the charge to mass ratios and electroneutrality, Eq. (5.1.25) for γ can be rewritten

$$\begin{aligned}\gamma &= \frac{2k_B T q_+ n_+}{l^2}(r_+\zeta_+/m_+ + r_-\zeta_-/m_-) \\ &= \frac{2k_B T q_+ n_+ \lambda(r_+ + r_-)}{l^2}, \end{aligned} \tag{5.1.30}$$

so that the right-hand side of Eq. (5.1.27) becomes

$$\gamma/\lambda^2 2\pi = k_B T q_+ n_+ (r_+ + r_-)/\pi\lambda l^2. \tag{5.1.31}$$

Finally noting that in this notation the electric conductivity in Eq. (5.1.12) becomes

$$\sigma = q_+ n_+ (r_+ + r_-)/\lambda l, \tag{5.1.32}$$

we obtain

$$\langle \Delta i^2(\omega) \rangle \Delta\omega = k_B T \sigma/\pi l = k_B T/\pi R, \tag{5.1.33}$$

where $R = l/\sigma$ is the resistance defined by Ohm's law, i.e.,

$$R i^{ss} = \phi \tag{5.1.34}$$

with $\phi = |\mathbf{E}| l$ the voltage drop across the length l parallel to the current.

The traditional form of the Nyquist formula involves the voltage fluctuations. Relating the current and voltage by Ohm's law, one has $R\Delta i = \Delta\phi$. Thus Eq. (5.1.33) can be rewritten

$$\langle \Delta\phi^2(\omega) \rangle \Delta\omega = k_B T R/\pi. \tag{5.1.35}$$

In this form the magnitude of the voltage fluctuations in the frequency band ω are seen to be proportional to the resistance of the solution, the absolute temperature, and Boltzmann's constant. These observations were first made experimentally by Johnson. His experiments, and Nyquist's interpretation of them, were published in the same issue of the *Physical Review* in 1928. Some of Johnson's data relating the proportionality of the voltage fluctuations to the resistance are shown in Fig. 5.2. Using these data Johnson deduced that

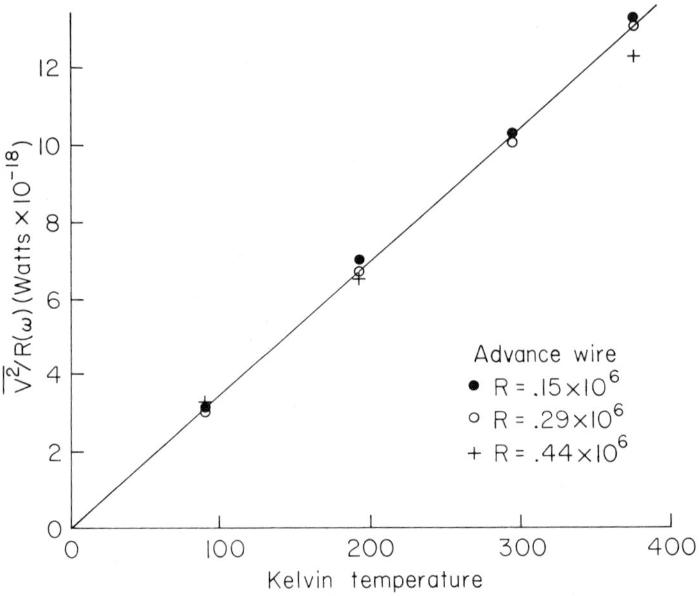

FIGURE 5.2. Johnson's measurements of voltage fluctuations at low frequency as a function of the Kelvin temperature. From J.B. Johnson, *Phys. Rev.* **32**, 97 (1928).

Boltzmann's constant was $(1.27 \pm .17) \times 10^{-16}\,\mathrm{erg} \cdot \mathrm{K}^{-1}$, which is within experimental error of the accepted value of $1.38 \times 10^{-16}\,\mathrm{erg} \cdot \mathrm{K}^{-1}$.

5.2. The Feher–Weissman Experiment

Johnson noise, as we have seen, is caused by the Brownian motion of charge carriers. This noise has a flat spectrum and can be observed even in an equilibrium ensemble in which there is no current on the average. In an equilibrium ensemble current fluctuations have no component due to fluctuations in the number of charge carriers. Indeed, the expressions for the current noise obtained in Section 5.1 are conditional on the number of charge carriers being fixed. However, Eqs. (5.1.32) and (5.1.33) show that the spectral density, conditioned on fixed n_+ and n_-, is proportional to n_+. Thus after averaging over n_+, the value of n_+ can be replaced by its equilibrium average. Consequently, fluctuations in the number of charge carriers do not contribute to the current noise at equilibrium.

The situation is different when a net current is flowing. If we consider only current in the direction of the applied field, then the current can be written as

$$i = \Delta i + (\sigma/l)\phi, \tag{5.2.1}$$

where Δi is the Johnson noise and $\sigma/l = R^{-1}$. Using Eq. (5.1.32) the conductivity per unit length can be expressed as

$$\sigma/l = n_+/r, \tag{5.2.2}$$

where r^{-1} is the conductivity for a pair of charge carriers per unit length and is independent of n_+. Hence when there is an applied voltage, the frequency dependence of the current fluctuations is given by

$$\delta i(\omega) = \Delta i(\omega) + \delta n_+(\omega)\phi/r. \tag{5.2.3}$$

Since the Johnson noise and fluctuations in the number of charge carriers are statistically independent, Eq. (5.2.3) implies that

$$\langle \delta i^2(\omega) \rangle \Delta\omega = \langle \Delta i^2(\omega) \rangle \Delta\omega + (\phi/r)^2 \langle \delta n_+^2(\omega) \rangle \Delta\omega. \tag{5.2.4}$$

In other words, when a net current is flowing, fluctuations in the number of charge carriers contribute to the power spectrum independently of the Johnson noise.

Feher and Weissman used this independence to study the charge carrier fluctuations caused by the association of Be^{2+} and SO_4^{2-} ions in aqueous solution. As discussed in Section 5.1, this process is governed by the elementary complex reaction

$$Be^{2+}_{aq} + SO^{2-}_{4aq} \rightleftarrows BeSO_{4aq}. \tag{5.2.5}$$

However, when a steady net current is flowing, the resulting ensemble will not be an equilibrium ensemble. Moreover, the steady current involves electrolysis of the electrolyte solution which leads to a steady increase of OH^- ions and depletion of water. Yet for a current small enough, this does not create a significant change in the solution. Since the Brownian motion of the Be^{2+} and SO_4^{2-} ions is only weakly coupled to their number fluctuations, the ensemble appears to be in equilibrium as far as the number fluctuations are concerned. Thus the charge carrier fluctuations can be analyzed using the equilibrium version of the canonical theory.

The canonical equations that govern the charge carrier fluctuations are easily written down. The stoichiometry of the reaction implies that $n_+ + n = n_0$, where n is the number of $BeSO_4$ molecules in solution and n_0 is the initial amount of $BeSO_4$ from which the solution is made. Since electroneutrality implies that $n_+ = n_-$, it follows that there is only a single independent extensive variable. As we are interested in fluctuations in n_+, we choose it as the independent variable. Thus the canonical equation is

$$d\bar{n}_+/dt = -\Omega\{\exp[(\bar{\mu}_+ + \bar{\mu}_-)/k_B T] - \exp(\bar{\mu}/k_B T)\}, \tag{5.2.6}$$

where μ is the chemical potential of $BeSO_4$ in solution and

$$d\delta n_+/dt = H\delta n_+ + \tilde{f} \tag{5.2.7a}$$

$$\langle \tilde{f}(t)\tilde{f}(t')\rangle = \Omega\{\exp[(\bar{\mu}_+ + \bar{\mu}_-)/k_B T] + \exp(\bar{\mu}/k_B T)\}\delta(t - t'). \tag{5.2.7b}$$

The average equilibrium state is determined by the Lewis equation (Eq. 3.8.7) which, for this reaction, involves the activities of the ions, $a_\pm$, and the neutral molecule, a:

$$\frac{a_+^e a_-^e}{a^e} = \exp(-\Delta G^0/k_B T). \tag{5.2.8}$$

Introducing the *molar* density-based activity coefficients, γ_i, defined by

$$\mu_i = \mu_i^0 + k_B T \ln(\gamma_i \rho_i), \tag{5.2.9}$$

Eq. (5.2.8) can be rewritten

$$\rho_+^e \rho_-^e/\rho^e = (\gamma^e/\gamma_+^e \gamma_-^e)\exp(-\Delta G^0/k_B T) \equiv K. \tag{5.2.10}$$

The solution used by Feher and Weissman was made up from 0.03 moles of beryllium sulfate per liter. Using the measured value of $K = 0.45\ M$, Eq. (5.2.10) and the conservation conditions imply that $\rho_+^e = \rho_-^e = 0.029\ M$ and $\rho^e = 0.002\ M$. Thus under these conditions beryllium sulfate is mostly dissociated.

Fluctuations in the number of Be^{+2} ions are determined by Eq. (5.2.7a). The relaxation rate H is obtained by linearizing Eq. (5.2.6) around equilibrium. To do so we rewrite it in the form

$$d\bar{n}_+/dt = -V(k^+\bar{\rho}_+\bar{\rho}_- - k^-\bar{\rho}). \tag{5.2.11}$$

where k^+ and k^- are the molar density-based kinetic constants defined by

$$k^+ = \Omega \exp[(\mu_+^0 + \mu_-^0)/k_B T]\gamma_+\gamma_-$$
$$k^- = \Omega \exp(\mu^0/k_B T)\gamma. \tag{5.2.12}$$

Neglecting the dependence of k^+ and k^- on ρ_+ (through the activity coefficients), linearization of Eq. (5.2.11) gives

$$H = -(2k^+\rho_+^e + k^-) \tag{5.2.13}$$

since the conservation relations give $\delta n_+ = \delta n_-$ and $\delta n_+ = -\delta n$. Equations (5.2.12) show that $k^-/k^+ = K$. Thus Eq. (5.2.13) can be rewritten as

$$H = -k^-[(2\rho_+^e/K) + 1]. \tag{5.2.14}$$

Detailed balance at equilibrium permits the variance of the random force in Eq. (5.2.7b) to be written

$$\langle \tilde{f}(t)\tilde{f}(t')\rangle = 2Vk^+\rho_+^{e2}\delta(t - t'). \tag{5.2.15}$$

The equilibrium fluctuation–dissipation theorem in Eq. (4.6.21) provides a way of determining the equilibrium variance of the Be^{2+} ions, σ, namely,

$$\sigma = -\gamma/2H, \tag{5.2.16}$$

where γ is the strength of the random term given in Eq. (5.2.15). Combining the preceding three equations yields

$$\langle(\delta n_+)^2\rangle = V\rho_+^{e2}/(2\rho_+^e + K). \tag{5.2.17}$$

The frequency dependence of the number fluctuations in Be^{2+} can be obtained from the version of the Wiener–Khintchine theorem in Eq. (1.8.44), i.e.,

$$\begin{aligned}\langle \delta n_+^2(\omega)\rangle\Delta\omega &= S(\omega)/2\pi \\ &= \gamma/2\pi(\omega^2 + H^2)\end{aligned} \tag{5.2.18}$$

where in the second equality the form of the spectral density for a scalar Ornstein–Uhlenbeck process obtained in Eq. (1.8.37) was used. Substituting the explicit expression for γ from Eq. (5.2.15) into Eq. (5.2.18) gives

$$\langle \delta n_+^2(\omega)\rangle\Delta\omega = Vk^+\rho_+^{e2}/\pi(\omega^2 + H^2). \tag{5.2.19}$$

Since Feher and Weissman measured voltage fluctuations, it is convenient to reexpress the results for current fluctuations in terms of the voltage. Using the expression $R\delta i = \delta\phi$ and the results in Eqs. (5.1.35) and (5.2.19), Eq. (5.2.4) can be rewritten

$$\langle \delta\phi^2(\omega)\rangle\Delta\omega = (k_B \mathrm{Tr}/V\pi\rho_+^e) + \phi^2 k^+/V\pi(\omega^2 + H^2), \tag{5.2.20}$$

where we have used the fact that $R = r/n_+$. Equation (5.2.20) predicts that the voltage fluctuations in a beryllium sulfate solution involve additive contributions from the Johnson noise and the number fluctuations. The Johnson noise is a frequency-independent background, which is also independent of the applied voltage and inversely proportional to the volume. The magnitude of fluctuations caused by the chemical reaction, on the other hand, is proportional to the square of the voltage and inversely proportional to the volume. If H is comparable to ω, the number fluctuations should have a measurable Lorentzian frequency spectrum. In fact, the relaxation rate, $-H$, as obtained by conventional techniques is 375 s^{-1} at a temperature of 298 K. This corresponds to a Lorentzian half-width at half-height of $\nu_{1/2} = H/2\pi = 60\ s^{-1}$, which is well within the range of experimentally observable frequencies.

To enhance the voltage fluctuations, Feher and Weissman utilized a conduction cell in which two large vessels were separated by a capillary 580 μm in length and 60 μm in diameter. Equation (5.2.20) shows that a small volume increases the magnitude of both the Johnson noise and the noise due to number fluctuations. Except for small corrections at the end of the capillary, the resistance of the conduction cell is caused by ionic conduction through the capillary. After carefully eliminating other sources of noise, the frequency spectra shown in Fig. 5.3 were obtained. The spectra were recorded electronically using an amplifier and frequency analyzer after subtracting off the average cell voltage. The lower spectrum was taken with zero applied voltage and is typical of Johnson noise. When a voltage of the order of 30 V was applied, the noise due to number fluctuations becomes visible as a Lorentzian peak superimposed on the Johnson noise, as predicted by Eq. (5.2.20). The half-width at half-height, $\nu_{1/2}$, is a function of both the temperature and the concentration. The temperature dependence of $\nu_{1/2}$ was measured in the range of 283–313 K. The characteristic Arrhenius dependence on the temperature is exhibited with an activation energy $E_a = 8.2 \pm 0.4$ kcal$\cdot$mol^{-1}, in good agreement with the value obtained from Eq. (5.2.20) using kinetic constants obtained by conventional techniques.

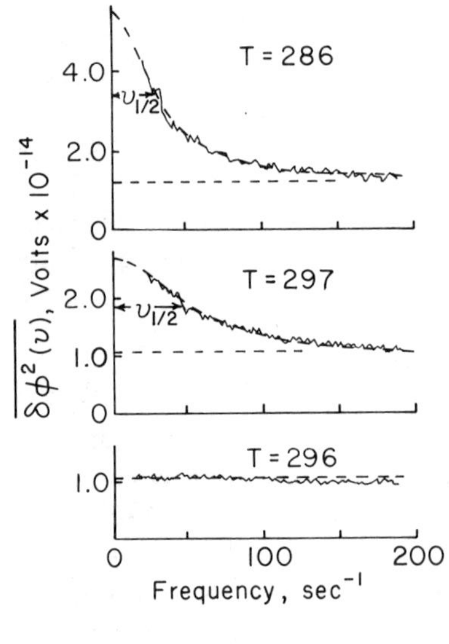

FIGURE 5.3. Voltage fluctuations for the association reaction of beryllium sulfate in water as measured by Feher and Weissman. The lower panel is the power spectrum in the absence of current and represents the Johnson noise. The upper two panels show the Lorentzian power spectrum in the presence of current with the half-width at half-height indicated by $\nu_{1/2}$. Data taken from G. Feher and M. Weissman, *Proc. Nat. Acad. Sci. U.S.A.* **70**, 870 (1973).

In addition to showing that number fluctuations caused by chemical reactions are accessible to measurement, the Feher–Weissman experiment validates the application of the canonical fluctuation theory to chemical reaction fluctuations at equilibrium. In the canonical theory, which reduces to the Onsager theory at equilibrium, fluctuations are small except near critical points. Indeed, the magnitude of the voltage fluctuations can be judged from the power spectra in Fig. 5.3. According to the Wiener–Khintchine theorem, the power spectrum gives the variance of the voltage fluctuations in a frequency band $\Delta\omega$. Thus at a given frequency its square root measures the width of the Gaussian probability density of voltage fluctuations in the frequency band $\Delta\omega$. Even at an applied voltage of 30 V, Fig. 5.3 shows that this width is of order 2×10^{-7} V, i.e., about one-hundred-millionth of the size of the applied voltage itself. Another lesson from the Feher–Weissman experiment is that elementary processes which occur on different time scales are effectively uncoupled from one another. Thus the Johnson noise, which is caused by the Brownian motion of the charge carriers and relaxes on the picosecond time scale, is independent of the number fluctuations which occur on a millisecond time scale. This clear separation of time scales occurs for many elementary processes and is the physical reason behind the success of the highly contracted thermodynamic level of description introduced in Chapter 2.

5.3. The General Linear Mechanism

The simplest class of elementary processes for which number fluctuations can be characterized involves linear reaction mechanisms. These reactions generalize the isomerization reaction, A ⇄ B, and involve only unimolecular elementary reactions. A typical example is the isomerization of *n*-propyl fluoride, which has three rotational isomers, the *gauche* and *trans* forms drawn below:

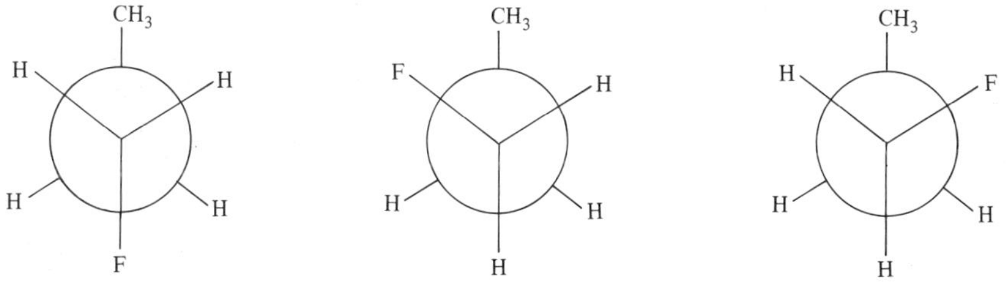

trans *gauche*

These isomers interconvert by passing over a barrier to internal rotation of about 0.05 eV. The energy required for this rotational motion is provided by collisions with other molecules and may be funneled into internal rotation from other degrees of freedom. Thinking of rotational isomers as different molecules is sensible because experimental techniques, such as nuclear magnetic resonance and microwave spectroscopy, allow the *gauche* and *trans* isomers to be distinguished. Extending this idea, it is possible to consider other internal states to provide a label for identifying molecules. For example, the diatomic molecule Br_2 has a vibrational level spacing which is 60% large than $k_B T$ at room temperature. Thus the first three vibrational levels are appreciably populated at room temperature and the vibrational state can be used as a label for differentiating Br_2 molecules. Many linear reaction mechanisms involve changes which are less subtle than a change in internal state. For example, the α and β anomers of D-glucose are six-membered rings of five carbons and one oxygen, whereas the aldehyde form of D-glucose is a linear molecule with an oxygen doubly-bonded to a carbon atom. In aqueous solution these molecules interconvert and are easily distinguished since only the aldehyde form can be reduced at a mercury electrode.

Despite these differences, linear reaction mechanisms share the characteristic that only a single molecule is involved in any elementary reaction, κ. In terms of the numbers $n_{\kappa i}^{\pm}$, this means that for each reaction $n_{\kappa i}^{+}$ is one for only one value of i and zero for other values; similarly $n_{\kappa j}^{-}$ is one for only one value of j and zero for other values or, symbolically, $i_\kappa \rightleftarrows j_\kappa$. Thus the net changes $\omega_{\kappa i} = n_\kappa^- - n_\kappa^+$ have the values -1, $+1$, or 0 depending on whether or not the molecule i is lost, created, or doesn't participate in the reaction κ. Using the canonical equations in Section 4.2 the rate of change in the average number of molecules of kind i is

$$d\bar{n}_i/dt = \sum_\kappa \omega_{\kappa i}\Omega_\kappa[\exp(\bar{\mu}_{i_\kappa}/k_B T) - \exp(\bar{\mu}_{j_\kappa}/k_B T)]. \tag{5.3.1}$$

There is a more convenient way to write Eq. (5.3.1) which involves pairs of molecules i and j. This is possible since each elementary reaction κ involves only two molecules. Making this explicit in the sum, Eq. (5.3.1) can be reexpressed as

$$d\bar{n}_i/dt = \sum_j \Omega_{ij}[\exp(\bar{\mu}_j/k_B T) - \exp(\bar{\mu}_i/k_B T)], \tag{5.3.2}$$

where $\Omega_{ij} = \Omega_{ji} = \sum_\kappa \Omega_\kappa$ with the sum restricted to those elementary processes κ of the form $i \rightleftarrows j$. Using Eq. (5.2.9) to introduce activity coefficients, γ_i, and number densities, ρ_i, the canonical equation, Eq. (5.3.2), can be written more conventionally as

$$d\bar{\rho}_i/dt = \sum_{j \neq i} [k_{ij}\bar{\rho}_j - k_{ji}\bar{\rho}_i], \tag{5.3.3}$$

where the sum over j excludes i,

$$k_{ij} = V^{-1}\Omega_{ij}\gamma_j \exp(\mu_j^0/k_B T), \tag{5.3.4}$$

and V is the volume.

Linear mechanisms conserve the total number of molecules, $n_T = \sum_i n_i$. This can be seen using Eq. (5.3.1) by summing over the subscript i on both sides of the equation and noticing the exact cancellation of terms on the right-hand side. This can also be seen from the canonical equation (5.3.1) using the fact that the number of atoms does not change in any of the elementary processes. Thus if ψ_{Ai} is the number of A atoms in molecule i, $\psi_{Ai} = \psi_{Aj}$ and the atom conservation condition discussed in Eq. (3.7.2) reduces to

$$\sum_i \omega_{\kappa i} = 0. \tag{5.3.5}$$

Consequently, summing over i in Eq. (5.3.1) also leads to conservation of the total number of molecules. This conservation condition is responsible for the simple gain–loss form of Eq. (5.3.3). It is convenient to recognize this explicitly and define the matrix

$$W_{ij} = \begin{cases} k_{ij} & i \neq j \\ -\sum_{l \neq i} k_{li} & i = j, \end{cases} \tag{5.3.6}$$

so that Eq. (5.3.3) can be written in matrix notation as

$$d\overline{\boldsymbol{\rho}}/dt = W\overline{\boldsymbol{\rho}}. \tag{5.3.7}$$

Assuming for simplicity that the activity coefficients in Eq. (5.3.4) are constant, the solution of Eq. (5.3.7) is

$$\overline{\boldsymbol{\rho}}(t) = \exp(Wt)\boldsymbol{\rho}^0. \tag{5.3.8}$$

The physical interpretation of the matrix elements W_{ij} depends on the elementary processes which make up the linear mechanism. If the reactions are overtly chemical, as in the isomerization of D-glucose, then W involves a linear combination of reaction rate constants. On the other hand, if the elementary processes involve the relaxation of internal states, then W is the matrix of transition rates. In any case, it follows from Eqs. (5.3.4) and (5.3.6) that the matrix elements W_{ij} are positive for $i \neq j$ and that

$$\sum_{i \neq j} W_{ij} = -W_{jj}. \tag{5.3.9}$$

Matrices of this form have nice mathematical properties, especially if they are *strong connected*. Strong connectivity implies for any i and j that either there exists a nonzero matrix element W_{ij} connecting i and j or that a sequence of nonzero matrix elements $W_{ij'}, W_{j'j''}, \ldots, W_{j'''j}$ exists which connects i to j. In the language of internal state relaxation this means that states i and j are connected either directly by a single transition or by a sequence of multiple transitions. If this is not the case, then the problem can always be broken down into groups of substates that have these properties. Thus it suffices to consider only the strongly connected case. Strong connectivity implies that the matrix W has a single zero eigenvalue and that all other eigenvalues have negative real parts. The single zero eigenvector of W solves

$$W\boldsymbol{\rho}_0 = \mathbf{0}.$$

Comparing this with Eq. (5.3.7), it follows that $\boldsymbol{\rho}_0$ is the equilibrium value of the molecule densities, $\boldsymbol{\rho}^e$. The condition of detailed balance at equilibrium in Eq. (4.5.3) applied to the canonical equation (5.3.1) yields

$$\mu_j^e = \mu_i^e$$

or in terms of activity coefficients

$$\gamma_j^e \exp(\mu_j^0/k_B T)\rho_j^e = \gamma_i^e \exp(\mu_i^0/k_B T)\rho_i^e. \tag{5.3.10}$$

Detailed balance can be expressed in terms of the matrix W_{ij} by combining Eq. (5.3.10) with Eqs. (5.3.4) and (5.3.6) which, since $\Omega_{ij} = \Omega_{ji}$, gives

$$W_{ij}\rho_j^e = W_{ji}\rho_i^e. \tag{5.3.11}$$

For the case of transitions between internal states of a molecule, the equilibrium densities are $\rho_j^e = \rho \exp(-\varepsilon_j/k_B T)$, where ε_j is the energy of the jth internal state. Thus Eq. (5.3.11) reduces to

$$W_{ij}\exp(-\varepsilon_j/k_B T) = W_{ji}\exp(-\varepsilon_i/k_B T)$$

This equality is often referred to as the condition of detailed balance, although we see that it is only a special case of the more general condition in Eq. (4.5.3).

A slight generalization of the linear mechanism occurs when we consider elementary processes which exchange molecules with outside reservoirs. For a linear exchange mechanism these processes add a term of the form

$$-\bar{\Omega}_i[\exp(\bar{\mu}_i/k_B T) - \exp(\mu_i^{ex}/k_B T)] \tag{5.3.12}$$

to the right-hand side of Eq. (5.3.1), where μ_i^{ex} is the chemical potential of the ith molecule in the external reservoir and is a constant. Changing to the density as a variable, these terms lead to the following modification of Eq. (5.3.7)

$$d\bar{\boldsymbol{\rho}}/dt = W\bar{\boldsymbol{\rho}} - K(\bar{\boldsymbol{\rho}} - \boldsymbol{\rho}^{ex}), \tag{5.3.13}$$

where K is the diagonal matrix

$$K_{ij} = V^{-1}\bar{\Omega}_i \exp(\mu_i^0/k_B T)\delta_{ij}, \tag{5.3.14}$$

and the activity coefficients have been set equal to one. Unless the densities in the external reservoirs, $\boldsymbol{\rho}^{ex}$, are precisely the equilibrium densities, $\boldsymbol{\rho}^e$, Eq. (5.3.13) no longer has an equilibrium solution. Instead, if the densities $\boldsymbol{\rho}^{ex}$ are time independent there will be a *steady-state* solution satisfying $d\boldsymbol{\rho}^{ss}/dt = \mathbf{0}$ or

$$W\boldsymbol{\rho}^{ss} = K(\boldsymbol{\rho}^{ss} - \boldsymbol{\rho}^{ex}). \tag{5.3.15}$$

In the steady state there is no net change in the number of molecules of kind i, although Eq. (5.3.15) shows that there is a steady flow of matter into or out of the reservoirs. It is possible to set up conditions which lead to steady state in a variety of linear systems. In fact, Eq. (5.3.13) has important applications in describing linear elementary processes in biophysical systems, including membrane transport and muscle contraction. The numbers, n_i, no longer represent molecule numbers in those cases, but rather the populations of certain well-defined molecular states. For example, proteins which form ion channels in membranes can exist in different internal states, each with different conductivities. If n_i represents the number of channels in the membrane in the ith state, then the matrix W describes transitions among these states. Some explicit examples are treated in Section 5.8. Alternatively, the n_i's may represent the state of protein receptor molecules bound to a cell membrane. In this case the states represent the number of receptors which are unoccupied, singlely occupied, or multiply occupied by ligands. There are molecular mechanisms which both remove and add receptors to the membrane, and the matrix K represents these elementary processes.

Fluctuations in the general linear system of Eq. (5.3.13) are described by the canonical theory in Sections 4.2–4.4. To simplify matters so that the resulting equations can be solved, we consider the special case of molecules that are in a dilute phase. In this case the activity coefficients, γ_i, are all unity and Eqs. (5.3.4) and (5.3.14) show that the transition matrices W and K are independent of the densities. Thus Eq. (5.3.13) is truly linear in the densities. Nonetheless, to be as general as possible we consider the reservoir concentrations to be time dependent, i.e., $\boldsymbol{\rho}^{ex} = \boldsymbol{\rho}^{ex}(t)$, which will include the special case of constant reservoir densities. To conform to the notation in earlier chapters, we write Eq. (5.3.13) as

$$d\bar{\boldsymbol{\rho}}/dt = H\bar{\boldsymbol{\rho}} + K\bar{\boldsymbol{\rho}}^{ex} \tag{5.3.16}$$

with

$$H = W - K. \tag{5.3.17}$$

By linearizing Eq. (5.3.16) the equation for the conditional density fluctuations is found to be

$$d\delta\boldsymbol{\rho}/dt = H\delta\boldsymbol{\rho} + \tilde{\mathbf{f}}. \tag{5.3.18}$$

The random term vanishes on the conditional average and its covariance can be determined using Eqs. (4.4.6), (5.3.1), and (5.3.12). Defining

$$\langle \tilde{f}_i(t)\tilde{f}_j(t')\rangle = \gamma_{ij}\delta(t - t'), \tag{5.3.19}$$

we find that

$$\gamma_{ij} = V^{-1}\left[\delta_{ij}\sum_k W_{ik}\bar{\rho}_k - (W_{ij}\bar{\rho}_j + W_{ji}\bar{\rho}_i) + K_{ij}(\bar{\rho}_j + \bar{\rho}_j^{ex})\right], \tag{5.3.20}$$

where the factor of V^{-1} arises because Eq. (5.3.18) is written for density, not number, fluctuations. It is easy to check that the terms in Eq. (5.3.20) are correct. For example, when $i \neq j$ Eq. (5.3.20) gives γ_{ij} equal to $-V^{-1}(W_{ij}\bar{\rho}_j + W_{ji}\bar{\rho}_i)$, which involves the sum of the rates of forward and reverse i,j transitions as required by Eq. (4.4.6) and a minus sign corresponding to $\omega_{\kappa i}\omega_{\kappa j}$. Using

Eqs. (5.3.9) and (5.3.14) the expression for the diagonal terms in (5.3.20) can be verified in a similar manner. Notice that γ_{ij} depends explicitly on time through the average densities and the time-dependent densities of the external reservoirs.

Equations (5.3.16)–(5.3.20) give the canonical stochastic description of linear elementary reactions. As in the general case described in Sections 4.2 and 4.3, the conditional fluctuations depend explicitly on the solution to the average equation (5.3.16). Consequently, that equation must be solved first. Since it is a first-order differential equation in the time and linear, it has the solution

$$\bar{\boldsymbol{\rho}}(\boldsymbol{\rho}^0, t) = \exp(Ht)\boldsymbol{\rho}^0 + \int_0^t \exp[H(t-s)]K\boldsymbol{\rho}^{ex}(s)\,ds. \tag{5.3.21}$$

As noticed earlier, the matrix W has at most one zero eigenvalue since it is strongly connected. Using this fact, one can show that all the eigenvalues of $H = W - K$ have negative real parts. Consequently, if $\lim_{t\to\infty} \boldsymbol{\rho}^{ex}(t) = \boldsymbol{\rho}^\infty$ exists, that is, if the densities of the reservoirs ultimately become constant, then Eq. (5.3.21) asymptotically becomes

$$\lim_{t\to\infty} \bar{\boldsymbol{\rho}}(\boldsymbol{\rho}^0, t) = \boldsymbol{\rho}^{ss} = -H^{-1}K\boldsymbol{\rho}^\infty \tag{5.3.22}$$

This state will be a nonequilibrium steady state unless $\boldsymbol{\rho}^\infty = \boldsymbol{\rho}^e$.

To see what the fluctuations are for the general linear reaction, we can use the fact obtained in Section 4.3 that the conditional fluctuations are Gaussian with zero mean. Furthermore, the conditional covariance matrix $\sigma_{ij} = \langle \delta\rho_i(t)\delta\rho_j(t)\rangle^0$ satisfies Eq. (4.3.9), i.e.,

$$d\sigma/dt = H\sigma + \sigma H^T + \gamma. \tag{5.3.23}$$

The formal solution to this equation is given by Eq. (4.3.2). However, since H is independent of time, there is an alternative expression for the solution in which the dependence on the average values is explicit. This form of the solution is obtained by examining the symmetric matrix

$$\alpha_{ij} \equiv (\bar{\rho}_i\delta_{ij} - \bar{\rho}_i\bar{\rho}_j c)V^{-1}, \tag{5.3.24}$$

where c is a constant. Differentiating the right-hand side of Eq. (5.3.24) with respect to time, using Eq. (5.3.16), and rearranging, it can be shown that

$$d\alpha/dt = H\alpha + \alpha H^T + \gamma - V^{-1}c[(K\boldsymbol{\rho}^{ex})\bar{\boldsymbol{\rho}}^T + \boldsymbol{\rho}(K\boldsymbol{\rho}^{ex})^T]. \tag{5.3.25}$$

Combining this equation with Eq. (5.3.23), it follows that the matrix

$$\xi_{ij} \equiv \sigma_{ij} - \alpha_{ij} \tag{5.3.26}$$

satisfies

$$d\xi/dt = H\xi + \xi H^T - V^{-1}c[(K\boldsymbol{\rho}^{ex})\bar{\boldsymbol{\rho}}^T + \boldsymbol{\rho}(K\boldsymbol{\rho}^{ex})^T]. \tag{5.3.27}$$

Taking $c = 0$, we obtain the equation

$$d\xi/dt = H\xi + \xi H^T, \tag{5.3.28}$$

which is solved by

$$\xi(t) = \exp(Ht)\xi(0)\exp(H^T t). \tag{5.3.29}$$

Consider, now, the conditional fluctuations for which $\sigma_{ij}(\boldsymbol{\rho}^0, 0) = 0$. Equations (5.3.26) and (5.3.24) imply the initial conditions

$$\xi_{ij}(0) = -V^{-1}\rho_i^0\delta_{ij}, \tag{5.3.30}$$

and Eqs. (5.3.26), (5.3.29), and (5.3.30) imply that

$$\sigma_{ij}(\boldsymbol{\rho}^0, t) = V^{-1}\bar{\rho}_i(\boldsymbol{\rho}^0, t)\delta_{ij} - V^{-1}\sum_l [\exp(Ht)]_{il}\rho_l^0[\exp(H^T t)]_{lj}. \tag{5.3.31}$$

According to Eq. (5.3.31) the dominant component of the conditional covariance will be proportional to the conditional averages since the exponentials will damp out the second term over time. Indeed, asymptotically,

$$\sigma_{ij}(\boldsymbol{\rho}^0, t) = V^{-1}\bar{\rho}_i(\boldsymbol{\rho}^0, t)\delta_{ij}. \tag{5.3.32}$$

Furthermore, if the external densities lose their transients and become constant, it follows that $\bar{\rho}_i$ approaches ρ_i^{ss} in Eq. (5.3.22) and that

$$\lim_{t\to\infty} \sigma_{ij}(\boldsymbol{\rho}^0, t) = V^{-1}\rho_i^{ss}\delta_{ij}. \tag{5.3.33}$$

These results can be extended to linear mechanisms for which external interactions with reservoirs do not explicitly appear. For these mechanisms $H_{ij} = W_{ij}$ and H will have a single zero eigenvector even if the states are strongly connected. Because the matrix K is zero for such mechanisms, the steady state is no longer determined by Eq. (5.3.22), but rather by $W\boldsymbol{\rho}^{ss} = \mathbf{0}$. As we have seen, if the reactions are linear, then detailed balance implies that this steady state is an equilibrium state. The time dependence of the conditional average values for this case are given by Eq. (5.3.8) and the total density $\rho_T \equiv \sum_i \bar{\rho}_i$ is a constant. The covariance matrix solves exactly the same equations as the general linear mechanism, except the matrix K now vanishes. Thus Eq. (5.3.27) implies that

$$d\xi/dt = W\xi + \xi W^T \tag{5.3.34}$$

where, from Eq. (5.3.24),

$$\xi_{ij} = \sigma_{ij} - V^{-1}(\bar{\rho}_i\delta_{ij} - \bar{\rho}_i\bar{\rho}_j c). \tag{5.3.35}$$

We are at liberty to choose a value of the constant c in this equation. The correct choice turns out to be $c = \rho_T^{-1}$. Since the asymptotic state is equilibrium, it follows from Eq. (5.3.34) that at equilibrium

$$W\xi^e + \xi^e W^T = 0. \tag{5.3.36}$$

This is a linear equation, which can be solved for ξ^e. Using the fact that W is strongly connected, it can be shown that the only solution to Eq. (5.3.36) is

$\xi_{ij} = 0$. Thus Eq. (5.3.35) implies that

$$\sigma_{ij}^e = V^{-1}(\rho_i^e \delta_{ij} - \rho_i^e \rho_j^e / \rho_T). \tag{5.3.37}$$

In analogy to Eq. (5.3.31), the complete time-dependent solution to Eq. (5.3.34) is

$$V\sigma_{ij}(\boldsymbol{\rho}^0, t) = \bar{\rho}_i(\boldsymbol{\rho}^0, t)\delta_{ij} - \bar{\rho}_i(\boldsymbol{\rho}^0, t)\bar{\rho}_j(\boldsymbol{\rho}^0, t)/\rho_T$$
$$- \sum_l \sum_m [\exp(Wt)]_{il}[\rho^0 \delta_{lm} - (\rho_l^0 \rho_m^0/\rho_T)][\exp(Wt)]_{lm}. \tag{5.3.38}$$

Because $\sigma_{ij}(\boldsymbol{\rho}^0, t)$ approaches σ_{ij}^e asymptotically in time, it follows that the terms in this expression involving $\exp(Wt)$ vanish as $t \to \infty$.

The simplest linear system is the isomerization reaction $A \rightleftarrows B$. The complete solution of the conditional probability for this reaction in the absence of external reservoirs was given at the end of Section 3.6. Due to the constraint of molecule conservation, fluctuations are described by a single progress variable. In the presence of external reservoirs, molecule conservation no longer applies and the fluctuations in the density of both A and B need to be considered. The formal solution to this problem is given in Eqs. (5.3.21) and (5.3.31) where the matrix H is

$$H = \begin{pmatrix} -(k^+ + K_A) & k^- \\ k^+ & -(k^- + K_B) \end{pmatrix}. \tag{5.3.39}$$

We can check the general results for linear systems against the results obtained for the isomerization reaction $A \rightleftarrows B$ in Chapter 3. For example, Eq. (5.3.37) gives the equilibrium variance of the number of B molecules, $n_B = V\rho_B$, as

$$\langle(\delta n_B)^2\rangle^e = n_B - n_B^2/n = n_A n_B/n, \tag{5.3.40}$$

which agrees with the result obtained in Eq. (3.6.26). Other results, including the conditional average of n_B and its variance, are easily checked against the specific results obtained in Eqs. (3.6.17) and (3.6.23).

5.4. Bimolecular Isomerization

Linear molecular processes underlie the isomerization reactions described in the previous section. The linearity of these mechanisms leads to linear stochastic differential equations in dilute solution. A nonlinear chemical reaction involves the coming together of several molecules, and the effects of the nonlinearity remain even in dilute solution. The simplest such reaction is the bimolecular isomerization, written symbolically as

$$A + B \rightleftarrows 2B. \tag{5.4.1}$$

The net effect of this reaction is $A = B$, so that the net changes are $\omega_A = -1$ and $\omega_B = 1$, just as for the linear mechanism $A \rightleftarrows B$. Although reactions with this mechanism are not of much importance chemically, they do occur in the gas phase. For example, the rotational conformers of methyl nitrite are separated by a significant energy barrier. Thus intermolecular energy transfer via bimolecular collisions is a dominant form of equilibration of the two conformational states. Be this as it may, the bimolecular isomerization reaction is interesting because it illustrates how the mechanism—as opposed to the overall reaction—determines the stochastic process.

For simplicity, we will consider the bimolecular isomerization to occur only in dilute solution or in the gas phase. In this case the chemical potentials of A or B have the form

$$\mu_i = \mu_i^0 + k_B T \ln \rho_i \tag{5.4.2}$$

with ρ_i the number density. Using these expressions, the canonical form of the average rate equations become

$$d\bar{\rho}_{\mathrm{A}}/dt = -d\bar{\rho}_{\mathrm{B}}/dt = -k^+ \bar{\rho}_{\mathrm{A}}\bar{\rho}_{\mathrm{B}} + k^- \bar{\rho}_{\mathrm{B}}^2 \tag{5.4.3}$$

where

$$\begin{aligned} k^+ &= V^{-1}\Omega \exp[(\mu_{\mathrm{A}}^0 + \mu_{\mathrm{B}}^0)/k_B T] \\ k^- &= V^{-1}\Omega \exp[2\mu_{\mathrm{B}}^0/k_B T] \end{aligned} \tag{5.4.4}$$

are the mass action rate constants and Ω is the intrinsic rate of the elementary process as defined by Eq. (4.2.3). Conservation of the total number of isomers is reflected by the fact that $d(\bar{\rho}_{\mathrm{A}} + \bar{\rho}_{\mathrm{B}})/dt = 0$, which follows from Eq. (5.4.3). We must therefore introduce a single progress variable for this reaction. To maintain a close parallel with the linear isomerization reaction which was treated in Section 3.7, we select ρ_{B} as the progress variable. Thus writing $\rho = \bar{\rho}_{\mathrm{A}} + \bar{\rho}_{\mathrm{B}}$, Eq. (5.4.3) reduces to

$$d\bar{\rho}_{\mathrm{B}}/dt = k^+ \bar{\rho}_{\mathrm{B}}(\rho - \bar{\rho}_{\mathrm{B}}) - k^- \bar{\rho}_{\mathrm{B}}^2. \tag{5.4.5}$$

In the canonical theory $\bar{\rho}_{\mathrm{B}}$ is the conditional-average number density, and Eq. (5.4.5) must be solved with the initial condition $\bar{\rho}_{\mathrm{B}}(0) = \rho_{\mathrm{B}}^0$.

In solving Eq. (5.4.5) it is convenient to rearrange the right-hand side as

$$d\bar{\rho}_{\mathrm{B}}/dt = -(\lambda\bar{\rho}_{\mathrm{B}}/\rho_{\mathrm{B}}^e)(\rho_{\mathrm{B}} - \rho_{\mathrm{B}}^e), \tag{5.4.6}$$

where $\lambda = (k^+ + k^-)\rho_{\mathrm{B}}^e$ and $\rho_{\mathrm{B}}^e = k^+\rho/(k^+ + k^-)$. It is obvious from Eq. (5.4.6) that ρ_{B}^e is the average equilibrium density of B. Rearranging the differentials Eq. (5.4.6) can be written

$$d\bar{\rho}_{\mathrm{B}}/\bar{\rho}_{\mathrm{B}}(\bar{\rho}_{\mathrm{B}} - \rho_{\mathrm{B}}^e) = -(\lambda/\rho_{\mathrm{B}}^e)\,dt \tag{5.4.7}$$

which can be integrated directly to obtain

$$\ln[(\bar{\rho}_{\mathrm{B}} - \rho_{\mathrm{B}}^e)/\bar{\rho}_{\mathrm{B}}] = -\lambda t + C, \tag{5.4.8}$$

with C a constant determined by the initial condition $\bar{\rho}_{\mathrm{B}}(0) = \rho_{\mathrm{B}}^0$. Evaluating C and rearranging leads to

$$\bar{\rho}_B(\rho_B^0, t) = \rho_B^e[1 - (1 - \rho_B^e/\rho_B^0)\exp(-\lambda t)]^{-1}. \tag{5.4.9}$$

We have set the initial time $t^0 = 0$, since time does not occur explicitly in the differential equation. From Eq. (5.4.9) it is easy to see that as long as $0 < \rho_B^0$, the asymptotic solution to Eq. (5.4.5) is ρ_B^e, the average equilibrium density. If the initial condition is close to equilibrium, we can expand the function $(1 - x)^{-1}$ on the right-hand side of Eq. (5.4.9) in the Taylor series $1 + x + x^2 + \cdots$. Keeping only the lowest-order terms gives

$$\bar{\rho}_B(\rho_B^0, t) = \rho_B^e + \exp[-\lambda t](\rho_B^0 - \rho_B^e). \tag{5.4.10}$$

This expression is identical in form to Eq. (3.6.17) which is satisfied by the linear isomerization reaction. The relaxation rate λ, however, equals $k^+\rho$, which is different from λ for the isomerization reaction in Eq. (3.6.16).

To calculate the conditional fluctuations in ρ_B, we follow the canonical prescription given in Eqs. (4.4.2)–(4.4.6). Linearizing Eq. (5.4.5) and adding the random component of the time derivative gives

$$d\delta\rho_B/dt = [k^+\rho - 2(k^+ + k^-)\bar{\rho}_B]\delta\rho_B + \tilde{f}. \tag{5.4.11}$$

Using Eqs. (5.4.5), (4.4.5), and (4.4.6) the variance of $\tilde{f}$ is found to be

$$\langle \tilde{f}(t)\tilde{f}(t')\rangle = V^{-1}[k^+\bar{\rho}_B(\rho - \bar{\rho}_B) + k^-\bar{\rho}_B^2]\delta(t - t'). \tag{5.4.12}$$

The factor V^{-1} on the right-hand side of Eq. (5.4.12) arises since $\omega_B = V^{-1}$ for the density and $(V_\kappa^+ + V_\kappa^-) = V[k^+\bar{\rho}_B(\rho - \bar{\rho}_B) + k^-\bar{\rho}_B^2]$. The stochastic differential equation (5.4.11) depends on the time through $\bar{\rho}_B(\rho_B^0, t)$, which is given explicitly in Eq. (5.4.9). We have already seen in Section 4.4 that the conditional fluctuation is Gaussian. Thus to obtain the conditional probability density it suffices to calculate the conditional variance of ρ_B. The formal solution to this problem is given in Eq. (4.3.2), which for this reaction becomes

$$\begin{aligned}\sigma(\rho_B^0, t) \equiv \langle[\delta\rho_B(t)]^2\rangle = \int_0^t \exp\left[2\int_{\tau'}^t [k^+\rho - 2(k^+ + k^-)\bar{\rho}_B(\rho_B^0, \tau)]\,d\tau\right] \\ \times V^{-1}\{k^+\bar{\rho}_B(\rho_B^0, \tau')[\rho - \bar{\rho}_B(\rho_B^0, \tau')] \\ + k^-\bar{\rho}_B^2(\rho_B^0, \tau')\}\,d\tau'\end{aligned} \tag{5.4.13}$$

since $t^0 = 0$.

The fact that only a single independent variable is involved in Eq. (5.4.11) implies that the time-ordered exponential in Eq. (4.3.2) reduces to the ordinary exponential in Eq. (5.4.13). This simplifies the integrations considerably. For example, consider the integrals inside the exponential function. The first one is trivial and the second involves

$$\begin{aligned}\int_{\tau'}^t \bar{\rho}_B(\rho_B^0, \tau)\,d\tau &= \int_{\tau'}^t \rho_B^e e^{\lambda\tau}(e^{\lambda\tau} - \alpha)^{-1}\,d\tau \\ &= \frac{\rho_B^e}{\lambda}\ln[(e^{\lambda t} - \alpha)/(e^{\lambda\tau'} - \alpha)],\end{aligned} \tag{5.4.14}$$

where $\alpha \equiv 1 - \rho_B^e/\rho_B^0$ and the explicit expression for $\bar{\rho}_B$ in Eq. (5.4.9) was used. Thus the exponential factor in the integrand becomes

$$\exp[\cdot] = \exp[2k^+\rho(t - \tau')](e^{\lambda\tau'} - \alpha)^4(e^{\lambda t} - \alpha)^{-4}, \tag{5.4.15}$$

since the factor $(k^+ + k^-)\rho_B^e/\lambda = 1$ as is seen from the definitions below Eq. (5.4.6). Substituting Eqs. (5.4.9) and (5.4.15) into Eq. (5.4.13), the remaining integrals are found to be straightforward. After a bit of calculation one obtains

$$\sigma(\rho_B^0, t) = V^{-1}(1 - \alpha e^{-\lambda t})^{-4}\left\{\rho_B^e\left[\frac{(1 - e^{-2\lambda t})}{2} - 3\alpha(e^{-\lambda t} - e^{-2\lambda t}) + 3\alpha^2\lambda t e^{-2\lambda t} + \alpha^3(e^{-3\lambda t} - e^{-2\lambda t})\right] + \frac{(k^- - k^{\prime})}{\lambda}\rho_B^{e2}\left[\frac{(1 - e^{-2\lambda t})}{2} - 2\alpha(e^{-\lambda t} - e^{-2\lambda t}) + t\alpha^2\lambda e^{-2\lambda t}\right]\right\}. \tag{5.4.16}$$

The explicit dependence on ρ_B^0 comes through $\alpha = 1 - \rho_B^e/\rho_B^0$.

Although the formula for the conditional variance of ρ_B in Eq. (5.4.16) is complicated, it simplifies for several special cases. If the initial condition ρ_B^0 is near to ρ_B^e, then α can be set equal to zero. This gives

$$\sigma(t) = \frac{k^+\rho_B^e(\rho - \rho_B^e)}{\lambda V}(1 - e^{-2\lambda t}). \tag{5.4.17}$$

Equation (5.4.17) is independent of ρ_B^0, and it is easy to verify that Eq. (5.4.17) is the result expected from the Onsager theory. Indeed Eq. (1.8.12), which is valid for a stationary, Gaussian, Markov process, reduces for a scalar random variable to

$$\sigma(t) = \sigma^e(1 - e^{-2\lambda t}), \tag{5.4.18}$$

where $\lambda = -H$, the relaxation rate. Equation (5.4.10) shows that near equilibrium λ is the relaxation rate for the bimolecular isomerization reaction. The variance at equilibrium is σ^e, which is independent of the mechanism of the isomerization reaction. It is given by the Einstein formula in Eq. (3.6.26), namely,

$$\sigma^e = \rho_A^e\rho_B^e/V\rho. \tag{5.4.19}$$

Using the facts that $\lambda = k^+\rho$ and that $\rho_A^e = \rho - \rho_B^e$, it is easy to check that the first factor in Eq. (5.4.17) equals σ^e. Thus we have another specific example of the equivalence of the mechanistic theory and the Onsager theory near equilibrium.

The behavior of the stochastic process far from equilibrium is much more interesting. Besides the equilibrium state at ρ_B^e, the average equation (Eq. 5.4.5) has an unstable stationary solution at $\bar{\rho}_B = 0$. The lack of stability of this state can be seen by examining Eq. (5.4.9) when ρ_B^0 is close to zero. For short times this yields

$$\bar{\rho}_B(t) \approx \rho_B^0 \exp(\lambda t), \tag{5.4.20}$$

that is, on the average an exponential escape from the region near zero. The initial behavior of the variance can be obtained from Eq. (5.4.16) by noticing that near $\rho_B^0 = 0$, $\alpha = 1 - \rho_B^e/\rho_B^0$ becomes a large negative number. Keeping only the dominant terms in α leads to

$$\sigma(t) = (\rho_B^0/V)e^{2\lambda t}(1 - e^{-\lambda t}). \tag{5.4.21}$$

Thus not only does the conditional average grow exponentially but there is an abnormally rapid exponential growth in the variance. This abnormal growth, of course, is bounded since both the average and variance ultimately attain the equilibrium values predicted by Eqs. (5.4.9) and (5.4.16).

The approach to equilibrium for the special case that $k^+ = k^-$ is shown in Fig. 5.4. There the conditional average and variance are plotted for several values of ρ_B^0. The largest possible initial value of ρ_B^0 when $k^+ = k^-$ is $2\rho_B^e$, and for that initial value the variance approaches its equilibrium value in much the same way as when $\rho_B^0 = \rho_B^e$, which is the Onsager result in Eq. (5.4.18). At

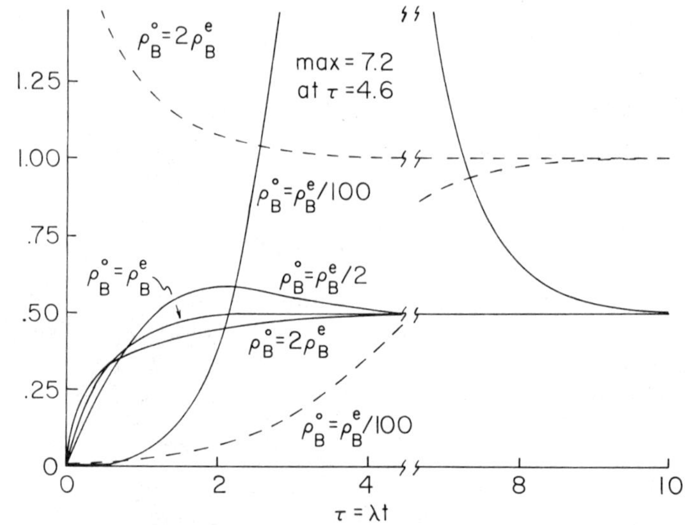

FIGURE 5.4. Time-dependent parameters for the Gaussian conditional probability density for the bimolecular isomerization reaction in Eq. (5.4.1). The rate constants are equal ($k^+ = k^-$) and the dashed lines represent $\bar{\rho}_B(\rho_B^0, t)/\rho_B^e$ and the full lines represent $V\sigma(\rho_B^0, t)/\rho_B^e$ for the indicated values of ρ_B^0. Calculations taken from J. Keizer, *J. Chem. Phys.* **63**, 398 (1975).

small values of ρ_B^0 the initial exponential growth predicted by Eqs. (5.4.20) and (5.4.21) is seen. Indeed if ρ_B^0 is small enough, the conditional variance overshoots its equilibrium value and achieves a maximum before its final approach to equilibrium. The reason for this can be seen by examining the time-dependent equation (Eq. 4.3.9) satisfied by σ. For the scalar variable ρ_B it can be written

$$d\sigma/dt = 2H\sigma + \gamma, \tag{5.4.22}$$

where from Eq. (5.4.11)

$$H(\rho_B^0, t) = [k^+\rho - 2(k^+ + k^-)\bar{\rho}_B(\rho_B^0, t)]. \tag{5.4.23}$$

According to Eq. (5.4.22), $\sigma(\rho_B^0, t)$ will reach a maximum if

$$\sigma(\rho_B^0, t) = -\gamma(\rho_B^0, t)/2H(\rho_B^0, t). \tag{5.4.24}$$

Since σ and γ are positive, this can only occur after H has become negative. When ρ_B^0 is small, Eq. (5.4.23) shows that $H = k^+\rho = \lambda$ is positive. At equilibrium, on the other hand, $H = -k^+\rho = -\lambda$. Indeed when $\bar{\rho}_B(\rho_B^0, t) = k^+\rho/2(k^+ + k^-) = \rho_B^e/2$, H vanishes and thereafter becomes negative. As H^{-1} changes from $+\infty$ at $\rho_B^e/2$ to its equilibrium value of $(-1/\lambda)$, it is therefore possible for the equality in Eq. (5.4.24) to be achieved. The size of the maximum in the variance depends on how far the initial state is from equilibrium, and for the condition $\rho_B^0 = \rho_B^e/100$ the maximum is more than an order of magnitude larger than the equilibrium variance.

These results for the bimolecular isomerization are in strong contrast to the results obtained for the linear isomerization reaction in Section 3.6. For $k^- = k^+$, the variance obtained for the linear mechanism in Eq. (3.6.23) has the time dependence

$$\sigma(\rho_B^0, t) = (\rho_A^e\rho_B^e/V\rho)(1 - e^{-2\lambda t}), \tag{5.4.25}$$

no matter how far ρ_B^0 is from its equilibrium value. Only near equilibrium is this the same as that for the nonlinear mechanism. The distinctive features of the stochastic process for the nonlinear isomerization mechanism—for example, exponential growth of the conditional average and variance—are a function of the details of the nonlinearities associated with the elementary process. As will become evident in the remainder of this book, ensembles that are maintained far from equilibrium will exhibit behavior that is strongly dependent on the molecular mechanism taking place.

5.5. Continuously Stirred Tank Reactors and Molecule Reservoirs

The bimolecular isomerization reaction treated in the previous section demonstrates some of the interesting statistical and dynamic features that occur when a nonlinear system is far from equilibrium. There are a variety of ways in

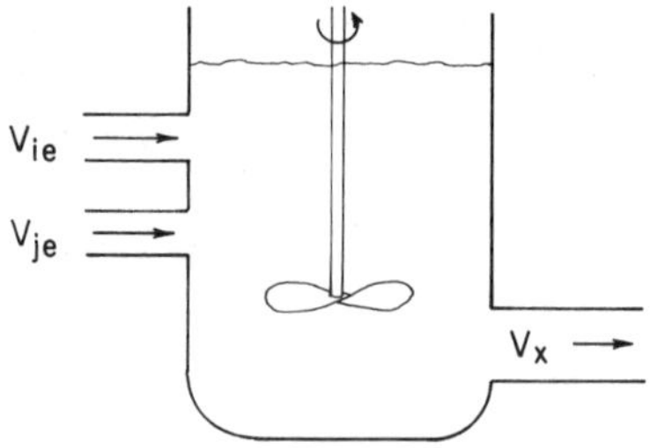

FIGURE 5.5. Diagram of a continuously stirred tank reactor (CSTR) showing the entrance ports with flow velocities v_{ie}, v_{je}, and the exit port with velocity v_x. Rapid stirring maintains densities that are effectively uniform throughout the tank.

which a system of chemical reactants and products can be maintained in a nonequilibrium state. One of the simplest is to use a continuously stirred tank reactor or CSTR. The schematic arrangement of a CSTR is shown in Fig. 5.5. Entrance ports carry a flow of reactants into the main tank where the incoming solutions are mixed rapidly. A single exit port carries the flow of mixed reactants and products out of the tank. The system considered here is the solution of reactants and products occupying the tank, which, since the system is open, may vary as a function of time. The condition of the system can be controlled by several operational variables. For example, the experimentalist can control the concentration of reactants coming into the tank through the entrance ports. By controlling the flow rate out of the exit port, it is possible to control the amount of time that the reactants spend in the tank together and, thus, the extent to which the reactions proceed.

A detailed analysis at the hydrodynamic level of the processes occurring in a CSTR is complicated by the entrance, exit, and mixing flows of the solutions. The purpose of rapid mixing, however, is to circumvent these difficulties by making concentrations in the system effectively uniform. In practice this can be done quite efficiently using a magnetic stirrer, and it is customary to analyze reactions in a CSTR as if the solution were completely homogeneous. Thus, as a first approximation, it is possible to describe the statistical thermodynamics of chemical reactions in a CSTR using the thermodynamic level of description. We defer until Chapters 6 and 8, where the hydrodynamic level of description is treated in more detail, a discussion of how to include the length scale of the mixing process into the present picture.

Besides chemical reactions there are other dissipative processes that occur in the CSTR, for example, the mixing of the incoming solutions of reactants with the existing contents of the tank. Although this mixing process has associated with it viscous dissipation, this is not included at the thermodynamic level of description. Dissipation is also involved in the process of dilution of the entrance solutions as they flow through the reactor. The characteristic time for this process, τ, is called the residence time. An expression for the residence time can be obtained from a mechanical description of the mixing process. Consider the rate of change of n_i, the number of molecules of kind i in the CSTR, due to mixing. This rate is given by the difference between the number of molecules that enter through the entrance port per unit time and the number that flow out the exit port per unit time. The entrance flow is characterized by a fixed density of i, ρ_{ie}, and a fixed velocity of magnitude v_{ie} directed into the tank. If the cross-sectional area of the entrance port is A_{ie}, then on the average

$$(d\bar{n}_i/dt)_{\text{entrance}} = \rho_{ie} v_{ie} A_{ie}. \tag{5.5.1}$$

The exit port removes the solution in the CSTR at a velocity v_x and with densities $\bar{\rho}_i = \bar{n}_i/V$, where V is the volume of the CSTR. Thus if the cross-sectional area of the exit port is A_x, it follows that

$$(d\bar{n}_i/dt)_{\text{mix}} = \rho_{ie} v_{ie} A_{ie} - \bar{\rho}_i v_x A_x. \tag{5.5.2}$$

Equation (5.5.2) can be written in terms of the average density as

$$(d\bar{\rho}_i/dt)_{\text{mix}} = -\tau^{-1}(\bar{\rho}_i - \rho_i^d) \tag{5.5.3}$$

where

$$\tau \equiv 1/v_x A_x \tag{5.5.4a}$$

$$\rho_i^d = v_{ie} A_i \rho_{ie}/v_x A_x. \tag{5.5.4b}$$

The relaxation time, τ, in Eq. (5.5.3) is the *residence time*. Since $v_x A_x$ is the volume removed at the exit in unit time, it is evident from Eq. (5.5.4a) that τ is the time required for the entire volume of the tank to be emptied by the exit port. Assuming an incompressible flow, the volume of solution containing i that enters per unit time, i.e., $V_e = v_{ie} A_{ie}$, is smaller than the overall volume leaving per unit time, $V_x = v_x A_x$. The ratio (V_e/V_x) in Eq. (5.5.4b) gives the fractional dilution of the entrance solution that would occur by mixing the volume V in the absence of reactions. Thus at the thermodynamic level of description, Eq. (5.5.3) can be characterized as the rate of the elementary process associated with mixing.

This derivation of the rate of the elementary process is limited to dilute, incompressible solutions that are mixed rapidly enough to be characterized by uniform densities. For nonideal solutions the elementary mixing process can be described by the canonical form. For example, one might write

$$(d\bar{n}_i/dt)_{\text{mix}} = -\Omega[\exp(\bar{\mu}_i/k_B T) - \exp(\mu_i^d/k_B T)], \tag{5.5.5}$$

where μ_i is the chemical potential of i and μ_i^d is the chemical potential that would result upon dilution of the entrance flows in the absence of reaction. The molecular mechanism of mixing described by Eq. (5.5.5) involves one molecule at a time ($n_i^+ = 1, n_i^- = 0$). It should be clear from this example that the conception of an elementary molecular process for mixing is very far removed from the molecular details of the mixing process, itself, which is described much more completely at the hydrodynamic level of description. Nonetheless, the analysis which leads to Eq. (5.5.3) is clearly valid for rapid mixing of ideal solutions. Moreover, Eq. (5.5.3) shows that the mixing process is dissipative in the system variables, ρ_i. Thus it is consistent at the thermodynamic level of description to treat mixing as an elementary process described by the canonical form. In practice, the effects of nonideality are difficult to calculate for any elementary process, and in most situations Eq. (5.5.3) will be used to describe the mixing process.

The mixing process in Eq. (5.5.3) eliminates the conservation conditions, which for chemical reactions couple changes in the density of reactant and product molecules. Consider, for example, the bimolecular reaction A + B $\rightleftarrows$ C + D. For an ideal solution the rate of this process can be expressed as

$$V^+ - V^- = V(k^+ \rho_A \rho_B - k^- \rho_C \rho_D). \tag{5.5.6}$$

If the reaction occurs in an isolated system, changes in the number of molecules of A, B, C, and D are determined by a single progress variable, as described in Section 3.7. If the reaction is carried out in a CSTR, on the other hand, the time rates of change of the density satisfy, on the average, the four coupled equations

$$\begin{aligned} d\bar{\rho}_A/dt &= -k^+ \bar{\rho}_A \bar{\rho}_B + k^- \bar{\rho}_C \bar{\rho}_D - \tau^{-1}(\bar{\rho}_A - \rho_A^d) \\ d\bar{\rho}_B/dt &= -k^+ \bar{\rho}_A \bar{\rho}_B + k^- \bar{\rho}_C \bar{\rho}_D - \tau^{-1}(\bar{\rho}_B - \rho_B^d) \\ d\bar{\rho}_C/dt &= k^+ \bar{\rho}_A \bar{\rho}_B - k^- \bar{\rho}_C \bar{\rho}_D - \tau^{-1}(\bar{\rho}_C - \rho_C^d) \\ d\bar{\rho}_D/dt &= k^+ \bar{\rho}_A \bar{\rho}_B - k^- \bar{\rho}_C \bar{\rho}_D - \tau^{-1}(\bar{\rho}_D - \rho_D^d). \end{aligned} \tag{5.5.7}$$

The mixing terms allow the densities to vary independently, and unless there is some additional constraint on the preparation of the system, variations in the densities will be independent on the average. The effect of the mixing terms is to introduce a new elementary process for each variable, in this case four new elementary processes. Since there are then five independent elementary processes, the four densities become independent variables.

Because the densities in a CSTR are independent variables, there are no conservation conditions that connect fluctuations in the densities. According to the canonical theory, conditional fluctuations in the density of A, B, C, and D in the CSTR described by Eq. (5.5.7) satisfy the stochastic differential equations

$$d\delta\boldsymbol{\rho}/dt = H(\bar{\boldsymbol{\rho}}(\boldsymbol{\rho}^0, t))\delta\boldsymbol{\rho} + \tilde{\mathbf{f}}, \tag{5.5.8}$$

where $\delta\boldsymbol{\rho}$ is the column vector $(\delta\rho_A, \delta\rho_B, \delta\rho_C, \delta\rho_D)^T$ and the matrix H is

$$H(\bar{\boldsymbol{\rho}}(\boldsymbol{\rho}^0, t)) = \begin{pmatrix} -[k^+\bar{\rho}_B + \tau^{-1}] & -k^+\bar{\rho}_A & k^-\bar{\rho}_D & k^-\bar{\rho}_C \\ -k^+\bar{\rho}_B & -[k^+\bar{\rho}_A + \tau^{-1}] & k^-\bar{\rho}_D & k^-\bar{\rho}_C \\ k^+\bar{\rho}_B & k^+\bar{\rho}_A & -[k^-\bar{\rho}_D + \tau^{-1}] & -k^-\bar{\rho}_C \\ k^+\bar{\rho}_B & k^+\bar{\rho}_A & -k^-\bar{\rho}_D & -[k^-\bar{\rho}_C + \tau^{-1}] \end{pmatrix} \tag{5.5.9}$$

The explicit dependence of H on time comes from the solution of Eqs. (5.5.7) with the initial condition $\bar{\boldsymbol{\rho}}(t) = \boldsymbol{\rho}^0$. The covariance of the random terms in Eq. (5.5.8) follows directly from the general results in Chapter 4. Equations

(4.4.5) and (4.4.6) give the canonical form of the covariance matrix in terms of the parameters of the five elementary processes in Eq. (5.5.7). Explicitly one finds that

$$\langle \tilde{\mathbf{f}}(t)\tilde{\mathbf{f}}^T(t')\rangle = \gamma(\bar{\boldsymbol{\rho}}(\boldsymbol{\rho}^0, t))\delta(t - t'), \tag{5.5.10}$$

where

$$\gamma(\bar{\boldsymbol{\rho}}(\boldsymbol{\rho}^0, t)) = V^{-1}\begin{pmatrix} (\bar{V}^+ + \bar{V}^-) + \tau^{-1}(\bar{\rho}_A + \rho_A^d) & (\bar{V}^+ + \bar{V}^-) & -(\bar{V}^+ + \bar{V}^-) & -(\bar{V}^+ + \bar{V}^-) \\ (\bar{V}^+ + \bar{V}^-) & (\bar{V}^+ + \bar{V}^-) + \tau^{-1}(\bar{\rho}_B + \rho_B^d) & -(\bar{V}^+ + \bar{V}^-) & -(\bar{V}^+ + \bar{V}^-) \\ -(\bar{V}^+ + \bar{V}^-) & -(\bar{V}^+ + \bar{V}^-) & (\bar{V}^+ + \bar{V}^-) + \tau^{-1}(\bar{\rho}_C + \rho_C^d) & (\bar{V}^+ + \bar{V}^-) \\ -(\bar{V}^+ + \bar{V}^-) & -(\bar{V}^+ + \bar{V}^-) & (\bar{V}^+ + \bar{V}^-) & (\bar{V}^+ + \bar{V}^-) + \tau^{-1}(\bar{\rho}_C + \rho_D^d) \end{pmatrix} \tag{5.5.11}$$

where $\bar{V}^\pm$ are the averages of the rates given in Eq. (5.5.6) and the volume factor, V^{-1}, comes from the fact that these are fluctuations in molecular number densities, not molecule numbers.

The stochastic description of the bimolecular reaction $A + B \rightleftarrows C + D$ in a CSTR is given by Eqs. (5.5.7)–(5.5.11). In principle, the solution of these equations would follow the same path used in obtaining the stochastic description of the bimolecular isomerization reaction in the previous section. Thus it is necessary first to solve the nonlinear conditional average equations, (5.5.7), and then solve the resulting linear stochastic differential equations, (5.5.8), using the explicit time dependence of matrices H and γ. Because four independent variables are involved in the solution of the average equations, it is pretty much hopeless to expect an explicit solution—like the one provided for the bimolecular isomerization reaction in Eq. (5.4.9). Moreover, even were such a solution available, one would still need to carry out the complicated time-ordered exponentials which through Eq. (4.4.8) define the solution to the stochastic differential equations (5.5.8). There are, nonetheless, special circumstances under which explicit solutions to these equations can be obtained, for example, at stable steady states. The statistical properties of stable steady states are treated in Chapter 7.

Another way to maintain a state of nonequilibrium in a coupled chemical system is using external molecule reservoirs. Indeed, there is a great similarity between the elementary process description of a CSTR and the description of external molecule reservoirs at the thermodynamic level. A molecule reservoir is permeable to a single kind of molecule and is taken to be so large that its chemical potential is constant. Because of this, a reservoir can act as a source or sink of chemical energy depending on the chemical potential of the corresponding molecule in the system. The simplest elementary process describing exchange of molecules with a reservoir has the form $i_S \rightleftarrows i_R$, where i_S represents a molecule of kind i in the system and i_R is the same kind of molecule in the reservoir. Since the system and reservoir are separated spatially, this elementary process represents a spatial transport process akin to diffusion.

This analogy is made more precise at the hydrodynamic level in Chapter 6. Although at the thermodynamic level of description spatial effects are ignored, molecule exchange with reservoirs can be used to provide a compartmentalized model of diffusion effects.

According to the canonical theory, the rate of the elementary process $i_S \rightleftarrows i_R$ is given by

$$V^+ - V^- = \Omega[\exp(\mu_i/k_B T) - \exp(\mu_i^R/k_B T)], \tag{5.5.12}$$

where μ_i^R is the chemical potential of i in the reservoir. This elementary process involves a single system molecule of i in the forward direction and a single reservoir molecule of i in the reverse direction. Thus $n_j^+ = \delta_{ij}$ and $n_j^- = 0$ for this process, where the $n_j^\pm$ represent the involvement only of system molecules, and so the net changes of system molecule numbers are $\omega_j = -\delta_{ij}$. The canonical form in Eq. (4.4.1) then implies that

$$(d\bar{n}_i/dt)_{\text{res}} = -\Omega[\exp(\bar{\mu}_i/k_B T) - \exp(\mu_i^R/k_B T)]. \tag{5.5.13}$$

The similarity of this equation to Eq. (5.5.5) which describes the mixing process for the CSTR is obvious.

It follows from Eq. (5.5.13) that if the chemical potential in the system exceeds the chemical potential in the reservoir, then on the average there is a loss of system molecules to the reservoir and *vice versa.* Associated with this change is a net increase in the local equilibrium entropy of the reservoir plus the system, i.e.,

$$\begin{aligned} dS/dt &= (-\bar{\mu}_i/T)(d\bar{n}_i/dt)_{\text{res}} + (-\mu_i^R/T)(d\bar{n}_i^R/dt)_{\text{res}} \\ &= [(\bar{\mu}_i - \mu_i^R)/T]\Omega[\exp(\bar{\mu}_i/k_B T) - \exp(\mu_i^R/k_B T)]. \end{aligned} \tag{5.5.14}$$

The fact that the change in the entropy is non-negative is guaranteed by the inequality in Eq. (4.5.6), which holds for any elementary process.

The contribution to fluctuations in n_i due to molecule exchange with a reservoir can be obtained from the fluctuation–dissipation theory. Just as for the mixing process in a CSTR, one finds that

$$d\delta n_i/dt = -(\Omega/k_B T)\exp(\bar{\mu}_i/k_B T)\sum_j (\partial\bar{\mu}_i/\partial\bar{n}_j)\delta n_j + \tilde{f}_i, \tag{5.5.15}$$

where in linearizing Eq. (5.5.13) temperature fluctuations have been ignored. The covariance of the random Gaussian terms $\tilde{f}_i$ can be obtained from Eqs. (5.5.12) and (5.5.13) using Eqs. (4.4.5) and (4.4.6). The result is

$$\langle \tilde{f}_k(t)\tilde{f}_j(t')\rangle = \Omega\sum_i \delta_{ik}\delta_{ij}[\exp(\bar{\mu}_i/k_B T) + \exp(\mu_i^R/k_B T)], \tag{5.5.16}$$

which has the same form as the CSTR mixing terms in Eq. (5.5.11). Because the CSTR mixing process and molecule exchange with a reservoir enter into the stochastic equations in the same fashion, a single calculation can be used to describe coupled chemical reactions that are driven away from equilibrium by either of these mechanisms.

5.6. Electrode Processes

The passage of current through electrolyte solutions is accompanied by oxidation and reduction processes occurring at solid electrodes. A pair of electrodes is required and each electrode provides an interface at which electrons can be taken up or given off by molecules in solution. When a molecule takes up electrons from an electrode, the molecule is said to be *reduced* and this elementary molecular process is called a *reduction* or *cathodic* process. The giving up of electrons, on the other hand, is called an *oxidation* or *anodic* process and the molecule is said to be *oxidized*. These oxidation–reduction processes involve identifiable molecular events and are the forward and reverse steps of elementary molecular processes. The kinetics of oxidation–reduction reactions at electrodes follow rate laws that were first identified by Butler and later explored in detail by Volmer. These rate laws are called the Butler–Volmer equations, and we will derive them here using the canonical form.

A rich variety of oxidation–reduction processes can occur at electrode surfaces. For example, indium cations, In^{3+}, can be reduced at a mercury electrode in aqueous solution. If the solution contains thiocyanate anions, SCN^-, the reduction process does not involve In^{3+} directly, but rather appears to go through a intermediate layer of $In(SCN)_2^-$ adsorbed at the mercury. Oxidation–reduction at a zinc electrode in acidic solution, on the other hand, can occur by the process

$$Zn_{(s)} \rightleftarrows 2e^- + Zn^{2+}_{(aq)}, \tag{5.6.1}$$

where e^- represents an electron. In this process zinc atoms are removed or deposited directly on the electrode surface. Other electrodes, however, like platinum are fairly inert. These electrodes act by providing a clean interface where ions and molecules in solution can be oxidized or reduced by electrons at the surface. This is the simplest type of oxidation–reduction process occurring at an electrode and is the one we examine here.

For definiteness consider the following reactions

$$\begin{aligned} Ce^{4+} + e^- &\rightleftarrows Ce^{3+} \\ Fe^{3+} + e^- &\rightleftarrows Fe^{2+} \end{aligned} \tag{5.6.2}$$

where, following the standard convention, the reactions are written as cathodic in the forward direction. These reactions can be coupled to produce electric current if the iron and cerium ions are kept separated in solution. Such an arrangement is shown in Fig. 5.6. There the two solutions are in electrical contact through a salt bridge. The circuit is completed by the platinum electrodes, which may be connected to an external voltage that provides an electrical force. When current flows, the anodic or reverse process in Eq. (5.6.2) dominates at one electrode and the cathodic or forward reaction dominates at the other. By convention, the electrode at which the cathodic process

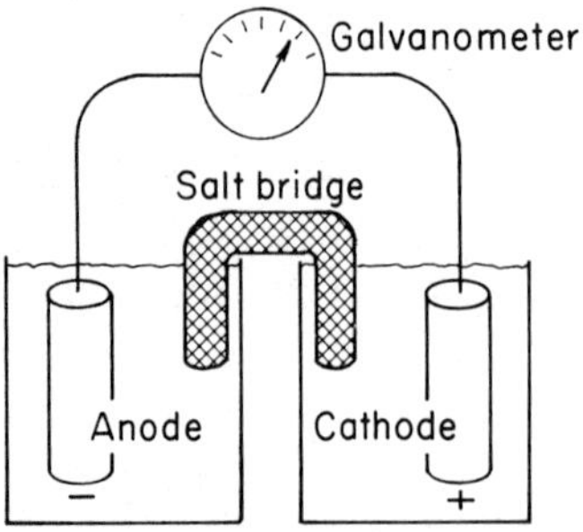

FIGURE 5.6. Diagram of an electrochemical cell consisting of two half-cells connected by a salt bridge. Electrons flow in the external circuit from the anode (−), where oxidation dominates, to the cathode (+), where reduction takes place.

dominates is called the *cathode* and is assigned a positive voltage. The negative electrode is called the *anode*. The entire apparatus is called an electrochemical cell.

To analyze the flow of electricity through such a cell we need to examine the rates of the elementary oxidation–reduction processes at the cathode and anode. Either process can be written symbolically as

$$A^{z+} + ne^- \rightleftarrows A^{z-n}, \tag{5.6.3}$$

where reduction has been chosen as the forward direction. This process occurs within a few Angstroms of the electrode surface, as diagramed in Fig. 5.7. The *Galvani potential*, ϕ, is the difference in electrical potential between the electrode and solution and is involved in determining the rate of the oxidation–reduction process. As is evident in Eq. (5.6.3), the forward step of this elementary process involves a single ion from solution, A^{z+}, as well as n electrons from the electrode, whereas the reverse step involves only the reduced species A^{z-n}. Since there is a variable electrical potential, the electrochemical potentials appear in the canonical form for the rates. Thus

$$\partial S/\partial N_i = -\tilde{\mu}_i/T, \tag{5.6.4}$$

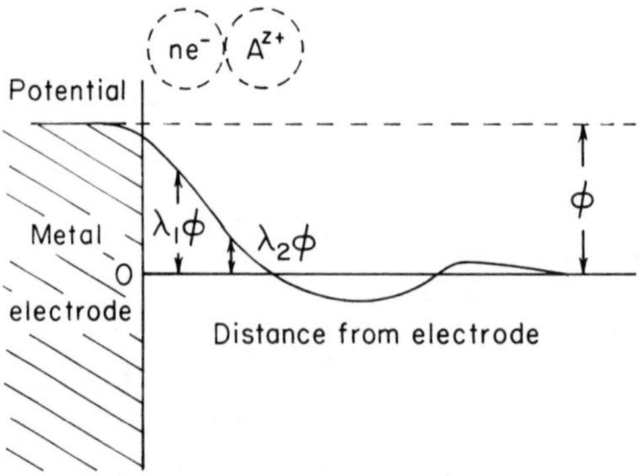

FIGURE 5.7. Diagram of the electrical potential as a function of distance from the surface of an inert metal electrode. The Galvani potential, ϕ, is the potential difference between the bulk metal and the bulk electrolyte. The electrical potentials of the reacting electrons and oxidized species A^{z+} are indicated by $\lambda_1\phi$ and $\lambda_2\phi$.

where $\tilde{\mu}_i$ is the *electrochemical potential.* Defining the activities of the oxidized and reduced species, A^{z+} and A^{z-n}, as a_o and a_r, respectively, one has

$$\tilde{\mu}_o = \mu_o^0 + k_B T \ln a_o \tag{5.6.5}$$

$$\tilde{\mu}_r = \mu_r^0 + k_B T \ln a_r \tag{5.6.6}$$

$$\tilde{\mu}_e = \mu_e - e\phi, \tag{5.6.7}$$

where for reference the potential of the solution is taken as zero. In addition to assembling the ions and electrons, it is necessary to have an elementary amount of internal energy, $\varepsilon^{\pm}$, if the forward or reverse direction of the process is to occur. Since internal energy is conserved in a chemical reaction, it follows that $\omega_E = \varepsilon^- - \varepsilon^+ = 0$. Thus $\varepsilon^{\pm} = \varepsilon$, which is called the *activation energy* for the process. According to the Arrhenius interpretation, the activation energy is a barrier that must be surmounted in order for reaction to occur. Since the reactants are charged, this barrier depends on the potential. We write it in the form

$$\varepsilon = \mu^{\ddagger} - \lambda_1 en\phi + \lambda_2 ez\phi, \tag{5.6.8}$$

where $\mu^{\ddagger}$ is the barrier when the Galvani potential vanishes. The constants λ_1 and λ_2 are between zero and one, which can be seen from Fig. 5.7, since at the barrier neither the electrons nor the ion experience the full potential difference. Defining α by

$$\alpha \equiv \lambda_1 - \lambda_2(z/n), \tag{5.6.9}$$

Eq. (5.6.8) can be written

$$\varepsilon = \mu^{\ddagger} - \alpha en\phi. \tag{5.6.10}$$

These effects of the electrical potential on the energy and chemical potentials are shown in Fig. 5.8.

To obtain the rate of the elementary process we recall that

$$(\partial S/\partial E)_{V,\mathbf{N}} = 1/T. \tag{5.6.11}$$

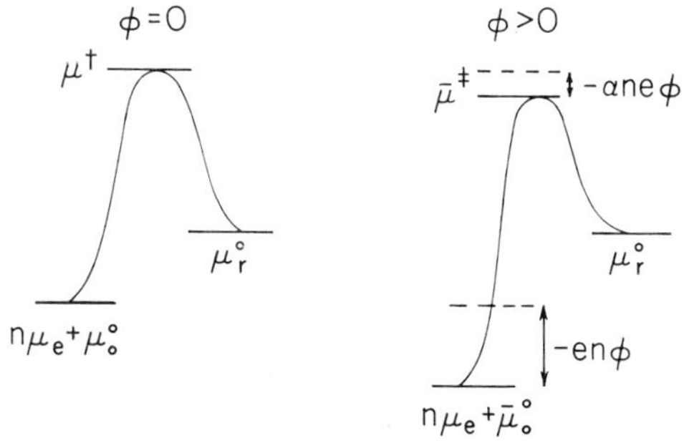

FIGURE 5.8. Schematic representation of the effect of a change in the Galvani potential, ϕ, on the relative free energies for the oxidation–reduction reaction $ne^- + A^{z+} = A^{z-n}$. An increase of ϕ decreases the barrier height, $\mu^{\ddagger}$, by $-\alpha ne\phi$, decreases the energy of the electrons and oxidized species, $n\mu_e + \mu_o^0$, by $-en\phi$, and leaves the energy of the reduced species in solution, μ_r^0, unchanged.

Thus applying the canonical formulas (4.2.3) and (4.2.4) for this process we find that

$$\begin{aligned} V^+ &= \Omega \exp(-\varepsilon/k_B T) \exp(\tilde{\mu}_0/k_B T) \exp(n\tilde{\mu}_e/k_B T) \\ V^- &= \Omega \exp(-\varepsilon/k_B T) \exp(\tilde{\mu}_r/k_B T). \end{aligned} \tag{5.6.12}$$

Substituting into these expressions the explicit functions for the chemical potentials and activation energy in Eqs. (5.6.5)–(5.6.7) and (5.6.10) gives

$$\begin{aligned} V^+ &= \Omega \exp(-\Delta\mu_+^{0\ddagger}/k_B T) \exp[-(1-\alpha)en\phi/k_B T] a_o \\ V^- &= \Omega \exp(-\Delta\mu_-^{0\ddagger}/k_B T) \exp[\alpha en\phi/k_B T] a_r \end{aligned} \tag{5.6.13}$$

where

$$\begin{aligned} \Delta\mu_+^{0\ddagger} &= \mu^\ddagger - n\mu_e - \mu_o^0 \\ \Delta\mu_-^{0\ddagger} &= \mu^\ddagger - \mu_r^0. \end{aligned} \tag{5.6.14}$$

These are the differences between the standard-state chemical potentials and the barrier height, as illustrated in Fig. 5.8. Equations (5.6.13) and (5.6.14) give the rates of the elementary oxidation–reduction process in Eq. (5.6.3).

If the concentrations of the oxidized and reduced species, A^{z+} and A^{z-n}, are held fixed in solution, there is a unique Galvani potential, ϕ^e, at which the rates of the cathodic process and the anodic process just balance one another. At this Galvani potential there is no net oxidation or reduction and the half-cell is in equilibrium. This potential is determined by the equation

$$\begin{aligned} &\Omega \exp(-\Delta\mu_+^{0\ddagger}/k_B T) \exp[-(1-\alpha)en\phi^e/k_B T] a_o \\ &\quad = \Omega \exp(-\Delta\mu_-^{0\ddagger}/k_B T) \exp[\alpha en\phi^e/k_B T] a_r. \end{aligned} \tag{5.6.15}$$

Canceling common terms, rearranging, and taking the logarithm gives

$$en\phi^e = (n\mu_e + \mu_o^0 - \mu_r^0) + k_B T \ln(a_o/a_r). \tag{5.6.16}$$

Although this characterizes the equilibrium state of an electrode in solution, this potential cannot be measured. Indeed, the properties of a test charge in two different phases always involve a chemical component and only the electrochemical potential difference can actually be measured. The operational solution to this dilemma is to insert a second reference electrode into the solution. By attaching electrical leads with a common composition to each electrode, it is possible to measure the electrical potential between the two leads. When the potential is adjusted so that no current flows in this cell, both electrodes, as well as the whole cell, are in equilibrium. This potential is called the *electromotive force*, or EMF, of the cell.

The Nernst expression for the EMF of an electrochemical cell can be derived from Eq. (5.6.16). To do so, consider the cell reaction

$$B^{z'+} + n'e^- \rightleftarrows B^{z'-n'} \tag{5.6.17}$$

which occurs at the reference electrode. According to Eq. (5.6.16), the equilibrium Galvani potential at this electrode is

$$n'e\phi^{e'} = (n'\mu_e + \mu_o^{0'} - \mu_r^{0'}) + k_B T \ln(a'_o/a'_r). \tag{5.6.18}$$

In general, n and n' may be the same or different. In either case let ν and ν' be the smallest whole numbers such that $\nu n = \nu' n' \equiv m$. Multiplying Eq. (5.6.16) by ν and Eq. (5.6.18) by ν' and subtracting the resulting expressions gives

$$em(\phi^e - \phi^{e'}) = (\nu\mu_o^0 + \nu'\mu_o^{0'} - \nu\mu_r^0 - \nu'\mu_r^{0'}) + k_B T \ln(a_o^\nu a_r'^{\nu'}/a_r^\nu a_o'^{\nu'}). \tag{5.6.19}$$

Since the two electrodes share a common electrolyte solution, the difference between their Galvani potentials is the electrical potential difference between the two electrodes. Moreover, since no current is flowing, the potential of each electrode is the same as that of the lead attached to it. Hence, $\phi^e - \phi^{e'} = \mathscr{E}$, the EMF of the cell. Thus Eq. (5.6.19) can be rewritten as

$$em\mathscr{E} = -\Delta G^0 - k_B T \ln[(a_r^\nu)(a_o'^{\nu'})/(a_o^\nu)(a_r'^{\nu'})] \tag{5.6.20}$$

or

$$em\mathscr{E} = -\Delta G, \tag{5.6.21}$$

where ΔG is the Gibbs free energy change defined in Eq. (3.7.34) for the net chemical reaction

$$\nu A^{z+} + \nu' B^{z'-n'} = \nu A^{z-n} + \nu' B^{z'+}. \tag{5.6.22}$$

Equation (5.6.21) is one form of the *Nernst equation* for the *balanced cell reaction* in Eq. (5.6.22). The Nernst equation can be expressed in terms of molar free energy difference by multiplying Eq. (5.6.20) by Avogadro's number. This gives

$$\mathscr{E} = \mathscr{E}^0 - \frac{RT}{mF} \ln[(a_r^\nu)(a_0'^{\nu'})/(a_0^\nu)(a_r'^{\nu'})], \tag{5.6.23}$$

where $F = 9.65 \times 10^4$ C mol^{-1} is Faraday's constant, R is the gas constant, and

$$\mathscr{E}^0 = -\Delta\bar{G}^0/mF, \tag{5.6.24}$$

with $\bar{G}$ the molar free energy. Clearly $\mathscr{E}^0$ is the EMF of the cell when reactants and products are in their standard states, i.e., when all the activities are unity. We recall from Section 3.8 that the equilibrium constant for a net chemical reaction is given by Eq. (3.8.20) or, in terms of molar quantities, by

$$K(T) = \exp(-\Delta\bar{G}^0/RT). \tag{5.6.25}$$

Combining Eqs. (5.6.24) and (5.6.25), the equilibrium constant can be written

$$K(T) = \exp(mF\mathscr{E}^0/RT). \tag{5.6.26}$$

This is the classical result of equilibrium thermodynamics, which relates an electrical measurement to an equilibrium constant. Notice that we have derived it directly from the canonical form rather than using the Gibbs free energy as a potential for reversible work, as is usually done.

The phenomenological equation that governs the current which flows through an electrode was discovered by Tafel. A more complete expression of this empirical result is embodied in the Butler–Volmer equation. For the elementary process in Eq. (5.6.3) the Butler–Volmer equation is

$$i = i_0\{\exp[\alpha nFn/RT] - \exp[-(1-\alpha)nF\eta/RT]\}, \tag{5.6.27}$$

where η is the *overpotential* defined by

$$\eta \equiv \phi - \phi^e. \tag{5.6.28}$$

In practice, the overpotential is measured with respect to a reference electrode, although the difference potential is the same as that in Eq. (5.6.28).

To obtain the Butler–Volmer equation from the canonical rate expressions, we need to use the rates $V^{\pm}$ to obtain the electrode current. For the elementary process in Eq. (5.6.3), the net change of charge on the electrode due to the process is $\omega = 0 - (-en) = en$. Thus using Eq. (4.2.10) we find that

$$(dq/dt)_{\mathrm{rx}} = en(V^+ - V^-), \tag{5.6.29}$$

where q is the charge on the electrode. Because of electroneutrality, current of opposite sign flows into the electrode from the external circuit. It is this current that is measured. Thus $i = -en(V^+ - V^-)$, so that, using the explicit expressions for $V^{\pm}$ in Eq. (5.6.13), the current can be written in molar units as

$$\begin{aligned} i = -Fn\{&\Omega\exp(-\Delta\mu_+^{0\ddagger}/k_BT)\exp[-(1-\alpha)en\phi/k_BT]a_o \\ &- \Omega\exp(-\Delta\mu_-^{0\ddagger}/k_BT)\exp[\alpha en\phi/k_BT]a_r\}. \end{aligned} \tag{5.6.30}$$

We now use Eq. (5.6.28) to rewrite ϕ in this equation as $\phi = \eta + \phi^e$ and then apply the condition of detailed balance in Eq. (5.6.15) to get

$$\begin{aligned} i = Fn\Omega\exp(-\Delta\mu_+^{0\ddagger}/k_BT)&\exp[-(1-\alpha)en\phi^e/k_BT]a_o\{\exp[\alpha en\eta/k_BT] \\ &- \exp[-(1-\alpha)en\eta/k_BT]\}. \end{aligned} \tag{5.6.31}$$

Defining the *exchange current* to be

$$i_0 = Fn\Omega\exp(-\Delta\mu_+^{0\ddagger}/k_BT)\exp[-(1-\alpha)en\phi^e/k_BT]a_o \tag{5.6.32}$$

and using molar instead of molecular quantities, Eq. (5.6.31) becomes

$$i = i_0\{\exp[\alpha nF\eta/RT] - \exp[-(1-\alpha)nF\eta/RT]\}, \tag{5.6.33}$$

which is the Butler–Volmer equation.

If the overpotential is large and negative, the reduction process in Eq. (5.6.33) is favored. In this case

$$i = -i_0\exp[-(1-\alpha)nF\eta/RT) \tag{5.6.34}$$

or, taking logarithms,

$$\ln|i| = \ln i_0 - (1 - \alpha)nF\eta/RT. \tag{5.6.35}$$

This can be rearranged to give

$$\eta = [RT/(1 - \alpha)nF]\ln i_0 - [RT/(1 - \alpha)nF]\ln|i|. \tag{5.6.36}$$

Thus at high negative overpotentials there is a linear relationship between η and the logarithm of the magnitude of the current. This relationship was discovered experimentally by Tafel, and Eq. (5.6.36) is known as *Tafel's equation.* A comparable equation holds when the anodic process dominates.

Written in the form of Eq. (5.6.29), the Butler–Volmer equation provides another example of the canonical form. Using the notation introduced in Eq. (5.6.32), Eq. (5.6.29) becomes

$$(dq/dt)_{rx} = en\bar{\Omega}\{\exp[-(1 - \alpha)en\eta/k_BT] - \exp[\alpha en\eta/k_BT]\}, \tag{5.6.37}$$

where $\bar{\Omega} \equiv (i_0/Fn)$. The extensive variable changing in Eq. (5.6.37) is the electrode charge, q. The intensive variable that is thermodynamically conjugate to the charge is

$$F_q = (\partial S/\partial q) = -\eta/T. \tag{5.6.38}$$

Thus, according to the canonical interpretation, Eq. (5.6.37) represents an elementary process involving a charge of $-\alpha en$ in the forward direction and a charge of $(1 - \alpha)en$ in the reverse direction. The difference of these elementary charges is $\omega = (1 - \alpha)en - (-\alpha en) = en$, the net change of electrode charge. Thought of in this way, Eq. (5.6.37) is the canonical formula for the rate of change of charge on the electrode due to the electrochemical reaction, written in terms of the electrode potential. This is a contracted version of the complete thermodynamic-level description provided by Eq. (5.6.30).

5.7. Fluctuations Caused by Electrochemical Reactions

As we have seen in the previous section, the equations that govern the kinetics of electrode processes have the canonical form. For a macroscopic system the canonical form dictates both the average behavior and fluctuations. In this section we continue our analysis of the statistical nature of electrode processes, focusing on the equations governing the time rate of change of the electrode potential, ϕ. Because both the average behavior and the fluctuations depend on the details of the kinetic mechanism, we concentrate our attention on the simple oxidation–reduction reaction in Eq. (5.6.3), i.e.,

$$A^{z+} + ne^- \rightleftarrows A^{z-n}. \tag{5.7.1}$$

For simplicity the ions A^{z+} and A^{z-n} are supposed dissolved in an inert solvent and electron transfer occurs from an inert electrode such as platinum. The canonical form which describes the rate of this elementary process is given in Eq. (5.6.13).

The electrode process in Eq. (5.7.1) actually occurs in conjunction with an external circuit. This circuit will supply a current i_{ext}. Because of conservation of charge

$$i_{\text{ext}} = i_T, \tag{5.7.2}$$

where i_T is the current flowing through the electrode. Part of this current is the oxidation–reduction current given by Eq. (5.6.30). The other contribution to the current comes from the layer of charges that accumulates around the electrode. Whenever the voltage of the electrode changes, the relative positions of these charges change, leading to a capacitive current, $i_C = C\, d\phi/dt$, where C is the capacitance of the electrode. Thus the condition of charge conservation can be written

$$i_{\text{ext}} = C\, d\phi/dt + i \tag{5.7.3}$$

with i given in Eq. (5.6.31). To use the conservation equation, we consider the external current to be an explicit function of time, $i_{\text{ext}}(t)$, and recall that the left-hand side of Eq. (5.7.3) will be valid on conditional average. Thus

$$\begin{aligned} \bar{C}\, d\bar{\phi}/dt = {} & i_c \exp[-(1-\alpha)en\bar{\phi}/k_B T]\bar{a}_o \\ & - i_a \exp[\alpha en\bar{\phi}/k_B T]\bar{a}_r + i_{\text{ext}}(t) \end{aligned} \tag{5.7.4}$$

where from Eq. (5.6.30) i_c and i_a are defined by

$$\begin{aligned} i_c &= Fn\Omega \exp(-\Delta\mu_+^{0\ddagger}/k_B T) \\ i_a &= Fn\Omega \exp(-\Delta\mu_-^{0\ddagger}/k_B T) \end{aligned} \tag{5.7.5}$$

and depend only on the temperature.

Equation (5.7.4) describes the average change in the electrode voltage. Because of the electrode reaction, the numbers of ions A^{z+} and A^{z-n} are also changing. Conservation of atoms for the process in Eq. (5.7.1) shows that the sum of the molecule numbers, $n_o + n_r$, is constant. Thus it suffices to take only one of the molecule numbers as an independent variable, say n_o. According to our basic *ansatz*, the conditional average density $\bar{\rho}_o = \bar{n}_o/V$ satisfies the equation

$$\begin{aligned} d\bar{\rho}_o/dt = {} & -(i_c/nFV)\exp[-(1-\alpha)en\bar{\phi}/k_B T]\bar{a}_o \\ & + (i_a/nFV)\exp[\alpha en\bar{\phi}/k_B T]\bar{a}_r. \end{aligned} \tag{5.7.6}$$

Because the activity a_r can be expressed as a function of $\bar{\rho}_o$ and the total density of A^{z+} and A^{z-n}, Eq. (5.7.6) is the remaining dynamical equation needed to describe the conditional average changes of ϕ and ρ_o.

Equations (5.7.4) and (5.7.6) are coupled nonlinear equations whose solutions describe the conditional average. This average will have a persistent transient component if the external current has persistent time dependence.

The description of the average equations simplifies considerably, however, if the volume of the electrolyte solution, V, is sufficiently large. This can be seen by comparing Eqs. (5.7.5) and (5.7.6), which shows that the time rate of change of $\bar{\rho}_o$ is proportional to Ω/V, where Ω is the intrinsic rate of the electrode process. Now Ω is proportional to the area of the electrode, A. Thus it follows from Eq. (5.7.6) that $d\bar{\rho}_o/dt$ vanishes in proportion to (A/V) as $V \to \infty$. This means that $\bar{\rho}_o$ as well as the activities $\bar{a}_o$ and $\bar{a}_r$ will be constant in this limit. Consequently, Eq. (5.7.4) can be written

$$\bar{C}\, d\bar{\eta}/dt = i_0\{\exp[-(1-\alpha)en\bar{\eta}/k_BT] - \exp[\alpha en\bar{\eta}/k_BT]\} + i_{\text{ext}}(t), \quad (5.7.7)$$

where η is the overpotential and i_0 is the exchange current defined in Eqs. (5.6.28) and (5.6.32). For a large reservoir of electrolyte, Eq. (5.7.7) provides the average dynamical description of the electrode process in Eq. (5.7.1).

Equation (5.7.7) is a highly nonlinear equation and is difficult to solve in general. It can be cast into a somewhat simpler form by making the change of variable

$$X = \exp[(1-\alpha)en\bar{\eta}/k_BT], \quad (5.7.8)$$

which gives

$$dX/dt = [(1-\alpha)/R\bar{C}][1 - X^{1/(1-\alpha)} + (i_{\text{ext}}/i_0)X] \quad (5.7.9)$$

where

$$R = k_BT/eni_0. \quad (5.7.10)$$

By linearizing Eq. (5.5.7) around the equilibrium state $\eta^e = 0$, the constant R is seen to satisfy

$$C^e d\bar{\eta}/dt = -\bar{\eta}/R + i_{\text{ext}}. \quad (5.7.11)$$

Thus if i_{ext} is constant, Eq. (5.7.11) shows that after the transient capacitive current has damped out the following relationship holds:

$$Ri_{\text{ext}} = \eta^{ss}. \quad (5.7.12)$$

Equation (5.7.12) is Ohm's law expressed for this electrode process and R is the *electrode resistance.*

The natural time scale in Eq. (5.7.9) is $R\bar{C}/(1-\alpha)$. Changing variables to the dimensionless time $\tau = t(1-\alpha)/R\bar{C}$ and the dimensionless current $i' = i_{\text{ext}}/i_0$, the equation becomes

$$dX/d\tau = 1 - X^{1/(1-\alpha)} + i'X. \quad (5.7.13)$$

Assuming a constant external current, the steady states of Eq. (5.7.13) are given by

$$0 = 1 - X^{ss\,1/(1-\alpha)} + i'X^{ss}. \quad (5.7.14)$$

Equation (5.7.13) is difficult to solve for arbitrary $0 \leq \alpha \leq 1$, even if the external current is constant. However, the symmetric case $\alpha = \frac{1}{2}$, which is relatively common experimentally, can solved analytically if i' is constant. For $\alpha = \frac{1}{2}$, Eq. (5.7.13) becomes

$$dX/d\tau = 1 - X^2 + i'X \tag{5.7.15}$$

or

$$dX/(1 - X^2 + i'X) = d\tau. \tag{5.7.16}$$

Integrating both sides of Eq. (5.7.16) gives

$$\ln\left[\frac{-2X + i' - (i'^2 + 4)^{1/2}}{-2X + i' + (i'^2 + 4)^{1/2}}\right] = (i'^2 + 4)^{1/2}\tau + C, \tag{5.7.17}$$

with C an integration constant. Exponentiating both sides, writing $X(0) = X^0$, and defining

$$A = \frac{-2X^0 + i' + (i'^2 + 4)^{1/2}}{-2X^0 + i' - (i'^2 + 4)^{1/2}}, \tag{5.7.18}$$

the solution for X is found to be

$$X(X^0, \tau) = \frac{i'[1 - A\exp(\beta\tau)] - \beta[1 + A\exp(\beta\tau)]}{2[1 - A\exp(\beta\tau)]} \tag{5.7.19}$$

where

$$\beta \equiv (i'^2 + 4)^{1/2}. \tag{5.7.20}$$

Notice that as $\tau \to \infty$,

$$X \to X^{ss} = [i' + (i'^2 + 4)^{1/2}]/2, \tag{5.7.21}$$

which is the appropriate solution of the steady-state equation (5.7.14) for $\alpha = \frac{1}{2}$. Thus, using Eq. (5.7.8), the time dependence of the conditional overvoltage for $\alpha = \frac{1}{2}$ is

$$\bar{\eta}(\eta^0, t) = (2k_B T/en)\ln X(X^0, \tau) \tag{5.7.22}$$

with $X(X^0, \tau)$ given in Eq. (5.7.19).

As given by Eqs. (5.7.19) and (5.7.22), the transient time behavior of the overvoltage is observable on the time scale dictated by the dimensionless time $\tau = t(1 - \alpha)/R\bar{C}$. Typical values of the electrode resistance are 10–10^4 ohm, whereas the capacitance of the double layer is of the order of 10^{-6} farad. Thus the characteristic time scale for relaxation to the steady state, $R\bar{C}/(1 - \alpha)$, is of the order of 10^{-5}–10^{-2} s. Consequently, except for very rapid measurements, the average dynamical state of the electrode will be a stationary state with a voltage determined by the solution to Eq. (5.7.14).

Fluctuations in the voltage around the conditional average are described by the general theory outlined in Chapter 4. Following the prescription given there, the fluctuations satisfy linearized versions of the average equation, with time-dependent coefficients which depend on the conditional average value,

$\bar{\eta}(\eta^0, t)$. Taking the capacitance as a constant, Eq. (5.7.7) then gives

$$\bar{C}d\delta\eta/dt = -(i_0 en/k_B T)\{(1-\alpha)\exp[-(1-\alpha)cn\bar{\eta}/k_B T] + \alpha\exp[\alpha en\bar{\eta}/k_B T]\}\delta\eta + \tilde{f}. \quad (5.7.23)$$

The random term, $\tilde{f}$, vanishes on the average and its variance is given by comparing the canonical form of Eq. (5.7.7), as given in Eqs. (5.6.37) and (5.6.38), to the canonical expression in Eq. (4.4.6). This gives

$$\langle \tilde{f}(t)\tilde{f}(t')\rangle = eni_0\{\exp[-(1-\alpha)en\bar{\eta}/k_B T] + \exp[\alpha en\bar{\eta}/k_B T]\}\delta(t-t'). \quad (5.7.24)$$

Analysis of the conditional fluctuations in the most general case involves complicated integrations similar to those carried out for the bimolecular isomerization reaction in Section 5.4. However, since the time course of these transients is short, it is sufficient to examine fluctuations close to the asymptotic steady state. In this case the coefficient of $\delta\eta$ on the right-hand side of Eq. (5.7.23), as well as the bracketed terms in Eq. (5.7.24), can be evaluated at steady state. This restricts our analysis to small fluctuations around steady state. This is justifiable as long as the single-time probability density, $W_1(\eta)$, for the steady-state ensemble is narrowly peaked around its steady-state value, η^{ss}. Using this approximation in Eq. (5.7.23) and (5.7.24) leads to the following stationary, Gaussian, Markov process:

$$d\delta\eta/dt = -\frac{1}{R\bar{C}}[(1-\alpha)X^{ss-1} + \alpha X^{ss(\alpha/1-\alpha)}]\delta\eta + \tilde{f} \quad (5.7.25)$$

$$\langle \tilde{f}(t)\tilde{f}(t')\rangle = 2k_B T[(X^{ss-1} + X^{ss(\alpha/1-\alpha)})/2\bar{C}^2 R]\delta(t-t'), \quad (5.7.26)$$

where X^{ss} and the resistance, R, as defined in Eqs. (5.7.8), (5.7.10), and (5.7.14), have been introduced. The factors $\bar{C}^{-1}$ in these equations come from dividing both sides of Eq. (5.7.23) by the capacitance.

As discussed in Section 5.1, voltage and current fluctuations are examined experimentally using the power spectrum, $S(\omega)$. For a stationary, Gaussian, Markov process we can use the result obtained in Eq. (1.8.45), which relates the frequency correlation function to the power spectrum, to write

$$\langle \delta\eta^2(\omega)\rangle^{ss}\Delta\omega = S(\omega)/2\pi. \quad (5.7.27)$$

Using the explicit result in Eq. (1.8.37) and the formulas in Eqs. (5.7.25) and (5.7.26) gives

$$\langle \delta\eta^2(\omega)\rangle\Delta\omega = \frac{k_B T(X^{ss-1} + X^{ss(\alpha/1-\alpha)})/2\bar{C}^2 R\pi}{\left[\dfrac{(1-\alpha)X^{ss-1} + \alpha X^{ss(\alpha/1-\alpha)}}{R\bar{C}}\right]^2 + \omega^2}. \quad (5.7.28)$$

Equation (5.7.28) simplifies considerably if the steady state is equilibrium, that is, no current is flowing. In this case Eq. (5.7.14) shows that $X^{ss} = 1$ and

Eq. (5.7.28) reduces to

$$\langle \delta\eta^2(\omega)\rangle^e \Delta\omega = \frac{k_B T/\bar{C}^2 R\pi}{\left(\frac{1}{R\bar{C}}\right)^2 + \omega^2}. \tag{5.7.29}$$

Recalling that typical values of the relaxation time $R\bar{C}$ are in the range 10^{-2}–10^{-5} s, the term $(1/R\bar{C})^2$ can be neglected for frequencies considerably less than 10^2–10^5 Hz. In this low-frequency range Eq. (5.7.29) takes the form

$$\langle \delta\eta^2(\omega)\rangle = k_B TR/\pi. \tag{5.7.30}$$

This is identical to the Nyquist formula for the Johnson noise of an electrolyte solution in Eq. (5.1.35). The only difference is the cause of the resistance, which in Eq. (5.7.30) is the rate of the electrochemical processes at the electrode.

At nonequilibrium steady states a net current is passing through the electrode, and Eq. (5.7.28) deviates from the Nyquist result. At low frequencies it reduces to

$$\langle \delta\eta^2(\omega)\rangle \Delta\omega = (k_B TR^*/\pi)\frac{(1 + X^{ss\ 1/1-\alpha})}{2([1-\alpha] + \alpha X^{ss\ 1/1-\alpha})}, \tag{5.7.31}$$

where $R^* = R/([1-\alpha] + \alpha X^{ss\ 1/1-\alpha})$ is the differential resistance obtained from $di/d\eta$ [cf. Eq. (5.7.25)]. The second factor in Eq. (5.7.31) depends on the current through the solution X^{ss}, in Eq. (5.7.14). The special case of the symmetric oxidation–reduction process, i.e. $\alpha = \frac{1}{2}$, is interesting since then the second factor is easily seen to equal unity. For the oxidation–reduction reaction of H^+ and H_2 on mercury the measured value of α is 0.50, and thus Eq. (5.7.31) predicts that the spectrum of voltage fluctuations should be independent of the applied current. The oxidation–reduction reaction of Ce^{4+} and Ce^{3+} on platinum, on the other hand, has a measured value of $\alpha = 0.75$. For this electrode process, the correction factor in Eq. (5.7.31) to the Nyquist formula can be written

$$C^* = 2(2 + i'X^{ss})/(4 + 3i'X^{ss}), \tag{5.7.32}$$

where α has been set equal to $\frac{3}{4}$ and Eq. (5.7.14) has been used to rewrite $X^{ss\ 1/1-\alpha}$ in terms of X^{ss}. The value of X^{ss} for a given value of i' can be obtained by rewriting Eq. (5.7.14) in the form

$$X^{ss} = (1 + i'X^{ss})^{(1-\alpha)} \tag{5.7.33}$$

and solving by iteration. For $\alpha = \frac{3}{4}$, the solutions are given for a range of i' values in Fig. 5.9 along with the value of the correction factor C^*. For this case the corrections to the Nyquist result become significant when the dimensionless current i' is of order unity. Since $i' = i_{ext}/i_0$, this means that when the external current is of the order of the exchange current, the voltage fluctuations become measurably different from those encountered at equilibrium. The exchange current for the Ce^{4+}, Ce^{3+} platinum electrode is about

FIGURE 5.9. Calculated values of $X^{ss} = \exp[(1-\alpha)en\eta^{ss}/k_BT]$ (Eq. 5.7.33) and the correction factor, C^*, to the Nyquist formula (Eq. 5.7.32) for the half-reaction $Ce^{4+} + e^- = Ce^{3+}$ as a function of the scaled external current $i' = i/i_0$.

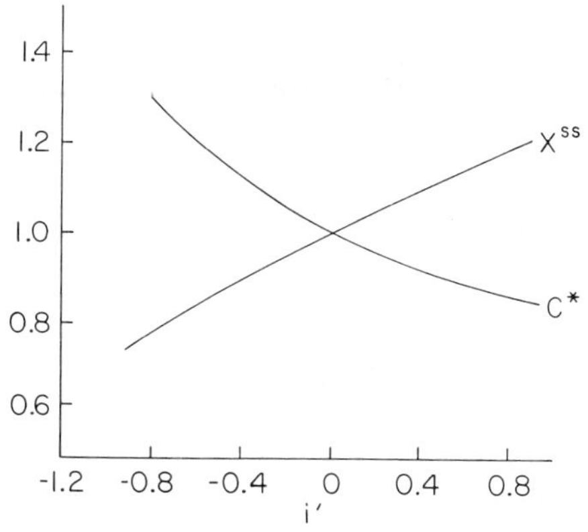

0.1 mA cm^{-2} under standard conditions. Thus it takes only a small current to significantly alter the power spectrum of the voltage. This type of effect, namely, an external flux producing an altered state of matter, turns out to be quite general. In Chapter 8 we examine the thermodynamic consequences of this phenomenon.

5.8. Ion Transport through Biological Membranes

Membranes in biological systems provide a rich source of mechanisms for the transport of electrical current. Two general mechanisms, one involving ion carriers and the other involving ion channels, seem to predominate. Indeed, living cells are filled with and surrounded by aqueous solutions, and ions are the current carriers in biological systems. The aqueous solution in a cell is separated by membranes that divide the cell into compartments. In animal cells the outer or *plasma membrane* provides an external permeability barrier for the whole cell. Biological membranes are composed primarily of lipids, protein, and carbohydrates. The phospholipids form a structural bilayer, with their terminal phosphate groups pointing outward into the two aqueous phases and their terminal hydrocarbon groups forming the interior of the bilayer. Permeability of the bilayer to ionic species is conveyed by proteins, which form membrane channels, or by small polypeptides dissolved in the lipid bilayer, which can act as carriers of specific ions. Protein channels usually exhibit a fair degree of specificity. For example, the plasma membrane of rat muscle contains a channel which efficiently transports Cl^- anions but not K^+ cations. The conductance of protein channels can also be modulated by binding ions that they do not transport, e.g., certain K^+-channels in rat muscle are activated by binding Ca^{2+}, as well as by changes in the electrical potential across the membrane.

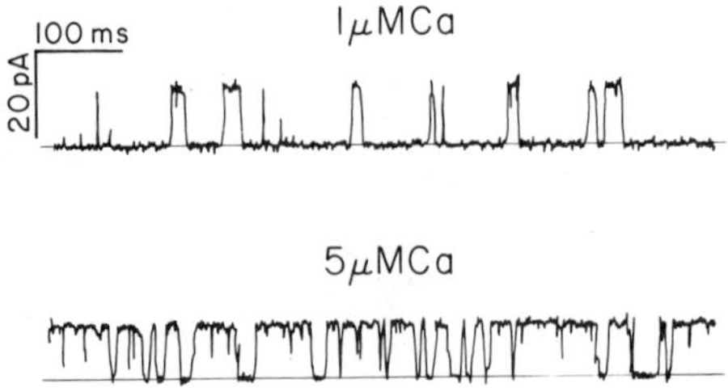

FIGURE 5.10. Single-channel records of the current through the Ca^{2+}-activated potassium channel from rat skeletal muscle using the patch clamp technique. The top record was recorded with an external calcium ion concentration of 1 μM, which was increased to 5 μM in the lower record. Data taken from E. Moczydlowski and R. Latorre, *J. Gen. Physiol.* **82**, 511 (1983).

Because membranes can pass current, it is possible to measure their electrical properties. Indeed, the electrophysiology of membranes is a highly developed discipline. From the vantage point of physical science, membranes provide an interesting testing ground for nonlinear effects caused by molecular mechanisms. The earliest experiments on membranes involved measuring the resistance or power spectrum of macroscopic patches of membrane. Recently, using a technique called the *patch clamp*, it has been possible to isolate a small portion of membrane containing one or just a few channels. Using sensitive amplifiers, this technique allows the conductance of single channels to be determined and permits one to examine the properties of one kind of channel in isolation from others in the membrane. A typical current record of such a patch clamp measurement is shown in Fig. 5.10. There one sees the random opening and closing of a single Ca^{2+}-activated K^{+}-channel from rat muscle. The baseline represents the background current. Each time the channel is conducting there is a burst of current corresponding to the conductance of the single channel. The effect of Ca^{2+} on the channel is to increase the amount of time the channel is open. By systematically studying channels in this way, it has been possible to determine the elementary molecular processes which govern their electrical properties.

To illustrate the application of the mechanistic theory to membrane transport, three mechanisms for membrane transport will be considered. The first involves *symmetric* channels. For a symmetric channel there is no difference between the kinetic parameters of the inside and the outside. As an example, we consider a channel that can be in a variety of internal states numbered $\alpha = 1, 2, \ldots, k$. Each state transmits an ion of interest at the intrinsic rate Ω_α. The elementary processes of conduction, thus, involve a single channel in the state α and a single ion, A^z, where z is a small whole positive or negative integer. Symbolically

$$\alpha + A_i^z \rightleftarrows \alpha + A_o^z, \tag{5.8.1}$$

where A_i^z represents the ion inside the membrane and A_o^z represents the ion outside. According to the canonical theory, the average time rate of change of the number of ions in the inside compartment, $\bar{n}_i$, has the form

$$d\bar{n}_i/dt = -\sum_\alpha \Omega_\alpha \exp(\bar{\mu}_\alpha/k_B T)[\exp(\tilde{\bar{\mu}}_i/k_B T) - \exp(\tilde{\bar{\mu}}_o/k_B T)], \tag{5.8.2}$$

where $\bar{\mu}_\alpha$ is the chemical potential of the channels in the state α, $\tilde{\bar{\mu}}_i$ is the electrochemical potential of the ions inside, and $\tilde{\bar{\mu}}_o$ is the electrochemical potential outside, all evaluated at the average conditions.

Since the membrane is symmetric, $\bar{\mu}_\alpha$ does not depend on the potential difference across the membrane, ϕ, which is defined by convention as the difference of the electrical potential inside minus outside, i.e.,

$$\phi_i - \phi_o = \phi. \tag{5.8.3}$$

A reference electrode taken as $\phi = 0$ is usually placed in the outside medium. Thus it is consistent to take $\phi_i = \phi$, $\phi_o = 0$, and to write the electrochemical potentials as

$$\begin{aligned} \tilde{\bar{\mu}}_i &= \mu^0 + k_B T \ln \bar{a}_i + ez\phi \\ \tilde{\bar{\mu}}_o &= \mu^0 + k_B T \ln \bar{a}_o. \end{aligned} \tag{5.8.4}$$

Membrane channels are present at rather low densities in membranes, usually no more than 50 or 100 per cell. This implies that, on the average, channels are about 3 μm apart and that they can be treated as a dilute two-dimensional solution. Consequently, chemical potentials can be written

$$\bar{\mu}_\alpha = \mu_\alpha^0 + k_B T \ln(\bar{n}_\alpha / A), \tag{5.8.5}$$

where A is the area of the membrane and n_α is the number of channels in state α. Substituting Eqs. (5.8.4) and (5.8.5) into Eq. (5.8.2) yields

$$d\bar{n}_i/dt = \sum_\alpha \bar{\Omega}_\alpha \exp[(\mu_\alpha^0 + \mu^0)/k_B T]\bar{n}_\alpha[\bar{a}_i \exp(ez\phi/k_B T) - \bar{a}_o], \tag{5.8.6}$$

where $\bar{\Omega}_\alpha \equiv \Omega_\alpha / A$.

To complete the kinetic description of the symmetric channel it is necessary to describe the mechanism by which the channel states change. Although for a symmetric channel the rate of transport of the ion depends on the states of the channel, the states of the channel are independent of the transport process. This is because the conduction process in Eq. (5.8.1) does not alter the state of the channel. The simplest scheme of this type involves one open and one closed channel state, O and C, and is described by the linear elementary process

$$\mathrm{O} \rightleftarrows \mathrm{C}.$$

This elementary process is identical to the isomerization reaction $\mathrm{A} \rightleftarrows \mathrm{B}$ whose statistical thermodynamics were examined in Sections 3.6 and 3.7. The only difference is that this process occurs in a two-dimensional solution rather than a three-dimensional solution so that number densities are per unit area, not volume. A nonlinear mechanism is required to explain the current characteristics of the gramicidin channel. Gramicidin is a polypeptide consisting of 15 amino acid residues. It is lipid soluble and in the membrane undergoes the dimerization reaction

$$G + G \rightleftarrows G_2. \tag{5.8.7}$$

The dimer has an elongated donut-like structure and is capable of spanning the membrane. Thus only the species G_2 conducts current and its rate of ion transport, as described by Eq. (5.8.6), is proportional to $\bar{n}_{G_2}$.

Other more complicated mechanisms have also been uncovered. Experiments on the Ca^{2+}-activated K^+ channel, like those shown in Fig. 5.10, have led to the following mechanism:

$$\begin{aligned} C + Ca_i^{2+} &\rightleftarrows C \cdot Ca^{2+} \\ C \cdot Ca^{2+} &\rightleftarrows O \cdot Ca^{2+} \\ O \cdot Ca^{2+} + Ca_i^{2+} &\rightleftarrows O \cdot (Ca^{2+})_2. \end{aligned} \tag{5.8.8}$$

This sequence of elementary processes involves a closed state, C, which can bind calcium ions on the inside of the membrane. This bound state is represented by $C \cdot Ca^{2+}$. The open channel state appears only in conjunction with one or two bound calcium ions, $O \cdot Ca^{2+}$ or $O \cdot (Ca^{2+})_2$. The elementary processes in Eq. (5.8.8) are nonlinear, which means that a complete stochastic description of these processes is complicated.

There are circumstances under which nonlinear mechanisms like that in Eq. (5.8.8) become effectively linear. Indeed, if the concentration of Ca^{2+} ions is maintained by a large reservoir, then its concentration fluctuations will be negligible. To see this, consider the first elementary binding process in Eq. (5.8.8). It contribution to the rate of change of the number of calcium ions inside the membrane is

$$\begin{aligned} d\bar{n}_{Ca^{2+}}/dt = &-\Omega_1[\exp(\bar{\mu}_C/k_B T)\exp(\bar{\mu}_{Ca^{+2}}/k_B T) \\ &- \exp(\bar{\mu}_{C \cdot Ca^{2+}}/k_B T)]. \end{aligned} \tag{5.8.9}$$

The intrinsic rate of this process, Ω_1, is proportional to the area of the membrane, A. Thus dividing by the volume of the reservoir, V, shows that the time rate of change of the density of Ca^{2+} ions is proportional to (A/V), which vanishes for a large reservoir. This means that the chemical potential of Ca^{2+} is fixed, so that the rate of the elementary process of binding takes the form

$$\bar{V}^+ - \bar{V}^- = \Omega_1^+ \exp(\bar{\mu}_C/k_B T) - \Omega_1^- \exp(\bar{\mu}_{C \cdot Ca^{2+}}/k_B T), \tag{5.8.10}$$

where the new intrinsic rates are

$$\begin{aligned} \Omega_1^+ &= \Omega_1 \exp(\bar{\mu}_{Ca^{2+}}/k_B T) \\ \Omega_1^- &= \Omega_1. \end{aligned} \tag{5.8.11}$$

Equation (5.8.10) corresponds to the linear mechanism

$$C \rightleftarrows C \cdot Ca^{2+}. \tag{5.8.12}$$

A somewhat more complicated argument, using the stochastic equations implied by Eq. (5.8.9), shows that this linear process can also be used to describe fluctuations in the densities of C and $C \cdot Ca^{2+}$. If the reservoir of Ca^{2+} ions is not large, as might be the case within a living cell, then the nonlinear elementary process in Eq. (5.8.8) must be utilized.

The current carried by the ion A^z through a symmetric membrane channel of the sort described by Eq. (5.8.1) can be obtained from Eq. (5.8.2). Indeed

$$i = -ezd\bar{n}_i/dt, \tag{5.8.13}$$

where the sign is chosen so that positive charges flowing outward give rise to a positive current. Just as for the electrochemical processes discussed in Section 5.8, it is natural to introduce the Nernst potential of the ionic species A^z. It is given by

$$\phi_N = (k_B T/ez)\ln(\bar{a}_o/\bar{a}_i), \tag{5.8.14}$$

where the activities of the ion inside and outside are defined by Eq. (5.8.4). Substituting this expression into Eq. (5.8.2) and using the definition of the current in Eq. (5.8.13) gives

$$i = ez \sum_{\alpha} \Omega'_\alpha n_\alpha \bar{a}_o \{\exp[ez(\phi - \phi_N)/k_B T] - 1\} \tag{5.8.15}$$

where

$$\Omega'_\alpha = \bar{\Omega}_\alpha \exp[(\mu^0_\alpha + \mu^0)/k_B T]. \tag{5.8.16}$$

The characteristic voltage in this equation is $k_B T/e$, which at room temperature is about 26 mV. Thus when the membrane potential is of the order of $\phi \approx \phi_N \pm 26$ mV, it is possible to linearize by expanding the exponential in Eq. (5.8.15) around the Nernst potential to get

$$\begin{aligned} i &= [(ez)^2/k_B T]\bar{a}_o\left(\sum_{\alpha} \Omega'_\alpha n_\alpha\right)(\phi - \phi_N) \\ &\equiv G(\phi - \phi_N). \end{aligned} \tag{5.8.17}$$

This is Ohm's law for the symmetric channel and G is called the conductance of the A^z-channel. When the concentration difference of A^z across the membrane vanishes, then the Nernst potential is zero and $\bar{a}_o$ in Eq. (5.8.17) is an equilibrium value. Otherwise G and ϕ_N are constant only insofar as $\bar{a}_i$ and $\bar{a}_o$ are maintained at constant values.

Other mechanisms in which the membrane channel has an asymmetric structure are easy to imagine. Consider, again, that the channel can be described by states, but that several states are involved in the transport of an ion. A simple example is the sequence of elementary processes

$$\begin{aligned} \alpha + A^z_i &\rightleftarrows \alpha \cdot A^z \\ \alpha \cdot A^z &\rightleftarrows \alpha' \cdot A^z \\ \alpha' \cdot A^z &\rightleftarrows \alpha' + A^z_o. \end{aligned} \tag{5.8.18}$$

In this sequence an A^z ion on the inside binds to the channel state α. This is followed by a change of conformation of the channel to state α', which in turn releases the ion to the outside. An abbreviated form of this sequence, which is still asymmetric, is obtained by adding the three elementary processes in Eq. (5.8.18). The resulting elementary complex process can be written symbolically as

$$\alpha + A_i^z \rightleftarrows \alpha' + A_o^z. \tag{5.8.19}$$

Any process like Eq. (5.8.18), which involves spatial locations of the ion other than inside and outside, adds a new complexity to the description of the electrical current. Indeed, the current of A^z through the first step in Eq. (5.8.18) does not have to equal that through the final step. Thus there are inward and outward currents in the membrane that may differ. Although these differences will be transient, they complicate the description of the current. For this reason, Eq. (5.8.19) provides a simpler description and we treat it as if it were an elementary process.

The molecular process in Eq. (5.8.19) is characterized by the two states α and α'. To remind us of their asymmetry with respect to the membrane, we write the intrinsic rate of this elementary process as $\Omega_{\alpha,\alpha'}$, where the first subscript represents the inside and the second the outside. Using the canonical form, the rate of change of the number of A^z ions inside is given by

$$\begin{aligned} d\bar{n}_i/dt = &-\sum_{\alpha,\alpha'} \Omega_{\alpha,\alpha'} \exp(\mu^0/k_B T)[\bar{a}_i \exp(ez\phi/k_B T)\exp(\bar{\mu}_\alpha/k_B T) \\ &- \bar{a}_o \exp(\bar{\mu}_{\alpha'}/k_B T)]. \end{aligned} \tag{5.8.20}$$

The mathematical asymmetry of this sort of process becomes apparent if the explicit expressions for the chemical potentials of the channels in Eq. (5.8.5) are introduced in Eq. (5.8.20). This lead to

$$d\bar{n}_i/dt = -\sum_{\alpha,\alpha'} \bar{\Omega}^+_{\alpha,\alpha'} \bar{n}_\alpha \bar{a}_i \exp(ez\phi/k_B T) - \bar{\Omega}^-_{\alpha,\alpha'} \bar{n}_{\alpha'} \bar{a}_o, \tag{5.8.21}$$

where

$$\begin{aligned} \bar{\Omega}^+_{\alpha,\alpha'} &= \Omega_{\alpha,\alpha'} \exp[(\mu^0_\alpha + \mu^0)/k_B T]/A \\ \bar{\Omega}^-_{\alpha,\alpha'} &= \Omega_{\alpha,\alpha'} \exp[(\mu^0_{\alpha'} + \mu^0)/k_B T]/A \end{aligned} \tag{5.8.22}$$

The asymmetry in Eq. (5.8.21) is expressed by the fact that the modified intrinsic rates, $\bar{\Omega}^\pm_{\alpha,\alpha'}$, are different, in contrast to what was found for the symmetric channel in Eq. (5.8.6).

Another important difference between symmetric and asymmetric channels is that the asymmetric mechanism couples changes of channel state to ion transport. Indeed, the elementary process in Eq. (5.8.19), which is responsible for conduction, also involves a change of state from α to α'. Thus, in addition to any other kinetic processes that are occurring one needs to consider contributions to the rate of change of $\bar{n}_\alpha$ like

$$(d\bar{n}_\alpha/dt)_{\alpha,\alpha'} = -\bar{\Omega}^+_{\alpha,\alpha'}\bar{a}_i \exp(ez\phi/k_B T)\bar{n}_\alpha + \bar{\Omega}^-_{\alpha,\alpha'}\bar{a}_o\bar{n}_{\alpha'}. \tag{5.8.23}$$

If conditions on both sides of the membrane keep the membrane potential and activities constant, then Eq. (5.8.23) is simply a linear process contributing to the state changes. Notice, however, that even in this case the apparent rate constants in Eq. (5.8.23), e.g.,

$$k^+_{\alpha,\alpha'} \equiv \bar{\Omega}^+_{\alpha,\alpha'}\bar{a}_i \exp(ez\phi/k_B T), \tag{5.8.24}$$

depend explicitly on the membrane potential.

Another difference between symmetric and asymmetric models is that the usual Nernst potential no longer locates the equilibrium potential for the elementary process in Eq. (5.8.19). This can be seen by rearranging Eq. (5.8.21) into the form

$$(d\bar{n}_i/dt) = -\sum_{\alpha,\alpha'} \bar{\Omega}^-_{\alpha,\alpha'}\bar{a}_o(\exp\{[ez(\phi - \phi_N) + (\mu^0_\alpha - \mu^0_{\alpha'})]/k_B T\}\bar{n}_\alpha - \bar{n}_{\alpha'}). \tag{5.8.25}$$

The terms in the brackets do not vanish when $\phi = \phi_N$, and it is not in general possible to write Eq. (5.8.25) in the form of Ohm's law. According to the canonical form in Eq. (5.8.20), the driving force is more appropriately expressed as the electrochemical difference

$$\mathscr{A}_{\alpha,\alpha'} = -(\bar{\mu}_{\alpha'} + \tilde{\bar{\mu}}_o) + (\bar{\mu}_\alpha + \tilde{\bar{\mu}}_i), \tag{5.8.26}$$

which is the analogue of the chemical affinity for the elementary process in Eq. (5.8.19). Using explicit expressions on the right-hand side of Eq. (5.8.26), the affinity can be written

$$\mathscr{A}_{\alpha,\alpha'} = \mu^0_\alpha - \mu^0_{\alpha'} + k_B T \ln(\bar{a}_\alpha\bar{a}_i/\bar{a}_{\alpha'}\bar{a}_o) + ez\phi. \tag{5.8.27}$$

One might think to define a modified Nernst potential by

$$ez\phi'_N = \mu^0_{\alpha'} - \mu^0_\alpha + k_B T \ln(\bar{a}_{\alpha'}\bar{a}_o/\bar{a}_\alpha\bar{a}_i), \tag{5.8.28}$$

so that the driving force becomes

$$\mathscr{A}_{\alpha,\alpha'} = ez(\phi - \phi'_N). \tag{5.8.29}$$

This modified Nernst potential, however, depends on α and α' and differs for each elementary process. Nonetheless, if there is only a single crossing process, then Eq. (5.8.25) can be linearized around ϕ'_N to give the current equation

$$i = -ez(dn_i/dt)_{\alpha,\alpha'} = G_{\alpha,\alpha'}(\phi - \phi'_N) \tag{5.8.30}$$

where

$$G_{\alpha,\alpha'} \equiv [(ez)^2/k_B T]\bar{\Omega}^-_{\alpha,\alpha'}\bar{a}_o\bar{n}_{\alpha'}. \tag{5.8.31}$$

If there are several elementary transport processes, it still may be possible to write an approximate Ohm's law expression like Eq. (5.8.20). For example, if $\bar{a}_\alpha \approx \bar{a}_{\alpha'}$ and $\mu^0_\alpha \approx \mu^0_{\alpha'}$, then the modified Nernst potential ϕ'_N is essentially the same as the usual Nernst potential in Eq. (5.8.14). In this special case, we can sum over the elementary processes α, α' in Eqs. (5.8.30) to obtain

$$i \approx [(ez)^2/k_B T] \sum_{\alpha,\alpha'} \bar{\Omega}^-_{\alpha,\alpha'} a_o n_{\alpha'} (\phi - \phi_N)$$
$$\equiv G(\phi - \phi_N), \qquad (5.8.32)$$

which is like the expression in Eq. (5.8.17) for the symmetric channel.

The second important class of membrane transport processes involves molecular carriers. These molecules posses ionic groups as well as aliphatic, aromatic, or lipid residues. These groups provide ion binding sites and convey solubility in both aqueous and lipid solutions. A simple example is the tetraphenylboron ion with the structure

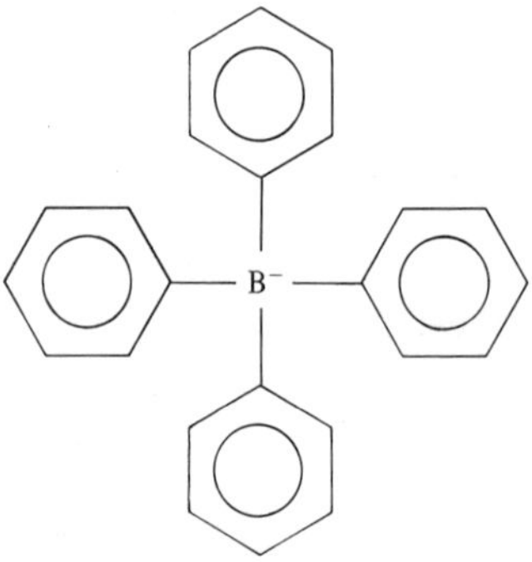

(5.8.33)

Tetraphenylboron binds Na^+ ions and when added to 0.1 *M* NaCl solutions, allow Na^+ ions to pass across phospholipid bilayers. Other nonionic carrier molecules, such as the dodecapeptide valinomycin, are macrocycles. By virtue of an appropriate-sized internal cavity, macrocycles can cage ions and, thence, transport them across bilayers. Because carriers are soluble in both bilayer and solution, they are constantly hopping on and off the membrane. This turns out to be a relatively slow process. Thus the basic processes involved in ion transport are the binding of ions to the carrier and the movement of the carrier from one side of the membrane to the other.

Using C_i to represent a carrier at the inside of the membrane while C_o represents carriers at the outside, these processes can be written

$$C_i \rightleftarrows C_o$$
$$A^z + C_i \rightleftarrows A^z \cdot C_i$$
$$A^z \cdot C_i \rightleftarrows A^z \cdot C_o$$
$$A^z \cdot C_o \rightleftarrows A^z + C_o. \qquad (5.8.34)$$

The ion A^z can accumulate in the membrane, and there are several ion currents in the carrier scheme. It is customary to use the third membrane crossing step to define the current. The canonical form for the rate change caused by this process is

$$\bar{V}^+ - \bar{V}^- = \Omega_3[\exp(\tilde{\bar{\mu}}_{iA}/k_BT) - \exp(\tilde{\bar{\mu}}_{oA}/k_BT)], \tag{5.8.35}$$

where the electrochemical potentials for $A^z \cdot C_i$ and $A^z \cdot C_o$ are

$$\begin{aligned} \tilde{\bar{\mu}}_{iA} &= \mu^0_{CB} + k_BT\ln(\bar{n}_{iA}/A) + ez'\phi \\ \tilde{\bar{\mu}}_{oA} &= \mu^0_{CB} + k_BT\ln(\bar{n}_{oA}/A). \end{aligned} \tag{5.8.36}$$

In Eq. (5.8.36) ez' is the total charge of carrier plus ion, the subscript CB stands for "carrier-bound" and for simplicity we assume that the electrical potential at the locations of the species $A^z \cdot C_i$ and $A^z \cdot C_o$ is the same as the inside or outside the membrane. Hence

$$\bar{V}^+ - \bar{V}^- = \Omega_3 \exp(\mu^0_{CB}/k_BT)[(\bar{n}_{iA}/A)\exp(ez'\phi/k_BT) - (\bar{n}_{oA}/A)]. \tag{5.8.37}$$

The rate of this process can be used to describe the current for the overall transport mechanism either at steady state or under the condition that the binding processes in Eq. (5.8.34) involve rapid equilibria. When binding is rapid, the conditions of equilibrium applied to the second and fourth elementary processes in Eq. (5.8.34) give

$$\begin{aligned} \bar{n}_{iA} &= \bar{n}_i \bar{a}_i K_D \\ \bar{n}_{oA} &= \bar{n}_o \bar{a}_o K_D, \end{aligned} \tag{5.8.38}$$

where n_i and n_o are the number of carrier molecules at the inside and outside faces of the membrane and the dissociation constant is

$$K_D = \exp[(\mu^0 + \mu^0_C - \mu^0_{CB})/k_BT] \tag{5.8.39}$$

with μ^0_C the standard-state chemical potential of either C_i or C_o. Substituting these equalities into Eq. (5.8.37) gives the current

$$i = ez\bar{\Omega}_3[\bar{n}_i\bar{a}_i\exp(ez'\phi/k_BT) - \bar{n}_o\bar{a}_o], \tag{5.8.40}$$

with $\bar{\Omega}_3 \equiv (\Omega_3 K_D/A)$. Comparing this expression to Eqs. (5.8.6) and (5.8.23), the carrier mechanism is seen to provide a third distinct kind of transport equation.

To complete the description of the kinetics of the carrier mechanism, it is necessary to describe changes in n_i and $\bar{n}_o$. Since the binding processes in Eq. (5.8.34) are assumed to be in equilibrium, they do not contribute to these changes. Only the first elementary process does, and it gives

$$d\bar{n}_i/dt = -k^+\bar{n}_i + k^-\bar{n}_o \tag{5.8.41}$$

The rate constants $k^\pm$ depend on the membrane potential according to

$$\begin{aligned} k^+ &= \Omega_1 \exp(\mu^0_C/k_BT)\exp(ez_C\phi/k_BT) \\ k^- &= \Omega_1 \exp(\mu^0_C/k_BT), \end{aligned} \tag{5.8.42}$$

with ez_C the charge of the carrier molecule. If this process is rapid, the condition of equilibrium gives $\bar{n}_i = \bar{n}_o \exp(-ez_C\phi/k_B T)$ and the current equation (5.8.40) reduces to

$$i = ez\Omega_3'\bar{n}_o\bar{a}_o\{\exp[ez(\phi - \phi_N)/k_B T] - 1\}, \tag{5.8.43}$$

with ϕ_N the Nernst potential. Under these conditions, the expression for the current due to carriers is identical to the expression for the current of a symmetrical membrane in Eq. (5.8.15).

We close this section with an application of the symmetric channel model to a problem of importance in nerve physiology. There are many ion-conducting channels in the membranes of nervous tissue. Of these, Hodgkin, Huxley, and others working after World War II recognized the importance of two channels, a Na^+-channel which carries an inward current and a K^+-channel which carries an outward current. According to Hodgkin and Huxley, the current in these channels can be described by the following equations:

$$i_{K^+} = \bar{G}_K n^4(\phi - \phi_{K^+}) \tag{5.8.44}$$

$$i_{Na^+} = \bar{G}_{Na} m^3 h(\phi - \phi_{Na^+}) \tag{5.8.45}$$

where

$$dn/dt = -\beta_n n + \alpha_n(1 - n) \tag{5.8.46}$$

$$dm/dt = -\beta_m m + \alpha_m(1 - m) \tag{5.8.47}$$

$$dh/dt = -\beta_h h + \alpha_h(1 - h). \tag{5.8.48}$$

The kinetic constants β_i and α_i all depend explicitly on the exponential of the voltage. The form of the conductances in Eqs. (5.8.44) and (5.8.45) is reminiscent of the symmetric channel conductance in Eq. (5.8.17). The difference is that the Hodgkin–Huxley equations are proportional to powers of parameters such as m, whose time dependence is given by linear equations such as Eq. (5.8.46).

To show that the Hodgkin–Huxley equations are just a special case of the symmetric channel mechanism described above, we first consider a simpler model. Here each channel will be made up of two identical, noninteracting subunits. The subunits can be in one of two states called open, o, and closed, c. Thus each channel has four states (o, o), (o, c), (c, o), and (c, c). Assume now that only the doubly open state can actually conduct ions. Then Eq. (5.8.17) implies that the current is

$$i = G'\bar{n}_{o,o}(\phi - \phi_N), \tag{5.8.49}$$

where $\bar{n}_{o,o}$ is the average number of doubly open channels. The kinetics of the four channel states and their rate constants are summarized in the diagram below:

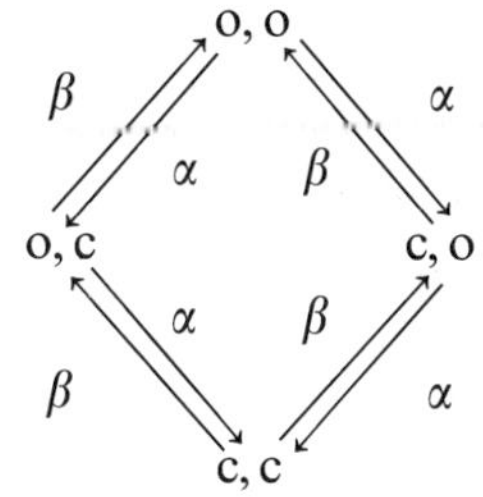

(5.8.50)

All the elementary processes arising from the change in the state of the subunits are included in the diagram. Because the subunits are identical and noninteracting, the rate constants for the processes o, o ⇄ o, c and o, o ⇄ c, o are the same. The average kinetic equations for these elementary processes are a specific case of the general linear mechanism in Section 5.3:

$$\begin{aligned} d\bar{n}_{o,o}/dt &= -2\alpha\bar{n}_{o,o} + \beta\bar{n}_{o,c} + \beta\bar{n}_{c,o} \\ d\bar{n}_{o,c}/dt &= -(\alpha+\beta)\bar{n}_{o,c} + \alpha\bar{n}_{o,o} + \beta\bar{n}_{c,c} \\ d\bar{n}_{c,o}/dt &= -(\alpha+\beta)\bar{n}_{c,o} + \alpha\bar{n}_{o,o} + \beta\bar{n}_{c,c} \\ d\bar{n}_{c,c}/dt &= -2\beta\bar{n}_{c,c} + \alpha\bar{n}_{o,c} + \alpha\bar{n}_{c,o}. \end{aligned} \tag{5.8.51}$$

Because the two subunits of a channel are identical and independent, the solution to Eqs. (5.8.51) can be expressed rather simply. At equilibrium the solution is easily verified to be

$$n^e_{o,o} = n\beta^2/(\alpha+\beta)^2,\ n^e_{o,c} = n^e_{c,o} = n\alpha\beta/(\alpha+\beta)^2,\ n^e_{c,c} = n\alpha^2/(\alpha+\beta)^2, \tag{5.8.52}$$

where n is the total number of channels in the membrane. Thus near equilibrium the number of states of the membrane has the form:

$$\bar{n}_{o,o} = np^2,\ \bar{n}_{o,c} = \bar{n}_{c,o} = np(1-p),\ \bar{n}_{c,c} = n(1-p)^2. \tag{5.8.53}$$

where $0 \le p \le 1$. If this form holds at an arbitrary initial time $t = 0$, then it also holds at a later $t > 0$ if p solves the equation

$$dp/dt = -\alpha p + \beta(1-p), \tag{5.8.54}$$

as can be verified by substituting Eqs. (5.8.53) into Eqs. (5.8.51). Thus, close to equilibrium the time dependence of the total number of open channels follows the power law

$$\bar{n}_{o,o} = np^2 \tag{5.8.55}$$

with p solving the linear equation (5.8.54). The current takes the form

$$i = \bar{G}p^2(\phi - \phi_N). \tag{5.8.56}$$

A similar argument can be used to derive the Hodgkin–Huxley equations. Examining Eqs. (5.8.44)–(5.8.48) the mechanism underlying the Na^+-channel is seen to involve four subunits, three identical and one different. Assume that all four have one open and one closed state and that only the completely open state is conducting. Then by analogy to the argument that leads to Eq. (5.8.56), the Na^+ conductance must be $\bar{G}_{Na}m^3h$, as found by Hodgkin and Huxley. The K^+-channel also involves four subunits, each with an open and closed state. These, however, are all identical, which gives rise to the K^+-channel conductance $\bar{G}_K n^4$. A typical elementary process for the K^+-channel involves the following change of state

$$\text{o, o, o, o} \underset{\beta_n}{\overset{\alpha_n}{\rightleftharpoons}} \text{c, o, o, o.} \tag{5.8.57}$$

The other channel states are combinations of open and closed subunits with transitions among them represented by the same rate constants, α_n and β_n. According to the canonical theory these rate constants can depend on the voltage through the electrochemical potentials of the closed and open states. Thus

$$\begin{aligned} \alpha_n &= \Omega_n \exp(\mu_o^0/k_B T)\exp(e\alpha\phi/k_B T) \\ \beta_n &= \Omega_n \exp(\mu_c^0/k_B T)\exp(e\beta\phi/k_B T), \end{aligned} \tag{5.8.58}$$

where $e\alpha\phi$ and $e\beta\phi$ are electrical potential energies associated with the open and closed states. These electrical terms give rise to the membrane potential dependence of the rate constants in the Hodgkin–Huxley model.

These various mechanisms of ion transport through biological membranes provide a clear example of how the canonical theory can be used to describe transport processes. The noise associated with these mechanisms can also be treated by the general methods described in Chapter 4. In the final section of this chapter, current fluctuations for biological membranes and a method for simulating these fluctuations are explored.

5.9. Simulation of Fluctuations

Chapter 4 outlines how the canonical formalism describes the stochastic process associated with molecular events. The basic assumption of the formalism is contained in Eq. (4.2.6), which relates the rates of the elementary molecular process to transition probabilities. By considering the thermodynamic limit in which the extensive variables reflect the collective behavior of many molecules, it was shown that the *ansatz* leads to the fluctuation–dissipation theory in Section 4.4. In that form we verified that the Second Law is true on the conditional average and we have obtained in this chapter some analytical results about fluctuations away from equilibrium. This is possible only when the canonical form is linear in the extensive variables or if no more than one or two extensive variables are involved. If a process involves nonlinear transition probabilities and a number of extensive variables as, for example, the Hodgkin–Huxley mechanism in Eqs. (5.8.44)–(5.8.48), then analytical techniques are unlikely to make much progress. In this case numerical simulation of the stochastic process on a computer offers an alternative to analytical techniques.

There are a variety of ways one might think to simulate the stochastic process associated with the basic *ansatz*. If the system is large enough, one could simulate the stochastic differential equations using the fluctuation–dissipation theory. As discussed previously, the conditional average equation (4.4.1), e.g.,

$$d\bar{n}_i/dt = R_i(\bar{\mathbf{n}}, t) \tag{5.9.1}$$

with $\bar{n}_i(0) = n_i^0$, needs to be solved first. Even if this equation is nonlinear, it can be solved using numerical techniques, such as the Gear algorithm, which are highly reliable. This, of course, produces only the average trajectory originating at the initial point $\mathbf{n}^0$. To describe conditional fluctuations about this trajectory, it is necessary to obtain solutions to Eq. (4.4.2), i.e.,

$$d\delta n_i/dt = \sum_j H_{ij}(\bar{\mathbf{n}}(\mathbf{n}^0, t))\delta n_j + \tilde{f}_i(t). \tag{5.9.2}$$

Because $\tilde{f}_i(t)$ is a Gaussian random variable, Eq. (5.9.2) has an uncountable infinity of solutions, each one representing a possible realization of the stochastic process.

A convenient way to simulate Eq. (5.9.2) begins with the equation

$$d\delta n_i = \sum_j [H_{ij}(\bar{\mathbf{n}}(\mathbf{n}^0, t))\delta n_j + \gamma_{ij}^{1/2}(\bar{\mathbf{n}}(\mathbf{n}^0, t))\tilde{g}_j(t)]\, dt, \tag{5.9.3}$$

where $\gamma_{ij}^{1/2}$ is the square-root matrix of the covariance of $\tilde{\mathbf{f}}(t)$. If the functions $\tilde{g}_j(t)$ are purely random Gaussians, each with unit variance and mutually uncorrelated, then it is easy to verify that

$$\left\langle \sum_j \gamma_{ij}^{1/2}\tilde{g}_j(t) \sum_k \gamma_{lk}^{1/2}\tilde{g}_k(t') \right\rangle = \gamma_{il}\delta(t - t'), \tag{5.9.4}$$

since $\gamma_{lk}^{1/2}$ is symmetric. Thus Eq. (5.9.3) is equivalent to Eq. (5.9.2). The simulation proceeds through the approximation

$$\delta n_i(t + dt) = \delta n_i(t) + \sum_j [H_{ij}(\bar{\mathbf{n}}^0(\mathbf{n}^0, t))\delta n_j(t)\, dt] + [\gamma_{ii}^{1/2}(\bar{\mathbf{n}}(\mathbf{n}^0, t))\tilde{g}_j(t)]\, dt. \tag{5.9.5}$$

To carry out the simulation one chooses the initial conditional fluctuation $\delta n_i(0) \equiv 0$ and a small time step dt. To use Eq. (5.9.5) to generate the fluctuations $\delta n_i(dt)$ after the first time step, it is necessary to calculate the sum on the right-hand side. The first term comes from the numerical solution of the average equation (5.9.1) since $H_{ij} \equiv \partial R_i/\partial \bar{n}_j$. Similarly, the matrix $\gamma_{ij}^{1/2}$ in the second term of the sum can be obtained from Eq. (4.4.6) and $\bar{\mathbf{n}}(\mathbf{n}^0, t)$. To actually simulate the second term in Eq. (5.9.5) we first write it as

$$\gamma_{ij}^{1/2}(\bar{\mathbf{n}}(\mathbf{n}^0, t))\tilde{g}_j(t)\,dt = \gamma_{ij}^{1/2}(\bar{\mathbf{n}}(\mathbf{n}^0, t))dw_j, \tag{5.9.6}$$

where $dw_j \equiv w_j(t + dt) - w_j(t)$ are the increments of independent, normalized Wiener processes, w_j. The equality in Eq. (5.9.6) follows from the relationship between white noise and the Wiener process discussed in Sections 1.5 and 1.6. To use Eq. (5.9.6) in the simulation, one chooses at each time t a value from $-\infty < dw_j(t) < +\infty$ according to a Gaussian probability distribution with a variance appropriate to a normalized Wiener process, i.e., dt. This is repeated for each $dw_j(t)$ to generate the right-hand side of Eq. (5.9.6), which is then used in the final term of Eq. (5.9.5).

A more direct method of simulating the canonical stochastic theory utilizes the basic *ansatz* in Eq. (4.2.6) itself. We recall that the *ansatz* is

$$P_2(\mathbf{n}, t|\mathbf{n}' = \mathbf{n} + d\mathbf{n}_s + d\mathbf{n}, t + dt) = \begin{cases} V_\kappa^\pm\, dt + \mathcal{O}(dt), \text{ if } d\mathbf{n} = \pm\omega_\kappa \\ 1 - \sum\limits_\kappa (V_\kappa^+ + V_\kappa^-)\,dt + \mathcal{O}(dt), \text{ if } d\mathbf{n} = \mathbf{0} \\ 0 \text{ otherwise} \end{cases} \tag{5.9.7}$$

where $V_\kappa^\pm$ are the forward and reverse rates of the elementary process κ, $\boldsymbol{\omega}_\kappa$ is the net change of $\mathbf{n}$ in the process, and $d\mathbf{n}_s$ is the nondissipative change of $\mathbf{n}$ in dt. To carry out the simulation using Eq. (5.9.7), the nondissipative change in $\mathbf{n}$ should be provided in the form of a function $d\mathbf{n}_s = \mathbf{S}(\mathbf{n}, t)\,dt$. The simulation begins at a fixed value $\mathbf{n}$. To determine $\mathbf{n}(dt)$ at a time step dt later, one first uses the second equality in Eq. (5.9.7). This shows that in the absence of transitions caused by the elementary processes $\mathbf{n}(dt)$ equals $\mathbf{n} + \mathbf{S}(\mathbf{n}, t)\,dt$. At each time step, however, any of the transitions $\pm\boldsymbol{\omega}_\kappa$ are possible. Thus the value of $\mathbf{n}$ at dt will be

$$\mathbf{n}(dt) = \mathbf{n} + \mathbf{S}(\mathbf{n}, t)\,dt + \sum_\kappa \boldsymbol{\omega}_\kappa(\chi_\kappa^+ - \chi_\kappa^-) \tag{5.9.8}$$

where, for example,

$$\chi_\kappa^+ = \begin{cases} 1 & \text{if the forward process of } \kappa \text{ occurs in } dt \\ 0 & \text{if it does not occur} \end{cases} \tag{5.9.9}$$

with a comparable definition for χ_κ^-. To determine whether or not the forward step or the reverse step of an elementary process occurs in dt, the following device is used. First calculate

$$P_\kappa^\pm(\mathbf{n}) = V_\kappa^\pm(\mathbf{n})\,dt. \tag{5.9.10}$$

According to Eq. (5.9.7), these are valid approximations to the jump probabilities as $dt \to 0$. For consistency, the size of the time step must certainly be chosen so that $P_\kappa^\pm(\mathbf{n})$ is less than one. A value of dt such that $P_\kappa^\pm(\mathbf{n})$ is less than a tenth is a reasonable compromise. After calculating $P_\kappa^\pm$ the next step is to generate a random number x_κ^+ between zero and one from a uniform distribution. If $0 \le x_\kappa^+ \le P_\kappa^+$, then the forward process occurs and χ_κ^+ is set equal to one. If $P_\kappa^+ < x_\kappa^+ < 1$, then the forward process does not occur and $\chi_\kappa^+ = 0$. A second random number x_κ^- is then generated. If $0 \le x_\kappa^- \le P_\kappa$, then the reverse process occurs and $\chi_\kappa^- = 1$. Otherwise the reverse process does not occur and $\chi_\kappa^- = 0$. Repeating this procedure for each elementary process provides the data to calculate the sum on the right-hand side of Eq. (5.9.8) and, thus, gives $\mathbf{n}(dt)$. To obtain the value of $\mathbf{n}$ after the next time step, one repeats these operations, beginning with $\mathbf{n}(dt)$ rather than $\mathbf{n}$.

All simulation methods, the present ones included, have inherent problems of accuracy. For example, random number generators on computers produce strings of numbers that are not completely random and the round-off errors in computer addition and multiplication can lead to cummulative errors. Finally some reasonable compromise must be made in the size of the time step; otherwise simulations take too much computer time. This limitation is particularly significant when nonstationary ensembles are examined, since each simulation represents only one possible trajectory, that is, the behavior of only one member of the ensemble. For a stationary ensemble, on the other hand, statistical analysis of a single, very long trajectory can be used to obtain information about ensemble averages, as described in Fig. 5.1.

Simulation techiques are particularly interesting for examining fluctuations in small systems. For small systems the thermodynamic limiting behavior is irrelevant and fluctuations become quite large. A compelling instance of small-system behavior is the current fluctuations of individual ion channels obtained using the patch clamp technique. Figure 5.10 shows typical results for a single Ca^{2+}-activated K^+-channel from rat muscle. Single openings and closings of the channel are easily recognizable. The symmetric mechanism which describes the state changes of this channel is given in Eq. (5.8.8). As was shown in Section 5.8, fluctuations in the Ca^{2+} ion concentration can be neglected when the Ca^{2+} reservoirs are large. Thus the channel dynamics can be described by the following rates:

$$\begin{aligned} V_1^+ &= k_1^+ \rho_{Ca} n_C, & V_1^- &= k_1^- n_{C \cdot Ca^{2+}} \\ V_2^+ &= k_2^+ n_{C \cdot Ca^{2+}}, & V_2^- &= k_2^- n_{O \cdot Ca^{2+}} \\ V_3^+ &= k_3^+ \rho_{Ca} n_{O \cdot Ca^{2+}}, & V_3^- &= k_3^- n_{O \cdot (Ca^{2+})_2}. \end{aligned} \tag{5.9.11}$$

The n's represent number of channels in the indicated state, the k's are rate constants, and the number density of Ca^{2+} ions is taken as a constant. The first and third of the elementary processes in Eqs. (5.8.8) and (5.9.11) represent Ca^{2+} binding to and dissociation from the channel. Since these are rapid steps, they rapidly achieve equilibrium. Thus, to a good approximation, $V_1^+ = V_1^-$ and $V_3^+ = V_3^-$. As a consequence, Eq. (5.9.11) implies that

$$\begin{aligned} n_C &= (k_1^-/k_1^+ \rho_{Ca^{2+}}) n_{C \cdot Ca^{2+}} \\ n_{O \cdot (Ca^{2+})_2} &= (k_3^+ \rho_{Ca^{2+}}/k_3^-) n_{O \cdot Ca^{2+}}. \end{aligned} \tag{5.9.12}$$

Analysis of single channel records like those in Fig. 5.10 reveals only a single conductance step height. Thus it is safe to assume that the two open states have the same conductance. If this is so, then according to Eqs. (5.8.13)–(5.8.17) the current will be proportional to the total number of open states, $n_{O\cdot Ca^{2+}} + n_{O\cdot (Ca^{2+})_2} \equiv n_{OT}$. For this channel Eq. (5.8.17) becomes

$$i = \bar{G} n_{OT}(\phi - \phi_N), \tag{5.9.13}$$

where $\bar{G}$ is the single K^+-channel conductance. Using Eq. (5.9.12), the total number of open and closed states can be expressed as

$$\begin{aligned} n_{CT} &= (1 + k_1^-/k_1^+ \rho_{Ca^{2+}}) n_{C\cdot Ca^{2+}} \\ n_{OT} &= (1 + k_3^+ \rho_{Ca^{2+}}/k_3^-) n_{O\cdot Ca^{2+}}. \end{aligned} \tag{5.9.14}$$

From these expressions the rates of the second elementary process in Eq. (5.9.11) become

$$\begin{aligned} V_2^+ &= k_2^+ (1 + k_1^-/k_1^+ \rho_{Ca^{2+}})^{-1} n_{CT} \\ V_2^- &= k_2^- (1 + k_3^+ \rho_{Ca^{2+}}/k_3^-)^{-1} n_{OT}. \end{aligned} \tag{5.9.15}$$

The assumption of rapid equillibrium in the binding processes, therefore, leaves only the total number of closed and open states as the relevant extensive variables. This is another example of a contraction of the stochastic description, which is explored in more detail in Chapter 9.

The rate constants associated with binding are strongly dependent on the membrane potential, ϕ. An exponential dependence on ϕ is predicted from the canonical form because charges are involved in the binding process. Experimentally Moczydlowski and Latorre found that the dissociation constants are

$$\begin{aligned} k_1^-/k_1^+ &= 180 \exp(-1.68 e\phi/k_B T) \\ k_3^-/k_3^+ &= 11 \exp(-2e\phi/k_B T), \end{aligned} \tag{5.9.16}$$

where the units are in μmoles/liter. The rate constants k_2^+ and k_2^- are found not to depend on the membrane potential and

$$k_2^+ = 480 \text{ s}^{-1}, \quad k_2^- = 280 \text{ s}^{-1}. \tag{5.9.17}$$

The lack of dependence of $k_2^\pm$ on the membrane potential suggest that the states $C\cdot Ca^{2+}$ and $O\cdot Ca^{2+}$ involve no significant charge at the inside of the membrane.

Using the contracted description in Eq. (5.9.15) only a single elementary process is required to describe the channel kinetics. For a large number n of channels, the conditional average equation can be written using Eq. (5.9.15) in the form

$$d\bar{n}_{OT}/dt = \bar{n}_{CT}/\tau_C - \bar{n}_{OT}/\tau_O, \tag{5.9.18}$$

where $\bar{n}_{CT} + \bar{n}_{OT} = n$ and

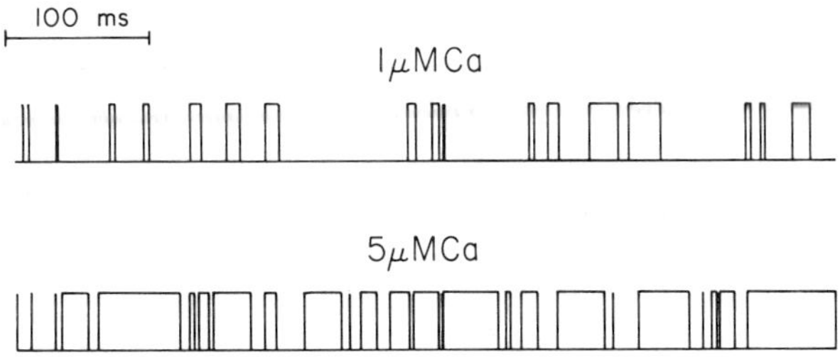

FIGURE 5.11. Simulated current fluctuations for the Ca^{2+}-activated potassium channels described in the legend of Fig. 5.10. The method of simulation is outlined in the text.

$$\begin{aligned} \tau_O &= (1 + k_3^+ \rho_{Ca^{2+}}/k_3^-)/k_2^- \\ \tau_C &= (1 + k_3^- \rho_{Ca^{2+}}/k_3^+)/k_2^+. \end{aligned} \tag{5.9.19}$$

For a single channel $n_{CT} + n_{OT} = 1$. Because the transition rates are linear, the conditional average still satisfies the linear Eq. (5.9.18). The fluctuations, however, are no longer Gaussian. An approximation to the stochastic process for a single Ca^{2+}-gated K^+-channel can be obtained by simulating the transition rates as described below Eq. (5.9.7). Typical results for a single channel with a membrane potential $\phi = +40$ mV are shown in Fig. 5.11. The two Ca^{2+} concentrations, 1 μM and 5 μM, correspond to the measured values $\tau_O = 12.5$ ms, $\tau_C = 33$ ms and $\tau_O = 40$ ms, $\tau_C = 10$ ms, respectively. The time step in the simulations was chosen to be 1 ms and the number of open channels, either 1 or 0, was plotted every 3 ms.

The calculated curves in Fig. 5.11 should be compared with the experimental results in Fig. 5.10. The only significant difference between the two is the high-frequency noise present in the experimental results which is superimposed on the conductance-step noise. Several sources contribute to this high-frequency noise, including measuring instruments and the movement of the K^+ ions through the open channel. Recall that Fig. 5.11 records only whether the channel is open or closed. To convert the abscissa to current, it is necessary to use the single-channel conductance in Eq. (5.9.13).

Although the simulated currents in Fig. 5.11 capture the essence of the opening and closing of the Ca^{2+}-gated K^+-channel, the simulation is incomplete. As stressed in Section 4.2, the basic stochastic *ansatz* in Eq. (4.2.6) holds only for large systems. If a system is composed of a single molecule, such as the Ca^{2+}-gated K^+-channel, then the intrinsic rates, Ω_κ, of the elementary processes are not well defined. Not only will these rates depend on internal molecular states, but the notions of entropy, chemical potential, and temperature which appear in the canonical form no longer have a macroscopic meaning. A portion of the high-frequency noise in Fig. 5.11 can probably be attributed to the breakdown of the canonical form for small systems. On the other hand, the agreement with the gross analysis of the statistics of open and closed channels shows that the *ansatz* in Eq. (4.2.6) can be useful even for systems of molecular size.

References

Ionic Conduction Noise

G. Feher and M. Weissman, Fluctuation spectroscopy: determination of chemical reaction kinetics from the frequency spectrum of fluctuations, *Proc. Nat. Acad. Sci. U.S.A.* **70**, 870–875 (1973).

J.B. Johnson, Thermal agitation of electricity in conductors, *Phys. Rev.* **32**, 97–109 (1928).

H. Nyquist, Thermal agitation of electric charge in conductors, *Phys. Rev.* **32**, 109–113 (1928).

Chemical Reaction Fluctuations

J. Keizer, A theory of spontaneous fluctuations in macroscopic systems, *J. Chem. Phys.* **63**, 398–403 (1975).

J. Keizer, Fluctuations stability and generalized state functions at nonequilibrium steady states, *J. Chem. Phys.* **65**, 4431–4444 (1976)

D.A. McQuarrie and J. Keizer, Fluctuations in chemically reacting systems, in *Theoretical Chemistry: Advances and Perspectives*, Vol. 6A (Academic Press, New York, 1981), pp. 165–213.

T.L. Hill, *Free Energy Transduction in Biology* (Academic, New York, 1977).

Electrochemical Kinetics

K.J. Velter, *Electrochemical Kinetics* (Academic Press, New York, 1967).

T. Erdey-Grúz, *Kinetics of Electrode Processes* (Wiley-Interscience, New York, 1972).

Membrane Conduction

B. Hille, *Ionic Channels of Excitable Membranes* (Sinauer Associates, Sunderland, 1984).

E. Moczydlowski and R. Latorre, Gating kinetics of Ca^{2+}-activated K^+ channels from rat muscle incorporated into planar lipid bilayers, *J. Gen. Physiol.* **82**, 511–542 (1983).

S.B. Hladky, The carrier mechanism, *Curr. Topics Memb. Trans.* **12**, 53–153 (1979).

L.J. De Felice, *Introduction to Membrane Noise* (Plenum, New York, 1981).

CHAPTER 6

The Hydrodynamic Level of Description

6.1. Diffusion in an Isotropic Medium

The hydrodynamic level of description is an extension of the thermodynamic level that takes into account the dependence of the extensive variables on spatial coordinates. Even an ensemble in which the systems are spatially uniform on the average involves fluctuations that differ from one position to another. Consider, for example, the particle mass density, $\rho(\mathbf{r}, t)$, in a simple fluid like water. In an equilibrium ensemble the average density, ρ^e, will be constant in the absence of an external field. Because of the molecular nature of water it is clear that at a given time t and position $\mathbf{r}$ different members of the ensemble will possess different values of the number density. We have already encountered this at the Boltzmann level of description in Sections 3.2 and 3.3. There it was necessary to keep track of the number of particles with a given range of positions and momenta. The hydrodynamic level is intermediate between the Boltzmann and thermodynamic levels and adds the momentum to the basic extensive thermodynamic variables. At the hydrodynamic level one has a closed description of the spatial dependence of the densities of extensive variables throughout a system.

Extensive variables, such as mass and momentum, are properties of clumps of matter, not points. Thus to examine the change in these quantities at the hydrodynamic level we decompose the volume occupied by a system into nonoverlapping regions. As in our treatment of the Boltzmann level, these regions fill up the entire volume of a system and, for convenience, we pick them all to be of the same size. Since the volume of these regions ultimately will be shrunk to zero, we write it as $d\mathbf{r}$. In each such volume we identify the internal energy, the mass, and the momentum as the extensive variables. To distinguish the volume elements from one another each will be designated by a spatial location, $\mathbf{r}$. Thus the extensive variables can be represented as $\mathbf{n}(\mathbf{r}, t)$.

To develop the stochastic description at the hydrodynamic level, it is necessary to delineate the elementary processes which cause the extensive variables to change. We begin in this section by considering the elementary process of diffusion. For the sake of illustration, diffusion in an isotropic, binary solution is treated first. To do so it is necessary to extend the hydrodynamic conservation laws discussed for a single-component system in Section 2.4. In particular, two mass conservation equations are necessary in a binary solution. Calling the mass density of the solvent, ρ_s, and the mass density of the solute, ρ_1, the local conservation laws in Eq. (2.4.6) become

$$\begin{aligned} \partial \rho_s/\partial t &= -\nabla \cdot \mathbf{j}_s \\ \partial \rho_1/\partial t &= -\nabla \cdot \mathbf{j}_1, \end{aligned} \tag{6.1.1}$$

with $\mathbf{j}_s$ and $\mathbf{j}_1$ the mass flux densities. According to Eq. (2.4.7) these can be written

$$\begin{aligned} \mathbf{j}_s &= \rho_s \mathbf{v}_s, \\ \mathbf{j}_1 &= \rho_1 \mathbf{v}_1, \end{aligned} \tag{6.1.2}$$

where $\mathbf{v}_s$ is the velocity of the solvent molecules at $\mathbf{r}$ and $\mathbf{v}_1$ is the velocity of the solute. By definition, the center of mass velocity is

$$\mathbf{v} = (\rho_s \mathbf{v}_s + \rho_1 \mathbf{v}_1)/(\rho_s + \rho_1). \tag{6.1.3}$$

From the kinematic point of view, diffusion can be defined in several ways. For our purposes it is convenient to think of diffusion as the motion of both solvent and solute respect to the center of mass motion. In general, neither the solvent nor the solute will move with the center of mass velocity. Consequently the diffusion fluxes are defined by

$$\begin{aligned} \mathbf{j}_{DS} &= \rho_s(\mathbf{v}_s - \mathbf{v}) \\ \mathbf{j}_{D1} &= \rho_1(\mathbf{v}_1 - \mathbf{v}). \end{aligned} \tag{6.1.4}$$

In terms of these diffusion fluxes, the conservation equations (6.1.1) become

$$\begin{aligned} \partial \rho_s/\partial t &= -\nabla \cdot \mathbf{j}_{DS} - \nabla \cdot \rho_s \mathbf{v} \\ \partial \rho_1/\partial t &= -\nabla \cdot \mathbf{j}_{D1} - \nabla \cdot \rho_1 \mathbf{v}. \end{aligned} \tag{6.1.5}$$

The diffusion fluxes are not independent of one another. Indeed Eq. (6.1.4) and the definition of the center of mass velocity in Eq. (6.1.3) imply that

$$\mathbf{j}_{DS} + \mathbf{j}_{D1} = \mathbf{0}. \tag{6.1.6}$$

This permits us to define a single a single diffusion flux, $\mathbf{j}_D \equiv \mathbf{j}_{D1} = -\mathbf{J}_{DS}$, in terms of which the mass conservation equations become

$$\begin{aligned} \partial \rho_s/\partial t &= \nabla \cdot \mathbf{j}_D - \nabla \cdot (\rho_s \mathbf{v}) \\ \partial \rho_1/\partial t &= -\nabla \cdot \mathbf{j}_D - \nabla \cdot (\rho_1 \mathbf{v}). \end{aligned} \tag{6.1.7}$$

The flux of solute mass due to diffusion is the only dissipative expression in Eq. (6.1.7). Indeed, diffusion is a result of molecular motion and can be described as an elementary process. To do this we decompose the solution into discrete cubic volume elements with sides of length $dx = dy = dz \equiv \delta$ and volume $d\mathbf{r} = \delta^3$, as shown in Fig. 6.1. In this picture, diffusion is represented

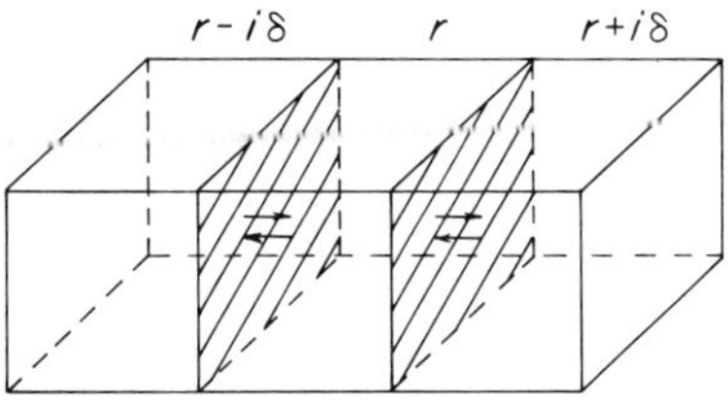

FIGURE 6.1. The cubic cells used to define the rate of change of particle number at the spatial position **r** for mass diffusion. The elementary processes that affect the cell at **r** occur through the six faces of the cell. The location of the processes that occur along the x-direction are illustrated by the cross-hatching.

as the elementary process which causes mass to move from one volume element to another. Although the volume of the boxes will ultimately be shrunk to zero, they should be regarded as having a length which is large with respect to the mean free path. In liquids the mean free path is of the order of a few tenths of an angstrom, whereas in gases at room temperature and pressure the mean free path is several hundred angstroms. Thus in liquids we except that the hydrodynamic-level description of diffusion will be accurate down to a distance scale the order of molecular size, whereas in gases it should be accurate only on a distance scale the order of 10^3–10^4 angstroms. In the limit of very low pressure gases, the mean free path is comparable to the size of the container and collisions with the wall are the most important molecular process. These so-called Knudsen gases do not have a hydrodynamic-level description.

Since the size of the cubic cells is large compared to the mean free path, diffusion can cause mass transport only to neighboring cells. Assuming that the solution is isotropic, diffusion will be unaffected by the orientation of the faces across which the solvent and solute move. Moreover, the elementary process, itself, will be localized at the faces of the volume element, as indicated by the arrows on the faces perpendicular to the x-axis in Fig. 6.1.

The simplest elementary diffusion process involves solvent and solute mass separately in neighboring cells. The mass must be transferred in such a way that the momentum of the center of mass is conserved. This process can occur at any of the faces of any of the cells, whose differing spatial locations allows these processes to be distinguished. Thus the index κ that labels elementary processes involves the spatial location of the face through which transport occurs, e.g., $\mathbf{r} \pm \mathbf{i}\delta/2$, as well as the masses of molecules involved in the two cells. To represent this aspect of the elementary process, consider the diffusion process at the faces $\mathbf{r} \pm \mathbf{i}\delta/2$ in Fig. 6.1. We adopt the convention that the forward direction of the elementary process occurs from left to right. The simplest diffusion process involves a molecular size amount of mass m in both cells. At the face of the box located at $\mathbf{r} + \mathbf{i}\delta/2$ it requires the participation of the following elementary amounts of mass:

$$\begin{aligned} n_1^+(\mathbf{r}) &= m, \qquad n_s^+(\mathbf{r} + \mathbf{i}\delta) = m, \\ n_1^-(\mathbf{r} + \mathbf{i}\delta) &= m, \qquad n_s^-(\mathbf{r}) = m \end{aligned} \tag{6.1.8}$$

and no mass in any of the other cells. In this process a mass, m, of solute moves from the volume at $\mathbf{r}$ to the volume at $\mathbf{r} + \mathbf{i}\delta$ and is replaced by an equivalent mass of solvent. The same type of process occurring at the face of the box at $\mathbf{r} - \mathbf{i}\delta/2$ involves

$$\begin{aligned} n_1^+(\mathbf{r} - \mathbf{i}\delta) &= m, \qquad n_s^+(\mathbf{r}) = m, \\ n_1^-(\mathbf{r}) &= m, \qquad n_s^-(\mathbf{r} - \mathbf{i}\delta) = m \end{aligned} \tag{6.1.9}$$

and no mass in any of the other cells.

The rates of these elementary process are given by the canonical formula in Eq. (4.2.3). The intensive variables conjugate to the masses are

$$F_s = (\partial S/\partial n_s) = -\mu_s/T$$

$$F_1 = (\partial S/\partial n_1) = -\mu_1/T,$$

where the μ's are the *mass-based* chemical potentials of solvent and solute. Since each volume element is characterized by its own chemical potential, we write $F_i(\mathbf{r}) = -\mu_i/T(\mathbf{r})$. Denoting the two elementary processes in Eqs. (6.1.8) and (6.1.9) by $+\delta$ and $-\delta$, respectively, the canonical formulas for their rates become

$$\begin{aligned} V_{+\delta}^+ &= \Omega \exp[m\mu_1/T(\mathbf{r}) + m\mu_s/T(\mathbf{r} + \mathbf{i}\delta)] \\ V_{+\delta}^- &= \Omega \exp[m\mu_1/T(\mathbf{r} + \mathbf{i}\delta) + m\mu_s/T(\mathbf{r})] \\ V_{-\delta}^+ &= \Omega \exp[m\mu_1/T(\mathbf{r} - \mathbf{i}\delta) + m\mu_s/T(\mathbf{r})] \\ V_{-\delta}^- &= \Omega \exp[m\mu_1/T(\mathbf{r}) + m\mu_s/T(\mathbf{r} - \mathbf{i}\delta)]. \end{aligned} \tag{6.1.10}$$

These processes affect the mass of solvent and solute in the cells at $\mathbf{r} - \mathbf{i}\delta$, $\mathbf{r}$, and $\mathbf{r} + \mathbf{i}\delta$. For the cell at $\mathbf{r}$, the net changes of solvent mass are

$$\begin{aligned} \omega_{+\delta,s}(\mathbf{r}) &= n_s^-(\mathbf{r}) - n_s^+(\mathbf{r}) = m \\ \omega_{-\delta,s}(\mathbf{r}) &= n_s^-(\mathbf{r}) - n_s^+(\mathbf{r}) = -m, \end{aligned} \tag{6.1.11}$$

as follows from Eqs. (6.1.8) and (6.1.9). For the solute the net changes are

$$\omega_{+\delta,1}(\mathbf{r}) = -m, \qquad \omega_{-\delta,1}(\mathbf{r}) = m. \tag{6.1.12}$$

Using the rates and the connection between fluctuations and dissipation, the average transport equations and the stochastic equations governing the fluctuations can be derived. The contribution to the rate of change of the conditional average due to these processes can be obtained from Eq. (4.4.1). The rate expressions in Eq. (6.1.10) and the net changes in Eq. (6.1.12) show that the average rate of change of solute mass in the volume element at $\mathbf{r}$ is

$$\begin{aligned} (\partial \bar{n}_1/\partial t)_x = {} & -m\Omega\{\exp[m\bar{\mu}_1/k_B T + m\bar{\mu}_s/k_B T(+)] \\ & - \exp[m\bar{\mu}_1/k_B T(+) + m\bar{\mu}_s/k_B T]\} \\ & + m\Omega\{\exp[m\bar{\mu}_1/k_B T(-) + m\bar{\mu}_s/k_B T] \\ & - \exp[m\bar{\mu}_1/k_B T + m\bar{\mu}_s/k_B T(-)]\}, \end{aligned} \tag{6.1.13}$$

where for compactness we have written $\mu_s/T(+) = \mu_s/T(\mathbf{r} + \mathbf{i}\delta)$, etc., and the subscript x denotes the elementary process in the x-direction. Comparing the net transfers in Eqs. (6.1.11) and (6.1.12) we see, in addition, that

$$(\partial \bar{n}_s/\partial t)_x = -(\partial \bar{n}_1/\partial t)_x. \tag{6.1.14}$$

Before taking the continuum limit of Eq. (6.1.13), it is instructive to examine the special case in which the solute is dilute. In this case the chemical potential of the solvent can be taken as the pure-solvent chemical potential. If the temperature is also uniform, μ_s/T can be replaced by a constant in Eq. (6.1.13). Under these conditions

$$(\partial \bar{n}_1/\partial t)_x = m\Omega'\{\exp[m\bar{\mu}_1/k_BT(+)] - 2\exp[m\bar{\mu}_1/k_BT] + \exp[m\bar{\mu}_1/k_BT(-)]\}, \tag{6.1.15}$$

where

$$\Omega' \equiv \Omega \exp(m\mu_s/k_BT)$$

is a constant. It is easy to take the continuum limit of Eq. (6.1.15). This requires using the formula

$$\lim_{\delta\to 0}\,[f(x+\delta) - 2f(x) + f(x-\delta)]/\delta^2 = \partial^2 f/\partial x^2, \tag{6.1.16}$$

which can be derived from the Taylor series expansions of $f(x \pm \delta)$ around $\delta = 0$. Dividing both sides of Eq. (6.1.15) by the volume of a cell, δ^3, and taking the continuum limit $\delta \to 0$ then gives

$$[\partial \bar{\rho}_1(\mathbf{r},t)/\partial t]_x = m\bar{\Omega}\partial^2 \exp(m\bar{\mu}_1/k_BT)/\partial x^2, \tag{6.1.17}$$

where

$$\bar{\Omega} = \lim_{\delta\to 0}\,(\Omega'/\delta). \tag{6.1.18}$$

It is not hard to verify that the limit in Eq. (6.1.18) exists. This follows from Eq. (6.1.15): since the mass of solute on the left-hand side is proportional to the volume, δ^3, the right-hand side is also proportional to δ^3. According to Eq. (6.1.16), the term in curly brackets in (6.1.15) is proportional to δ^2 when δ is small. Thus, it follows that Ω' is proportional to δ and that the limit in Eq. (6.1.18) exists.

The argument leading to Eq. (6.1.17) does not depend in any way on our choice of the x-direction. Indeed, the faces in the y-direction and the z-direction participate in mass diffusion in an identical fashion. Adding the elementary processes for these faces of the cell at $\mathbf{r}$ to those for the x-faces and taking the continuum limit gives the contribution of diffusion to the average change of density, namely,

$$[\partial \bar{\rho}_1(\mathbf{r},t)/\partial t]_{\text{diff}} = m\bar{\Omega}\nabla^2 \exp(m\bar{\mu}_1/k_BT). \tag{6.1.19}$$

The right-hand side can be rewritten in a more familiar form as

$$[\partial\bar{\rho}_1(\mathbf{r},t)/\partial t]_{\text{diff}} = -\boldsymbol{\nabla}\cdot[-m\bar{\Omega}\boldsymbol{\nabla}\exp(m\bar{\mu}_1/k_BT)]. \tag{6.1.20}$$

Comparing Eqs. (6.1.7) and (6.1.20), the average diffusion flux can be identified as

$$\bar{\mathbf{j}}_{D1} = \bar{\mathbf{j}}_D = -m\bar{\Omega}\boldsymbol{\nabla}\exp(m\bar{\mu}_1/k_BT). \tag{6.1.21}$$

If the mass that is transferred in the elementary process is that of a single solute molecule, then m is the mass of a solute molecule and $m\bar{\mu}_1$ equals the usual molecule-number-based chemical potential. In general, a variety of masses may be involved in the elementary diffusion processes and the resulting flux would involve a sum over these masses. It should be noted that in the continuum limit Eq. (6.1.14) generalizes to the result that

$$(\partial\bar{\rho}_s/\partial t)_{\text{diff}} = -(\partial\bar{\rho}_1/\partial t)_{\text{diff}}. \tag{6.1.22}$$

Thus if we write

$$(\partial\bar{\rho}_s/\partial t)_{\text{diff}} = -\boldsymbol{\nabla}\cdot\bar{\mathbf{j}}_{S1}, \tag{6.1.23}$$

it follows from Eqs. (6.1.20)–(6.1.22) that

$$\bar{\mathbf{j}}_{DS} = -\bar{\mathbf{j}}_{D1}, \tag{6.1.24}$$

as required by Eq. (6.1.6).

A similar limiting procedure can be carried out when solvent effects are not neglected. Returning to Eq. (6.1.13) and dividing by δ^3, we see that it is necessary to evaluate a limit of the form

$$\lim_{\delta\to 0}\{[f(x+\delta)g(x) - f(x)g(x+\delta) + f(x-\delta)g(x) - f(x)g(x-\delta)]/\delta^2\}$$
$$= g\partial^2 f/\partial x^2 - f\partial^2 g/\partial x^2, \tag{6.1.25}$$

where the equality can be verified using Taylor series expansions of f and g. Making the identifications $f = \exp(m\bar{\mu}_1/k_BT)$ and $g = \exp(m\bar{\mu}_s/k_BT)$ and using Eq. (6.1.25), the continuum limit of Eq. (6.1.13) becomes

$$(\partial\bar{\rho}_1(\mathbf{r},t)/\partial t)_x = m\bar{\Omega}'[\exp(m\bar{\mu}_s/k_BT)\partial^2\exp(m\bar{\mu}_1/k_BT)/\partial x^2$$
$$- \exp(m\bar{\mu}_1/k_BT)\partial^2\exp(m\bar{\mu}_s/k_BT)/\partial x^2], \tag{6.1.26}$$

with $\bar{\Omega}' \equiv \lim_{\delta\to 0}(\Omega'/\delta)$. Again adding the contributions from the y- and z-faces gives the complete effect of diffusion:

$$(\partial\bar{\rho}_1(\mathbf{r},t)/\partial t)_{\text{diff}} = m\bar{\Omega}'[\exp(m\bar{\mu}_s/k_BT)\boldsymbol{\nabla}^2\exp(m\bar{\mu}_1/k_BT)$$
$$- \exp(m\bar{\mu}_1/k_BT)\boldsymbol{\nabla}^2\exp(m\bar{\mu}_s/k_BT)]. \tag{6.1.27}$$

Using the identity $\boldsymbol{\nabla}\cdot(g\boldsymbol{\nabla}f - f\boldsymbol{\nabla}g) = g\boldsymbol{\nabla}^2 f - f\boldsymbol{\nabla}^2 g$, Eq. (6.1.27) can be rewritten in the form

$$[\partial\bar{\rho}_1(\mathbf{r},t)/\partial t]_{\text{diff}} = -\boldsymbol{\nabla}\cdot\bar{\mathbf{j}}_D, \tag{6.1.28}$$

where

$$\bar{\mathbf{j}}_D = -m\bar{\Omega}'[\exp(m\bar{\mu}_s/k_B T)\nabla \exp(m\bar{\mu}_1/k_B T) - \exp(m\bar{\mu}_1/k_B T)\nabla \exp(m\bar{\mu}_s/k_B T)]. \quad (6.1.29)$$

In the case that the chemical potential of the solvent is uniform, Eq. (6.1.29) reduces to the diffusion flux for dilute solution given in Eq. (6.1.21).

According to *Fick's law*, in terms of which most experiments on binary diffusion are interpreted, the diffusion flux has the form

$$\bar{\mathbf{j}}_D = -D\nabla\bar{\rho}_1. \quad (6.1.30)$$

D is a transport coefficient called the *diffusion constant*. The flux can also be written in the form

$$\bar{\mathbf{j}}_D = -L'\nabla\bar{\mu}_1 = L\nabla(-\bar{\mu}_1/T). \quad (6.1.31)$$

This is the continuum expression of Onsager's relationship in Eq. (2.3.3) between the mass flux, $\bar{\mathbf{j}}_D$, and the thermodynamic force, $\nabla(-\bar{\mu}_1/T)$. By rearranging Eq. (6.1.29) it is possible to establish the connections among the intrinsic rate constant $\bar{\Omega}'$, the diffusion constant, and the transport coefficient L. Taking the gradients of the exponentials in Eq. (6.1.29) gives

$$\bar{\mathbf{j}}_D = (m^2\bar{\Omega}'/k_B)\exp[m(\bar{\mu}_s + \bar{\mu}_1)/k_B T][\nabla(-\bar{\mu}_1/T) - \nabla(-\bar{\mu}_s/T)]. \quad (6.1.32)$$

The term involving the gradients can be simplified using the Gibbs–Duhem relationship between the two chemical potentials. The generalization of the single-component result, derived earlier in Eq. (2.4.30), is

$$\nabla p = -T\rho_1\nabla(-\bar{\mu}_1/T) - T\rho_s\nabla(-\mu_s/T) - Th\nabla(1/T), \quad (6.1.33)$$

where p is the pressure and h is the enthalpy density. For a system in mechanical and thermal equilibrium this reduces to

$$-\nabla(-\mu_s/T) = (\rho_1/\rho_s)\nabla(-\mu_1/T). \quad (6.1.34)$$

Since the gradients in Eq. (6.1.34) are carried out at constant temperature and pressure, the notation $(\nabla)_{T,p}$ is sometimes used for these derivatives. For ease in writing, we do not follow that notation. Under conditions of constant temperature and pressure, Eq. (6.1.34) allows the chemical potential of the solvent to be eliminated from Eq. (6.1.32), yielding

$$\bar{\mathbf{j}}_D = (m^2\bar{\Omega}'/k_B X_s)\exp[m(\bar{\mu}_s + \bar{\mu}_1)/k_B T]\nabla(-\bar{\mu}_1/T), \quad (6.1.35)$$

where $X_s = \rho_s/(\rho_1 + \rho_s)$ is the local mass fraction of the solvent. Comparing Eqs. (6.1.31) and (6.1.35), the Onsager coefficient L is

$$L = (m^2\bar{\Omega}'/k_B X_s)\exp[m(\bar{\mu}_s + \bar{\mu}_1)/k_B T]. \quad (6.1.36)$$

To obtain Fick's law, we recall that the Gibbs–Duhem relationship or, alternatively, Gibbs' phase rule implies that there are only three independent intensive variables in a binary solution. Thus at constant temperature and pressure

$$\nabla\bar{\mu}_1 = (\partial\bar{\mu}_1/\partial\bar{\rho}_1)_{T,p}\nabla\bar{\rho}_1. \tag{6.1.37}$$

Combining Eqs. (6.1.35) and (6.1.37) we see that the diffusion constant is

$$D = (m^2\bar{\Omega}'/k_BTX_s)(\partial\bar{\mu}_1/\partial\bar{\rho}_1)_{T,p}\exp[m(\bar{\mu}_s + \bar{\mu}_1)/k_BT]. \tag{6.1.38}$$

In dilute solution X_s approaches unity and the expression for the diffusion constant is identical to that obtained using the simpler expression in Eq. (6.1.21).

It is straightforward to generalize these results to multicomponent systems. To do so, we reconsider the diagram in Fig. 6.1. Each of the two faces perpendicular to the x-direction has one cell to the left and one cell to the right. At each face the same type of elementary process occurs. If the n_i's represent mass numbers, then symbolically the elementary processes are

$$(n_{\kappa 1}^{+L}, n_{\kappa 2}^{+L}, \ldots; n_{\kappa 1}^{+R}, n_{\kappa 2}^{+R}, \ldots) \rightleftarrows (n_{\kappa 1}^{-L}, n_{\kappa 2}^{-L}, \ldots; n_{\kappa 2}^{-R}, n_{\kappa 2}^{-R}, \ldots) \tag{6.1.39}$$

Here $n_{\kappa i}^{+L}$ represents the mass of molecule of type i in the left-hand cell required in the forward direction, $n_{\kappa i}^{+R}$ is the mass in the right-hand cell, and $n_{\kappa i}^{-L}$, $n_{\kappa i}^{-R}$ are the final values. These quantities are not arbitrary since the diffusion fluxes are defined with respect to the center of mass velocity, i.e.,

$$\mathbf{j}_{Di} = \rho_i(\mathbf{v}_i - \mathbf{v}), \tag{6.1.40}$$

where the velocity field of the molecules of kind i is $\mathbf{v}_i$ and the center of mass velocity is

$$\mathbf{v} = \sum_i \rho_i\mathbf{v}_i/\sum_i \rho_i, \tag{6.1.41}$$

with the ρ_i's mass densities. As a consequence, it follows that

$$\sum_i \mathbf{j}_{Di} = \mathbf{0}, \tag{6.1.42}$$

which is the statement that diffusion conserves the momentum of the center of mass. To satisfy this condition the net transfers in the elementary process, i.e.,

$$\begin{aligned} \omega_{\kappa i}^L &= n_{\kappa i}^{-L} - n_{\kappa i}^{+L} \\ \omega_{\kappa i}^R &= n_{\kappa i}^{-R} - n_{\kappa i}^{+R} \end{aligned} \tag{6.1.43}$$

must satisfy

$$\sum_i \omega_{\kappa i}^L = \sum_i \omega_{\kappa i}^R = 0. \tag{6.1.44}$$

They also must satisfy conservation of mass. This means that $n_{\kappa i}^{+L} + n_{\kappa i}^{+R} = n_{\kappa i}^{-L} + n_{\kappa i}^{-R}$, which rearranges to

$$\omega_{\kappa i}^L = -\omega_{\kappa i}^R. \tag{6.1.45}$$

It is easy to check that the simple binary elementary process in Eq. (6.1.8) satisfies both conditions.

The rates of the elementary processes in Eq. (6.1.39) have the canonical form. For the central cell shown in Fig. 6.1 the rates can be written at the $\mathbf{r} + \mathbf{i}\delta/2$ face as

$$V_\kappa^\pm(\mathbf{r}, \mathbf{r} + \mathbf{i}\delta) = \Omega_\kappa \exp\left[\sum_i n_{\kappa i}^{\pm L} \mu_i(\mathbf{r})/k_B T\right] \exp\left[\sum_i n_{\kappa i}^{\pm R} \mu_i(\mathbf{r} + \mathbf{i}\delta)/k_B T\right]. \tag{6.1.46}$$

At the $\mathbf{r} - \mathbf{i}\delta/2$ face they are

$$V_\kappa^\pm(\mathbf{r} - \mathbf{i}\delta, \mathbf{r}) = \Omega_\kappa \exp\left[\sum_i n_i^{\pm L} \mu_i(\mathbf{r} - \mathbf{i}\delta)/k_B T\right] \exp\left[\sum_i n_i^{\pm R} \mu_i(\mathbf{r})/k_B T\right]. \tag{6.1.47}$$

The conditional average rate of change of the mass of i in the cell at $\mathbf{r}$ due to these processes is, therefore,

$$(\partial \bar{n}_i(\mathbf{r}, t)/\partial t)_x = \sum_\kappa \omega_\kappa^L [\bar{V}_\kappa^+(\mathbf{r}, \mathbf{r} + \mathbf{i}\delta) - \bar{V}_\kappa^-(\mathbf{r}, \mathbf{r} + \mathbf{i}\delta) - \bar{V}_\kappa^+(\mathbf{r} - \mathbf{i}\delta, \mathbf{r}) + \bar{V}_\kappa^-(\mathbf{r} - \mathbf{i}\delta, \mathbf{r})], \tag{6.1.48}$$

where Eq. (6.1.45) was used to eliminate ω_κ^R. To take the continuum limit, we divide both sides of Eq. (6.1.48) by the volume δ^3 and take the limit that $\delta \to 0$. The left-hand side becomes the partial derivative of the mass density with respect to time. The limit of the right-hand side can be established using Taylor series expansions similar to those used in verifying Eqs. (6.1.16) and (6.1.25). To do this we first define the functions

$$\begin{aligned} f_\kappa^\pm(\mathbf{r}) &= \exp\left[\sum_i n_{\kappa i}^{\pm L} \bar{\mu}_i(\mathbf{r})/k_B T\right] \\ g_\kappa^\pm(\mathbf{r}) &= \exp\left[\sum_i n_{\kappa i}^{\pm R} \bar{\mu}_i(\mathbf{r})/k_B T\right]. \end{aligned} \tag{6.1.49}$$

Using these definitions the terms inside the square brackets in Eq. (6.1.48) can be written

$$\begin{aligned} [\cdot] = [&f_\kappa^+(\mathbf{r}) g_\kappa^+(\mathbf{r} + \mathbf{i}\delta) - f_\kappa^-(\mathbf{r}) g_\kappa^-(\mathbf{r} + \mathbf{i}\delta) \\ &- f_\kappa^+(\mathbf{r} - \mathbf{i}\delta) g_\kappa^+(\mathbf{r}) + f_\kappa^-(\mathbf{r} - \mathbf{i}\delta) g_\kappa^-(\mathbf{r})]. \end{aligned} \tag{6.1.50}$$

Next we carry out Taylor series expansions of each term around $\mathbf{r}$. The constant terms cancel and, using the identity above Eq. (6.1.45), the terms of order δ are seen to vanish. Thus the leading term is

$$[\cdot] = (\delta^2/2)[f_\kappa^+ \partial^2 g_\kappa^+/\partial x^2 - g_\kappa^+ \partial^2 f_\kappa^+/\partial x^2 - f_\kappa^- \partial^2 g_\kappa^-/\partial x^2 + g_\kappa^- \partial^2 f_\kappa^-/\partial x^2]. \tag{6.1.51}$$

Again defining

$$\bar{\Omega}_\kappa = \lim_{\delta \to 0} (\Omega_\kappa/\delta), \tag{6.1.52}$$

the continuum limit of Eq. (6.1.48) can be written

$$[\partial\bar{\rho}_i(\mathbf{r},t)/\partial t]_x = \frac{1}{2}\sum_\kappa \omega_\kappa^L\bar{\Omega}_\kappa[f_\kappa^+\partial^2 g_\kappa^+/\partial x^2 - g_\kappa^+\partial^2 f_\kappa^+/\partial x^2$$
$$- f_\kappa^-\partial^2 g_\kappa^-/\partial x^2 + g_\kappa^-\partial^2 f_\kappa^-/\partial x^2]. \tag{6.1.53}$$

Finally, including the elementary processes affecting the y- and z-directions yields

$$[\partial\bar{\rho}_i(\mathbf{r},t)/\partial t]_{\text{diff}} = \frac{1}{2}\sum_\kappa \omega_\kappa^L\bar{\Omega}_\kappa[f_\kappa^+\nabla^2 g_\kappa^+ - g_\kappa^+\nabla^2 f_\kappa^+ - f_\kappa^-\nabla^2 g_\kappa^- + g_\kappa^-\nabla^2 f_\kappa^-]. \tag{6.1.54}$$

The form of the diffusion equation in Eq. (6.1.54) can be simplified considerably. It is easy to see that it can be written as a conservation equation

$$[\partial\bar{\rho}_i(\mathbf{r},t)/\partial t]_{\text{diff}} = -\nabla\cdot\mathbf{j}_{Di} \tag{6.1.55}$$

if we define

$$\mathbf{j}_{Di} = -\frac{1}{2}\sum_\kappa \omega_\kappa^L\bar{\Omega}_\kappa[f_\kappa^+\nabla g_\kappa^+ - g_\kappa^+\nabla f_\kappa^+ - f_\kappa^-\nabla g_\kappa^- + g_\kappa^-\nabla f_\kappa^-]. \tag{6.1.56}$$

Furthermore, from the identity above Eq. (6.1.45) and Eq. (6.1.49) it follows that $f_\kappa^+ g_\kappa^+ = f_\kappa^- g_\kappa^-$. Carrying out the differentiations indicated in Eq. (6.1.56) and using this fact, it is easy to show that

$$\mathbf{J}_{Di} = \sum_\kappa\sum_j (\omega_{\kappa i}^L\bar{\Omega}_\kappa\omega_{\kappa j}^L/k_B)\exp\left[\sum_l (n_{\kappa l}^{+L} + n_{\kappa l}^{+R})\bar{\mu}_l/k_B T\right]\nabla(-\bar{\mu}_j/T).$$
$$= \sum_j L_{ij}\nabla(-\bar{\mu}_j/T) \tag{6.1.57}$$

The explicit form of the flux–force coupling matrix is

$$L_{ij} \equiv \sum_\kappa (\omega_{\kappa i}^L\bar{\Omega}_\kappa\omega_{\kappa j}^L/k_B)\prod_l \bar{z}_l^{m_{\kappa l}} \tag{6.1.58}$$

where $\bar{z}_l$ is the *absolute activity* defined by

$$\bar{z}_l = \exp(\bar{\mu}_l/k_B T) \tag{6.1.59}$$

and $m_{\kappa l} \equiv (n_{\kappa l}^{+L} + n_{\kappa l}^{+R})$.

Equations (6.1.55)–(6.1.58) govern the average effects of diffusion at the hydrodynamic level of description. Notice that the coupling matrix L_{ij} is symmetric, even if the ensemble is not close to equilibrium. As in the case of the binary solution treated already, the chemical potential and density of the solvent depend on those of all the other components through the Gibbs–Duhem equation. At constant temperature and pressure this implies that

$$\sum_{i\neq s} -\rho_i\nabla(-\mu_i/T) = \rho_s\nabla(-\mu_s/T). \tag{6.1.60}$$

Eliminating the solvent in this way, the expression for the diffusion fluxes in Eq. (6.1.58) can be written in a more symmetrical way if the thermodynamic forces are defined as

$$\bar{\mathbf{X}}_j = \nabla(-\bar{\mu}_j/T) + \sum_{k \neq s} (\bar{\rho}_k/\bar{\rho}_s)\nabla(-\bar{\mu}_k/T) = \nabla(-\bar{\mu}_j/T) - \nabla(-\bar{\mu}_s/T), \qquad (6.1.61)$$

where the second equality follows from Eq. (6.1.60). For these variables we verify that

$$\begin{aligned} \sum_{j \neq s} L_{ij}\bar{\mathbf{X}}_j &= \sum_{j \neq s} L_{ij}\nabla(-\bar{\mu}_j/T) - \left(\sum_{j \neq s} L_{ij}\right)\nabla(-\bar{\mu}_s/T) \\ &= \sum_{j} L_{ij}\nabla(-\bar{\mu}_j/T), \end{aligned} \qquad (6.1.62)$$

since

$$\sum_j L_{ij} = 0,$$

as Eqs. (6.1.44) and (6.1.58) show. This allows us to write the symmetric flux–force relationship for the solute molecules as

$$\bar{\mathbf{j}}_{Di} = \sum_{j \neq s} L_{ij}\bar{\mathbf{X}}_j \qquad (6.1.63)$$

with independent thermodynamic forces, $\mathbf{X}_j$. In an equilibrium ensemble the coupling matrix is a constant matrix and Eq. (6.1.63) becomes Onsager's linear law. In a nonequilibrium ensemble the explicit dependence of L_{ij} on the activities given in Eq. (6.1.58) becomes significant. Because Eq. (6.1.63) is linear, transforming the fluxes to new frames of reference, e.g., solvent velocity rather than center of mass velocity, can be carried out just as described by deGroot and Mazur or Fitts for the linear theory (see the reference list at the end of this chapter).

6.2. Density Fluctuations Caused by Diffusion

At the hydrodynamic level of description the mass density fluctuates because molecules enter or leave volume elements due to random molecular events. In the previous section we considered the elementary processes which are responsible for mass diffusion. This led to nonlinear average equations, like Eq. (6.1.57), which describe the conditional average diffusion fluxes. Fluctuations around the conditional average are dictated by the fluctuation–dissipation relationships in Section 4.4. As we have seen in numerous examples, all the quantities required to describe these fluctuations are contained in the parameters of the elementary process. The only new features of this correspondence that enter at the hydrodynamic level are the dependence of the elementary processes on spatial location. As preparation for the description of heat conduction, thermal diffusion, and momentum transport, the way in which diffusion causes fluctuations in the mass density is examined in detail in this section.

To illustrate the basic relationships, consider first a binary solution and the elementary process which describes diffusion when the solvent is neglected. The expression for the diffusion flux of the solute in this case is given by Eq. (6.1.21). Combining the flux with the conservation equation (6.1.5) gives the conditional average equation

$$\partial \bar{\rho}_1/\partial t = m\bar{\Omega}\nabla^2 \exp(m\bar{\mu}_1/k_B T) - \nabla\cdot(\bar{\rho}_1\bar{\mathbf{v}}). \tag{6.2.1}$$

According to the fluctuation–dissipation relationships in Eqs. (4.4.1) and (4.4.2), the equation which describes fluctuations around the conditional average is obtained by linearizing Eq. (6.2.1) and adding the appropriate random term. Doing so gives

$$\partial \delta\rho_1/\partial t = (\bar{\Omega}m^2/k_B T)\nabla^2[\exp(m\bar{\mu}_1/k_B T)\delta\mu_1] + \nabla\cdot(\bar{\rho}_1\,\delta\mathbf{v}) + \nabla\cdot(\bar{\mathbf{v}}\delta\rho_1) + \tilde{f}_1, \tag{6.2.2}$$

where the temperature has been taken as a constant. The only dissipative term effecting the mass density is the diffusion flux. The streaming term, $\nabla\cdot(\bar{\rho}_1\bar{\mathbf{v}})$, in Eq. (6.2.1) is purely kinematic and corresponds to molecules of the solute convecting with the center of mass velocity. Thus to characterize the term $\tilde{f}_1$ it is only necessary to consider the elementary diffusion process.

At the hydrodynamic level of description the extensive variables are replaced by their densities, e.g., the mass density. Thus the extensive variables appear as scalar fields which depend on the position $\mathbf{r}$. In applying the canonical formulas for properties of elementary processes, this is taken into account by writing the density of an extensive variable as $\rho(\mathbf{r}, t)$, rather than $\boldsymbol{\rho}(t)$. This makes it clear that the index of the vector $\boldsymbol{\rho}(t)$ is continuous. Furthermore, it reminds us that the density at point $\mathbf{r}$ may be correlated with the density at point $\mathbf{r}'$, since the density, $\rho_i(t)$, in a volume element at $\mathbf{r}_i$ may be correlated with the density, $\rho_j(t)$, in the volume element at $\mathbf{r}_j$. When several mass densities are involved, they will be written collectively as the column of vector fields, $\boldsymbol{\rho}(\mathbf{r}, t)$.

Using this notation, the random term in Eq. (6.2.2) becomes the scalar field $\tilde{f}_1(\mathbf{r}, t)$. As we know, it is a purely random, nonstationary Gaussian process which vanishes on the conditional average, i.e.,

$$\langle \tilde{f}_1(\mathbf{r}, t)\rangle = 0. \tag{6.2.3}$$

Its covariance is dictated by the canonical formulas in Eqs. (4.4.5) and (4.4.6). In the notation of scalar fields these equations become

$$\langle \tilde{f}_1(\mathbf{r}, t)\tilde{f}_1(\mathbf{r}', t)\rangle = \bar{\gamma}(\mathbf{r}, \mathbf{r}'; \rho^0(t))\delta(t - t'), \tag{6.2.4}$$

$$\bar{\gamma}(\mathbf{r}, \mathbf{r}'; \rho^0(t)) = \lim_{\delta\to 0}\sum_{\kappa} \omega_{\kappa i}(\bar{V}_\kappa^+ + \bar{V}_\kappa^-)\omega_{\kappa j}. \tag{6.2.5}$$

The covariance of $\tilde{f}_1$ is no longer a matrix but a continuous function of the two variables $\mathbf{r}$ and $\mathbf{r}'$. The correspondence here between the discrete thermodynamic and continuous hydrodynamic descriptions is that the linear operator, which is a matrix for thermodynamic variables, becomes a continuous linear operator acting on continuous density fields in hydrodynamics. The covariance matrix, as we shall see, becomes the kernel of a linear integral operator. As Eq. (6.2.5) shows, the covariance matrix can be obtained by applying the continuum limit $\delta \to 0$ to the elementary diffusion processes.

To obtain the covariance of $\tilde{f}_1(\mathbf{r}, t)$, we need to write out the sum of the left-hand side of Eq. (6.2.4) explicitly. That requires using all the elementary diffusion processes at all the faces of all the volume elements depicted in Fig. 6.1. To simplify this, we consider first only the faces perpendicular to the x-axis and neglect the participation of the solvent. For the cell at $\mathbf{r}$, only the processes at the faces $\mathbf{r} \pm \mathbf{i}\delta/2$ involve net changes $\boldsymbol{\omega}_\kappa$ that affect that cell. These processes affect the cells at $\mathbf{r} \pm \mathbf{i}\delta$ and no others. The processes which describe diffusion of a single solute, neglecting the solvent, can be gleaned from Eqs. (6.1.8) and (6.1.9). At the $\mathbf{r} + \mathbf{i}\delta/2$ face one has

$$\begin{gathered} n^+_{+\delta}(\mathbf{r}') = m\delta(\mathbf{r}, \mathbf{r}'), \quad n^-_{+\delta}(\mathbf{r}') = m\delta(\mathbf{r} + \mathbf{i}\delta, \mathbf{r}') \\ \omega_{+\delta}(\mathbf{r}') = m[\delta(\mathbf{r} + \mathbf{i}\delta, \mathbf{r}') - \delta(\mathbf{r}, \mathbf{r}')] \\ \bar{V}^+_{+\delta} = \Omega' \exp[m\bar{\mu}_1(\mathbf{r})/k_B T], \quad \bar{V}^-_{+\delta} = \Omega' \exp[m\bar{\mu}_1(\mathbf{r} + \mathbf{i}\delta)k_B T], \end{gathered} \tag{6.2.6}$$

where $\mathbf{r}' = \mathbf{r} \pm \mathbf{i}k\delta$ with k an integer and $\delta(\mathbf{r}, \mathbf{r}')$ is the Kronecker delta. In this process only solute mass is being followed and an amount m is transported. The description of the elementary process at the $\mathbf{r} - \mathbf{i}\delta/2$ face is similar. Thus

$$\begin{gathered} n^+_{-\delta}(\mathbf{r}') = m\delta(\mathbf{r} - \mathbf{i}\delta, \mathbf{r}'), \quad n^-_{-\delta}(\mathbf{r}') = m\delta(\mathbf{r}, \mathbf{r}') \\ \omega_{-\delta}(\mathbf{r}') = m[\delta(\mathbf{r}, \mathbf{r}') - \delta(\mathbf{r} - \mathbf{i}\delta, \mathbf{r}')] \\ \bar{V}^+_{-\delta} = \Omega' \exp[m\bar{\mu}_1(\mathbf{r} - \mathbf{i}\delta)/k_B T], \quad \bar{V}^-_{-\delta} = \Omega' \exp[m\bar{\mu}_1(\mathbf{r})/k_B T], \end{gathered} \tag{6.2.7}$$

These expressions need to be substituted in Eq. (6.2.5) to get the contribution of the x − axis faces to the covariance of $\tilde{f}_1(\mathbf{r}, t)$. Since we are treating fluctuations in mass densities, we need to divide by the volume of a cell, δ^3, and thus make the identifications $\omega_{\kappa i} = \omega_{\pm\delta}(\mathbf{r})\delta^{-3}$ and $\omega_{\kappa j} = \omega_{\pm\delta}(\mathbf{r}')\delta^{-3}$. Noticing that $\omega_{+\delta}(\mathbf{r}) = -m$ and $\omega_{-\delta}(\mathbf{r}) = m$ and using Eqs. (6.2.6) and (6.2.7) gives

$$\begin{aligned} \gamma[\mathbf{r}, \mathbf{r}'; \rho^0(t)]_x = \lim_{\delta \to 0} (m^2\Omega'/\delta^3)\{ & [\exp(m\bar{\mu}_1(-)/k_B T) + \exp(m\bar{\mu}_1/k_B T)] \\ & \times [\delta(\mathbf{r}, \mathbf{r}') - \delta(\mathbf{r} - \mathbf{i}\delta, \mathbf{r}')]/\delta^3 \\ & - [\exp(m\bar{\mu}_1/k_B T) + \exp(m\bar{\mu}_1(+)/k_B T)] \\ & \times [\delta(\mathbf{r} + \mathbf{i}\delta, \mathbf{r}') - \delta(\mathbf{r}, \mathbf{r}')]/\delta^3 \}. \end{aligned} \tag{6.2.8}$$

The first term comes from the face at $\mathbf{r} - \mathbf{i}\delta/2$ and the second comes from the face at $\mathbf{r} + \mathbf{i}\delta/2$.

To aid in carrying out the continuum limit in Eq. (6.2.8), we note that $\delta(\mathbf{r}, \mathbf{r}')/\delta^3$ approaches the Dirac delta function. This follows from the properties of the Kronecker delta, which show that

$$\delta(\mathbf{r}, \mathbf{r}') \equiv \lim_{\delta \to 0} \delta(\mathbf{r}, \mathbf{r}')/\delta^3 = \begin{cases} \infty & \mathbf{r} = \mathbf{r}' \\ 0 & \mathbf{r} \neq \mathbf{r}'. \end{cases} \tag{6.2.9}$$

Furthermore, since δ^3 is the volume of a cell and

$$\sum_{\mathbf{r}'} [\delta(\mathbf{r}, \mathbf{r}')/\delta^3]\delta^3 = 1, \tag{6.2.10}$$

it follows that

$$\lim_{\delta \to 0} \sum_{\mathbf{r}'} [\delta(\mathbf{r}, \mathbf{r}')/\delta^3]\delta^3 = \int \delta(\mathbf{r} - \mathbf{r}')\, d\mathbf{r}' = 1. \tag{6.2.11}$$

Thus the delta function is properly normalized. As a consequence, in Eq. (6.2.8) the terms in brackets involving the Kronecker deltas can be replaced by Dirac delta functions. Writing $f(\mathbf{r}) = \exp[m\bar{\mu}_1(\mathbf{r})/k_B T]$ and $\hat{\delta}(\mathbf{r}) \equiv \delta(\mathbf{r} - \mathbf{r}')$, the Taylor series expansion of the curly-bracketed term in Eq. (6.2.8) is

$$\begin{aligned} \{\cdot\} &= \left[2f - \delta\frac{\partial f}{\partial x}\right]\left[\delta\frac{\partial \hat{\delta}}{\partial x} - \frac{\delta^2}{2}\frac{\partial^2 \hat{\delta}}{\partial x^2}\right] - \left[2f + \delta\frac{\partial f}{\partial x}\right] \\ &\quad \times \left[\delta\frac{\partial \hat{\delta}}{\partial x} + \frac{\delta^2}{2}\frac{\partial^2 \hat{\delta}}{\partial x^2}\right] + \mathcal{O}(\delta^2) \\ &= -2\delta^2\left[\frac{\partial f}{\partial x}\frac{\partial \hat{\delta}}{\partial x} + f\frac{\partial^2 \hat{\delta}}{\partial x^2}\right] + \mathcal{O}(\delta)^2 \\ &= -2\delta^2\frac{\partial}{\partial x}[f\partial\hat{\delta}/\partial x] + \mathcal{O}(\delta^2). \end{aligned} \tag{6.2.12}$$

Substituting this expression into Eq. (6.2.8) and defining

$$\lim_{\delta \to 0} \Omega'/\delta = \bar{\Omega} \tag{6.2.13}$$

gives

$$\gamma(\mathbf{r}, \mathbf{r}'; \rho^0(t))_x = -2m^2\bar{\Omega}\frac{\partial}{\partial x}\exp(m\bar{\mu}_1/k_B T)\frac{\partial}{\partial x}\delta(\mathbf{r} - \mathbf{r}'). \tag{6.2.14}$$

To obtain the complete expression for the covariance of $\tilde{f}_1$ we add to Eq. (6.2.14) similar contributions coming from faces along the y- and the z-axis. This gives

$$\gamma(\mathbf{r}, \mathbf{r}'; \rho^0(t)) = -2m^2\bar{\Omega}\nabla_{\mathbf{r}}\cdot\exp(m\bar{\mu}_1/k_B T)\nabla_{\mathbf{r}}\delta(\mathbf{r} - \mathbf{r}'), \tag{6.2.15}$$

where the subscript on the gradient emphasizes that it is taken with respect to $\mathbf{r}$. Equation (6.2.15) also can be written in terms of the diffusion constant.

Neglecting solvent, the diffusion constant is [cf. Eq. (6.1.38) with $X_s = 1$]

$$D = (m^2\bar{\Omega}/k_B T)(\partial \bar{\mu}_1/\partial \bar{\rho}_1)_{T,p} \exp(m\bar{\mu}_1/k_B T).$$

Consequently Eq. (6.2.15) can be written as

$$\gamma(\mathbf{r}, \mathbf{r}'; \rho^0(t)) = -2k_B T \nabla_{\mathbf{r}} \cdot D(\partial \bar{\rho}_1/\partial \bar{\mu}_1)_{T,p} \nabla_{\mathbf{r}} \delta(\mathbf{r} - \mathbf{r}'). \tag{6.2.16}$$

Another way to write the random term involves the so-called random diffusion flux, $\tilde{\mathbf{j}}_D$, defined by

$$-\nabla \cdot \tilde{\mathbf{j}}_D = \tilde{f}_1. \tag{6.2.17}$$

In terms of $\tilde{\mathbf{j}}_D$ Eq. (6.2.2) for the conditional fluctuations becomes

$$\partial \delta\rho_1/\partial t = \nabla^2(D\delta\rho_1) + \nabla \cdot (\bar{\rho}_1 \delta\mathbf{v}) + \nabla \cdot (\bar{\mathbf{v}}\delta\rho_1) - \nabla \cdot \tilde{\mathbf{j}}_D. \tag{6.2.18}$$

Since $\tilde{\mathbf{j}}_D$ and $\tilde{f}_1$ are related by a linear operator, $\tilde{\mathbf{j}}_D$ inherits the Gaussian property from $\tilde{f}_1$. It also can be taken to vanish on the conditional average since it follows from Eq. (6.2.17) that

$$0 = \langle -\nabla \cdot \tilde{\mathbf{j}}_D \rangle = -\nabla \cdot \langle \tilde{\mathbf{j}}_D \rangle. \tag{6.2.19}$$

To find the covariance of $\tilde{\mathbf{j}}_D$ we need the help of the following *lemma*: If

$$\langle \tilde{h}(\mathbf{r}, t)\tilde{g}(\mathbf{r}', t) \rangle = \lambda(\mathbf{r}, t)\delta(\mathbf{r} - \mathbf{r}')\delta(t - t'), \tag{6.2.20}$$

then

$$\langle [\partial \tilde{h}(\mathbf{r}, t)/\partial x_i][\partial \tilde{g}(\mathbf{r}', t)/\partial x_j'] \rangle = -\frac{\partial}{\partial x_i}\lambda(\mathbf{r}, t)\frac{\partial}{\partial x_j}\delta(\mathbf{r} - \mathbf{r}')\delta(t - t') \tag{6.2.21}$$

where $x_i, x_j = x, y$, or z. The proof of this lemma follows from the identities

$$\langle [\partial \tilde{h}(\mathbf{r}, t)/\partial x_i]\tilde{g}(\mathbf{r}', t) \rangle = \frac{\partial}{\partial x_i}\lambda(\mathbf{r}, t)\delta(\mathbf{r} - \mathbf{r}')\delta(t - t') \tag{6.2.22}$$

and

$$\langle \tilde{h}(\mathbf{r}, t)\partial \tilde{g}(\mathbf{r}', t)/\partial x_j' \rangle = -\lambda(\mathbf{r}, t)\frac{\partial}{\partial x_j}\delta(\mathbf{r} - \mathbf{r}')\delta(t - t') \tag{6.2.23}$$

Indeed, applying the identity in Eq. (6.2.23) to the left-hand side of Eq. (6.2.21) followed by an application of Eq. (6.2.22) to the resulting expression obviously gives the right-hand side of Eq. (6.2.21).

To prove these identities, we use the fact that the covariance is the kernel of a linear, integral operator. Thus for $F(\mathbf{r})$, a differentiable function, Eq. (6.2.22) should be interpreted as

$$\begin{aligned}\int \langle [\partial \tilde{h}(\mathbf{r}, t)/\partial x_i]\tilde{g}(\mathbf{r}', t) \rangle F(\mathbf{r}')\, d\mathbf{r}' &= \int \frac{\partial}{\partial x_i}\lambda(\mathbf{r}, t)\delta(\mathbf{r} - \mathbf{r}')\delta(t - t')F(\mathbf{r}')\, d\mathbf{r}' \\ &= \frac{\partial}{\partial x_i}[\lambda(\mathbf{r}, t)F(\mathbf{r})]\delta(t - t'). \end{aligned} \tag{6.2.24}$$

But this is obvious since taking the derivative out of the integral on the left-hand side of Eq. (6.2.24) and using Eq. (6.2.20), we have

$$\int \langle [\partial \tilde{h}(\mathbf{r},t)/\partial x_i] \tilde{g}(\mathbf{r}',t) \rangle F(\mathbf{r}')\, d\mathbf{r}' = \frac{\partial}{\partial x_i} \int \langle \tilde{h}(\mathbf{r},t) \tilde{g}(\mathbf{r}',t) \rangle F(\mathbf{r}')\, d\mathbf{r}'$$

$$= \frac{\partial}{\partial x_i} [\lambda(\mathbf{r},t) F(\mathbf{r})] \delta(t-t'). \qquad (6.2.25)$$

The integral form of the left-hand side of the identity in Eq. (6.2.23) is

$$\int \langle \tilde{h}(\mathbf{r},t) [\partial \tilde{g}(\mathbf{r}',t)/\partial x_j'] \rangle F(\mathbf{r}')\, d\mathbf{r}'$$

$$= \lim_{\delta \to 0} \int \left\langle \tilde{h}(\mathbf{r},t) \left[\frac{\tilde{g}(\mathbf{r}' + \boldsymbol{\delta}_j, t') - \tilde{g}(\mathbf{r}',t)}{\delta} \right] F(\mathbf{r}') \right\rangle d\mathbf{r}' \qquad (6.2.26)$$

where $\boldsymbol{\delta}_j$ is a vector of length δ along the x_j-axis. Using Eq. (6.2.20), this can be written

$$\lim_{\delta \to 0} \int \lambda(\mathbf{r},t) [\delta(\mathbf{r} - \mathbf{r}' - \boldsymbol{\delta}_j) - \delta(\mathbf{r} - \mathbf{r}')] \frac{F(\mathbf{r}')}{\delta} d\mathbf{r}' \delta(t-t')$$

$$= \lim_{\delta \to 0} \{\lambda(\mathbf{r},t)[F(\mathbf{r} - \boldsymbol{\delta}_j) - F(\mathbf{r})]/\delta\} \delta(t-t')$$

$$= -\int \gamma(\mathbf{r},t) \frac{\partial}{\partial x_j} \delta(\mathbf{r} - \mathbf{r}') F(\mathbf{r}') \delta(t-t')\, d\mathbf{r}'. \qquad (6.2.27)$$

Since the last expression is the integral form of the right-hand side of Eq. (6.2.23), the equalities in Eq. (6.2.27) prove that identity, too.

To use the lemma in Eqs. (6.2.20) and (6.2.21) to obtain the covariance of the random diffusion flux, we write out the covariance expression which follows from the definition of $\tilde{\mathbf{j}}_D$ in Eq. (6.2.17),

$$\langle [\partial \tilde{j}_{Di}(\mathbf{r},t)/\partial x_i][\partial \tilde{j}_{Dj}(\mathbf{r}',t)/\partial x_j'] \rangle = \langle \tilde{f}_1(\mathbf{r},t) \tilde{f}_1(\mathbf{r}',t') \rangle, \qquad (6.2.28)$$

where the summation convention on repreated indices has been invoked. According to the lemma in Eq. (6.2.21) the left-hand side of Eq. (6.2.28) can be written

$$\langle [\partial \tilde{j}_{Di}(\mathbf{r},t)/\partial x_i] \partial \tilde{j}_{Dj}(\mathbf{r}',t')/\partial x_j' \rangle = -\frac{\partial}{\partial x_i} \gamma_{ij}(\mathbf{r},t) \frac{\partial}{\partial x_j} \delta(\mathbf{r} - \mathbf{r}') \delta(t-t') \quad (6.2.29)$$

if

$$\langle \tilde{j}_{Di}(\mathbf{r},t) \tilde{j}_{Dj}(\mathbf{r}',t') \rangle \equiv \gamma_{ij}(\mathbf{r},t) \delta(\mathbf{r} - \mathbf{r}') \delta(t-t'). \qquad (6.2.30)$$

On the other hand, from Eq. (6.2.16) it follows that the right-hand side of Eq. (6.2.28) is

$$\langle \tilde{f}_1(\mathbf{r},t) \tilde{f}_1(\mathbf{r}',t') \rangle = -\frac{\partial}{\partial x_i} \{2k_B T D (\partial \bar{\rho}_1/\partial \bar{\mu}_1)_{T,p}\} \frac{\partial}{\partial x_i} \delta(\mathbf{r} - \mathbf{r}') \delta(t-t'). \quad (6.2.31)$$

Hence, comparing Eqs. (6.2.29) and (6.2.31), it is consistent to take

$$\gamma_{ij}(\mathbf{r}, t) = 2k_B TD(\partial\bar{\rho}_1/\partial\bar{\mu}_1)_{T,p}\delta_{ij}. \qquad (6.2.32)$$

In dilute solution where the mass-based chemical potential, $\bar{\mu}_1$, is related to the density by

$$m_1\bar{\mu}_1 = \mu_1^0 + k_B T \ln \bar{\rho}_1, \qquad (6.2.33)$$

with m_1 the molecular mass of the solute, the covariance formula reduces to

$$\gamma_{ij}(\mathbf{r}, t) = 2D\bar{\rho}_1 m_1 \delta_{ij}. \qquad (6.2.34)$$

This makes it evident that all the dependence of the random flux on space and time is contained in the conditional average density.

It is convenient in many applications, especially involving chemical reactions, to consider number densities instead of mass densities as the basic variables. The corresponding formulas for the effects of diffusion are only slightly modified by this change. For the example treated in this section, the solute number density is $\rho = (\rho_1/m_1)$ and the number-density-based chemical potential is $\mu = m_1\mu_1$. With these changes the average equation (6.2.1) becomes

$$\partial\bar{\rho}/\partial t = \omega\bar{\Omega}\nabla^2 \exp(\omega\bar{\mu}/k_B T) - \nabla\cdot(\bar{\rho}\,\bar{\mathbf{v}}). \qquad (6.2.35)$$

where $\omega \equiv (m/m_1)$. Similarly, Eq. (6.2.2) becomes

$$\partial\delta\rho/\partial t = \nabla^2(D\delta\rho) + \nabla\cdot(\bar{\rho}\delta\mathbf{v}) + \nabla\cdot(\bar{\mathbf{v}}\delta\rho) - \nabla\cdot\tilde{\mathbf{j}}_D' \qquad (6.2.36)$$

where $\tilde{\mathbf{j}}_D' = \tilde{\mathbf{j}}_D/m_1$, the number-based random diffusion flux. The random flux vanishes on the average and using Eq. (6.2.32) its covariance becomes

$$\langle \tilde{j}_{Di}'(\mathbf{r}, t)\tilde{j}_{Dj}'(\mathbf{r}', t')\rangle = 2k_B TD(\partial\bar{\rho}/\partial\bar{\mu})_{T,p}\delta_{ij}. \qquad (6.2.37)$$

If, instead, the random term in Eq. (6.2.36) is defined through the flux, then

$$-\nabla\cdot\tilde{\mathbf{j}}_D' \equiv \tilde{f}, \qquad (6.2.38)$$

and the identity implies that

$$\langle \tilde{f}(\mathbf{r}, t)\tilde{f}(\mathbf{r}', t')\rangle = -2k_B T\nabla_{\mathbf{r}}\cdot D(\partial\bar{\rho}/\partial\bar{\mu})_{T,p}\nabla_{\mathbf{r}}\delta(\mathbf{r} - \mathbf{r}')\delta(t - t'). \quad (6.2.39)$$

Diffusion provides a good example of how fluctuations depend on spatial coordinates at the hydrodynamic level of description. As we have seen, the density $\rho(\mathbf{r}, t)$ is a random field. Since the dynamical equations for both the average and the fluctuations in the density couple one point in space to another, it is necessary to consider the entire density field, not just the density at an isolated point. Thus in denoting an initial condition for a conditional fluctuation, the initial density field throughout the entire system, $\rho^0(\mathbf{r})$, must be specified. The conditional equations, e.g. Eqs. (6.2.35) and (6.2.36), then must be solved for each different initial condition.

To illustrate how this is done, we examine solute number density fluctuations as described by Eqs. (6.2.35)–(6.2.39). Since we have not yet considered velocity fluctuations, we neglect these terms. For simplicity we consider the case $\omega = 1$, which is linear in the density. With these restrictions the conditional average equation (6.2.35) becomes

$$\partial \bar{\rho}/\partial t = D\nabla^2 \bar{\rho}, \tag{6.2.40}$$

where using the definition of D above Eq. (6.2.16), Eq. (6.2.33), and the fact that $\omega = m/m_1 = 1$ we have noticed that

$$D = \bar{\Omega} \exp(\mu_1^0/k_B T). \tag{6.2.41}$$

In this case D is independent of the density. Thus after neglect of the velocity Eq. (6.2.36) become

$$\partial \delta\rho/\partial t = D\nabla^2 \delta\rho + \tilde{f}. \tag{6.2.42}$$

The covariance of $\tilde{f}$ is given by Eq. (6.2.39), which in dilute solution reduces to

$$\langle \tilde{f}(\mathbf{r}, t)\tilde{f}(\mathbf{r}', t')\rangle = -2D\nabla_{\mathbf{r}}\cdot \bar{\rho}\nabla_{\mathbf{r}}\delta(\mathbf{r} - \mathbf{r}')\delta(t - t'). \tag{6.2.43}$$

As we know, Eqs. like (6.2.40), (6.2.41), and (6.2.43) have conditional probability densities that are Gaussian. In this case since the density is a field variable, the corresponding probability density is a *functional*. This means that the conditional average is also a density field. Indeed, it is determined by the solution of Eq. (6.2.40) using the initial condition $\rho^0(\mathbf{r})$. Moreover, the covariance of the Gaussian depends on two field variables, $\mathbf{r}$ and $\mathbf{r}'$. Like the covariance of the random fields $\tilde{f}$ in Eqs. (6.2.42) and (6.2.43), the conditional covariance of $\delta\rho(\mathbf{r}, t)$ can be written

$$\sigma(\mathbf{r}, \mathbf{r}', t) = \langle \delta\rho(\mathbf{r}, t)\delta\rho(\mathbf{r}', t)\rangle^0. \tag{6.2.44}$$

It is also called the conditional *density–density correlation function*. According to the general results obtained in Chapter 4, the covariance function solves Eq. (4.3.9), which has the formal structure

$$d\sigma/dt = H\sigma + \sigma H^T + \gamma. \tag{6.2.45}$$

Equation (6.2.45) is a linear equation. As we have seen, for discrete variables σ and H are matrices, whereas for continuous variables σ and H are continuous linear operators. The form of H follows from Eq. (6.2.42), i.e.

$$H[\delta\rho] \equiv \int H(\mathbf{r}, \mathbf{r}')\delta\rho(\mathbf{r}')\,d\mathbf{r}' = D\nabla^2 \delta\rho. \tag{6.2.46}$$

Thus the kernel of the linear functional must be

$$H(\mathbf{r}, \mathbf{r}') = D\nabla_{\mathbf{r}}^2 \delta(\mathbf{r} - \mathbf{r}'), \tag{6.2.47}$$

which should be compared to the integral representation of the Onsager functional in Section 2.4 at the hydrodynamic level of description. Using this notation, a more explicit representation of Eq. (6.2.45) is

$$d\sigma(\mathbf{r},\mathbf{r}',t)/dt = \int [H(\mathbf{r},\mathbf{r}'')\sigma(\mathbf{r}'',\mathbf{r}',t) + \sigma(\mathbf{r},\mathbf{r}'',t)H(\mathbf{r}',\mathbf{r}'')]\,d\mathbf{r}'' + \gamma(\mathbf{r},\mathbf{r}',t). \tag{6.2.48}$$

It is often convenient to couch statistical problems involving field variables in another representation. For a field that is defined throughout all space, we can use the Fourier transform to define

$$\delta\hat{\rho}(\mathbf{k},t) \equiv (2\pi)^{-3}\int_{-\infty}^{+\infty} \exp(i\mathbf{k}\cdot\mathbf{r})\delta\rho(\mathbf{r},t)\,d\mathbf{r}. \tag{6.2.49}$$

One advantage of this transformation is that for any function f which vanishes as $|\mathbf{r}| \to \infty$, one has

$$\widehat{\nabla\cdot f} = -i\mathbf{k}\hat{f}, \tag{6.2.50}$$

which follows by integrating by parts. Consequently, if we interchange the integral over space with the time derivative, the Fourier transform of Eq. (6.2.42) becomes

$$\partial\delta\hat{\rho}/\partial t = -k^2 D\delta\hat{\rho} + \hat{\tilde{f}}. \tag{6.2.51}$$

The correlation function of the Fourier-transformed term, $\hat{\tilde{f}}$, can be obtained by using its definition, i.e.,

$$\begin{aligned}
&\langle\hat{\tilde{f}}(\mathbf{k},t)\hat{\tilde{f}}(\mathbf{k}',t')\rangle \\
&= (2\pi)^{-6}\int d\mathbf{r}\int d\mathbf{r}' \exp(i\mathbf{k}\cdot\mathbf{r})[-2D\nabla_{\mathbf{r}}\cdot\bar{\rho}\nabla_{\mathbf{r}}\delta(\mathbf{r}-\mathbf{r}')]\exp(i\mathbf{k}'\cdot\mathbf{r}')\delta(t-t') \\
&= (2\pi)^{-6}\int d\mathbf{r}\exp(i\mathbf{k}\cdot\mathbf{r})[-2D\nabla_{\mathbf{r}}\cdot\bar{\rho}\nabla_{\mathbf{r}}\exp(i\mathbf{k}'\cdot\mathbf{r})]\delta(t-t') \\
&= (2\pi)^{-6}(-2D)\int d\mathbf{r}\exp[i(\mathbf{k}+\mathbf{k}')\cdot\mathbf{r}][i\mathbf{k}'\cdot\nabla_{\mathbf{r}}\bar{\rho} - k'^2\bar{\rho}]\delta(t-t').
\end{aligned}$$

Simplifying the first term in the final integral using Eq. (6.2.50) and introducing the Fourier transform of the average density, $\hat{\bar{\rho}}$, gives

$$\begin{aligned}
\langle\hat{\tilde{f}}(\mathbf{k},t)\hat{\tilde{f}}(\mathbf{k}',t)\rangle &= 2D[k'^2 - \mathbf{k}'\cdot(\mathbf{k}+\mathbf{k}')]\hat{\bar{\rho}}(\mathbf{k}+\mathbf{k}',t)\delta(t-t')/(2\pi)^3 \\
&\equiv \hat{\gamma}(\mathbf{k},\mathbf{k}',t)\delta(t-t').
\end{aligned} \tag{6.2.52}$$

In the Fourier transform representation, $\delta\hat{\rho}$ is still a Gaussian process and its conditional covariance solves the analogue of Eq. (6.2.48). Equation (6.5.51) shows that $\hat{H}(\mathbf{k},\mathbf{k}') = -k^2 D\delta(\mathbf{k}-\mathbf{k}')$, so this equation has the form

$$d\hat{\sigma}(\mathbf{k},\mathbf{k}',t)/dt = -D(k^2 + k'^2)\hat{\sigma}(\mathbf{k},\mathbf{k}',t) + \hat{\gamma}(\mathbf{k},\mathbf{k}',t). \tag{6.2.53}$$

Since Eq. (6.2.53) is linear and first order in the time, it has the formal solution

$$\hat{\sigma}(\mathbf{k},\mathbf{k}',t) = \int_0^t \exp[-D(k^2 + k'^2)(t-\tau)]\hat{\gamma}(\mathbf{k},\mathbf{k}',\tau)\,d\tau, \tag{6.2.54}$$

where we have used the initial condition $\hat{\sigma}(\mathbf{k},\mathbf{k}',0) = 0$, appropriate for a conditional covariance.

The simplest case in which Eq. (6.2.54) can be solved involves the uniform initial conditional $\rho^0(\mathbf{r}) = \rho^0$, a constant. In this case the solution to the average equation (6.2.40) is also a constant, i.e., $\bar{\rho}(\mathbf{r},t) = \rho^0$. Thus

$$\hat{\bar{\rho}}(\mathbf{k}+\mathbf{k}',t) = (2\pi)^{-3}\int_{-\infty}^{+\infty} \rho^0 \exp[i(\mathbf{k}+\mathbf{k}')\cdot\mathbf{r}]\,d\mathbf{r} = \rho^0\delta(\mathbf{k}+\mathbf{k}'), \tag{6.2.55}$$

as follows from the well-known representation of the Dirac delta function. Consequently, for an initial density field which is uniform, Eq. (6.2.54) can be written

$$\begin{aligned}\hat{\sigma}(\mathbf{k},\mathbf{k}',t) &= 2D\rho^0 k^2 \int_0^t \exp[-2Dk^2(t-\tau)]\,d\tau\,\delta(\mathbf{k}+\mathbf{k}')/(2\pi)^3 \\ &= \rho^0[1 - \exp(-2Dk^2 t)]\delta(\mathbf{k}+\mathbf{k}')/(2\pi)^3. \end{aligned}\tag{6.2.56}$$

The relative ease with which we have obtained the Fourier transform of the density–density covariance function is a consequence of the fact that Eq. (6.2.53) is a linear ordinary differential equation. This contrasts with the analogous equation (6.2.48) in the position representation which is a linear integral, partial differential equation. We can, however, obtain the density–density covariance function in the position representation by using the Fourier inversion theorem,

$$f(\mathbf{r},t) = \int \exp(-i\mathbf{k}\cdot\mathbf{r})\hat{f}(\mathbf{k},t)\,d\mathbf{k}. \tag{6.2.57}$$

Applying this formula twice to $\hat{\sigma}$ in Eq. (6.2.56) gives

$$\begin{aligned}&\sigma(\mathbf{r},\mathbf{r}',t) \\ &= (2\pi)^{-3}\rho^0 \int d\mathbf{k} \int d\mathbf{k}'\,\delta(\mathbf{k}+\mathbf{k}')\exp[-i(\mathbf{k}\cdot\mathbf{r} + \mathbf{k}'\cdot\mathbf{r}')][1 - \exp(-2Dk^2 t)] \\ &= (2\pi)^{-3}\rho^0 \int d\mathbf{k}\exp[-i\mathbf{k}\cdot(\mathbf{r}-\mathbf{r}')][1 - \exp(-2Dk^2 t)]. \end{aligned}\tag{6.2.58}$$

Changing variables in the integrand to spherical coordinates so that $d\mathbf{k} = k^2 \sin\theta\,dk\,d\theta\,d\phi$ and choosing the z-axis to lie along $\mathbf{r} - \mathbf{r}'$, the integral in Eq. (6.2.58) can be rewritten

$$\begin{aligned}\sigma(\mathbf{r},\mathbf{r}',t) = \rho^0\delta(\mathbf{r}-\mathbf{r}') - \rho^0(2\pi)^{-2}\int_0^{\infty}\int_0^{\pi} \exp(-ik|\mathbf{r}-\mathbf{r}'|\cos\theta) \\ \times \sin\theta\,d\theta\,k^2\exp(-2Dk^2 t)\,dk, \end{aligned}\tag{6.2.59}$$

where in the first term we have used Eq. (6.2.55). The θ integral can be performed by changing variables to $y = -\cos\theta$, so that $dy = \sin\theta\, d\theta$. This gives

$$\sigma(\mathbf{r}, \mathbf{r}', t) = \rho^0 \delta(\mathbf{r} - \mathbf{r}') - 2\rho^0 (2\pi)^{-2} \int_0^\infty \frac{k \sin(k|\mathbf{r} - \mathbf{r}'|)}{|\mathbf{r} - \mathbf{r}'|} \exp(-2Dk^2 t)\, dk. \tag{6.2.60}$$

Finally, the integral over k can be carried out by contour integration or found in tables. Substituting its value into Eq. (6.2.60) yields

$$\sigma(\mathbf{r}, \mathbf{r}', t) = \rho^0 \{\delta(\mathbf{r} - \mathbf{r}') - \exp[-|\mathbf{r} - \mathbf{r}'|^2/8Dt]/(8\pi Dt)^{3/2}\}. \tag{6.2.61}$$

Notice that as t approaches infinity the density–density correlation function approaches

$$\sigma^e(\mathbf{r}, \mathbf{r}') = \rho^0 \delta(\mathbf{r} - \mathbf{r}'). \tag{6.2.62}$$

This is the equilibrium density–density correlation function for an ideal solution. In Section 6.9 this result is obtained in another way from the functional derivatives of thermodynamic quantities.

There is another initial density profile for which the conditional covariance is easily obtained, namely,

$$\rho^0(\mathbf{r}) = \rho^0 \delta(\mathbf{r}). \tag{6.2.63}$$

This initial profile is far from equilibrium and leads to the conditional average solution

$$\bar{\rho}(\mathbf{r}, t) = \rho^0 (4\pi Dt)^{-3/2} \exp(-r^2/4Dt). \tag{6.2.64}$$

The Fourier transform of $\bar{\rho}$ is easily carried out using manipulations similar to those in Eqs. (6.2.58)–(6.2.60). Application of the formula in Eq. (6.2.54) then gives, after considerable manipulation,

$$\begin{aligned}\sigma(\mathbf{r}, \mathbf{r}', t) = \rho^0 \{&\delta(\mathbf{r} - \mathbf{r}') \exp[-|\mathbf{r} + \mathbf{r}'|^2/4Dt]/(4\pi Dt)^{3/2} \\ &- \exp[-|\mathbf{r} - \mathbf{r}'|^2/8Dt] \exp[-|\mathbf{r} + \mathbf{r}'|^2/8Dt]/(8\pi Dt)^3\}.\end{aligned} \tag{6.2.65}$$

This covariance function vanishes as t approaches infinity because the initial density profile does not contain a sufficient amount of solute to fill up an infinite container with a nonzero density.

For the sake of completeness we close this section with a discussion of density fluctuations caused by diffusion in multicomponent solutions. At the end of Section 6.1 we described the canonical form for the most general elementary processes associated with molecular diffusion in an isotropic medium. These equations suffice to determine the stochastic equations for the density fluctuations. The calculations are analogous to the calculations used to obtain Eqs. (6.2.15) and (6.2.32) for the simple linear diffusion process for a single dilute solute. The conditional average equations can be obtained by combining Eqs. (6.1.56) and (6.1.58) with the condition of mass conservation. Thus

$$\partial \bar{\rho}_i / \partial \tau = -\nabla \cdot (\bar{\rho}_i \bar{\mathbf{v}}) - \nabla \cdot \bar{L}_{ij} \nabla(-\bar{\mu}_j / T), \tag{6.2.66}$$

where the form L_{ij} is given in Eqs. (6.1.58) and (6.1.59). The fluctuations around the conditional average satisfy a linearized version of Eq. (6.2.66), i.e.,

$$\begin{aligned} \partial \delta\rho_i / \partial t = &-\nabla \cdot (\delta\rho_i \bar{\mathbf{v}}) - \nabla \cdot (\bar{\rho}_i \delta\mathbf{v}) - \nabla \cdot (\delta L_{ij} \nabla(-\bar{\mu}_j / \bar{T})) \\ &- \nabla \cdot \bar{L}_{ij} \nabla(-\delta(\mu_j / T)) + \tilde{f}_i. \end{aligned} \tag{6.2.67}$$

The random term, $\tilde{f}_i(\mathbf{r}, t)$, vanishes on the average and its covariance can be calculated from the description of the elementary processes in Section 6.1. One finds that

$$\langle \tilde{f}_i(\mathbf{r}, t) \tilde{f}_j(\mathbf{r}', t') \rangle = -2k_B \nabla_{\mathbf{r}} \cdot \bar{L}_{ij} \nabla_{\mathbf{r}} \delta(\mathbf{r} - \mathbf{r}') \delta(t - t'). \tag{6.2.68}$$

Equation (6.2.67) can also be written in terms of random diffusion fluxes, $\tilde{\mathbf{j}}_i$, as

$$\begin{aligned} \partial \delta\rho_i / \partial t = &-\nabla \cdot (\delta\rho_i \bar{\mathbf{v}}) - \nabla \cdot (\bar{\rho}_i \delta\mathbf{v}) - \nabla \cdot [\delta L_{ij} \nabla(-\bar{\mu}_j / T)] \\ &- \nabla \cdot \bar{L}_{ij} \nabla[-\delta(\mu_j / T)] - \nabla \cdot \mathbf{j}_1. \end{aligned} \tag{6.2.69}$$

The form of the covariance of $\tilde{\mathbf{j}}_i$ can be obtained using the lemma in Eqs. (6.2.20) and (6.2.21) and Eq. (6.2.67). This gives

$$\langle \tilde{\mathbf{j}}_i(\mathbf{r}, t) \tilde{\mathbf{j}}_j^T(\mathbf{r}', t') \rangle = 2k_B I \bar{L}_{ij} \delta(\mathbf{r} - \mathbf{r}'), \tag{6.2.70}$$

where I is the 3×3 identity matrix. The absence of correlations between the x-, y-, and z-components of $\tilde{\mathbf{j}}_i$ reflects the fact that the solution has been assumed to be isotropic.

6.3. Heat Conduction and Thermal Diffusion

The phenomenological relationship which describes the conduction of heat is Fourier's law, discussed in Section 2.4. Fourier's law is analogous to Fick's law for mass diffusion and both describe fluxes in terms of transport coefficients and gradients. Although this analogy is often useful in relating solutions of the average differential equation for diffusion to solutions of the corresponding heat equation, the elementary molecular processes underlying Fourier's law are different from those underlying Fick's law. In this section we describe the elementary processes which give rise to heat conduction as well as those which give rise to the phenomenon of thermal diffusion. Thermal diffusion is a coupled transport process, involving both mass and energy. As we shall see, it leads to the possibilities of heat being transported via a gradient in chemical potential and mass being transported via a temperature gradient. Since heat conduction in anisotropic media, such as a liquid crystal or a solid, is of interest experimentally, we begin by considering a simple case of an anisotropic medium.

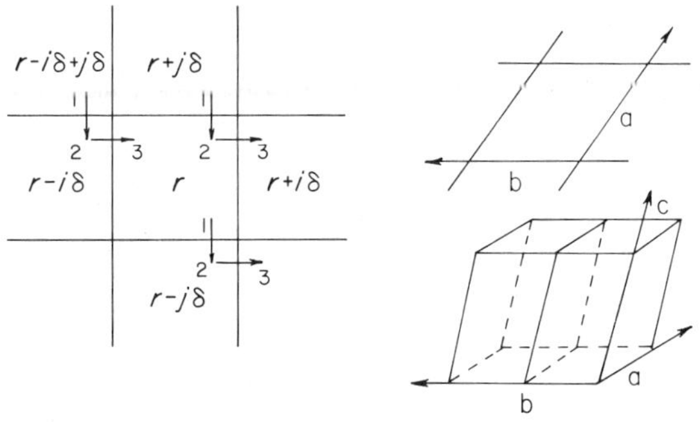

FIGURE 6.2. An illustration of a possible set of elementary processes for heat conduction in a monoclinic crystal, shown in perspective and from the top at the right-hand side of the diagram. The elementary processses, shown at the left, correspond to those described by Eq. (6.3.1).

The description of the elementary molecular processes which operate in an anisotropic medium is considerably more complicated than the description of the elementary processes which operate in an isotropic substance. In an anisotropic medium the physical properties depend on orientation of the substance with respect to particular axes. Consider, for example, a monoclinic crystal which has the skewed lattice structure shown from a top view in Fig. 6.2. That plane is spanned by the nonorthogonal a and b axes. The c-axis of the crystal is perpendicular to this plane and can be taken as the z-axis. When we describe the conduction of heat within the a–b plane using the orthogonal x and y axes, energy which is transported across a face perpendicular to the y-axis will be coupled by molecular motions along the crystal axes to energy that is transported across a face perpendicular to the x-axis. Because the c-axis (z-axis) is perpendicular to the plane, there will be no coupling of heat transport between the a–b plane and the z-direction. The degree of coupling within the a–b plane will depend on the particular elementary processes which occur in the crystal.

One possible set of elementary processes that would affect a cubic volume element located at position $\mathbf{r}$ are shown schematically in Fig. 6.2. Each process involves three adjoining volume elements in the x–y plane arranged in the L-shaped fashion shown in that figure. The arrows indicate the forward direction of the elementary process which involves simultaneous energy transport across two faces of the volume element at $\mathbf{r}$. All faces of the volume element which are parallel to the z-axis can be involved in this process. The energy of the volume element $\mathbf{r}$ is affected only when the process occurs at the three locations illustrated in Fig. 6.2. At each of these locations the elementary process can be written in terms of the molecular-sized amounts of energy, $\varepsilon^{\pm}$, involved in the process:

$$(\varepsilon_1^+, \varepsilon_2^+, \varepsilon_3^+) \rightleftharpoons (\varepsilon_1^-, \varepsilon_2^-, \varepsilon_3^-) \tag{6.3.1}$$

where the subscripts 1, 2, and 3 label adjacent cells as shown in the figure. The net transfers of energy are $\omega_i = \varepsilon_i^- - \varepsilon_i^+$ and the intrinsic rate of the elementary process is Ω. Since the elementary process conserves energy, it follows that $\varepsilon_1^+ + \varepsilon_2^+ + \varepsilon_3^+ = \varepsilon_1^- + \varepsilon_2^- + \varepsilon_3^-$ or

$$\omega_1 + \omega_2 + \omega_3 = 0. \tag{6.3.2}$$

Since the intensive variable conjugate to the energy is $\partial S/\partial E = 1/T$, the canonical form for the rate of change of energy in the cell at $\mathbf{r}$ becomes

$$\begin{aligned}
\partial E/\partial t = \omega_1\Omega\Bigg\{&\exp\left[-\left(\frac{\varepsilon_1^+}{k_BT} + \frac{\varepsilon_2^+}{k_BT(-j)} + \frac{\varepsilon_3^+}{k_BT(i-j)}\right)\right] \\
&- \exp\left[-\left(\frac{\varepsilon_1^-}{k_BT} + \frac{\varepsilon_2^-}{k_BT(-j)} + \frac{\varepsilon_3^-}{k_BT(i-j)}\right)\right]\Bigg\} \\
+ \omega_2\Omega\Bigg\{&\exp\left[-\left(\frac{\varepsilon_1^+}{k_BT(j)} + \frac{\varepsilon_2^+}{k_BT} + \frac{\varepsilon_3^+}{k_BT(i)}\right)\right] \\
&- \exp\left[-\left(\frac{\varepsilon_1^-}{k_BT(j)} + \frac{\varepsilon_2^-}{k_BT} + \frac{\varepsilon_3^-}{k_BT(i)}\right)\right]\Bigg\} \\
+ \omega_3\Omega\Bigg\{&\exp\left[-\left(\frac{\varepsilon_1^+}{k_BT(j-i)} + \frac{\varepsilon_2^+}{k_BT(-i)} + \frac{\varepsilon_3^+}{k_BT}\right)\right] \\
&- \exp\left[-\left(\frac{\varepsilon_1^-}{k_BT(j-i)} + \frac{\varepsilon_2^-}{k_BT(-i)} + \frac{\varepsilon_3^-}{k_BT}\right)\right]\Bigg\},
\end{aligned} \tag{6.3.3}$$

where, for example, we have used the shorthand notation $T(\mathbf{r} + \mathbf{i}\delta) = T(i)$ to represent the Kelvin temperature in the cell at $\mathbf{r} + \mathbf{i}\delta$.

If we carry out the continuum limit of Eq. (6.3.3), we can obtain the transport equation for the energy density

$$e(\mathbf{r}, t) \equiv \lim_{\delta\to 0} E(\mathbf{r}, t)/\delta^3. \tag{6.3.4}$$

The details of this are similar to what was required in taking the continuum limit for the canonical equations of mass diffusion [cf. Eqs. (6.1.15) or (6.1.48)]. The canonical form in Eq. (6.3.3), however, involves three adjacent cells, rather than two, and the calculation is considerably more laborious. To simplify the analysis, we consider the specific elementary process

$$(\varepsilon, \varepsilon, 0) \rightleftarrows (0, 0, 2\varepsilon), \tag{6.3.5}$$

which satisfies conservation of energy as can be checked using Eq. (6.3.2). For this process Eq. (6.3.3) can be written in the form

$$\begin{aligned}
(\varepsilon\Omega/\delta)^{-1}\frac{\partial E/\delta^3}{\partial t} &= \delta^{-2}\{-[RR(-j) - R^2(i-j)] - [R(j)R - R^2(i)] \\
&\quad + 2[R(j-i)R(-i) - R^2]\}
\end{aligned} \tag{6.3.6}$$

where $R = \exp(-\varepsilon/k_BT(\mathbf{r}))$ and we have used a shorthand notation $R(j)$, etc., like that in Eq. (6.3.3). The continuum limit of Eq. (6.3.6) can be obtained by Taylor expanding the functions R on the right-hand side around the point $\mathbf{r}$. The constant and linear terms vanish. Thus the limit $\delta \to 0$ leaves only the contribution from the quadratic terms. After some rearrangement the right-hand side of Eq. (6.3.6) becomes

$$-\frac{\partial}{\partial x}(2R\partial R/\partial y - 4R\partial R/\partial x) - \frac{\partial}{\partial y}(2R\partial R/\partial x - R\partial R/\partial y). \tag{6.3.7}$$

Combining this with the limit of the right-hand side of Eq. (6.3.6) yields

$$\partial e(\mathbf{r}, t)/\partial t = -\nabla \cdot \mathbf{j}_q, \tag{6.3.8}$$

where e is the internal energy density defined in Eq. (6.3.4) and the heat flux vector has components

$$\begin{aligned} j_{qx} &= \varepsilon\bar{\Omega}(-4R\partial R/\partial x + 2R\partial R/\partial y) \\ j_{qy} &= \varepsilon\bar{\Omega}(2R\partial R/\partial x - R\partial R/\partial y) \\ j_{qz} &= 0, \end{aligned} \tag{6.3.9}$$

with

$$\bar{\Omega} \equiv \lim_{\delta \to 0} (\Omega/\delta).$$

Using the definition of R given below Eq. (6.3.6), the heat fluxes can be written explicitly in terms of the gradient of T^{-1}, the intensive variable conjugate to the internal energy. This gives

$$\begin{aligned} j_{qx} &= L_{xx}\partial T^{-1}/\partial x + L_{xy}\partial T^{-1}/\partial y \\ j_{qy} &= L_{yx}\partial T^{-1}/\partial x + L_{yy}\partial T^{-1}/\partial y \\ j_{qz} &= 0 \end{aligned} \tag{6.3.10}$$

with

$$L_{xx} = (4\varepsilon^2\bar{\Omega}/k_B)\exp(-2\varepsilon/k_B T) \tag{6.3.11}$$

and $L_{xy} = L_{yx} = L_{xx}/2$, $L_{yy} = L_{xx}/4$.

This simple elementary process illustrates the sort of coupling between the components of the temperature gradient that can occur in isotropic crystals. The form is a generalization of Fourier's law for isotropic media. Notice that the coupling matrix L, which depends explicitly on the temperature, is symmetric and satisfies the conditions L_{xx}, $L_{yy} > 0$ and $L_{xx}L_{yy} - L_{xy}^2 = 0$. These properties reflect the fact that the continuum limit of the dissipation function is non-negative. Based on its definition in Eq. (4.5.5), the continuum limit of the dissipation function for this elementary process is

$$\Phi(t) = \int (T^{-1} - T^{e-1})\partial e/\partial t \, d\mathbf{r}, \tag{6.3.12}$$

where T^e is the equilibrium temperature, which is independent of **r**, and the integral is over the volume of the system. Using Eqs. (6.3.8) and (6.3.10), Eq. (6.3.12) can be written

$$\phi(t) = \int (T^{-1} - T^{e-1})(-\nabla \cdot L\nabla T^{-1})\, d\mathbf{r}$$

$$= \int \nabla T^{-1} \cdot L\nabla T^{-1}\, d\mathbf{r} \geq 0. \quad (6.3.13)$$

The second equality comes from integrating by parts and noting that to achieve an equilibrium temperature, T^e, one of two conditions must hold: Either the system is closed so that the heat flux vanishes on the boundary or the system is open and the constant temperature on the boundary equals the final equilibrium temperature, T^e.

It is difficult in the case of an anisotropic substance to describe the most general elementary processes that are responsible for heat conduction. This is due to the fact that a laboratory system of Cartesian coordinates will generally be unrelated to the system of molecular axes. Because of the symmetry of the heat conductivity tensor, however, an orthogonal coordinate system exists in which the conductivity tensor is diagonal. These axes lie on the principal axes of conductivity. In an equilibrium ensemble the orientation of these axes will not depend on one's location, **r**, in the crystal. In a nonequilibrium ensemble the temperature and the components of the conductivity tensor, L_{ij}, will depend on **r** through the average local temperature, $\bar{T}(\mathbf{r})$. Thus in a nonequilibrium ensemble it is not generally possible to select a fixed orthogonal system of axes for which heat conduction across perpendicular faces will be uncoupled.

Certain noncrystalline materials, which are termed *orthotropic*, have independent heat fluxes of the form

$$j_x = L_{xx}\partial T^{-1}/\partial x, \quad j_y = L_{yy}\partial T^{-1}/\partial y, \quad j_z = L_{zz}\partial T^{-1}/\partial z \quad (6.3.14)$$

along three orthogonal axes. An orthotropic material is isotropic in the case $L_{xx} = L_{yy} = L_{zz}$. For orthotropic materials one can derive the heat equation using independent elementary processes along the x, y, and z directions. The same methods that lead to the diffusion equation (6.1.17) can be used to obtain the conditional averaged heat equation:

$$\partial \bar{e}(\mathbf{r}, t)/\partial t = -\nabla \cdot L\nabla \bar{T}^{-1} \quad (6.3.15)$$

where L is the diagonal matrix of the form

$$L_{ij} = (\omega_i^2 \bar{\Omega}_i/k_B)\exp[-(\varepsilon_1^+ + \varepsilon_2^+)/k_B T]\delta_{ij}, \quad (6.3.16)$$

and $i, j = x$, y, or z. The equations which govern fluctuations in the energy density can also be derived easily using the methods described in Section 6.2. By analogy to Eqs. (6.2.2) and (6.2.14) we see that

$$\partial \delta e/\partial t = -\nabla \cdot \bar{L}\nabla \delta T^{-1} - \nabla \cdot \delta L\nabla \bar{T}^{-1} + \tilde{f}, \quad (6.3.17)$$

and by analogy to Eqs. (6.1.36) and (6.2.14),

$$\langle \tilde{f}(\mathbf{r},t)\tilde{f}(\mathbf{r}',t')\rangle = -2k_B \nabla_{\mathbf{r}} \cdot \bar{L} \nabla_{\mathbf{r}} \delta(\mathbf{r}-\mathbf{r}')\delta(t-t'). \qquad (6.3.18)$$

Equations (6.3.15)–(6.3.18) govern the molecular fluctuations caused by heat conduction in orthotropic materials, which are the most general anisotropic materials we will consider.

The differential equations that govern the coupled transfer of mass and heat are similar to those which describe the uncoupled transport processes. These equations are easily derived using the canonical form. The development is quite similar to that used in Section 6.1 to obtain the equation for multicomponent mass diffusion. The major difference is that both mass and energy are simultaneously transported. For simplicity we consider only an isotropic medium. Thus the elementary processes which occur at the x (or y or z) faces of the volume element in Fig. 6.1 can be written symbolically as

$$(n_{\kappa 1}^{+L}, n_{\kappa 2}^{+L}, \ldots, \varepsilon_{\kappa}^{+L}; n_{\kappa 1}^{+R}, n_{\kappa 2}^{+R}, \ldots, \varepsilon_{\kappa}^{+R}) \rightleftarrows (n_{\kappa 1}^{-L}, n_{\kappa 2}^{-L}, \ldots, \varepsilon_{\kappa}^{-L}; n_{\kappa 1}^{-R}, n_{\kappa 2}^{-R}, \ldots, \varepsilon_{\kappa}^{-R}). \qquad (6.3.19)$$

Just as in the case of mass diffusion, the superscripts L and R refer to the volume elements at the left (L) or right (R) of the face at which the elementary process occurs. Mass conservation and energy conservation imply that

$$\begin{aligned} n_{\kappa i}^{+L} + n_{\kappa i}^{+R} &= n_{\kappa i}^{-L} + n_{\kappa i}^{-R} \\ \varepsilon_{\kappa}^{+L} + \varepsilon_{\kappa}^{+R} &= \varepsilon_{\kappa}^{-L} + \varepsilon_{\kappa}^{-R}. \end{aligned} \qquad (6.3.20)$$

The first of these equations is identical to that solved by mass diffusion alone [cf. Eq. (6.1.45)]. Moreover, the second equation has the same formal structure as the first. Consequently, we can use our previous work on mass diffusion to see that the number densities must satisfy equations like (6.1.46)–(6.1.48). The final term in the sum in the exponential, however, is now $-\varepsilon_{\kappa}^{\pm}/k_B T$. Thus, to obtain the partial differential equations satisfied by the mass density, we can use Eq. (6.1.48) with rate expressions

$$\begin{aligned} V_{\kappa}^{\pm}(\mathbf{r}, \mathbf{r}+\mathbf{i}\delta) &= \Omega_{\kappa} \exp\left[-\sum_{i}^{k+1} m_{\kappa i}^{\pm L} F_i(\mathbf{r})/k_B\right] \exp\left[-\sum_{i}^{k+1} m_{\kappa i}^{\pm R} F_i(\mathbf{r}+\mathbf{i}\delta)/k_B\right] \\ V_{\kappa}^{\pm}(\mathbf{r}-\mathbf{i}\delta, \mathbf{r}) &= \Omega_{\kappa} \exp\left[-\sum_{i}^{k+1} m_{\kappa i}^{\pm L} F_i(\mathbf{r}-\mathbf{i}\delta)/k_B\right] \exp\left[-\sum_{i}^{k+1} m_{\kappa i}^{\pm R} F_i(\mathbf{r})/k_B\right] \end{aligned} \qquad (6.3.21)$$

where, for example, $m_{\kappa i}^{\pm L} = n_{\kappa i}^{\pm L}$ and $F_i = -\mu_i/T$ for $i = 1, 2, \ldots, k$ and $m_{\kappa, k+1}^{\pm L} = \varepsilon_{\kappa}^{+L}$ and $F_{k+1} = 1/T$. Proceeding now as with the derivation of Eqs. (6.1.55)–(6.1.57), we find that

$$[\partial \bar{\rho}_i(\mathbf{r},t)/\partial t]_{\text{therm.diff.}} = -\nabla \cdot \mathbf{j}_{TDi} \qquad (6.3.22a)$$

$$\mathbf{j}_{TDi} = \sum_{\kappa} \sum_{j}^{k+1} (\omega_{\kappa i}^{L} \bar{\Omega}_{\kappa} \omega_{\kappa j}^{L}/k_B) \exp\left[-\sum_{l}^{k+1} (m_{\kappa l}^{+L} + m_{\kappa l}^{+R})\bar{F}_l/k_B\right] \nabla \bar{F}_j, \qquad (6.3.22b)$$

where $\omega^L_{\kappa i} = n^{-L}_{\kappa i} - n^{+L}_{\kappa i}$ for $i = 1, 2, \ldots, k$ and $\omega^L_{\kappa,k+1} = \varepsilon^{-L}_\kappa - \varepsilon^{+L}_\kappa$, etc. Similarly, the equation governing the internal energy density is

$$[\partial \bar{e}(\mathbf{r}, t)/\partial t]_{\text{therm. diff.}} = -\nabla \cdot \mathbf{j}_{TD} \tag{6.3.23}$$

$$\mathbf{j}_{TD} = \sum_\kappa \sum_j^{k+1} (\omega^L_\kappa \bar{\Omega}_\kappa \omega^L_{\kappa j}/k_B) \exp\left[-\sum_l^{k+1} (m^{+L}_{\kappa l} + m^{+R}_{\kappa l})\bar{F}_l/k_B\right] \nabla \bar{F}_j. \tag{6.3.24}$$

These equations have a rather simple formal structure that becomes evident if we write them in terms of the $(k + 1)$-element column vector of the mass and internal energy density, $\bar{\mathbf{p}}(\mathbf{r},t)$. Then Eqs. (6.3.21)–(6.3.24) can be written

$$[\partial \bar{\mathbf{p}}(\mathbf{r}, t)/\partial t]_{\text{therm. diff.}} = -\nabla \cdot \bar{L} \nabla \mathbf{F} \tag{6.3.25}$$

with the $k + 1$ by $k + 1$ matrix L defined by

$$\bar{L}_{ij} \equiv \sum_\kappa (\omega^L_{\kappa i} \bar{\Omega}_\kappa \omega^L_{\kappa j}/k_B)\left(\prod_{l=1}^{k} \bar{z}_l^{m_{\kappa l}}\right) \exp[-(\varepsilon^{+L}_\kappa + \varepsilon^{+R}_\kappa)/k_B T], \tag{6.3.26}$$

where $z_l = \exp(\mu_l/k_B T)$, the activity, and $m_{\kappa l} = n^{+L}_{\kappa l} + n^{+R}_{\kappa l}$. Notice that temperature gradients appear in two places in Eq. (6.3.25)—first through the terms $\nabla(-\bar{\mu}_j/T)$ and second through the terms $\nabla(1/T)$ associated with energy transport. Notice also that the expressions in Eqs. (6.3.25) and (6.3.26) for thermal diffusion encompass mass diffusion and heat conduction. For example, if $n^{+L}_{\kappa i} = n^{+R}_{\kappa i}$ for $i = 1, 2, \ldots, k$, then Eq. (6.3.25) reduces to the heat equation (6.3.15) in an isotropic medium.

The equations which describe the molecular fluctuations caused by thermal diffusion can also be written down by analogy to multicomponent mass diffusion. Linearizing Eq. (6.3.25), the conditional fluctuations solve the stochastic differential equation

$$[\partial \delta \mathbf{p}(\mathbf{r}, t)/\partial t]_{\text{therm. diff.}} = -\nabla \cdot \bar{L} \nabla \delta \mathbf{F} - \nabla \cdot \delta L \nabla \bar{\mathbf{F}} + \tilde{\mathbf{f}}(\mathbf{r}, t). \tag{6.3.27}$$

As usual the random term $\tilde{\mathbf{f}}$ vanishes on the conditional average. Referring to the manipulations for mass diffusion in Eqs. (6.2.4)–(6.2.14) we see that their covariance is given by

$$\langle \tilde{f}_i(\mathbf{r}, t) \tilde{f}_j(\mathbf{r}', t) \rangle = -2k_B \nabla_\mathbf{r} \cdot \bar{L}_{ij} \nabla_\mathbf{r} \delta(\mathbf{r} - \mathbf{r}') \delta(t - t'). \tag{6.3.28}$$

It is possible to describe the random terms in fluctuation equations using random thermal diffusion flux vectors, $\tilde{\mathbf{j}}_i$. We leave the details of this up to the reader.

6.4. Viscous Fluids: The Canonical Form

At the hydrodynamic level of description, transport of momentum is caused by viscosity. In Section 2.4 we discussed the phenomenological Newtonian form of the stress tensor, which is related to the shear and bulk viscosity coefficients, η and ζ. In an isotropic medium the relationship is

$$\sigma'_{ij} = 2\eta\mathring{e}_{ij} + \zeta e_{ii}\delta_{ij}, \tag{6.4.1}$$

where $\mathring{e}_{ij}$ is the traceless part of the rate of strain tensor and e_{ii} is its trace. Denoting the components of the center of mass velocity, $\mathbf{v}$, by v_i and the components of the position vector, $\mathbf{r}$, by $x = x_1$, $y = x_2$, and $z = x_3$, the rate of strain tensor is defined by

$$e_{ij} = \frac{1}{2}\left(\frac{\partial v_i}{\partial x_j} + \frac{\partial v_j}{\partial x_i}\right). \tag{6.4.2}$$

According to Eq. (2.4.14), which governs conservation of momentum, the contributions of the shear and bulk viscosity to momentum transport are given by

$$(\partial \mathbf{p}/\partial t)_{\text{vis}} = 2\nabla \cdot \eta\mathring{e} + \nabla \cdot \zeta e_{ii}, \tag{6.4.3}$$

where $\mathbf{p} = \rho\mathbf{v}$ is the local momentum density. These contributions are dissipative and are due to molecular processes occurring in the fluid. Like the molecular events which are responsible for diffusion, heat conduction, and thermal diffusion, these processes can be described as elementary processes with rates having the canonical form. Because the momentum density is a vector, the description involves some features of elementary processes which are new. One feature is that the transport of momentum also entails the transport of energy. Indeed, Eq. (2.4.25) shows that the shear and bulk viscosity give rise to changes in the total energy density, ε, of the form

$$(\partial \varepsilon/\partial t)_{\text{vis}} = \nabla \cdot (\zeta \mathbf{v} \nabla \cdot \mathbf{v}) + 2\nabla \cdot (\eta \mathbf{v} \cdot \mathring{e}). \tag{6.4.4}$$

In this section we obtain the canonical forms for the rate of transport of momentum and energy by the elementary processes of shear and bulk viscosity. In the following section we combine these with heat conduction to obtain the fluctuating hydrodynamic description of an isotropic, single-component fluid.

The shear viscosity and the bulk viscosity arise from distinctly different elementary processes. The shear viscosity is caused by interactions that transfer momentum between molecules in one fluid volume element to molecules in a neighboring volume element. In a dilute gas these events can be described in terms of isolated binary collisions and, as we show in Chapter 9, the shear viscosity coefficient can be expressed in terms of the collision cross section. The phenomenon of bulk viscosity, however, does not exist in a dilute gas of point particles (cf. Chapter 9, Sections 9.4 and 9.5). Indeed, in the gas phase the bulk viscosity is caused by molecular processes involving disequilibrium in internal states of a molecule, for example, *cis–trans* isomerization. This disequilibrium produces a nonequilibrium contribution to the diagonal part of the pressure tensor which is proportional to the trace of the stress tensor. Although in a condensed phase the bulk viscosity of point particles can be nonzero, the origin of the bulk viscosity can be thought of in similar terms.

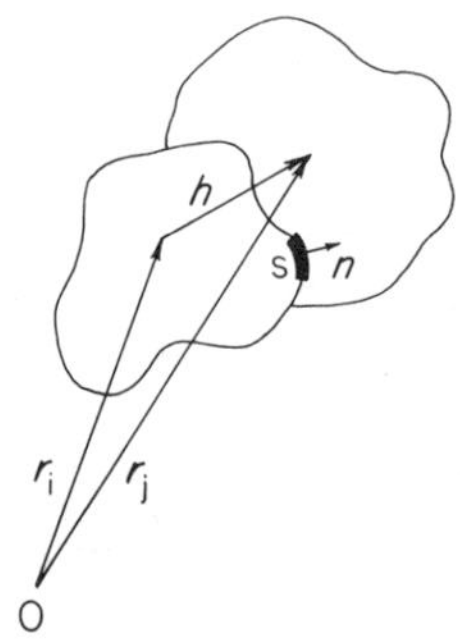

FIGURE 6.3. A schematic illustration of the neighboring volume elements used in the construction of the canonical form for the shear viscosity as described by Eqs. (6.4.5)–(6.4.7). The vector **h** is the difference of the locations of the centers of mass and the vector **n** is perpendicular to the surface area s where the elementary process is localized.

There the motion of the fluid modifies the radial distribution function, which in turn changes the pressure tensor. The radial distribution function is a quantitative measure of the arrangement of particles and vacancies in the fluid. Thus the motion of the fluid can be thought of as disturbing the spatial equilibrium between the particles and vacancies. Because the two viscosities have such different origins, we treat them separately.

To obtain the canonical form for the momentum transport due to the shear viscosity, we again divide up the fluid into volume elements. Although the fluid is taken to be isotropic, the fact that the quantity being transported, i.e., momentum, is a vectorial quantity requires that we consider volume elements of arbitrary shape. Figure 6.3 shows a typical situation with two neighboring volume elements whose centers of mass are located at $\mathbf{r}_i$ and $\mathbf{r}_j$. We will consider the elementary processes, $\boldsymbol{\kappa}$, which transport momentum between these cells at the element of surface area indicated by the heavy line at s in Fig. 6.3. Thus the elementary process $\boldsymbol{\kappa}$ actually indexes three variables: the location of the adjacent cell, j; the orientation of the surface element, s; and the molecular event, κ. The volume elements are fixed in space and have volumes V_i which in the continuum limit shrink to zero. Since the fluid may be compressible, the mass in the volume element, M_i, as well as the total energy, E_i, and the momenta, $\mathbf{P}_i$, can change. For the volume elements i and j the forward direction of the elementary process $\boldsymbol{\kappa}$ involves a momentum $\mathbf{p}_\kappa^+$ in cell i and $\overline{\mathbf{p}}_\kappa^+$ in cell j, while the reverse process involves the momenta $\mathbf{p}_\kappa^-$ and $\overline{\mathbf{p}}_\kappa^-$, respectively. Conservation of momentum implies that

$$\mathbf{p}_\kappa^+ + \overline{\mathbf{p}}_\kappa^+ = \mathbf{p}_\kappa^- + \overline{\mathbf{p}}_\kappa^-. \tag{6.4.5a}$$

In addition, the total energies involved in the elementary process are $\varepsilon_\kappa^\pm$ in cell i and $\overline{\varepsilon}_\kappa^\pm$ in cell j. Energy conservation implies that

$$\varepsilon_\kappa^+ + \overline{\varepsilon}_\kappa^+ = \varepsilon_\kappa^- + \overline{\varepsilon}_\kappa^-. \tag{6.4.5b}$$

The rates of the shear viscosity process associated with this elementary event have the canonical form

$$V_{\boldsymbol{\kappa}}^\pm = \Omega_\kappa \exp\{\mathbf{v}_i \cdot \mathbf{p}_\kappa^\pm / k_B T_i + \mathbf{v}_j \cdot \overline{\mathbf{p}}_\kappa^\pm / k_B T_j - \varepsilon_\kappa^\pm / k_B T_i - \overline{\varepsilon}_\kappa^\pm / k_B T_j\}, \tag{6.4.6}$$

since, as we noted in Eq. (2.4.33), $\partial S/\partial \mathbf{p} = -\mathbf{v}/T$ and $\partial S/\partial \varepsilon = 1/T$. The change of energy and momentum in the ith cell for this elementary process are

$$\begin{aligned}\boldsymbol{\omega}_\kappa &= \mathbf{p}_\kappa^- - \mathbf{p}_\kappa^+ = \overline{\mathbf{p}}_\kappa^+ - \overline{\mathbf{p}}_\kappa^- \\ \omega_\kappa &= \varepsilon_\kappa^- - \varepsilon_\kappa^+ = \overline{\varepsilon}_\kappa^+ - \overline{\varepsilon}_\kappa^-,\end{aligned} \tag{6.4.7}$$

where the second equalities follow from the conservation equations (6.4.5a) and (6.4.5b).

To determine the quantities $\varepsilon_\kappa^\pm$ and $\overline{\varepsilon}_\kappa^\pm$, we recall that the total energy E_i has a contribution from the local momentum through the kinetic energy, $\mathbf{P}_i^2/2M_i$. The only energy involved in the elementary process is kinetic energy since it comes from the momenta $\mathbf{p}_\kappa^\pm$ and $\overline{\mathbf{p}}_\kappa^\pm$. Thus it follows that the values of $\varepsilon_\kappa^\pm$ and $\overline{\varepsilon}_\kappa^\pm$ are determined by $\mathbf{p}_\kappa^\pm$ and $\overline{\mathbf{p}}_\kappa^\pm$. A consistent relationship can be determined by writing $\mathbf{P}_i^2/2M_i + \varepsilon_\kappa^\pm = (\mathbf{P}_i + \mathbf{p}_\kappa^\pm)^2/2M_i$ and identifying $\varepsilon_\kappa^\pm$ with the dominant term. This gives $\varepsilon_\kappa^\pm = \mathbf{v}_i \cdot \mathbf{p}_\kappa^\pm$. This value for $\varepsilon_\kappa^\pm$ combined with the conservation relations in Eqs. (6.4.5) and (6.4.6) implies that $\overline{\varepsilon}_\kappa^\pm = \mathbf{v}_i \cdot \overline{\mathbf{p}}_\kappa^\pm$. It is also consistent to use $\mathbf{v}_j$ or $\mathbf{v}^* = (\mathbf{v}_i + \mathbf{v}_j)/2$ instead of $\mathbf{v}_i$ in these definitions. Since the elementary process actually occurs at the surface between the two cells, we will adopt the choice

$$\varepsilon_\kappa^\pm = \mathbf{v}^* \cdot \mathbf{p}_\kappa^\pm, \qquad \overline{\varepsilon}_\kappa^\pm = \mathbf{v}^* \cdot \overline{\mathbf{p}}_\kappa^\pm. \tag{6.4.8}$$

All three choices, however, give the same result in the continuum limit. Substituting these expressions into Eq. (6.4.6), the canonical form becomes

$$V_\kappa^\pm = \Omega_\kappa \exp\left\{\frac{(\mathbf{v}_i - \mathbf{v}_j)\cdot \mathbf{p}_\kappa^\pm}{2k_B T_i} + \frac{(\mathbf{v}_j - \mathbf{v}_i)\cdot \overline{\mathbf{p}}_\kappa^\pm}{2k_B T_j}\right\}. \tag{6.4.9}$$

To obtain the average rate at which these elementary processes dissipate momentum and energy we need to take the continuum limit of the expressions

$$\begin{aligned} d\mathbf{P}_i/dt &= \sum_\kappa \boldsymbol{\omega}_\kappa (V_\kappa^+ - V_\kappa^-) \\ dE_i/dt &= \sum_\kappa \omega_\kappa (V_\kappa^+ - V_\kappa^-), \end{aligned} \tag{6.4.10}$$

where the sum is over all adjacent cells, j, all surface elements, s, and all processes, κ. We begin by expanding the difference of the rates in terms of $\mathbf{h} = \mathbf{r}_j - \mathbf{r}_i$, the separation between cells j and i. Using Eq. (6.4.9) one finds that

$$\begin{aligned} &V_\kappa^+ - V_\kappa^- \\ &= \Omega_\kappa\left[\frac{(\mathbf{v}_i - \mathbf{v}_j)\cdot \mathbf{p}_\kappa^+}{2k_B T_i} + \frac{(\mathbf{v}_j - \mathbf{v}_i)\cdot \overline{\mathbf{p}}_\kappa^+}{2k_B T_j} - \frac{(\mathbf{v}_i - \mathbf{v}_j)\cdot \mathbf{p}_\kappa^-}{2k_B T_i} - \frac{(\mathbf{v}_j - \mathbf{v}_i)\cdot \overline{\mathbf{p}}_\kappa^-}{2k_B T_j} + \mathcal{O}(\mathbf{v}_i - \mathbf{v}_j)\right] \\ &= \Omega_\kappa\left[\frac{-\mathbf{p}_\kappa^+ \cdot (\nabla \mathbf{v})_i \cdot \mathbf{h}}{2k_B T_i} + \frac{\overline{\mathbf{p}}_\kappa^+ \cdot (\nabla \mathbf{v})_i \cdot \mathbf{h}}{2k_B T_j} + \frac{\mathbf{p}_\kappa^- \cdot (\nabla \mathbf{v})_i \cdot \mathbf{h}}{2k_B T_i} - \frac{\overline{\mathbf{p}}_\kappa^- \cdot (\nabla \mathbf{v})_i \cdot \mathbf{h}}{2k_B T_j} + \mathcal{O}(h)\right] \\ &= \Omega_\kappa \boldsymbol{\omega}_\kappa \cdot (\nabla \mathbf{v})_i \mathbf{h}/k_B T_i + \mathcal{O}(h), \end{aligned} \tag{6.4.11}$$

where $(\nabla \mathbf{v})_i$ is the rate of strain tensor evaluated at cell i. Thus to dominant order in the separation $\mathbf{h}$ we can write

$$V_\kappa^+ - V_\kappa^- = \bar{\Omega}_{\kappa i} \boldsymbol{\omega}_\kappa \cdot (\nabla \mathbf{v})_i \cdot \mathbf{h}, \tag{6.4.12}$$

where $\bar{\Omega}_{\kappa i} \equiv \Omega_\kappa / k_B T_i$. Substituting Eq. (6.4.12) into Eq. (6.4.10) gives to lowest order in $\mathbf{h}$

$$\begin{aligned} d\mathbf{P}_i/dt &= \sum_s \sum_j \left[\sum_\kappa \boldsymbol{\omega}_\kappa \bar{\Omega}_{\kappa i} \boldsymbol{\omega}_\kappa^T \right] \cdot (\nabla \mathbf{v})_i \mathbf{h} \\ dE_i/dt &= \sum_s \sum_j \mathbf{v}_i \cdot \left[\sum_\kappa \boldsymbol{\omega}_\kappa \bar{\Omega}_{\kappa i} \boldsymbol{\omega}_\kappa^T \right] \cdot (\nabla \mathbf{v})_i \cdot \mathbf{h}. \end{aligned} \tag{6.4.13}$$

To complete the continuum limit, we introduce into Eq. (6.4.13) the matrix $\Gamma_j(s)$ which transforms the unit outward normal at s, $\mathbf{n}_{js}$, which is shown in Fig. 6.3, into the separation vector $\mathbf{h}$. Since $\Gamma_j(s)\mathbf{n}_{js} = \mathbf{h}$, Eqs. (6.4.13) can be rewritten in terms of the momentum density, $\mathbf{p}_i$, and total energy density, ε_i, as

$$\begin{aligned} V_i\, d\mathbf{p}_i/dt &= \sum_{j,s} 2\hat{\eta}_j(s)(\nabla \mathbf{v})_i \cdot \mathbf{n}_{js} A_{js} \\ V_i\, d\varepsilon_i/dt &= \sum_{j,s} \mathbf{v}_i \cdot 2\hat{\eta}_j(s)(\nabla \mathbf{v})_i \cdot \mathbf{n}_{js} A_{js}, \end{aligned} \tag{6.4.14}$$

where $\hat{\eta}_i$ is the rank-4 shear viscosity tensor defined by

$$2\hat{\eta}_j(s) \equiv \sum_\kappa (\boldsymbol{\omega}_\kappa \bar{\Omega}_{\kappa i} \boldsymbol{\omega}_\kappa) \Gamma_j(s)/A_{js} \tag{6.4.15}$$

and A_{js} is the area of the surface element at s. The right-hand sides of Eqs. (6.4.14) approximate surface integrals over the cell i of the indicated summands. The surface integrals can be transformed using the divergence theorem to obtain

$$\begin{aligned} V_i\, d\mathbf{p}_i/dt &= 2\nabla \cdot [\hat{\eta}(\nabla \mathbf{v})_i] V_i + \mathcal{O}(V_i) \\ V_i\, d\varepsilon_i/dt &= 2\nabla \cdot [\mathbf{v}_i \cdot \hat{\eta}(\nabla \mathbf{v})_i] V_i + \mathcal{O}(V_i). \end{aligned} \tag{6.4.16}$$

Finally shrinking the volume V_i to zero gives

$$\begin{aligned} (\partial \mathbf{p}/\partial t)_{\text{shear}} &= 2\nabla \cdot (\hat{\eta} \nabla \mathbf{v}) \\ (\partial \varepsilon/\partial t)_{\text{shear}} &= 2\nabla \cdot \mathbf{v} \cdot (\hat{\eta} \nabla \mathbf{v}). \end{aligned} \tag{6.4.17}$$

Because we have assumed an isotropic fluid the viscosity tensor must be symmetric, isotropic, and traceless. These properties imply that

$$\hat{\eta} \nabla \mathbf{v} = \eta \mathring{\mathbf{e}}, \tag{6.4.18}$$

where η is the shear viscosity coefficient. Thus we deduce from Eqs. (6.4.17) and (6.4.18) that

$$
\begin{aligned}
(\partial \mathbf{p}/\partial t)_{\text{shear}} &= 2\nabla \cdot \eta \mathring{\mathbf{e}} \\
(\partial \varepsilon/\partial t)_{\text{shear}} &= 2\nabla \cdot (\eta \mathbf{v} \cdot \mathring{\mathbf{e}}).
\end{aligned}
\tag{6.4.19}
$$

This agrees with the Netwonian expressions in Eqs. (6.4.3) and (6.4.4).

The canonical form for the bulk viscosity can be derived on the assumption that the fluid consists of two molecular states, 1 and 2, which are rapidly interconverting. As a result of this reaction, the equation of mass conservation (2.4.9) for the mass densities, ρ_1 and ρ_2, must be supplemented with source and sink terms. Since this process conserves mass one can write

$$
\begin{aligned}
\partial \rho_1/\partial t &= -\nabla \cdot \rho_1 \mathbf{v} - R \\
\partial \rho/\partial t &= -\nabla \cdot \rho \mathbf{v},
\end{aligned}
\tag{6.4.20}
$$

where the total mass density is $\rho = \rho_1 + \rho_2$ and we have written $\mathbf{v}_1 = \mathbf{v}_2 = \mathbf{v}$ since the molecules only differ by an internal state. For definiteness we assume the elementary reaction

$$
(n_1^+, n_2^+) \rightleftarrows (n_1^-, n_2^-) \tag{6.4.21}
$$

with $n_1^- - n_1^+ = \omega_1$. The sink term, R, thus takes the canonical form for chemical reactions described in Section 3.6:

$$
\begin{aligned}
-R = \omega_1(V^+ - V^-) = \omega_1 \Omega \{&\exp[(n_1^+ \mu_1 + n_2^+ \mu_2)/k_B T] \\
&- \exp[(n_1^- \mu_1 + n_2^- \mu_2)/k_B T]\}.
\end{aligned}
\tag{6.4.22}
$$

Equations (6.4.20) couple the motion of the fluid to the rate of the internal relaxation process. When the density ρ_1 relaxes rapidly, this coupling can be expressed in terms of the bulk viscosity. To see this, we write the first of Eqs. (6.4.20) in terms of the convective derivative,

$$
D\rho_1/Dt \equiv \partial \rho_1/\partial t + \mathbf{v} \cdot \nabla \rho_1, \tag{6.4.23}
$$

which gives the density change with respect to an observer moving along with the flow of the fluid. Using the identity

$$
\nabla \cdot (\rho_1 \mathbf{v}) = \rho_1 \nabla \cdot \mathbf{v} + \mathbf{v} \cdot \nabla \rho_1 \tag{6.4.24}
$$

allows the first of Eqs. (6.4.20) to be written

$$
D\rho_1/Dt = -\rho_1 \nabla \cdot \mathbf{v} - R. \tag{6.4.25}
$$

If ρ_1 relaxes rapidly within the moving fluid element, Eq. (6.4.25) implies that

$$
R = -\rho_1 \nabla \cdot \mathbf{v}. \tag{6.4.26}
$$

Thus as long as the fluid is flowing, the two internal states cannot be in equilibrium, unless the flow is incompressible and $\nabla \cdot \mathbf{v} = 0$. This affects, in turn, the local equilibrium pressure, p, which, as we have seen, is the diagonal part of the pressure tensor in the hydrodynamic picture. Indeed, the local equilibrium pressure is a function of ρ_1, ρ_2, and T. When the two states are in local equilibrium the condition $R = 0$ implies a relationship $\rho_2 = \rho_2^e(\rho_1, T)$ between the two mass densities so that the pressure actually depends on only one mass density. Expanding the local equilibrium equation of state in a Taylor series with respect to the internal state disequilibrium, R, implied by Eq. (6.4.26) gives

$$p(\rho_1, \rho_2, T) = p^e(T, \rho) - (\partial p/\partial R)^e \rho_1 \nabla \cdot \mathbf{v} + \cdots. \tag{6.4.27}$$

According to this result we can write the local equilibrium part of the pressure in the form

$$p = p^e - \zeta \nabla \cdot \mathbf{v}, \tag{6.4.28}$$

where the term p^e implies complete internal equilibrium and

$$\zeta \equiv (\partial p/\partial R)^e \rho_1. \tag{6.4.29}$$

Thus disequilibrium in the internal states leads to a contribution to the pressure tensor of the form

$$P_{ij} = p(T, \rho)\delta_{ij} - \zeta(T, \rho)\nabla \cdot \mathbf{v}\delta_{ij}. \tag{6.4.30}$$

Comparing Eq. (6.4.29) with Eqs. (2.4.21) and (2.4.22), we see that $\zeta(T, \rho)$ is the coefficient of bulk viscosity.

Although it is not evident from its definition in Eq. (6.4.29), the bulk viscosity is a positive quantity. This can be seen by comparing the local value of the dissipation function for the reaction in Eq. (6.4.21) with that for the bulk viscosity. According to Eq. (2.3.15), the dissipation function has the form

$$\Phi = \sum_i \sum_j L_{ij} X_i X_j, \tag{6.4.31}$$

where L_{ij} is the Onsager matrix and X_j is the thermodynamic force. For the bulk viscosity the form of L_{ij} is given in the final term in Eq. (2.4.25), and the thermodynamic forces are $-v_i/T^e$. Substituting these into Eq. (6.4.31) gives the bulk viscosity contribution

$$T^e \Phi = -\int (\mathbf{v} \cdot \nabla)(\zeta^e \nabla \cdot \mathbf{v})\, d\mathbf{r} = \int \zeta^e (\nabla \cdot \mathbf{v})^2\, d\mathbf{r}, \tag{6.4.32}$$

where the second equality follows from the first by parts integration assuming that $\mathbf{v}$ vanishes on the boundaries. Thus, locally in space, the dissipation function is

$$T^e \phi(\mathbf{r}) = \zeta^e (\nabla \cdot \mathbf{v})^2. \tag{6.4.33}$$

For the internal relaxation process, on the other hand, the local dissipation function can be expressed in terms of the affinity, $\mathscr{A} = -(\mu_1 \omega_1 + \mu_2 \omega_2)/T$. Using the general expression for the dissipation function derived in Eq. (4.5.15) we find to the linear order used in deriving Eq. (6.4.29) that

$$\phi_{\text{rx}}(\mathbf{r}) = \Omega \exp[(n_1^- \mu_1^e + n_2^- \mu_2^e)/k_B T] \mathscr{A}^2/k_B. \tag{6.4.34}$$

To linear order the affinity can be related to the reaction rate $(V^+ - V^-)$ by Eq. (3.7.39a). Thus linearizing Eq. (6.4.22) we deduced that

$$\mathscr{A} = -\{\omega_1 \Omega \exp[(n_1^- \mu_1^e + n_2^- \mu_2^e)/k_B T^e]/k_B\}^{-1} R. \tag{6.4.35}$$

We can use this equation and Eqs. (6.4.22) and (6.4.26) to rewrite Eq. (6.4.34) in the form

$$T\phi_{\text{rx}}(\mathbf{r}) = k_B T \rho_1^2 \Omega^{-1} \omega_1^{-2} \exp[-(n_1^- \mu_1^e + n_2^- \mu_2^e)/k_B T](\nabla \cdot \mathbf{v})^2. \tag{6.4.36}$$

Since the origin of the dissipation due to the bulk viscosity is in the chemical reaction, we can compare ϕ_{rx} in Eq. (6.4.36) with ϕ in Eq. (6.4.33) and conclude that

$$\zeta^e = k_B T^e \rho_1^2 \Omega^{-1} \omega_1^{-2} \exp[-(n_1^- \mu_1^e + n_2^- \mu_2^e)/k_B T^e]. \tag{6.4.37}$$

As all the quantities on the right-hand side of this equation are positive, the coefficient of bulk viscosity is positive. Finally, combining Eqs. (6.4.29) and (6.4.37) we find that

$$(\partial p/\partial R)^{e2} = k_B T \zeta / \omega_1^2 \Omega \exp[(n_1^- \mu_1^e + n_2^- \mu_2^e)/k_B T], \tag{6.4.38}$$

a relationship that will be useful in discussing fluctuations.

Reviewing the derivation of the expressions for the bulk viscosity, it is seen that we have contracted our hydrodynamic description from four variables $(\rho_1, \rho_2, \mathbf{p}, \text{and } \varepsilon)$ to three variables $(\rho, \mathbf{p}, \text{and } \varepsilon)$. After the process of contraction, a new elementary process—the bulk viscosity—appeared with a transport coefficient ζ that could be expressed in terms of the rate of another elementary process. This exemplifies the general notion of *contraction of the description*, which we have discussed in passing several times already. In the next section we show how the stochastic process associated with the bulk viscosity can be contracted, too. A more thorough treatment of contracted descriptions is given in Chapter 9.

6.5. Fluctuating Hydrodynamics

Having seen how to write down the canonical form for the shear and bulk viscosity, we can use the general theory in Chapter 4 to obtain the canonical stochastic description at the hydrodynamic level. For simplicity we continue to consider a single-component fluid in this section. The combined effects of chemical reaction and diffusion, which can be important in multicomponent solutions, are treated separately in Section 6.6.

For a single-component, isotropic fluid the densities of extensive variables used in the hydrodynamic description are the mass density, ρ, the momentum density, $\mathbf{p}$, and the total internal energy, $\varepsilon = e + p^2/2\rho + \rho\phi$, where e is the internal energy density and ϕ is the potential energy per unit mass. Using the canonical form the rate of dissipation of energy and momentum were shown to be given by Eq. (6.3.14), which is Fourier's law, Eq. (6.4.19), which is Newton's law, and Eq. (6.4.30), which describes the bulk viscosity. Combining these expressions with the nondissipative terms in the hydrodynamic-level conservation laws in Section 2.4, we can write out the conditional-average hydrodynamic equations as

$$\partial\bar{\rho}/\partial t = -\nabla\cdot\bar{\mathbf{p}} \tag{6.5.1}$$

$$\partial\bar{\varepsilon}/\partial t = -\nabla\cdot(\bar{\varepsilon}\bar{\mathbf{v}}) - \nabla\cdot(\bar{p}^e\bar{\mathbf{v}} - \bar{\zeta}\bar{\mathbf{v}}\nabla\cdot\bar{\mathbf{v}}) + 2\nabla\cdot(\bar{\eta}\bar{\mathbf{v}}\cdot\bar{\mathring{e}}) - \nabla\cdot(\bar{K}\nabla\bar{T}^{-1}) + \bar{\rho}\bar{\mathbf{v}}\cdot\mathbf{F} \tag{6.5.2}$$

$$\partial\bar{\mathbf{p}}/\partial t = -\nabla\cdot(\bar{\mathbf{v}}\bar{\mathbf{p}}^T) - \nabla\bar{p}^e + 2\nabla\cdot\bar{\eta}\bar{\mathring{e}} + \nabla(\bar{\zeta}\nabla\cdot\bar{\mathbf{v}}) + \bar{\rho}\mathbf{F}. \tag{6.5.3}$$

These equations are identical to Eqs. (2.4.24)–(2.4.26) of phenomenological hydrodynamics, except now the overbars emphasize that the equations describe the conditional average.

Fluctuations around the conditional average satisfy linearized versions of Eqs. (6.5.1)–(6.5.3) with the appropriate random components of the time derivatives added. Because these equations involve a good many nonlinear terms, it is cumbersome to write out their linearization explicitly. Instead we introduce the linearization operator, $\hat{\delta}$, defined by $\hat{\delta}G(\rho, \varepsilon, \mathbf{p}) = (\partial\bar{G}/\partial\bar{\rho})\delta\rho + (\partial\bar{G}/\partial\bar{\varepsilon})\delta\varepsilon + (\partial\bar{G}/\partial\bar{p}_i)\delta p_i$. Using this operator the equations satisfied by the fluctuations can be written in compact notation as

$$\partial\delta\rho/\partial t = -\nabla\cdot\delta\mathbf{p} \tag{6.5.4}$$

$$\partial\delta\varepsilon/\partial t = -\nabla\cdot\hat{\delta}(h\mathbf{v}) + \nabla\cdot\hat{\delta}(\mathbf{v}\cdot\sigma') - \nabla\cdot\hat{\delta}(K\nabla T^{-1}) + \mathbf{F}\cdot\hat{\delta}(\rho\mathbf{v}) + \tilde{f}_\varepsilon \tag{6.5.5}$$

$$\partial\delta\mathbf{p}/\partial t = -\nabla\cdot\hat{\delta}(\mathbf{v}\mathbf{p}) - \nabla\cdot\hat{\delta}p^e + \nabla\cdot\hat{\delta}\sigma' + \mathbf{F}\delta\rho + \tilde{f}_m, \tag{6.5.6}$$

where $h = \varepsilon + p^e$ is the *total* enthalpy density and σ' is the nonequilibrium portion of the stress tensor, i.e.,

$$\sigma'_{ij} = 2\eta\mathring{e}_{ij} + \zeta\nabla\cdot\mathbf{v}\delta_{ij}. \tag{6.5.7}$$

The random components of the time derivative are Gaussian, purely random processes with covariance matrices determined by the canonical form. No random term appears in the continuity equation since in a single-component fluid there are no elementary processes which change the mass density. To obtain the covariance matrix of the random terms, we need to use the canonical formula

$$\langle\tilde{\mathbf{f}}(t)\tilde{\mathbf{f}}^T(t')\rangle = \sum_\kappa \boldsymbol{\omega}_\kappa(V_\kappa^+ + V_\kappa^-)\boldsymbol{\omega}_\kappa^T\delta(t-t'). \tag{6.5.8}$$

The right-hand side involves a sum over elementary processes, and since heat conduction, the shear viscosity, and the bulk viscosity are independent molecular processes, there are no correlations between them.

The derivation of the covariance formula for $\tilde{\mathbf{f}}$ using the canonical form—with the exception of the process involving bulk viscosity—is almost identical to the derivation encountered for mass diffusion in Section 6.2. To keep track of components it is convenient to introduce the fourth-rank tensor

$$\Delta_{il,kj} = \delta_{il}\delta_{kj} + \delta_{ik}\delta_{lj} - (2/3)\delta_{ij}\delta_{lk}. \tag{6.5.9}$$

In terms of this tensor the elements of the covariance matrix are

$$\begin{aligned}\langle \tilde{f}_\varepsilon(\mathbf{r},t)\tilde{f}_\varepsilon(\mathbf{r}',t')\rangle = &-2k_B\frac{\partial}{\partial x_j}\bar{K}\frac{\partial}{\partial x_j}\delta(\mathbf{r}-\mathbf{r}')\delta(t-t')\\ &+2k_B\frac{\partial}{\partial x_j}\bar{T}\bar{v}_l\bar{v}_i[\bar{\eta}\Delta_{il,jk}+\bar{\zeta}\delta_{lj}\delta_{ik}]\frac{\partial}{\partial x_k}\delta(\mathbf{r}-\mathbf{r}')\delta(t-t')\end{aligned} \tag{6.5.10}$$

$$\langle \tilde{f}_{ml}(\mathbf{r},t)\tilde{f}_\varepsilon(\mathbf{r}',t')\rangle = -2k_B\frac{\partial}{\partial x_j}\bar{T}\bar{v}_i[\bar{\eta}\Delta_{il,jk}+\bar{\zeta}\delta_{ik}\delta_{jl}]\frac{\partial}{\partial x_k}\delta(\mathbf{r}-\mathbf{r}')\delta(t-t') \tag{6.5.11}$$

$$\langle \tilde{f}_\varepsilon(\mathbf{r},t)\tilde{f}_{ml}(\mathbf{r}',t')\rangle = -2k_B\frac{\partial}{\partial x_j}\bar{T}\bar{v}_i[\bar{\eta}\Delta_{il,kj}+\bar{\zeta}\delta_{ij}\delta_{kl}]\frac{\partial}{\partial x_k}\delta(\mathbf{r}-\mathbf{r}')\delta(t-t') \tag{6.5.12}$$

$$\langle \tilde{f}_{mi}(\mathbf{r},t)\tilde{f}_{ml}(\mathbf{r}',t)\rangle = -2k_B\frac{\partial}{\partial x_j}\bar{T}[\bar{\eta}\Delta_{il,kj}+\bar{\zeta}\delta_{ij}\delta_{kl}]\frac{\partial}{\partial x_k}\delta(\mathbf{r}-\mathbf{r}')\delta(t-t'). \tag{6.5.13}$$

Rather than deriving these equations in detail, we simply call attention to the analogy with similar formulas which have been derived in previous sections. For example, the first term in Eq. (6.5.10) is due to heat conduction. Its form was obtained in Eq. (6.3.18) for an orthotropic substance. For an isotropic substance $L_{xx} = L_{yy} = L_{zz} \equiv K$ and Eq. (6.3.18) reduces to the first term in Eq. (6.5.10). All the terms involving the shear viscosity in Eqs. (6.5.10)–(6.5.13) can be derived from Eq. (6.5.8) for the rates of the shear elementary processes. The derivation is analogous to the derivation of Eq. (6.2.14) for mass diffusion. The only additional identity that one needs is an expression for the symmetric, traceless part of the gradient tensor, $\nabla\mathbf{G}$. Using the definition of the tensor Δ in Eq. (6.5.8) one verifies easily that

$$2(\overset{\circ}{\nabla\mathbf{G}})_{ij} = \Delta_{il,kj}(\partial G_l/\partial x_k), \tag{6.5.14}$$

where the degree sign above the tensor implies the symmetric, traceless part. On the other hand, some of the terms in Eqs. (6.5.10)–(6.5.13) require novel analysis since they arise from a contraction of an elementary process. Thus we describe in detail how those terms are obtained.

Recall that we treated the bulk viscosity as arising from internal relaxation processes using Eq. (6.4.25). By making the steady-state assumption in Eq. (6.4.26) we eliminated the mass density ρ_1 from the hydrodynamic-level description and obtained the expression for the bulk viscosity in Eq. (6.4.28).

To eliminate fluctuations in ρ_1 something similar is required. When ρ_1 is included in the description, the elementary process in Eq. (6.4.21) gives rise to a random term in the fluctuation equation for $\delta\rho_1$. Indeed, linearizing Eq. (6.4.25) and following the familiar procedure we find that

$$\hat{\delta}(D\rho_1/Dt) = -\hat{\delta}(\rho_1\nabla\cdot\mathbf{v}) - \hat{\delta}R + \tilde{f}_1. \tag{6.5.15}$$

The conditional average equation, on the other hand, is

$$\bar{D}\bar{\rho}_1/\bar{D}t = -\bar{\rho}_1\nabla\cdot\bar{\mathbf{v}} - \bar{R}, \tag{6.5.16}$$

where $\bar{D}/\bar{D}t = \partial/\partial t + \bar{\mathbf{v}}\cdot\nabla$ is the average material derivative. The stochastic analogue of the steady-state assumption introduced in Eq. (6.4.26) is

$$\hat{\delta}(D\rho_1/Dt) = \bar{D}\rho_1/\bar{D}t = 0. \tag{6.5.17}$$

These assumptions, which are justified in a more general context in Chapter 9, can be used to eliminate both $\bar{\rho}_1$ and $\delta\rho_1$ from the stochastic description. To do this we combine Eqs. (6.5.15)–(6.5.17) to obtain

$$\bar{R} = -\bar{\rho}_1\nabla\cdot\bar{\mathbf{v}} \tag{6.5.18}$$

and

$$\hat{\delta}R = \tilde{f}_1 - \hat{\delta}(\rho_1\nabla\cdot\mathbf{v}). \tag{6.5.19}$$

Just as in Section 6.4, Eq. (6.5.18) can be used to show that Eq. (6.4.28) holds on the conditional average, i.e.,

$$\bar{p} = \bar{p}^e - \bar{\zeta}\nabla\cdot\bar{\mathbf{v}}. \tag{6.5.20}$$

Just as the average effect of the elimination of ρ_1 affects only the average pressure, the elimination of $\delta\rho_1$ using Eq. (6.5.19) affects only fluctuations in the pressure. To see what their effect is, we linearize the Taylor series expansion of the pressure, i.e.,

$$p = p^e + (\partial p/\partial R)^e R + \ldots, \tag{6.5.21}$$

to get

$$\hat{\delta}p = \delta p^e + \bar{R}\hat{\delta}(\partial p/\partial R)^e + \hat{\delta}R(\partial\bar{p}/\partial R)^e + \ldots. \tag{6.5.22}$$

Using Eqs. (6.5.18) and (6.5.19) in this expression then implies that

$$\delta\hat{p} = \hat{\delta}p^e - \hat{\delta}\{\rho_1\nabla\cdot\mathbf{v}(\partial p/\partial R)^e\} + (\partial\bar{p}/\partial R)^e\tilde{f}_1. \tag{6.5.23}$$

Finally recalling that Eq. (6.4.29) defines the coefficient of bulk viscosity as $\zeta = \rho_1(\partial p/\partial R)^e$, we see that

$$\delta\hat{p} = \hat{\delta}p^e - \hat{\delta}(\zeta\nabla\cdot\mathbf{v}) + (\partial\bar{p}/\partial R)^e\tilde{f}_1. \tag{6.5.24}$$

The first two terms in Eq. (6.5.24) appear in the linearization of the local equilibrium pressure and the bulk viscosity part of the stress tensor in Eqs. (6.5.5) and (6.5.6). The third term also contributes to the fluctuation equations, and has been absorbed into the random terms $\tilde{f}_\varepsilon$ and $\tilde{\mathbf{f}}_m$. Using Eq. (6.4.38) this contribution to Eq. (6.5.24) can be written

$$\delta\tilde{p} \equiv (\partial\bar{p}/\partial R)^e\tilde{f}_1 = [k_B\bar{T}\bar{\zeta}/\omega_1^2\Omega\exp[(n_1^-\mu_1^e + n_2^-\mu_2^e)/k_B\bar{T}]]^{1/2}\tilde{f}_1. \quad (6.5.25)$$

The random part of the pressure fluctuation appears in both the momentum and energy equations. In the fluctuating momentum equation it gives rise to the term

$$\tilde{\mathbf{f}}_{\text{bulk}} = -\nabla\cdot(\delta\tilde{p}) = -\nabla\cdot\tilde{f}_1[(k_B\bar{T}\bar{\zeta}/\omega_1^2\Omega\exp[(n_1^-\mu_1^e + n_2^-\mu_2^e)/k_B\bar{T})]]^{1/2}, \quad (6.5.26)$$

whereas in the energy equation it contributes the term

$$\tilde{f}_{\text{bulk}} = -\nabla\cdot\bar{\mathbf{v}}\tilde{f}_1[(k_B\bar{T}\bar{\zeta}/\omega_1^2\Omega\exp[(n_1^-\mu_1^e + n_2^-\mu_2^e/k_B\bar{T})]]^{1/2}. \quad (6.5.27)$$

We can calculate the covariances of these terms by using the lemma in Eqs. (6.2.20) and (6.2.21). To apply the lemma we need to know the covariance of the internal state relaxation term, $\tilde{f}_1$. Its form follows from Eq. (6.4.22) and the general expressions in Chapter 4, which give

$$\begin{aligned}\langle\tilde{f}_1(\mathbf{r},t)\tilde{f}_1(\mathbf{r}',t')\rangle &= 2\omega_1^2\Omega\exp[(n_1^-\mu_1^e + n_2^-\mu_2^e)/k_BT]\delta(\mathbf{r}-\mathbf{r}')\delta(t-t')\\ &\equiv \gamma_1\delta(\mathbf{r}-\mathbf{r}')\delta(t-t'). \end{aligned} \quad (6.5.28)$$

For the term $\tilde{f}_{\text{bulk}}$, Eq. (6.5.28) and lemma (6.2.20) imply that

$$\langle\tilde{f}_{\text{bulk}}(\mathbf{r},t)\tilde{f}_{\text{bulk}}(\mathbf{r}',t')\rangle = -\frac{\partial}{\partial x_i}\bar{v}_i\bar{v}_jG\frac{\partial}{\partial x_j}\delta(\mathbf{r}-\mathbf{r}')\delta(t-t'), \quad (6.5.29)$$

where

$$\begin{aligned}G &\equiv -\gamma_1\{k_B\bar{T}\bar{\zeta}/\omega_1^2\Omega\exp[n_1^-\mu_1^e + n_2^-\mu_2^e)/k_BT)]\}\\ &= -2k_B\bar{T}\bar{\zeta},\end{aligned} \quad (6.5.30)$$

with the equality coming from the identity in Eq. (6.5.28). Thus we conclude that the contribution of the bulk viscosity to the covariance in Eq. (6.5.10) should be

$$\langle\tilde{f}_{\text{bulk}}(\mathbf{r},t)\tilde{f}_{\text{bulk}}(\mathbf{r}',t')\rangle = 2k_B\frac{\partial}{\partial x_j}\bar{T}\bar{\zeta}\bar{v}_j\bar{v}_k\frac{\partial}{\partial x_k}\delta(\mathbf{r}-\mathbf{r}')\delta(t-t'). \quad (6.5.31)$$

A glance at Eq. (6.5.10) shows that Eq. (6.5.31) agrees with the result given there. A similar argument can be used to verify that the other contributions of the bulk viscosity to the covariances in Eqs. (6.5.10)–(6.5.13) are correct. No cross terms involving the bulk viscosity appear there since the internal state relaxation process is independent of the shear viscosity and heat conduction processes.

It is often useful to use the internal energy density, e, rather than the total energy density, ε, as the energy variable. On the conditional average the two are related by the expression

$$\bar{\varepsilon} = \bar{e} + \bar{p}^2/2\bar{\rho} + \bar{\rho}\phi \tag{6.5.32}$$

whereas the conditional fluctuations are related by

$$\delta\varepsilon = \delta e + \bar{\mathbf{v}}\cdot\delta\mathbf{p} - (\bar{v}^2/2)\delta\rho + \phi\delta\rho. \tag{6.5.33}$$

Using Eq. (6.5.32) the conditional average equation satisfied by $\bar{e}$ can be deduced to be

$$\partial\bar{e}/\partial t = \partial\bar{\varepsilon}/\partial t - \bar{\mathbf{v}}\cdot\partial\bar{\mathbf{p}}/\partial t + (\bar{v}^2/2)\partial\bar{\rho}/\partial t - \phi(\partial\bar{\rho}/\partial t). \tag{6.5.34}$$

Substituting the expressions for the partial derivatives of $\bar{\rho}$, $\bar{\varepsilon}$, and $\bar{\mathbf{p}}$ in Eqs. (6.5.1)–(6.5.3) into Eq. (6.5.34) one finds that

$$\partial\bar{e}/\partial t = -\nabla\cdot(\bar{\mathbf{v}}\bar{e}) - \bar{p}\nabla\cdot\bar{\mathbf{v}} + \bar{\zeta}(\nabla\cdot\bar{\mathbf{v}})^2 + 2\bar{\eta}\,\mathring{\bar{e}}_{ij}\mathring{\bar{e}}_{ji} - \nabla\cdot(\bar{K}\nabla\bar{T}^{-1}). \tag{6.5.35}$$

The equation satisfied by conditional fluctuations in the internal energy density, on the other hand, can be obtained using Eq. (6.5.33). It is

$$\partial\delta e/\partial t = \partial\delta\varepsilon/\partial t - (\partial\bar{\mathbf{v}}/\partial t)\cdot\delta\mathbf{p} - \bar{\mathbf{v}}\cdot\partial\delta\mathbf{p}/\partial t + \frac{1}{2}(\partial\bar{v}^2/\partial t)\delta\rho$$
$$+ (\bar{v}^2/2)\partial\delta\rho/\partial t - \phi\partial\delta\rho/\partial t. \tag{6.5.36}$$

Except for a random term, $\tilde{f}_e$, the right-hand side of this equation is the same as the linearized form of Eq. (6.5.35). This follows from the fact that the linearization operator, $\hat{\delta}$, and the partial derivative operator, $\partial/\partial t$, commute. The random term, $\tilde{f}_e$, however, is a linear combination of the random terms which appear in Eqs. (6.5.5) and (6.5.6). Only the first and third terms on the right-hand side of Eq. (6.5.36) contribute, so that one has the transformation formula

$$\tilde{f}_e = \tilde{f}_\varepsilon - \bar{\mathbf{v}}\cdot\tilde{\mathbf{f}}_m. \tag{6.5.37}$$

The covariance formula for $\tilde{f}_e$ and $\tilde{\mathbf{f}}_m$ can be calculated without much difficulty using Eq. (6.5.37) and the covariance formulas in Eqs. (6.5.10)–(6.5.13). One finds that

$$\langle\tilde{f}_e(\mathbf{r},t)\tilde{f}_e(\mathbf{r}',t')\rangle = 2k_B\bar{T}\left[\frac{\partial\bar{v}_l}{\partial x_j}\right]\left[\frac{\partial\bar{v}_i}{\partial x_k}\right](\bar{\eta}\Delta_{il,jk} + \bar{\zeta}\delta_{lj}\delta_{ik})\delta(\mathbf{r}-\mathbf{r}')\delta(t-t')$$
$$- 2k_B\frac{\partial}{\partial x_j}\bar{K}\frac{\partial}{\partial x_j}\delta(\mathbf{r}-\mathbf{r}')\delta(t-t') \tag{6.5.38}$$

$$\langle\tilde{f}_e(\mathbf{r},t)\tilde{f}_{ml}(\mathbf{r}',t')\rangle = -2k_B\bar{T}\left[\frac{\partial\bar{v}_i}{\partial x_j}\right]\{\bar{\eta}\Delta_{il,kj} + \bar{\zeta}\delta_{ij}\delta_{kl}\}\frac{\partial}{\partial x_k}\delta(\mathbf{r}-\mathbf{r}')\delta(t-t') \tag{6.5.39}$$

$$\langle\tilde{f}_{ml}(\mathbf{r},t)\tilde{f}_e(\mathbf{r}',t')\rangle = 2k_B\frac{\partial}{\partial x_j}\bar{T}(\partial\bar{v}_i/\partial x_k)(\bar{\eta}\Delta_{il,jk} + \bar{\zeta}\delta_{ik}\delta_{jl})\delta(\mathbf{r}-\mathbf{r}')\delta(t-t'). \tag{6.5.40}$$

The expression for the momentum–momentum covariance is unchanged from Eq. (6.5.13).

For the sake of completeness, we note that the random terms $\tilde{\mathbf{f}}_m$ and $\tilde{f}_e$ can be expressed in terms of a random stress tensor, $\tilde{\sigma}$, and a random heat flux vector, $\tilde{\mathbf{q}}$. Both of these random functions vanish on the average and are defined by

$$\tilde{f}_e = \tilde{\sigma}_{ij}\partial\bar{v}_i/\partial x_j + \partial\tilde{q}_j/\partial x_j \tag{6.5.41}$$

and

$$\tilde{f}_{ml} = \partial\tilde{\sigma}_{lj}/\partial x_j. \tag{6.5.42}$$

These expressions are similar to the relation defined in Eq. (6.2.17) between the random term for diffusion and the random diffusion flux. In a similar way one can use the lemmas in Eqs. (6.2.20)–(6.2.23) to obtain the covariance of $\tilde{\sigma}$ and $\tilde{\mathbf{q}}$. With the help of Eqs. (6.5.13) and (6.5.38)–(6.5.40) one finds that

$$\langle\tilde{q}_k\rangle = \langle\tilde{\sigma}_{ij}\rangle = \langle\tilde{\sigma}_{ij}\tilde{q}_k\rangle = 0, \tag{6.5.43}$$

$$\langle\tilde{\sigma}_{ij}(\mathbf{r},t)\tilde{\sigma}_{kl}(\mathbf{r}',t')\rangle = 2k_B\bar{T}[\bar{\eta}\Delta_{il,jk} + \bar{\zeta}\delta_{ij}\delta_{lk}]\delta(\mathbf{r}-\mathbf{r}')\delta(t-t'), \tag{6.5.44}$$

and

$$\langle\tilde{q}_i(\mathbf{r},t)\tilde{q}_j(\mathbf{r}',t')\rangle = 2k_B\bar{K}\delta_{ij}\delta(\mathbf{r}-\mathbf{r}')\delta(t-t'). \tag{6.5.45}$$

These expressions are sometimes useful in applications.

6.6. Chemical Reactions and Diffusion

In the preceding sections of this chapter we have set down the basic statistical, nonequilibrium thermodynamic theory at the hydrodynamic level. This theory has a wide range of applicability and includes the thermodynamic-level theory as a special case. In this and following sections we show how the theory can be applied to several simple physical and chemical systems. We begin with the calculation of density–density correlation functions for chemically reacting systems.

The hydrodynamic level of description involves local densities of the extensive variables. As we saw in Section 6.2, this implies that the covariance functions depend on the spatial coordinates. In this section we focus our attention on chemical systems in which chemical reaction and diffusion are taken as the only significant molecular processes. The molecule numbers, N_i, will be used as extensive variables, and their densities, $\rho_i(\mathbf{r},t)$, will be the only variables included in the hydrodynamic description. For a multicomponent chemical system the average diffusion flux, after elimination of the solvent, was obtained in Eq. (6.1.63). Expressed in terms of chemical potential gradients, this gives

$$\bar{\mathbf{j}}_{Di} = \sum_j L_{ij} \nabla[-(\bar{\mu}_j - \bar{\mu}_s)/T)], \tag{6.6.1}$$

where the sum is over the solute molecules and $\bar{\mu}_s$ is the chemical potential of the solvent. If the solvent is chemically inert and the solution is dilute, the chemical potential of the solvent can be treated as a constant. In dilute solution the coupling matrix, L_{ij}, also simplifies. According to Eq. (6.1.58) it is proportional to the absolute activities, $\bar{z}_l$, taken to powers which depend on the number of molecules participating in the elementary processes. Since in dilute solution $\bar{z}_l$ is proportional to $\bar{\rho}_l$, the dominant terms in dilute solution are due to molecules diffusing in isolation. Consequently, to dominant order in the density

$$L_{ij} = L_{ii}\delta_{ij}. \tag{6.6.2}$$

Thus in dilute solution the diffusion flux can be written

$$\bar{\mathbf{j}}_{Di} = -(L_{ii}/T)\nabla\bar{\mu}_i, \tag{6.6.3}$$

where the temperature has been taken as constant. Finally noting that $\bar{\mu}_i = \mu_i^0(T) + k_B T \ln \bar{\rho}_i$ in dilute solution, we obtain Fick's law for independent components in the form

$$\bar{\mathbf{j}}_{Di} = -D_i \nabla\bar{\rho}_i, \tag{6.6.4}$$

with the diffusion constants given by

$$D_i = k_B L_{ii}/\bar{\rho}_i. \tag{6.6.5}$$

For the isolated molecule diffusion mechanism, Eq. (6.1.58) shows that L_{ii} is proportional to $\bar{z}_i = \exp(\mu_i^0/k_B T)\bar{\rho}_i$. Thus the diffusion constants in Eq. (6.6.5) are independent of density.

To obtain the conditional average equations for diffusion and chemical reaction, we need to generalize Eq. (6.2.66), which describes mass conservation in the absence of chemical reactions. Chemical reactions occur locally in space and contribute source terms to the mass conservation equation. In general, then, one has

$$\partial\bar{\rho}_i/\partial t = -\nabla\cdot\bar{\mathbf{j}}_{Di} - \nabla\cdot(\bar{\rho}_i\bar{\mathbf{v}}) + \sum_\kappa \omega_{\kappa i}(\bar{v}_\kappa^+ - \bar{v}_\kappa^-), \tag{6.6.6}$$

where $\bar{v}_\kappa^\pm$ are the per unit volume conditional average rates of the chemical reaction, κ. To apply Eq. (6.6.6) to the present problem, we note that in dilute solution $\bar{\mathbf{v}}$ will be the average solvent velocity, which we can take equal to zero. Thus, substituting the explicit expression for $\bar{\mathbf{j}}_{Di}$ from Eq. (6.6.4), we find that

$$\partial\bar{\rho}_i/\partial t = D_i\nabla^2\bar{\rho}_i + \sum_\kappa \omega_{\kappa i}(\bar{v}_\kappa^+ - \bar{v}_\kappa^-). \tag{6.6.7}$$

Equation (6.6.7) describes the conditional average density in a dilute, chemically reacting system in which energy and momentum effects are neglected.

Fluctuations around the conditional average are described by linearizing Eq. (6.6.7) to get

$$\partial\delta\rho_i/\partial t = D_i\nabla^2\delta\rho_i + \sum_\kappa \omega_{\kappa i}\sum_j (\partial\bar{v}_\kappa^+ - \bar{v}_\kappa^-/\partial\bar{\rho}_j)\delta\rho_j + \tilde{f}_i. \tag{6.6.8}$$

The covariance of the random terms involves independent contributions from reaction and diffusion. Thus the chemical terms can be added to the diffusion terms, which are obtained by simplifying Eq. (6.2.68) using Eqs. (6.6.2) and (6.6.5). This yields

$$\langle\tilde{f}_i(\mathbf{r},t)\tilde{f}_j(\mathbf{r}',t')\rangle$$
$$= \left[\sum_\kappa \omega_{\kappa i}(\bar{v}_\kappa^+ + \bar{v}_\kappa^-)\omega_{\kappa j}\delta(\mathbf{r}-\mathbf{r}') - 2D_i\nabla_\mathbf{r}\cdot\bar{\rho}_i\delta_{ij}\nabla_\mathbf{r}\delta(\mathbf{r}-\mathbf{r}')\right]\delta(t-t'). \tag{6.6.9}$$

Notice that the chemical terms are localized in space by the delta function. If the chemical terms are integrated over $\mathbf{r}$ and $\mathbf{r}'$, it is easy to check that they give the correct expression for the covariance of $\tilde{f}_i$ for the overall molecule numbers, $N_j(t) = \int d\mathbf{r}\,\rho_j(\mathbf{r},t)$. Equations (6.6.7)–(6.6.9) provide the hydrodynamic-level stochastic description of number densities for a dilute, chemically reacting system. Within the restrictions of the hydrodynamic description, they are applicable close to or far from equilibrium.

A simple, solvable example that illustrates the way in which diffusion and reaction are coupled is provided by the dimerization reaction

$$A + A \rightleftarrows A_2. \tag{6.6.10}$$

To simplify the calculation further, we neglect the reverse reaction and consider only

$$A + A \rightarrow A_2, \tag{6.6.11}$$

but imagine that A is added to solution at a constant rate, $2F$. In this case the average equation for ρ_A takes the form

$$\partial\bar{\rho}_A/\partial t = -2k^+\bar{\rho}_A^2 + D_A\nabla^2\bar{\rho}_A + 2F. \tag{6.6.12}$$

This equation depends only on $\bar{\rho}_A$ and, for the time being, the dimer A_2 will be ignored. According to Eqs. (6.6.8) and (6.6.9), fluctuations in ρ_A satisfy

$$\partial\delta\rho_A/\partial t = -4k^+\bar{\rho}_A\delta\rho_A + D_A\nabla^2\delta\rho_A + \tilde{f}_A \tag{6.6.13}$$

where $\langle\tilde{f}_A(\mathbf{r},t)\rangle \equiv 0$ and

$$\langle\tilde{f}_A(\mathbf{r},t)\tilde{f}_A(\mathbf{r}',t')\rangle = [4k^+\bar{\rho}_A^2 - 2D_A\nabla\cdot\bar{\rho}_A\nabla]\delta(\mathbf{r}-\mathbf{r}')\delta(t-t'), \tag{6.6.14}$$

with the gradient operator acting on $\mathbf{r}$. Equation (6.6.12), which describes the average, is a nonlinear, partial differential equation whose solution requires the specification of boundary conditions. We will take the *Neumann* boundary condition at infinity, namely that

$$\lim_{|\mathbf{r}|\to\infty} \nabla\bar{\rho}_A = \mathbf{0}.$$

This insures that a uniform steady state exists which satisfies $\partial\rho_A^{ss}/\partial t = \nabla^2\rho_A^{ss} = 0$. Using these conditions in Eq. (6.6.12) give

$$\rho_A^{ss} = (F/k^+)^{1/2}. \tag{6.6.15}$$

This uniform steady state is unchanged over time and we will use it as the average value around which we examine the conditional fluctuations. For this average state Eqs. (6.6.13) and (6.6.14) become

$$\partial\delta\rho_A/\partial t = -2k^+\rho_A^{ss}\delta\rho_A + D_A\nabla^2\delta\rho_A + \tilde{f}_A \tag{6.6.16}$$

$$\langle\tilde{f}_A(\mathbf{r},t)\tilde{f}_A(\mathbf{r}',t')\rangle = [4k^+\rho_A^{ss2} - 2D_A\rho_A^{ss}\nabla^2]\delta(\mathbf{r}-\mathbf{r}')\delta(t-t'). \tag{6.6.17}$$

Equation (6.6.16) is a linear partial differential equation and also requires the specification of a boundary condition. Since we are interested primarily in the correlation function, we will assume the boundary conditions

$$\lim_{|\mathbf{r}|\to\infty} \langle\delta\rho_A(\mathbf{r},t)\delta\rho_A(\mathbf{r}',t)\rangle^0 = \lim_{|\mathbf{r}'|\to 0} \langle\delta\rho_A(\mathbf{r},t)\delta\rho_A(\mathbf{r}',t)\rangle^0 = 0. \tag{6.6.18}$$

These boundary conditions allow Fourier transform techniques to be used. To solve for the probability distribution of the fluctuations, we recall that the probability density functional is Gaussian. Thus, since the average is ρ_A^{ss}, it suffices to obtain the covariance of the fluctuations,

$$\sigma(\mathbf{r},\mathbf{r},t) \equiv \langle\delta\rho_A(\mathbf{r},t)\delta\rho_A(\mathbf{r}',t)\rangle^0. \tag{6.6.19}$$

As we have seen, this function satisfies Eq. (6.2.48), which can be solved in a similar way by introducing the Fourier transform

$$\hat{\sigma}(\mathbf{k},\mathbf{k}',t) = (2\pi)^{-6}\int_{-\infty}^{+\infty} d\mathbf{r} \int_{-\infty}^{+\infty} d\mathbf{r}' \exp[i(\mathbf{k}\cdot\mathbf{r} + \mathbf{k}'\cdot\mathbf{r}')]\sigma(\mathbf{r},\mathbf{r}',t). \tag{6.6.20}$$

The equation satisfied by $\hat{\sigma}$ can be obtained from Eq. (6.2.48) by Fourier transforms in the same way that Eq. (6.2.53) was obtained for diffusion alone. For the present problem the matrix $\hat{H}(\mathbf{k},\mathbf{k}')$ comes from Eq. (6.6.17). It is diagonal and has the form

$$\hat{H}(\mathbf{k},\mathbf{k}') = -[4k^+\rho_A^{ss} + k^2 D_A]\delta(\mathbf{k}-\mathbf{k}'). \tag{6.6.21}$$

The double Fourier transformation of the covariance of the random term in Eq. (6.6.17) is also straightforward and gives

$$\begin{aligned}\langle\hat{\tilde{f}}_A(\mathbf{k},t)\hat{\tilde{f}}_A(\mathbf{k}',t)\rangle &\equiv \hat{\gamma}(\mathbf{k},\mathbf{k}')\delta(t-t') \\ &= [4k^+\rho_A^{ss2} + 2D_A\rho_A^{ss}k^2](2\pi)^{-3}\delta(\mathbf{k}+\mathbf{k}')\delta(t-t'). \end{aligned} \tag{6.6.22}$$

Substituting these expressions into the Fourier-transformed version of Eq. (6.2.48), i.e.,

$$\partial\hat{\sigma}(\mathbf{k},\mathbf{k}',t)/\partial t = \int [\hat{H}(\mathbf{k},\mathbf{k}'',t)\hat{\sigma}(\mathbf{k}'',\mathbf{k}',t) + \hat{\sigma}(\mathbf{k},\mathbf{k}'',t)\hat{H}^T(\mathbf{k}'',\mathbf{k}',t)]\,d\mathbf{k}'' + \hat{\gamma}(\mathbf{k},\mathbf{k}'), \tag{6.6.23}$$

then gives

$$\begin{aligned}\partial\hat{\sigma}(\mathbf{k},\mathbf{k}',t)/\partial \mathbf{t} = &-[8k^+\rho_A^{ss} + D_A(k^2 + k^2)]\sigma(\mathbf{k},\mathbf{k}',t)\\ &+ [4k^+\rho_A^{ss2} + 2D_A\rho_A^{ss}k^2](2\pi)^{-3}\delta(\mathbf{k}+\mathbf{k}').\end{aligned} \tag{6.6.24}$$

Equation (6.6.24) must be solved with the initial condition $\hat{\sigma} \equiv 0$, corresponding to the fact that $\hat{\sigma}$ is the covariance of the conditional fluctuation. The solution is obviously

$$\begin{aligned}\hat{\sigma}(\mathbf{k},\mathbf{k}',t) = 2\int_0^t &\exp[-2(4k^+\rho_A^{ss} + D_A k^2)(t-\tau)]\\ &\times \{2k^+\rho_A^{ss2} + D_A\rho_A^{ss}k^2\}\,d\tau(2\pi)^{-3}\sigma(\mathbf{k}+\mathbf{k}'),\end{aligned}$$

which can be integrated to give:

$$\begin{aligned}&\hat{\sigma}(\mathbf{k},\mathbf{k}',t)\\ &= \left[\frac{2k^+\rho_A^{ss2} + D_A\rho_A^{ss}k^2}{4k^+\rho_A^{ss} + D_A k^2}\right](2\pi)^{-3}\delta(\mathbf{k}+\mathbf{k}')\{1 - \exp[-2(4k^+\rho_A^{ss} + D_A k^2)t]\}.\end{aligned}$$

Although it is possible to invert the Fourier transform of $\hat{\sigma}(\mathbf{k},\mathbf{k}',t)$ analytically, the resulting expression is long and unwieldy. Thus we content ourselves with writing down the asymptotic expression for $t \to \infty$, i.e.,

$$\hat{\sigma}(\mathbf{k},\mathbf{k}') = \left[\rho_A^{ss} - \frac{2k^+\rho_A^{ss2}}{(4k^+\rho_A^{ss} + D_A k^2)}\right]\delta(\mathbf{k}+\mathbf{k}')/(2\pi)^3 \tag{6.6.25}$$

As is discussed in Chapter 7, this is the covariance for the stationary probability functional, W_1, in the steady-state ensemble. The inverse transform of Eq. (6.6.25) can be found in tables, which give

$$\langle\delta\rho_A(\mathbf{r})\delta\rho_A(\mathbf{r}')\rangle^{ss} \equiv \sigma(\mathbf{r},\mathbf{r}') = \rho_A^{ss}\delta(\mathbf{r}-\mathbf{r}') - \frac{k^+\rho_A^{ss2}}{2\pi D_A|\mathbf{r}-\mathbf{r}'|}\exp(-\xi|\mathbf{r}-\mathbf{r}'|). \tag{6.6.26}$$

The factor ξ^{-1} is called the *correlation length* and is defined by

$$\xi^{-1} = (D_A/4k^+\rho_A^{ss})^{1/2}. \tag{6.6.27}$$

To appreciate the effect that the coupling of reaction and diffusion has on the density fluctuations, we need to refer back to Eq. (6.2.62), which shows that the first term in Eq. (6.6.26) has the same form as the equilibrium density–density correlation function for a dilute solution. Then second term, which involves long-range correlations through the exponential, is due to the coupling of reaction and diffusion. This term vanishes in proportion to $k^+\rho_A^{ss2}$.

According to Eq. (6.6.15) this will occur when the rate, F, at which A is added to solution vanishes. Thus we see that in an ensemble with inputs the coupling of diffusion and chemical reactions is reflected in the stationary probability density. This persistent correlation can lead to measurable changes in the reaction rate constant, a phenomenon discussed further in Chapter 9.

It is easy to check that the density correlation function in Eq. (6.6.26) is compatible with the thermodynamic-level description of the dimerization reaction. To show this we need to integrate over $\mathbf{r}$ and $\mathbf{r}'$ in Eq. (6.6.26) to obtain the number density fluctuations, $\delta N_A \equiv \int d\mathbf{r}\, \delta\rho_A$. This yields

$$\langle \delta N_A^2 \rangle^{ss} = N_A^{ss} - \frac{2k^+ \rho_A^{ss2}}{D_A} V \int_0^\infty x \exp(-\xi x)\, dx, \tag{6.6.28}$$

where V is the volume and integration variables were changed to $\mathbf{x} = \mathbf{r} - \mathbf{r}'$ and $(\mathbf{r} + \mathbf{r}')/2$. Performing the integral and using the definition of ξ in Eq. (6.6.27) leads to

$$\langle \delta N_A^2 \rangle^{ss} = N_A^{ss}/2. \tag{6.6.29}$$

The comparable thermodynamic-level theory of number fluctuations is given by the equations

$$d\delta N_A/dt = -4k^+ \rho^{ss} \delta N_A + \tilde{f}_A \tag{6.6.30a}$$

$$\langle \tilde{f}_A(t) \tilde{f}_A(t') \rangle = V 4k^+ \rho^{ss2} \delta(t - t'). \tag{6.6.30b}$$

To calculate the conditional variance, σ, we recall that it satisfies Eq. (4.3.9), which for this problem becomes

$$d\sigma/dt = -8k^+ \rho^{ss} \sigma + 4k^+ V \rho^{ss2}. \tag{6.6.31}$$

We need the asymptotic solution of Eq. (6.6.31) as $t \to \infty$ to compare with Eq. (6.6.29). In this limit the stationary state is reached and the left-hand side of Eq. (6.6.31) vanishes. Solving the resulting equation gives

$$\sigma^{ss} = N^{ss}/2, \tag{6.6.32}$$

in agreement with the integrated hydrodynamic-level result in Eq. (6.6.29).

To simplify the calculations of the density–density correlation functions for the dimerization reaction in Eq. (6.6.10), we neglected the back reaction and assumed that it occurred in a solution at low density. The inclusion of the back reaction complicates the calculation because it couples the density fluctuations of A to density fluctuations of the dimer, A_2. To treat this problem we enrich the example somewhat by including a photochemical mechanism for generation of the monomer, e.g.,

$$h\nu + A_2 \to 2A \tag{6.6.33}$$

If the light involved in this process is generated by a monochromatic laser, then Eq. (6.6.33) does not describe an elementary process and its rate does not have the canonical form. Nonetheless, it does introduce a systematic variation into the average rate equation, which we write as $2k^*\bar{\rho}_{A_2}$. The constant k^* is proportional to the intensity of the laser light. Combining this expression with the rate expressions for the forward and reverse reaction in Eq. (6.6.10) gives the average equations

$$\begin{aligned} \partial\bar{\rho}_A/\partial t &= 2k^*\bar{\rho}_{A_2} - 2k^+\bar{\rho}_A^2 + 2k_u\bar{\rho}_{A_2} + D_A\nabla^2\bar{\rho}_A \\ \partial\bar{\rho}_{A_2}/\partial t &= -k^*\bar{\rho}_{A_2} - k_u\bar{\rho}_{A_2} + k^+\bar{\rho}_A^2 + D_{A_2}\nabla^2\bar{\rho}_{A_2}, \end{aligned} \tag{6.6.34}$$

where k_u is the rate constant for the back reaction $A_2 \rightarrow 2A$. At the uniform steady state, Eqs. (6.6.34) have the solution

$$\rho_{A_2}^{ss} = [k^+/(k^* + k_u)]\rho_A^{ss2}. \tag{6.6.35}$$

Two partial stochastic differential equations are needed to characterize the fluctuations around this steady state. Linearizing Eqs. (6.6.34) we find

$$\begin{aligned} \partial\delta\rho_A/\partial t &= -4k^+\rho_A^{ss}\delta\rho_A + 2(k^* + k_u)\delta\rho_{A_2} + D_A\nabla^2\delta\rho_A + \hat{f}_A \\ \partial\delta\rho_{A_2}/\partial t &= 2k^+\rho_A^{ss}\delta\rho_A - (k^* + k_u)\delta\rho_{A_2} + D_{A_2}\nabla^2\delta\rho_{A_2} + \hat{f}_{A_2}. \end{aligned} \tag{6.6.36}$$

The covariance matrix of the random terms has the structure

$$\langle\hat{f}_i(\mathbf{r}, t)\hat{f}_j(\mathbf{r}', t')\rangle = \gamma_{ij}(\mathbf{r}, \mathbf{r}')\delta(t - t'),$$

where, defining $K = k^+\rho_A^{ss2} + k_u\rho_{A_2}^{ss}$,

$$\gamma(\mathbf{r}, \mathbf{r}') = \begin{pmatrix} 4K - 2D_A\rho_A^{ss}\nabla_{\mathbf{r}}^2 & -2K \\ -2K & K - 2D_{A_2}\rho_{A_2}^{ss}\nabla_{\mathbf{r}}^2 \end{pmatrix}\delta(\mathbf{r} - \mathbf{r}'). \tag{6.6.37}$$

The form of γ follows from the canonical result for diffusion and reaction given in Eq. (6.6.9).

The simplest way to find the covariance matrix of the density fluctuations for A and A_2 is to Fourier transform Eqs. (6.6.36) and (6.6.37). Defining the column vector of transformed density fluctuations by $\delta\hat{\boldsymbol{\rho}}^T \equiv (\delta\hat{\rho}_A(\mathbf{k}, t), \delta\hat{\rho}_{A_2}(\mathbf{k}, t))$, Eq. (6.6.36) shows that $\delta\hat{\boldsymbol{\rho}}$ solves the equation

$$d\delta\hat{\boldsymbol{\rho}}(\mathbf{k}, t)/dt = \int \hat{H}(\mathbf{k}, \mathbf{k}')\delta\hat{\boldsymbol{\rho}}(\mathbf{k}', t)\,d\mathbf{k}' + \hat{\hat{\mathbf{f}}}(\mathbf{k}, t)$$

where

$$\begin{aligned} \hat{H}(\mathbf{k}, \mathbf{k}') &= \begin{Bmatrix} -4k^+\rho_A^{ss} - D_A k^2 & 2(k^* + k_u) \\ 2k^+\rho_A^{ss} & -(k^* + k_u) - D_{A_2}k^2 \end{Bmatrix}\delta(\mathbf{k} - \mathbf{k}') \\ &\equiv \hat{H}(k)\delta(\mathbf{k} - \mathbf{k}'). \end{aligned} \tag{6.6.38}$$

Fourier transformation of Eq. (6.6.37) gives

$$\langle\hat{\hat{\mathbf{f}}}(\mathbf{k}, t)\hat{\hat{\mathbf{f}}}^T(\mathbf{k}', t')\rangle = \hat{\gamma}(\mathbf{k}, \mathbf{k}')\delta(t - t')$$

with

$$\hat{\gamma}(\mathbf{k}, \mathbf{k}') = \begin{Bmatrix} 4K + 2D_{\mathrm{A}}\rho_{\mathrm{A}}^{ss}k^2 & -2K \\ -2K & K + 2D_{\mathrm{A}_2}\rho_{\mathrm{A}_2}^{ss}k^2 \end{Bmatrix} \frac{\delta(\mathbf{k} + \mathbf{k}')}{(2\pi)^3}$$
$$\equiv \gamma(k)\delta(\mathbf{k} + \mathbf{k}')/(2\pi)^3. \tag{6.6.39}$$

We are interested in the conditional covariance of the density fluctuations. For two variables this is a matrix, i.e.,

$$\sigma_{ij}(\mathbf{r}, \mathbf{r}', t) \equiv \langle \delta\rho_i(\mathbf{r}, t)\delta\rho_j(\mathbf{r}', t)\rangle^0. \tag{6.6.40}$$

Its Fourier transform, $\hat{\sigma}(\mathbf{k}, \mathbf{k}', \mathbf{t})$, satisfies Eq. (6.6.23), which as $t \to \infty$ asymptotically becomes

$$\hat{H}(k)\hat{\sigma}(\mathbf{k}, \mathbf{k}') + \hat{\sigma}(\mathbf{k}, \mathbf{k}')\hat{H}^T(k) = -\hat{\gamma}(k)\delta(\mathbf{k} + \mathbf{k}')/(2\pi)^3. \tag{6.6.41}$$

This equation has a unique solution of the form

$$\hat{\sigma}(\mathbf{k}, \mathbf{k}') = \hat{\sigma}(k)\delta(\mathbf{k} + \mathbf{k}')/(2\pi)^3, \tag{6.6.42}$$

where

$$\hat{H}(k)\hat{\sigma}(k) + \hat{\sigma}(k)\hat{H}(k)^T = -\hat{\gamma}(k). \tag{6.6.43}$$

This is a linear matrix equation for $\hat{\sigma}$. Since $\hat{\sigma}(k)$ and $\hat{\gamma}(k)$ are symmetric matrices, this equation can be broken down into three linear equations for $\hat{\sigma}_{11}(k)$, $\hat{\sigma}_{12}(k) = \hat{\sigma}_{21}(k)$, and $\hat{\sigma}_{22}(k)$. As such it can be solved by standard methods of linear algebra, say, Cramer's rule. After some lengthy, but straightforward, algebraic manipulations we find that

$$\hat{\sigma}_{11}(\mathbf{k}, \mathbf{k}') = \left[\rho_{\mathrm{A}}^{ss} - \frac{2k^*\rho_{\mathrm{A}_2}^{ss}}{(D_{\mathrm{A}} + D_{\mathrm{A}_2})(k^2 + \xi_2^2)} - \frac{k^*\rho_{\mathrm{A}_2}^{ss}(D_{\mathrm{A}_2}/D_{\mathrm{A}})k^2}{(D_{\mathrm{A}} + D_{\mathrm{A}_2})(k^2 + \xi_1^2)(k^2 + \xi_2^2)}\right]$$
$$\times\ \delta(\mathbf{k} + \mathbf{k}')/(2\pi)^3 \tag{6.6.44}$$

$$\begin{aligned} \xi_1^2 &= [(4D_{\mathrm{A}_2}/\rho_{\mathrm{A}}^{ss}) + (D_{\mathrm{A}}/\rho_{\mathrm{A}_2}^{ss})]K/D_{\mathrm{A}}D_{\mathrm{A}_2} \\ \xi_2^2 &= [(4/\rho_{\mathrm{A}}^{ss}) + (1/\rho_{\mathrm{A}_2}^{ss})]K/(D_{\mathrm{A}} + D_{\mathrm{A}_2}). \end{aligned} \tag{6.6.45}$$

The expression for $\hat{\sigma}_{11}$ in Eq. (6.6.44) is similar to the one obtained in Eq. (6.6.25) based on the neglect of the back reaction. Indeed, if both D_{A_2} and k_u, the rate constant for the back reaction, are set equal to zero in Eq. (6.6.44), it reduces to

$$\hat{\sigma}_{11}(\mathbf{k}, \mathbf{k}') = \left[\rho_{\mathrm{A}}^{ss} - \frac{2k^+\rho_{\mathrm{A}}^{ss2}}{D_{\mathrm{A}}k^2 + (4k^+\rho_{\mathrm{A}}^{ss} + k^*)}\right]\delta(\mathbf{k} + \mathbf{k}')/(2\pi)^3. \tag{6.6.46}$$

Except for the term k^* in the denominator, this is identical to Eq. (6.6.25). In the limit that the illumination intensity vanishes, k^* also vanishes and the two expressions are the same. This provides another illustration of the hierarchical nature of the fluctuation theory, which reduces to a coarser version of the theory when appropriate elementary processes are neglected.

The application of the hydrodynamic-level theory to reaction–diffusion systems problems typifies some of the problems encountered at the hydrodynamic level of description. The real challenge to calculations, however, comes about when systems with spatial gradients are treated. We take up this complication in Section 6.8.

6.7. Quasi-elastic Scattering Theory

An excellent way to get information about the space–time structure of fluctuations is using quasi-elastic scattering techniques. Sources of widely varying wavelength have proven to be useful. Thermal neutrons, for example, have wavelengths the order of one angstrom and can be used to explore a spatial domain of molecular size. The wavelengths of visible light, on the other hand, are the order of 5×10^3 angstroms, which is in the domain of the hydrodynamic theory. In this section we outline some of the important features of quasi-elastic scattering theory and relate the results to quantities which can be calculated using fluctuation theory.

A schematic diagram of a scattering experiment is illustrated in Fig. 6.4. The incoming wave is a plane wave characterized by a wave vector, $\mathbf{k}_i$, and a frequency, ω_i. The wave vector is oriented perpendicular to the plane wave and has a magnitude

$$k = 2\pi/\lambda, \tag{6.7.1}$$

where λ is the wavelength. The frequency and wavelength are related by $\omega\lambda/2\pi = c$, the speed of light, or

$$\omega/k = c, \tag{6.7.2}$$

which is the dispersion relationship for light in vacuum. When the plane wave encounters a molecular medium, some of its energy is absorbed by the medium and reemitted as spherical waves. If the medium were completely homogeneous, the reemitted wave fronts would add up coherently to form a reconstructed plane wave without loss of intensity. Because of molecular

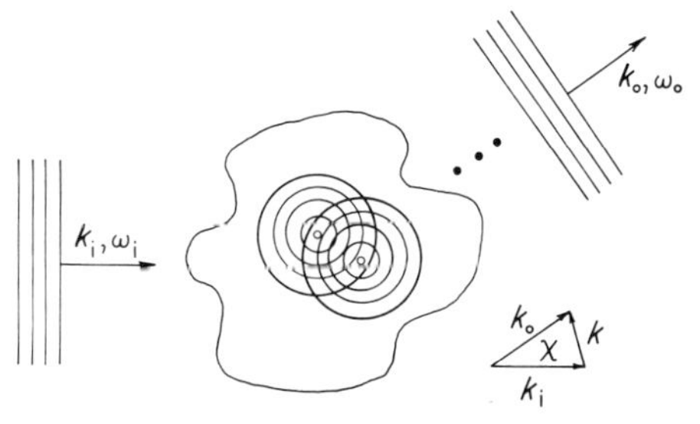

FIGURE 6.4. An illustration of the scattering of an incoming plane wave with wave vector $\mathbf{k}_i$ and frequency ω_i and the emerging outgoing plane wave with wave vector $\mathbf{k}_0$ and frequency ω_0. The scattered wave vector, $\mathbf{k}$, and its connection to the scattering angle, χ, are shown by the vector addition diagram.

fluctuations in density, however, the medium is not homogeneous on the scale of the wavelength of the incident wave and light is reemitted incoherently. This leads to scattered light which, far from the source, appears again as plane waves. Since the molecular processes involved in absorption may be inelastic, the energy of the incident wave, $\hbar\omega_i$, may be changed. Thus the frequency of the observed wave, ω_0, may differ from that of the incoming wave. According to Eq. (6.7.2), this means that the magnitude of the observed wave vector is also changed. The angle χ between the incoming and observed wave vectors is called the *scattering angle.*

In quasi-elastic scattering the energy change $\hbar(\omega_0 - \omega_i) \equiv \hbar\omega$ is assumed to be small, that is, the scattering is almost elastic. In this case the magnitude of the observed scattering vector $\mathbf{k}_0$ will be essentially identical to that of the incoming wave. The so-called *scattering vector* is defined by

$$\mathbf{k} = \mathbf{k}_i - \mathbf{k}_0. \tag{6.7.3}$$

If $\mathbf{k}_i$ and $\mathbf{k}_0$ have essentially the same magnitude, then it follows from the vector diagram in Fig. 6.4 that

$$k = 2k_i \sin(\chi/2). \tag{6.7.4}$$

Thus by changing the angle at which the scattered wave is observed, one changes the magnitude of the scattering vector.

This description of a quasi-elastic scattering experiment is appropriate for light scattering. However, only minor changes need be made to describe neutron scattering. Since neutrons are particle waves, their wave vectors are given by the de Broglie formula

$$\mathbf{k} = \mathbf{p}/\hbar, \tag{6.7.5}$$

where $\mathbf{p}$ is the momentum of the neutron. The associated frequency is given by the dispersion relationship

$$\hbar\omega = p^2/2m = \hbar^2 k^2/2m. \tag{6.7.6}$$

For quasi-elastic neutron scattering Eqs. (6.7.3) and (6.7.4) are obviously still applicable.

As we saw in our treatment of the Boltzmann equation in Section 2.7, the scattering of particles can be interpreted in terms of a differential scattering cross section. The same is true for waves or particle-waves. In the case of a plane wave beam of neutrons, the cross section depends both on their energy, $\hbar\omega$, and their wave vector. For a monochromatic beam, we use the double differential cross section, $\Sigma(\mathbf{k}, \omega)$, to describe the scattering. By definition, the number of neutrons scattered in time dt with a scattered wave vector $\mathbf{k}$ and in a frequency range $d\omega$ is given by $dN(\mathbf{k}, \omega) = I\Sigma(\mathbf{k}, \omega)\, d\Omega\, d\omega\, dt$, where I is the magnitude of the flux density in the initial beam and $d\Omega = \sin\chi\, d\chi\, d\theta$ is the differential solid angle of scattering. For quasi-elastic neutron scattering the cross section can be calculated from quantum mechanics. In the Born approximation this gives

$$\Sigma(\mathbf{k}, \omega) = C \sum_i \sum_j \langle \exp(i\mathbf{k}\cdot\mathbf{r}_i)\exp(-i\mathbf{k}\cdot\mathbf{r}_j)\rangle \delta(\omega + \omega_i - \omega_0), \quad (6.7.7)$$

where C depends on the magnitude of $\mathbf{k}_i$, $\mathbf{k}$, and the number of molecules in the scattering region, and $\mathbf{r}_i$ is the position of the center of mass of the ith molecule in the scattering volume. The average indicated in Eq. (6.7.7) is over an ensemble representing the scattering medium. Equation (6.7.7) can be rewritten using the representation of the delta function

$$2\pi\delta(\omega + \omega_i - \omega_0) = \int_{-\infty}^{+\infty} \exp[i(\omega + \omega_i - \omega_0)t]\, dt \quad (6.7.8)$$

If we then take advantage of the fact that the ensemble average in Eq. (6.7.7) can be written in terms of the eigenvectors of the Hamiltonian of the unperturbed medium, it is possible to rewrite Eq. (6.7.7) as a two-time average. This leads to the expression

$$S(\mathbf{k}, \omega) \equiv \frac{\Sigma(\mathbf{k}, \omega)}{C} = \frac{1}{2\pi}\int_{-\infty}^{+\infty} dt \exp(i\omega t) F(\mathbf{k}, t), \quad (6.7.9)$$

where

$$F(\mathbf{k}, t - t') = \sum_{i,j} \langle \exp[i\mathbf{k}\cdot\mathbf{r}_i(t)] \exp[-i\mathbf{k}\cdot\mathbf{r}_j(t')]\rangle. \quad (6.7.10)$$

The dependence of F on $t - t'$ is appropriate for a stationary ensemble. $S(\mathbf{k}, \omega)$ is called the *scattering function* and $F(\mathbf{k}, t - t')$ is called the *intermediate scattering* function. To a good approximation, the motions of the molecules can be treated by classical mechanics. Thus the operators in Eq. (6.7.10) commute and one can simplify this expression by writing

$$\sum_i \exp(i\mathbf{k}\cdot\mathbf{r}_i) = \int_{-\infty}^{+\infty} d\mathbf{r} \exp(i\mathbf{k}\cdot\mathbf{r}) \sum_i \delta(\mathbf{r}_i - \mathbf{r}). \quad (6.7.11)$$

Now in any system in the ensemble the number density of molecules at point $\mathbf{r}$ is given by

$$\rho(\mathbf{r}, t) = \sum_i \delta(\mathbf{r}_i(t) - \mathbf{r}). \quad (6.7.12)$$

Making these two substitutions, Eq. (6.7.10) can be written

$$F(\mathbf{k}, t - t') = \iint \exp(i\mathbf{k}\cdot\mathbf{r})\langle\rho(\mathbf{r}, t)\rho(\mathbf{r}', t')\rangle \exp(-i\mathbf{k}\cdot\mathbf{r}')\, d\mathbf{r}\, d\mathbf{r}'. \quad (6.7.13)$$

Finally, multiplying both sides of Eq. (6.7.13) by $\delta(\mathbf{k} + \mathbf{k}')$ and introducing the Fourier transform of the density, $\hat{\rho}$, Eq. (6.7.13) can be written

$$\frac{F(\mathbf{k}, t - t')}{(2\pi)^3}\delta(\mathbf{k} + \mathbf{k}') = \langle\hat{\rho}(\mathbf{k}, t)\hat{\rho}(\mathbf{k}', t')\rangle. \quad (6.7.14)$$

Equation (6.7.14) shows that the intermediate scattering function is related to the density–density correlation function. Fourier transforming Eq. (6.7.14) twice with respect to time and comparing with Eq. (6.7.9) gives

$$S(\mathbf{k},\omega)\delta(\mathbf{k}+\mathbf{k}')\delta(\omega+\omega')/(2\pi)^4 = \langle\hat{\rho}(\mathbf{k},\omega)\hat{\rho}(\mathbf{k}',\omega')\rangle. \qquad (6.7.15)$$

If we forget about forward scattering, i.e., $\mathbf{k} = \mathbf{0}$, then S can be reexpressed in terms of the density fluctuations. For a uniform system

$$\delta\hat{\rho}(\mathbf{k},t) = \hat{\rho}(\mathbf{k},t) - \bar{\rho}\delta(\mathbf{k}), \qquad (6.7.16)$$

and for $\mathbf{k} \neq \mathbf{0}$ Eq. (6.7.15) becomes

$$S(\mathbf{k},\omega)\delta(\mathbf{k}+\mathbf{k}')\delta(\omega+\omega)/(2\pi)^4 = \langle\delta\hat{\rho}(\mathbf{k},\omega)\delta\hat{\rho}(\mathbf{k}',\omega')\rangle. \qquad (6.7.17)$$

Equation (6.7.17) is the fundamental relationship between the neutron scattering function, S, and the density correlation function in a stationary ensemble.

For quasi-elastic light scattering experiments one can derive a relationship comparable to Eq. (6.7.17). For light scattering one also has an incoming intensity and a scattered intensity. The scattering, however, can be followed either by using the electric field or the intensity of the field, i.e., its square magnitude. If one follows the electric field, then the spectrum of the scattered light is determined by fluctuations in the dielectric tensor of the medium, $\delta\varepsilon$. Using an approach similar to that used to derive Eq. (6.7.17), one can express the electric field correlation function, $I(\mathbf{k},\omega)$, of the light scattered with frequency $\omega = \omega_i - \omega_0$ and scattering vector $\mathbf{k}$ in terms of fluctuations in the dielectric constant. In particular, in an isotropic medium one finds that

$$I(\mathbf{k},\omega)\delta(\mathbf{k}+\mathbf{k}')\delta(\omega+\omega')/(2\pi)^4 = C\langle\delta\hat{\varepsilon}(\mathbf{k},\omega)\delta\hat{\varepsilon}(\mathbf{k}',\omega')\rangle, \qquad (6.7.18)$$

where C is proportional to the intensity of the incoming beam and ε is the scalar dielectric constant. $I(\mathbf{k},\omega)$ is analogous to the scattering function for neutron scattering and is the quantity measured using heterodyne detection methods; see Berne and Pecora in the list of references at the end of the chapter.

The dielectric function in Eq. (6.7.18) depends on the temperature, density, and composition of the scattering medium. Thus fluctuations in the dielectric constant can be written as

$$\delta\varepsilon = (\partial\varepsilon/\partial T)_{\boldsymbol{\rho}}\delta T + \sum_{i=1}^{k} (\partial\varepsilon/\partial\rho_i)_{T,\boldsymbol{\rho}'}\delta\rho_i. \qquad (6.7.19)$$

This implies that the dielectric constant correlation function can be reexpressed in terms of the temperature and density correlation functions. In this way one can compare the results of heterodyne light scattering experiments with theoretical calculations based on hydrodynamic fluctuation theory. In most substances the dependence of the dielectric constant on temperature is much weaker than its dependence on the density. Thus in a single-component fluid one can write $\delta\varepsilon = (\partial\varepsilon/\partial\rho)_T\delta\rho$ and Eq. (6.7.18) simplifies to

$$I(\mathbf{k},\omega)\delta(\mathbf{k}+\mathbf{k}')\delta(\omega+\omega')/(2\pi)^4 = C(\partial\varepsilon/\partial\rho)_T^2\langle\delta\hat{\rho}(\mathbf{k},\omega)\delta\hat{\rho}(\mathbf{k}',\omega')\rangle. \quad (6.7.20)$$

In this case the light scattering spectrum is a measure of the density–density correlation function.

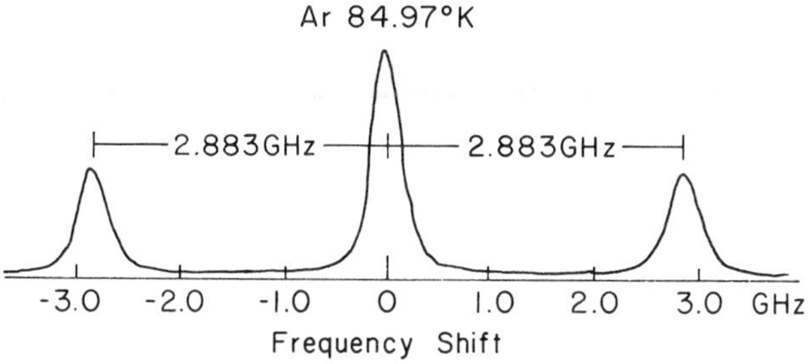

FIGURE 6.5. A typical quasi-elastic light scattering spectrum for liquid argon at 84.97 K. The central Rayleigh peak and the two Brillouin peaks, symmetrically disposed around the incident frequency with shifts of $2.883 \times 10^9\ s^{-1}$, are plainly visible. Taken from R.A. Fleury and J.P. Boon, *Phys. Rev.* **186**, 244 (1969).

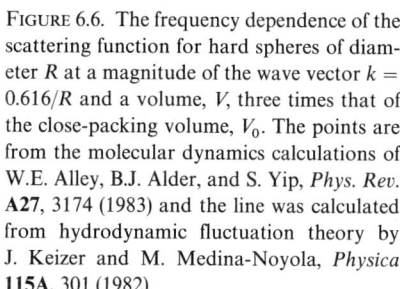

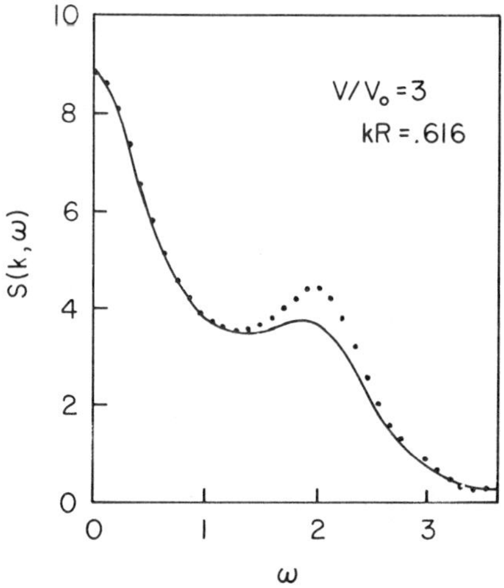

FIGURE 6.6. The frequency dependence of the scattering function for hard spheres of diameter R at a magnitude of the wave vector $k = 0.616/R$ and a volume, V, three times that of the close-packing volume, V_0. The points are from the molecular dynamics calculations of W.E. Alley, B.J. Alder, and S. Yip, *Phys. Rev.* **A27**, 3174 (1983) and the line was calculated from hydrodynamic fluctuation theory by J. Keizer and M. Medina-Noyola, *Physica* **115A**, 301 (1982).

An example of the sort of spectra that are obtained from quasi-elastic light scattering experiments is shown in Fig. 6.5. There the frequency dependence of light scattered from liquid argon is shown. The incident wavelength is 5145 Å and the scattering angle is $\chi = 90°14'$. The argon was at equilibrium with a thermal reservoir at 84.97 K. The central peak at $\omega = 0$ is called the *Rayleigh peak* and the smaller side bands are called the *Brillouin peaks*. In this spectrum they are located at $\omega = \pm 2.883 \times 10^9$ Hz and have a shape that is approximately Lorentzian. Light scattering spectra with similar features are also found for molecules like CCl_4 that have internal degrees of freedom.

Neutron scattering spectra tend to have fewer features than light scattering spectra because the wavelength of neutrons is of molecular dimensions. Recent molecular dynamics experiments, however, have been used to investigate wave vectors somewhat larger than are accessible by thermal neutrons. Figure 6.6

gives an example of the scattering function, $S(k, \omega)$, taken from molecular dynamics calculations by Alley, Alder, and Yip on hard spheres. The density is one-third the close-packing density and the magnitude of the scattering vector is $k = 0.616/R$, where R is the hard-sphere diameter. Notice that a central Rayleigh peak still appears at these short wavelengths, but that the Brillouin peak has moved closer to the center. As only positive frequencies are shown, the second Brillouin peak does not appear. Both this spectrum and the spectrum for argon in Fig. 6.5. can be explained on the basis of hydrodynamic fluctuation theory.

6.8. Light Scattering in a Thermal Gradient

As an example of how to use hydrodynamic fluctuation theory, we examine the problem of light scattering from a fluid with a thermal gradient. The idea of the experiment is a simple extension of the argon scattering experiment discussed in the previous section. Instead of equilibrating the liquid with a thermal reservoir, two researvoirs at temperatures T_1 and T_2 are arranged at opposite ends of the vessel containing the liquid. As heat flows from the high-temperature to the low-temperature reservoir a temperature gradient is set up. In the absence of convection, the gradient is determined by the steady-state condition based on Fourier's law, i.e.,

$$\nabla \cdot \bar{\kappa} \nabla \bar{T} = 0. \tag{6.8.1}$$

If the temperature dependence of the thermal conductivity is ignored, then solving Eq. (6.8.1) using the boundary conditions $\bar{T} = T_1$, $\bar{T} = T_2$ at the ends of the vessel leads to a constant temperature gradient of magnitude

$$G = |\nabla \bar{T}| = |T_1 - T_2|/L, \tag{6.8.2}$$

where L is the length of the vessel. An ensemble of such systems represents a nonequilibrium steady state. The light scattering spectrum in this state is affected by the flux of energy through the systems in the ensemble. The magnitude of the flux is

$$f = \bar{\kappa}|T_1 - T_2|/L,$$

which is a measure of the distance the ensemble is away from equilibrium.

According to Eq. (6.7.20) the light scattering spectrum is proportional to the Fourier transform of the density–density correlation function. This correlation function can be calculated using the hydrodynamic fluctuation theory developed in Section 6.5. For an ensemble with the steady temperature gradient given by Eq. (6.8.2), the equations satisfied by the density, velocity, and internal energy fields are given by Eqs. (6.5.1), (6.5.2), and (6.5.3). At the steady state all these fields are time independent. To obtain the fixed temperature gradient we have already assumed that there is no convection, i.e., $\mathbf{v}^{aa} = 0$. This follows from Eq. (6.5.35) since we have utilized Eq. (6.8.1) to determine the size of the steady gradient. Setting the right-hand side of Eq. (6.5.3) equal to zero provides the final condition characterizing the steady state, namely,

$$\nabla \bar{p}^e = \bar{\rho}\mathbf{F}, \tag{6.8.3}$$

where p^e is the pressure and $\mathbf{F}$ the external field. Thus, neglecting gravity, the pressure will be constant. The effect of temperature on the equation of state leads to an average density gradient in a system of magnitude

$$\begin{aligned} |\nabla \bar{\rho}| &= (\partial \bar{\rho}/\partial \bar{T})_p |\nabla T| \\ &= (\partial \bar{\rho}/\partial \bar{T})_p |T_1 - T_2|/L. \end{aligned} \tag{6.8.4}$$

Thus the relative magnitude of the gradient, i.e., $|\nabla \bar{\rho}/\bar{\rho}|$, is proportional to $(\partial \bar{\rho}/\partial \bar{T})_p/\bar{\rho}$, which is the thermal expansivity. As thermal expansivities are the order of $10^{-3}\ \mathrm{K}^{-1}$ for normal liquids away from their critical points, it is a good approximation to take the density at steady state to be a constant, ρ^0. Thus in the steady state ensemble one has, on the average, $\mathbf{v}^{ss} = \mathbf{0}$ and $\rho^{ss} = \rho^0$. The temperature profile is linear and if the high-temperature reservoir (T_1) is located at $x = -L/2$ and the low-temperature reservoir (T_2) at $x = +L/2$, then

$$T^{ss}(\mathbf{r}) = -Gx + T_2 + GL/2. \tag{6.8.5}$$

The average internal energy density is $e^{ss}(\mathbf{r}) = e(\rho^0, T^{ss}(\mathbf{r}))$.

Fluctuations around these steady-state values can be determined using Eqs. (6.5.4), (6.5.5), (6.5.6), and (6.5.36). Linearizing around the values $\bar{\rho} = \rho^0$, $\bar{\mathbf{v}} = \mathbf{v}^{ss} = 0$, $\bar{T} = T^{ss}$, and $\bar{e} = e^{ss}$ leads to

$$\partial \delta\rho/\partial t = -\nabla \cdot \delta\mathbf{p} \tag{6.8.6}$$

$$\partial \delta e/\partial t = -h^{ss}\nabla \cdot \delta\mathbf{v} + \delta v_x c_v^{ss} G + \bar{\kappa}\nabla^2 \delta T + \tilde{f}_e \tag{6.8.7}$$

$$\partial \delta\mathbf{p}/\partial t = -\nabla \delta p^e + 2\bar{\eta}\nabla \cdot \delta\mathring{e} + \bar{\zeta}\nabla^2 \delta\mathbf{v} + \tilde{\mathbf{f}}_m, \tag{6.8.8}$$

where $c_v^{ss} = (\partial e^{ss}/\partial T^{ss})_\rho$ is the constant-volume heat capacity. The covariance of the random terms is given in Eqs. (6.5.13) and (6.5.38)–(6.5.40), which for the special case of a constant thermal gradient are

$$\begin{aligned} \langle \tilde{f}_e(\mathbf{r}, t)\tilde{f}_e(\mathbf{r}', t')\rangle = &-2k_B\bar{\kappa}T^{ss2}\frac{\partial^2}{\partial x_j^2}\delta(\mathbf{r} - \mathbf{r}')\delta(t - t') \\ &+ 4k_B\bar{\kappa}T^{ss}G\frac{\partial}{\partial x}\delta(\mathbf{r} - \mathbf{r}')\delta(t - t') \end{aligned} \tag{6.8.9}$$

$$\langle \tilde{f}_e(\mathbf{r}, t)\tilde{f}_{ml}(\mathbf{r}', t')\rangle = \langle \tilde{f}_{ml}(\mathbf{r}, t)\tilde{f}_e(\mathbf{r}', t')\rangle = 0 \tag{6.8.10}$$

$$\langle \tilde{f}_{mi}(\mathbf{r},t)\tilde{f}_{ml}(\mathbf{r}',t')\rangle = -2k_B T^{ss}[\bar{\eta}\Delta_{il,kj} + \bar{\zeta}\delta_{ij}\delta_{kl}]\frac{\partial^2}{\partial x_j \partial x_k}\delta(\mathbf{r}-\mathbf{r}')\delta(t-t')$$
$$+ 2k_B G[\bar{\eta}\Delta_{il,k1} + \bar{\zeta}\delta_{i1}\delta_{kl}]\frac{\partial}{\partial x_k}\delta(\mathbf{r}-\mathbf{r}')\delta(t-t'), \tag{6.8.11}$$

where in the final expression $x_1 = x$. The terms that depend on the existence of the gradient are the second terms on the right-hand side of Eqs. (6.8.7), (6.8.9), and (6.8.11). The remaining terms involve contributions that are present in an equilibrium ensemble where G vanishes.

The fluctuation equations (6.8.6)–(6.8.8) are strongly coupled. They can be solved analytically for the structure factor, however, in the special case that the thermodynamic derivative $(\partial\bar{\rho}/\partial T)_p$ vanishes. In this special case, which occurs for water a few degrees above its melting temperature, the fluctuation in the pressure can be written

$$\delta p^e = (\partial\bar{p}/\partial\bar{\rho})_e \delta\rho. \tag{6.8.12}$$

This follows from the fact that

$$\delta p^e = (\partial\bar{p}/\partial\bar{e})_\rho \delta e + (\partial\bar{p}/\partial\bar{\rho})_e \delta\rho \tag{6.8.13}$$

and the standard thermodynamic identities

$$(\partial\bar{p}/\partial\bar{e})_\rho = -(\partial\bar{p}/\partial\bar{\rho})_e(\partial\bar{\rho}/\partial\bar{e})_p$$
$$= -(\partial\bar{p}/\partial\bar{\rho})_e(\partial\bar{\rho}/\partial\bar{T})_p(\partial\bar{T}/\partial\bar{e})_p, \tag{6.8.14}$$

which show that if $(\partial\bar{\rho}/\partial\bar{T})_p$ vanishes, then the first term in Eq. (6.8.13) also vanishes. Using Eq. (6.8.12), Eqs. (6.8.6) and (6.8.8) can be written

$$\partial\delta\rho/\partial t = -\nabla\cdot\delta\mathbf{p} \tag{6.8.15}$$

$$\partial\delta\mathbf{p}/\partial t = -(\partial\bar{p}/\partial\bar{\rho})_{\bar{e}}\nabla\delta\rho + 2\bar{\eta}\nabla\cdot\delta\mathring{e} + \bar{\zeta}\nabla^2\delta\mathbf{v} + \tilde{\mathbf{f}}_m. \tag{6.8.16}$$

Since $\delta\mathring{e}$ depends only on the velocities and $\delta\mathbf{v} = \delta\mathbf{p}/\rho^0$, these equations are uncoupled from the internal energy equation (6.8.7).

To make further progress in obtaining the structure factor, we need to Fourier transform Eqs. (6.8.15) and (6.8.16), as well as Eq. (6.8.11), with respect to space and time. The time transformation is straightforward, whereas the spatial transformation requires transforming the term $\nabla\cdot\delta\mathring{e}$, which is a bit tricky. To do this, it is helpful to use the identity

$$2(\nabla\cdot\delta\mathring{e})_l = \frac{\partial}{\partial x_i}\frac{\partial}{\partial x_k}\Delta_{ik,lj}\delta v_j, \tag{6.8.17}$$

which can be verified using the definitions of the tensor Δ in Eq. (6.5.9) and the fact that

$$\mathring{e}_{il} = \frac{1}{2}\left(\frac{\partial v_i}{\partial x_l} + \frac{\partial v_l}{\partial x_i}\right) - \frac{1}{3}\frac{\partial v_k}{\partial x_k}\delta_{il}. \tag{6.8.18}$$

FIGURE 6.7. A vector diagram illustrating the decomposition of the momentum fluctuation vector, $\delta\hat{\mathbf{p}}$, into its longitudinal, $\delta\hat{\mathbf{p}}_l$, and transverse, $\delta\hat{\mathbf{p}}_t$, components. The longitudinal component lies along the wave vector $\mathbf{k}$, while the transverse component is in a plane perpendicular to $\mathbf{k}$.

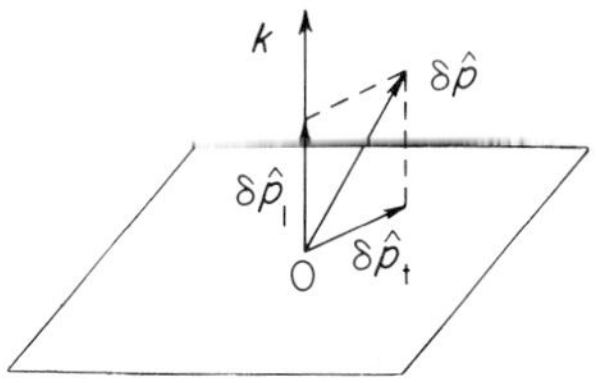

Using the correspondence $\partial/\partial x_l \rightarrow -ik_l$, the spatial Fourier transform of Eq. (6.8.17) is

$$(\nabla \cdot \delta\hat{e})_l = -k_i k_k \Delta_{ik,lj} \delta\hat{v}_j$$
$$= -k^2 \delta\hat{v}_l - k_l \mathbf{k} \cdot \delta\hat{\mathbf{v}}, \tag{6.8.19}$$

where in the final equality the definition of Δ in Eq. (6.5.9) was used. This result allows us to take the spatial Fourier transform of Eqs. (6.8.15) and (6.8.16) in a straightforward manner. One finds that

$$\partial\delta\hat{\rho}/\partial t = i\mathbf{k} \cdot \delta\hat{\mathbf{p}}$$
$$\partial\delta\hat{\mathbf{p}}/\partial t = i(\partial\bar{p}/\partial\bar{\rho})_e \mathbf{k}\delta\hat{\rho} - (\bar{\eta}/\rho^0)\left[k^2\delta\hat{\mathbf{p}} + \frac{1}{3}\mathbf{k}\mathbf{k}^T \cdot \delta\hat{\mathbf{p}}\right] - (\bar{\zeta}/\rho^0)k^2\delta\hat{\mathbf{p}} + \hat{\tilde{\mathbf{f}}}_m. \tag{6.8.20}$$

The problem of calculating the density–density correlation function simplifies even further if we introduce the longitudinal and transverse components of the momentum fluctuations. These are the components of $\delta\hat{\mathbf{p}}$ which are in the direction of the wave vector $\mathbf{k}$ or in a plane perpendicular to it, as illustrated in Fig. 6.7. Thus the *longitudinal* momentum fluctuations is

$$\delta\hat{\mathbf{p}}_l = (\mathbf{k}/k) \cdot \delta\hat{\mathbf{p}}(\mathbf{k}/k) \equiv \delta\hat{p}_l \mathbf{e}_\parallel, \tag{6.8.21}$$

where $\mathbf{e}_\parallel = \mathbf{k}/k$ is the unit vector parallel to $\mathbf{k}$, and the *transverse* momentum fluctuation is

$$\delta\hat{\mathbf{p}}_t = |(\mathbf{k}/k) \times \delta\hat{\mathbf{p}}|\,\mathbf{e}_\perp, \tag{6.8.22}$$

where $\mathbf{e}_\perp$ is a unit vector in the plane perpendicular to $\mathbf{k}$ with direction such that

$$\delta\hat{\mathbf{p}} = \delta\hat{\mathbf{p}}_l + \delta\hat{\mathbf{p}}_t. \tag{6.8.23}$$

Taking the dot product of $\mathbf{k}$ on both sides of the second of Eqs. (6.8.20) and using Eq. (6.8.21), we find that

$$\partial\delta\hat{\rho}/\partial t = ik\delta\hat{p}_l \tag{6.8.24}$$

$$\partial\delta\hat{p}_l/\partial t = i(\partial\bar{p}/\partial\bar{\rho})_e k\delta\hat{\rho} - k^2\bar{D}\delta\hat{p}_l + (\mathbf{k}/k)\cdot\tilde{\mathbf{f}}_m, \tag{6.8.25}$$

where $D \equiv 4\bar{\eta}/3\rho^0 + \bar{\zeta}/\rho^0$. Thus by introducing the longitudinal component of the momentum density, the fluctuation problem has been reduced to one involving two variables, the density and the longitudinal momentum.

The first term in Eq. (6.8.25) corresponds to the coupling between the density and momentum caused by sound waves. This can be seen by taking the second partial time derivative of Eq. (6.8.24) and substituting Eq. (6.8.25) for the right-hand side. Ignoring, for the time being, the other terms gives

$$\partial^2 \delta\hat{\rho}/\partial t^2 = -(\partial \bar{p}/\partial \bar{\rho})_e k^2 \delta\hat{\rho}. \tag{6.8.26}$$

We recognize this as the Fourier-transformed version of the wave equation,

$$\partial^2 \delta\rho/\partial t = c^2 \nabla^2 \delta\rho, \tag{6.8.27}$$

where the speed of the wave is

$$c = (\partial \bar{p}/\partial \bar{\rho})_e^{1/2}. \tag{6.8.28}$$

Equation (6.8.27) describes a density wave that represents the propagation of sound. Thus c is the sound speed.

The scattering function is related to the density–density correlation function, which can be obtained from the power spectrum of the stochastic process in Eqs. (6.8.24) and (6.8.25). Indeed, the power spectrum is related to the time Fourier transform of the time correlation function, as Eq. (1.8.33) shows. More explicitly, we want to find $\hat{s}$, defined by

$$\hat{s}(\mathbf{k}, \mathbf{k}', \omega)\delta(\omega + \omega')/2\pi \equiv \langle \delta\mathbf{a}(\mathbf{k}, \omega)\delta\mathbf{a}^T(\mathbf{k}', \omega')\rangle, \tag{6.8.29}$$

where $\delta a_1 = \delta\rho$ and $\delta a_2 = \delta p_l$. According to Eqs. (1.8.41), $\hat{s}$ is the power spectrum and satisfies Eq. (1.8.36), i.e.,

$$(i\omega + \hat{H})^{-1}\hat{\gamma}(-i\omega + \hat{H}^T)^{-1} = \hat{s}. \tag{6.8.30}$$

Remembering that the streaming terms are *anti*symmetric, the relaxation matrices $\hat{H}$ and $\hat{H}^T$ can be read off Eqs. (6.8.24) and (6.8.25):

$$\hat{H} = \begin{pmatrix} 0 & ik \\ ic^2 k & -k^2 \bar{D} \end{pmatrix}, \qquad \hat{H}^T = \begin{pmatrix} 0 & -ic^2 k \\ -ik & -k^2 \bar{D} \end{pmatrix} \tag{6.8.31}$$

while the matrix $\hat{\gamma}$ has only a single nonzero element arising from the term $\mathbf{k} \cdot \tilde{\mathbf{f}}_m$ in Eq. (6.8.25). Indeed,

$$\hat{\gamma}(k, k') = (1/kk')\begin{pmatrix} 0 & 0 \\ 0 & k_i \hat{\gamma}_{il} k_l' \end{pmatrix} \tag{6.8.32}$$

where the summation convention on repreated indices is used and $\hat{\gamma}_{il}$ can be found using Eq. (6.8.11). That equation contains two terms, the first of which depends explicitly on space through the temperature profile, $T^{ss}(x) = -Gx + T_2 + GL/2$. A convenient way of dealing with this term is to substitute the steady-state profile

$$T^{ss}(x) = \left(T_2 + \frac{GL}{2}\right) - (G/q)\sin qx, \tag{6.8.33}$$

which in the limit that $q \to 0$ becomes identical to the true profile. This facilitates Fourier transforming Eq. (6.8.11). Using Eq. (6.8.32), the analogue of Eq. (6.8.11) becomes

$$\begin{aligned}\langle \tilde{f}_{mi}(\mathbf{r}, t)\tilde{f}_{ml}(\mathbf{r}', t')\rangle = &- 2k_B T^0[\bar{\eta}\Delta_{il,kj} + \bar{\zeta}\delta_{ij}\delta_{kl}]\frac{\partial^2}{\partial x_j \partial x_k}\delta(\mathbf{r} - \mathbf{r}')\delta(t - t')\\ &- ik_B(G/q)\frac{\partial}{\partial x_j}(e^{iqx} - e^{-iqx})[\bar{\eta}\Delta_{il,kj} + \bar{\zeta}\delta_{ij}\delta_{kl}]\frac{\partial}{\partial x_k}\delta(\mathbf{r} - \mathbf{r}')\delta(t - t'),\end{aligned} \tag{6.8.34}$$

where $T^0 \equiv T_2 + GL/2$.

The Fourier transform of Eq. (6.8.34) is required for the calculation of the correlation function $\hat{s}$ in Eq. (6.8.30). The transform of the first term in Eq. (6.8.34) can be carried out by inspection and gives for the wave-vector-dependent factor

$$2k_B T^0 k_j k_k'[\bar{\eta}\Delta_{il,kj} + \bar{\zeta}\delta_{ij}\delta_{kl}]\delta(\mathbf{k} + \mathbf{k}'). \tag{6.8.35}$$

This is the part of $\hat{\gamma}_{il}$ that does not depend on the gradient and its contribution to $\hat{\gamma}$ in Eq. (6.8.32) is

$$2k_B T^0(k_i/k)k_j k_k'(k_l'/k')[\bar{\eta}\Delta_{il,kj} + \bar{\zeta}\delta_{ij}\delta_{kl}]\delta(\mathbf{k} + \mathbf{k}'). \tag{6.8.36}$$

It is easy to evaluate the repeated sums in Eq. (6.8.36) using the definition of Δ in Eq. (6.5.9). This gives

$$2k_B T^0(1/kk')[2\bar{\eta}(\mathbf{k}\cdot\mathbf{k}')^2 + (\bar{\zeta} - 2\bar{\eta}/3)k^2 k'^2]\delta(\mathbf{k} + \mathbf{k}'). \tag{6.8.37}$$

To Fourier transform the second term in Eq. (6.8.34) we need to use the fact that the Fourier transform of

$$\frac{\partial}{\partial x_j} f(\mathbf{r})\frac{\partial}{\partial x_k}\delta(\mathbf{r} - \mathbf{r}') \tag{6.8.38}$$

is

$$k_j k_k' \hat{f}(\mathbf{k} + \mathbf{k}')/(2\pi)^3, \tag{6.8.39}$$

a fact which can be proved by writing the delta function in Eq. (6.8.38) as

$$\delta(\mathbf{r} - \mathbf{r}') = \frac{1}{(2\pi)^3}\int e^{-i\mathbf{k}\cdot(\mathbf{r}-\mathbf{r}')}\,d\mathbf{k}$$

and expressing f in terms of its Fourier transform, $\hat{f}$. Applying this identity to the second term in Eq. (6.8.34) then gives for the gradient-dependent contribution to $\hat{\gamma}_{il}$

$$-ik_B(G/q)k_j' k_k[\bar{\eta}\Delta_{il,kj} + \bar{\zeta}\delta_{ij}\delta_{kl}]\{\delta(\mathbf{k} + \mathbf{k}' + \mathbf{q}) - \delta(\mathbf{k} + \mathbf{k}' - \mathbf{q})\}, \tag{6.8.40}$$

where $\mathbf{q} = q\mathbf{i}$. Multiplying this by $k_i k_l'/kk'$, summing as indicated in the definition of $\hat{\gamma}$ in Eq. (6.8.32), and adding in Eq. (6.8.37) finally yields

$$\hat{\gamma}_{22}(\mathbf{k}, \mathbf{k}') = k_B[2T^0\delta(\mathbf{k} + \mathbf{k}') - i(G/q)\{\delta(\mathbf{k} + \mathbf{k}' + \mathbf{q}) - \delta(\mathbf{k} + \mathbf{k}' - \mathbf{q})\}] \times (1/kk')[2\bar{\eta}(\mathbf{k}\cdot\mathbf{k}')^2 + (\bar{\zeta} - 2\bar{\eta}/3)k^2k'^2]. \tag{6.8.41}$$

Before completing the analysis of the fluctuation spectrum for the temperature gradient, it is instructive to examine the situation at equilibrium where the gradient vanishes. In that case Eq. (6.8.41) reduces to

$$\hat{\gamma}_{22}(\mathbf{k}, \mathbf{k}') = (k_i/k)\hat{\gamma}_{il}(k_l'/k') = 2k_B\rho^0\bar{D}T^ek^2\delta(\mathbf{k} + \mathbf{k}')/(2\pi)^3. \tag{6.8.42}$$

To obtain the density–density correlation function we need to take the $1-1$ element of the matrix $\hat{s}$ given by Eq. (6.8.30). Thus

$$\hat{s}_{11}(\mathbf{k}, \mathbf{k}', \omega) = (i\omega + \hat{H})^{-1}_{12}\hat{\gamma}_{22}(-i\omega + \hat{H}^T)^{-1}_{21}, \tag{6.8.43}$$

since Eq. (6.8.32) shows that the only nonvanishing element of $\hat{\gamma}$ is $\hat{\gamma}_{22}$. To obtain the inverse of the matrices in Eq. (6.8.43), we use the formula

$$A^{-1} = \begin{pmatrix} A_{22} & -A_{12} \\ -A_{21} & A_{11} \end{pmatrix}(\det A)^{-1}, \tag{6.8.44}$$

a fact that is easily verified by matrix multiplication. Combining this result with the expression for $\hat{H}$ in Eq. (6.8.31) gives

$$\begin{aligned} (i\omega + \hat{H})^{-1}_{12} &= -ik/[(c^2k^2 - \omega^2) - i\omega k^2\bar{D}] \\ (-i\omega + \hat{H}^T)^{-1}_{21} &= ik'/[(c^2k^2 - \omega^2) + i\omega k^2\bar{D}]. \end{aligned} \tag{6.8.45}$$

Equations (6.8.42) and (6.8.45) give all the factors needed to evaluate $\hat{s}_{11}$ in Eq. (6.8.43) and yield

$$\hat{s}_{11}(\mathbf{k}, \mathbf{k}', \omega) = \frac{2k_B\rho^0\bar{D}T^ek^4\delta(\mathbf{k} + \mathbf{k}')/(2\pi)^3}{(\omega^2 - c^2k^2)^2 + \omega^2k^4\bar{D}^2}. \tag{6.8.46}$$

For frequencies close to $\omega = \pm ck$ and for small values of k, the expression in Eq. (6.8.46) can be written approximately as the sum of two Lorentzians. Indeed, it can be verified using straightforward algebra that for small k and $\omega \simeq \pm ck$

$$\frac{4c^2k^2}{(\omega^2 - c^2k^2)^2 + \omega^2k^4\bar{D}^2} \simeq \frac{1}{(\omega + ck)^2 + (k^2\bar{D}/2)^2} + \frac{1}{(\omega - ck)^2 + (k^2\bar{D}/2)^2}. \tag{6.8.47}$$

This identity combined with Eqs. (6.7.17), (6.8.29), and (6.8.46) allow us to write the scattering function approximately as

$$S(k, \omega) = k_B\rho^0(\bar{D}/2c^2)T^ek^2\left[\frac{1}{(\omega + ck)^2 + (k^2\bar{D}/2)^2} + \frac{1}{(\omega - ck)^2 + (k^2\bar{D}/2)^2}\right]. \tag{6.8.48}$$

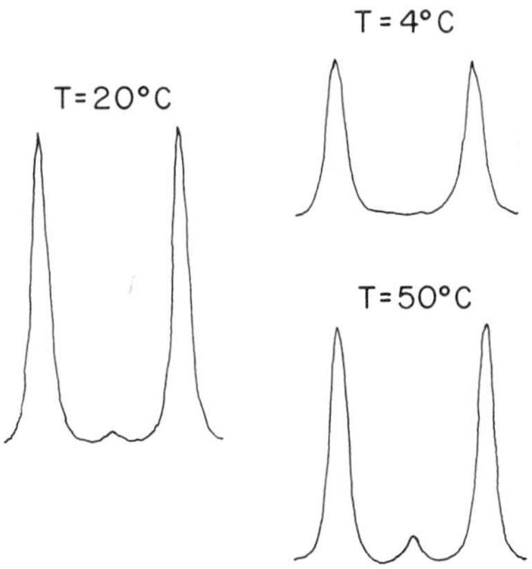

FIGURE 6.8. The light scattering spectrum of water at three different temperatures. Notice the absence of the Rayleigh peak at 4°C, where the isothermal compressibility of water vanishes. Data taken form C.L. O'Connor and J.P. Schlupf, *J. Chem. Phys.* **47**, 31 (1967).

In this form the spectrum is seen to have two Lorentzian peaks [cf. Eq. (1.8.38)] centered at the frequencies $\omega = \pm ck$. These are the Brillouin peaks and are due to the propagating sound modes in the fluid. The half-width at half-maximum of the peaks is given by $\omega_{1/2} = k^2\bar{D}/2$. These peaks are easily identified in the experimental argon spectrum in Fig. 6.5. The central Rayleigh peak present in the argon spectrum is missing from Eq. (6.8.48). A more complete calculation shows that its amplitude is proportional to the isothermal compressibility, $(\partial\bar{p}/\partial\bar{\rho})_T$, which we have taken to be zero and which is why the Rayleigh peak is missing from our calculation. Figure 6.8 shows an experimental light scattering spectrum for water near 4°C, where the isothermal compressibility is close to zero. There one sees only the Brillouin peaks, as predicted by Eq. (6.8.48). At 20°C and 50°C, where the isothermal compressibility is no longer zero, the central Rayleigh peak, though small, is plainly visible.

Returning to the problem of light scattering in a temperature gradient, we expect that the spectrum should resemble that in Eq. (6.8.48) with some modifications which depend on the gradient, G. In the presence of the sinusoidal temperature profile in Eq. (6.8.33) the experimental light scattering spectrum involves two additive contributions from the wave vector $\mathbf{q}$. Indeed, by taking the finite size of the scattering volume into account one deduces (see Tremblay, Siggia, and Arai in the list of references at the end of this chapter) that

$$S(\mathbf{k},\omega)\delta(\mathbf{k}+\mathbf{k}')/(2\pi)^3 = \lim_{q\to 0}\,[s_{11}(\mathbf{k},\mathbf{k}',\omega) + s_{11}(\mathbf{k}-\mathbf{q}/2,\mathbf{k}'-\mathbf{q}/2,\omega) + s_{11}(\mathbf{k}-\mathbf{q}/2,\mathbf{k}'-\mathbf{q}/2,\omega)]. \tag{6.8.49}$$

The procedure for solving Eq. (6.8.30) for $s_{11}(\mathbf{k},\mathbf{k}',\omega)$ is the same as when the gradient vanishes, that is, Eq. (6.8.43) still holds. Now, however, Eq. (6.8.41) must be used for $\hat{\gamma}_{22}$. Thus it follows that

$$s_{11}(\mathbf{k}, \mathbf{k}', \omega) = k_B[2T^0\delta(\mathbf{k} + \mathbf{k}') - i(G/q)\{\delta(\mathbf{k} + \mathbf{k}' - \mathbf{q}) - \delta(\mathbf{k} + \mathbf{k}' + \mathbf{q}\}]$$
$$\times [2\bar{\eta}(\mathbf{k}\cdot\mathbf{k}')^2 + (\bar{\zeta} - 2\bar{\eta}/3)k^2k'^2]$$
$$\times [c^2k^2 - \omega^2 - i\omega k^2\bar{D}]^{-1}[c^2k^2 - \omega^2 + i\omega k^2\bar{D}]^{-1}. \tag{6.8.50}$$

Substitution of this expression in Eq. (6.8.49) to obtain $S(\mathbf{k}, \omega)$ is straightforward although the algebraic manipulations in taking the limit $q \to 0$ are tedious and require using the identity

$$\lim_{q\to 0} f(x)\left[\frac{\delta(x - q) - \delta(x + q)}{q}\right] = 2f'(x)\delta(x). \tag{6.8.51}$$

After a good deal of algebra one obtains

$$S(\mathbf{k}, \omega) = k_B T^0 \rho^0 k^2 \left\{\frac{2\bar{D}k^2}{(\omega^2 - c^2k^2)^2 + (\omega\bar{D}k^2)^2} - \frac{4\omega^2(\omega\bar{D}k^2)\bar{D}k_xG/T^0}{[(\omega^2 - c^2k^2)^2 + (\omega\bar{D}k^2)^2]^2}\right\}. \tag{6.8.52}$$

Finally, if it is assumed that the time required for a sound wave to traverse the length of the container, i.e., $t_c \equiv T^0/Gc$, is long compared to the time it takes momentum to diffuse over a wavelength k^{-1}, i.e., $t_m = 1/\bar{D}k^2$, then Eq. (6.8.52) can be simplified further. Indeed, following Tremblay and colleagues one finds, as at equilibrium, two Brillouin peaks, but modified because of the flow of heat along the gradient:

$$S(\mathbf{k}, \omega) = (k_B T^0 \rho^0/c^2)\left[\frac{(\bar{D}k^2/2)[1 - \varepsilon(\mathbf{k}, \omega)]}{(\omega - ck)^2 + \omega(\bar{D}k^2/2)^2} + \frac{(\bar{D}k^2/2)[1 + \varepsilon(\mathbf{k}, \omega)]}{(\omega + ck)^2 + (\bar{D}k^2/2)^2}\right] \tag{6.8.53}$$

with

$$\varepsilon(\mathbf{k}, \omega) \equiv (c/\bar{D}k^2)(k_xG/T^0)\left[\frac{2(\omega\bar{D}k^2)^2}{(\omega^2 - c^2k^2)^2 + (\omega\bar{D}k^2)^2}\right]. \tag{6.8.54}$$

The Brillouin peaks in the presence of a temperature gradient are asymmetric. The peak that corresponds to sound absorption, i.e., $\omega = -ck$, has a larger amplitude if $k_x > 0$, that is, if the scattering vector $\mathbf{k}$ has a positive component in the direction of the heat flux. This asymmetry is reversed if the scattering vector opposes the flux of heat. No effect is predicted if $\mathbf{k}$ is perpendicular to the gradient. These qualitative features of the spectrum have been observed in the light scattering experiments of Beysens and coworkers using samples of H_2O with $T^0 \simeq 4°C$. Spectra which show the dependence of the asymmetry of the peaks on the orientation of the scattering vector are shown in Fig. 6.9. Actual experimental data at a scattered wave vector of $k = 2110\ \text{cm}^{-1}$ are illustrated in the top panel. The large Rayleigh peak, which

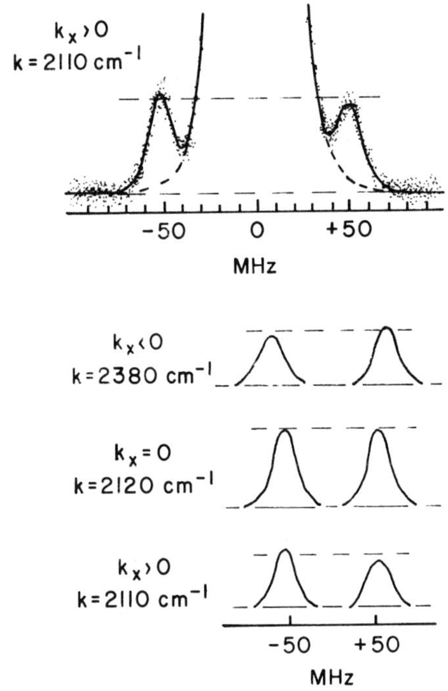

FIGURE 6.9. The light scattering spectrum of water in a temperature gradient of 59 K cm^{-1}, taken from D. Beysens, Y. Garrabos, and G. Zalczer, *Phys. Rev. Letters* **45**, 403 (1980). The upper panel shows the measured spectrum at the indicated wave vector in the direction of the gradient. The central peak is due to scattering by the windows of the apparatus and dust particles and has been subtracted out in the lower panels. The asymmetry of the Brillouin peaks when the wave vector is either along ($k_x > 0$), opposed ($k_x < 0$), or perpendicular to ($k_x = 0$) the gradient agrees qualitatively with Eq. (6.8.53).

is subtracted out in the lower spectra, is due to dust particles and scattering from the windows of the scattering chamber. These experiments involve a thin sample (0.5 cm), a temperature gradient $G = 59$ K cm^{-1}, $c = 1.4 \times 10^5$ cm s^{-1}, and $\bar{D} = 1 \times 10^{-2}$ cm^2 s^{-1}. Thus the assumption that $T^0/Gc \gg 1/\bar{D}k^2$ that allowed Eq. (6.8.53) to be deduced from Eq. (6.8.52) is not valid. The presence of the walls, which was not taken into account in the above calculation, is important in the experiment and partially explains why Eq. (6.8.52) fits the data only qualitatively.

6.9. Local versus Nonlocal Fluctuations

In this chapter the hydrodynamic level of the canonical theory has been applied to problems involving diffusion, diffusion coupled to chemical reaction, and heat and momentum transport. In these applications we have tacitly assumed that fluctuations in the intensive variables are spatially local. For example, in deriving Eq. (6.2.18), which describes the density fluctuations of a solute in a two-component solution, we linearized the chemical potential in Eq. (6.2.2) in the following way:

$$\delta\mu_1(\mathbf{r}, t) = (\partial\bar{\mu}_1/\partial\bar{\rho}_1(\mathbf{r}, t))_T \delta\rho_1(\mathbf{r}, t). \tag{6.9.1}$$

In other words, fluctuations in the chemical potential at position $\mathbf{r}$ were assumed to depend only on fluctuations in the density at the same position. This reflects the fact that we have implicitly assumed a *local* description of the thermodynamic quantities.

In spatially distributed systems, such as encountered in the hydrodynamic description, the entropy is no longer simply a function of the densities of the extensive variables. It is, instead, a *functional.* For example, the entropy, S, of a single-component system should be written

$$S = \int s(\rho, \varepsilon, \mathbf{p})\, d\mathbf{r}, \tag{6.9.2}$$

where ρ is the mass density, $\varepsilon = e + p^2/2\rho + \rho\phi$ is the total energy density, and $\mathbf{p}$ is the momentum density. As in Section 2.4, s denotes the entropy density. Equation (6.9.2) makes it clear that a change in the distribution of ρ, ε, or $\mathbf{p}$ over the volume will affect the entropy, even if the total mass, energy, and momentum are unchanged. Thus S is a functional. Its functional derivatives are determined by the functional variation of S,

$$\delta S = \int [(\partial s/\partial \rho)_{\varepsilon,\mathbf{p}}\delta\rho + (\partial s/\partial \varepsilon)_{\rho,\mathbf{p}}\delta\varepsilon + (\partial s/\partial p_i)_{\rho,\varepsilon}\delta p_i]\, d\mathbf{r}. \tag{6.9.3}$$

Recalling the identities in Eqs. (2.4.33), we can write the first *functional derivatives* of S as

$$\begin{aligned} (\delta s/\delta \rho(\mathbf{r}))_{\varepsilon,\mathbf{p}} &= -[\mu(\mathbf{r}) + \phi(\mathbf{r})]/T(\mathbf{r}) + v^2(\mathbf{r})/T(\mathbf{r}) \\ (\delta s/\delta \varepsilon(\mathbf{r}))_{\rho,\mathbf{p}} &= 1/T(\mathbf{r}) \\ (\delta s/\delta p_i(\mathbf{r}))_{\rho,\varepsilon,p_j} &= -v_i(\mathbf{r})/T(\mathbf{r}). \end{aligned} \tag{6.9.4}$$

The intensive variables on the right-hand side of Eqs. (6.9.4) have their customary meaning. However, in a system where the extensive variables vary from position to position, the intensive variables—which depend explicitly on ρ, ε, and $\mathbf{p}$—are also functionals. Indeed, by analogy by Eq. (6.9.3), a variation in $1/T(\mathbf{r})$ takes the form

$$\delta T^{-1}(\mathbf{r}) = \int [S_{21}(\mathbf{r},\mathbf{r}')\delta\rho(\mathbf{r}') + S_{22}(\mathbf{r},\mathbf{r}')\delta\varepsilon(\mathbf{r}')]\, d\mathbf{r}' \tag{6.9.5}$$

where

$$S_{ij}(\mathbf{r},\mathbf{r}') \equiv \delta^2 S/\delta n_i(\mathbf{r})\delta n_j(\mathbf{r}') \tag{6.9.6}$$

with $n_i = \rho$ or ε. In a similar fashion, using Eq. (6.9.4), a variation in $\delta S/\delta\rho(\mathbf{r})$ can be written

$$\delta[\delta S/\delta\rho(\mathbf{r})] = -\delta(\mu/T)(\mathbf{r}) + 2\mathbf{v}(\mathbf{r})\cdot\delta\mathbf{v}(\mathbf{r})T^{-1}(\mathbf{r}) + \{v^2(\mathbf{r}) + \phi(\mathbf{r})\}\delta T^{-1}(\mathbf{r}). \tag{6.9.7}$$

The first term in this expression involves the variation of the chemical potential, which like T^{-1} is independent of δp_i. Thus

$$-\delta(\mu/T)(\mathbf{r}) = \int [S_{11}(\mathbf{r},\mathbf{r}')\delta\rho(\mathbf{r}') + S_{12}(\mathbf{r},\mathbf{r}')\delta\varepsilon(\mathbf{r}')]\,d\mathbf{r}'. \tag{6.9.8}$$

Since S is the local equilibrium entropy, these nonlocal expressions for variations in the intensive variables can be evaluated using equilibrium statistical mechanics. All variations of the intensive variables in Eqs. (6.9.4) can be expressed in terms of the variations of $v_i(\mathbf{r})$, $T^{-1}(\mathbf{r})$, and $\mu/T(\mathbf{r})$, and the symmetric matrix

$$S^{(2)}(\mathbf{r},\mathbf{r}') = \begin{pmatrix} S_{11}(\mathbf{r},\mathbf{r}') & S_{12}(\mathbf{r},\mathbf{r}') \\ S_{21}(\mathbf{r},\mathbf{r}') & S_{22}(\mathbf{r},\mathbf{r}') \end{pmatrix}, \tag{6.9.9}$$

which contains all the relevant nonlocal thermodynamic information. According to Eqs. (6.9.6) and (6.9.8), $S^{(2)}$ is the matrix of second functional derivatives of the local equilibrium entropy with respect to mass and energy density.

The entropy functional enters into the hydrodynamic-level theory in the same manner as the local equilibrium entropy function does at the thermodynamic level. For an equilibrium ensemble the thermodynamic forces, defined at the hydrodynamic level in Section 2.4, can be written

$$X_i = (\delta S/\delta a_i)(\mathbf{r}) - (\delta S/\delta a_i)^e(\mathbf{r}) \equiv F_i(\mathbf{r}) - F_i^e(\mathbf{r}), \tag{6.9.10}$$

where $a_i(\mathbf{r}) = n_i(\mathbf{r}) - n_i^e(\mathbf{r})$ is the deviation of the density n_i from its equilibrium value. The variation in the entropy, up to second order, is

$$\Delta S = \int F_i^e a_i(\mathbf{r})\,d\mathbf{r} + \frac{1}{2}\iint \mathbf{a}^T(\mathbf{r})S(\mathbf{r},\mathbf{r}')\mathbf{a}(\mathbf{r}')\,d\mathbf{r}\,d\mathbf{r}', \tag{6.9.11}$$

In the absence of an external field the 5×5 matrix S takes the form

$$S(\mathbf{r},\mathbf{r}') = \begin{pmatrix} S^{(2)}(\mathbf{r},\mathbf{r}') & 0 \\ 0 & \dfrac{-I_3\delta(\mathbf{r}-\mathbf{r}')}{\rho^e T^e} \end{pmatrix}, \tag{6.9.12}$$

where I_3 is the 3×3 identity. Equation (6.9.12) is easily verified using Eqs. (6.9.4), (6.9.5), and (6.9.8) since in the absence of an external field, $\phi(\mathbf{r})$ as well as $\mathbf{v}^e(\mathbf{r})$ vanish and $\rho^e(\mathbf{r})$ is independent of $\mathbf{r}$. For an isolated system, the variations $a_i(\mathbf{r})$ in the densities must satisfy

$$\int a_i(\mathbf{r})\,d\mathbf{r} = \Delta(N_i - N_i^e) = 0 \tag{6.9.13}$$

where $N_i \equiv \int n_i(\mathbf{r})\,d\mathbf{r}$, the total amount of the extensive variable. Since the intensive variables are constant in the equilibrium ensemble, it follows from Eqs. (6.9.11) and (6.9.13) that

$$\Delta S = \frac{1}{2}\iint \mathbf{a}^T(\mathbf{r})S(\mathbf{r},\mathbf{r}')\mathbf{a}(\mathbf{r}')\,d\mathbf{r}\,d\mathbf{r}' \leq 0, \tag{6.9.14}$$

with the inequality implied by the entropy maximum principle. As at the thermodynamic level of description, Eq. (6.9.14) shows that $S(\mathbf{r},\mathbf{r}')$ is negative semi-definite.

According to the definition in Eq. (6.9.10), the thermodynamic forces can be expressed nonlocally as

$$\mathbf{X}(\mathbf{r}) = \int S(\mathbf{r},\mathbf{r}')\mathbf{a}(\mathbf{r}')\,d\mathbf{r}', \tag{6.9.15}$$

so that Eq. (6.9.14) implies that

$$\Delta S = \frac{1}{2}\int \mathbf{a}^T(\mathbf{r})\mathbf{X}(\mathbf{r})\,d\mathbf{r}. \tag{6.9.16}$$

Taking the time derivative of Eq. (6.9.14) gives an expression for the entropy production, namely,

$$dS/dt = \int \mathbf{X}^T(\mathbf{r})\partial\mathbf{a}(\mathbf{r})/\partial t\,d\mathbf{r} \geq 0. \tag{6.9.17}$$

Thus Eq. (6.9.15) shows that the entropy production density has the nonlocal form

$$\mathbf{X}^T(\mathbf{r})\partial\mathbf{a}(\mathbf{r})/\partial t = \int \mathbf{a}^T(\mathbf{r}')S(\mathbf{r},\mathbf{r}')\partial\mathbf{a}(\mathbf{r})/\partial t\,d\mathbf{r}'. \tag{6.9.18}$$

For an isotropic material at equilibrium the second differential of the entropy can be expressed rather simply. In this case the second functional derivatives in $S^{(2)}(\mathbf{r},\mathbf{r}')$ depend only on the magnitude of the separation $|\mathbf{r}-\mathbf{r}'|$. Thus in component form

$$\delta F_i(\mathbf{r}) = \int S_{ij}(|\mathbf{r}-\mathbf{r}'|)\delta n_j(\mathbf{r}')\,d\mathbf{r}', \tag{6.9.19}$$

where the repeated index is summed. This expression can be reexpressed conveniently in terms of Fourier transforms. Recalling the convolution theorem, the transform of Eq. (6.9.19) becomes

$$\delta\hat{F}_i(\mathbf{k}) = \hat{S}_{ij}(k)\delta\hat{n}_j(\mathbf{k}), \tag{6.9.20}$$

with the Fourier transform defined as in Eq. (6.2.49). Because of the structure of Eq. (6.9.20), the matrix $\hat{S}(\mathbf{k})$ can be thought of as consisting of *wave-vector-dependent thermodynamic derivatives.* Consequently, we will write Eq. (6.9.20) in the equivalent form

$$\delta\hat{F}_i(\mathbf{k}) = [\partial\hat{F}_i/\partial\hat{n}_j(\mathbf{k})]\delta\hat{n}_j(\mathbf{k}), \tag{6.9.21}$$

where

$$\partial \hat{F}_i / \partial \hat{n}_j(\mathbf{k}) \equiv \hat{S}_{ij}(k). \tag{6.9.22}$$

The wave-vector-dependent thermodynamic functions are an analogue of familiar spatially local functions at the thermodynamic level. For example, Eq. (6.9.21) is the wave-vector-dependent version of

$$dF_i = (\partial^2 S/\partial N_i \partial N_j)\, dN_j, \tag{6.9.23}$$

where the N_j's are the extensive variables defined by

$$N_j = \int n_j(\mathbf{r})\, d\mathbf{r}. \tag{6.9.24}$$

Accordingly, we notice that

$$\delta N_j = \lim_{\mathbf{k}\to\mathbf{0}} \int \delta n_j(\mathbf{r}) \exp(i\mathbf{k}\cdot\mathbf{r})\, d\mathbf{r} = \lim_{\mathbf{k}\to\mathbf{0}} \delta \hat{n}_j(\mathbf{k})(2\pi)^3. \tag{6.9.25}$$

This relationship states that variations in the overall extensive variables of the system are proportional to the zero-wave-vector variations of the densities, $\delta\hat{n}_j(\mathbf{0})$. The relationship between wave-vector-dependent and local thermodynamics can be appreciated using Eq. (6.9.20). In the limit $\mathbf{k} \to \mathbf{0}$ that equation becomes

$$\lim_{\mathbf{k}\to\mathbf{0}} \int \delta F_i(\mathbf{r}) \exp(i\mathbf{k}\cdot\mathbf{r})\, d\mathbf{r} = \hat{S}_{ij}(0)\delta N_j, \tag{6.9.26}$$

where we have used Eq. (6.9.25) and

$$\hat{S}_{ij}(0) = \int S_{ij}(\mathbf{r})\, d\mathbf{r}. \tag{6.9.27}$$

In the thermodynamic-level theory the intensive variables are independent of the spatial coordinate. Thus taking δF_i independent of $\mathbf{r}$ in Eq. (6.9.26) gives

$$\delta F_i = [\hat{S}_{ij}(0)/V]\delta N_j, \tag{6.9.28}$$

where V is the volume. Comparing this to Eq. (6.9.23), we can make the identification

$$\partial^2 S/\partial N_i \partial N_j = \int S_{ij}(\mathbf{r})\, d\mathbf{r}/V, \tag{6.9.29}$$

or, equivalently, in terms of the density n_j,

$$\partial F_i/\partial n_j = \lim_{\mathbf{k}\to\mathbf{0}} \partial \hat{F}_i/\partial \hat{n}_j(\mathbf{k}). \tag{6.9.30}$$

Thus the usual thermodynamic derivatives of the intensive variables are nothing more than the wave-vector-dependent thermodynamic derivatives in the limit of zero wave vector. For example, using Eq. (6.9.30) and Eq. (6.9.8) we can write

$$\partial(-\mu/T)/\partial\rho = \int S_{11}(\mathbf{r})\,d\mathbf{r}. \tag{6.9.31}$$

The second functional differential martix, $S^{(2)}(\mathbf{r}, \mathbf{r}')$, can be related to spatial correlation functions that are commonly used in equilibrium thermodynamics. The most important of these are the radial distribution function, $g^{(2)}(r)$, and the direct correlation function, $c^{(2)}(r)$. The radial distribution function is determined by the density–density correlation function. For an isotropic substance the relationship can be written

$$\langle \delta\rho(\mathbf{r})\delta\rho(\mathbf{r}')\rangle = m\rho\delta(\mathbf{r} - \mathbf{r}') + \rho^2[g^{(2)}(|\mathbf{r} - \mathbf{r}'|) - 1]. \tag{6.9.32}$$

Because of this relationship it is often convenient to use the equivalent function $h^{(2)}(r) \equiv g^{(2)}(r) - 1$. The radial distribution gives the number of molecules, $N[r, r + dr]$, lying in a spherical shell of width dr around a central molecule. The precise relationship is

$$N[r, r + dr] = (\rho/m)4\pi r^2 g^{(2)}(r)\,dr. \tag{6.9.33}$$

The direct correlation function is related to $h^{(2)}(r)$ through the Ornstein–Zernicke integral equation,

$$h^{(2)}(r) = c^{(2)}(r) + (\rho/m)\int c^{(2)}(|\mathbf{r} - \mathbf{r}'|)h^{(2)}(r')\,d\mathbf{r}. \tag{6.9.34}$$

The direct correlation function vanishes when r is greater than the order of a molecular diameter and is useful in describing the structure of simple fluids. Fourier transforming Eq. (6.9.34) we can explicitly solve the resulting algebraic equation to obtain

$$\hat{c}^{(2)}(k) = \hat{h}^{(2)}(k)/[1 + (\rho/m)\hat{h}^{(2)}(k)]. \tag{6.9.35}$$

The relationship between the spatially nonlocal thermodynamic derivatives $S^{(2)}$ and the spatial correlation functions comes from the expression

$$\langle \delta n_i(\mathbf{r})\delta n_j(\mathbf{r}')\rangle^e = -k_B(S^{-1})_{ij}(\mathbf{r}, \mathbf{r}'). \tag{6.9.36}$$

This relationship follows from the Einstein formula in Eq. (2.5.16) applied at the hydrodynamic level of description. Note that the inverse must be interpreted as the functional inverse of the matrix S. Thus, for example, $S^{(2)-1}$ satisfies

$$\int S^{(2)}(\mathbf{r}, \mathbf{r}'')S^{(2)-1}(\mathbf{r}'', \mathbf{r}')\,d\mathbf{r}'' = I_2\delta(\mathbf{r} - \mathbf{r}'), \tag{6.9.37}$$

where I_2 is the 2×2 identity matrix. In particular, Eq. (6.9.36) implies that

$$\langle \delta\rho(\mathbf{r})\delta\rho(\mathbf{r}')\rangle = -k_B(S^{(2)-1})_{11}(\mathbf{r}, \mathbf{r}'). \tag{6.9.38}$$

By comparing equations like (6.9.32) and (6.9.38) it is possible to obtain explicit expressions for $S^{(2)}$ in terms of the spatial correlation functions.

The expressions for $S^{(2)}$ in terms of correlation functions are rather complicated, except for hard spheres. Nonetheless, there are a few wave-vector-dependent thermodynamic derivatives for which general expressions can be obtained. For example, one can show that

$$(k_B T/m)(\partial\hat{\rho}/\partial\hat{p})_{\hat{T}} = 1 + (\rho/m)\hat{h}^{(2)}(k). \tag{6.9.39}$$

This is the wave-vector-dependent generalization of the compressibility equation

$$(k_B T/m)(\partial\rho/\partial p)_T = 1 + (\rho/m)\int h^{(2)}(r)\,d\mathbf{r}, \tag{6.9.40}$$

which, in fact, follows from (6.9.39) by taking the limit $\mathbf{k} \to \mathbf{0}$.

The derivation of Eq. (6.9.39) follows a line of reasoning similar to that used in local thermodynamics. In fact, dropping the explicit dependence on $\mathbf{k}$ in Eq. (6.9.21) and using the usual differential notation, we can write

$$d\hat{F}_i = (\partial\hat{F}_i/\partial\hat{n}_j)\,d\hat{n}_j. \tag{6.9.41}$$

This is formally identical to the spatially local expression $dF_i = (\partial F_i/\partial n_j)\,dn_j$ and, thus, it can be manipulated in the same fashion. This implies, for example, the wave-vector-dependent Gibbs–Duhem relationship

$$d\hat{p} = -Thd(1/\hat{T}) - T\rho d(-\hat{\mu}/T). \tag{6.9.42}$$

By eliminating the differential $d(-\hat{\mu}/T)$ in the Gibbs–Duhem equation in favor of $d\hat{\rho}$, we can derive Eq. (6.9.39). To do this we first write out Eq. (6.9.41) explicitly as

$$\begin{aligned} d(-\hat{\mu}/T) &= \hat{S}^{(2)}_{11}\,d\hat{\rho} + S^{(2)}_{12}\,d\hat{\varepsilon} \\ d(\hat{1}/T) &= \hat{S}^{(2)}_{21}\,d\hat{\rho} + \hat{S}^{(2)}_{22}\,d\hat{\varepsilon}. \end{aligned} \tag{6.9.43}$$

Solving these equations for $d(-\hat{\mu}/T)$ in terms of $d(1/\hat{T})$ and $d\hat{\rho}$ gives

$$d(-\hat{\mu}/T) = (\hat{S}^{(2)}_{12}/\hat{S}^{(2)}_{22})d(1/\hat{T}) + [\det\hat{S}^{(2)}/\hat{S}^{(2)}_{22}]\,d\hat{\rho}. \tag{6.9.44}$$

Substituting this into the Gibbs–Duhem equation (6.9.42) we deduce that

$$(\partial\hat{p}/\partial\hat{\rho})_{\hat{T}} = -T\rho/(\hat{S}^{(2)-1})_{11}, \tag{6.9.45}$$

where we have used the fact that for a 2×2 matrix $(\hat{S}^{(2)-1})_{11} = \hat{S}^{(2)}_{22}/\det\hat{S}^{(2)}$. The right-hand side of (6.9.45) can be simplified if we combine Eqs. (6.9.32) and (6.9.38) to obtain

$$-k_B(\hat{S}^{(2)-1})_{11}(k) = m\rho + \rho^2\hat{h}^{(2)}(k). \tag{6.9.46}$$

This allows (6.9.45) to be rewritten as

$$(\partial\hat{p}/\partial\hat{\rho})_{\hat{T}} = k_B T/m[1 + (\rho/m)\hat{h}^{(2)}(k)], \tag{6.9.47}$$

which is equivalent to Eq. (6.9.39).

The spatially nonlocal formulation of thermodynamics is an essential feature of the hydrodynamic-level theory. This nonlocality enters only in the way that variations in the intensive variables depend on variations in the densities of the extensive variables. Thus the canonical equations, which depend on the average values of the intensive variables, but not on their variations, are spatially local. This can be seen clearly by inspection of the average hydrodynamic equations (6.5.1)–(6.5.3). The equations describing fluctuations in the extensive variables, on the other hand, explicitly involve variations, δF_i, of the intensive variables. Since these variations depend in a spatially nonlocal way on the densities of the extensive variables, the fluctuation equations are nonlocal in space.

The manner in which this nonlocality enters into the theory can be appreciated by examining the fluctuation equations for a simple fluid, Eqs. (6.5.4)–(6.5.6). Equation (6.5.6), for example, involves the term $\nabla\hat{\delta}p^e$, the gradient of the fluctuation in the local equilibrium pressure. According to definition, the fluctuation, $\hat{\delta}p^e$, is the variation in the pressure induced by linear variations in the densities of the extensive variables. Prior to this section we have written this in the local approximation

$$\hat{\delta}p^e = (\partial\bar{p}^e/\partial\bar{\rho})\delta\rho + (\partial\bar{p}^e/\partial\bar{\varepsilon})\delta\varepsilon, \tag{6.9.48}$$

where, for example, $\partial\bar{p}^e/\partial\bar{\rho} = \lim_{\mathbf{k}\to\mathbf{0}} \partial\hat{\bar{p}}/\partial\hat{\bar{\rho}}(\mathbf{k})$. To emphasize the spatially nonlocal character of the variations, this expression should be interpreted as a *functional* variation. In its spatially nonlocal form Eq. (6.9.43) should read

$$\hat{\delta}p^e = \delta\bar{p}^e/\delta\bar{\rho}[\delta\rho] + \delta\bar{p}^e/\delta\bar{\varepsilon}[\delta\varepsilon], \tag{6.9.49}$$

which emphasizes that p^e is a functional of ρ and ε. The only other terms in Eqs. (6.5.4)–(6.5.6) which involve a nonlocal functional variation come from Fourier's and Newton's laws. These terms explicitly involve the temperature, either as a thermodynamic force (i.e., ∇T^{-1}) or as a temperature-dependent transport coefficient. All the other terms are local.

The local approximation, which we have used prior to this section, consists in replacing equations like (6.9.49) by those like (6.9.48). This is often an excellent approximation since the range of the functional derivatives is usually short, that is, of the order of a molecular diameter R. For example, in light scattering experiments the magnitude of the wave vector is in the range 10^3–10^5 cm^{-1}, so that $k \ll R^{-1}$. Consequently, it is a good approximation to set $\mathbf{k} = \mathbf{0}$, a fact that is borne out by good agreement of calculations with experiment. Near equilibrium critical points, on the other hand, correlation lengths increase tremendously and the range of nonlocal thermodynamic derivatives can be comparable to the scattering vectors of visible light. This leads to critical opalescence, which cannot be explained without including these nonlocal effects. Nonlocal effects are also important in treating neutron scattering where was vectors of magnitude 10^8–10^{10} cm^{-1} are routinely used.

References

Diffusion and Heat Transport

S.R. deGroot and P. Mazur, *Non-equilibrium Thermodynamics* (North Holland, Amsterdam, 1962; reprinted by Dover, 1984), Chapter XI.

D.D. Fitts, *Nonequilibrium Thermodynamics* (McGraw-Hill, New York, 1962).

D.G. Miller, Thermodynamics of irreversible processes: Experimental verification of the Onsager reciprocal relations, *Chem. Rev.* **60**, 15–37 (1960).

D.G. Miller, V. Vitagliano, and R. Sartorio, Some comments on multicomponent diffusion: Negative main term diffusion coefficients, Second Law constraints, solvent choices, and reference frame transformations, *J. Phys. Chem.* **90**, 1509–1519 (1986).

The Hydrodynamic Level of Description

G.K. Batchelor, *An Introduction to Fluid Mechanics* (Cambridge University Press, Cambridge, 1970).

R.F. Fox and G.E. Uhlenbeck, Contributions to nonequilibrium thermodynamics. I. Theory of hydrodynamical fluctuations, *Phys. Fluids* **13**, 1893–1902 (1970).

D. Forster, *Hydrodynamic Fluctuations, Broken Symmetry, and Correlation Functions* (W.A. Benjamin, New York, 1975).

Fluctuating Hydrodynamics

L.D. Landau and E.M. Lifshitz, *Fluid Mechanics* (Pergamon, London, 1959), Chapter XVII.

R.F. Fox, Gaussian stochastic processes in physics, *Phys. Reports* **48**, 179–283 (1978).

J. Keizer, A theory of spontaneous fluctuations in viscous fluids far from equilibrium, *Phys. Fluids* **21**, 198–208 (1978).

Nonlocal Fluctuating Hydrodynamics

P. Schofield, in *Physics of Simple Liquids*, H.N.V. Temperely, J.S. Rowlinson, and G.S. Rushbrooke, eds. (North-Holland, Amsterdam, 1968), p. 564–592.

M. Medina-Noyola and J. Keizer, Spatial correlations in nonequilibrium systems: the effect of diffusion, *Physica* **107A**, 437–463 (1981).

J. Keizer and M. Medina-Noyola, Spatially nonlocal fluctuation theories: hydrodynamic fluctuations for simple liquids, *Physica* **115A**, 301–339 (1982).

E. Peacock-Lopez and J. Keizer, Hydrodynamic calculation of static correlation functions for homogeneous shear, *Physics Letters* **108A**, 85–90 (1985).

Chemical Reactions and Diffusion

C.W. Gardiner, K.J. McNeil, D.F. Walls, and I.S. Matheson, Correlations in stochastic theories of chemical reactions, *J. Stat. Phys.* **14**, 307–331 (1976).

J. Keizer, Master equations, Langevin equations, and the effect of diffusion on concentration fluctuations, *J. Chem. Phys.* **67**, 1473–1476 (1977).

D. McQuarrie and J. Keizer, Fluctuations in chemically reacting systems, in *Theoretical Chemistry: Advances and Perspectives*, Vol. 6A, D. Henderson, ed. (Academic Press, New York, 1981), pp. 165–213.

N.G. van Kampen, *Stochastic Processes in Physics and Chemistry* (North-Holland, Amsterdam, 1981), Chapter XII.

Light and Neutron Scattering

B.J. Berne and R. Pecora, *Dynamic Light Scattering* (Wiley, New York, 1976).
J.R.D. Copley and S.W. Lovesey, The dynamic properties of monatomic liquids, *Rep. Prog. Phys.* **38**, 461–563 (1975).
D. McQuarrie, *Statistical Mechanics* (Harper and Row, New York, 1976), Chapter 21.
R.A. Fleury and J.P. Boon, Brillouin scattering in simple liquids: argon and neon, *Phys. Rev.* **186**, 244–254 (1969).
W.E. Alley, B.J. Alder, and S. Yip, The neutron scattering function for hard spheres, *Phys. Rev. A* **27**, 3174–3186 (1983).
C.L. O'Connor and J.P. Schlupf, Brillouin scattering in water: The Landau–Placzek ratio, *J. Chem. Phys.* **47**, 31–38 (1967).

Light Scattering in Thermal Gradients

R. Fox, Testing theories of nonequilibrium processes with light-scattering techniques, *J. Phys. Chem.* **86**, 2812–2818 (1982).
D. Beysens, Y. Garrabos, and G. Zalczer, Experimental evidence for Brillouin asymmetry induced by a temperature gradient, *Phys. Rev. Letters* **45**, 403–406 (1980).
A.M.S. Tremblay, E. Siggia, and M. Arai, Fluctuations about simple nonequilibrium steady states, *Phys. Rev. A* **23**, 1451–1480 (1981).
T.R. Kirkpatrick, E.G.D. Cohen, and J.R. Dorfman, Light scattering by a fluid in a nonequilibrium steady state. I. Small gradients, *Phys. Rev. A* **26**, 972–994 (1982).
I. Procaccia, D. Ronis, and I. Oppenheim, Light scattering from nonequilibrium stationary states: The implications of broken time-reversal symmetry, *Phys. Rev. Letters A* **19**, 287–291 (1979).

CHAPTER 7

Nonequilibrium Steady States

7.1. Steady-State Ensembles

In several places in this book we have encountered the notion of a nonequilibrium steady state. Because of statistical fluctuations a steady state, like an equilibrium state, should not be thought of as the state of a single system, but rather as the state of an ensemble. For example, in Section 4.7 we examined the coupled chemical reactions $A + X \to 2X$ and $2X \to E$. Using the canonical theory we discovered that, on the average, there are two densities of X which do not change as a function of time. One of these was $\rho_x^{(1)} = 0$, which was found to be unstable to small perturbations, and the other was $\rho_x^{(2)} = \bar{k}_1/2k_2$, which is stable. The stable density is like an equilibrium density in that it supports a stationary probability distribution. In other words, associated with the time-independent average density $\rho_x^{(2)}$ is a unique, stationary probability distribution that characterizes single-time averages in the steady-state ensemble. This situation turns out to be relatively common. Indeed, it has already arisen in our treatment of electrochemical reactions in Section 5.7, in our discussion of reaction–diffusion fluctuations in Section 6.6, and in the calculation of the light scattering spectrum from a temperature gradient in Section 6.8. In this chapter we consider the statistical thermodynamic description of stable nonequilibrium steady states in a more general setting. We begin in this section by characterizing the average statistical state.

Because we are interested in hierarchies of levels of description—thermodynamic, hydrodynamic, Boltzmann, etc.—it is useful to introduce a notation that applies symbolically to all these descriptions. It is convenient, therefore, to consider an abstract vector-valued stochastic process $\mathbf{n}(t)$ of extensive variables that satisfies the canonical statistical theory discussed in Chapter 4. This, of course, corresponds exactly to the discrete thermodynamic level of description with $\mathbf{n}$ the vector of extensive variables. However, even if we are thinking of variables with spatial variation as described in Chapter 6, $\mathbf{n}$ can be interpreted as a vector-valued field of densities of the extensive variables. In that case sums must be interpreted as integrals and partial derivatives as functional derivatives, as outlined in Section 6.9. At the Boltzmann level of description, which was described for dilute gases in Chapter 3, $\mathbf{n}$ becomes a density field defined on a space of three coordinates and three momenta and the nonlinear function that gives the rate of change of the density becomes a nonlinear functional, i.e., the collision operator.

Keeping these facts in mind, we can write the conditional average equations in the canonical theory symbolically as

$$\partial\bar{\mathbf{n}}/\partial t = \mathbf{V}(\bar{\mathbf{n}}) - \mathbf{S}(\bar{\mathbf{n}}) + \mathbf{F}(\bar{\mathbf{n}}) \equiv \mathbf{R}(\bar{\mathbf{n}}). \tag{7.1.1}$$

In deference to applications involving field variables we use the partial derivative notation for the time variable. The other terms in Eq. (7.1.1) are supposed to represent dissipative contributions (**V**), or mechanical (nondissipative) contributions (**S**), and the effect of external fluxes of the extensive variables (**F**). The dissipative terms arise from elementary processes and have the canonical form discussed in Chapter 4. As we have seen, these terms sometimes involve parameters related to the flux of external variables. Indeed, for the continuously stirred tank reactor described in Section 5.5, the inverse of the residence time, τ, appears in the canonical equations and serves as a measure of the fluxes of chemicals into the system. The mechanical term, **S**, may involve external fields or may simply reflect the streaming motion or Euler terms at the hydrodynamic level. The external fluxes, **F**, will often be taken to be independent of $\bar{\mathbf{n}}$, as was the case in our description of the dimerization reaction $A + A \rightarrow A_2$ in Section 6.6. That is not, however, always the case. Indeed, if we think of applying Eq. (7.1.1) to hydrodynamics, then the external flux terms generally will be localized near the boundaries and may depend on gradients of the intensive variables. The external heat flux, for example, might take the form $\mathbf{F} = -K\delta(\mathbf{r} - \mathbf{r}_B)\nabla 1/T \cdot \hat{\mathbf{n}}(\mathbf{r}_B)$, where $\mathbf{r}_B$ is a point on the boundary and $\hat{\mathbf{n}}(\mathbf{r}_B)$ is the outward normal. Such terms, of course, are often more conveniently treated as a boundary condition on the differential equation. In all cases that we consider in this chapter, **V**, **S**, and **F** will be explicitly independent of time. Thus Eq. (7.1.1) is an autonomous, ordinary differential equation which is first order in the time.

Solutions to Eq. (7.1.1) that are independent of time will be referred to as "steady states," although it should be remembered that these are only the average states of the ensemble. Mathematically, the steady states, $\mathbf{n}^{ss}$, are determined by the equations

$$\partial\mathbf{n}^{ss}/\partial t = \mathbf{R}(\mathbf{n}^{ss}) = \mathbf{0}. \tag{7.1.2}$$

Equation (7.1.2) can have more than one solution, even if **R** is a linear function. This case was treated in Section 5.3, where $\mathbf{R}(\mathbf{n}) = W\mathbf{n}$, with W a matrix. If **R** is linear, the steady states are determined by the condition

$$W\mathbf{n}^{ss} = \mathbf{0}, \tag{7.1.3}$$

that is, the steady states are eigenvectors with eigenvalue zero. As discussed in Section 5.3, only when W is a strongly connected matrix will Eq. (7.1.3) have a unique solution. The absence of strong connectivity implies that subsets of the extensive variables n_1, n_2, n_3, ... are uncoupled from one another dynamically. Thus this sort of multiplicity of steady states is a trivial consequence of the variables being dynamically independent. In what follows we will imagine that only dynamically coupled variables are included in the vector **n**.

It is often convenient to think of the variables n_1, n_2, n_3, ... as being functionally independent. In other words, no conditions of the form $g_i(\mathbf{n}) = 0$ exist. Although we have seen that conservation equations like this are contained in the canonical formulas, such conditions will often be destroyed by the presence of external fluxes in an open system. For example, using the canonical equation (6.1.15) associated with diffusion, it is easy to verify that the total mass of the solute is conserved. External fluxes of mass at the boundary, however, will break this condition. Similarly, elementary chemical reactions involve atom conservation. However, if the effect of diffusion is included, atom conservation no longer applies locally, but only for the total numbers of atoms. In any case, we will assume that any dependent variables have been eliminated from the set $n_1, n_2, n_3, \ldots$ so that Eq. (7.1.2) can be solved without considering subsidiary equations.

Although we have seen examples of nonequilibrium steady states in Chapters 5 and 6, it is instructive to examine another simple example. To make things easy we look at a set of coupled, reversible chemical reactions in which only one species, X, is allowed to vary, i.e.,

$$\begin{gathered} \mathrm{A} + 2\mathrm{X} \rightleftarrows 3\mathrm{X} + \mathrm{G} \\ \mathrm{Y} + \mathrm{A} \rightleftarrows \mathrm{X} + \mathrm{F} \\ \mathrm{A} \rightleftarrows \mathrm{Q}. \end{gathered} \tag{7.1.4}$$

We have in mind fixing the concentrations of A, G, Y, F, and Q by some feedback control mechanism. Thus if the n_i's represent number densities, then

$$d\bar{n}_\mathrm{A}/dt = d\bar{n}_\mathrm{G}/dt = d\bar{n}_\mathrm{Y}/dt = d\bar{n}_\mathrm{F}/dt = d\bar{n}_\mathrm{Q}/dt = 0 \tag{7.1.5}$$

is a condition that holds for all time. The time derivative of $\bar{n}_\mathrm{X}$ involves no external inputs; these affect only the other densities and insure that Eq. (7.1.5) holds. In dilute solution the canonical equations for the reaction rates imply that $\bar{n} \equiv \bar{n}_\mathrm{X}$ satisfies

$$d\bar{n}/dt \equiv k_1 A\bar{n}^2 - k_2\bar{n}^3 + k_3 A - k_4\bar{n}, \tag{7.1.6}$$

where k_1 is the forward rate constant for the first reaction in Eq. (7.1.4), $A \equiv \bar{n}_\mathrm{A}$, and the other constants are combinations of rate parameters and concentrations. In Eq. (7.1.6) $\bar{n}$ is the only time-dependent quantity and the steady states are determined by

$$R(n^{ss}) \equiv k_1 A n^{ss2} - k_2 n^{ss3} + k_3 A - k_4 n^{ss} = 0. \tag{7.1.7}$$

Because of the particular elementary reactions involved, the constants in Eq. (7.1.7) can be chosen so that the steady state is not an equilibrium state.

If we had chosen to write, instead, the reactions

$$\begin{gathered} A + 2X \rightleftarrows 3X \\ A \rightleftarrows X, \end{gathered} \tag{7.1.8}$$

then this would not be the case. Although the canonical equation satisfied by $\bar{n}_X$ for these reactions is identical in form to Eq. (7.1.6), the canonical form restricts the values of the constants in that equation. In Section 3.8 we discovered that microscopic reversibility implies that the ratio of the forward and reverse rate constants for an elementary reaction depends only on the standard Gibbs' free energy difference for the reaction. Since both reactions in Eq. (7.1.8) have the same net stoichiometry, i.e., A = X, it follows that the ratio of the forward and reverse rate constants for the reactions must be identical. This implies, in turn, that the constants in Eq. (7.1.7) must satisfy $k_1/k_2 = k_3/k_4$. As we shall see, this condition implies the existence of a unique time-independent state at $n^e = k_1 A/k_2$. This is the equilibrium state.

The reactions in Eqs. (7.1.4) also satisfy microscopic reversibility. Because none of the three reactions have the same stoichiometry, this does not, however, restrict the allowable values of k_1, k_2, k_3, and k_4 in Eq. (7.1.7). Thus we can treat these as non-negative, but otherwise disposable, parameters and write Eq. (7.1.7) as the cubic equation

$$n^{ss3} - Bn^{ss2} + Rn^{ss} - PB = 0, \tag{7.1.9}$$

where $B \equiv k_1 A/k_2$, $R \equiv k_4/k_2$, $P \equiv k_3/k_1$. In this form it is easy to check that when $R = P$, that is, when $k_4/k_3 = k_2/k_1$ holds, as it does for reactions (7.1.8), then the left-hand side of Eq. (7.1.9) factors as

$$(n^{ss2} + R)(n^{ss} - B) = 0. \tag{7.1.10}$$

This implies a unique positive solution $n^{ss} = B = k_1 A/k_2$, which, as mentioned above, is an equilibrium state. Thus to obtain a nonequilibrium steady state we must pick $R \neq P$. A simple way to visualize these solutions is to solve Eq. (7.1.9) for the parameter B in terms of n^{ss}, i.e.,

$$B = n^{ss}(n^{ss2} + R)/(n^{ss2} + P). \tag{7.1.11}$$

Because the numerator of this expression is cubic, there can be one, two, or three positive solutions for appropriate values of B. To illustrate this, Eq. (7.1.11) has been plotted in Fig. 7.1 for several fixed values of R and P. When $P = 1 \times 10^3$ and $R = 2.5 \times 10^4$ one obtains a region of three steady states, whereas for $P = 1 \times 10^3$ and R in the range $10^3 \leq R \leq 5.001 \times 10^3$ only one steady state exists for a given value of B. The curve for $R = 5.001 \times 10^3$ is called a critical curve and represents a parameter boundary that separates regions of multiple and unique steady states. The line $n^{ss} = B$ represents the equilibrium states that occur when $R = P$.

It is not clear from what we have seen so far whether or not the steady states

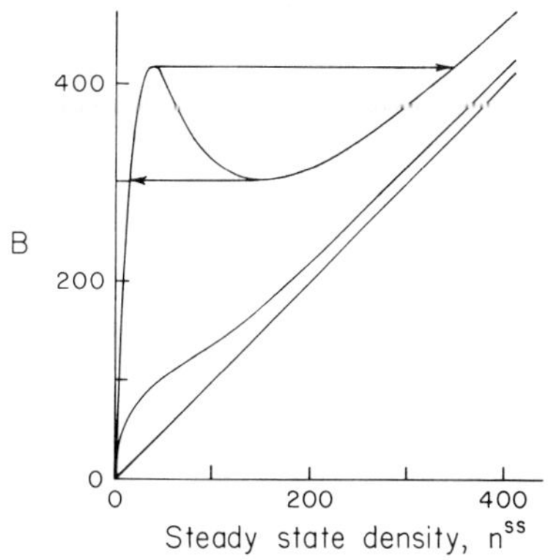

FIGURE 7.1. A graph of the parameter B as a function of the steady-state density of X molecules for the chemical reaction mechanism in Eq. (7.1.4). The upper curve corresponds to $P = 1 \times 10^3$ and $R = 2.5 \times 10^4$, the middle curve is the critical curve with the same value of P and $R = 5.001 \times 10^3$, and the straight line $B = n^{ss}$ is the equilibrium curve. The arrows represent hysteresis after traversing the region of multiple steady states.

defined by Eq. (7.1.2) are actually "steady" in the sense that equilibrium is "steady." In other words, it is not clear that these average states can support a stable, stationary probability distribution like equilibrium. We turn our attention to this question, in an average sense, in the next section.

7.2. Stability of Steady States

The mathematical theory of the stability of differential equations was developed in response to the question: what happens when the solution to an equation is perturbed slightly? Does the solution stay close, in some sense, to the original solution or does it move away? This is a rather broad question and attempts to answer it have led to a vast mathematical literature. For example, one might ask if the solutions, $\mathbf{n}(\mathbf{n}^0, t)$, to a chemical rate equation are stable with respect to small changes in the value of a rate constant, k. In other words, starting from the same initial condition, $\mathbf{n}^0$, do the solutions to the rate equation with the rate constants k and $k + \varepsilon$ stay close to one another if ε is small? This reflects the so-called structural stability of the rate equation and is not the sort of stability that interests us here. Rather we are concerned with the situation in which the kinetic constants, transport coefficients, and so forth are held fixed, but the initial condition is changed.

To be more precise we will consider autonomous equations such as Eq. (7.1.1), i.e.,

$$d\mathbf{n}/dt = \mathbf{R}(\mathbf{n}). \tag{7.2.1}$$

As we have seen in numerous examples, the solutions to this equation depend parametrically on the initial condition $\mathbf{n}^0 = \mathbf{n}(0)$. We wish to compare the time course of two solutions to Eq. (7.2.1) that begin near to one another, say, at $\mathbf{n}^0$ and at $\mathbf{n}^{0\prime}$. To make the idea of "near" definite we use the Euclidean distance

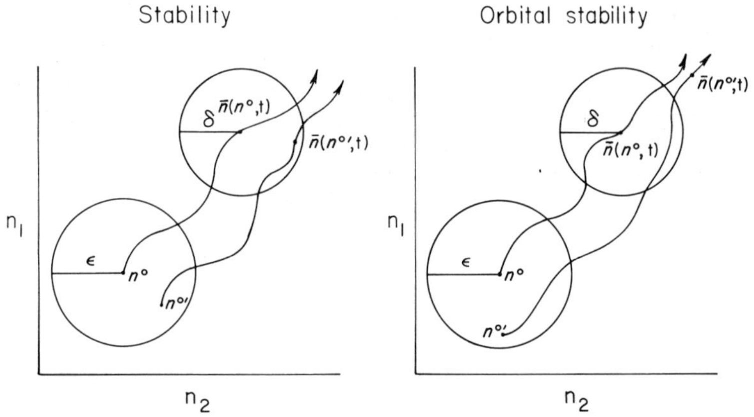

FIGURE 7.2. The left panel represents trajectories of a stable system, which when initially confined to a circle of radius ε remain within a circle of preassigned radius δ at time t. The right panel represents orbital stability for which it suffices that points on the trajectory, perhaps at different times, be within the preassigned distance, δ.

in the space of the variables $\mathbf{n}$. Thus the distance between $\mathbf{n}^0$ and $\mathbf{n}^{0\prime}$ will be defined by

$$|\mathbf{n}^0 - \mathbf{n}^{0\prime}| \equiv \left[\sum_i (n_i^0 - n_i^{0\prime})^2\right]^{1/2}. \tag{7.2.2}$$

Using this distance the solutions to Eq. (7.2.1) are said to be *Liapunov stable* at $\mathbf{n}^0$ if for a preassigned $\delta > 0$ one can always find an $\varepsilon > 0$ such that for all $\mathbf{n}^{0\prime}$ satisfying

$$|\mathbf{n}^0 - \mathbf{n}^{0\prime}| < \varepsilon, \tag{7.2.3}$$

it follows for all $t > 0$ that

$$|\mathbf{n}(\mathbf{n}^0, t) - \mathbf{n}(\mathbf{n}^{0\prime}, t)| < \delta. \tag{7.2.4}$$

In other words, the solutions can always be made to stay close to each other (Eq. (7.2.4)) if they start out close enough (Eq. (7.2.3)). This is illustrated schematically in Fig. 7.2. This condition is rather strict since it requires closeness for all times t. A less stringent condition, which is useful in discussing limit cycles and chaotic trajectories, is that of *orbital stability*. For orbital stability, trajectories do not necessarily need to be close at each time t, but over the course of time the trajectories themselves should stay close. In this case the condition in Eq. (7.2.4) is replaced by

$$\inf_{t' \geq 0} |\mathbf{n}(\mathbf{n}^0, t) - \mathbf{n}(\mathbf{n}^{0\prime}, t')| < \delta, \tag{7.2.5}$$

where inf means the infinum or greatest lower bound. In this chapter we shall be mainly concerned with Liapunov stability.

A knowledge of the stability of the steady-state solutions, $\mathbf{n}^{ss}$, to Eq. (7.2.1) is important physically. Indeed, if a steady state is not stable, then trajectories which start out close to $\mathbf{n}^{ss}$ will not stay close. Thus it is not possible even to prepare the system in the steady state. A slightly stronger condition of stability, called asymptotic stability, guarantees that the steady state can be prepared. A steady state is said to be *asymptotically stable* if it is stable and, furthermore, if one can always find an $\varepsilon > 0$ such that whenever $|\mathbf{n}^{ss} - \mathbf{n}^{0\prime}| < \varepsilon$, it follows that

$$\lim_{t\to\infty} \mathbf{n}(\mathbf{n}^{0\prime}, t) = \mathbf{n}^{ss}. \tag{7.2.6}$$

Thus there exists a neighborhood of the steady state for which all trajectories that start in a neighborhood of the steady state stay close to it and lead asymptotically to the steady state. For this reason asymptotically stable steady states are referred to as *attractors* and a steady state satisfying Eq. (7.2.6) is said to be *attractive*. The entire set of points that are "attracted" to $\mathbf{n}^{ss}$ is called its *domain of attraction*.

All states of thermodynamic equilibrium which are not critical points are asymptotically stable. This turns out to be a consequence of the Second Law. Although we defer the proof of this to Chapter 8, we have already worked through an example that illustrates this in Section 5.4. There we looked at the bimolecular isomerization reaction $A + B \rightleftarrows 2B$ and found that the conditional-averaged density of B solved the differential equation

$$d\bar{\rho}_B/dt = k^+\bar{\rho}_B(\rho - \bar{\rho}_B) - k^-\bar{\rho}_B^2 \equiv R(\rho_B). \tag{7.2.7}$$

This has the form of Eq. (7.2.1) with the explicit solution for $0 < \rho_B^0 \le \rho$ given by Eq. (5.4.9), i.e.,

$$\bar{\rho}_B(\rho_B^0, t) = \rho_B^e[1 - (1 - \rho_B^e/\rho_B^0)\exp(-\lambda t)]^{-1} \tag{7.2.8}$$

with $\lambda = (k^+ + k^-)\rho_B^e$. The initial condition $\rho_B^0 = \rho_B^e$ is easily seen to be a steady state since the time dependence in Eq. (7.2.8) vanishes when $\rho_B^0 = \rho_B^e$. This state is also asymptotically stable since for all $0 < \rho_B^0 \le \rho$, it follows from Eq. (7.2.8) that

$$\lim_{t\to\infty} \bar{\rho}_B(\rho_B^0, t) = \rho_B^e. \tag{7.2.9}$$

Thus all initial conditions, except $\rho_B^0 = 0$, end up asymptotically at ρ_B^e, and the domain of attraction of ρ_B^e is $0 < \rho_B^0 \le \rho$. The state $\rho_B = 0$ is also a steady state, since Eq. (5.4.5) shows that $R(0) = 0$. Since all other values of ρ_B^0 are in the domain of attraction of ρ_B^e, $\rho_B = 0$ is said to be an *unstable state*.

There are several criteria that are sufficient to guarantee that a steady state is asymptotically stable. One is based on linear stability analysis and the other is based on the idea of *Liapunov function*. A Liapunov function is a kind of distance function that can be used, in essence, to replace the Euclidean distance defined in Eq. (7.2.2). In particular, a Liapunov function is a real-valued function, Δ, on the state space that satisfies

$$\Delta(\mathbf{n} - \mathbf{n}^{ss}) \geq 0, \tag{7.2.10}$$

with equality only when $\mathbf{n} = \mathbf{n}^{ss}$, and that also satisfies

$$d\Delta(\mathbf{n} - \mathbf{n}^{ss})/dt \leq 0. \tag{7.2.11}$$

A function Δ will be called a *negative Liapunov function* if it satisfies Eqs. (7.2.10) and (7.2.11) with the inequalities reversed. [In that case $-\Delta$ satisfies Eqs. (7.2.10) and (7.2.11) as written. Thus it turns out that all the properties associated with a positive L iapunov function are also valid for a negative Liapunov function.] In Eq. (7.2.11) the derivative is taken along the trajectory determined by Eq. (7.2.1). Thus

$$d\Delta(\mathbf{n} - \mathbf{n}^{ss})/dt \equiv \sum_i (\partial\Delta/\partial n_i) R_i(\mathbf{n} - \mathbf{n}^{ss}). \tag{7.2.12}$$

Thinking of the function Δ as a distance, Eqs. (7.2.10) and (7.2.11) state that any point which is a positive distance away from the steady state gets closer to the steady state as time proceeds.

Given this interpretation, it is not surprising that the existence of a Liapunov function is a sufficient condition for stability. In fact, Liapunov proved that if a Liapunov function exists for all $\mathbf{n}$ in a neighborhood of $\mathbf{n}^{ss}$, then $\mathbf{n}^{ss}$ is stable. Furthermore, in the case that the inequality in Eq. (7.2.11) is replaced by the condition

$$\Delta(\mathbf{n} - \mathbf{n}^{ss}) \geq \phi_1(|\mathbf{n} - \mathbf{n}^{ss}|) \tag{7.2.13}$$

and Eq. (7.2.12) is replaced by the condition

$$d\Delta/dt \leq -\phi_2(|\mathbf{n} - \mathbf{n}^{ss}|), \tag{7.2.14}$$

where ϕ_1 and ϕ_2 are smooth, increasing functions, then the steady state is asymptotically stable.

The first part of Liapunov's theorem is easy to prove in the case that the Euclidean distance, which is positive and so satisfies Eq. (7.2.10), is a Liapunov function, that is, it satisfies

$$d|\mathbf{n} - \mathbf{n}^{ss}|/dt \leq 0. \tag{7.2.15}$$

In this case Eq. (7.2.15) implies that

$$|\mathbf{n} - \mathbf{n}^{ss}| < |\mathbf{n}^0 - \mathbf{n}^{ss}|. \tag{7.2.16}$$

Consequently, if the Euclidean distance is a Liapunov function for all $\mathbf{n}^0$ in a neighborhood of $\mathbf{n}^{ss}$, then for any $\delta > 0$ the condition $|\mathbf{n}^0 - \mathbf{n}^{ss}| < \varepsilon \equiv \delta$ and Eq. (7.2.16) imply that

$$|\mathbf{n} - \mathbf{n}^{ss}| < \delta. \tag{7.2.17}$$

Since this was our original definition of stability, the state must be stable. The beauty of Liapunov's theorem is that one is not restricted to using the Euclidean distance in this way. Indeed, any Liapunov function works just as well.

Although the general proof of Liapunov's theorem is straightforward, we will restrict ourselves to proving it when Δ is a quadratic form. This case has a simple geometric interpretation and is the one of interest in nonequilibrium thermodynamics. Consider, first, a positive definite quadratic form

$$v(\mathbf{a}) = \mathbf{a}^T V \mathbf{a} \geq 0, \tag{7.2.18}$$

where $\mathbf{a} \equiv \mathbf{n} - \mathbf{n}^{ss}$. If the matrix V_{ij} is not symmetric, it can be arranged to be so by defining a new matrix $V_{ij} = \frac{1}{2}(V_{ij} + V_{ji})$, which leaves the value of $v(\mathbf{a})$ unchanged. Because V is positive definite, the equality in Eq. (7.2.18) holds only for $\mathbf{a} = \mathbf{0}$. Thus, according to the definition in Eq. (7.2.10), v is a possible candidate for a Liapunov function. To actually be a Liapunov function, v in addition must satisfy

$$dv/dt \leq 0, \tag{7.2.19}$$

which implies that

$$v(\mathbf{a}(\mathbf{n}^0, t)) \leq v(\mathbf{a}(\mathbf{n}^0, 0)). \tag{7.2.20}$$

The fact that v is symmetric and positive definite, however, implies that there is an orthogonal transformation $\mathbf{a}' = O_V \mathbf{a}$ which diagonalizes the quadratic form of V. This is the so-called principal axis transformation, illustrated for two variables in Fig. 7.3. This transformation rotates the coordinate axes to coincide with the principal axes of the elliptical surface $v =$ constant. As a result of this transformation one has

$$v(\mathbf{a}) = v(\mathbf{a}') = \sum_{i=1}^{k} \lambda_i a_i'^2, \tag{7.2.21}$$

where the λ_i are the k eigenvalues of V, all of which are positive. Thus if λ_M is the largest of the λ_i and λ_m is the smallest, we can write

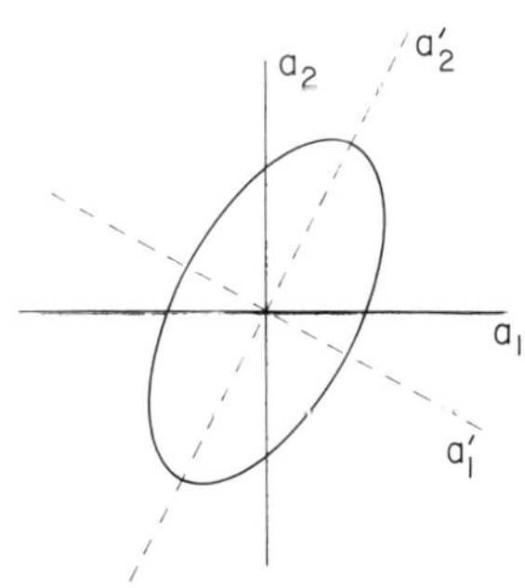

FIGURE 7.3. Illustration of the principal axis transformation, which rotates the coordinate frame to the primed axes and thereby represents the elliptical quadratic form as a sum of squares.

$$0 \leq \lambda_m |\mathbf{a}| \leq v(\mathbf{a}) \leq \lambda_M |\mathbf{a}|, \tag{7.2.22}$$

where we have used the fact that $|\mathbf{a}'| = |\mathbf{a}|$ since the transformation only rotates the coordinate system. Equation (7.2.22) is the key result. When it is combined with Eq. (7.2.20), we obtain

$$\lambda_m |\mathbf{a}(\mathbf{n}^0, t)| \leq v(\mathbf{a}(\mathbf{n}^0, t)) \leq v(\mathbf{a}(\mathbf{n}^0, 0)) \leq \lambda_M |\mathbf{a}(\mathbf{n}^0, 0)|. \tag{7.2.23}$$

The first and last terms in this string of inequalities imply that

$$|\mathbf{n}(\mathbf{n}^0, t) - \mathbf{n}^{ss}| \leq (\lambda_M/\lambda_m)|\mathbf{n}^0 - \mathbf{n}^{ss}|. \tag{7.2.24}$$

For a given $\delta > 0$, if we pick $|\mathbf{n}^0 - \mathbf{n}^{ss}| < \varepsilon \equiv \delta(\lambda_m/\lambda_M)$, it follows from Eq. (7.2.24) that $|\mathbf{n}(\mathbf{n}^0, t) - \mathbf{n}^{ss}| < \delta$ for all $t > 0$. Hence, according to our definition, the steady is stable. If instead of being positive definite, v were a negative definite Liapunov function, the same proof—now applied to $-v$—would still be valid. Thus the existence of either a quadratic positive definite or negative definite Liapunov function is a sufficient condition for stability.

There is another fact about Liapunov functions that is particularly relevant in nonequilibrium thermodynamics. Consider the special case that the kinetic equation (7.2.1) is linear, i.e.,

$$d\mathbf{n}/dt = H\mathbf{n} \tag{7.2.25}$$

with a unique steady state $\mathbf{n}^{ss} = \mathbf{0}$. Then the deviation from steady state, $\mathbf{a}$, satisfies

$$d\mathbf{a}/dt = H\mathbf{a}. \tag{7.2.26}$$

Again we will assume that the quadratic form $v(\mathbf{a}) = \mathbf{a}^T V\mathbf{a}$ is a Liapunov function in the sense of Eqs. (7.2.10) and (7.2.11). Thus v is positive definite. We will further assume that dv/dt is negative definite, i.e.,

$$\begin{aligned} dv(\mathbf{a})/dt &= (d\mathbf{a}^T/dt)V\mathbf{a} + \mathbf{a}^T V(d\mathbf{a}/dt) \\ &= \mathbf{a}^T H^T V\mathbf{a} + \mathbf{a}^T VH\mathbf{a} \\ &= \mathbf{a}^T(H^T V + VH)\mathbf{a} \leq 0. \end{aligned} \tag{7.2.27}$$

Thus the matrix $B = H^T V + VH$ is a negative definite matrix. Since the matrix V is symmetric and positive definite, it has an inverse V^{-1} that is also symmetric and positive definite, i.e., $V^{-1} = V^{-1T}$. Hence if we carry out an orthogonal transformation of B with V^{-1}, we obtain the matrix

$$V^{-1}BV^{-1T} = HV^{-1} + V^{-1}H^T = G. \tag{7.2.28}$$

The matrix G is symmetric and negative definite because B is. Thus defining $\mathbf{a}' = V^{-1}\mathbf{a}$, we can write

$$\begin{aligned} dv(\mathbf{a})/dt &= \mathbf{a}'^T G\mathbf{a}' \\ &= \sum_{i=1}^{k} \bar{\lambda}_i \bar{a}_i^2, \end{aligned}$$

where $\bar{\mathbf{a}} = O_G \mathbf{a}'$ is the orthogonal transformation which diagonalizes G. Since G is negative definite, all the $\bar{\lambda}_i$ are negative. Thus, arguing as before, we can write

$$\bar{\lambda}_m |\mathbf{a}| \leq dv(\mathbf{a})/dt \leq \bar{\lambda}_M |\mathbf{a}| \leq 0. \tag{7.2.29}$$

Using this inequality and the inequality in Eq. (7.2.22), we prove that if v is positive definite and dv/dt is negative definite, then the steady state of the linear Eq. (7.2.26) is *asymptotically stable.* To do this, notice from Eq. (7.2.22) that $0 \leq v(\mathbf{a})/\lambda_M \leq |\mathbf{a}|$, so that the central inequality in Eq. (7.2.29) can be written

$$dv(\mathbf{a})/dt \leq (\bar{\lambda}_M/\lambda_M) v(\mathbf{a}) \leq 0 \tag{7.2.30}$$

or

$$dv/dt \leq -bv \leq 0, \tag{7.2.31}$$

where $b \equiv |\bar{\lambda}_M/\lambda_m|$. Thus dividing by v, which is positive, and integrating Eq. (7.2.31) gives

$$v(\mathbf{a}(\mathbf{n}^0, t)) \leq \exp(-bt) v(\mathbf{a}(\mathbf{n}^0, 0)). \tag{7.2.32}$$

Applying the inequalities in Eq. (7.2.22) to each end of Eq. (7.2.32) gives

$$\begin{aligned} \lambda_m |\mathbf{a}(\mathbf{n}^0, t)| \leq v(\mathbf{a}(\mathbf{n}^0, t)) &\leq \exp(-bt) v(\mathbf{a}(\mathbf{n}^0, 0)) \\ &\leq \exp(-bt) \lambda_M |\mathbf{a}(\mathbf{n}^0, 0| \end{aligned} \tag{7.2.33}$$

or

$$|\mathbf{n}(\mathbf{n}^0, t) - \mathbf{n}^{ss}| \leq \exp(-bt)(\lambda_M/\lambda_m)|\mathbf{n}^0 - \mathbf{n}^{ss}|. \tag{7.2.34}$$

Equation (7.2.34) implies both that $\mathbf{n}^{ss}$ is stable and that

$$\lim_{t \to 0} \mathbf{n}(\mathbf{n}^0, t) = \mathbf{n}^{ss}. \tag{7.2.35}$$

Hence the steady state is asymptotically stable. Thus the existence of a positive definite quadratic form with a negative definite time derivative, i.e., a Liapunov function, is a *sufficient* condition for asymptotic stability of the linear equation (7.2.26). By introducing a minus sign, the same proof shows that the existence of a negative definite quadratic Liapunov function also implies asymptotic stability.

The asymptotic stability of a linear equation can be assessed in another fashion. This method uses the fact that Eq. (7.2.26) has the formal solution

$$\mathbf{a}(t) = \exp(Ht)\mathbf{a}(0). \tag{7.2.36}$$

As is well known, when H can be diagonalized, the exponential of a matrix can be expressed in terms of projection operators, P_i, on the space of eigenvectors with eigenvalues λ_i. In this case one has

$$\exp(Ht) = \sum_{j=1}^{k} \exp(\lambda_j t) P_i. \tag{7.2.37}$$

In general, the eigenvalues λ_j will be complex with real part α_j, i.e., $\lambda_j = \alpha_j + i\beta_j$. Combining Eqs. (7.2.36) and (7.2.37) it follows that

$$|\mathbf{n}(\mathbf{n}^0, t) - \mathbf{n}^{ss}| \le \sum_{j=1}^{k} \exp(\alpha_j t)|P_i \mathbf{a}(0)|$$

$$\le \sum_{j=1}^{k} \exp(\alpha_j t)|\mathbf{a}(0)| \tag{7.2.38}$$

or

$$|\mathbf{n}(\mathbf{n}^0, t) - \mathbf{n}^{ss}| \le k \exp(\alpha_M t)|\mathbf{n}^0 - \mathbf{n}^{ss}|, \tag{7.2.39}$$

where α_M is the largest value of the real parts α_j. If all the real parts of the eigenvalues of H are negative, then Eq. (7.2.39) implies that $\mathbf{n}^{ss}$ is asymptotically stable. Although in the general case the matrix H cannot be diagonalized, using the Jordan canonical form one can show that the exponential can be written

$$\exp(Ht) = \sum_{j=1}^{m} \sum_{n=1}^{m_j} Z_{jn} t^{n-1} \exp(\lambda_j t). \tag{7.2.40}$$

In Eq. (7.2.40) the λ_j are the m distinct eigenvalues of H of degeneracy m_j ($\sum_{j=1}^{m} m_j = k$) and the matrices Z_{jn} are the linearly independent component matrices of H. Using this result it is again easy to see in general that *if the real parts of the eigenvalues of H are negative, then the steady state is asymptotically stable.* Such a matrix H will be called stable.

Using this criterion for asymptotic stability for linear systems one can prove that asymptotic stability implies the existence of a positive definite quadratic form which has a negative definite time derivative. This is the converse of our earlier result, which means that the existence of a positive definite quadratic form with a negative definite time derivative is a *necessary and sufficient* condition for stability of a linear equation. The proof is based on the fact that when H is a stable matrix, the equation

$$HV^{-1} + V^{-1}H^T = G, \tag{7.2.41}$$

with G a negative definite matrix, has the unique positive definite solution

$$V^{-1} = -\int_0^\infty \exp(Ht) G \exp(Ht)\, dt. \tag{7.2.42}$$

However, as we saw earlier in Eqs. (7.2.27) and (7.2.28), Eq. (7.2.41) is precisely the condition which guarantees that $v(\mathbf{a}) = \mathbf{a}^T V \mathbf{a}$ has a negative definite time derivative. Consequently, asymptotic stability implies the existence of a positive definite quadratic, V^{-1}, form whose time derivative is negative definite.

Analysis of the linear problem is important because near a steady state it is possible to linearize the nonlinear kinetic equations and obtain information about stability in a neighborhood of the steady state. This is called *linear stability analysis*, and frequently it can be used to assess whether or not a steady state is asymptotically stable. To see why this procedure works, consider a steady state, $\mathbf{n}^{ss}$, of Eq. (7.2.1), that is,

$$\mathbf{R}(\mathbf{n}^{ss}) = \mathbf{0}. \tag{7.2.43}$$

If this steady state is isolated from other states, we can linearize Eq. (7.2.1) using the Taylor series expansion in $\mathbf{a} = \mathbf{n} - \mathbf{n}^{ss}$ to obtain

$$d\mathbf{n}/dt = \mathbf{R}(\mathbf{n}^{ss} + \mathbf{a}) = \mathbf{R}(\mathbf{n}^{ss}) + (\partial \mathbf{R}/\partial \mathbf{n})^{ss}\mathbf{a} + \mathcal{O}(\mathbf{a}). \tag{7.2.44}$$

Using Eq. (7.2.43) and the definition $H_{ij} = (\partial R_i/\partial n_j)^{ss}$, Eq. (7.2.44) becomes

$$d\mathbf{a}/dt = H\mathbf{a} + \mathcal{O}(\mathbf{a}). \tag{7.2.45}$$

Linear stability analysis consists in checking the asymptotic stability of the linear system in Eq. (7.2.45) by neglecting the higher-order terms. As we have seen, a necessary and sufficient condition for H to be stable is the existence of a positive definite quadratic form, $v(\mathbf{a})$, that has a negative definite derivative. However, according to Eqs. (7.2.22) and (7.2.29) v and dv/dt satisfy the conditions

$$\begin{aligned} v &\geq \lambda_m|\mathbf{n} - \mathbf{n}^{ss}| \\ dv/dt &< -\alpha|\mathbf{n} - \mathbf{n}^{ss}|, \end{aligned} \tag{7.2.46}$$

where $0 < \alpha < |\bar{\lambda}_M|$. The time derivative in Eq. (7.2.46) is taken using only the linear term in Eq. (7.2.45). However, if we choose a neighborhood of $\mathbf{n}^{ss}$ that is very small, the inequality of dv/dt will also be satisfied if we evaluate the time derivative using the full Eq. (7.2.46). This follows from the fact that the corrections introduced to Eq. (7.2.46) will be $\mathcal{O}(\mathbf{a})$. This implies that v satisfies conditions (7.2.13) and (7.2.14) for the full Eq. (7.2.45) and, thus, that the full nonlinear equation is asymptotically stable near $\mathbf{n}^{ss}$. Thus asymptotic stability of the linearized equation at $\mathbf{n}^{ss}$ implies asymptotic stability of the full nonlinear equation. This is the basis of linear stability analysis.

There are two other useful facts about linear stability analysis that we will need, but not prove. The first concerns the vanishing of the real part of one or more of the eigenvalues of H when the other real parts are negative. This is called *marginal stability*. The name is a good one since linear stability analysis gives no information about the stability of a nonlinear equation in this situation. If the stability of a steady state is marginal, the steady state is called a *critical point*. On the other hand, if the real part of even one of the eigenvalues of H is positive, the linear system is unstable. This is clear from Eq. (7.2.40), which shows that the solution blows up exponentially in time. One can show that instability of a linearized equation is sufficient to imply instability of the full nonlinear equation. Although such an instability may imply that the solution to the nonlinear equation also blows up, it is frequently the case that the nonlinear terms lead trajectories to another steady state or to some type of bounded periodic, quasi-periodic, or aperiodic motion. Some examples of this are discussed in Chapter 10.

The reader may have noticed several parallels between the discussion of linear stability in this section and the discussion of stationary, Gaussian, Markov processes in Section 1.8. Indeed, the fluctuation–dissipation theorem in Eq. (1.8.14) is identical in form to Eq. (7.2.41), which was used to obtain the Liapunov function for H. Moreover, Eq. (1.8.10) expresses the stationary covariance matrix, σ, in terms of an integral comparable to the integral in Eq. (7.2.42). This is no accident. Indeed, in the thermodynamic limit these relationships between fluctuations and stability turn out to be rather general. This connection is explored in more detail in the next section.

7.3. Fluctuations at Steady States

To treat the statistical aspects of steady states requires specifying an initial ensemble. If we are concerned with a particular steady state, it is necessary that the initial states represented by the ensemble be concentrated in the domain of attraction of that steady state. Otherwise, the equations that govern the conditional average behavior will not end up at the steady state. This suggests that unstable steady states, which have no domain of attraction, cannot be characterized by a stationary statistical ensemble.

The correctness of this conjecture is illustrated by the reaction scheme treated in Section 7.1. There we found that Eqs. (7.1.4) could support three steady states if the rate constants and concentrations were adjusted correctly. A specific case is shown in Fig. 7.1. To test if these steady states are asymptotically stable and, therefore, possess domains of attraction, we can carry out linear stability analysis. This is trivial for this model, which involves only a single variable n. Thus the eigenvalue, λ, and the relaxation matrix, H, coincide. To facilitate linearization around the steady state, we write Eq. (7.1.6) in the form

$$d\bar{n}/dt = -k_2(\bar{n}^3 - B\bar{n}^2 + R\bar{n} - PB), \qquad (7.3.1)$$

which can be linearized to give

$$da/dt = -k_2(3n^{ss2} - 2Bn^{ss} + R)a \equiv Ha. \qquad (7.3.2)$$

Thus a steady state is asymptotically stable if

$$3n^{ss2} - 2Bn^{ss} + R > 0. \qquad (7.3.3)$$

However, differentiating Eq. (7.1.9), which defines the steady state with respect to n^{ss}, we find that for fixed values of R and P

$$3n^{ss2} - 2Bn^{ss} + R = (n^{ss2} + P)(\partial B/\partial n^{ss}) \qquad (7.3.4)$$

Comparing this to the stability criterion in Eq. (7.3.3), we see that the steady state will be asymptotically stable if

$$(\partial B/\partial n^{ss}) > 0. \tag{7.3.5}$$

Referring to Fig. 7.1, it is clear that the steady states to the left of the maximum (n_1^{ss}) or the right of the minimum (n_3^{ss}) in the curve B versus n^{ss} are stable whereas those between the maximum and minimum (n_2^{ss}) are unstable. Since the kinetics involve only a single variable, it is also clear that the domain of attraction of n_1^{ss} is the half-open set $[0, n_2^{ss})$ while the domain of attraction of n_3^{ss} is (n_2^{ss}, ∞). The fact that n_3^{ss} is an attractor for $n > n_3^{ss}$ follows from the fact that there are no other steady states for $n > n_3^{ss}$.

It obviously makes no sense to look for a stationary ensemble to describe the unstable state, n_2^{ss}. For the other two steady states, which are asymptotically stable, the fluctuation–dissipation theory gives rise to characteristic stationary probability densities. Consider, for example, an initial condition, n^0, in the domain of attraction of n_1^{ss}, i.e., $0 \le n^0 < n_2^{ss}$. Then the conditional average $\bar{n}(n^0, t)$ satisfies Eq. (7.3.1) and since n_1^{ss} is an attractor,

$$\lim_{t\to\infty} \bar{n}(n^0, t) = n_1^{ss}. \tag{7.3.6}$$

Fluctuations around this trajectory are Gaussian and satisfy the stochastic equation

$$d\delta n/dt = -k_2(3\bar{n}^2 - 2B\bar{n} + R)\delta n + \tilde{f}. \tag{7.3.7}$$

As usual, $\tilde{f}$ vanishes on the average and, using the elementary processes in Eq. (7.1.4) and the canonical formula in Eq. (4.4.6), its variance is found to be

$$\langle \tilde{f}(t)\tilde{f}(t')\rangle = (k_2/V)(\bar{n}^3 + B\bar{n}^2 + R\bar{n} + PB)\delta(t - t') \tag{7.3.8}$$

where the factor, V, the volume of the system, occurs because n is a density. The conditional covariance of the Gaussian, $\sigma(n^0, t) = \langle \delta n(t)^2 \rangle$, satisfies Eq. (4.3.9). Because H is a scalar, it takes the form

$$d\sigma/dt = -2k_2(3\bar{n}^2 - 2B\bar{n} + R)\sigma + (k_2/V)(\bar{n}^3 + B\bar{n}^2 + R\bar{n} + PB). \tag{7.3.9}$$

As time proceeds, the conditional average approaches n_1^{ss}. Thus, asymptotically, this equation becomes

$$\begin{aligned} d\sigma/dt = &-2k_2(3n_1^{ss2} - 2Bn_1^{ss} + R)\sigma + (k_2/V) \\ &\times (n_1^{ss3} + Bn_1^{ss2} + Rn_1^{ss} + PB). \end{aligned} \tag{7.3.10}$$

Notice that this is a linear equation with time-independent coefficients. It is a stable equation because the relaxation rate is the same as that for the linearized average equation (7.3.2). The unique steady state of Eq. (7.3.10) is determined by the condition $d\sigma^{ss}/dt = 0$, which gives

$$\sigma^{ss}(n_1^{ss}) = \frac{(n_1^{ss3} + Bn_1^{ss2} + Rn_1^{ss} + PB)}{2V(3n_1^{ss2} - 2Bn_1^{ss} + R)}. \tag{7.3.11}$$

From this we see that any initial condition that falls within the domain of attraction of n_1^{ss} leads asymptotically to the Gaussian probability density

$$W_1^{(1)}(n) = (2\pi\sigma_1^{ss})^{-1/2}\exp[-(n - n_1^{ss})^2/2\sigma^{ss}(n_1^{ss})]. \tag{7.3.12}$$

Since there was nothing special about the steady state n_1^{ss}, except its asymptotic stability, a similar result holds for n_3^{ss}. Thus conditional ensembles initially within the domain of attraction n_3^{ss} asymptotically approach

$$W_1^{(2)}(n) = (2\pi\sigma_3^{ss})^{-1/2}\exp[-(n - n_3^{ss})^2/2\sigma^{ss}(n_3^{ss})]. \tag{7.3.13}$$

These asymptotic probability densities are extremely narrow with respect to the size of the average. Indeed Eq. (7.3.11) shows that their variances scale inversely with respect to the volume. Consequently, in the thermodynamic limit the Gaussians become delta functions centered at n_1^{ss} or n_3^{ss}. Hence in this limit fluctuations in the density are so small that they are negligible. This, in fact, is why a state n^0 that starts out in the domain of attraction n_1^{ss} is unaffected by the attractor at n_3^{ss}. Fluctuations are so small that they simply do not detect the presence of another attractor when the system is large. If the system is small, on the other hand, a state that starts out in one domain of attraction can penetrate into the other domain. We discuss this situation in more detail in Section 7.4.

Equations (7.3.12) and (7.3.13) show that, locally, there is a time-independent single-time Gaussian probability density associated with each of the asymptotically stable stationary states. These probability densities are approached asymptotically by all conditional probability densities that are in the domain of attraction of the steady states. Consequently, we can predict the asymptotic behavior of an arbitrary initial ensemble corresponding to an initial preparation, $W_1(n^0, t^0)$. The evolution equation for W_1 is given in Eq. (1.3.20) and is

$$W_1(n, t) = \int W_1(n^0, t^0)P_2(n^0, t^0|n, t)\,dn^0. \tag{7.3.14}$$

Taking the limit that $t \to \infty$ on both sides, we find that

$$W_1(n) \equiv \lim_{t\to\infty} W(n, t) = \int_0^{n_2^{ss}} W_1(n^0, t^0)W_1^{(1)}(n)\,dn^0 + \int_{n_2^{ss}}^{\infty} W_1^{(2)}(n^0, t^0)W_1^{(2)}(n)\,dn^0, \tag{7.3.15}$$

where the integrals extend over the domains of attraction of the steady states n_1^{ss} and n_3^{ss}. Now the fractions of the initial ensemble in the two domains of attraction are

$$\int_0^{n_2^{ss}} W_1(n^0, t)\,dn^0 \equiv f_1^0, \quad \int_{n_2^{ss}}^{\infty} W_1(n^0, t)\,dn^0 \equiv f_3^0 = 1 - f_1^0. \tag{7.3.16}$$

Consequently, Eq. (7.3.15) can be rewritten as

$$W_1(n) = f_1^0 W_1^{(1)}(n) + (1 - f_1^0) W_1^{(3)}(n). \tag{7.3.17}$$

In other words, an arbitrary initial ensemble separates asymptotically into two subensembles based on the initial fraction in the domain of attraction of n_1^{ss} and n_3^{ss}.

These results clearly apply to any system that has stable steady states. In the general case, however, the conditional average equation (7.1.1) will have the form

$$\partial\bar{\mathbf{n}}/\partial t = \mathbf{R}(\bar{\mathbf{n}}) \tag{7.3.18}$$

and the steady states are determined by

$$\mathbf{R}(\mathbf{n}^{ss}) = \mathbf{0}. \tag{7.3.19}$$

Let us assume that m of these steady states, $\mathbf{n}_1^{ss}, \mathbf{n}_2^{ss}, \ldots, \mathbf{n}_m^{ss}$, are asymptotically stable while the remainder are unstable. As in the example above, it follows that if n^0 is in the domain of attraction of $\mathbf{n}_j^{ss}$, then

$$\lim_{t\to\infty} \bar{\mathbf{n}}(n^0, t) = \mathbf{n}_j^{ss}. \tag{7.3.20}$$

Furthermore, based on the fluctuation–dissipation theory, the conditional fluctuations satisfy the stochastic differential equation

$$\partial\delta\mathbf{n}/\partial t = H(\mathbf{n}^0, t)\delta\mathbf{n} + \tilde{\mathbf{f}} \tag{7.3.21}$$

with $H_{ij} = \partial R_i/\partial \bar{n}_j$, $\langle \tilde{\mathbf{f}}(t)\rangle = \mathbf{0}$, and

$$\langle \tilde{\mathbf{f}}(t)\tilde{\mathbf{f}}^T(t')\rangle = \gamma(\mathbf{n}^0, t)\delta(t - t'). \tag{7.3.22}$$

Although the explicit form for γ depends on the elementary processes involved, the resulting stochastic process $\delta\mathbf{n}(t)$ is Gaussian and Markovian. As usual the conditional covariance, $\sigma(\mathbf{n}^0, t) \equiv \langle \delta\mathbf{n}(t)\delta\mathbf{n}^T(t)\rangle$, satisfies

$$d\sigma/dt = H\sigma + \sigma H^T + \gamma. \tag{7.3.23}$$

For the steady states that are attractors we can conclude, as before, that if $\mathbf{n}^0$ is in the domain of attraction of $\mathbf{n}_j^{ss}$, then

$$\lim_{t\to\infty} P_2(\mathbf{n}^0|\mathbf{n}, t) \equiv W_1^{(j)}(\mathbf{n}) = [(2\pi)^k \det \sigma(\mathbf{n}_j^{ss})]^{-1/2} \exp[-(\mathbf{n} - \mathbf{n}_j^{ss})^T \times \sigma^{-1}(\mathbf{n}_j^{ss})(\mathbf{n} - \mathbf{n}_j^{ss})/2], \tag{7.3.24}$$

where the steady-state covariance solves the equation

$$H(\mathbf{n}_j^{ss})\sigma(\mathbf{n}_j^{ss}) + \sigma(\mathbf{n}_j^{ss})H^T(\mathbf{n}_j^{ss}) = -\gamma(\mathbf{n}_j^{ss}). \tag{7.3.25}$$

Equation (7.3.25) is the generalized fluctuation–dissipation theorem at the steady state $\mathbf{n}_j^{ss}$. For an arbitrary initial preparation, $W_1(\mathbf{n}^0, t^0)$, that is concentrated within the domains of attraction of the steady states, it follows from Eqs. (7.3.14) and (7.3.24) that

$$W_1(\mathbf{n}) \equiv \lim_{t\to\infty} W_1(\mathbf{n}, t) = \sum_{j=1}^{m} f_j^0 W_1^{(j)}(\mathbf{n}), \tag{7.3.26a}$$

where

$$f_j^0 \equiv \int_{D_j} W_1(\mathbf{n}^0, t^0)\, d\mathbf{n}^0, \tag{7.3.26b}$$

with D_j the domain of attraction of $\mathbf{n}_j^{ss}$. Thus we conclude rather generally that any asymptotically stable steady state has an asymptotic Gaussian probability density associated with it.

As long as the asymptotic probability density in Eq. (7.3.24) is sharply peaked around the steady state, it is possible to assign a unique stationary, Gaussian, Markov process to an asymptotically stable steady state. As we have seen in Section 4.7, the fact that the extensive variables scale in proportion to the size of a system, Ω, means that fluctuations in the densities of extensive variables vanish like $\Omega^{-1/2}$ for large systems. Expressed in terms of the extensive variables, this implies that the covariance matrix in Eq. (7.3.24) is proportional to Ω for a large system. It is easy to see that the half-width of a Gaussian at half-height, $w_{1/2}$, is proportional to the square root of its variance. Indeed, equating $\exp(-w_{1/2}^2/2\sigma) = 1/2$, shows that

$$w_{1/2} = (2\sigma \ln 2)^{1/2}. \tag{7.3.27}$$

Consequently, for the extensive variables the width of the Gaussian is proportional to $\Omega^{-1/2}$ while the extensive variables themselves are proportional to Ω. In other words, for a large system the error inherent in a measurement of an extensive variable, n, will be negligible in comparison to the variable itself since $w_{1/2}/\bar{n}$ vanishes like $\Omega^{-1/2}$.

An exception to this situation occurs at critical points. Recall that in Section 7.2 a critical point was defined by the vanishing of the real part of some eigenvalues of the matrix H while the remaining real parts are negative. For example, in the chemical reaction model treated at the beginning of this section, we derived the condition for asymptotic stability $\partial B/\partial n^{ss} > 0$. Referring to Eqs. (7.3.3) and (7.3.4), we see that a critical point is determined by

$$(\partial B/\partial n^{ss}) = 0. \tag{7.3.28}$$

Thus in Fig. 7.1 maximum and minimum of the upper curve correspond to critical points. These correspond to values of B for which $n_1^{ss} = n_2^{ss}$ or $n_3^{ss} = n_2^{ss}$ where the stable branches merge with the unstable branch. Figure 7.1 also shows the curve $B(n^{ss})$ at critical values of the parameters P and R for which there are no unstable points, but only a single critical point, where $n_1^{ss} = n_2^{ss} = n_3^{ss}$. To see what effect critical points have on the variance of the fluctuations in Eq. (7.3.11), we use the steady-state condition (7.1.9) and Eq. (7.3.11) to rewrite the variance as

$$\sigma^{ss}(n^{ss}) = B/V(\partial B/n^{ss}). \tag{7.3.29}$$

Combining this with Eq. (7.3.28) we see that the variance becomes infinite at the critical point. This is a general phenomenon at critical points, as we will see in Sections 7.5 and 7.6. Thus near a critical point the scaling of fluctuations is not sufficient to insure that fluctuations are small.

Excluding states that are near critical points, it is possible to associate a stationary ensemble with asymptotically stable steady states. To do this we begin by considering the asymptotic probability density, $W_1^{(j)}(\mathbf{n})$, in Eq. (7.3.24). Since the steady state $\mathbf{n}_j^{ss}$ is not close to a critical point, this probability density will be sharp. Thus, if we consider $W_1^{(j)}$ as describing an initial ensemble, the probability of a member of the ensemble having a value of $\mathbf{n}$ outside the domain of attraction of $\mathbf{n}_j^{ss}$ will be negligible. In fact, the probability of $\mathbf{n}$ being appreciably different from $\mathbf{n}_j^{ss}$ is also negligible. This is easy to see using $W_1^{(1)}(n)$ for the chemical reaction in Eq. (7.3.12). Since the distribution is Gaussian, the probability that $a = n - n_1^{ss}$ is outside the interval $[-b, b]$ is

$$\begin{aligned} \text{Prob}(a > |b|) &= 2(2\pi\sigma^{ss})^{-1/2} \int_b^\infty \exp(-y^2/2\sigma^{ss})\, dy \\ &= (4/\pi)^{1/2} \int_{b/(2\sigma^{ss})^{1/2}}^\infty \exp(-x^2)\, dx \\ &= \operatorname{erfc}[b/(2\sigma^{ss})^{1/2}], \end{aligned} \tag{7.3.30}$$

where erfc is the complementary error function. Away from critical points Eq. (7.3.29) shows that σ^{ss} is the order of V^{-1}. Since $\lim_{x\to\infty} \operatorname{erfc}(x) = 0$, the only deviations $\mathbf{n} - \mathbf{n}_1^{ss}$ with an appreciable probability of being detected are small, in fact the order of $V^{-1/2}$.

This sort of reasoning assures us that the only states that have much probability of being found in the ensemble represented by $W_1^{(j)}(\mathbf{n})$ are $\mathbf{n} \simeq \mathbf{n}^{ss}$. Consequently, when we consider the dynamics of the ensemble [using Eqs. (7.3.18), (7.3.21), and (7.3.22)] it is a good approximation to consider only states $\mathbf{n}^0$ which are close to $\mathbf{n}_j^{ss}$. This implies that we can substitute $\bar{\mathbf{n}} = \mathbf{n}_j^{ss} + \bar{\mathbf{a}}$ into Eq. (7.1.1) and treat $\bar{\mathbf{a}}$ as small, i.e.,

$$\partial\bar{\mathbf{a}}/dt = \partial\bar{\mathbf{n}}/dt = \mathbf{R}(\mathbf{n}_j^{ss} + \bar{\mathbf{a}}) = H(\mathbf{n}^{ss})\bar{\mathbf{a}}. \tag{7.3.31}$$

Similarly, to the same degree of approximation we can substitute $\mathbf{n}_j^{ss}$ for $\mathbf{n}^0$ in Eqs. (7.3.21) and (7.3.22) to obtain

$$\partial\delta\mathbf{a}/dt = H(\mathbf{n}_j^{ss})\delta\mathbf{a} + \tilde{\mathbf{f}} \tag{7.3.32}$$

$$\langle\tilde{\mathbf{f}}(t)\tilde{\mathbf{f}}^T(t')\rangle = \gamma(\mathbf{n}_j^{ss})\delta(t - t'). \tag{7.3.33}$$

Notice that since $\mathbf{n}_j^{ss}$ is a steady state, setting $\mathbf{n}^0 = \mathbf{n}_j^{ss}$ eliminates the explicit time dependence in $H(\mathbf{n}^0, t)$ and $\gamma(\mathbf{n}^0, t)$. Recalling that $\mathbf{a} = \bar{\mathbf{a}} + \delta\mathbf{a}$, we can combine Eqs. (7.3.31) and (7.3.32) to get

$$\partial\mathbf{a}/dt = H(\mathbf{n}_j^{ss})\mathbf{a} + \tilde{\mathbf{f}}. \tag{7.3.34}$$

As we have seen, Eqs. (7.3.33) and (7.3.34) describe an Ornstein–Uhlenbeck process. According to the results in Section 1.8, an Ornstein–Uhlenbeck process is a stationary, Gaussian, Markov process, whose single-time probability density satisfies Eq. (1.8.14). This, however, is just the fluctuation–dissipation theorem obtained in Eq. (7.3.25) for the asymptotic probability density $W_1^{(j)}(\mathbf{n})$. Hence we can conclude that $W_1^{(j)}(\mathbf{n})$ is the stationary, single-time probability density for the ensemble.

Equations (7.3.33) and (7.3.34) describe the stationary ensemble locally in the neighborhood of an asymptotically stable stationary state. Indeed, this is the asymptotic behavior of any initial ensemble that is concentrated on the domain of attraction of $\mathbf{n}_j^{ss}$. More generally, if there are multiple steady states that are asymptotically stable, then aging an ensemble that is initially concentrated on domains of attraction leads to a composite of these local ensembles. Indeed, we have already obtained the single-time probability density for such a composite ensemble in Eq. (7.3.26a). The reason that the local subensembles act independently is that fluctuations are small. Consequently, a system initially in the ensemble that was in the domain of attraction of $\mathbf{n}_j^{ss}$ will have no mechanism to explore states in any other domain of attraction. This is easy to visualize for an ensemble of macroscopic spheres rolling in a smooth double-well potential. As long as the spheres and the wells are of macroscopic size, Brownian motion is negligible and friction will cause the spheres to come to rest in one well or the other. Indeed, if a sphere starts out in one well without sufficient energy to cross the barrier, the probability that a momentum fluctuation will be sufficient to take it across the barrier is negligible. The results in this section show that a comparable result holds for other macroscopic systems at asymptotically stable steady states.

The stationary ensemble associated with an asymptotically stable steady state represents what happens when an ensemble, originally concentrated in the domain of attraction of $\mathbf{n}^{ss}$, is aged for a long time. The stochastic process ultimately becomes stationary, Gaussian, and Markov with the probability densities

$$W_1(\mathbf{n}) = \left[(2\pi)^k \det \sigma^{ss}\right]^{-1/2} \exp\left[-(\mathbf{n} - \mathbf{n}^{ss})^T \sigma^{ss-1} (\mathbf{n} - \mathbf{n}^{ss})/2\right] \quad (7.3.35)$$

$$P_2(\mathbf{n}^0|\mathbf{n}, t) = \left[(2\pi)^k \det \sigma(t)\right]^{-1/2}$$

$$\times \exp\left[-[\mathbf{n} - \exp(Ht)\mathbf{n}^0]^T \sigma^{-1}(t)[\mathbf{n} - \exp(Ht)\mathbf{n}^0]/2\right], \quad (7.3.36)$$

where $H \equiv H(\mathbf{n}^{ss})$. Using the results in Section 1.8, the conditional covariance matrix, $\sigma(t)$, satisfies the equation

$$d\sigma/dt = H\sigma + \sigma H^T + \gamma^{ss} \quad (7.3.37)$$

with $\gamma^{ss} \equiv \gamma(\mathbf{n}^{ss})$ and $\sigma_{ij}(0) = 0$. The formal solution of Eq. (7.3.37) is given in

Eq. (1.8.10), i.e.,

$$\sigma(t) = \int_0^t \exp(Hs)\gamma^{ss}\exp(H^T s)\,ds. \tag{7.3.38}$$

This expression gives an integral representation for the stationary covariance, σ^{ss}. Indeed, since $\mathbf{n}^{ss}$ is stable, it follows from Eq. (7.3.36) that as $t \to \infty$, P_2 becomes independent of $\mathbf{n}^0$. Combining this with the fact that the ensemble is stationary, i.e.,

$$W_1(\mathbf{n}) = \int W_1(\mathbf{n}^0)P_2(\mathbf{n}^0|\mathbf{n}, t)\,d\mathbf{n}^0, \tag{7.3.39}$$

and that $W_1(\mathbf{n}^0)$ is normalized, we see that

$$W_1(\mathbf{n}) = \lim_{t\to\infty} P_2(\mathbf{n}^0|\mathbf{n}, t). \tag{7.3.40}$$

Since both W_1 and P_2 are Gaussian, this implies that

$$\sigma^{ss} = \lim_{t\to\infty} \sigma(t) = \int_0^\infty \exp(Hs)\gamma^{ss}\exp(Hs)\,ds. \tag{7.3.41}$$

Evaluation of this integral is sometimes a useful way of obtaining an explicit expression for σ^{ss}.

Another way to obtain σ^{ss} is to treat the fluctuation–dissipation theorem in Eq. (7.3.25) as a linear equation to be solved for the elements of σ_{ij}^{ss}. Since σ^{ss} is symmetric, we can solve for the $k(k+1)/2$ independent components σ_{ij}^{ss} $(1 \le i \le j \le k)$. For example, for two variables the independent components of σ_{ij}^{ss} are σ_{11}^{ss}, σ_{12}^{ss}, and σ_{22}^{ss}. According to Eq. (7.3.25), they solve the linear equations

$$\begin{aligned} 2H_{11}\sigma_{11}^{ss} + 2H_{12}\sigma_{12}^{ss} &= -\gamma_{11}^{ss} \\ H_{21}\sigma_{12}^{ss} + (H_{11} + H_{22})\sigma_{12}^{ss} + H_{12}\sigma_{22}^{ss} &= -\gamma_{12}^{ss} \\ 2H_{21}\sigma_{12}^{ss} + 2H_{22}\sigma_{22}^{ss} &= -\gamma_{22}^{ss}. \end{aligned} \tag{7.3.42}$$

As long as the matrix H has eigenvalues with negative real parts, then the fluctuation–dissipation theorem will have a unique solution. Quite frequently, posing the problem as a linear problem in $k(k+1)/2$ unknowns, as done in Eq. (7.3.42), provides the simplest route to obtaining σ^{ss}.

7.4. Multiple Steady States in Chemically Reactive Systems

In this section we examine the statistical properties of steady states that can develop in two types of chemically reacting systems. The first involves the dimerization reaction $2A \rightleftarrows A_2$. This reaction, including the effect of diffusion, was treated in Section 6.7. There the reaction was driven to a nonequilibrium steady state by a steady input of radiation, which photolyzed the dimer into the monomer. Since the temperature of that system was held fixed, a unqiue nonequilibrium steady state occurs for each value of the intensity of the incident radiation. Here we look at a related situation in which the radiation is absorbed by the monomer. In the absence of fluorescence, this energy causes an increase in the temperature of the system. If the process of heat transport to the outside is slow, the internal energy of the system increases until the rate of input of energy from the radiation equals the loss by heat transport.

A nice example of this process has been developed by Zimmermann and Ross. Using a gas-phase reaction cell to contain their system, they irradiated peroxydisulfuryl difluoride ($S_2O_6F_2$) with laser light of wavelength 488 nm. Peroxydisulfuryl difluoride is a dimer consisting of two fluorosulfate radicals ($FSO_3^{\cdot}$), which react according to the elementary reaction

$$S_2O_6F_2 \rightleftarrows 2FSO_3^{\cdot}. \tag{7.4.1}$$

At room temperature the equilibrium constant for this reaction is about 5×10^{-13} mol cm^{-3}. Thus at equilibrium the dimer is overwhelmingly the favored species. The 488-nm wavelength laser light excites only the fluorosulfate radical, in a process which can be represented by

$$FSO_3^{\cdot} \xrightarrow{h\nu} FSO_3^{\cdot *}. \tag{7.4.2}$$

One way for the excited state to decay is by fluorescence,

$$FSO_3^{\cdot *} \rightarrow FSO_3^{\cdot} + h\nu', \tag{7.4.3}$$

where the emitted frequency ν' is smaller than ν. As long as there is a high concentration of the dimer, fluorescence is a minor process and the excited state decays predominantly by the quenching reaction

$$FSO_3^{\cdot *} + S_2O_6F_2 \rightarrow FSO_3^{\cdot} + S_2O_6F_2. \tag{7.4.4}$$

Only the forward steps in reactions (7.4.2)–(7.4.4) have been indicated because the rates of the back reactions are negligible under typical experimental conditions.

The quenching reaction is rapid enough that the number of fluorosulfate radicals in the excited state is always small compared to the number in the ground state. Thus the excited state can be neglected in analyzing the dynamics of the system. If we also neglect the effect of diffusion, then the elementary reaction in Eq. (7.4.1) is the only process affecting the total number, N, of the radical. On the conditional average it satisfies the canonical equation

$$d\bar{N}/dt = 2\Omega[\exp(\bar{\mu}'/k_BT) - \exp(2\bar{\mu}/k_BT)] \tag{7.4.5}$$

with μ and μ' the chemical potentials of the radical and dimer, respectively. In the experimental arrangement used by Zimmermann and Ross, the total pressure of radical and dimer is held fixed by keeping the vessel in contact with liquid peroxydisulfuryl difluoride. If the vapor pressure is p_0, then the ideal gas law gives $p_0 = (\rho + \rho')k_B T$, with ρ and ρ' the density of radical and dimer, respectively. Using this and the fact that the chemical potentials of dilute gases have the form $\mu = \mu^0 + k_B T \ln \rho$, Eq. (7.4.5) can be written as

$$d\bar{N}/dt = \Omega \exp(2\mu^0/k_B\bar{T})\{K[(p_0/k_B\bar{T}) - \bar{\rho}] - \bar{\rho}^2\}, \tag{7.4.6}$$

where $K \equiv \exp(-[2\mu^0 - \mu^{0\prime}]/k_B T)$. From Section 3.8 we recall that K is the equilibrium constant for the reaction and that the factor $\Omega \exp(2\mu^0/k_B T)$ is proportional to the rate constant, k^-, for the reverse reaction. In this notation

$$d\bar{N}/dt = 2Vk^-\{K[(p_0/k_B\bar{T}) - \bar{\rho}] - \bar{\rho}^2\}, \tag{7.4.7}$$

where V is the volume of the system.

The internal energy of the system changes because of the light absorbed by the system and the heat given off to a reservoir at temperature T_R. In keeping with the neglect of spatial effects on the concentration, we will treat heat transport only at the thermodynamic level of description. The canonical form for heat transport is given in Eq. (4.5.30). Using this expression, the average rate of change of the internal energy of the system is

$$d\bar{E}/dt = f(\bar{\rho}) - \omega\bar{\Omega}\left\{\exp\left[-\left(\frac{\varepsilon_s^+}{k_B\bar{T}} + \frac{\varepsilon_R^+}{k_B T_R}\right)\right] - \exp\left[-\left(\frac{\varepsilon_s^-}{k_B\bar{T}} + \frac{\varepsilon_R^-}{k_B T_R}\right)\right]\right\} \tag{7.4.8}$$

with $\omega \equiv \varepsilon_s^- - \varepsilon_s^+$ and $f(\bar{\rho})$ representing the energy absorbed by the light. As long as the temperature of the system differs only slightly from that of the reservoir, the heat conduction term can be linearized around $\bar{T} = T_R$. It is also convenient to introduce the absorbance, $A(\bar{\rho})$, of the fluorosulfate radical, which is determined by Beer's law to be

$$A(\bar{\rho}) \equiv f(\bar{\rho})/I_0 = 1 - \exp(-\varepsilon\bar{\rho}l). \tag{7.4.9}$$

In this expressional I_0 is the incoming power of the laser beam (intensity times area), ε is the molecular extinction coefficient, and l is the pathlength of the laser beam within the system. In terms of these expressions we have

$$d\bar{E}/dt = A(\bar{\rho})I_0 - \beta(\bar{T} - T_R) \tag{7.4.10}$$

with $\beta \equiv (\omega^2\bar{\Omega}/k_B T_R^2)\exp[-(\varepsilon_s^+ + \varepsilon_R^+)/k_B T_R]$. Equations (7.4.7) and (7.4.10) determine the average behavior of this reaction system at the thermodynamic level of description.

The steady states, which interest us here, are determined by the condition $d\bar{E}/dt = d\bar{N}/dt = 0$. Since the absorbance of the radical and the temperature are directly accessible to measurement, the steady states will be described in terms of these variables. At steady state Eq. (7.4.10) gives that

$$A = (\beta/I_0)(T - T_R), \tag{7.4.11}$$

while Eq. (7.4.7) yields a single positive root,

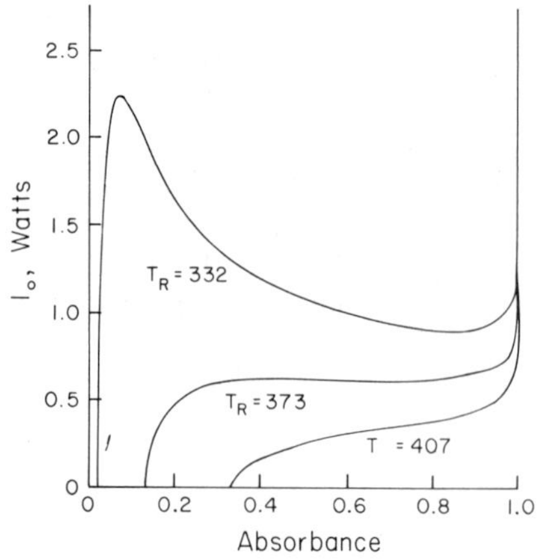

FIGURE 7.4. Multiple steady states in the peroxydisulfuryl difluoride system studied by E.C. Zimmermann and J. Ross, *J. Chem. Phys.* **80**, 720 (1984). Depending on the temperature of the thermal reservoir, a given intensity of the laser, I_0, can correspond to steady states of one or three different absorbances. The curves were calculated using Eq. (7.4.14).

$$\rho^{ss} = -(K/2)[1 - (1 + 4p_0/k_B TK)^{1/2}]. \tag{7.4.12}$$

Substituting this into the definition of the absorbance gives

$$A = 1 - \exp\{(\varepsilon l K/2)[1 - (1 + 4p_0/k_B TK)^{1/2}]\} \tag{7.4.13}$$

The simultaneous solutions to Eqs. (7.4.11) and (7.4.13) yield the steady states. Equating the right-hand sides of the two equations we can solve for the parameter I_0 in terms of T, i.e.,

$$I_0 = \beta(T - T_R)[1 - \exp\{(\varepsilon l K/2)[1 - (1 + 4p_0/k_B TK)^{1/2}]\}]^{-1}. \tag{7.4.14}$$

Since the temperature dependence of K is known, we can plot $I_0(T)$ for different values of T_R and p_0. A more convenient scale for the abscissa is obtained if I_0 is plotted versus $\beta(I_0)(T - T_R)$, i.e., versus A. This has been done in Fig. 7.4 for the experimental values $\beta = 5.5 \times 10^{-3}$ J K^{-1} s^{-1}, $p_0 = 0.261$ atm, and $l = 1.0$ cm used by Zimmermann and Ross. When the power of the laser vanishes, the gas is at equilibrium and the absorbance of the radical increases with temperature as the dimer dissociates. For bath temperatures above $T_R = 373$ K, only a single steady state exists for each value of I_0, whereas below $T_R = 373$ K, three steady states are possible. Thus $T_R = 373$ K is a critical curve.

For bath temperatures above 373 K, Zimmermann and Ross verified experimentally that a single steady state exists. Below 373 K, they found that only the high- and low-absorbance states are stable. As we have seen in the previous section, the stability of a state can be understood on the basis of fluctuations in the extensive variables. For the heat transport process and chemical reaction described by Eqs. (7.4.7) and (7.4.10), fluctuations in molecular number and internal energy satisfy the stochastic differential equations

$$d\delta N/dt = -2Vk^-(K + 2\bar{\rho})\delta\rho + 2Vk^-\left\{\frac{dK}{dT}[(p_0/k_B\bar{T}) - \bar{\rho}] - Kp_0/k_B\bar{T}^2\right\}$$
$$\times\, \delta T + \tilde{f}_1 \tag{7.4.15}$$
$$d\delta E/dt = I_0(dA/d\bar{\rho})\delta\rho - \beta\delta T + \tilde{f}_2,$$

where k^- has been treated as a constant. Using the canonical formula one also finds

$$\langle \tilde{f}_1(t)\rangle = \langle \tilde{f}_2(t)\rangle = \langle \tilde{f}_1(t)\tilde{f}_2(t')\rangle = \langle \tilde{f}_2(t)\tilde{f}_1(t')\rangle = 0 \tag{7.4.16}$$

and

$$\begin{aligned}\langle \tilde{f}_1(t)\tilde{f}_1(t')\rangle &= 4Vk^-\{K(\bar{T})[(p_0/k_B\bar{T}) - \bar{\rho}] + \bar{\rho}^2\}\delta(t - t')\\ \langle \tilde{f}_2(t)\tilde{f}_2(t')\rangle &= 2\beta k_B T_R^2\,\delta(t - t'),\end{aligned} \tag{7.4.17}$$

where in the second equality the expression for β below Eq. (7.4.10) was used.

The asymptotic stability of the steady states can be determined, as we have seen, by examining the eigenvalues of the matrix of the relaxation equation of the fluctuations. Equations (7.4.15) express the time derivative of δN and δE in terms of $\delta\rho$ and δT. Thus they cannot be used for this purpose until we have changed variables on one side of the equations to agree with those on the other side. It is most convenient to eliminate δE in favor of δT. This change requires that we linearize E in terms of δT and δN. This gives

$$\delta E = (\partial E/\partial T)_{V,N,N'}\delta T + (\partial E/\partial N)_{V,T,N'}\delta N + (\partial E/\partial N')_{V,T,N}\delta N',$$

where N' is the number of $S_2O_6F_2$ molecules and the volume is held fixed. Since the monomer and dimer change only due to the reaction in Eq. (7.4.1), $\delta N' = -\delta N/2$. Thus

$$\delta E = C_V\delta T + \Delta E\delta\rho, \tag{7.4.18}$$

where $V\Delta E \equiv (\partial E/\partial N)_{V,T,N'} - \frac{1}{2}(\partial E/\partial N')_{V,T,N}$ is the internal energy of reaction per radical of $FSO_3^{\cdot}$ produced and C_V is the constant-volume heat capacity. Using this change of variable on the left-hand side of Eqs. (7.4.15) and dividing the first of those equations by V gives

$$d\delta\rho/dt = -2k^-(K + 2\bar{\rho})\delta\rho + 2k^-\left\{\frac{dK}{dT}[(p_0/k_B\bar{T}) - \bar{\rho}] - Kp_0/k_B\bar{T}^2\right\}$$
$$\times\, \delta T + \tilde{f}_1 \tag{7.4.19}$$
$$d\delta T/dt = (I_0/C_V)\frac{dA}{d\bar{\rho}}\delta\rho - (\beta/C_V)\delta T - (\Delta E/C_V)\,d\delta\rho/dt + \tilde{f}_2,$$

where for simplicity ΔE and C_V have to be treated as constants. Notice that this change of variable adds the term $-(\Delta E/C_V)\,d\delta\rho/dt$ to the right-hand side of the second of Eqs. (7.4.19). This not only changes the relaxation matrix by adding terms such as $2(\Delta E/C_V)k^-(K + 2\bar{\rho})$ to the second equation, but adds an additional random term $-(\Delta E/C_V)\tilde{f}_1$ to that equation. The fluctuation equations for $\delta\rho$ and δT can, thus, be written symbolically as

$$d\delta\mathbf{n}'/dt = H'\delta\mathbf{n}' + \tilde{\mathbf{f}}' \tag{7.4.20}$$

where $\delta n_1' = \delta\rho$, $\delta n_2' = \delta T$, $\tilde{f}_1' = \tilde{f}_1$, and $\tilde{f}_2' = \tilde{f}_2 + (\Delta E/C_V)\tilde{f}_1$.

The asymptotic stability of the steady states can be examined using the eigenvalues of the matrix H evaluated at steady state. The eigenvalues of H are the roots to the characteristic equation $\det(H' - \lambda I) = 0$. For this problem H is a 2×2 matrix, and the characteristic equation takes the form

$$\lambda^2 - (\mathrm{tr}\, H')\lambda + \det H' = 0 \tag{7.4.21}$$

with roots

$$\lambda_\pm = \frac{1}{2}(\mathrm{tr}\, H' \pm [(\mathrm{tr}\, H')^2 - 4\det H']^{1/2}). \tag{7.4.22}$$

By inspection we see that the real parts of both $\lambda_\pm$ are negative if and only if

$$\mathrm{tr}\, H' < 0 \quad \text{and} \quad \det H' > 0. \tag{7.4.23}$$

Thus it sufficies to verify these inequalities at each steady state to determine that the steady states asymptotically stable. Zimmermann and Ross have analyzed the stability of the steady states for these equations using this criterion and have found that the middle branch of steady states is unstable.

The thermodynamic level of description turns out to be only of limited validity in formulating a picture of this reaction system. The reason for this is that as the laser beam passes through the gas, radiation is absorbed, which diminishes the intensity of the laser. Thus the temperature will decrease along the path of the laser. This implies that the system is inhomogeneous and that the spatial dependence of the energy and particle number must be taken into account. Indeed, because of the attenuation of the laser beam one expects to see the high-absorbance state only at the end of the cell where the laser enters. This is, indeed, the case as is seen in the photographs in Fig. 7.5. The light recorded in the photographs represents the weak fluorescence from the excited state of $FSO_3^{\cdot}$, which is proportional to the absorbance. The two photographs represent two states of differing absorbance, which can be prepared using conditions that are otherwise identical. The high-absorbance state is not uniform in space, but is localized near the end of the cell where the laser enters, as expected. Zimmermann and Ross have analyzed these spatial effects using a hydrodynamic theory that includes the processes of heat transport, mass diffusion, and chemical reaction and obtain a relatively accurate description of this phenomenon.

The same type of bistability is exhibited by an illuminated aqueous solution of *o*-cresolphthalein (OCP). This weak acid dissociates in aqueous solution according to the elementary process

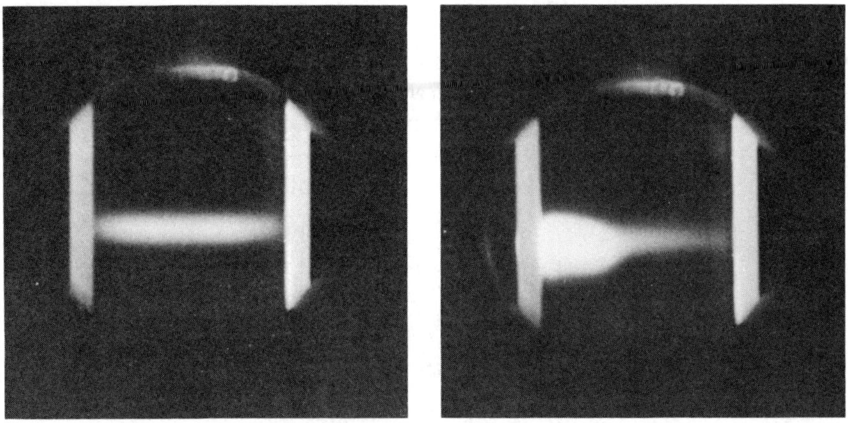

FIGURE 7.5. Photographs courtesy of E.C. Zimmermann and J. Ross showing the two stable steady states for the peroxydisulfuryl difluoride system as visualized by the fluoresence of the fluorosulfate radical. The panel on the left is the low-absorbance state and the panel on the right is the high-absorbance state, both produced at the same laser intensity.

$$\text{OCP} \rightleftarrows \text{OCP}^- + \text{H}^+, \tag{7.4.24}$$

where H^+ represents a hydrated proton. The basic form, OCP^-, absorbs light of wavelength 514 nm, yielding an excited state OCP^{-*}. In solution the excited state is quenched almost instantaneously with a net increase in the temperature of the solution. The differential equations that describe these events are analogous to those for the dimerization of $FSO_3^{\cdot}$. Thus, on the average

$$d\bar{N}/dt = Vk^-[K(\rho_0 - \bar{\rho}) - \bar{\rho}_H\bar{\rho}] \tag{7.4.25}$$

$$d\bar{E}/dt = A(\bar{\rho})I_0 - \beta(\bar{T} - T_R), \tag{7.4.26}$$

where ρ_0 is the total number density of *o*-creosolphthalein and ρ and ρ_H are the number densities of OCP^- and H^+, respectively. Since pH buffers are used, $\bar{\rho}_H$ can be treated as a constant.

This system is considerably simpler to analyze than the dimerization reaction since the relaxation time for the chemical reaction is about six orders of magnitude shorter than that for the energy. Indeed, from Eqs. (7.4.25) and (7.4.26) we can estimate that $\tau_{RX} = 1/k^-\rho_0 = 2 \times 10^{-8}$ s and $\tau_E = C_V/\beta = 4 \times 10^{-2}$ s, where we have used the experimental values $k^- = 5 \times 10^{13}$ cm^3 mol^{-1} s^{-1}, $\rho_0 = 8.7 \times 10^{-7}$ mol cm^{-3}, $C_V = 4.3 \times 10^{-5}$ J K^{-1}, $\beta = 1.2 \times 10^{-3}$ J K^{-1} s^{-1}. As a consequence, we can neglect the dynamics of the chemical reaction on the time scale of the energy relaxation processes. We do this by setting $d\bar{N}/dt = 0$ and solving Eq. (7.2.25) for $\bar{\rho}$. Since the equilibrium constant, K, is a function of temperature, this gives $\bar{\rho}$ as a function, $\bar{\rho}(T)$, of the temperature. Substituting this into Eq. (7.4.26) we can deduce that

$$d\bar{E}/dt = A(\bar{T})I_0 - \beta(\bar{T} - T_R). \tag{7.4.27}$$

This is a contraction of the description given by Eqs. (7.4.25) and (7.4.26). Although we defer to Chapter 9 a complete justification of this sort of contraction, on the time scale of heat transport we can treat Eq. (7.4.27) as the thermodynamic-level description of this system.

The steady states of Eq. (7.4.27) are easy to analyze. Using the assumption $d\bar{N}/dt = 0$ and Eq. (7.4.9), $A(\bar{T})$ can be written

$$A(\bar{T}) = 1 - \exp[-\varepsilon l K(\bar{T})\rho_0/K(\bar{T}) + \bar{\rho}_H]. \tag{7.4.28}$$

Thus the steady states are given by the transcendental equation

$$(\beta/I_0)(\bar{T} - T_R) = 1 - \exp[-\varepsilon l K(\bar{T})\rho_0/K(\bar{T}) + \bar{\rho}_H]. \tag{7.4.29}$$

The equilibrium constant has been measured to be $K(\bar{T}) = 2.3 \times 10^{-9} \times \exp(-36 \times 10^3/\bar{T})$ mol cm^{-3}, $\varepsilon l \rho_0 = 7.3$, and $\beta = 1.2 \times 10^{-3}$ watts K^{-1}.The steady states can be found graphically by finding the intersections of the linear curve of slope β/I_0 on the left-hand side of Eq. (7.4.29) with the nonlinear curve on the right-hand side. Alternatively, given values of $\bar{\rho}_H$ and T_R one can solve Eq. (7.4.29) for I_0 and plot this parameter as a function of $\bar{T}$, as we have done before, i.e.,

$$I_0 = \beta(\bar{T} - T_R)/A(\bar{T}). \tag{7.4.30}$$

For T_R near 300 K and pH below 9, this leads to curves for absorption as a function of intensity that contain regions of three steady states, just as for the isomerization reaction. This is shown in Fig. 7.6 for $T_R = 292$ K and pH $= 8.8$.

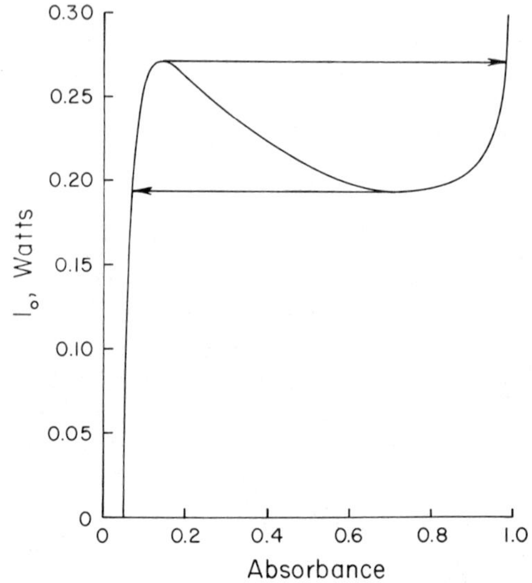

FIGURE 7.6. The multiple steady states of absorbance that develop as a function of the laser intensity, I_0, in the *o*-cresolphthalein system studied by Kramer and Ross. The theoretical curve was calculated using Eqs. (7.4.28) and (7.4.29) for $T_R = 292$ K and pH $=$ 8.8.

For this system it is easy to see that the steady states lying between the maximum and minimum in Fig. 7.6 are unstable. According to Eq. (7.4.27), fluctuations in the internal energy satisfy

$$d\delta E/dt = [I_0(dA/d\bar{T}) - \beta]C_V^{-1}\delta E + \tilde{f}, \tag{7.4.31}$$

where we have used the fact that $\delta E = C_V \delta T$. The covariance of the random term can be obtained from Eqs. (7.4.15)–(7.4.17) and is

$$\langle \tilde{f}(t)\tilde{f}(t')\rangle = 2\beta k_B T_R^2 \delta(t - t'). \tag{7.4.32}$$

The single eigenvalue of the linearized matrix in Eq. (7.4.31) is $H = [I_0(dA/d\bar{T}) - \beta]C_V$. Thus since the heat capacity is positive, the criterion for asymptotic stability of a stationary state is

$$I_0(dA/d\bar{T}) < \beta. \tag{7.4.33}$$

The steady states which violate this condition can be ascertained using Eq. (7.4.30). Indeed, differentiating $I_0(\bar{T})$ in that equation with respect to A using the chain rule we find that

$$dI_0/dA = (\beta\, d\bar{T}/dA - I_0)/A(\bar{T}). \tag{7.4.34}$$

Since the absorbance is positive, Eq. (7.4.34) shows that whenever the slope of the plot of I_0 versus A is positive one has

$$\beta\, d\bar{T}/dA - I_0 > 0. \tag{7.4.35}$$

This is equivalent to Eq. (7.4.33). The only steady states which violate this condition and are, thus, unstable lie between the maximum and minimum of the curve in Fig. 7.6.

The variance of the energy fluctuations at the asymptotically stable steady states can be obtained from the fluctuation–dissipation theorem in Eq. (7.3.25). Using Eqs. (7.4.31) and (7.4.32) to evaulate H and γ one obtains

$$\sigma^{ss} = k_B T_R^2 C_V/[1 - (dA/d\bar{T})I_0/\beta]. \tag{7.4.36}$$

Notice that for $I_0 = 0$, which is the equilibrium state, $\sigma^{ss} = k_B T_R^2 C_V$. This is just the result expected on the basis of the Einstein formula applied to this problem. As long as one stays away from the maximum and minimum in Fig. 7.6, the denominator in Eq. (7.4.36) is different from zero and σ^{ss} is proportional to the system size through the heat capacity, C_V. Thus, except at these critical points, fluctuations will be negligible.

It has been possible to test one of the basic assumptions of the fluctuation–dissipation theory using the *o*-cresolphthalein acid–base equilibrium system, namely, that the conditional averages satisfy the deterministic equations. For this system the deterministic equation is Eq. (7.4.27) which, when reexpressed in terms of the temperature alone, becomes

$$C_V d\bar{T}/dt = A(\bar{T})I_0 - \beta(\bar{T} - T_R). \tag{7.4.37}$$

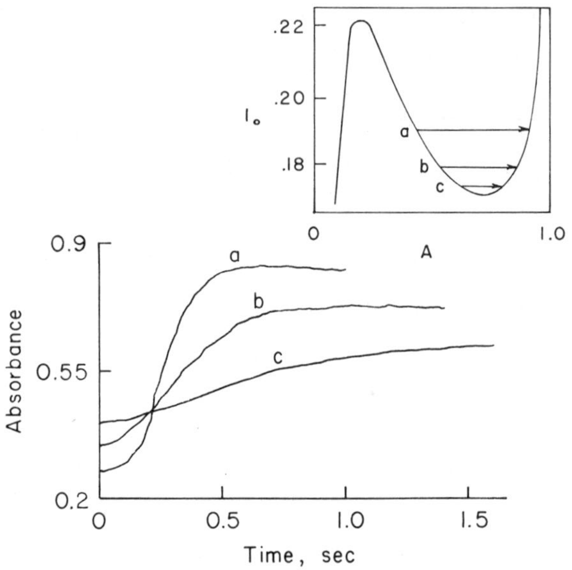

FIGURE 7.7. The experimentally determined time course of the absorbance for the *o*-cresolphthalein system at $T_R = 292$ K and pH $= 8.85$. The trajectories begin near the unstable steady states *a*, *b*, and *c* indicated by the inset. Data taken from J. Kramer and J. Ross, *J. Chem. Phys.* **83**, 6234 (1985).

A stringent test of this equation was made by preparing the initial temperature, T^0, very close to an unstable steady state. Since near the unstable steady Eq. (7.4.31) shows that the fluctuations grow rather than decay, one might anticipate a large variance in the behavior of repeated experiments and doubt the applicability of Eq. (7.4.37). Typical average experimental trajectories, expressed as absorbance versus time, are shown in Fig. 7.7. The three distinct initial points *a*, *b*, *c* are on the unstable manifold and correspond to three different laser intensities, I_0. Experimentally, it is impossible to specify the initial temperature with complete precision, and this leads to a slight variance among repetitions of these experiments. Nonetheless, it is possible to explain all the measured variaton in terms of the deterministic equation (7.4.37) and the experimental inaccuracies in reproducing T^0. In other words, to within the accuracy of these experiments, Eq. (7.4.37) holds.

Although a complete discussion of the transients in this system is deferred to Section 10.2, the size of the molecular fluctuations to be expected in this experiment can be judged from Eq. (7.4.31). Written in terms of the temperature, it becomes

$$d\delta T/dt = [I_0(dA/d\bar{T}) - \beta]C_V^{-1}\delta T + \tilde{f}/C_V. \tag{7.4.38}$$

Near the unstable steady state $[I_0(dA/d\bar{T}) - \beta]C_V^{-1} \equiv \lambda$ is positive, as the remarks below Eq. (7.4.35) show. According to the fluctuation–dissipation theory, the conditional variance of the temperature fluctuations, $\langle \delta T^2(t) \rangle^0 \equiv \sigma(t)$, satisfies Eq. (4.3.9), which in this notation can be written

$$d\sigma/dt = 2\lambda\sigma + 2k_B\bar{T}^2\beta C_V^{-2}. \tag{7.4.39}$$

Since the conditional covariance vanishes at $t = 0$, for short times one has

$$\sigma(t) = k_B \bar{T}^2 \beta C_V^{-1}(e^{2\lambda t} - 1)/[I_0(dA/d\bar{T}) - \beta], \qquad (7.4.40)$$

where λ is evaluated at the unstable state. It is evident from this result that the size of the variance is proportional to the inverse size of the system, i.e., to C_V^{-1}. The initial exponential growth is stabilized, ultimately, by the presence of the stable state that is the attractor for T^0. Thus, as we saw for the bimolecular dimerization reaction in Section 5.4, the variance is bounded and always stays of the order of the inverse size of the system. This shows that even for transient states in this system fluctuations are negligible. The sole exceptions to this are states initially near critical points, where the denominator in Eq. (7.4.40) is seen to vanish. We take up these exceptional states in the following section.

7.5. Critical Points

Steady states that are critical are characterized by the eigenvalues of their relaxation matrix, H, some of which have negative real parts and the remainder which have vanishing real parts. The eigenvectors of the relaxation matrix represent deviations from steady state that relax with a single exponential. In analogy to normal mode analysis, the eigenvectors are sometimes called the *relaxation modes*. In this terminology, a state is critical if there is at least one mode which either is constant in time or at worst oscillates with a frequency equal to the imaginary part of its eigenvalue. Thus a critical state is not asymptotically stable, although it may, in fact, be stable in the general sense of Liapunov.

Critical points can be classified into two broad categories depending on what happens to the manifold of steady states when one of the operational parameters is changed. Referring to Fig. 7.1 we see that the critical points occur at definite values of the operational parameter B. For parameter values just above (or just below) the critical value, the steady state is stable. If as the operational parameter is increased (or decreased) a discontinuous change occurs, such as a change in the number of steady states, the critical state is said to be a *bifurcation point*. This occurs for the values of B at the maximum and minimum of the upper curve in Fig. 7.1. A critical point does not have to be a bifurcation point, as is illustrated by the critical curve in Fig. 7.1. On that curve there is a unique point for which

$$(\partial B/\partial n^{ss}) = (\partial^2 B/\partial n^{ss2}) = 0. \qquad (7.5.1)$$

Otherwise on that curve $\partial B/\partial n^{ss} > 0$. Thus we can deduce from the previous analysis in Eqs. (7.3.2)–(7.3.5) that only the critical state lacks asymptotic stability. (It is, nonetheless, stable since the average trajectory is bounded and not oscillatory.) This critical point is like that encountered on the critical isotherm of a gas–liquid phase transition and has led to the bistable systems, like those discussed in the previous sections, being considered as analogous to first-order phase transitions. This analogy is incomplete, however, since as we have seen these states coexist only in the ensemble and not as separate phases in physical contact.

The critical bifurcation points in bistable systems lead to *hysteresis*. Imagine, for example, starting with the *o*-cresolphthalein system described in the previous section in a steady state with the laser turned off. According to Fig. 7.6, this is a state of negligible absorbance and $T = T_R$. As the laser intensity increases, the region of bistability is reached. As the intensity is increased further, the system stays on the low-absorbance branch until the critical bifurcation point is reached. Any further increase in laser intensity causes a discontinuous jump in absorbance indicated by the arrow in Fig. 7.6. If the laser intensity is gradually decreased, the system does not traverse this path in reverse since the system is now on the stable high-absorbance branch of steady states. This leads to a loop as indicated by the lower arrow as the system returns discontinuously to the low-absorbance branch at sufficiently low intensity. This sort of hysteresis has been observed experimentally for the *o*-cresolpthalein system and is modeled by the behavior in Fig. 7.6 rather well.

Another property of critical points is the lengthening of the overall time scale for relaxation as the critical point is approached. This is called *critical slowing down*. At a critical point at least one relaxation mode has an eigenvalue whose real part vanishes. Thus just before the critical point is reached, that mode relaxes very slowly. This behavior will also be reflected in the dynamics of the fluctuations. Indeed, the time course of the conditional covariance is given by Eq. (7.3.38). If we substitute the expression for $\exp(Ht)$ given in Eq. (7.2.40) into this integral and integrate the resulting expression we find that

$$\sigma(t) = \sum_{j,j'} Z_j \gamma Z_j \{\exp[(\lambda_j + \lambda_j')t] - 1\}/(\lambda_j + \lambda_j'), \tag{7.5.2}$$

where for ease in writing we have assumed that all the eigenvalues are distinct. This result shows that the variance depends on the eigenvalues of H. In particular, if H is a real matrix, then its eigenvalues are either real or come in complex conjugate pairs. If the jth mode is real and becomes critical, then the j-j term in Eq. (7.5.2) will diverge as the critical point is approached. If the jth mode is complex, then the term involving the jth mode and its complex conjugate will diverge since $\lambda_j + \lambda_j^*$ becomes zero at the critical point. Thus for real relaxation matrices, the conditional variance will grow in an unbounded way as one approaches the critical point. It does so, however, extremely slowly since the exponential factor in Eq. (7.5.2) has a characteristic time, $\tau = 1/(\lambda_j + \lambda_j^*)$, that approaches infinity. Thus, at critical points fluctuations are greatly enhanced, although the time course of their growth is greatly retarded.

To obtain an estimate of how the variance depends on the distance from the critical point, we must have an explicit solution for $\sigma(t)$. The simplest cases involve only a single variable and, for the sake of illustration, we return to the chemical reaction model examined in Sections 7.1 and 7.3. The explicit expression for $\sigma(t)$, obtained by solving Eq. (7.3.10), is

$$\sigma(t) = (B/V\partial B/\partial n^{ss})\{1 - \exp[-2k_2 t(3n^{ss2} - 2Bn^{ss} + R)]\}, \quad (7.5.3)$$

where we have used the expression for σ^{ss} in Eq. (7.3.29). We consider first critical points not on the critical curve. In particular, consider the lower critical point, n_c, at the maximum of the curve in Fig. 7.1. The value of n_c can be obtained by setting the relaxation rate in the exponential of Eq. (7.5.3) equal to zero. This gives

$$n_c = \frac{1}{3}[B_c - (B_c^2 - 3R)^{1/2}], \quad (7.5.4)$$

with B_c the value of B at the maximum. Just below the critical point we can write $n^{ss} = n_c - \delta$, with δ small and positive, and expand the relaxation rate to find that

$$3n^{ss2} - 2Bn^{ss} + R \simeq -6n_c\delta + 2B_c\delta = 2(B_c^2 - 3R)^{1/2}\delta, \quad (7.5.5)$$

where the second equality uses Eqs. (7.5.4). The steady-state variance, on the other hand, is

$$\sigma^{ss} = B/V\partial B/\partial n^{ss}. \quad (7.5.6)$$

Since $\partial B/\partial n^{ss}$ vanishes at the critical point, it follows from a Taylor series expansion that, to lowest order in δ,

$$\partial B/\partial n^{ss} \simeq -(\partial^2 B/\partial n^{ss2})^c \delta. \quad (7.5.7)$$

Moreover, the coefficient of δ is positive because n_c is at the maximum. Consequently, we can combine Eqs. (7.5.3) and (7.5.5)–(7.5.7) to see that near the critical point

$$\sigma(t) = -B_c\{V(\partial^2 B/\partial n^{ss2})^c\delta\}^{-1}\{1 - \exp[-4k_2 t(B_c^2 - 3R)^{1/2}\delta]\}. \quad (7.5.8)$$

According to Eq. (7.5.8), the variance of the density fluctuations increases like δ^{-1} as the critical point is approached from below. To judge just how large the fluctuations become, two criteria can be used. One is to compare the square root of the variance to the size of the average density at the steady state. They will be comparable, and thus fluctuations will be large, when

$$(\sigma^{ss})^{1/2}/n^c \simeq 1, \quad (7.5.9)$$

i.e., when

$$\delta \simeq -B_c/(\partial^2 B/\partial n^{ss2})^c n_c^2 V. \quad (7.5.10)$$

Since the right-hand side vanishes in the thermodynamic limit, one must be extremely close to the critical point for fluctuations to be large by this criterion.

Another measure of the size of the fluctuations compares the variance to the distance of the steady state from the boundary of the domain of attraction, i.e., from the unstable state n_2^{ss} which lies between the extrema in Fig. 7.1. From Eq. (7.3.27) we know that the width at half-height of the steady-state probability density is proportional to the square root of σ^{ss}. Since near the critical point the curve B versus n^{ss} is a parabola, it follows that $n_1^{ss} - n_2^{ss} \simeq 2(n^c - n_1^{ss}) = 2\delta$. Thus for a fluctuation to have an appreciable probability of leaving the domain of attraction,

$$(\sigma^{ss})^{1/2} \simeq 2\delta. \tag{7.5.11}$$

Using Eq. (7.5.8) this criterion can be written

$$\delta \simeq [-B_c/4V(\partial^2 B/\partial n^{ss2})^c]^{1/3}. \tag{7.5.12}$$

Equation (7.5.12) provides a much more stringent condition since δ only need be the order of $V^{-1/3}$ for it to be satisfied. When this condition is satisfied, it is likely that a spontaneous fluctuation will occur which takes a system into the domain of attraction of the upper branch of steady states. Because of critical slowing down, however, Eq. (7.5.8) shows that the expected time for such a fluctuation to develop is of order $V^{1/3}$. Thus it is likely that there will be an appreciable wait before such a fluctuation occurs. A formula like Eq. (7.5.12) can be obtained in a similar manner for the *o*-cresolpthalein reaction. Using Eqs. (7.4.34) and (7.4.40) one finds that the criteria for fluctuations to be large near the maximum in Fig. 7.6 is

$$\delta \simeq \{k_B T_c^2 \beta/[-2C_V(d^2 I_0/dT^2)^c A^c]\}^{1/3}. \tag{7.5.13}$$

For the critical piont at the maximum in Fig. 7.6 one can estimate that $(d^2 I_0/dT^2)^c A^c \simeq -2.5 \times 10^{-5}$ watts K^{-2}. Using the experimental values of $\beta = 1.2 \times 10^{-3}$ watts K^{-1}, $T_c = 324$ K, and $C_V = 4.3 \times 10^{-5}$ J K^{-1}, it follows that $\delta \simeq 10^{-4}$ K. Since it is not possible to control any of the parameters in this experiment to this degree of accuracy, fluctuations generated by the chemical reaction will always be negligible operationally.

Because a critical state like n_c has a domain of attraction consisting of a single point, the only stationary probability density for the steady state is the delta function, $\delta(n - n_c)$. This is not so for the critical point lying on the critical curve in Fig. 7.1. That point is an inflection point and is defined by the conditions in Eq. (7.5.1). As we noted earlier, that critical point is stable. Thus it can support a nontrivial probability density. Nonetheless, since it is not asymptotically stable, there will be critical slowing down and a divergence in the variance as the critical state is approached, either from above or below. Differentiating Eqs. (7.1.9) and using (7.5.1), it is easy to see that on the critical curve

$$n_c = B_c/3 = (R/3)^{1/3}, \tag{7.5.14}$$

where the second equality comes from Eq. (7.5.4). We can use the same method to estimate critical slowing down and divergence of the variance as was used in obtaining Eq. (7.5.8). This gives

$$\sigma(t) = 2B_c[V(\partial^3 B/\partial n^{ss3})^c \delta^2]^{-1}[1 - \exp(-6k_2\delta^2 t)]. \qquad (7.5.15)$$

Thus the critical slowing down is more pronounced on the critical curve. Since on the critical curve the critical point is stable, the only criterion for assessing when the variance is large is the comparison of $\sigma^{ss1/2}$ to n_c. Using Eqs. (7.5.9), (7.5.14), and (7.5.15), this gives

$$\delta \simeq [18VB_c(\partial^3 B/\partial n^{ss3})^c]^{-1/2}. \qquad (7.5.16)$$

Equation (7.5.16) implies that δ need only be of order $V^{-1/2}$ for fluctuations to be significant. Comparing this to Eq. (7.5.13) shows that fluctuations are not as important on the critical curve as they are near the unstable critical points.

According to Eq. (7.5.15), the probability density at the critical point is Gaussian with an infinite variance. If we trace back this result to the basic assumptions in Chapter 4, we see that it arises because we have scaled fluctuations in the densities of intensive variables using $\delta n = qV^{-1/2}$. This scaling makes sense only if the variance of q is of order unity. According to Eq. (7.5.15), however, $\langle q^2 \rangle$ diverges like δ^{-2} as the critical point is approached. Thus to see what the probability density looks like at the critical point, it is necessary to scale variables somewhat differently.

To see what is required, we use the master equation, which in Section 4.7 was shown to be equivalent to the basic postulate in Eq. (4.2.6). For the chemical reactions in Eqs. (7.1.4) the master equation expressed for the density $n = N/V$ is

$$\begin{aligned} dW_1(n,t)/dt = {} & V[k_2 B(n - V^{-1})^2 + k_2 PB]W_1(n - V^{-1}, t) \\ & + V[k_2(n + V^{-1})^3 + k_2 R(n + V^{-1})]W_1(n + V^{-1}, t) \\ & - V[k_2 n^3 + k_2 Bn^2 + k_2 Rn + k_2 PB]W_1(n,t), \end{aligned} \qquad (7.5.17)$$

where we have used the notation for the rate constants given below Eq. (7.1.9). In deriving the fluctuation–dissipation theory from a master equation in Section 4.7, we used the Kramers–Moyal expansion, Eq. (4.7.28). A Kramers–Moyal expansion for this reaction can be obtained in a similar fashion. Considering only the conditional probability density and using the Taylor series expansion of Eq. (7.5.17) gives

$$\partial P_2/\partial t = \sum_{j=1}^{\infty} \frac{V^{-j+1}}{j!}\frac{\partial^j}{\partial n^j}\{[k_2 n^3 + k_2 Rn + (-1)^j(k_2 Bn^2 + k_2 PB)]P_2\}, \qquad (7.5.18)$$

where P_2 is the solution to Eq. (7.5.17) conditional on $W_1(n,0) = \delta_{n^0 n}$. As in Chapter 4, we are interested only in deviations around the solutions to the deterministic solution. Thus we write $P_2(n^0|n,t) = P_2(n^0|\bar{n} + \delta n, t) = \hat{P}(\delta n, t)$

where $\bar{n}$ satisfies the deterministic equation

$$d\bar{n}/dt = k_2(B\bar{n}^2 + PB - \bar{n}^3 - R\bar{n}) \equiv R(\bar{n}). \tag{7.5.19}$$

Clearly, $\hat{P}$ gives the probability of a density fluctuation around $\bar{n}(n^0, t)$. Using the chain rule we can write

$$\partial\hat{P}/\partial t = (\partial P_2/\partial n)\, d\bar{n}/dt + \partial P_2/\partial t. \tag{7.5.20}$$

Thus substituting Eq. (7.5.18) into Eq. (7.5.20) yields

$$\partial\hat{P}/\partial t = (\partial\hat{P}/\partial\delta n)\, d\bar{n}/dt + \sum_{j=1}^{\infty} \frac{V^{-j+1}}{j!}\frac{\partial^j}{\partial\delta n^j}\{[k_2 n^3 + k_2 R n + (-1)^j(k_2 B n^2 + k_2 PB)]\hat{P}\}. \tag{7.5.21}$$

Close to the critical point on the critical curve we can no longer assume that density fluctuations scale in proportion to $V^{-1/2}$ [c.f. Eq. (4.7.29)]. For this reaction the correct scaling is

$$\delta n = qV^{-1/4}. \tag{7.5.22}$$

Writing $\hat{P}(q, t) = \hat{P}(\delta n, t)$ and making the substitution $n = \bar{n} + qV^{-1/4}$ into Eq. (7.5.21) leads to

$$\begin{aligned}\partial\hat{P}/\partial t = {} & (\partial\hat{P}/\partial q)R(\bar{n})V^{1/4} + \sum_{j=1}^{\infty} \frac{V^{-(3j-4)/4}}{j!}\frac{\partial^j}{\partial q^j} \\ & \times \{[k_2(\bar{n} + qV^{-1/4})^3 + k_2 R(\bar{n} + qV^{-1/4}) \\ & + (-1)^j(B(\bar{n} + qV^{-1/4})^2 + k_2 PB)]\hat{P}\}.\end{aligned} \tag{7.5.23}$$

Separating the remaining terms by powers of $V^{-1/4}$ and redefining the time scale according to $\tau = V^{1/2}t$, we find that

$$\begin{aligned}\partial\hat{P}/\partial\tau = {} & \frac{\partial}{\partial q}\{-q^3\hat{P}[R(\bar{n} + qV^{-1/4}) - R(\bar{n})]/(qV^{-1/4})^3\} + \frac{1}{2} \\ & \times \frac{\partial^2}{\partial q^2}\{k_2 B\bar{n}^2 + k_2 PB + k_2\bar{n}^3 + k_2 R\bar{n}\}\hat{P} + \mathcal{O}(V^{-1/4}).\end{aligned} \tag{7.5.24}$$

To carry out the limit $V \to \infty$ in Eq. (7.5.24) we must examine the first term with some care. By differentiating the condition of steady state $R(n^{ss}) = 0$, we easily verify that

$$\begin{aligned}\partial R/\partial n^{ss} &= -k_2(n^{ss2} + P)\partial B/\partial n^{ss} \\ \partial^2 R/\partial n^{ss} &= -k_2[2n^{ss}(\partial B/\partial n^{ss}) + (n^{ss2} + P)\partial^2 B/\partial n^{ss2}]\end{aligned} \tag{7.5.25}$$

Recalling that the critical point of the critical curve occurs when $(\partial B/\partial n^{ss})^c \equiv 0$, we see from Eqs. (7.5.25) that

$$R(n_c) = (\partial R/\partial n^{ss})^c = (\partial^2 R/\partial n^{ss2})^c = 0. \tag{7.5.26}$$

On the other hand, Eq. (7.5.19) shows that $\partial^3 R/\partial n^3 = -6k_2$. Thus if we take $\bar{n} = n_c$, the limit $V \to \infty$ of Eq. (7.5.24) becomes

$$\partial \hat{P}/\partial \tau = \frac{\partial}{\partial q}[k_2 q^3 \hat{P}] + \frac{1}{2}\frac{\partial^2}{\partial q^2}[\gamma(n_c)\hat{P}]. \tag{7.5.27}$$

The rescaling of the time in Eq. (7.5.27) is a manifestation of critical slowing down, and to keep $t = V^{-1/2}\tau$ fixed, τ must go to infinity. In the limit $\tau \to \infty$, the stationary solution of Eq. (7.2.27) satisfies the equation

$$dq^3 \hat{P}/dq = -(\gamma/2k_2)d^2\hat{P}/dq^2. \tag{7.5.28}$$

The normalized solution of Eq. (7.5.28) is easily verified to be

$$\hat{P}(q) = \frac{2(\gamma/8k_2)^{1/4}}{\Gamma(1/4)}\exp[-\gamma q^4/8k_2], \tag{7.5.29}$$

where Γ is the gamma function. Reexpressing this in terms of the probability density for n gives

$$W_1(n) = \frac{2(\gamma/8Vk_2)^{1/4}}{\Gamma(1/4)}\exp[-\gamma(n-n_c)^4/8Vk_2]. \tag{7.5.30}$$

Thus were it possible to adjust the reaction parameters precisely enough to actually sit on the critical point, the postulates in Chapter 4 predict that the non-Gaussian probability density in Eq. (7.5.30) would be observed.

7.6. The Gunn Effect

The critical points discussed in the previous sections are associated with the margin of stability between regions of multiple steady states. Another kind of critical point occurs at the boundary between stable steady states and stable oscillations. If this involves a pair of complex conjugate eigenvalues whose real part vanishes at the critical point, then the critical point is called a *Hopf bifurcation.* According to the results in Section 7.3, the covariance matrix of the fluctuations will have some unbounded components at any critical point. Thus we anticipate that steady states close to a Hopf bifurcation point will exhibit anomalously large fluctuations.

An experimental system in which this phenomenon has been observed is the Gunn diode. The Gunn diode is a semiconductor device consisting of a thin wafer of gallium arsenide (GaAs) with metal contacts attached to its two largest faces. When a small external voltage is placed across the crystal, it exhibits an ohmic resistance of the sort discussed in Section 5.1 for ionic solutions. In GaAs the charge carriers are electrons occupying two states of differing mobility, μ, in the conduction band, For electrons in the high-

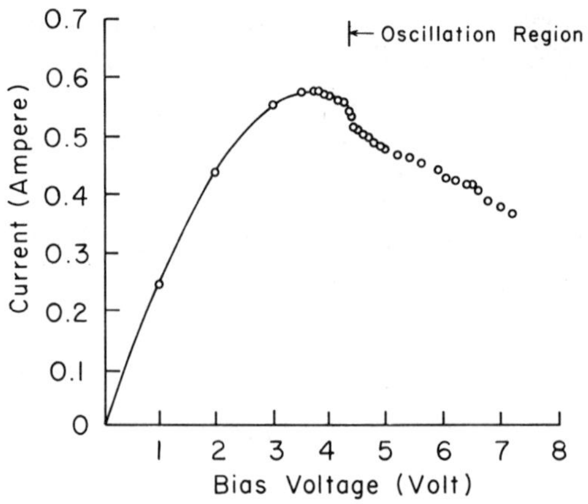

FIGURE 7.8. The steady-state current versus voltage curve for a thin wafer of GaAs, showing the region of negative differential resistance and the region of spontaneous oscillations, called the Gunn effect, in which the steady states are unstable. From the experiments of S. Kabashimi, H. Yamazaki, and T. Kawakubo, *J. Phys. Soc. Japan* **40**, 921 (1976).

mobility state $\mu_1 \simeq 8 \times 10^3$ cm^2 V^{-1} s^{-1}, while $\mu_2 \simeq 200$ cm^2 V^{-1} s^{-1} for electrons in the low-mobility state. Because the low-mobility state has an energy 0.36 eV greater than the high-mobility state, the joule heating that accompanies the electric current tends to populate the low-mobility state. This leads to an increase in the differential resistance at higher external voltages and above about 3.5 V the differential resistance actually becomes negative. Finally, as shown by the data in Fig. 7.8, a critical voltage is reached beyond which the current oscillates with a frequency of about 10^{10} s^{-1}. This is the so-called Gunn effect, which has become a technologically important way of generating microwave radiation.

Molecular fluctuations in the voltage across the GaAs wafer in the Gunn diode have been measured by Kabashimi, Yamazaki, and Kawakubo. Using electronic equipment they Fourier transformed the time record of the voltage in steady states below the critical point. Although they restricted their analysis to the components of the fluctuations in a small range of frequencies around 50 kHz, they were able to obtain the complete probability density for these components. To within the accuracy of their experiment, they found that the probability density was Gaussian. Furthermore, as the external voltage was increased toward the critical point, they found an increase in the variance of nearly six orders of magnitude.

These results are qualitatively what we would expect based on our knowledge of fluctuation theory. It is of some interest, nonetheless, to see if we can make a quantitative comparison of the theory with these experiments. To do so, it is necessary to enumerate the elementary molecular processes involved in the Gunn effect. There are five that are of primary importance. First, there is the spontaneous transition between the two states of the electron in the conduction band, $e_1 \rightleftarrows e_2$. There is also the process of current conduction by the two states, which involves the local momentum of the charge carriers as discussed in Section 5.1. Finally, there are the processes of diffusion of the electrons and heat transport to the external environment. To include all these processes we must use a hydrodynamic-level description. The appropriate densities of the extensive variables that are changing are the number densities of the two electron states $\rho_1(r, t)$ and $\rho_2(r, t)$, the two momentum densities, and the energy density.

Fortunately for the sake of our analysis, the energy and momentum densities relax on a rapid time scale, of the order of 10^{-12}–10^{-13} s. This is also true of the transitions between the two electron states for which the characteristic time is $\tau = 2 \times 10^{-12}$ s. Consequently, we can assume that these processes rapidly reach steady state and that only the dynamics of diffusion need to be considered. This is a good approximation for these experiments since measurements at 50 kHz will be sensitive only to dynamical processes with a time scale greater than about 10^{-4} s. This implies that locally in space the populations of the two states will be determined by the temperature dependence of their equilibrium constant. The local temperature, in turn, will be determined by the local electric field through the joule heating.

The relationship between the mobility and the local current, i, carried by the two electron states is

$$\begin{aligned} i_1 &= eA\rho_1\mu_1 E \\ i_2 &= eA\rho_2\mu_2 E, \end{aligned} \tag{7.6.1}$$

where E is the local electrical field and A is the cross-sectional area of the GaAs. Thus the total charge flux density due to the convective current flow is

$$j = e(\rho_1\mu_1 + \rho_2\mu_2)E. \tag{7.6.2}$$

Assuming that there is rapid equilibration at the local temperature T, Eq. (7.6.2) can be rewritten using the fact that $\rho_1/\rho_2 = \exp(\Delta/k_B T)$, where $\Delta = 0.36$ eV is the energy separating the two states. Thus

$$j = e\rho\mu(T)E \tag{7.6.3}$$

with $\rho = \rho_1 + \rho_2$, the total density of charge carriers, and

$$\mu(T) = [\mu_1 \exp(\Delta/k_B T) + \mu_2]/[1 + \exp(\Delta/k_B T)], \tag{7.6.4}$$

the average mobility of the two charge carriers. This mobility depends implicitly on the electric field through the temperature because of the steady state maintained by the joule heating. Recall that the rate of production of joule heat equals current times voltage, or in our notation, $(Aj) \times (El)$. Thus if the heat loss to the surroundings is given by Newton's law of cooling [c.f. Eq. (7.4.10)], then the steady state for the internal energy is determined by

$$\beta(T - T_R) = jEV, \tag{7.6.5}$$

where $V = Al$ is the volume. Substituting Eqs. (7.6.3) and (7.6.4) into the right-hand side of Eq. (7.6.5), then gives

$$(\beta/\rho V)(T - T_R) = eE^2[\mu_1 \exp(\Delta/k_B T) + \mu_2]/[1 + \exp(\Delta/k_B T)]. \quad (7.6.6)$$

The solution of Eq. (7.6.6) gives the local temperature as a function of the electric field. If we substitute this result into Eq. (7.6.3) and express the current in terms of the velocity $v \equiv \mu E$ of the charge carriers, we find that

$$j = e\rho v(E). \quad (7.6.7)$$

Since the populations of the two states are in equilibrium and the temperature and current have their steady-state values, the only variables whose time dependence need be considered are the total density of charge carriers and the electric field. Indeed, a graph of the current versus applied voltage based on Eq. (7.6.7) exhibits a negative differential resistance like that in Fig. 7.8 and agrees with experiment, except that it does not exhibit a critical point. To understand the critical behavior, the spatial dependence of the density and electric field must be taken into account. To treat spatial variations in the current, one needs a Maxwell equation, which in rationalized MKS units is

$$\varepsilon \partial E/\partial t = -j_T + I/A, \quad (7.6.8)$$

where $\varepsilon = 1.1 \times 10^{-10}$ C N^{-1} m^{-1} is the permittivity of GaAs, I is the applied current, and j_T is the total charge current density. Adding the effect of diffusion to Eq. (7.6.7) we have

$$j_T = eD\partial\rho/\partial x + e\rho v(E). \quad (7.6.9)$$

If we use this expression in Maxwell's equation, we must interpret it as involving the conditional average. Hence

$$\varepsilon \partial \bar{E}/\partial t = -eD\partial\bar{\rho}/\partial x - e\bar{\rho}v(E) + I/A. \quad (7.6.10)$$

Poisson's equation provides an additional relationship between the density and electric field, namely,

$$\partial \bar{E}/\partial x = (\rho_b - e\bar{\rho})/\varepsilon, \quad (7.6.11)$$

where ρ_b is the background positive charge density.

Equations (7.6.10) and (7.6.11) can be used to describe the steady states and their stability. At steady state one has

$$\begin{aligned} eDd\rho^{ss}/dx &= -e\rho^{ss}v(E^{ss}) + I/A \\ dE^{ss}/dx &= (\rho_b - e\rho^{ss})/\varepsilon. \end{aligned} \quad (7.6.12)$$

If we utilize the empirical form of the charge carrier velocity, $v(E)$, which can be obtained from Fig. 7.8, these coupled nonlinear partial differential equations can be solved numerically. This has been done by McCumber and Chynoweth and their result for the electric field at an applied voltage below the critical point is shown in Fig. 7.9. Notice that the electric field has a

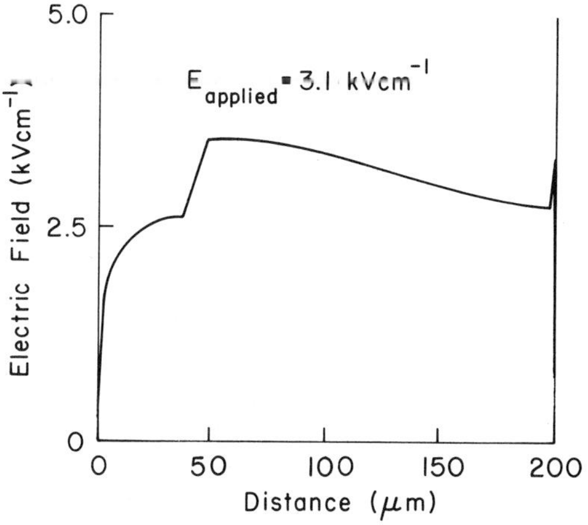

FIGURE 7.9. The electric field calculated as a function of distance for the Gunn diode at an applied field strength below the Gunn instability. The abrupt change in the field near 40 μm is due to a change in the density of positive background charge used in the calculation. Redrawn from D.E. McCumber and A.G. Chynoweth, *IEEE Trans. Elect. Devices* **13**, 4 (1966).

maximum near the anode and changes appreciably over the thickness of the GaAs wafer. The abrupt changes in the field near $x = 40$ μm are due to the fact that the background density, ρ_b, was taken to be a constant, except in a 10-μm notch between 40 and 50 μm.

To describe the dynamic character of the fluctuations around these steady states, we need to linearize Eqs. (7.6.10) and (7.6.11) and add the appropriate random terms. Since diffusion is the only dissipative process, the fluctuation–dissipation postulates imply that

$$\varepsilon \partial \delta E/\partial t = -eD\partial \delta \rho/\partial x - e\rho^{ss}(x)(dv/dE)^{ss}\delta E - ev^{ss}\delta \rho + \tilde{j}, \quad (7.6.13)$$

where $\tilde{j}$ is the random diffusion flux and, for simplicity, we have treated D as a constant. Using the results in Section 6.2 and the chemical potential $\mu = \mu^0 + k_B T \ln \rho$, the variance of $\tilde{j}$ is found to be

$$\langle \tilde{j}(x,t)\tilde{j}(x',t')\rangle = 2D\rho^{ss}e^2\delta(\mathbf{r}-\mathbf{r}')\delta(t-t'). \quad (7.6.14)$$

Because we are interested only in the voltage fluctuations, we need to find the stochastic equations solved by

$$V(t) = -\int_0^l E(x,t)\,dx. \quad (7.6.15)$$

The steady-state voltage, V^{ss}, can be obtained by integrating the solution $E^{ss}(x)$, while the time derivative of fluctuations in the voltage comes from integrating Eq. (7.6.13) over x. Doing this gives

$$d\delta V/dt = (eD/\varepsilon)(\delta\rho_l - \delta\rho_0) + (e/\varepsilon)\int_0^l \rho^{ss}(x)(dv/dE)^{ss}\delta E(x,t)\,dx$$
$$+ (e/\varepsilon)\int_0^l v^{ss}\delta\rho(x,t)\,dx + \tilde{f}(t), \quad (7.6.16)$$

where

$$\tilde{f}(t) \equiv -\int_0^l \tilde{j}(\mathbf{r}, t)\,dx. \tag{7.6.17}$$

Because $\langle \tilde{j} \rangle$ vanishes, it follows that $\langle \tilde{f} \rangle$ does also. Similarly, $\tilde{f}(t)$ is Gaussian, and using Eq. (7.6.14) we obtain

$$\langle \tilde{f}(t)\tilde{f}(t') \rangle = 2D(e/\varepsilon)^2 \rho_T l \delta(y - y')\delta(z - z')\delta(t - t'), \tag{7.6.18}$$

with $\rho_T \equiv \int_0^l \rho^{ss}(x, t)\,dx$, the overall density of charge carriers.

Equation (7.6.16) is not a closed equation for $\delta V(t)$. If, however, we neglect the dependence of the functions ρ^{ss} and v^{ss} on x in the two integral expressions, the equation simplifies considerably. Indeed, in that approximation Eq. (7.6.16) becomes

$$d\delta V/dt = (eD/\varepsilon)(\delta\rho_l - \delta\rho_0) - (e/\varepsilon)\rho_T(dv/dE)^{ss}\delta V + \tilde{f}(t), \tag{7.6.19}$$

where the third term has vanished since $\int_0^l \delta\rho\,dx = 0$ by electroneutrality. This is a homogeneous approximation to the fluctuation equation for the voltage that includes fluctuations in the charge density at the boundaries $x = l$ (the anode) and $x = 0$ (the cathode). These density fluctuations can be reexpressed using Poisson's equation (7.6.11). Thus

$$e\delta\rho/\varepsilon = -\partial\delta E/\partial x \simeq -\delta E/d^*, \tag{7.6.20}$$

where d^* represents a characteristic distance over which the electric field is changing. At the cathode, fluctuations in the electric field tend to vary only slightly in space. Thus d^* at the cathode is much larger than at the anode. Consequently, we introduce the further approximation that $\delta\rho_0 \simeq 0$ (the so-called virtual cathode assumption) and at the anode write

$$\delta\rho_l = -(\varepsilon/e)\delta E_l/d^* \simeq -(\varepsilon/e)\delta V/ld^*. \tag{7.6.21}$$

Using these expressions, Eq. (7.6.19) becomes

$$d\delta V/dt = -[(D/ld^*) + (e\rho_T/\varepsilon)(dv/dE)^{ss}]\delta V + \tilde{f}(t). \tag{7.6.22}$$

Thus in this approximation the voltage fluctuations around the steady state become a stationary, Gaussian, Markov process.

We can use Eq. (7.6.22) to assess the stability of the steady state as well as to evaluate the frequency spectrum of the voltage fluctuations. The steady state will be asymptotically stable as long as the relaxation rate in Eq. (7.6.22) is negative, i.e., as long as

$$-D\varepsilon/d^* e(dv/dE)^{ss} < l\rho_T. \tag{7.6.23}$$

This condition can be violated only when the differential resistance, as reflected by dv/dE in Eq. (7.6.23), is negative. In fact, it is known from experiment that the Gunn diode does not produce microwave oscillations if $l\rho_T \leq 6 \times 10^{11}\ \text{cm}^{-2}$. This fact allows us to estimate d^*. Notice first that equality of the two expressions in Eq. (7.6.23) defines the critical point for the Gunn diode. Using the value 6×10^{11} cm^{-2} for $l\rho_T$, we can estimate that in *CGS* units

$$d^* \simeq -1.7 \times 10^{-12}(D\varepsilon/d^* e)(dv/dE)^{ss}. \tag{7.6.24}$$

The experimental results in Fig. 7.8 show that the instability occurs when $dv/dE \simeq -3 \times 10^3$ cm^2 V^{-1} s^{-1}. Combining this with the experimental values of the diffusion and dielectric constants, we can use Eq. (7.6.24) to obtain an approximate value of d^* at the critical point. This gives $d^* \simeq 3 \times 10^{-6}$ cm, which is compatible with the distance over which the electric field changes at the anode in Fig. 7.9. Below the critical point, d^* will depend on the voltage, gradually increasing to a large value at equilibrium where the electric field vanishes. Using the value $d^* = 3 \times 10^{-6}$ cm in Eq. (7.6.22) and the experimental curve in Fig. 7.8 to evaluate dv/dE, we find that the critical point occurs at the correct experimental value of applied voltage, approximately 4.35 V.

Because the voltage fluctuations are stationary, Gaussian, and Markovian, the probability density of their frequency-dependent Fourier transform, i.e.,

$$\delta V(\omega) = \frac{1}{2\pi}\int_{-\infty}^{+\infty} \delta V(t) e^{i\omega t}\, dt, \tag{7.6.25}$$

will also be Gaussian. This is the qualitative result obtained by Kabashimi, Yamazaki, and Kawakubo and referred to at the beginning of this section. To make a quantitative comparison with their experiment, we need only calculate the variance of $\delta V(\omega)$ as a function of the applied voltage. As we saw in Section 1.8, this is given by the power spectrum, $S(\omega)$. Applying the expressions for $S(\omega)$ in Eqs. (1.8.37) and (1.8.45) to the stochastic process in Eq. (7.6.22), we find that

$$S(\omega) = \frac{D\rho_T(e/\varepsilon)^2 l\Delta\omega}{\pi A}\{[D/d^* l + (\rho_T e/\varepsilon)(dv/dE)^{ss}]^2 + \omega^2\}^{-1}. \tag{7.6.26}$$

All of the parameters in Eq. (7.6.26) are either known from experiment or, like d^*, can be estimated. Thus Eq. (7.6.26) provides an unambiguous expression for the variance of $\delta V(\omega)$ which can be compared to experiment.

At equilibrium Eq. (7.6.26) reduces to the Nyquist formula. Indeed, since at equilibrium d^* is very large, one has

$$S^e(\omega) = \frac{D\rho_T(e/\varepsilon)^2 l\Delta\omega}{\pi A}[(\rho_T e\mu/\varepsilon)^2 + \omega^2]^{-1}. \tag{7.6.27}$$

Furthermore, at equilibrium $\mu \simeq 0.8$ m^2 V^{-1} s^{-1} and $\rho_T \simeq 10^{21}$ m^{-3}, so that the characteristic Lorentzian frequency $\rho_T e\mu/\varepsilon$ in Eq. (7.6.27) is about 10^{12} Hz. Thus at frequencies less than 10^{10} Hz the dependence of the power spectrum on ω is negligible and

$$S^2(\omega) = Dl\Delta\omega/\pi A\mu^2\rho_T. \tag{7.6.28}$$

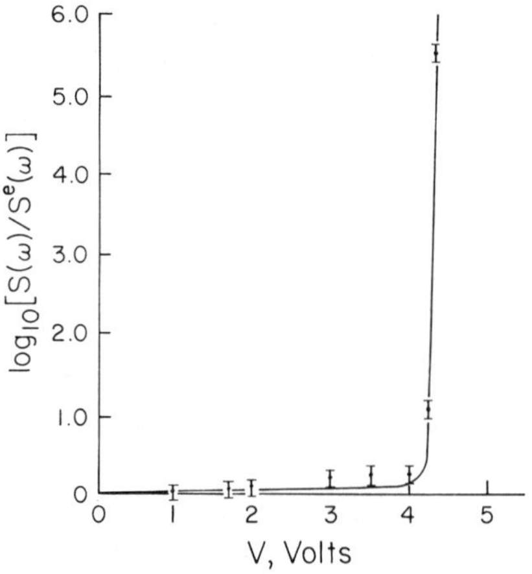

FIGURE 7.10. The power spectrum of a Gunn diode at 50 kHz with a 660-Hz band width as a function of the applied voltage. The curve is calculated on the basis of Eq. (7.6.26) and the points are taken from the experiments of Kabashimi, Yamazaki, and Kawakubo. From the points we have subtracted an additive constant corresponding to noise measured at equilibrium in excess of the Johnson noise. Taken from J. Keizer, *J. Chem. Phys.* **74**, 1350 (1981).

This expression can be shown to be identical with the Nyquist formula. To do so we need to use the Nernst–Einstein relationship, $D = k_B T\mu/e$, which holds since e/μ is the friction constant for the electron. Similarly, it is easy to show that the electrical resistance can be written as $R = l/eA\rho_T\mu$. Substituting these expressions into Eq. (7.6.28) then gives

$$S^e(\omega) = k_B TR\Delta\omega/\pi, \tag{7.6.29}$$

which is the Nyquist formula derived previously in Section 5.1. The resistance of the particular GaAs crystal used by Kabashimi, Yamazaki, and Kawakubo was 3.8 ohms. Thus, based on Eq. (7.6.29) one expects the variance of the equilibrium voltage fluctuations to be $3.2 \times 10^{-18}\ \mathrm{V}^2$.

The experimental value of the power spectrum at equilibrium observed by Kabashimi, Yamazaki, and Kawakubo was $3.5 \times 10^{-15}\ \mathrm{V}^2$. Since experiments on other systems have provided excellent confirmation of the Nyquist formula, it is sensible to assume that the additional noise arises from other sources. In this spirit, we have subtracted the excess equilibrium noise from the experimental data and plotted the result as the points in Fig. 7.10. The fit to the theoretical curve based on Eq. (7.6.26) is rather good, especially given the somewhat crude approximations that were made in deriving Eq. (7.6.22). The variance increases by almost six orders of magnitude when the applied voltage is changed from about 4 V to the threshold at 4.35 V. Although this is a tremendous increase, it is still rather small on the macroscopic scale. Indeed, in our calculation this gives a width at half-height of the probability density of the order of 10^{-6} V, which is still far below the value of the applied voltage. The quality of the agreement with experiment for the Gunn diode demonstrates the way in which the theory can be used to understand statistical effects in nonequilibrium systems.

References

Multiple Stationary States

I. Matheson, D.F. Walls, and C. Gardiner, Stochastic models of first order phase transitions in chemical reactions, *J. State. Phys.* **12**, 21–34 (1975).
J. Keizer, Maxwell-type constructions for multiple nonequilibrium steady states, *Proc. Nat. Acad. Sci. U.S.A.* **75**, 3023–3026 (1978).
I. Oppenheim, K. Shuler, and G. Weiss, Stochastic theory of nonlinear rate processes with multiple stationary states, *Physica* **88A**, 191–214 (1977).
E.C. Zimmermann and J. Ross, Light induced bistability in $S_2O_6F_2 \rightleftarrows SO_3F$: Theory and experiment, *J. Chem. Phys.* **80**, 720–729 (1984).
J. Kramer and J. Ross, Stabilization of unstable states, relaxation, and critical slowing down in a bistable system, *J. Chem. Phys.* **83**, 6234–6241 (1985).
J. Kramer and J. Ross, Thermochemical bistability in an illuminated liquid-phase system, *J. Phys. Chem.* **90**, 923–926 (1986).

Stability and Liapunov Functions

W. Hahn, *The Stability of Motion* (Springer-Verlag, Berlin, 1967).
L.S. Pontryagin, *Ordinary Differential Equations* (Addison-Wesley, Reading, MA, 1962), Chapter 5.

Fluctuation at Steady States

R. Kubo, K. Matsuo, and K. Kitahara, Fluctuation and relaxation of macrovariables, *J. Stat. Phys.* **9**, 51–96 (1973).
J. Keizer, Fluctuations, stability, and generalized state functions at nonequilibrium steady states, *J. Chem. Phys.* **65**, 4431–4444 (1976).
N.G. van Kampen, *Stochastic Processes in Physics and Chemistry* (North-Holland, Amsterdam, 1981).
D. Ronis, I. Procaccia, and I. Oppenheim, Statistical mechanics of stationary states III: Fluctuations in dense fluids with applications to light scattering, *Phys. Rev. A* **19**, 1324–1339 (1979).
T.R. Kirkpatrick, E.G.D. Cohen, and J.R. Dorfman, Fluctuations in a nonequilibrium steady state: Basic equations, *Phys. Rev. A* **26** 950–971 (1982).

Critical Points

M. Mangel, Simple theory of relaxation from instabilities, *Phys. Rev. A* **24**, 3226–3238 (1981).
R. Fox, Master equation derivation of Keizer's theory of nonequilibrium thermodynamics with critical fluctuations, *J. Chem. Phys.* **70**, 4660–4663 (1979).
D. McQuarrie and J. Keizer, Fluctuations in chemically reacting systems, in *Theoretical Chemistry: Advances and Perspectives*, Vol. 6A, D. Henderson, ed. (Academic Press, New York, 1981), pp. 165–213.
A. Nitzan, P. Ortoleva, J. Deutch, and J. Ross, Fluctuations and transitions at chemical instabilities: The analogy to phase transitions, *J. Chem. Phys.* **61**, 1056–1074 (1974).

Gunn Effect

J.B. Gunn, Microwave oscillations of current in III-V semiconductors, *Solid State Commun.* **1**, 88–91 (1963).
J.E. Carroll, *Hot Electron Microwave Devices* (Elsevier, New York, 1970).
S. Kabashimi, H. Yamazaki, and T. Kawakubo, Critical fluctuation near threshold of Gunn instability. *J. Phys. Soc. Japan* **40**, 921–924 (1976).

D.E. McCumber and A.G. Chynoweth, Theory of negative-conductance amplification of Gunn instabilities in "two-valley" semiconductors, *IEEE Trans. Elect. Devices* **13**, 4–21 (1966).
J. Keizer, Calculation of voltage fluctuations at the Gunn instability, *J. Chem. Phys.* **74**, 1350–1356 (1981).
A. Diaz-Guilera and J.M. Rubi, On fluctuations about nonequilibrium steady states near Gunn instability, *Physica* **135A**, 200–212 (1986).

CHAPTER 8

Thermodynamics and the Stability of Steady States

8.1. The Thermodynamic Stability of Equilibrium

In the preceding chapter we examined the relationship between fluctuations and the stability of steady states. That relationship depends on the dynamical character of the fluctuations, which incessantly explore nearby states testing for stability. At equilibrium there is another important connection between stability and macroscopic observations. This connection is provided by the Second Law of thermodynamics and leads to a quadratic Liapunov function for relaxation processes near equilibrium. This relationship is closely connected to the classical LeChatelier–Braun principle and, as we will see, actually follows in a simple fashion from what we learned in Chapter 7 regarding the dynamics of fluctuations. In the subsequent sections of this chapter, we show that a similar connection to thermodynamics can be made at nonequilibrium steady states by introducing a suitable generalization of the thermodynamic state functions.

To begin with, however, we shall assume that our steady state is an equilibrium state. Furthermore, we shall assume that the equilibrium state is far enough from a critical point that the aged equilibrium ensemble is stationary, Gaussian, and Markovian as described in Sections 4.6 and 7.3. Consequently, if $\mathbf{a} = \mathbf{n} - \mathbf{n}^e$ is the deviation of the extensive variables from their equilibrium value, then

$$d\mathbf{a}/dt = H\mathbf{a} + \tilde{\mathbf{f}} \tag{8.1.1}$$

and $\langle \tilde{\mathbf{f}}(t) \rangle = \mathbf{0}$ with

$$\langle \tilde{\mathbf{f}}(t)\tilde{\mathbf{f}}^T(t') \rangle = 2k_B L_S(t - t'), \tag{8.1.2}$$

where L_S is the symmetric part of the flux–force coupling matrix. The matrix L can be expressed in terms of the second derivative matrix of the entropy, S, and the relaxation matrix, H. According to Eq. (4.6.14) one has

$$L = HS^{-1}. \tag{8.1.3}$$

The connection between the Second Law and stability depends on the average behavior in the ensemble. On the conditional average, the deviations from equilibrium satisfy

$$d\bar{\mathbf{a}}/dt = H\bar{\mathbf{a}}, \tag{8.1.4}$$

where $\bar{\mathbf{a}}(\mathbf{a}^0, t) \equiv \int P_2(\mathbf{a}^0|\mathbf{a}, t)\mathbf{a}\, d\mathbf{a}$. For more general initial conditions, which nonetheless deviate little from the stationary ensemble, we can use the relationship

$$\begin{aligned}\langle \mathbf{a}(t)\rangle &= \int W_1(\mathbf{a}, t)\mathbf{a}\, d\mathbf{a} = \int\int W_1(\mathbf{a}^0, 0)P_2(\mathbf{a}^0|\mathbf{a}, t)\mathbf{a}\, d\mathbf{a}^0\, d\mathbf{a} \\ &= \int W_1(\mathbf{a}^0, 0)\bar{\mathbf{a}}(\mathbf{a}^0, t)\, d\mathbf{a}^0,\end{aligned} \tag{8.1.5}$$

to calculate unconditioned averages. Using Eq. (8.1.5) to average over both sides of Eq. (8.1.4) and recalling that H is independent of $\mathbf{a}^0$, we see that the equation satisfied by $\langle \mathbf{a}(t)\rangle$ is

$$d\langle \mathbf{a}(t)\rangle/dt = H\langle \mathbf{a}(t)\rangle. \tag{8.1.6}$$

Thus even in a nonequilibrium ensemble that is close to equilibrium an equation of the form of (8.1.4) gives the correct average time rate of change of the extensive variables.

According to the Second Law, which was derived in Section 4.5, the entropy of an isolated system increases on the average path. If the system is not isolated, but in contact with a reservoir that keeps the temperature of the system fixed, then one can also show that the Helmholtz free energy decreases on the average path. Similarly, at constant temperature and pressure, the Gibbs free energy decreases on the average path. These dynamical statements of the Second Law can be combined with the thermodynamic extremum principles to show that the second differential of the entropy is a Liapunov function.

Consider, first, the case of an isolated system, which is described by the so-called microcanonical ensemble. The Second Law then states that

$$dS(\bar{\mathbf{n}})/dt \geq 0 \tag{8.1.7}$$

with equality only at equilibrium. We have seen in Chapter 4 that this follows from the canonical form. In particular, near equilibrium we can expand the entropy in a Taylor series as in Section 2.3, to see that

$$d\delta^2 S/dt \geq 0, \tag{8.1.8}$$

where $\delta^2 S$ is the second differential of the entropy, i.e.,

$$\delta^2 S = \sum_{i,j} (\partial^2 S/\partial n_i \partial n_j)^e a_i a_j. \tag{8.1.9}$$

The first differential of the entropy does not appear in Eq. (8.1.8) since it vanishes at equilibrium by the entropy maximum principle. Indeed, the entropy maximum principle implies for an isolated system that

$$S(\bar{\mathbf{n}}) - S(\mathbf{n}^e) = \frac{1}{2}\delta^2 S + \cdots \leq 0. \tag{8.1.10}$$

Since this inequality must hold when $\bar{\mathbf{n}}$ is close to equilibrium, it follows that

$$\delta^2 S \leq 0. \tag{8.1.11}$$

Thus the second differential of the entropy is a negative definite quadratic form. Equation (8.1.8) shows, however, that the time derivative of $\delta^2 S$ is positive definite. As we saw in Section 7.2, this implies that the second differential of the entropy is a negative definite Liapunov function. Indeed, since equality in both equations occurs only when $\bar{\mathbf{n}} = \mathbf{n}^e$, it follows that equilibrium states that are noncritical are asymptotically stable.

This connection between the Second Law and stability at equilibrium is independent of the sort of equilibrium ensemble one is considering. For example, the canonical ensemble involves systems in thermal contact with heat baths. In the canonical ensemble the Second Law implies that the Helmholtz free energy, $A = E - TS$, is monotone decreasing, i.e.,

$$[dA(\bar{\mathbf{n}})/dt]_T \leq 0, \tag{8.1.12}$$

where the subscript emphasizes that the temperature must be held constant. In this ensemble the thermodynamic extremum principle is a minimum principle. This means that the Helmholtz free energy is a minimum at equilibrium for changes at constant temperature. Thus from its Taylor expansion around $\mathbf{n}^e$ it follows that

$$[A(\bar{\mathbf{n}}) - A^e]_T = \frac{1}{2}\delta^2 A_T + \cdots \geq 0, \tag{8.1.13}$$

where the second differential of A, $\delta^2 A_T$, must be evaluated at constant temperature. Close to equilibrium Eq. (8.1.13) shows that

$$\delta^2 A_T \geq 0, \tag{8.1.14}$$

while Eqs. (8.1.12) and (8.1.13) show that

$$d\delta^2 A_T/dt \leq 0. \tag{8.1.15}$$

Thus in the canonical ensemble the second differential of the Helmholtz free energy taken at constant temperature is a Liapunov function.

It is not difficult to see that $\delta^2 A_T$ is proportional to $\delta^2 S$. To do so, we write the left-hand side of Eq. (8.1.13) as

$$\begin{aligned}[A(\bar{\mathbf{n}}) - A^e]_T &\equiv \bar{E} - TS(\bar{\mathbf{n}}) - [E^e - TS(\mathbf{n}^e)] \\ &= \delta E - T[S(\bar{\mathbf{n}}) - S(\mathbf{n}^e)]. \end{aligned} \tag{8.1.16}$$

Next we introduce the Taylor expansion of the entropy to obtain

$$[A(\bar{\mathbf{n}}) - A^e]_T = (\delta E - T\delta S) - \frac{1}{2}T\delta^2 S + \cdots. \tag{8.1.17}$$

The first term in parentheses on the right-hand side of Eq. (8.1.17) is the linear term in the Taylor expansion and it equals δA_T, while the next term is second order and it must be $\frac{1}{2}\delta^2 A_T$. This second identification implies that

$$\delta^2 A_T = -T\delta^2 S. \tag{8.1.18}$$

Since the temperature is constant in the canonical ensemble, Eq. (8.1.18) shows that the thermodynamic Liapunov function in this ensemble can still be taken to be $\delta^2 S$.

A similar argument can be made for other equilibrium ensembles to show that the second differential of the entropy is a Liapunov function. For example, in the isothermal, isobaric ensemble both T and p are fixed by reservoirs. In that ensemble, the Gibbs free energy is minimized at equilibrium and it is easy to verify that

$$\delta^2 G_{T,p} = -T\delta^2 S. \tag{8.1.19}$$

A somewhat more systematic set of state functions for our purpose are the Mathieu functions. These functions are Legendre transforms of the entropy, rather than the internal energy, and, like the entropy, satisfy maximum principles in the various ensembles. In the canonical ensemble, for example, the appropriate Mathieu function is

$$\psi \equiv S - E/T = -A/T. \tag{8.1.20}$$

This Mathieu function is monotone increasing and maximized at equilibrium under the condition of constant temperature. Under this latter restriction, Eq. (8.1.20) shows that its second differential is

$$\delta^2 \psi_T = -\delta^2 A_T/T = \delta^2 S, \tag{8.1.21}$$

where the second equality follows from Eq. (8.1.18). Similarly, in the isothermal, isobaric ensemble the Mathieu function is

$$\psi' \equiv S - E/T - pV/T = -G/T \tag{8.1.22}$$

and

$$\delta^2 \psi'_{T,p} = \delta^2 S. \tag{8.1.23}$$

In terms of the properties of these somewhat less familiar state functions, we see that the second differential of the entropy provides all the thermodynamic information about the stability of equilibrium ensembles.

The thermodynamic criterion of stability at equilibrium can be used to prove the LeChatelier–Braun principle. As originally formulated this principle states that for any system in equilibrium a variation in a parameter that affects the equilibrium will be accompanied by a change which tends to cancel the initial variation. This principle is quite general although we prove it here only for chemical reactions at constant temperature and pressure. In this ensemble, the Liapunov function is $\delta^2 G_{T,p}$. To obtain the form of this function we write out the differential of the Gibbs free energy as

$$dG = -S\,dT + V\,dp + \sum_i \mu_i\,dN_i. \tag{8.1.24}$$

Since the molecule numbers change only as a result of chemical reaction, we can use the progress variable $d\xi = dN_i/\nu_i$ introduced in Section 3.8 to write

$$\delta G = -S\delta T + V\delta p + \Delta G_{rx}\delta\xi, \tag{8.1.25}$$

where ΔG_{rx} is the Gibbs free energy of reaction. Thus at constant temperature and pressure the second differential of the Gibbs free energy is

$$\delta^2 G_{T,p} = (\partial \Delta G_{rx}/\partial\xi)_{T,p}(\delta\xi)^2. \tag{8.1.26}$$

Since $\delta^2 G_{T,p}$ is a Liapunov function, it follows that

$$(\partial \Delta G_{rx}/\partial\xi)_{T,p} > 0 \tag{8.1.27}$$

for a stable chemical equilibrium.

To prove the LeChatelier–Braun principle we consider the criterion of chemical equilibrium at constant temperature and pressure obtained in Section 3.8 using the canonical form, i.e.,

$$\Delta G_{rx} = 0. \tag{8.1.28}$$

Notice that this also follows from Eq. (8.1.25) and the fact that G is minimized at equilibrium when T and p are held constant. The LeChatelier–Braun principle deals with perturbations that change the chemical equilibrium from one equilibrium state to another. Thus the perturbation will involve changes of p, T, and the progress variable, ξ. Since the equilibrium condition in Eq. (8.1.28) is the same in any equilibrium state, it follows that

$$\begin{aligned}\delta\Delta G_{rx} &= (\partial \Delta G_{rx}/\partial T)_{\xi,p}\delta T + \partial(\Delta G_{rx}/\partial p)_{\xi,T}\delta p \\ &\quad + (\partial \Delta G_{rx}/\partial\xi)_{T,p}\delta\xi = 0.\end{aligned} \tag{8.1.29}$$

This can be rewritten using the fact that $\Delta G_{rx} = (\partial G/\partial\xi)_{T,p}$ and Maxwell relationships like $\partial^2 G/\partial\xi\partial T = \partial^2 G/\partial T\partial\xi = -(\partial S/\partial\xi)_{T,p}$ to obtain

$$-(\partial S/\partial\xi)_{T,p}\delta T + (\partial V/\partial\xi)_{T,p}\delta P + (\partial \Delta G_{rx}/\partial\xi)_{T,p}\delta\xi = 0. \tag{8.1.30}$$

The quantity $(\partial S/\partial\xi)_{T,p}$ is the change of entropy due to reaction, ΔS_{rx}, and $(\partial V/\partial\xi)_{T,p} = \Delta V_{rx}$, the change of volume due to reaction. Substituting these expressions into Eq. (8.1.30) we see that changes in T, p, and ξ are related by the expression

$$\delta\xi = (\Delta S_{rx}\delta T - \Delta V_{rx}\delta P)/(\partial \Delta G_{rx}/\partial\xi)_{T,p}. \tag{8.1.31}$$

For example, if the pressure is changed at constant temperature, then

$$(\partial\xi/\partial P)_T = -\Delta V_{rx}/(\partial \Delta G_{rx}/\partial\xi)_{T,p}. \tag{8.1.32}$$

As we found in Eq. (8.1.27), the stability of the equilibrium state implies that the denominator of the right-hand side of Eq. (8.1.32) is positive. Thus if the pressure is increased when the change in volume on reaction is positive, the chemical reaction compensates by decreasing the amount of product. On the other hand, if δP is positive and ΔV_{rx} is negative, then the reaction compensates by increasing the amount of product.

For temperature changes which occur at constant pressure the comparable expression obtained from Eq. (8.1.31) is

$$(\partial\xi/\partial T)_p = \Delta S_{rx}/(\partial \Delta G_{rx}/\partial\xi)_{T,p}. \tag{8.1.33}$$

This can be rewritten using the fact that $\Delta H_{rx} = T\Delta S_{rx}$, where ΔH_{rx} is the change in enthalpy or heat of reaction. This follows directly from the condition of chemical equilibrium in Eq. (8.1.28) and allows us to write

$$(\partial\xi/\partial T)_p(\partial \Delta G_{rx}/\partial\xi)_{T,p} = \Delta H_{rx}/T. \tag{8.1.34}$$

Thus, for example, if heat is released in a chemical reaction (i.e., $\Delta H_{rx} < 0$), the effect of an increase in temperature is to reverse the direction of the reaction. Since a reversal in the reaction causes the system to take up heat, this response tends to diminish the original increase in temperature. This is the basic content of the LeChatelier–Braun principle.

There are other useful equilibrium thermodynamic relationships that stem from the thermodynamic Liapunov function $\delta^2 S$. As an example, consider a single-component fluid for which the second differential of the entropy functional is given by Eq. (6.9.11), i.e.,

$$\begin{aligned}\delta^2 S = \iint &\{[\delta 1/T(\mathbf{r})/\delta e(\mathbf{r}')]\delta e(\mathbf{r})\delta e(\mathbf{r}') + 2[\delta 1/T(\mathbf{r})/\delta\rho(\mathbf{r}')]\delta e(\mathbf{r})\delta\rho(\mathbf{r}') \\ &+ [\delta - \mu/T(\mathbf{r})/\delta\rho(\mathbf{r}')]\delta\rho(\mathbf{r})\delta\rho(\mathbf{r}')\}\, d\mathbf{r}\, d\mathbf{r}'.\end{aligned} \tag{8.1.35}$$

In the spatially local approximation the functional derivatives in the integrand are approximated using a delta function spatial dependence as described in Section 6.9. Thus in this approximation

$$\begin{aligned}\delta^2 S = \int &\{[\partial(1/T)/\partial e][\partial e(\mathbf{r})]^2 + 2[\partial(1/T)/\partial\rho]\delta e(\mathbf{r})\delta\rho(\mathbf{r}) \\ &+ [\partial(-\mu/T)/\partial\rho][\delta\rho(\mathbf{r})]^2\}\, d\mathbf{r}.\end{aligned} \tag{8.1.36}$$

The Liapunov property of $\delta^2 S$, however, requires that $\delta^2 S \leq 0$ for all possible variations $\delta e(\mathbf{r})$ and $\delta\rho(\mathbf{r})$. This can only be true if the integrand in Eq. (8.1.36) is negative definite since the variations could be localized anywhere. Consequently we have

$$[\partial(1/T)/\partial e]\rho(\delta e)^2 + 2[\partial(1/T)/\partial\rho]_e\delta e\delta\rho + [\partial(-\mu/T)/\partial\rho]_e(\delta\rho)^2 \leq 0. \tag{8.1.37}$$

There are two inequalities which follow immediately from the definiteness of this quadratic form. Considering first variations for which $\delta\rho$ vanishes and then variations for which δe vanishes, it is clear that

$$[\partial(1/T)/\partial c]_\rho < 0 \quad \text{and} \quad [\partial(\ \ \mu/T)/\partial\rho]_e < 0. \tag{8.1.38}$$

Carrying out the indicated differentiation in the first of these inequalities we see that

$$1/T^2 c_V > 0, \tag{8.1.39}$$

where c_V is the heat capacity per unit volume. Equation (8.1.39) proves that away from an equilibrium critical point the constant-volume heat capacity is positive.

Other interesting inequalities can be obtained by changing variables from the internal energy to the temperature using the relationship

$$\delta e = c_V \delta T + (\partial e/\partial\rho)_T \delta\rho. \tag{8.1.40}$$

Substituting Eq. (8.1.40) for δe into Eq. (8.1.37) and collecting terms yields

$$-c_V T^{-2}(\delta T)^2 - \{T^{-2}(\partial T/\partial\rho)_e(\partial e/\partial\rho)_T + [\partial(\mu/T)/\partial\rho]_e\}(\delta\rho)^2 \le 0, \tag{8.1.41}$$

where we have used the thermodynamic identity $(\partial e/\partial\rho)_T = -(\partial e/\partial T)_\rho(\partial T/\partial\rho)_e$ to eliminate the cross term and the identity $(\partial T/\partial e)_\rho(\partial e/\partial\rho)_T = -(\partial T/\partial\rho)_e$ to simplify the coefficient of $(\delta\rho)^2$. Clearly, for the inequality to hold the coefficients of both terms in Eq. (8.1.41) must be negative. Applying this inequality to the coefficient of $(\delta T)^2$ gives the same result as Eq. (8.1.39), while for the coefficient of $(\delta\rho)^2$ one obtains

$$[\partial(-\mu/T)/\partial e]_\rho(\partial e/\partial\rho)_T + [\partial(-\mu/T)/\partial\rho]_e < 0, \tag{8.1.42}$$

where we have used a Maxwell relationship to write $[\partial(1/T)/\partial\rho] = [\partial(-\mu/T)/\partial e]$. Using a well-known thermodynamic identity, the left-hand side of Eq. (8.1.42) can be shown to equal $[\partial(-\mu/T)/\partial\rho]_T$. Thus we find that

$$[\partial(-\mu/T)/\partial\rho]_T < 0. \tag{8.1.43}$$

For a single-component system, this is the analogue of the multicomponent chemical reaction stability criterion given in Eq. (8.1.27).

The inequality in Eq. (8.1.43) can be written in a more transparent from using the Gibbs–Duhem relationship in Eq. (2.4.29), which shows that

$$(\partial p/\partial\rho)_T = \quad T\rho[\partial(\ \ \mu/T)/\partial\rho]_T. \tag{8.1.44}$$

Combining Eq. (8.1.44) with Eq. (8.1.43) it follows that

$$1/\kappa_T \equiv -\rho(\partial p/\partial\rho)_T < 0. \tag{8.1.45}$$

The function κ_T is the isothermal compressibility, and according to Eq. (8.1.45) κ_T is strictly negative at a stable, noncritical equilibrium state. This inequality becomes an equality at the critical point of a first-order phase transition. As is well known, the critical isotherm in a graph of pressure versus density has a point of inflection at the critical point. Thus at this point the pressure satisfies the conditions

$$(\partial p/\partial \rho)_T^c = (\partial^2 p/\partial \rho^2)_T^c = 0, \quad (\partial^3 p/\partial \rho^3)_T^c > 0, \tag{8.1.46}$$

and the isothermal compressibility is infinite.

The first of the equalities in Eq. (8.1.46) implies that $\delta^2 S$ actually vanishes at the critical point for nonzero values of $\delta\rho$. This has an important significance for the density fluctuations at the critical point. Recall that the Einstein formula in Eq. (2.5.14) relates the statistical fluctuations at equilibrium to the inverse of the second derivative matrix of the entropy. Since $\delta^2 S$ is only positive semi-definite at the critical point, the inverse of the second derivative matrix S does not exist there. This is signaled as one approaches the critical point by an anomalous increase in the density–density correlation function. Indeed, from Eq. (2.5.16) we can see that local fluctuations in the density are given by the equation

$$\langle(\delta\rho)^2\rangle = k_B T \rho^2 \kappa_T / V, \tag{8.1.47}$$

with V the volume of the system. At the critical point, Eq. (8.1.46) shows that κ_T diverges, which implies that the variance of the density fluctuations does also. In fact, the spatial correlations of the density fluctuations become extremely long-ranged close to the critical point. This can also be seen using Eq. (6.9.40), which relates the compressibility to the radial distribution function $g(r)$ by the formula

$$\kappa_T = -(V/k_B T)\left\{1 + 4\pi \int_0^\infty [g(r) - 1] r^2 \, dr\right\}. \tag{8.1.48}$$

Away from critical points $g(r)$ approaches its asymptotic value of unity within a few molecular diameters. At the critical point, however, κ_T becomes infinite and it follows from the integral expression in Eq. (8.1.48) that the radial distribution function must become extremely long-ranged there. Indeed, to make the integral infinite, $g(r)$ must have an asymptotic behavior at least as long-ranged as

$$[g(r) - 1] \sim 1/r^3. \tag{8.1.49}$$

It is interesting to note that the positive semi-definiteness of the second differential of the entropy is not limited to equilibrium states. Recall that to calculate $\delta^2 S$ we have used the local equilibrium entropy, that is, the same functional form that the entropy has at equilibrium, but with nonequilibrium values of the extensive variables in its arguments. Thus even when evaluated at a nonequilibrium state, the second derivatives of this entropy function will satisfy all the same inequalities as they do at equilibrium. Consequently, as long as a state $\mathbf{n}$ is noncritical, the second differential of the local equilibrium entropy evaluated at this state satisfies

$$\delta^2 S \leq 0, \tag{8.1.50}$$

with equality only when $\delta\mathbf{n}$ vanishes. Because the second differential is a definite quadratic form, we might hope that it is still a Liapunov function at nonequilibrium steady states. Indeed, if we could show that near the steady state

$$d\delta^2 S/dt \geq 0, \tag{8.1.51}$$

then $\delta^2 S$ would be a Liapunov function. Unfortunately, at nonequilibrium steady states we no longer have the Second Law and the maximum principle to guarantee that Eq. (8.1.51) holds. In fact, numerous examples are known for which $d\delta^2 S/dt$ has no definite sign at nonequilibrium steady states. This, however, conveys no information about the stability of the state, since as we saw in Section 7.2, the existence of a particular quadratic form as a Liapunov function is only a *sufficient* criterion for stability.

Nevertheless, Glansdorff and Prigogine adopted this thermodynamic approach and for some time referred to the possible Liapunov property of $\delta^2 S$ as the generalized evolution criterion. Because the existence of $\delta^2 S$ as a Liapunov function is not a necessary condition for stability, the criterion frequently fails. Although it may be useful for steady states quite close to equilibrium, it has no thermodynamic significance for steady states that are far from equilibrium.

There is another way of thinking about the relationship between $\delta^2 S$ and stability that extends in a natural way to steady states. As we know the inverse of the matrix of second derivatives of S is related to fluctuations by the Einstein formula, i.e.,

$$(S^{-1})_{ij} = -\langle \delta n_i \delta n_j \rangle^e / k_B. \tag{8.1.52}$$

Alternatively, if we know the equilibrium covariance matrix $\sigma^e_{ij} = \langle \delta n_i \delta n_j \rangle^e$, we can invert Eq. (8.1.52) to obtain the second derivatives of S in the form

$$S_{ij} = -k_B(\sigma^{e-1})_{ij}. \tag{8.1.53}$$

Recall that σ^e will be a positive definite matrix if the extensive variables are independent of one another and the equilibrium state is noncritical. Thus it follows that σ^{e-1} is positive definite, too, so that S is negative definite. Using the definition of $\delta^2 S$ in Eq. (8.1.9) we conclude immediately that $\delta^2 S$ is a negative definite quadratic form.

More than this can be proven if we consider the dynamical properties of the fluctuations. According to the fluctuation–dissipation theorem in Eqs. (8.1.2) and (8.1.3) we have

$$\gamma^e/k_B = HS^{-1} + S^{-1}H^T, \tag{8.1.54}$$

where $\gamma^e = 2k_B L_S$ is the covariance matrix of the random terms $\tilde{\mathbf{f}}$. Because the intensive variables are independent, γ^e is a positive definite matrix. Indeed, from its definition in Eq. (4.3.10) we see that

$$\sum_{i,j} \gamma_{ij}^e x_i x_j = 2 \sum_{i,j} \sum_{\kappa} \omega_{\kappa i} x_i x_j V_\kappa^{+e} \omega_{\kappa j}$$
$$= 2 \sum_{\kappa} V_\kappa^{+e} \left(\sum_i \omega_{\kappa i} x_k \right)^2 \geq 0. \tag{8.1.55}$$

Thus as long as any conservation conditions among the extensive variables have been removed (cf. Section 3.7), the quadratic form is positive definite. We have seen Eq. (8.1.54) before. Indeed in Eq. (7.2.28) we discovered that any positive definite symmetric matrix V^{-1} which satisfies

$$G = HV^{-1} + V^{-1}H^T \tag{8.1.56}$$

with G a negative definite matrix provides a Liapunov function for H. Thus multiplying Eq. (8.1.54) by minus one and making the identifications $G = -\gamma^e/k_B$ and $-S^{-1} = V^{-1}$, we see that $-\delta^2 S$ is a Liapunov function and, therefore, that $\delta^2 S$ is one, too.

Notice that we have not had to invoke the Second Law in this proof. All that we relied upon is the Einstein relationship in Eq. (8.1.52) to show that $\delta^2 S$ is negative definite and the fluctuation–dissipation theorem to show that $d\delta^2 S/dt$ is positive definite. In this way the fact that the entropy is a Liapunov function is completely contained in the statistical properties of the equilibrium ensemble.

8.2. Fluctuations and Stability at Steady States

The property that the entropy is a Liapunov function at equilibrium follows from the fluctuation–dissipation theorem. It is instructive to rewrite the proof a little, which will show how to obtain a comparable Liapunov function at nonequilibrium steady states. In Eq. (8.1.54) we have written the fluctuation–dissipation theorem in terms of the matrix of second derivatives of the entropy. Using Eq. (8.5.53) it can be written in the equivalent form

$$H\sigma^e + \sigma^e H^T = -\gamma^e. \tag{8.2.1}$$

It is easy to see that σ^e, and therefore σ^{e-1}, is a positive definite matrix. Indeed,

$$\sum_{i,j} \sigma_{ij}^e x_i x_j = \sum_{i,j} \langle \delta n_k \delta n_j \rangle^e x_i x_j = \left\langle \sum_i \delta n_i x_i \sum_j \delta n_j x_j \right\rangle^e$$
$$= \left\langle \left(\sum_i \delta n_i x_i \right)^2 \right\rangle^e \geq 0. \tag{8.2.2}$$

Since we have assumed that the extensive variables are independent, only when all the x_i vanish will $\sum_i \delta n_i x_i$ also vanish. Thus σ is positive definite.

According to Eq. (8.1.55), γ^e is also positive definite. Consequently $-\gamma^e$ is negative definite and Eq. (8.2.1) is equivalent to Eq. (8.1.56), which implies that the inverse of the covariance function, σ^{e-1}, is a stability matrix.

There is nothing about this proof that is restricted to equilibrium. Indeed, in Section 7.3 we showed that every asymptotically stable steady state that is noncritical has a stationary, Gaussian single-time probability density of the form

$$W_1(\mathbf{n}) = [(2\pi)^k \det \sigma^{ss}]^{-1/2} \exp[-(\mathbf{n} - \mathbf{n}^{ss})^T(\sigma^{ss})^{-1}(\mathbf{n} - \mathbf{n}^{ss})/2]. \quad (8.2.3)$$

The covariance matrix, σ^{ss}, solves the generalized fluctuation–dissipation theorem in Eq. (7.3.25), which we write as

$$H\sigma^{ss} + \sigma^{ss}H^T = -\gamma^{ss}. \quad (8.2.4)$$

As long as the extensive variables are chosen to be dynamically independent, it follows from comparing Eqs. (8.1.56) and (8.2.4) that σ^{ss-1} is a stability matrix for H. In other words, the positive definite quadratic form

$$\sum_{i,j} (\sigma^{ss-1})_{ij} a_i a_j, \quad (8.2.5)$$

where $a_i \equiv n_i - n_i^{ss}$, is a Liapunov function. Notice that this is true for all asymptotically stable steady states.

The quadratic form in Eq. (8.2.5) provides an important generalization of the stability property of the second differential of the entropy. In fact, when the steady state is an equilibrium state, the form in Eq. (8.2.5) reduces to the thermodynamic Liapunov function. This is easy to see using Eq. (8.1.53), which shows that $\sigma^{e-1} = -S/k_B$, and the quadratic form in Eq. (8.2.5) becomes $-\delta^2 S/k_B$. Thus the inverse of the covariance matrix gives rise to a Liapunov function that is a natural generalization of the equilibrium thermodynamic Liapunov function. This suggests that σ_{ij}^{ss-1} may be a state function. In Section 8.3 it is shown that this is, indeed, the case.

Fluctuations can be used in another way to assess the stability of a steady state. Consider a state, $\mathbf{n}^{ss}$, which is known to be a steady state, but whose stability is as yet undetermined. Accordingly, the extensive variables are all independent of time, or using the notation of Chapter 7,

$$\mathbf{R}(\mathbf{n}^{ss}) = \mathbf{0}. \quad (8.2.6)$$

Thus if we examine the trajectories that begin at $\mathbf{n}^{ss}$, the fluctuation–dissipation theory predicts that their conditional average value has a constant value $\bar{\mathbf{n}}(\mathbf{n}^{ss}, t) = \mathbf{n}^{ss}$. The fluctuations associated with these trajectories can be determined using the fluctuation–dissipation theory. They satisfy

$$d\delta\mathbf{n}/dt = H(\mathbf{n}^{ss})\delta\mathbf{n} + \tilde{\mathbf{f}}(t) \quad (8.2.7)$$

with

$$\langle \tilde{\mathbf{f}}(t)\tilde{\mathbf{f}}^T(t') \rangle = \gamma(\mathbf{n}^{ss})\delta(t - t'), \quad (8.2.8)$$

as we see from Eq. (4.4.2). Because $\mathbf{n}^{ss}$ is a fixed point, $H \equiv H(\mathbf{n}^{ss})$ and $\gamma^{ss} \equiv \gamma(\mathbf{n}^{ss})$ are independent of time. As a consequence, Eq. (4.4.8), which describes the time dependence of the conditional covariance matrix, takes the form

$$\sigma(\mathbf{n}^{ss}, t) = \int_0^t \exp(Hs)\gamma^{ss} \exp(H^T s)\, ds. \tag{8.2.9}$$

Whenever the matrix γ^{ss} is positive definite, the long-time behavior of the conditional covariance matrix provides a necessary and sufficient condition for the asymptotic stability of the steady state. Indeed, if the limit

$$\lim_{t\to\infty} \sigma(\mathbf{n}^{ss}, t) \equiv \sigma^{ss} \tag{8.2.10}$$

exists, then the steady state is asymptotically stable. Furthermore, if the steady state is asymptotically stable, then the limit in Eq. (8.2.10) exists.

The easy part of the proof is that asymptotic stability implies that the limit in Eq. (8.2.10) exists. Indeed, we have already implicitly used this fact in showing that the steady-state covariance can be calculated using Eq. (7.3.41). The formal proof requires the expansion of $\exp(Hs)$ in terms of its component matrices given in Eq. (7.2.40). Because of asymptotic stability, all the eigenvalues of H have negative real parts. Thus all the terms in the integrand of Eq. (8.2.9) are exponentially damped and the limit

$$\sigma^{ss} \equiv \int_0^\infty \exp(Hs)\gamma^{ss} \exp(Hs)\, ds \tag{8.2.11}$$

exists. On the other hand, if the limit exists, then the positive definiteness of γ^{ss} implies that σ^{ss} is positive definite. Furthermore, differentiating Eq. (8.2.9) with respect to the time we see that

$$d\sigma/dt = H\sigma + \sigma H^T + \gamma^{ss}. \tag{8.2.12}$$

Moreover, since σ^{ss} exists, it follows that $\lim_{t\to\infty} d\sigma/dt = 0$. Thus from Eq. (8.2.12)

$$H\sigma^{ss} + \sigma^{ss} H^T = -\gamma^{ss}.$$

This relationship is just the fluctuation–dissipation theorem and is what one requires to show that σ^{ss-1} is a stability matrix for $\mathbf{n}^{ss}$. Thus the steady state is asymptotically stable.

From the details of the proof, we see that there is another way of expressing this stability criterion. Since $\sigma(\mathbf{n}^{ss}, t)$ as defined in Eq. (8.2.9) is a positive definite matrix, it follows that the existence of a finite, positive definite solution to the fluctuation–dissipation theorem is a necessary condition for asymptotic stability. It is also sufficient, since the existence of such a solution implies that σ^{ss-1} is a Liapunov function.

To help understand these ideas, it is instructive to look at several simple examples. Consider first the *o*-cresolphthalein reaction treated in Section 7.4.

In that system the acid–base reaction in Eq. (7.4.24) is driven to a nonequilibrium steady state at a temperature T^{ss} greater than that of the surrounding thermal reservoir. Because of the rapid time scale on which the relaxation of the chemical reaction occurs, it was only necessary to treat internal energy fluctuations in that system. In this approximation only the conditional variance of δE is required to test the stability of the system. The integral in Eq. (8.2.9), which provides the necessary and sufficient condition for stability, can be obtained from Eqs. (7.4.31) and (7.4.32). One finds that

$$\sigma(E^{ss}, t) = 2\beta k_B T_R^2 \int_0^t \exp\{2[I_0(dA/d\bar{T}) - \beta]C_V^{-1}s\}\, ds. \tag{8.2.13}$$

This integral can be carried out explicitly since all the physical parameters are independent of time. As long as $I_0(dA/d\bar{T}) - \beta \neq 0$,

$$\sigma(E^{ss}, t) = (k_B T_R^2 C_V/[1 - (dA/d\bar{T})I_0/\beta]) \times (\exp\{-2[1 - (dA/d\bar{T})I_0/\beta]\beta C_V^{-1}t\} - 1) \tag{8.2.14}$$

whereas when $I_0(dA/d\bar{T}) - \beta = 0$,

$$\sigma(E^{ss}, t) = 2\beta k_B T_R^2 t.$$

From an examination of these expressions it is clear that the infinite time limit of $\sigma(E^{ss}, t)$ exists if and only if

$$(dA/d\bar{T})I_0/\beta < 1. \tag{8.2.15}$$

According to Eq. (8.2.10), this is a necessary and sufficient condition for asymptotic stability. Indeed, Eq. (8.2.15) is identical to the criterion in Eq. (7.4.33), which was derived on the basis of linear stability analysis. In the case that the steady state is stable, the limiting form of $\sigma(E^{ss}, t)$ is seen from Eq. (8.2.14) to be

$$\sigma^{ss} = k_B T_R^2 C_V/[1 - (dA/d\bar{T})I_0/\beta], \tag{8.2.16}$$

which agrees with the result in Eq. (7.4.36) based on the fluctuation–dissipation theorem. Notice that for values of the absorbance between the two critical points illustrated for this system in Fig. 7.6, Eq. (8.2.14) shows that the infinite time limit of $\sigma(E^{ss}, t)$ does not exist. Thus, as was established earlier, those steady states are unstable.

A more involved example of the stability criterion is provided by a two-state system that can be excited by radiation. We have in mind an organic molecule with a triplet state that is rapidly populated by intersystem crossing from an excited singlet state. Repopulation of the ground singlet state occurs by collisional energy transfer and phosphorescence from the triplet state. For molecules such as naphthalene and benzene, relaxation times for intersystem crossing are the order of 10^{-7}–10^{-8} s, while phosphorescene lifetimes are the order of 1–100 s. Thus it is a good approximation to treat these molecules as consisting of the ground singlet state and the excited triplet state. If N_0 and N_1 are the number of molecules in each of these states, then molecule conservation implies that $N_0 + N_1 = N$, the total number of molecules.

Assuming that the ground state absorbs radiation of frequency ω, then the absorbance at this frequency is given by Beer's law in Eq. (7.4.9) as

$$A = 1 - \exp(-\varepsilon\bar{\rho}_0 l), \tag{8.2.17}$$

where $\bar{\rho}_0$ is the number density of ground state molecules. For simplicity we can take the molecular extinction coefficient to be small, so that $A \simeq \varepsilon l\bar{\rho}_0$. Thus the rate of transitions to the excited state induced by the radiation can be written

$$(dN_0/dt)_{\text{rad}} = -I_0\varepsilon l\bar{\rho}_0 \equiv -f\bar{\rho}_0, \tag{8.2.18}$$

with I_0 the intensity of the incoming radiation. The average rate of change of the number of ground state molecules involves this term plus others due to molecular processes and has the form

$$d\bar{N}_0/dt = \Omega\exp(-\varepsilon/k_BT)[\exp(\bar{\mu}_1/k_BT) - \exp(\bar{\mu}_0/k_BT)] - f\bar{\rho}_0. \tag{8.2.19}$$

The elementary process in Eq. (8.2.19) represents unimolecular transitions between the two states. The factor $\exp(-\varepsilon/k_BT)$ comes from the internal energy required for these transitions, which is the same for both the forward and the reverse step, as discussed in Section 5.6. If the system is in contact with a thermal reservoir at temperature T_R, then the average rate of change of internal energy is

$$d\bar{E}/dt = \hbar\omega f\bar{\rho}_0 + \kappa(1/\bar{T} - 1/T_R), \tag{8.2.20}$$

where we have taken the linearized canonical form for heat transport from Eq. (4.5.30) expressed in terms of $1/T$, rather than T, and have assumed that radiation losses, e.g., from phosphorescence, can be neglected.

The steady states in this system are determined by the equation

$$\Omega\exp(-\varepsilon/k_BT^{ss})[\exp(\mu_1^{ss}/k_BT^{ss}) - \exp(\mu_0^{ss}/k_BT^{ss})] = f\rho_0^{ss} \tag{8.2.21a}$$

$$-\kappa(1/T^{ss} - 1/T_R) = \hbar\omega f\rho_0^{ss}, \tag{8.2.21b}$$

along with the conservation condition $\rho_0^{ss} + \rho_1^{ss} = \rho$. In order to use the fluctuations to assess the stability of the steady states, we need to obtain the relaxation matrix H and the covariance of the random terms γ, both evaluated at ρ^{ss}, T^{ss}. The covariance matrix involves only terms for the molecular excitation process and heat transfer to the reservoir. Applying Eq. (4.2.9) and the steady-state conditions in Eqs. (8.2.21a) and (8.2.21b) to this problem gives

$$\gamma^{ss} = \begin{pmatrix} 2\Omega\exp\left(\dfrac{\mu_0^{ss} - \varepsilon}{k_BT^{ss}}\right) + f\rho_0^{ss} & 0 \\ 0 & 2k_B\kappa \end{pmatrix}. \tag{8.2.22}$$

The relaxation matrix is most easily expressed in terms of the affinity, $\mathscr{A} = (\mu_1 - \mu_0)/T$. Linearization of Eqs. (8.2.19) and (8.2.20) gives

$$
\begin{aligned}
H_{11} &= (\Omega/k_B)\exp[(\mu_0^{ss} - \varepsilon)/k_B T^{ss}](\partial\mathscr{A}/\partial N)^{ss} \\
&\quad + (f\rho_0^{ss}/k_B)(\partial[\mu_1/T - \varepsilon/T]/\partial N_0)^{ss} - f/V \\
H_{12} &= (\Omega/k_B)\exp[(\mu_0^{ss} - \varepsilon)/k_B T^{ss}](\partial\mathscr{A}/\partial E)^{ss} \\
&\quad + (f\rho_0^{ss}/k_B)(\partial[\mu_1/T - \varepsilon/T]/\partial E)^{ss} \qquad (8.2.23) \\
H_{21} &= \kappa[(\partial(1/T)/\partial N_0]^{ss} + (\hbar\omega f/V) \\
H_{22} &= \kappa[(\partial(1/T)/\partial E]^{ss}.
\end{aligned}
$$

The necessary and sufficient condition for the stability of the steady states of this system is the existence of the infinite time limit of the integral in Eq. (8.2.11). Equivalently one can check for the existence of a finite, positive definite solution, σ^{ss}, to the fluctuation–dissipation theorem

$$H\sigma^{ss} + \sigma^{ss}H^T = -\gamma^{ss}. \qquad (8.2.24)$$

As discussed at the end of Section 7.3, Eq. (8.2.24) is a set of linear equations which can be solved for the independent quantities σ_{ij}^{ss} with $i < j$. For the present problem these variables are $\sigma_{11}^{ss} = \langle(\delta N_0)^2\rangle^{ss}$, $\sigma_{12}^{ss} = \langle\delta N_0 \delta E\rangle^{ss}$, and $\sigma_{22}^{ss} = \langle(\delta E)^2\rangle^{ss}$. The form of Eq. (8.2.24) is written out for this special case in Eq. (7.3.42). Since $\gamma_{12}^{ss} = 0$, the formal solution can be obtained rather easily using Cramer's rule. After considerable algebra one obtains

$$
\begin{aligned}
\sigma_{11}^{ss} &= [(\det H + H_{22}^2)\gamma_{11} + H_{12}^2\gamma_{22}]/2[(\det H + H_{22}^2)H_{11} + H_{12}H_{22}H_{21}] \\
\sigma_{12}^{ss} &= -\gamma_{11}/2H_{12} - H_{11}\sigma_{11}^{ss}/H_{12} \qquad (8.2.25) \\
\sigma_{22}^{ss} &= [(\det H + H_{11}^2)\gamma_{22} + H_{21}^2\gamma_{11}]/2[(\det H + H_{11}^2)H_{22} + H_{21}H_{11}H_{12}].
\end{aligned}
$$

Eq. (8.2.25) is only a formal solution to Eq. (8.2.24), and we still must check that σ^{ss} is finite and positive definite. Thus to determine the stability of a steady state, it is necessary to know the rate parameters and thermodynamic derivatives so that the matrices H_{ij} and γ_{ij}^{ss} can be evaluated. The simplest case is an ideal gas, for which

$$
\begin{aligned}
\mu_0 &= \mu^0(T) + k_B T \ln \rho_0 \\
\mu_1 &= \mu^0(T) + \hbar\omega + k_B T \ln(\rho - \rho_0) \qquad (8.2.26) \\
E &= \frac{3}{2} N k_B T + (N - N_0)\hbar\omega,
\end{aligned}
$$

since the two levels are separated by an energy $\hbar\omega$. We can reduce the number of parameters in the calculation by assuming that $\varepsilon = \hbar\omega$. In this case the steady-state condition in Eqs. (8.2.21) can be written

$$y^{ss} = [1 + \exp(Kry^{ss} - x_R) + r]^{-1}$$
$$x^{ss} = x_R - Kry^{ss}, \tag{8.2.27}$$

where the nondimensional variables x, y, K, and r are defined by

$$x = \hbar\omega/k_B T, \quad y = N_0/N$$
$$r = f/Vk, \qquad K = (\hbar\omega)^2 Nk/\kappa k_B. \tag{8.2.28}$$

The parameter k in Eq. (8.2.28) is the reciprocal lifetime of the excited state, i.e.,

$$k = (\Omega/V)\exp(\mu_{ss}^0/k_B T),$$

as can be seen by substituting the expression for μ_1 in Eq. (8.2.26) into the canonical form in Eq. (8.2.19). The parameter r is the ratio of the excitation rate to the spontaneous emission rate, and K is the ratio of the spontaneous emission rate to the cooling rate. When $K = 0$, heat is removed by the thermal reservoir much faster than it is absorbed from the light source and Eq. (8.2.27) shows that $T^{ss} = T_R$.

The assumptions that have gone into linearizing Eq. (8.2.17) and the heat transport term can be expressed in the form $Kry^{ss} \ll x_R$. If we make the further assumption that $x_R = \hbar\omega/k_B T \gg 1$, then Eqs. (8.2.27) take the simple form

$$y^{ss} = [1 + \exp(-x_R) + r]^{-1}$$
$$x^{ss} = x_R - Kr/(1 + r). \tag{8.2.29}$$

Using these expressions and Eqs. (8.2.25), the following explicit formulas for H_{ij} and γ_{ij}^{ss} can be obtained:

$$H_{11} = -k\left[\frac{(1+r)^2}{(1+\alpha)} + \frac{rx^{ss}}{(1+r)} + \frac{2x^{ss2}}{3(1+r)}\right]\exp(-x^{ss})$$
$$H_{12} = -(k/\hbar\omega)2x^{ss2}\exp(-x^{ss})/3(1+r)$$
$$H_{21} = k\hbar\omega[r - 2x^{ss2}/3K]$$
$$H_{22} = -2kx^{ss2}/3K \tag{8.2.30}$$
$$\gamma_{11} = kN[2\exp(-x^{ss}) + r]/(1+r)$$
$$\gamma_{12} = \gamma_{21} = 0$$
$$\gamma_{22} = 2(\hbar\omega)^2 kN/K,$$

where $\alpha \equiv \exp(-x_R) \ll 1$ and x^{ss} is given in Eq. (8.2.29).

Substituting these expressions into Eqs. (8.2.25), the covariance matrix can be evaluated as a function of the nondimensional excitation rate. In evaluating σ_{ij}^{ss} we have used a value of $\kappa = 8 \times 10^2$ J K s^{-1}, which is comparable to that obtained from the value of β measured by Zimmermann and Ross in Section 7.4 using the relationship $\kappa = T_R^2\beta$. The value $K = 5$ corresponds to a lifetime of the excited state of 10^{-2} s–10 s and gas pressures of 1–100 torr.

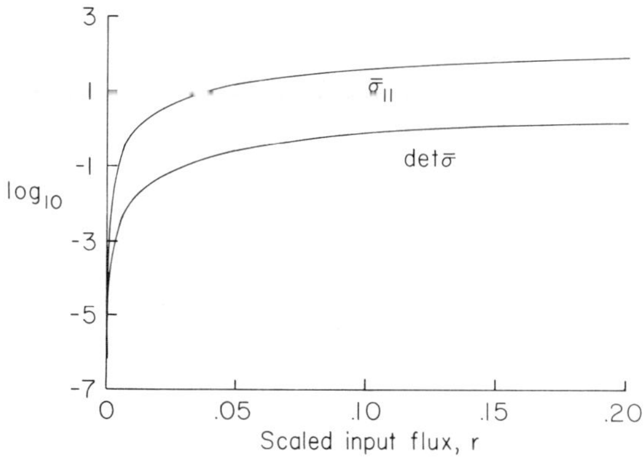

FIGURE 8.1. The determinant of the scaled covariance matrix (det $\bar{\sigma}$) and the scaled variance of the ground state occupancy number ($\bar{\sigma}_{11}$) as a function of the scaled input flux, $r = f/Vk$, for the steady states of the two-state model in Eqs. (8.2.19) and (8.2.20). The fact that both quantities are nonzero implies that the steady states at these input fluxes are asymptotically stable.

The value of x_R was chosen to be 10, which is compatible with the assumption $Kry^{ss} \ll x_R$ as long as r is less than or about 0.2. The results of this calculation are shown in Fig. 8.1. There we have graphed the scaled variance $\bar{\sigma}_{11} = \sigma_{11}/N$ as well as the determinant of the covariance matrix, with the additional scalings $\bar{\sigma}_{12} = \sigma_{12}/N\hbar\omega$ and $\bar{\sigma}_{22} = \sigma_{22}/N(\hbar\omega)^2$. At $r = 0$ the laser is turned off and one has equilibrium, where $\bar{\sigma}_{11}$ is very small (about 4.4×10^{-5}). As r increases, $\bar{\sigma}_{11}$ increases by about six orders of magnitude until r is about 0.3. Similarly, the determinant of $\bar{\sigma}$ increases dramatically over this range of values of r.

Notice that both $\bar{\sigma}_{11}$ and det $\bar{\sigma}$ are positive over the entire range of values of r. This is an important fact since, according to Sylvester's theorem, a matrix is positive definite if and only if all its main determinants are positive. Thus $\bar{\sigma}$ is a positive definite solution to the fluctuation–dissipation theorem, and it follows that these steady states are all stable. This result can be checked easily using linear stability analysis. Indeed, recall that in Section 7.4 we found that a necessary and sufficient condition for asymptotic stability for a two-variable system is

$$\operatorname{tr} H < 0 \quad \text{and} \quad \det H > 0. \tag{8.2.31}$$

Equations (8.2.30) show that the first of these conditions holds, and an easy calculation gives

$$\det H = \frac{2x^{ss2}k^2}{2K}\exp(-x^{ss})\left[\frac{(1+r)^2}{(1+\alpha)} + \frac{r(x^{ss}+R)}{(1+r)}\right],$$

which verifies the second condition.

These calculations emphasize that there is nothing in the Liapunov property of the covariance matrix that cannot also be obtained from linear stability analysis. On the other hand, it shows that the stability of a steady state is intimately associated with the molecular fluctuations. In the following section, we use this fact together with our knowledge of the properties of the equilibrium entropy function to define a generalized entropy that is useful at steady states.

The fluctuation–dissipation theorem is at the heart of the Liapunov criterion, and we close this section with some identities that are useful for solving that equation. The chief difficulty encountered in solving the fluctuation–dissipation theorem is that the matrix H couples together all or most of the elements of σ_{ij}^{ss}. Sometimes this coupling can be reduced by an appropriate change of variables. The fundamental variables are, of course, the extensive variables $\mathbf{n}$. Thus any invertible change of variables can be represented by

$$\mathbf{n}' = \mathbf{G}(\mathbf{n}), \tag{8.2.32}$$

where $\mathbf{G}$ may be a nonlinear function. This, however, always induces a *linear* change of variables for the conditional fluctuations, since we have

$$\delta n_i' = \hat{\delta} G_i = \sum_j (\partial G_i/\partial \bar{n}_j)\delta n_j. \tag{8.2.33}$$

In particular, at steady state we can write

$$\delta \mathbf{n}' = G\delta \mathbf{n}, \tag{8.2.34}$$

where G is the invertible matrix $G_{ij} \equiv (\partial G_i/\partial n_j)^{ss}$. Operating on Eq. (8.2.7) with G from the left, it follows that fluctuations in the new variables satisfy

$$d\delta \mathbf{n}'/dt = GHG^{-1}\delta \mathbf{n}' + G\tilde{\mathbf{f}}. \tag{8.2.35}$$

Consequently, we can write

$$d\delta \mathbf{n}'/dt = H'\delta \mathbf{n}' + \tilde{\mathbf{f}}' \tag{8.2.36}$$

with

$$\begin{aligned} H' &\equiv GHG^{-1} \\ \tilde{\mathbf{f}}' &\equiv G\tilde{\mathbf{f}}. \end{aligned} \tag{8.2.37}$$

Combining this second equality with Eq. (8.2.8), it follows that $\langle \tilde{\mathbf{f}}'(t) \rangle = \mathbf{0}$ and

$$\langle \tilde{\mathbf{f}}'(t)\tilde{\mathbf{f}}'^T(t') \rangle = \gamma' \delta(t - t'), \tag{8.2.38}$$

where

$$\gamma' \equiv G\gamma G^T. \tag{8.2.39}$$

The transformation law for σ^{ss} is also easy to obtain since

$$\sigma^{ss'} \equiv \langle \delta \mathbf{n}' \delta \mathbf{n}'^T \rangle = \langle G\delta \mathbf{n} \delta \mathbf{n}^T G^T \rangle = G\sigma^{ss}G^T. \tag{8.2.40}$$

Using Eqs. (8.2.37), (8.2.39) and (8.2.40) we can now write the fluctuation–dissipation theorem in terms of $\sigma^{ss'}$. Indeed, operating on Eq. (8.2.24), first from the left with G and then from the right with G^T, gives

$$GH\sigma^{ss}G^T + G\sigma^{ss}H^TG^T = -\gamma'. \tag{8.2.41}$$

Inserting the identity matrix between H and σ^{ss} in the form of $G^{-1}G$ and between σ^{ss} and H^T in the form of $G^T(G^{-1})^T$ gives

$$H'\sigma^{ss\prime} + \sigma^{ss\prime}H' = -\gamma'. \tag{8.2.42}$$

Equation (8.2.42) shows that the form of the fluctuation–dissipation theorem is invariant to a change in variables of this sort.

Transformations which diagonalize H give a particularly simple solution to the fluctuation–dissipation theorem. When H is diagonal the solution to the two-variable fluctuation–dissipation theorem, which is written out in detail in Eq. (7.3.42), has the same form as the multivariable case. The explicit solution is easily seen to be

$$\sigma_{ij}^{ss\prime} = -\gamma_{ij}^{ss\prime}/(\lambda_i + \lambda_j), \tag{8.2.43}$$

where the λ's are the eigenvalues of H', since it is diagonal. The solution of Eq. (8.2.42) is also facilitated if the matrix γ' is diagonal. For two variables, the explicit solution for σ^{ss} when γ' is diagonal is given in Eq. (8.2.25).

8.3. Thermodynamic Functions at Steady State

In the previous section the inverse of the single-time covariance matrix in a steady-state ensemble was shown to be a Liapunov function. At equilibrium this function reduces to the thermodynamic Liapunov function $-\delta^2 S/k_B$ and, thus, it is natural to seek a state function that has the same property at steady state. Formally, we need to find a function, Σ, for which

$$-\delta^2\Sigma/k_B = \sum_{i,j} (\sigma^{ss-1})_{ij} a_i a_j, \tag{8.3.1}$$

where $a_i = n_i - n_i^{ss}$. A more useful way to express this relationship is

$$\partial^2\Sigma/\partial n_i \partial n_j = -k_B(\sigma^{ss-1})_{ij}, \tag{8.3.2}$$

which provides a natural generalization of the Einstein formula at equilibrium. In fact, if such a function Σ could be found, then the single-time probability density in Eq. (8.2.3) could be written

$$W_1(\mathbf{n}) = N\exp(\delta^2\Sigma/2k_B), \tag{8.3.3}$$

where N is the normalization factor. Thus Σ would also provide a natural generalization of the Einstein formulation of static fluctuation theory.

If such a function exists, what variables does it depend upon? Certainly it must depend on the extensive variables, since it must reduce to the local equilibrium entropy when the steady state is an equilibrium state. However, it must also depend on the operational variables that are used to prepare the steady state. As we have seen in specific examples in this and previous chapters, steady states are maintained by a steady input of mass, momentum, or energy into a system. For the two examples in Section 8.2, the input is energy in the form of photons from a laser. For the steady thermal gradient treated in Section 6.8, the input is heat from a thermal reservoir, and for the steady state in the reaction–diffusion system in Section 6.6 the inputs are mass and energy in the form of reactant molecules.

As we discussed below Eq. (7.1.1), these inputs can be treated as external fluxes of the extensive variables and written as $\mathbf{F}(\mathbf{n})$. In general, the average values of the extensive variables at steady state, i.e., $\mathbf{n}^{ss}$, will depend on these fluxes. For definiteness we will assume that the functions $F_j(\mathbf{n})$ are characterized by a set of l parameters f_i, which involve inputs of extensive variables, and a set of m parameters, R_i, which refer to the reservoirs. The reservoir parameters include reservoir temperatures and pressures, as for example in the fluorosulfate radical system discussed in Section 7.4, or the reservoir chemical potential, as appears for membrane transport in Section 5.8. Using this characterization we can write

$$\mathbf{n}^{ss} = \mathbf{n}^{ss}(\mathbf{f}, \mathbf{R}). \tag{8.3.4}$$

An explicit example of Eq. (8.3.4) is given in Eq. (8.2.29), where the scaled reservoir variable is $x_R = \hbar\omega/k_B T_R$ and the scaled flux is $r = f/Vk$.

Since Σ depends only on the variables $\mathbf{n}$, $\mathbf{f}$, and $\mathbf{R}$, we can write

$$\Sigma(\mathbf{n}; \mathbf{f}, \mathbf{R}) = S(\mathbf{n}) + \sum_j f_j v_j(\mathbf{n}; \mathbf{f}, \mathbf{R}). \tag{8.3.5}$$

Clearly the terms $f_j v_j$ must vanish as $|\mathbf{f}| \to 0$, since Σ reduces to the entropy in an equilibrium ensemble, where all the fluxes vanish. Because of this property, we will often refer to Σ as the generalized entropy. Using Eq. (8.3.5) we can also define intensive variables, conjugate to the extensive variables, by

$$\phi_j \equiv \partial\Sigma/\partial n_j = \partial S/\partial n_j + \sum_j f_j \partial v_j/\partial n_j. \tag{8.3.6}$$

For example, if n_1 is the internal energy, then

$$\phi_1 = 1/T + \sum_j f_j \partial v_j/\partial E, \tag{8.3.7}$$

with T the Kelvin temperature.

These properties, of course, are vacuous unless we can find a function Σ whose second partials satisfy Eq. (8.3.2). Notice that if we assign a functional dependence on $\mathbf{n}$ to the right-hand side of Eq. (8.3.2), the equation becomes a set of second-order partial differential equations for Σ. Not all such equations are well posed, since an arbitrary set of functions of several variables are not guaranteed to be the second partial derivatives of some function. Indeed, the integrability condition for Eq. (8.3.2) is that

$$\partial(\sigma^{ss-1})_{ij}/\partial n_k = \partial(\sigma^{ss-1})_{ik}/\partial n_j, \tag{8.3.8}$$

i.e., that the mixed third partials of Σ be equal. Since we know how to calculate σ_{ij}^{ss} using the fluctuation–dissipation theorem, we might think of testing this condition in special cases. This is not, however, a well-defined procedure since σ^{ss} calculated in this way depends explicitly on $\mathbf{n}^{ss}$ through the matrices H and γ^{ss} [cf. Eqs. (8.2.25)], and $\mathbf{n}^{ss}$ depends implicitly on $\mathbf{f}$ and $\mathbf{R}$ as in Eq. (8.3.4). In fact, it seems most natural to write σ_{ij}^{ss} as a function of $\mathbf{f}$ and $\mathbf{R}$ rather than $\mathbf{n}$.

One way out of this dilemma involves separating the covariance matrix into its local equilibrium and nonequilibrium components. The local equilibrium component is defined by

$$k_B \sigma_{ij}^{le-1} \equiv -\partial^2 S/\partial n_i \partial n_j. \tag{8.3.9}$$

At steady state σ_{ij}^{le-1} can be written as an explicit function of $\mathbf{n}$, i.e., $\sigma_{ij}^{le-1}(\mathbf{n}^{ss})$, while using Eq. (8.3.4) it can be written as an explicit function of $\mathbf{f}$ and $\mathbf{R}$, i.e., $\sigma_{ij}^{le-1}(\mathbf{n}^{ss}(\mathbf{f}, \mathbf{R}))$. Indeed, using the natural dependence of σ_{ij}^{ss-1} on $\mathbf{f}$ and $\mathbf{R}$ and adding and subtracting the two forms of σ_{ij}^{le-1} to and from the right-hand side of Eq. (8.3.2), we obtain

$$\partial^2 \Sigma/\partial n_i \partial n_j = \partial^2 S/\partial n_i \partial n_j + B_{ij}(\mathbf{f}, \mathbf{R}), \tag{8.3.10}$$

where

$$B_{ij}(\mathbf{f}, \mathbf{R}) \equiv -k_B[\sigma_{ij}^{-1}(\mathbf{f}, \mathbf{R}) - \sigma_{ij}^{le-1}(\mathbf{f}, \mathbf{R})]. \tag{8.3.11}$$

Thus B_{ij} is the nonequilibrium component of the second partials of Σ. With this interpretation, the integrability condition in Eq. (8.3.8) reduces to the equality of the mixed third partials of the local equilibrium entropy, which is obviously valid.

Equation (8.3.10) also provides a natural way of introducing the dependence of Σ on the operational variables $\mathbf{f}$ and $\mathbf{R}$. In fact, integrating Eq. (8.3.10) gives the expressions

$$\phi_i = 1/T + \sum_j B_{ij}(\mathbf{f}, \mathbf{R}) n_j + \alpha_i(\mathbf{f}, \mathbf{R}) \tag{8.3.12}$$

and

$$\Sigma = S + \frac{1}{2} \sum_{i,j} B_{ij}(\mathbf{f}, \mathbf{R}) n_i n_j + \sum_i a_i(\mathbf{f}, \mathbf{R}) n_i + \beta(\mathbf{f}, \mathbf{R}), \tag{8.3.13}$$

where α_i and β are integration constants that may depend on $\mathbf{f}$ and $\mathbf{R}$. The integration constants must vanish at equilibrium since Σ is supposed to reduce to the entropy there. Otherwise these functions appear to the arbitrary. Their explicit form does not affect the characteristic Liapunov property of the sigma function or its relationship to fluctuations in Eq. (8.3.3), although it will affect the values of the intensive variables and Σ itself. For lack of a convincing argument on which to base a choice for α_i and β, we take the simplest choice, namely, $\alpha_i = \beta = 0$. When these values of the integration constants are used, we will refer to Σ as the sigma function.

The generalized entropy, Σ, can be used to define generalized thermodynamic forces. The natural extension of Onsager's definition in Eq. (2.3.2) is

$$\begin{aligned}\chi_i &\equiv \partial\Sigma/\partial n_i - (\partial\Sigma/\partial n_i)^{ss}\\ &= \phi_i - \phi_i^{ss},\end{aligned} \tag{8.3.14}$$

where ϕ_i is the conjugative intensive variable. As long as we consider only small excursions around the steady state, Eq. (8.3.14) can be linearized to give

$$\begin{aligned}\chi_i &= \sum_j (\partial\phi_i/\partial n_j)^{ss} a_j\\ &= \sum_j (\partial^2\Sigma/\partial n_i \partial n_j)^{ss} a_j,\end{aligned} \tag{8.3.15}$$

where Eq. (8.3.6) was used in the second line. In vector notation this can be written

$$\Sigma^{-1}\chi = \mathbf{a}, \tag{8.3.16}$$

with the matrix $\Sigma_{ij} \equiv (\partial^2\Sigma/\partial n_i \partial n_j)^{ss}$. This allows us to write the stochastic differential equations (7.3.34), which are satisfied locally by $\mathbf{a}$, in the form

$$\partial\mathbf{a}/\partial t = H\Sigma^{-1}\chi + \tilde{\mathbf{f}} \tag{8.3.17a}$$

with

$$\langle\tilde{\mathbf{f}}(t)\tilde{\mathbf{f}}^T(t')\rangle = -(H\sigma^{ss} + \sigma^{ss}H^T)\delta(t - t'), \tag{8.3.17b}$$

where we have used the fluctuation–dissipation theorem to rewrite γ^{ss}. Because of the way in which Σ is defined by Eq. (8.3.2), it follows that

$$\sigma^{ss} = -k_B\Sigma^{-1}, \tag{8.3.18}$$

so that if the generalized Onsager matrix is defined by

$$L \equiv H\Sigma^{-1}, \tag{8.3.19}$$

we see that

$$d\mathbf{a}/dt = L\mathbf{a} + \tilde{\mathbf{f}} \tag{8.3.20a}$$

and

$$\langle\tilde{\mathbf{f}}(t)\tilde{\mathbf{f}}^T(t')\rangle = 2k_B L^s\delta(t - t'). \tag{8.3.20b}$$

Written this way, the stochastic theory of a steady state is seen to have the same thermodynamic structure as the Onsager theory in Eqs. (2.6.10)–(2.6.12). As a consequence, it is reasonable to inquire if there is a dissipation function associated with a steady state. We have already seen that the canonical form provides a nonlinear generalization of the Rayleigh–Onsager dissipation function, Φ. At a stable steady state this function is positive and independent of time. Indeed, combining its definition in Eq. (4.5.5) with the average evolution equation (7.1.1), we see that

$$\Phi^{ss} = \sum [(\partial S/\partial n_j)^{ss} - (\partial S/\partial n_j)^{e}][S_j^{ss} - F_j^{ss}] > 0, \tag{8.3.21}$$

where S_j^{ss} is the streaming terms and F_j^{ss} is the external flux term evaluated at steady state. This is not, however, the dissipation function that we have in mind. We are interested in a function which would measure the production of Σ as a system relaxes back to a steady state. Near equilibrium this function is $\mathbf{X}^T L \mathbf{X}$, where L is the Onsager matrix and $\mathbf{X}$ is the thermodynamic force. The comparable quantity at steady state is

$$\hat{\Phi} \equiv \chi^T L \chi = \chi^T L^s \chi. \tag{8.3.22}$$

Indeed, from Eq. (8.3.20b) we see that $L^s = \gamma^{ss}/2k_B$, which is positive definite since the extensive variables have been chosen to be independent. Thus it follows that

$$\hat{\Phi} > 0, \tag{8.3.23}$$

with equality only at steady state. This function differs from the Rayleigh–Onsager dissipation function, and we shall refer to $\hat{\Phi}$ as the steady-state dissipation function.

Using the average equations

$$d\bar{\mathbf{n}}/dt = H\bar{\mathbf{a}} \tag{8.3.24}$$

and the definition of χ in Eqs. (8.3.14) and (8.3.16), $\hat{\Phi}$ can be rewritten as

$$\hat{\Phi} = \sum_j (\phi_j - \phi_j^{ss})(d\bar{n}_j/dt) \geq 0, \tag{8.3.25}$$

which is analogous to Eq. (4.5.44) for the Rayleigh–Onsager dissipation function in an equilibrium ensemble. This analogy can be extended further to obtain the *generalized Clausius inequality*, namely,

$$d\Sigma/dt \geq \sum_j \phi_j^{ss}(d\bar{n}_j/dt), \tag{8.3.26}$$

where the time derivative is evaluated on the conditional average trajectory with $\mathbf{f}$ and $\mathbf{R}$ fixed. Notice that in an equilibrium ensemble Eq. (8.3.26) reduces to

$$dS/dt \geq \sum_j F_j^{e}(d\bar{n}_j/dt),$$

which is identical to the Clausius inequality in Eq. (4.5.45). As we saw in Eq.(8.3.24), these inequalities are a result of the positive definiteness of the strength of the random terms, γ^{ss}, which is also the essential ingredient in the proof that $\delta^2\Sigma$ is a Liapunov function. Thus it should not be surprising that Eq. (8.3.25) can also be derived from the identities

$$\delta^2\Sigma = \sum_j \chi_j a_j \leq 0 \tag{8.3.27}$$

and

$$d\delta^2\Sigma/dt = 2\sum_j \chi_j(d\bar{a}_j/dt) \geq 0. \tag{8.3.28}$$

The importance of the dissipation function $\hat{\Phi}$ is that it allows us to define a reversible process at steady state using the analogy to what is done at equilibrium. A *reversible process* is a change occurring at fixed values of **f** and **R** for which

$$\hat{\Phi} \equiv 0. \tag{8.3.29}$$

The requirement that the fluxes **f** and reservoir variables **R** be held fixed is analogous to what is required at equilibrium, although at equilibrium the condition $\mathbf{f} \equiv \mathbf{0}$ is left understood. Using Eq. (8.3.25) the vanishing of $\hat{\Phi}$ can be written as

$$\sum_j \phi_j(d\bar{n}_j/dt)_{\text{rev}} = \sum_j \phi_j^{ss}(d\bar{n}_j/dt), \tag{8.3.30}$$

which we will call the *generalized Clausius equality*. According to Eq. (8.3.5), the total differential of Σ is

$$d\Sigma = \sum_{j=1}^{k} \phi_j\, dn_j + \sum_{i=1}^{l} (\partial\Sigma/\partial f_i)\, df_i + \sum_{i=1}^{m} (\partial\Sigma/\partial R_i)\, dR_i. \tag{8.3.31}$$

Thus the left-hand side of Eq. (8.3.30) can be written in terms of $d\Sigma/dt$ at *fixed values of* **f** *and* **R**, i.e.,

$$(d\Sigma/dt)_{\text{rev}} \equiv \sum_j \bar{\phi}_j(d\bar{n}_j/dt)_{\mathbf{f},\mathbf{R}} = \sum_j \phi_j^{ss}(d\bar{n}_j/dt)_{\mathbf{f},\mathbf{R}}. \tag{8.3.32}$$

Equation (8.3.32) involves total time derivatives, and the differential expression

$$(d\Sigma)_{\text{rev}} = \sum_j \phi_j\, dn_j = \sum_j \phi_j^{ss}\, dn_j \tag{8.3.33}$$

represents the infinitesimal change of Σ in a reverible process. Notice that the second equality implies that in a reversible process **n** always takes on steady-state values with **f** and **R** fixed. Thus a reversible change is a path in the space of steady states with the quantities **f** and **R** held constant.

The time rate of change of Σ in an irreversible process at fixed **f** and **R** can be obtained by combining Eqs. (8.3.25), (8.3.31), and (8.3.32). This gives

$$(d\Sigma/dt)_{\mathbf{f},\mathbf{R}} = (d\Sigma/dt)_{\text{rev}} + \hat{\Phi}. \tag{8.3.34}$$

Equation (8.3.34) makes it clear that $\hat{\Phi}$ is the production of the generalized entropy associated with an irreversible process. Changes in the entropy can also occur because of changes in the external fluxes and reservoir variables. Using Eq. (8.3.31) the most general change in Σ can be written

$$d\Sigma = (d\Sigma)_{\text{rev}} + \hat{\Phi}\, dt + (d\Sigma)_{\text{out}}. \tag{8.3.35}$$

The term $(d\Sigma)_{\text{out}}$ is due to effects outside the system, corresponding to changes in **f** and **R**, and neither this change nor the reversible change have a definite sign. Combining them into a term $(d\Sigma)_{ext}$, we can write Eq. (8.3.35) in the form

$$d\Sigma = (d\Sigma)_{ext} + (d\Sigma)_{in} \tag{8.3.36a}$$

with

$$(d\Sigma)_{in} = \hat{\Phi}\, dt \geq 0. \tag{8.3.36b}$$

In an equilibrium ensemble, Eq. (8.3.36a) reduces to the Second Law of thermodynamics. At steady states Eq. (8.3.36a) provides a local generalization of the Second Law.

The reason that **f** and **R** must be held constant in the definition of a reversible process is that Σ is defined in terms of the static fluctuations and the static fluctuations sample states with fixed values of **f** and **R**. Because the steady state is asymptotically stable under these conditions, the generalized Clausius inequality holds. This inequality restricts the dynamics in such a way that $\hat{\Phi}$ is non-negative and leads to the definition of a reversible process. Since the fluctuations sample only states with **f** and **R** fixed, they cannot provide dynamical information about processes in which **f** and **R** change. That information can be obtained only from a knowledge of the external devices that control **f** and **R**, a knowledge which has been excluded from our consideration.

There are interesting extremum principles associated with Legendre transforms of the sigma function. All of them can be derived from the generalized Clausius inequality. Indeed, refering to Eq. (8.3.25), we see that if the ϕ_j are fixed at their steady state values, then

$$d\left(\Sigma - \sum_j \phi_j^{ss}\bar{n}_j\right)/dt \geq 0. \tag{8.3.37}$$

Since $\phi_j = \partial\Sigma/\partial n_j$, the quantity in brackets is by definition the Legendre transformation of Σ with respect to the n_j. However, the time derivative of some of the terms $\phi_j^{ss}\bar{n}_j$ often turns out to vanish. At equilibrium this occurs because of microscopic reversibility, as we saw in Section 4.5. In fact, at equilibrium we found in Eq. (4.5.34) that

$$d\sum_j F_j^e\bar{n}_j/dt = 0. \tag{8.3.38}$$

At steady state some subset of the extensive variables is often unaffected by the inputs on the system. These include variables which, because they relax rapidly, can be ignored in the dynamical treatment. Recall that this was the case for the *o*-cresolphthalein reaction in Eq. (7.4.27), where the chemical dynamics could be ignored. In that example the generalized entropy is not changed due to the effects of the chemical reaction. This implies that the intensive variables conjugate to the molecule numbers are unchanged, i.e., $(\partial\Sigma/\partial N_i)^{ss} = -\mu_i^e/T$. Consequently the chemical conservation theorems, discussed in Sections 3.7 and 4.5 and contained in Eq. (8.3.38), imply that the terms conjugate to the molecule numbers will not contribute to the left-hand side of Eq. (8.3.37). Indicating this explicitly by a prime on the sum, Eq. (8.3.37) becomes

$$d\left(\Sigma - \sum_j{}' \phi_j^{ss}\bar{n}_j\right)/dt \geq 0. \tag{8.3.39}$$

Equation (8.3.39) states that an appropriate Legendre transformation of Σ with respect to the extensive variables is a nondecreasing function of time when the conjugate intensive variables are held fixed. For the *o*-cresolphthalein reaction, for example, Eq. (8.3.39) can be written

$$d\hat{\psi}/dt \geq 0, \tag{8.3.40}$$

where $\hat{\psi} = \Sigma - (\partial\Sigma/\partial E)\bar{E}$ is the generalization of the Mathieu function defined in Eq. (8.1.20). This Mathieu function achieves its maximum at steady state and cannot decrease as a function of time if $(\partial\Sigma/\partial E)$ is held fixed. If the flux of energy from the laser, which is used to excite the *o*-cresolphthalein, vanishes, then the resulting steady state is an equilibrium state and $\hat{\psi} = \psi = S - \bar{E}/T$, the equilibrium Mathieu function.

The analogy to the local equilibrium state functions becomes more transparent if we explicitly introduce the internal energy, volume, and molecule numbers as variables. Equation (8.3.33) then becomes

$$d\Sigma = \phi_1\, dE + \phi_2\, dV + \sum_{j=3}^{k} \phi_j\, dN_j. \tag{8.3.41}$$

Comparing this formula to that for the local equilibrium entropy, i.e.,

$$dS = dE/T + p\, dV/T - \sum_{j=3}^{k} \mu_j\, dN_j/T, \tag{8.3.42}$$

suggests that we define the intensive variables

$$\begin{aligned} 1/\hat{T} &\equiv \phi_1 \\ \hat{p}/\hat{T} &\equiv \phi_2 \\ -\hat{\mu}_j/\hat{T} &\equiv \phi_j. \end{aligned} \tag{8.3.43}$$

In terms of these expressions Eq. (8.3.41) becomes

$$d\Sigma = dE/\hat{T} + \hat{p}\, dV/\hat{T} - \sum_{j=3}^{k} \hat{\mu}_j\, dN_j/\hat{T} \tag{8.3.44}$$

or

$$dE = \hat{T}\, d\Sigma - \hat{p}\, dV + \sum_{j=3}^{k} \hat{\mu}_j\, dN_j. \tag{8.3.45}$$

In Section 8.4 we show that $\hat{p}$ is equal to the pressure, while $\hat{T}$ and the $\hat{\mu}_j$ generalize the Kelvin temperature and the chemical potentials, respectively. Thus we will often refer to $\hat{T}$ as the *generalized thermodynamic temperature.*

It is not difficult to show that $\hat{T}$ must be positive. To see this we use the definition of $\hat{T}$ in Eq. (8.3.43) and Eqs. (8.3.2) and (8.3.6) to write

$$\frac{1}{\hat{T}^2}\left[\frac{\partial \hat{T}}{\partial E}\right]_{V,\mathbf{N},\mathbf{f},\mathbf{R}} = k_B(\sigma^{ss-1})_{EE}. \tag{8.3.46}$$

Since the steady state is asymptotically stable, it follows that σ^{ss-1} is a finite, positive definite matrix and that the right-hand side of Eq. (8.3.46) is positive. Thus, except when $(\partial\hat{T}/\partial E) = 0$, $\hat{T}$ cannot be zero. However, if we assume that $\hat{T}$ vanishes at E_0, we generate a contradiction. Indeed, if $\hat{T}$ is an analytic function of E and $\hat{T}(E_0) = \partial\hat{T}/\partial E(E_0) = 0$, then the Taylor expansion of $\hat{T}$ at E_0 is

$$\hat{T}(E) = (\partial^2\hat{T}/\partial E^2)^0(E - E_0)^2/2 + \cdots \tag{8.3.47a}$$

and

$$(\partial\hat{T}/\partial E) = (\partial^2\hat{T}/\partial E^2)^0(E - E_0) + \cdots. \tag{8.3.47b}$$

Recall that the right-hand side of Eq. (8.3.46) must be finite at $E = E_0$. It follows using Eqs. (8.3.47) in Eq. (8.3.46) that $(\partial^2\hat{T}/\partial E^2)^0 \equiv \partial^2\hat{T}/\partial E^2(E_0)$ must also vanish. Clearly this argument can be extended to show that $\partial^n\hat{T}/\partial E^n(E_0) = 0$ for all n, which implies that $\hat{T}$ is identically zero. This, however, contradicts the positive, finite character of σ^{ss-1}, and implies that $\hat{T}$ is never zero. Because we have constructed $\hat{T}$ so that it reduces to the Kelvin temperature at equilibrium, it follows that $\hat{T}$ must be strictly positive.

It is easy to obtain explicit expressions for the intensive variables using the steady states examined in the previous sections. For example, in Section 7.1 we analyzed the steady states for the coupled chemical reactions in Eq. (7.1.4), and in Section 7.3 we found that the variance of the fluctuations in the density of X molecules was

$$\sigma^{ss}(n^{ss}) = B/V\partial B/\partial n^{ss}. \tag{8.3.48}$$

Using the definition of Σ in Eq. (8.3.2), Eq. (8.3.48) implies that

$$\partial^2\Sigma/\partial N^2 = -k_B(\partial \ln B/\partial N), \tag{8.3.49}$$

where N is the number of X molecules. According to the integration procedure in Eq. (8.3.10) we must rewrite Eq. (8.3.49) in the form

$$\begin{aligned}\partial^2\Sigma/\partial N^2 &= [\partial(-\hat{\mu}/\hat{T})/\partial N]_E \\ &= [\partial(-\mu/T)/\partial N]_E - k_B V^{-1}[(\partial \ln B/\partial n)^{ss} - 1/n^{ss}],\end{aligned} \tag{8.3.50}$$

where the local equilibrium variance for this model is $\sigma^{le} = N$. To be consistent with Eq. (8.3.11) the quantity in square brackets must be treated as a function of the parameters P, B, and R. Using the condition of steady state in Eq. (7.1.9) and a little algebra this term can be written

$$B_{11}(P, B, R) = -k_B V^{-1} n^{ss}(n^{ss} - B)/(n^{ss2} + R). \tag{8.3.51}$$

The explicit dependence of n^{ss} on P, B, and R can be obtained graphically, as in Fig. 7.1. Recalling that equilibrium occurs when $n^{ss} = B$, we see that this correction term to Σ vanishes at equilibrium. Away from equilibrium it contributes a new term to the chemical potential of X which follows by integrating Eq. (8.3.50).

Integration of Eq. (8.3.50) is facilitated by use of the formula for the change of partial derivatives:

$$[\partial(-\hat{\mu}/\hat{T})/\partial N]_E = [\partial(-\hat{\mu}/\hat{T})/\partial N]_T + [\partial(-\hat{\mu}/\hat{T}/\partial(1/T)]_n[\partial(1/T)/\partial N]_E. \tag{8.3.52}$$

The fact that we have ignored fluctuations in the energy in this problem means that $\hat{T} = T$. Furthermore, it implies that the system must be dilute in X. This allows us to neglect the second term in Eq. (8.3.52). Indeed, the first factor in that term is an intensive quantity while, using a Maxwell relationship, the second factor can be written

$$[\partial(1/T)/\partial N]_E = [\partial(-\mu/T)/\partial E]_N. \tag{8.3.53}$$

The right-hand side of this equation vanishes as the number of solvent molecules goes to infinity since the solvent dominates the energy of the system. Hence to the same degree of approximation that energy fluctuations can be neglected, we can write Eq. (8.3.52) as

$$[\partial(-\hat{\mu}/\hat{T})/\partial N]_E \simeq [\partial(-\hat{\mu}/T)/\partial N]_T. \tag{8.3.54}$$

Since a similar approximation is valid for the local equilibrium term, Eq. (8.3.50) becomes

$$[\partial(-\hat{\mu}/T)/\partial N]_T = [\partial(-\mu/T)/\partial N]_T + B_{11}(P, B, R). \tag{8.3.55}$$

Thus integrating over N at constant T, we find that

$$\hat{\mu} = \mu - k_B T n n^{ss}(n^{ss} - B)/(n^{ss2} + R), \tag{8.3.56}$$

where the formula for B_{11} in Eq. (8.3.51) was used.

The nonequilibrium correction to the chemical potential is proportional to $k_B T$. As the local equilibrium chemical potential is also the order of $k_B T$, it follows that the correction to the local equilibrium chemical potential can be significant. Indeed, near the lower critical point in this reaction model one can use Eq. (7.5.4) to show that

$$2n^{ss}(B - n^{ss})/(n^{ss2} + R) \simeq 1. \tag{8.3.57}$$

Thus it follows that near this critical point the correction to the chemical potential is approximately $-k_B T n^{ss}/2$. For the parameter values and density units chosen in Fig. 7.1, n^{ss} is approximately 10. Thus the correction is about $-5k_B T$ or, in molar units, about -3 kcal mol^{-1} at room temperature.

In a similar manner one can obtain nonequilibrium corrections to the

Kelvin temperature for the *o*-cresolphthalein reaction in Section 7.4. In that system the chemical reaction is rapid so that only fluctuations in the energy were treated. In this approximation the chemical affinity is unchanged from its local equilibrium form, while the thermodynamic temperature is modified. Using Eq. (7.4.36) for the variance of the internal energy and the formulas for the nonequilibrium contribution to the sigma function in Eqs. (8.3.10) and (8.3.11), we find that

$$[\partial(1/\hat{T})/\partial E]_N = -\frac{1}{T^2 C_V} - \frac{[(1 - (dA/d\bar{T})^{ss} I_0/\beta) - (T_R/T^{ss})^2]}{T_R^2 C_V}. \quad (8.3.58)$$

Since the second term is taken to depend only on the energy flux through I_0 and the reservoir temperature, T_R, Eq. (8.3.58) can be integrated to obtain

$$1/\hat{T} = 1/T - E\{[1 - (dA/d\bar{T})^{ss} I_0/\beta] - (T_R/T^{ss})^2\}/T_R^2 C_V. \quad (8.3.59)$$

The size of the correction term can be calculated from a knowledge of the dependence of the absorbance and Kelvin temperature on I_0 and T_R, as given in Eqs. (7.4.28) and (7.4.30).

8.4. Thermodynamic Properties of Steady States

The sigma function introduced in Section 8.3 generalizes the local equilibrium entropy. As we have seen, it is related to stability, fluctuations, and dynamical changes at steady state in the same way as the local equilibrium entropy is at equilibrium. When the local equilibrium entropy is restricted to the manifold of equilibrium states, it becomes the fundamental state function for the extensive variables. From its dependence on the extensive variables one can obtain heat capacities and the mechanical and thermal equations of state and connect seemingly independent quantities like the isothermal compressibility and the bulk modulus. Indeed, one of the remarkable features of thermodynamics is that so many properties of matter at equilibrium can be obtained from a knowledge of the entropy.

We have seen from the calculations in previous sections that the statistical properties of matter are modified at steady states. This implies that on a fundamental, structural level, matter is changed as one forces a substance away from equilibrium. As a consequence, the physical and chemical properties of a substance are changed. For example, in Section 8.2 we found that the fraction of time a molecule spends in a given molecular state changes drastically when a gas is pumped with a laser. This is associated with a change in the measurable statistical properties of the gas and, according to the ideas in Section 8.3, a change in its thermodynamic properties. For these ideas to have significance, these changes in the entropy should be manifest in the thermal, mechanical, and chemical properties of a system. We pursue these questions in this section, focusing on the experimental significance of the intensive variables $\hat{T}$, $\hat{p}$, and $\hat{\mu}_i$.

To help understand the thermodynamic theory at steady state, it is useful to introduce a statistical mechanical description of the state of a system. Following the Gibbs' approach, which was outlined in Section 1.1, we recognize that our steady state ensemble has associated with it an underlying mechanical description. For reasons of generality, we will use quantum mechanics to describe the mechanical state of a system. The associated measure of probability of a quantum state is provided by a density matrix, $\hat{P}$. For an equilibrium ensemble of isolated systems, $\hat{P}$ can be taken as the microcanonical density matrix, while if the systems are in contact with a thermal reservoir, $\hat{P}$ is the canonical density matrix. Although at steady state the form of $\hat{P}$ is unknown, the basic paradigm of statistical mechanics involves the existence of a density matrix. The paradigm further asserts that average values can be calculated using the formula

$$\bar{O} = \mathrm{tr}(\hat{O}\hat{P}), \tag{8.4.1}$$

where $\hat{O}$ is an operator representing a measurable quantity. For example, if $\hat{H}$ is the Hamiltonian of a system, then the total energy is

$$\bar{E} = \mathrm{tr}(\hat{H}\hat{P}) = \sum_l E_l P_l, \tag{8.4.2}$$

where the E_l are the eigenvalues of $\hat{H}$ and P_l is the probability of finding state l in the ensemble.

To make a connection with the thermodynamic theory in the previous section, we consider an infinitesimal change in $\bar{E}$. Using Eq. (8.4.2), this can be written

$$d\bar{E} = \sum_l (dE_l)P_l + \sum_l E_l\, dP_l. \tag{8.4.3}$$

The first term in this equation involves changes in the energy levels of the systems. These are properties of the Hamiltonian and, thus, depend on quantities like the volume and the numbers of molecules in the system and such things as external fields, which contribute to the potential energy of the system. If we designate these variables, which are not extensive, by x_α, then we have

$$dE_l = (\partial E_l/\partial V)\, dV + \sum_i (\partial E_l/\partial N_i)\, dN_i + \sum_\alpha (\partial E_l/\partial x_\alpha)\, dx_\alpha. \tag{8.4.4}$$

The contribution of these terms to Eq. (8.4.3) is no different than at equilibrium, and we can write

$$d\bar{E} = \sum_l (\partial E_l/\partial V)P_l\, dV + \sum_i \left[\sum_l (\partial E_l/\partial N_i)P_l\right] dN_i$$

$$+ \sum_\alpha \left[\sum_l (\partial E_l/\partial x_\alpha)P_l\, dx_\alpha\right] + \sum_l E_l\, dP_l. \tag{8.4.5}$$

The first three terms in Eq. (8.4.5) represent average changes in the energy levels with respect to changes in V, N_i, and x_α. Just as at equilibrium, these are mechanical changes and, thus, they must represent work. In fact, if we contemplate only changes with the fluxes, **f**, and reservoir variables, **R**, fixed, these changes correspond to a reversible process since they stay on the manifold of steady states. Thus Eq. (8.4.5) can be written

$$d\bar{E} = đW_{\text{rev}} + \sum_l E_l \, dP_l \tag{8.4.6}$$

with the reversible work given by

$$đW_{\text{rev}} = -\bar{p}\, dV + \sum_i \bar{\mu}_i \, dN_i + \sum_\alpha \bar{e}_\alpha \, dx_\alpha, \tag{8.4.7}$$

where we have made the definitions

$$\begin{aligned} \bar{p} &\equiv -\sum_l (\partial E_l/\partial V) P_l \\ \bar{\mu}_i &= \sum_l (\partial E_l/\partial N_i) P_l \\ \bar{e}_\alpha &= \sum_l (\partial E_l/\partial x_\alpha) P_l. \end{aligned} \tag{8.4.8}$$

Since the work done on a system by changing the volume reversibly is $-p\, dV$, where p is the pressure of the system, we can identify $\bar{p}$ as the mechanical pressure. The other terms represent reversible work due to chemical and other mechanical changes.

To compare Eqs. (8.4.6)–(8.4.8) with the thermodynamic theory in the previous section, we need to add the variables x_α to that theory. That can be done by introducing the relationship between the internal energy, E, and the total energy, $\bar{E}$,

$$d\bar{E} = dE + \sum_\alpha e_\alpha \, dx_\alpha. \tag{8.4.9}$$

Substituting this into Eq. (8.3.45), which holds for a reversible process, gives the equivalent expression

$$d\bar{E} = \hat{T}\, d\Sigma - \hat{p}\, dV + \sum_i \hat{\mu}_i \, dN_i + \sum_\alpha e_\alpha \, dx_\alpha. \tag{8.4.10}$$

To compare Eq. (8.4.10) with Eq. (8.4.6), notice that only the second term in Eq. (8.4.6) involves changes in the statistical distribution of the ensemble. However, if the statistical distribution is held fixed, then it follows from the definition of Σ in terms of the covariance matrix that $d\Sigma = 0$. Thus from Eqs. (8.4.6) and (8.4.10) we deduce that

$$đW_{\text{rev}} = -\hat{p}\, dV + \sum_i \hat{\mu}_i \, dN_i + \sum_\alpha e_\alpha \, dx_\alpha. \tag{8.4.11}$$

Furthermore, since the variables V, N_i, and x_α are independent, we can compare Eqs. (8.4.11) and (8.4.7) term by term to see that

$$\hat{p} = \bar{p}, \quad \hat{\mu}_i = \bar{\mu}_i, \quad \text{and} \quad e_\alpha = \bar{e}_\alpha. \tag{8.4.12}$$

Thus since $\bar{p}$ is the mechanical pressure, can conclude that $\hat{p}$ is also.

To identify the second term in Eq. (8.4.6) we can use the First Law of thermodynamics, i.e.,

$$d\bar{E} = đW + đQ. \tag{8.4.13}$$

For a reversible process this allows us to conclude from Eq. (8.4.6) that

$$đQ_{\text{rev}} = \sum_l E_l \, dP_l \tag{8.4.14}$$

and from Eqs. (8.4.10) and (8.4.11) that

$$đQ_{\text{rev}} = \hat{T}\, d\Sigma. \tag{8.4.15}$$

As we know from equilibrium thermodynamics, the heat taken up or released in a process depends on the path and $đQ$ is, therefore, not the differential of any function. For reversible processes at equilibrium, $đQ_{\text{rev}}$ can be made an exact differential by dividing by the Kelvin temperature, i.e., $đQ_{\text{rev}}/T = dS$. According to Eq. (8.4.15), $\hat{T}$ serves a similar function at steady state. Indeed, since $\hat{T}$ reduces to T at equilibrium and Σ reduces to S, Eq. (8.4.15) is the natural generalization of the equilibrium result. This further justifies calling $\hat{T}$ the thermodynamic temperature at steady state.

Although we have seen that the intensive variable $\hat{p}$ equals the mechanical pressure, it is not generally true that $\hat{T}$ equals the Kelvin temperature. This might be expected from the fact that the pressure has a mechanical definition, while the temperature is a property of the statistical distribution even at equilibrium. This makes the thermodynamic temperature a difficult quantity to measure at a steady state. One can, of course, appeal to its definition in Eq. (8.3.46) in terms of fluctuations in the internal energy. At a steady state these fluctuations can in principle be measured by light scattering, as we saw in Section 6.8.

There is an important special case for which the thermodynamic temperature reduces to the Kelvin temperature. This involves systems that exchange heat with a reservoir. If the heat exchange is rapid with respect to other processes in the system, one can deduce that $\hat{T} = T$. For example, if we reconsider the two-state laser excitation problem in Section 8.2, we find using Eq. (8.2.25) and the inversion formula $\sigma_{22}^{-1} = \sigma_{11}/\det \sigma$ that

$$\sigma_{22}^{-1} = -\frac{2[\gamma_{11}(\det H + H_{22}^2) + \gamma_{22} H_{12}^2] \operatorname{tr} H}{\gamma_{11}\gamma_{22}(\operatorname{tr} H)^2 + (\gamma_{11} H_{21} - \gamma_{22} H_{12})^2}. \tag{8.4.16}$$

If heat transport to the reservoir is rapid, the scaled parameter K approaches zero and the formulas for H_{ij} and γ_{ij} in Eqs. (8.2.30) can be used to show that

$$\sigma_{22}^{-1} = -2\gamma_{11} H_{22}^3/\gamma_{11}\gamma_{22} H_{22}^2 = -2H_{22}/\gamma_{22}. \tag{8.4.17}$$

Equation (8.4.17) can be expressed in terms of the unscaled variables using

Eqs. (8.2.22) and (8.2.23). This gives

$$-k_B\sigma_{22}^{-1} = [\partial(1/T)/\partial E], \tag{8.4.18}$$

which is the local equilibrium result. Consequently from the definition of Σ it follows that $\hat{T} = T$, the Kelvin temperature.

Although it is often the case that thermal relaxation is rapid, this is not always so. For example, the fluorosulfate reaction treated in Section 7.4 involves heat exchange with a reservoir. However, because the relaxation time for heat transport in that system is comparable to the chemical relaxation time, the energy fluctuations are strongly coupled to the density fluctuations.

Explicit calculations of the fluctuations are easy to carry out when only two variables are involved. We have done this for the two-level laser model in Section 8.2 in order to illustrate the effect of this coupling on the thermodynamic temperature. The calculation requires only that we invert the covariance matrix, which was previously given in Fig. 8.1. Using the inversion formula for a 2×2 matrix in Eq. (6.8.44), this is straightforward. $\hat{T}$, in turn, can be obtained from Eq. (8.3.12). For this problem that formula becomes

$$1/\hat{T} = 1/T + B_{21}N_0 + B_{22}E. \tag{8.4.19}$$

Equation (8.4.19) has been evaluated at the steady-state values of N_0 and E given by Eq. (8.2.21). The results are given in Fig. 8.2 where the scaled inverse temperature $\hat{x}^{ss} = \hbar\omega/k_B\hat{T}^{ss}$ is plotted as a function of the scaled laser intensity, r. There are large differences between $\hat{T}^{ss}$ and T^{ss} when r is just a few tenths. Notice that T^{ss} increases as the intensity of the laser increases, while $\hat{T}^{ss}$ decreases. This reflects the fact that fluctuations in the internal energy and the number of ground state molecules are greatly increased at steady state. The increase in the Kelvin temperature, while a useful measure of the ability of the system to exchange heat with its surroundings, gives no information about the molecular statistics of the system, whereas the thermodynamic temperature does.

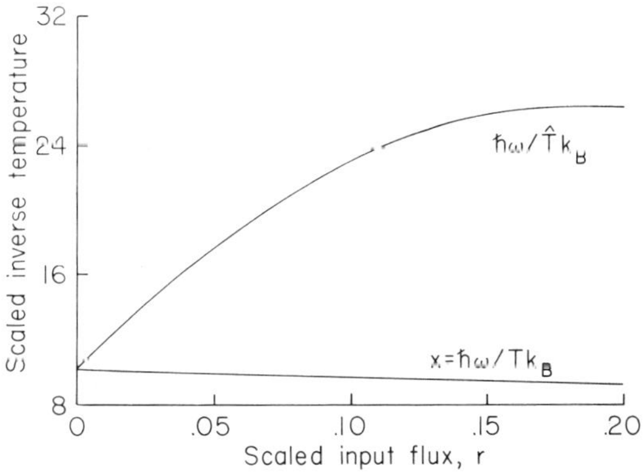

FIGURE 8.2. The scaled inverse Kelvin temperature, $1/T$, and the scaled thermodynamic temperature, $1/\hat{T}$, as a function of the scaled input flux, $r = f/Vk$, for the steady states of the two-state model in Eqs. (8.2.19) and (8.2.20). The difference between the two temperatures reflects the deviation from local equilibrium at the steady states.

Using classical mechanics it is possible to simulate steady states using computer calculations in which the kinetic energy is held fixed. Given the classical interpretation of the temperature as proportional to the kinetic energy per molecule, this is equivalent to rapid heat exchange with a reservoir. In one such simulation used by Hanley and Evans, a constant velocity gradient and a constant rate of strain are set up along one direction in the fluid. Using a slight generalization of periodic boundary conditions, the fluid appears to be homogeneous on the macroscopic scale, and this simulation is called *homogeneous shear*. Because of the rapid heat exchange with the reservoir, we can write $\hat{T} = T$ and analyze the thermodynamics of this system using Eq. (8.4.15), i.e.,

$$đQ_{\text{rev}} = T\,d\Sigma. \tag{8.4.20}$$

For example, integrating this expression gives

$$\Delta\Sigma = \int_1^2 đQ_{\text{rev}}/T, \tag{8.4.21}$$

where states 1 and 2 and those in between have identical values of the shear force, which in this system serves as the energy input. Equation (8.4.21) shows that the change in the generalized entropy that occurs on compressing the system with T fixed is simply

$$\Delta\Sigma = Q_{\text{rev}}/T. \tag{8.4.22}$$

Another way to analyze this system involves the total differential of the sigma function in Eq. (8.3.31). For homogeneous shear this can be written

$$T\,d\Sigma = dE + \hat{p}\,dV - \hat{\mu}\,dN + \theta\,df, \tag{8.4.23}$$

where $\theta = T\partial\Sigma/\partial f$ and f is the rate of input of energy. If we define the generalized Helmholtz free energy by

$$\hat{A} = E - T\Sigma, \tag{8.4.24}$$

then Eq. (8.4.23) implies that at constant N

$$d\hat{A} = -\Sigma\,dT - \hat{p}\,dV - \theta\,df. \tag{8.4.25}$$

Since $d\hat{A}$ is an exact differential, the equality of the mixed partials gives the *generalized Maxwell relationships*

$$\begin{aligned}(\partial\Sigma/\partial V)_{T,f,N} &= (\partial\hat{p}/\partial T)_{V,f,N}\\ (\partial\hat{p}/\partial f)_{T,V,N} &= (\partial\theta/\partial V)_{T,f,N}.\end{aligned} \tag{8.4.26}$$

On the other hand, taking f and N fixed, Eq. (8.4.23) implies that

$$(\partial E/\partial V)_{T,f,N} = T(\partial\Sigma/\partial V)_{T,f,N} - \hat{p}. \tag{8.4.27}$$

Differentiating this equation with respect to f at constant T, V, and N thus gives

$$\partial^2 E/\partial f \partial V = T\partial^2 \hat{p}/\partial f \partial T - \partial \hat{p}/\partial f, \tag{8.4.28}$$

where the first of Eqs. (8.4.26) was used.

The interesting feature of Eq. (8.4.28) is that it relates the thermal equation of state on the left-hand side to the mechanical equation of state on the right-hand side. Hanley and Evans have tested Eq. (8.4.28) against their molecular dynamics calculation of homogeneous shear. By measuring the internal energy and pressure as a function of the density, temperature, and rate of strain, they were able to calculate the derivatives in Eq. (8.4.28) numerically. To within the accuracy of their numerical data, they found that Eq. (8.4.28) was valid. Equation (8.4.28) illustrates an important use of the steady-state thermodynamic theory, namely, that it can be used to relate seemingly unconnected data on variables such as the energy, pressure, and volume.

8.5. Free Energy and the Electromotive Force

At equilibrium the Gibbs and Helmholtz free energies provide information about the work that can be extracted from a system. Similar properties hold for the generalized Helmholtz free energy,

$$\hat{A} = E - \hat{T}\Sigma = -\hat{T}\hat{\psi}, \tag{8.5.1}$$

where $\hat{\psi}$ is the generalized Mathieu function in Eq. (8.3.40), and for the generalized Gibbs free energy

$$\hat{G} = \hat{A} + \hat{p}V. \tag{8.5.2}$$

We will treat only $\hat{G}$ explicitly, since the properties of $\hat{A}$ are easy to derive by analogy.

To understand the work that can be obtained through external fields we must use the total energy in the definition of $\hat{G}$. Thus, in general, we will write

$$\hat{G} \equiv \bar{E} - \hat{T}\Sigma + \hat{p}V, \tag{8.5.3}$$

where $\bar{E}$ is the total energy. Using this definition, the total differential of $\hat{G}$ is

$$d\hat{G} = d\bar{E} - \hat{T}\,d\Sigma - \Sigma\,d\hat{T} + \hat{p}\,dV + V\,d\hat{p}. \tag{8.5.4}$$

Restricting ourselves to conditions of constant $\hat{T}$ and $\hat{p}$ and using Eq. (8.4.10) to eliminate $d\bar{E}$, Eq. (8.5.4) reduces to

$$d\hat{G} = \sum_i \hat{\mu}_i\,dN_i + \sum_\alpha e_\alpha\,dx_\alpha. \tag{8.5.5}$$

But according to Eq. (8.4.11), the right-hand side of Eq. (8.5.5) is the reversible work exclusive of the pressure–volume work. Thus at constant $\hat{T}$ and $\hat{p}$ we can write

$$d\hat{G} = đW_{\text{rev, non-}pV}. \tag{8.5.6}$$

This justifies thinking of $\hat{G}$ as a work potential since it gives the useful work that can be extracted from a steady-state system in a reversible process.

The generalized Clausius inequality in Eq. (8.3.26) provides another important property of $\hat{G}$. Multiplying both sides of that equation by $-\hat{T}^{ss}$ gives

$$d\left(E - \hat{T}^{ss}\Sigma + \hat{p}^{ss}V - \sum_i \hat{\mu}_i^{ss}N_i\right) \leq 0, \tag{8.5.7}$$

where the differential is taken on the conditional average path. Adding the differential of the work done by external forces, $đW_{\text{ext}} = \sum_\alpha e_\alpha\, dx_\alpha$, to both sides of Eq. (8.5.7) and rearranging gives

$$d\hat{G} \leq \sum_i \mu_i^{ss}\, dN_i + đW_{\text{ext}}. \tag{8.5.8}$$

Because $\hat{T}$ and $\hat{p}$ are held fixed at their steady-state values, we can combine Eq. (8.5.8) with Eq. (8.5.6) to write

$$đW_{\text{rev, chem}} + đW_{\text{rev, ext}} = d\hat{G} \leq đW_{\text{rev, chem}} + đW_{\text{ext}} \tag{8.5.9}$$

where we have written $\sum_i \mu_i^{ss}\, dN_i = đW_{\text{rev, chem}}$ for the reversible work from chemical processes. Canceling the common term in Eq. (8.5.9), we see that

$$đW_{\text{rev, ext}} \leq đW_{\text{ext}}. \tag{8.5.10}$$

In our convention for the First Law, work done by a system is negative. Thus Eq. (8.5.10) states that at constant $\hat{p}$ and $\hat{T}$ the work done by a steady-state system is maximized on a reversible path.

From its definition in Eq. (8.5.3) we can calculate the total differential of $\hat{G}$ to be

$$d\hat{G} = -\Sigma\, d\hat{T} + V\, d\hat{p} + \sum_i \hat{\mu}_i\, dN_i - \sum_i \theta_i\, df_i - \sum_i \lambda_i\, dR_i. \tag{8.5.11}$$

Thus at constant $\hat{T}$, $\hat{p}$, $\mathbf{f}$, and $\mathbf{R}$,

$$d\hat{G} = \sum_i \hat{\mu}_i\, dN_i. \tag{8.5.12}$$

According to Eq. (8.5.12) the chemical potentials contain all the information about changes in $\hat{G}$ for reversible processes at constant $\hat{T}$ and $\hat{p}$. From Eq. (8.5.11) we can also obtain Maxwell relationships. One of the most interesting is

$$(\partial\hat{\mu}_i/\partial\hat{p})_{\hat{T},\mathbf{N},\mathbf{f},\mathbf{R}} = (\partial V/\partial N_i)_{\hat{T},\hat{p},\mathbf{f},\mathbf{R}}. \tag{8.5.13}$$

The quantity on the right-hand side of Eq. (8.5.13) is a generalization of the partial molar volume. It is determined by the mechanical equation of state, while the chemical potential on the left-hand side comes from the chemical equation of state. Integration of Eq. (8.5.13) when the generalized partial molar volume is known will produce an expression for the generalized chemical potential.

There is another method for determining the generalized chemical potentials that relies on the fact that $\hat{G}$ is a work potential. In Section 5.6 we discussed the electromotive force, or *EMF*, of an electrochemical cell. The *EMF* is the electrical potential which evokes no current flow between the two electrodes in a cell. Since no current is flowing and all other processes are at equilibrium, the *EMF* is determined by the Nernst equation (5.6.21), i.e.,

$$em\mathscr{E} = -\Delta G, \tag{8.5.14}$$

with m the number of electrons involved in the cell reaction. At a nonequilibrium steady state there are at least some dissipative molecular processes occurring and the Nernst equation cannot be used to determine the electromotive force.

At steady state changes in the generalized Gibbs free energy for the cell reaction are related to the *EMF* at steady state. To see this, we begin with Eq. (8.5.6) written in the form

$$đW_{\text{rev, non-}pV} = \sum_i \hat{\mu}_i \, dN_i + \sum_\alpha e_\alpha \, dx_\alpha. \tag{8.5.15}$$

To apply Eq. (8.5.15) we must specify what the system is so that the work on the system is well defined. As the system we take the electrochemical cell, including the electrodes and the connection between them, but excluding other external circuits. In the measurement of the *EMF* no non-pV work is done by the system since charge is transferred from one electrode to another without dissipation. Thus $đW_{\text{rev, non-}pV} = 0$ and

$$-\sum_i \hat{\mu}_i \, dN_i = \sum_\alpha e_\alpha \, dx_\alpha. \tag{8.5.16}$$

The right-hand side of Eq. (8.5.16) gives the change in potential energy associated with the virtual changes dN_i. If these virtual changes are due to the electrochemical reaction, then

$$dN_i = \nu_i \, d\xi, \tag{8.5.17}$$

where the ν_i are the signed stoichiometric coefficients and ξ is the progress variable. The change in potential energy involves one term for the electrodes and another for the solution. In the absence of a liquid junction potential, the electrostatic potential, ϕ, is the same for all molecules in solution. Thus the potential energy of the solution is unchanged since electroneutrality gives

$$\sum_i \phi \, dq_i = 0, \tag{8.5.18}$$

where $dq_i = ez_i\nu_i \, d\xi$, with z_i the electron charge on species i. The contribution to the right-hand side of Eq. (8.5.16) due to the electrodes can be written

$$\sum_\alpha e_\alpha \, dx_\alpha = -\int \nabla\phi \cdot d\mathbf{s} \, q \tag{8.5.19}$$

with q the charge that is transferred and the line integral along the direction of motion of the charge. If $d\xi$ is positive, electrons move from the negative terminal (anode) to the positive terminal (cathode), and the electron charge that moves is $q = -em\,d\xi$. Thus the integral in Eq. (8.5.19) gives

$$\sum_\alpha e_\alpha dx_\alpha = em(\phi_c - \phi_a)\,d\xi, \tag{8.5.20}$$

with ϕ_c and ϕ_a the cathode and anode potentials. Substituting Eq. (8.5.20) into Eq. (8.5.16) using Eq. (8.5.17) gives

$$em\mathscr{E} = -\Delta\hat{G}, \tag{8.5.21}$$

with

$$\Delta\hat{G} \equiv \sum_i \hat{\mu}_i \nu_i. \tag{8.5.22}$$

Equation (8.5.21) is the generalized Nernst equation for the *EMF* at steady states.

We have seen that the intensive variables at steady state are usually modified from their local equilibrium form. This means that we can write

$$\Delta\hat{G} = -em\mathscr{E}^{le} + \Delta\hat{G}^{ne}, \tag{8.5.23}$$

where E^{le} is the *EMF* based on the Nernst equation (5.6.23). At steady state we anticipate a nonequilibrium contribution to the *EMF*

$$\mathscr{E}^{ne} = \mathscr{E} - \mathscr{E}^{le} = -\Delta\hat{G}^{ne}/em. \tag{8.5.24}$$

In looking over the derivation of these formulas, we see that it was assumed that $\hat{p}$ (the pressure) and $\hat{T}$ (the generalized thermodynamic temperature) as well as the fluxes and reservoir variables are held fixed. Although keeping the pressure and the other operational variables fixed is no problem, we have given no indication of how to keep $\hat{T}$ fixed. The sole exception to this is when the system rapidly exchanges heat with a thermal reservoir. In that case we noted in Section 8.4 that $\hat{T} = T_R$, the reservoir temperature. Thus for rapid heat exchange, fixing the Kelvin temperature of the reservoir fixes the thermodynamic temperature.

8.6. The Nonequilibrium EMF in a Stirred Tank Reactor

A continuously stirred tank reactor (CSTR) is a convenient device in which to maintain nonequilibrium steady states for chemical reactions. The operation and characteristics of CSTRs are described in Section 5.5. Because a CSTR is easy to thermostat, it provides an ideal system in which to measure the nonequilibrium electromotive force. As an example, consider the half-reaction

$$\mathrm{Fe^{3+}_{(aq)}} + \mathrm{e^-} = \mathrm{Fe^{2+}_{(aq)}}, \tag{8.6.1}$$

in which ferric ion is reduced to ferrous ion in aqueous solution. The kinetics of this reaction are rapid, and it is easy to measure the EMF of this half-cell at equilibrium. The measured EMF satisfies the Nernst equation over many orders of magnitude of change in concentration. For this reason the half-cell is a good candidate for examining the size of the nonequilibrium EMF.

Steady nonequilibrium concentrations of Fe^{2+} and Fe^{3+} can be achieved in a CSTR by mixing two inflows, one containing Fe^{2+} and Fe^{3+} and another containing the oxidant peroxydisulfate, $S_2O_8^{2-}$. Peroxydisulfate oxidizes ferrous ion via the two elementary reactions

$$Fe^{2+} + S_2O_8^{2-} \rightleftarrows Fe^{3+} + SO_4^{\cdot -} + SO_4^{2-}$$

$$Fe^{2+} + SO_4^{\cdot -} \rightleftarrows Fe^{3+} + SO_4^{2-}.$$

Neither reaction is very reversible and the second, which involves the free radical $SO_4^{\cdot -}$, is extremely rapid. The two reactions have the net effect of oxidizing two ferrous ions for each peroxydisulfate. Since the free radical is at such a low concentration, it can be eliminated by applying a steady-state approximation. This eliminates $SO_4^{\cdot -}$ from the rate equations, which have an overall stoichiometry

$$2Fe^{2+} + S_2O_8^{2-} = 2Fe^{3+} + 2SO_4^{2-}. \tag{8.6.2}$$

The process of the elimination of variables using the steady-state approximation is described in detail in Section 9.2. The basic idea is to set $d\bar{\rho}/dt = 0$, where $\bar{\rho}$ is the average density of $SO_4^{\cdot -}$ radicals, and to solve the resulting equation for $\bar{\rho}$ in terms of the remaining densities. This expression is then used to eliminate $\bar{\rho}$ in the remaining kinetic equations. When this is done for the reactions in Eq. (8.6.1), taking into account the slowness of the back reactions and the presence of excess SO_4^{2-}, one finds that the rate of the net reaction in Eq. (8.6.2) is given by

$$\bar{V}^+ - \bar{V}^- = k^+ \bar{\rho}_1 \bar{\rho}_2 - k^- \bar{\rho}_3, \tag{8.6.3}$$

where $\bar{\rho}_1$ is the molar density of Fe^{2+}, $\bar{\rho}_2$ is the density of peroxydisulfate, and $\bar{\rho}_3$ is the molar density of Fe^{3+}. The forward rate constant for this reaction is $k^+ = 3.7\ M^{-1}\ s^{-1}$, while the reverse rate constant is too small to measure. By adjusting the input concentrations of Fe^{2+}, Fe^{3+}, and $S_2O_8^{2-}$ along with the residence time, τ, it is easy to obtain stable nonequilibrium steady states using this reaction.

The analysis of reaction (8.6.2) in a CSTR is simpler than that of the four-component reaction treated in Section 5.5 since only three components need be considered, i.e., Fe^{2+}, $S_2O_8^{2-}$, and Fe^{3+}. The rate equations for the conditional average are determined by the stoichiometry in Eq. (8.6.2) and the rate law in Eq. (8.6.3) to be

$$d\bar{\boldsymbol{\rho}}/dt = \boldsymbol{\omega}(\bar{V}^+ - \bar{V}^-) - \tau^{-1}(\bar{\boldsymbol{\rho}} - \boldsymbol{\rho}^d), \tag{8.6.4}$$

where $\boldsymbol{\rho}$ is the column vector of densities, $\boldsymbol{\rho}^d$ is the column vector of diluted input densities, and $\boldsymbol{\omega}$ is the vector of stoichiometric coefficients. According to Eq. (8.6.2), $\omega_1 = -2$, $\omega_2 = -1$, and $\omega_3 = +2$, and we have assumed that the concentration of sulfate is held constant by use of a sulfate buffer. The steady-state densities in the CSTR are determined by setting $d\bar{\boldsymbol{\rho}}/dt = \mathbf{0}$, which gives

$$\boldsymbol{\omega}(k^+ \rho_1^{ss} \rho_2^{ss} - k^- \rho_3^{ss}) = \tau^{-1}(\boldsymbol{\rho}^{ss} - \boldsymbol{\rho}^d). \tag{8.6.5}$$

These equations are easily solved by making the substitution $\boldsymbol{\rho}^{ss} - \boldsymbol{\rho}^d = \boldsymbol{\omega}\xi$, which reduces the three equations to a single equation for ξ,

$$k^+(\rho_1^d - 2\xi)(\rho_2^d - \xi) - k^-(\rho_3^d + 2\xi) = \tau^{-1}\xi. \tag{8.6.6}$$

Equation (8.6.6) has a unique solution that corresponds to the reaction proceeding in the direction required by the diluted concentrations, $\boldsymbol{\rho}^d$, namely,

$$\xi = \frac{1}{2}\left(\rho_1^d + 2\rho_2^d + \frac{2k^- + \tau^{-1}}{2k^+}\right) - \frac{1}{2}\left[\left(\rho_1^d + 2\rho_2^d + \frac{2k^- + \tau^{-1}}{2k^+}\right)^2 + 2\left(\frac{k^- \rho_3^d}{k^+} - \rho_1^d \rho_2^d\right)\right]^{1/2}. \tag{8.6.7}$$

To estimate the nonequilibrium electromotive force for the half-cell in Eq. (8.6.1), we need to calculate the generalized chemical potentials, $\hat{\mu}_1$ and $\hat{\mu}_3$, of Fe^{2+} and Fe^{3+}, respectively. Indeed, since none of the other molecules involved in the CSTR reaction are involved in the electrode process, the nonequilibrium EMF reflects only the nonequilibrium contributions to $\hat{\mu}_1$ and $\hat{\mu}_3$. To find the nonequilibrium chemical potentials we first must calculate the covariance of the concentration fluctuations. In doing so, we need to account for the turbulent mixing in the CSTR. This can be done using diffusion by carrying out a calculation at the hydrodynamic level. If the three molecules have the diffusion constants D_i, then the density fluctuations satisfy

$$\partial \delta\boldsymbol{\rho}/\partial t = \boldsymbol{\omega}\hat{\delta}(V^+ - V^-) - \tau^{-1}\delta\boldsymbol{\rho} + D_i \nabla^2 \delta\boldsymbol{\rho} + \tilde{\mathbf{f}} \tag{8.6.8}$$

with

$$\hat{\delta}(V^+ - V^-) = k^+ \rho_1^{ss} \delta\rho_2 + k^+ \rho_2^{ss} \delta\rho_1 - k^- \delta\rho_3. \tag{8.6.9}$$

The random term, $\tilde{\mathbf{f}}$, results from the three processes of diffusion, chemical reaction, and dilution in the CSTR. The contributions of diffusion and the dilution process to the covariance of $\tilde{\mathbf{f}}$ were treated in Sections 6.6 and 5.5, respectively. The contribution of the chemical reaction requires additional calculations since the net reaction in Eq. (8.6.2) is the result of the contraction of elementary processes. In Section 9.2 the formula for the covariance for a steady-state contraction is given in Eq. (9.2.37). After a bit of computation one finds that the canonical formula $\gamma_{ij} = \omega_i(\bar{V}^+ + \bar{V}^-)\omega_j$ holds for this reaction even after the contraction. Thus the complete covariance can be written:

$$\begin{aligned}\mathcal{N}\langle \tilde{f}_i(\mathbf{r},t)\tilde{f}_j(\mathbf{r}',t')\rangle &= \{[\omega_i(V^{+ss} + V^{-ss})\omega_j + \tau^{-1}(\rho_i^{ss} + \rho_i^{d})\delta_{ij} - 2D_i\delta_{ij}\rho_i^{ss}\nabla_{\mathbf{r}}^2\} \\ &\quad\times \delta(\mathbf{r} - \mathbf{r}')\delta(t - t') \\ &\equiv \hat{\gamma}(\mathbf{r},\mathbf{r}')\delta(t - t'),\end{aligned} \tag{8.6.10}$$

where Avogadro's number, $\mathcal{N}$, appears since these are molar densities. These equations describe the stationary, Gaussian, Markov process of the concentration fluctuations on the length scale where mixing is complete. Based on turbulent mixing theory, this length scale is of the order of $l = (R_c/R)^{3/4}DL/v$, where v is the viscosity of the fluid, R is the Reynolds number, R_c is the Reynolds number at the onset of turbulence, L is the size of the CSTR, and D is a typical diffusion constant.

To obtain the generalized chemical potentials, we do not need to know the spatial dependence of the density–density correlation function. We need only the overall mole number fluctuations, which can be calculated using the Fourier transform, i.e.,

$$\begin{aligned}\delta n_j &= \lim_{\mathbf{k}\to\mathbf{0}} \int \delta\rho_j(\mathbf{r},t)\exp(i\mathbf{k}\cdot\mathbf{r})\,d\mathbf{r} \\ &= \lim_{\mathbf{k}\to\mathbf{0}} \delta\hat{\rho}_j(\mathbf{k},t)(2\pi)^3.\end{aligned} \tag{8.6.11}$$

This expression leads to equations similar to those found in Section 6.9 for the local approximation to thermodynamic derivatives at equilibrium. Thus we can carry out calculations of the fluctuations using the Fourier-transformed concentrations, $\hat{\rho}_j$, and take the limit $\mathbf{k} \to \mathbf{0}$. The basic quantity which we need to calculate, therefore, is the single-time covariance

$$\hat{\sigma}_{ij}(\mathbf{k},\mathbf{k}') = \langle \delta\hat{\rho}_i(\mathbf{k})\delta\hat{\rho}_j(\mathbf{k}')\rangle^{ss} = \hat{\sigma}_{ij}(k)\delta(\mathbf{k} + \mathbf{k}')/(2\pi)^3\mathcal{N}, \tag{8.6.12}$$

where the final equality follows from the work in Section 6.6. Referring to the details in that section, we see that the matrix $\hat{\sigma}(k)$ solves the fluctuation–dissipation theorem in Eq. (6.6.43), i.e.,

$$\hat{H}(k)\hat{\sigma}(k) + \hat{\sigma}(k)\hat{H}(k)^T = -\hat{\gamma}(k). \tag{8.6.13}$$

Here $\hat{H}(k)$ is the relaxation matrix, whose form can be obtained by Fourier transforming Eq. (8.6.8). One finds that

$$\hat{H}(k) = \begin{pmatrix} -(2k^+\rho_2^{ss}+k^2D_1+\tau^{-1}) & -2k^+\rho_1^{ss} & 2k^- \\ -k^+\rho_2^{ss} & -(k^+\rho_1^{ss}+k^2D_2+\tau^{-1}) & k^- \\ 2k^+\rho_2^{ss} & 2k^+\rho_1^{ss} & -(2k^-+k^2D_3+\tau^{-1}) \end{pmatrix}. \tag{8.6.14}$$

The form of $\hat{\gamma}$ comes from Fourier transforming Eq. (8.6.10),

$$\hat{\gamma}_{ij}(k) = \omega_i(V^{+ss} + V^{-ss})\omega_j + [\tau^{-1}(\rho_i^{ss} + \rho_i^{d}) + 2k^2D_i\rho_i^{ss}]\delta_{ij}. \tag{8.6.15}$$

Instead of solving Eq. (8.6.13) directly, it is easier to use the equation for the nonequilibrium portion of $\hat{\sigma}(k)$, i.e.,

$$\hat{s}_{ij}(k) \equiv \hat{\sigma}_{ij}(k) - \rho_i^{ss}\delta_{ij}, \tag{8.6.16}$$

where the local equilibrium form has been subtracted out. Substituting Eq. (8.6.16) into Eq. (8.6.13), it is easy to see that $\hat{s}$ satisfies an equation of the form

$$\hat{H}(k)\hat{s}(k) + \hat{s}(k)\hat{H}(k)^T = -\hat{g}(k). \tag{8.6.17}$$

The explicit formula for $\hat{H}(k)$ shows that all the nonzero elements of $\hat{g}(k)$ are proportional to $\tau^{-1}(\rho_1^{ss} - \rho_1^d)$. Explicitly one finds that

$$\hat{g} = \begin{pmatrix} -1 & \frac{1}{2} & 1 \\ \frac{1}{2} & 0 & 0 \\ 1 & 0 & -1 \end{pmatrix} \tau^{-1}(\rho_1 - \rho_1^d).$$

Using this formula and Eq. (8.6.14), one can solve Eq. (8.6.17) for the six linearly independent quantities $\hat{s}_{11}, \hat{s}_{12}, \hat{s}_{13}, \hat{s}_{22}, \hat{s}_{23}$, and $\hat{s}_{33}$.

To complete the calculation we need to evaluate the inverse of the covariance matrix of the mole numbers. Combining Eqs. (8.6.11) and (8.6.12) we find that

$$\sigma_{ij}^{-1} \equiv (\langle \delta\mathbf{n}\delta\mathbf{n}^T\rangle^{-1})_{ij} = \lim_{k\to 0} \mathcal{N}\hat{\sigma}(k)_{ij}^{-1}/V, \tag{8.6.18}$$

where V is the volume of the system. We must, however, be careful taking the limit $k \to 0$ in Eq. (8.6.18) since this corresponds to fluctuations of infinite wavelength. As we have already noted, turbulent mixing restricts the present calculation to the well-mixed region of linear dimension $l = (R_c/R)^{3/4}DL/v$. This corresponds to a minimum wave vector $k_m = 2\pi/l$. Thus the limit $k \to 0$ must be interpreted as $k \to k_m$. With this restriction, we can apply the basic formulas in Section 8.3 to calculate the nonequilibrium chemical potentials. Equations (8.3.10)–(8.3.12) can be written explicitly as

$$(\partial\hat{\mu}_i/\partial n_j) = RT\delta_{ij}/n_j + RT\bar{B}_{ij}/V, \tag{8.6.19}$$

where

$$\bar{B}_{ij} = [\hat{\sigma}(k_m)_{ij}^{-1} - \delta_{ij}/\rho_j^{ss}]. \tag{8.6.20}$$

The gas constant, rather than Boltzmann's constant, appears in Eq. (8.6.19) because the n_j have been taken as mole numbers, not molecule numbers. Integrating Eq. (8.6.19) and recalling that $\bar{B}_{ij}$ is treated as a function of ρ_j^d and τ only, gives

$$\hat{\mu}_i = \mu_i^0 + RT\ln\rho_i + RT\sum_{j=1}^{3}\bar{B}_{ij}(\tau, \boldsymbol{\rho}^d)\rho_j. \tag{8.6.21}$$

Again we see that the corrections to the local equilibrium formula for $\hat{\mu}_i$ are the order of RT.

The expression for the electromotive force for the half-reaction in Eq. (8.6.1) comes from the generalized Nernst equation (8.5.21). This is a one-electron oxidation so that $m = 1$. The signed stoichiometric coefficients for Fe^{2+} and

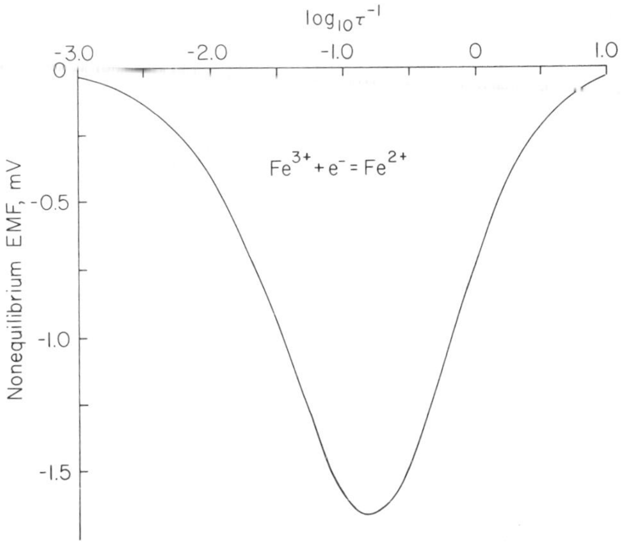

FIGURE 8.3. Calculated values of the nonequilibrium electromotive force (EMF) for the reduction of Fe^{3+} in a CSTR as a function of the inverse residence time, τ^{-1}. The CSTR is maintained in a nonequilibrium steady state by the reaction of ferrous ion with peroxydisulfate as described in the text.

Fe^{3+} in the half-reaction (8.6.1) are $\nu = +1$ and $\nu = -1$, respectively. Thus Eqs. (8.5.21) and (8.6.21) yield

$$\mathscr{E} = \mathscr{E}^{le} + (RT/F) \sum_{j=1}^{3} (\bar{B}_{3j} - \bar{B}_{1j})\rho_j^{ss}. \tag{8.6.22}$$

At room temperature RT/F is 25.7 mV, which is the order of the size of the nonequilibrium EMF. The actual size of this contribution is easy to obtain numerically when the calculation outlined above is carried out explicitly. The results of this, using the values $k_m^2 = 7.1 \times 10^4\ \text{cm}^{-2}$ and $D_1 = 7.0 \times 10^{-6}\ \text{cm}^2\ \text{s}^{-1}$, $D_2 = 3.5 \times 10^{-6}\ \text{cm}^2\ \text{s}^{-1}$, $D_3 = 2.1 \times 10^{-6}\ \text{cm}^2\ \text{s}^{-1}$ are given in Fig. 8.3 for a range of values of the residence time, τ. Notice that as $\tau \to \infty$ the correction to the local equilibrium EMF vanishes. This is due to the fact that long residence times correspond to equilibrium, as can be seen from the steady-state condition in Eq. (8.6.5). The correction also vanishes when $\tau \to 0$. In this limit Eq. (8.6.5) shows that $\boldsymbol{\rho}^{ss} = \boldsymbol{\rho}^d$ and there is no time for non-equilibrium correlations to build up at the steady state. Recent experiments on this system have measured corrections to the EMF that agree with the calculations in Fig. 8.3.

The maximum corrections to the local equilibrium electromotive force in Fig. 8.3 are the order of a few millivolts. By adjusting the rate constants k^+ and k^- and the concentrations $\boldsymbol{\rho}^d$ it can be shown that this is about as large as these corrections can become for a reaction of this stoichiometry.

References

Thermodynamic Stability of Equilibrium

H. Callen, *Thermodynamics* (Wiley, New York, 1960).

L. Tisza, On the general theory of phase transitions, in *Phase Transitions in Solids* (Wiley, New York), Chapter 1.

J. Kirkwood and I. Oppenheim, *Chemical Thermodynamics* (McGraw-Hill, New York, 1961).

Thermodynamic Stability of Steady States

J. Keizer and R. Fox, Qualms regarding the range of validity of the Glansdorff–Prigogine criterion for stability of nonequilibrium states, *Proc. Nat. Acad. Sci. U.S.A.* **71**, 192–196 (1974).

P. Glansdorff and I. Prigogine, *Thermodynamic Theory of Structure Stability, and Fluctuations* (Wiley-Interscience, New York, 1971).

J. Keizer, Fluctuations, stability, and generalized state functions at nonequilibrium steady states, *J. Chem. Phys.* **65**, 4431–4444 (1976).

J. Keizer, Nonequilibrium thermodynamics and the stability of states far from equilibrium, *Acc. Chem. Res.* **12**, 243–249 (1979).

R. Fox, The "excess entropy" around nonequilibrium steady states, $(\delta^2 S)_{ss}$, is not a Liapunov function, *Proc. Nat. Acad. Sci. U.S.A.* **77**, 3763–3766 (1980).

Thermodynamic Functions at Nonequilibrium Steady States

J. Keizer, Thermodynamics at nonequilibrium steady states, *J. Chem. Phys.* **69**, 2609–2620 (1978).

H.J.M. Hanley and D. Evans, A thermodynamics for a system under shear, *J. Chem. Phys.* **76**, 3225–3232 (1982).

J. Keizer, Reversibility, work, and heat at nonequilibrium steady states, *Phys. Rev. A* **30**, 1115–1117 (1984).

J. Keizer, Heat, work, and the thermodynamic temperature at nonequilibrium steady states, *J. Chem. Phys.* **82**, 2751–2771 (1985).

CHAPTER 9

Hierarchies and Contractions of the Description

9.1. Introduction

One of the characteristic features of the mechanistic statistical theory is that it can be used at a variety of levels of description. Indeed, we have seen that molecular processes ranging from momentum and heat transport to electrochemical reactions and molecular scattering events all can be described by the canonical equations in Chapter 4. The amount of molecular detail contained in these descriptions varies tremendously. At the thermodynamic level the extensive variables are characteristic of properties of the entire system, such as the total internal energy. At the hydrodynamic level, on the other hand, one keeps track of densities of extensive variables at each point in space. Finally, at the Boltzmann level the number density of particles in the molecular phase space is the fundamental quantity of interest.

In the preceding chapters these levels of description have been treated as separate even though they form a natural hierarchy in which the elementary dynamical events involve progressively more molecular detail. In this chapter we examine the sort of connections that can be made between the levels of the hierarchy. This leads to an extension of the kinetic theory of Maxwell and Boltzmann in which kinetic coefficients at one level of description are expressed in terms of kinetic coefficients at a more detailed level of description. The general process in which variables are eliminated will be referred to as *contraction of the description.*

The connection between levels of description is based on a separation of time scales among dynamical processes, and variables that relax rapidly often can be eliminated from the statistical description of a system. A case of particular interest involves elementary processes that are rapid enough to maintain a balance between their forward and reverse rates in the presence of other elementary processes that are out of equilibrium. On several occasions, e.g., in our discussion of the *o*-cresolphthalein reaction in Section 7.4, we have used this as justification for neglecting the dynamical effects of such a process. In Section 9.2 we examine this sort of contraction and find conditions under which it can be justified. Another kind of contraction is possible when several coupled elementary processes produce a slow time variation in certain variables.

This quasi-steady-state condition allows these variables to be eliminated although, as we will see, their elimination modifies the statistical properties of the variables that remain.

The notion of a contracted stochastic process provides another kind of connection between descriptions of dynamical processes. Recall that the fundamental hypothesis of our approach in Chapter 4 leads to a nonstationary, Gaussian, Markov description of conditional fluctuations. In this sense the fluctuations have no memory of their previous values, except as is inherited from the time dependence of the conditional average. This does not always provide a correct description of measurements in an ensemble. For example, at pressures corresponding to a dense gas, molecular dynamics calculations by Alley, Alder, and Yip (see Section 6.7) show that the hard-sphere scattering function does not agree well with that calculated using hydrodynamic fluctuation theory. What is missing is a detailed description of molecular collisions, such as contained in the Boltzmann level of description. This can, in principle, be included without adding additional variables to the hydrodynamic description by performing a contraction of the Boltzmann description. This leads to memory terms in the stochastic theory, which, in general, are no longer Markovian.

Using ideas inspired by the Chapman–Enskog expansion, Fox has found conditions under which this contraction procedure leads to a stationary, Gaussian, Markov process. In Sections 9.3 and 9.4 we show how this idea can be used to derive explicit expressions for transport coefficients using the Boltzmann equation. Similar problems arise in the theory of rapid chemical reactions, where reaction rate constants can become dependent on the diffusion constant. This also can be treated using a contraction of the hydrodynamic-level description, although the contraction differs from that between the Boltzmann and hydrodynamic levels of description. It is based on the occurrence of nonequilibrium corrections to spatial correlation functions rather than on a difference of time scales. This sort of contraction provides a generalization of the Smoluchowski theory of diffusion-controlled reactions and is treated in Section 9.5.

9.2. Contractions without Memory

The elimination of variables is implicit in all descriptions of molecular systems. Indeed, our statistical approach leads to the description of physical properties in terms of an ensemble in which all but a few variables are ignored. In statistical nonequilibrium thermodynamics the fundamental variables are extensive and reflect the collective properties of many molecules. In a particular system there are many possible choices of extensive variables, even at a particular level of description. At the thermodynamic level, for example, one might choose to describe a beaker of water using the energy, volume, and number of moles of H_2O or, instead of just moles of H_2O, one might add the number of protons, H^+, and hydroxyl ions, OH^-, or even the individual vibrational states of some or all of these species. The choice of which extensive variables to select depends chiefly on what variables one can measure, which in turn depends on the sophistication of one's instrumentation. Nonetheless, some choices clearly will not give rise to a simple self-contained description, whereas others will.

It is important to develop criteria that provide information about when a particular choice of extensive variables is self-contained. The single most useful criterion has to do with separation of time scales. A collection of extensive variables provide a self-contained statistical ensemble description, of the sort used in Chapter 4, if they vary on a time scale which is well separated from that of other extensive variables. For example, the rates of most chemical reactions that occur in aqueous solution are adequately described by a knowledge of chemical potentials of the molecules involved, while a knowledge of the internal molecular states, which relax to equilibrium on a picosecond time scale, is irrelevant for all but the most rapid chemical reactions.

To make this statement more precise, consider the $l + k$ elementary processes with rates $V_\kappa^\pm$. The final k of these processes will be assumed to be slow with respect to the first l processes. According to the canonical formula in Eq. (4.2.3)

$$V_\kappa^\pm = \Omega_\kappa^\pm \exp\left[-\sum_j F_j n_{\kappa j}^\pm / k_B\right]. \tag{9.2.1}$$

This means that the intrinsic rates, $\Omega_\kappa^\pm$, of the final k processes are much smaller than those for the first l processes. The jump probabilities for the extensive variables $\mathbf{n}$ are determined from these rates by Eq. (4.2.6), i.e.,

$$\begin{aligned} P_2(\mathbf{n}, t^0 | \mathbf{n} + d\mathbf{n}_s + d\mathbf{n}, t^0 + dt) &= V_\kappa^\pm(\mathbf{n})\,dt + \mathcal{O}(dt) \quad \text{if } d\mathbf{n} = \pm\boldsymbol{\omega}_\kappa \\ &= 1 - \sum_\kappa V_\kappa^\pm(\mathbf{n})\,dt + \mathcal{O}(dt) \quad \text{if } d\mathbf{n} = \mathbf{0} \\ &= 0 \quad \text{otherwise.} \end{aligned} \tag{9.2.2}$$

Imagine now that we are simulating this stochastic process on a computer. To do so, we would need to choose $dt \ll 1/\Omega_\kappa^* \ll 1/\Omega_\kappa^{**}$, where Ω_κ^* is the maximum of $\Omega_\kappa^\pm$ and Ω_κ^{**} is the minimum. In the simulation the fast processes will cause transitions on the time scale $1/\Omega_\kappa^*$, which is characteristic of the rapid processes, while the slow processes will change on the time scale $1/\Omega_\kappa^{**}$. As long as these two time scales are well separated, the rapid processes will lead to numerous transitions before the slow processes cause appreciable change. On the slow time scale the extensive variables will have adjusted themselves so that the number of forward transitions is approximately balanced by the number of reverse transitions for each rapid process. On the slow time scale this will be equivalent to setting

$$V_\kappa^+(\mathbf{n}) = V_\kappa^-(\mathbf{n}) \tag{9.2.3}$$

for the fast processes. In other words, on the slower time scale we can use the functional relationship in Eq. (9.2.3) to determine the values of the rapidly changing extensive variables in terms of the slowly changing variables. The way in which the slow variables change on the slow time scale will thus be determined by only the slow processes in Eq. (9.2.2), with the subsidiary condition that the values of the fast variables be determined by Eq. (9.2.3). Using Eq. (9.2.3) the fast variables are contracted out from the description and only the slow elementary processes are left in the stochastic process.

A specific example will help illustrate this sort of contraction. Consider the two elementary chemical reactions

$$\begin{aligned} &\mathrm{A} \overset{\alpha}{\rightleftarrows} \mathrm{B} \\ &\mathrm{B} + \mathrm{C} \overset{\beta}{\rightleftarrows} \mathrm{D}. \end{aligned} \tag{9.2.4}$$

We will assume that α is the rapid process and that β is the slow process. In the jargon of chemistry, α represents a rapid preequilibrium for reaction β. Because of rapid equilibration, we expect that the intermediate species B can be eliminated. This means that instead of four species, only three will occur in the stochastic theory. If the molecule numbers of B, A, C, and D are labeled n_1, n_2, n_3, and n_4, then the conditional average equations corresponding to Eq. (9.2.4) are

$$d\bar{n}_1/dt = \bar{V}_\alpha - \bar{V}_\beta, \quad d\bar{n}_2/dt = -\bar{V}_\alpha, \quad d\bar{n}_3/dt = -\bar{V}_\beta, \quad d\bar{n}_4/dt = \bar{V}_\beta, \tag{9.2.5a}$$

where the $\bar{V}_j$'s represent the reaction rates. As a first step, we eliminate B in favor of the total amount of A and B, i.e., $n_1 + n_2 \equiv n_1^*$. Combining the first two equations in Eq. (9.2.5a), this gives

$$d\bar{n}_1^*/dt = -\bar{V}_\beta, \quad d\bar{n}_3/dt = -\bar{V}_\beta, \quad d\bar{n}_4/dt = \bar{V}_\beta. \tag{9.2.5b}$$

These equations correspond to the stoichiometry of the β reaction alone. Indeed, if the reaction α rapidly achieves equilibrium we can write $d\bar{n}_2/dt = 0$. Using Eq. (9.2.5a) we see that this is equivalent to setting $\bar{V}_\alpha^+ = \bar{V}_\alpha^-$ or

$$k_\alpha^+ \bar{\rho}_2 = k_\alpha^- \bar{\rho}_1, \tag{9.2.6}$$

where the ρ_i's are number densities and the k's are rate constants. Solving this relationship for $\bar{\rho}_i$ in terms of $\bar{\rho}_1^* = \bar{\rho}_1 + \bar{\rho}_2$ gives

$$\bar{\rho}_1 = \bar{\rho}_1^* k_\alpha^+/(k_\alpha^+ + k_\alpha^-). \tag{9.2.7}$$

We can now eliminate $\bar{\rho}_2$ in the expression for $\bar{V}_\beta$ to obtain

$$\bar{V}_\beta = (\bar{k}_\beta^+ \bar{\rho}_1^* \bar{\rho}_3 - k_\beta^- \bar{\rho}_4) V \tag{9.2.8}$$

with V the volume and $\bar{k}_\beta^+ \equiv k_\beta^+ k_\alpha^+/(k_\alpha^+ + k_\alpha^-)$. The rate expression in Eq. (9.2.8) obviously corresponds to the reaction

$$B^* + C \overset{\beta}{\rightleftarrows} D, \tag{9.2.9}$$

where B* represents the total of A + B. Equations (9.2.5) and (9.2.8) show how the contraction affects the average behavior of the slow variables: the canonical form of the rate of reaction β is preserved while the rate constant is modified by the factor $k_\alpha^+/(k_\alpha^+ + k_\alpha^-)$.

Fluctuations in the uncontracted four-variable system satisfy the usual linearized equations. These can easily be expressed in terms of fluctuations in n_1^*, n_2, n_3, and n_4 since $n_1^* = n_1 + n_2$. If we write the fluctuation equations in terms of the densities of these quantities we obtain

$$\begin{aligned} d\delta\rho_2/dt &= -(k_\alpha^+ + k_\alpha^-)\delta\rho_2 + k_\alpha^-\delta\rho_1^* - \tilde{f}_\alpha \\ d\delta\boldsymbol{\rho}^*/dt &= H\delta\boldsymbol{\rho}^* + H'\delta\rho_2 + \tilde{\mathbf{f}}^*. \end{aligned} \tag{9.2.10}$$

In this equation $\delta\boldsymbol{\rho}^*$ is the column vector of density fluctuations of the slow variables, $\tilde{f}_\alpha$ is the random term associated with reaction α, and $\tilde{\mathbf{f}}^*$ and H' are the column vectors

$$\tilde{\mathbf{f}}^* = f_\beta \begin{pmatrix} 1 \\ 1 \\ -1 \end{pmatrix}, \quad H' = k_\beta^+\bar{\rho}_3 \begin{pmatrix} 1 \\ 1 \\ -1 \end{pmatrix}, \tag{9.2.11}$$

and H^* is the 3×3 matrix

$$H = \begin{pmatrix} -k_\beta^+\bar{\rho}_3 & -k_\beta^+\bar{\rho}_1 & k_\beta^- \\ -k_\beta^+\bar{\rho}_3 & -k_\beta^+\bar{\rho}_1 & k_\beta^- \\ k_\beta^+\bar{\rho}_3 & k_\beta^+\bar{\rho}_1 & -k_\beta^- \end{pmatrix}. \tag{9.2.12}$$

The value of $\bar{\rho}_1$ in Eq. (9.2.12) is given by Eq. (9.2.7).

Equations (9.2.9)–(9.2.12) exhibit the intrinsic coupling between the fast variable, $\delta\rho_2$, and the slow variables, $\delta\boldsymbol{\rho}^*$. However, because the α reaction is fast, the relaxation time $\tau = 1/(k_\alpha^+ + k_\alpha^-)$ in Eq. (9.2.10) is small. Indeed, solving that equation in the usual way we have

$$\delta\rho_2(t) = \int_0^t \exp[-(t-s)/\tau]\{k_\alpha^-\delta\rho_1^*(s) - \tilde{f}_\alpha(s)\}\,ds, \tag{9.2.13}$$

since $\delta\rho_2(0) = 0$. Because of the separation of time scales, it is appropriate to consider the limit that $\tau \to 0$. In this limit

$$\lim_{\tau\to 0} \tau^{-1}\exp[-|t-s|/\tau] = 2\delta(t-s). \tag{9.2.14}$$

Using this to integrate Eq. (9.2.13), we can approximate $\delta\rho_2(t)$ by the expression

$$\tau^{-1}\delta\rho_2(t) = k_\alpha^-\delta\rho_1^*(t) - \tilde{f}_\alpha(t). \tag{9.2.15}$$

Notice that this is equivalent to setting $d\delta\rho_2/dt = 0$ in Eq. (9.2.10), which is similar to the contraction of the average equation using $d\bar{n}_2/dt = 0$. Equation (9.2.15) allows us to eliminate $\delta\rho_2$ from Eq. (9.2.10) and, thus, uncouple $\delta\rho_2$ and $\delta\boldsymbol{\rho}^*$. After some algebra one finds

$$d\delta\,\boldsymbol{\rho}^*/dt = H^*\delta\boldsymbol{\rho}^* + \tilde{\mathbf{f}}^* - \tau H'\tilde{f}_\alpha \tag{9.2.16}$$

with

$$H^* = \begin{pmatrix} -\bar{k}_\beta^+\bar{\rho}_3 & -\bar{k}_\beta^+\bar{\rho}_1^* & k_\beta^- \\ -\bar{k}_\beta^+\bar{\rho}_3 & -\bar{k}_\beta^+\bar{\rho}_1^* & k_\beta^- \\ \bar{k}_\beta^+\bar{\rho}_3 & \bar{k}_\beta^+\bar{\rho}_1^* & -k_\beta^- \end{pmatrix}. \tag{9.2.17}$$

To obtain the covariance of the random term $\tilde{\mathbf{f}}^* - \tau H'\tilde{f}_\alpha$, we recall from the canonical formula in Eq. (3.2.25) that $\tilde{f}$'s for independent reactions are uncorrelated. Thus the covariance is

$$\langle\tilde{\mathbf{f}}^*(t)\tilde{\mathbf{f}}^{*T}(t')\rangle + \tau^2 H'H'^T\langle\tilde{f}_\alpha(t)\tilde{f}_\alpha(t')\rangle. \tag{9.2.18}$$

The second term, however, vanishes in the limit that the time scales are well separated. Indeed the covariance of $\tilde{f}_\alpha$ is proportional to $(\bar{V}_\alpha^+ + \bar{V}_\alpha^-)$, which is proportional to Ω_α [cf. Eq. (9.2.1)], which is proportional to $k_\alpha^+ + k_\alpha^- = \tau^{-1}$. Thus the second term in Eq. (9.2.18) is proportional to τ and vanishes in the limit that $\tau \to 0$.

Putting these results together, we find that on the time scale of the slow variables, the density fluctuations satisfy

$$d\delta\,\boldsymbol{\rho}^*/dt = H^*\delta\boldsymbol{\rho}^* + \tilde{\mathbf{f}}^*, \tag{9.2.19}$$

where from Eq. (9.2.11)

$$\langle\tilde{f}_i^*(t)\tilde{f}_j^*(t')\rangle = \omega_i(\bar{V}_\beta^+ + \bar{V}_\beta^-)\omega_j\delta(t-t') \tag{9.2.20}$$

with $\omega_1^* = -1, \omega_3 = -1, \omega_4 = +1$. Notice that Eq. (9.2.20) has the canonical form with the stoichiometric coefficients identical to those for B^*, C, and D based on the contracted elementary process in Eq. (9.2.9). Moreover, comparing the explicit expression for H^* given in (9.2.17) to the contracted form of $\bar{V}_\beta$ in Eq. (9.2.8), we see that H^* results from the linearization of the contracted average equations. In other words, when the time scales are well separated, the fast reaction α no longer appears in the canonical theory. Only the contracted elementary process remains, and on the contracted time scale it satisfies the canonical theory.

This is true for general schemes of elementary processes and is not restricted to chemical reactions. The proof follows the above example rather closely and is left as an exercise. This result is important because it allows us to ignore elementary processes that are rapid on the time scale that interests us. This justifies, for example, our neglect of density fluctuations associated with intermediate species $FSO_3^{\cdot *}$ in the dimerization of fluorosulfate radicals in Section 7.4. Similarly, it shows that fluctuations associated with the ionization of o-cresolphthalein in Eq. (7.4.24) can be ignored when treating thermal fluctuations, as was assumed there also.

We used a related notion of contraction in our treatment of fluctuations associated with the bulk viscosity in Section 6.5. In Eq. (6.5.17) we introduced a quasi-steady-state assumption to eliminate fluctuations in an internal state that was assumed to relax rapidly with respect to the hydrodynamic variables. This sort of contraction is commonly employed to treat problems in chemical kinetics. It differs from the rapid-equilibrium assumption since the quasi-steady state may be achieved by balancing rates of several reactions. An important example in biochemistry is the Michaelis–Menten scheme for the enzyme-catalyzed conversion of substrate (S) to product (P), e.g.,

$$\mathrm{E} + \mathrm{S} \overset{\alpha}{\rightleftarrows} \mathrm{ES} \overset{\beta}{\rightleftarrows} \mathrm{P} + \mathrm{E}. \tag{9.2.21}$$

When the enzyme, E, is present in catalytic amounts the intermediate enzyme–substrate complex, ES, achieves a quasi-steady, nonequilibrium concentration determined by a balancing of the two elementary reactions in Eq. (9.2.21).

The contracted stochastic process associated with variables that achieve a quasi-steady state is easy to derive. The contracted stochastic process is similar in structure to that which occurs when there is rapid equilibrium, although the coupling of the rapid variables via the elementary processes leads to new random terms in the statistical description. To see this, we begin with an arbitrary set of extensive variables, $\mathbf{n}$, which we partition into two sets. The first set will be written $\mathbf{n}_1$, and these will be the variables that rapidly achieve a quasi-steady state. The remaining variables, which relax slowly, will be written $\mathbf{n}_2$. Thus $\mathbf{n} = (\mathbf{n}_1, \mathbf{n}_2)$. The conditional averages of $\mathbf{n}_1$ and $\mathbf{n}_2$ are coupled by the average equations

$$\begin{aligned} d\bar{\mathbf{n}}_1/dt &= \mathbf{R}_1(\bar{\mathbf{n}}_1, \bar{\mathbf{n}}_2) \\ d\bar{\mathbf{n}}_2/dt &= \mathbf{R}_2(\bar{\mathbf{n}}_1, \bar{\mathbf{n}}_2, t), \end{aligned} \tag{9.2.22}$$

where it is assumed that rate equations for the fast variables do not depend explicitly on the time. The coupling of the conditional fluctuations can be written

$$\begin{aligned} d\delta\mathbf{n}_1/dt &= H_{11}\delta\mathbf{n}_1 + H_{12}\delta\mathbf{n}_2 + \tilde{\mathbf{f}}_1 \\ d\delta\mathbf{n}_2/dt &= H_{21}\delta\mathbf{n}_1 + H_{22}\delta\mathbf{n}_2 + \tilde{\mathbf{f}}_2. \end{aligned} \tag{9.2.23}$$

As usual, the matrices H_{ij} are obtained by linearizing Eq. (9.2.22), i.e.,

$$H_{ij} \equiv [\partial \mathbf{R}_i(\bar{\mathbf{n}}_1, \bar{\mathbf{n}}_2, t)/\partial \bar{\mathbf{n}}_j]_{\mathbf{n}_i}, \tag{9.2.24}$$

with $i, j = 1, 2$. The covariance of the random terms takes its canonical form, so that

$$\begin{aligned} \langle \tilde{\mathbf{f}}_i(t)\tilde{\mathbf{f}}_j^T(t')\rangle &= \sum_\kappa \boldsymbol{\omega}_{\kappa i}(\bar{V}_\kappa^+ + \bar{V}_\kappa^-)\boldsymbol{\omega}_{\kappa j}^T\delta(t - t') \\ &\equiv \gamma_{ij}\delta(t - t'), \end{aligned} \tag{9.2.25}$$

where, for example, $\boldsymbol{\omega}_{\kappa 1}$ is the vector of net changes in process κ for the fast variables. Thus if some of the fast variables are coupled by elementary processes to the slow variables, then the random terms $\tilde{\mathbf{f}}_1$ and $\tilde{\mathbf{f}}_2$ will be coupled by the matrices γ_{12} and γ_{21}. If there are l fast variables and k slow variables, the H_{11} and γ_{11} are $l \times l$ matrices, H_{12} and γ_{12}, are $l \times k$ matrices, etc. While it is not true that γ_{12} equals γ_{21}, it is true that

$$\gamma_{12} = \gamma_{21}^T, \tag{9.2.26}$$

where the superscript T means that the rows and columns are interchanged.

The assumption that a quasi-steady state is achieved by the variables $\mathbf{n}_1$ implies that

$$\mathbf{R}_1(\bar{\mathbf{n}}_1, \bar{\mathbf{n}}_2) = \mathbf{0}. \tag{9.2.27}$$

These are l equations, which we will assume can be solved for $\mathbf{n}_1$ in terms of $\mathbf{n}_2$. According to the inverse function theorem, this can be done if the matrix $H_{11} = \partial \mathbf{R}_1/\partial \bar{\mathbf{n}}_1$ is invertible. The solution to Eq. (9.2.27) will be written

$$\bar{\mathbf{n}}_1 = \mathbf{G}(\bar{\mathbf{n}}_2). \tag{9.2.28}$$

Substituting this expression into Eq. (9.2.22) gives the contracted average equation

$$d\bar{\mathbf{n}}_2/dt = \mathbf{R}_2[\mathbf{G}(\bar{\mathbf{n}}_2), \bar{\mathbf{n}}_2, t] \equiv \mathbf{R}_2(\bar{\mathbf{n}}_2, t), \tag{9.2.29}$$

which no longer depends explicitly on $\bar{\mathbf{n}}_1$. The quasi-steady state in Eq. (9.2.28) is supposed to occur rapidly and to be stable. Thus on the time scale of the rapid variables, it follows that a perturbation $\Delta\mathbf{n}_1$ around the quasi-steady state rapidly relaxes to it. Using the first of Eqs. (9.2.22) on the rapid time scale, this implies that

$$d\Delta\mathbf{n}_1/dt \simeq H_{11}\Delta\mathbf{n}_1, \tag{9.2.30}$$

and that H_{11} corresponds to a rapid relaxation process. Thus there exists a characteristic time scale, τ_1, such that $H_{11} = \bar{H}_{11}/\tau_1$ with τ_1 much smaller than the characteristic time, τ_2, for the slow variables.

To see how this relaxation affects the fluctuations, we need to use Eqs. (9.2.23). Recall that the matrices H_{ij} are time dependent. If we choose a time $t_0 > 0$ such that the quasi-steady state has been achieved, $H_{11}(t)$ will be slowly varying for $\tau_2 \gg (t - t_0) \gg \tau_1$. In such a time interval the H_{ij}'s are constants. Consequently, solving the first of Eqs. (9.2.23) for $\delta\mathbf{n}_1(t)$ gives

$$\delta\mathbf{n}_1(t) = \exp[\bar{H}_{11}(t)(t - t_0)/\tau_1]\delta\mathbf{n}_1(t_0) + \int_{t_0}^{t} \exp[\bar{H}_{11}(t)(t - t_0)/\tau_1]\{\bar{H}_{12}(s)\delta\mathbf{n}_2(s) + \tilde{\mathbf{f}}_1(s)\}\, ds. \tag{9.2.31}$$

Since $(t - t_0)/\tau_1 \gg 1$, the first term in Eq. (9.2.31) vanishes. Furthermore, using the matrix generalization of Eq. (9.2.14) we have

$$\lim_{\tau_1 \to 0} \tau_1^{-1} \exp[\bar{H}_{11}(t_0)(t - t_0)/\tau_1] = -2H_{11}^{-1}(t_0)\delta(t - t_0). \tag{9.2.32}$$

Using Eq. (9.2.32) in Eq. (9.2.31), we have the formula

$$\delta \mathbf{n}_1(t) = -H_{11}^{-1}[H_{12}\delta\mathbf{n}_2(t) + \tilde{\mathbf{f}}_1(t)], \tag{9.2.33}$$

which holds asymptotically as $\tau_1 \to 0$. Notice that this is precisely the formula which is obtained if we simply set $d\delta\,\mathbf{n}_1/dt = \mathbf{0}$ in Eq. (9.2.23). Equation (9.2.33) justifies the use of this assumption for the bulk viscosity in Sections 6.4 and 6.5.

Equation (9.2.33) allows us to eliminate $\delta\mathbf{n}_1$ from the second equation in (9.2.23) and obtain

$$d\delta\mathbf{n}_2/dt = (-H_{21}H_{11}^{-1}H_{12} + H_{22})\delta\mathbf{n}_2 - H_{21}H_{11}^{-1}\tilde{\mathbf{f}}_1 + \tilde{\mathbf{f}}_2. \tag{9.2.34}$$

This is the equation that describes conditional fluctuations for the contracted variables. We can write it as

$$d\delta\mathbf{n}_2/dt = H\delta\mathbf{n}_2 + \tilde{\mathbf{f}} \tag{9.2.35}$$

if we define

$$H = -H_{21}H_{11}^{-1}H_{12} + H_{22} \tag{9.2.36}$$

and take

$$\langle\tilde{\mathbf{f}}(t)\tilde{\mathbf{f}}^T(t')\rangle = (\gamma_{22} - \gamma_{21}H_{11}^{T-1}H_{21}^T - H_{21}H_{11}^{-1}\gamma_{12} + H_{21}H_{11}^{-1}\gamma_{11}H_{11}^{T-1}H_{21}^T) \times \delta(t - t'), \tag{9.2.37}$$

which follows from Eq. (9.2.34) and (9.2.25). It is not difficult to see that the matrix H is what one gets from the linearization of the contracted average equation (9.2.29). In fact, differentiating the contracted $\mathbf{R}_2$ as defined in Eq. (9.2.29), the chain rule gives

$$\begin{aligned}\partial\mathbf{R}_2/\partial\bar{\mathbf{n}}_2 &= (\partial\mathbf{R}_2/\partial\bar{\mathbf{n}}_1)_{\mathbf{n}_2}(\partial\bar{\mathbf{n}}_1/\partial\bar{\mathbf{n}}_2) + (\partial\mathbf{R}_2/\partial\bar{\mathbf{n}}_2)_{\mathbf{n}_1} \\ &= H_{21}(\partial\bar{\mathbf{n}}_1/\partial\bar{\mathbf{n}}_2) + H_{22},\end{aligned} \tag{9.2.38}$$

where we have used the definition of the H_{ij} in Eq. (9.2.24). However, if we use the fact that $\bar{\mathbf{n}}_1$ comes from solving Eq. (9.2.27), we can differentiate that expression to obtain

$$\partial\mathbf{R}_1/\partial\bar{\mathbf{n}}_2 = (\partial\mathbf{R}_1/\partial\bar{\mathbf{n}}_1)_{\mathbf{n}_2}(\partial\bar{\mathbf{n}}_1/\partial\bar{\mathbf{n}}_2) + (\partial\mathbf{R}_1/\partial\bar{\mathbf{n}}_2)_{\mathbf{n}_1} = \mathbf{0}, \tag{9.2.39}$$

or

$$\partial\bar{\mathbf{n}}_1/\partial\bar{\mathbf{n}}_2 = -H_{11}^{-1}H_{12}. \tag{9.2.40}$$

Substituting this into Eq. (9.2.37) yields

$$\partial\bar{\mathbf{R}}_2/\partial\bar{\mathbf{n}}_2 = -H_{21}H_{11}^{-1}H_{12} + H_{22} \tag{9.2.41}$$

for the linearized relaxation rate matrix based on Eq. (9.2.29), which is identical to the contracted matrix H in Eqs. (9.2.35) and (9.2.36). Thus we can use the contracted average equation to generate the relaxation matrix H for the fluctuations by differentiation in the usual way.

Except under special circumstances, it is not true that the random terms in the quasi-steady-state contraction are those expected based on the canonical formula. Indeed, the canonical form gives only the term γ_{22} in Eq. (9.2.37). Only two of the remaining terms will often be important, however, since as we have seen $H_{11}^{-1} = \tau_1 \bar{H}_{11}^{-1}$. Indeed, γ_{11} is at worst proportional to τ_1^{-1} (through terms like Ω_κ). Thus if H_{21} is independent of τ_1, it follows that the final term in Eq. (9.2.37) vanishes as $\tau_1 \to 0$. If the variables $\mathbf{n}_1$ and $\mathbf{n}_2$ are uncoupled by the elementary processes, then it follows from Eq. (9.2.25) that γ_{12} and γ_{21} vanish. This is the case, for example, for progress variables for elementary chemical reactions. If this is true, then only the term γ_{22} remains in the covariance formula for the random term $\tilde{\mathbf{f}}$ and the form of the canonical theory is preserved. Another important special case occurs when the slow variables are nondissipative quantities when considered by themselves, for example, total mass density at the hydrodynamic level of description. In this case γ_{22} as well as γ_{12} and γ_{21} vanish. This means that only the final term in Eq. (9.2.37) remains for the covariance, and it cannot be neglected. This was the case for the bulk viscosity in Section 6.5, and led us to the formula for the stress tensor correlation function in Eq. (6.5.31).

As an example of the steady-state contraction process, we work out the details for the Michaelis–Menten enzyme reaction mechanism in Eq. (9.2.21). The intermediate species is the enzyme–substrate complex, ES, and it will be assumed to come to a rapid steady state. We will label the molecular densities of the species E, ES, S, and P by ρ_1, ρ_2, ρ_s, and ρ_p. In a dilute solution the average rate equations corresponding to the elementary processes of this mechanism are

$$\begin{aligned} d\bar{\rho}_1/dt &= -k_\alpha^+ \bar{\rho}_1 \bar{\rho}_s + k_\alpha^- \bar{\rho}_2 + k_\beta^+ \bar{\rho}_2 - k_\beta^- \bar{\rho}_1 \bar{\rho}_p = -d\bar{\rho}_2/dt \\ d\bar{\rho}_s/dt &= -k_\alpha^+ \bar{\rho}_1 \bar{\rho}_s + k_\alpha^- \bar{\rho}_2 = -d\bar{\rho}_p/dt = -d\bar{\rho}_2/dt \end{aligned} \tag{9.2.42}$$

where α and β refer to the α and β reactions in Eq. (9.2.21). According to the steady-state assumption, $d\bar{\rho}_1/dt = -d\bar{\rho}_2/dt = 0$. Using this we can solve the first of Eqs. (9.2.42) to obtain $\bar{\rho}_1$ and $\bar{\rho}_2$ in terms of $\rho = \bar{\rho}_1 + \bar{\rho}_2$, $\bar{\rho}_s$, and $\bar{\rho}_p$. One obtains

$$\begin{aligned} \bar{\rho}_2 &= \frac{(k_\alpha^+ \bar{\rho}_s + k_\beta^- \bar{\rho}_p)\rho}{(k_\alpha^+ \bar{\rho}_s + k_\beta^- \bar{\rho}_p + k_\alpha^- + k_\beta^+)} \\ \bar{\rho}_1 &= \frac{(k_\alpha^- + k_\beta^+)\rho}{(k_\alpha^+ \bar{\rho}_s + k_\beta^- \bar{\rho}_p + k_\alpha^- + k_\beta^+)}. \end{aligned} \tag{9.2.43}$$

Using these expressions we can eliminate the intermediates from the average equations for substrate and product to obtain

$$d\bar{\rho}_s/dt = \frac{-\rho(k_\alpha^+ k_\beta^+ \bar{\rho}_s - k_\alpha^- k_\beta^- \bar{\rho}_p)}{(k_\alpha^+ \bar{\rho}_s + k_\beta^- \bar{\rho}_p + k_\alpha^- + k_\beta^+)} = -d\bar{\rho}_p/dt. \tag{9.2.44}$$

Equation (9.2.44) is the average equation satisfied by the slow variables, ρ_s and ρ_p.

According to the general contraction procedure, fluctuations in the slow variables satisfy the linearized average equation with a random term dictated by Eq. (9.2.37). Since $\delta\rho_s = -\delta\rho_p - \delta\rho_2$ and $\delta\rho_1 = -\delta\rho_2$, it suffices to consider fluctuations in the substrate ($\delta\rho_s$) and the enzyme ($\delta\rho_1$). Thus the matrices γ_{11}, γ_{12}, γ_{21}, and γ_{22} are all scalars. Using the canonical formula (9.2.25) for the Michaelis–Menten mechanism we find

$$\begin{aligned} \gamma_{11} &= V^{-2}[\bar{V}_\alpha^+ + \bar{V}_\alpha^- + \bar{V}_\beta^+ + \bar{V}_\beta] \\ \gamma_{12} = \gamma_{21} &= V^{-2}[\bar{V}_\alpha^+ + \bar{V}_\alpha^-] \\ \gamma_{22} &= V^{-2}[\bar{V}_\alpha^+ + \bar{V}_\alpha^-], \end{aligned} \tag{9.2.45}$$

where V is the volume and the $\bar{V}_\kappa^\pm$ represent the rates of the elementary reactions α an β. The linearization of Eq. (9.2.44) is straightforward and after a little algebra one finds

$$\begin{aligned} d\delta\rho_s/dt = &- \frac{k_\alpha^+ \bar{\rho}_1(k_\beta^- \bar{\rho}_p + k_\beta^+)\delta\rho_s}{(k_\alpha^+ \bar{\rho}_s + k_\beta^- \bar{\rho}_p + k_\alpha^- + k_\beta^+)} \\ &+ \frac{k_\beta^- \bar{\rho}_1(k_\alpha^+ \bar{\rho}_s + k_\alpha^-)\delta\rho_p}{(k_\alpha^+ \bar{\rho}_s + k_\beta^- \bar{\rho}_p + k_\alpha^- + k_\beta^+)} + \tilde{f}_s. \end{aligned} \tag{9.2.46}$$

To obtain the covariance of $\tilde{f}_s$ we use Eq. (9.2.37). Thus we need to know H_{11} and H_{21}, which are obtained from the linearization of the complete equations (9.2.42). Carrying out this linearization we find that

$$\begin{aligned} H_{11} &= -(k_\alpha^+ \bar{\rho}_s + k_\beta^- \bar{\rho}^* + k_\alpha^- + k_\beta^+) \\ H_{21} &= -(k_\alpha^+ \bar{\rho}_s + k_\alpha^-), \end{aligned} \tag{9.2.47}$$

with $\bar{\rho}^* = \bar{\rho}_p + \bar{\rho}_1$ in the Michaelis–Menten mechanism the rapid time scale, τ_1, is determined by $k_\alpha^+ \bar{\rho}_s$ and k_β^+. Thus both H_{11} and H_{12} are of the order τ_1^{-1} and all the terms in Eq. (9.2.37) contribute to the covariance. Substituting the expressions in Eqs. (9.2.47) and (9.2.45) into Eq. (9.2.37) gives the explicit result

$$\begin{aligned} &\langle \tilde{f}_s(t)\tilde{f}_s(t')\rangle \\ &= \frac{V^{-2}[(k_\beta^- \bar{\rho}^* + k_\beta^+)^2(\bar{V}_\alpha^+ + \bar{V}_\alpha^-) + (k_\alpha^+ \bar{\rho}_s + k_\alpha^-)^2(\bar{V}_\beta^+ + \bar{V}_\beta^-)]}{(k_\alpha^+ \bar{\rho}_s + k_\beta^- \bar{\rho}^* + k_\alpha^- + k_\beta^+)^2}\delta(t - t'). \end{aligned} \tag{9.2.48}$$

Equations (9.2.46) and (9.2.48) make it clear that the quasi-steady-state contraction explicitly involves both elementary reactions. This should be contrasted with contractions caused by rapid equilibrium, in which the rapid elementary processes no longer appear.

9.3. Contraction of Stationary, Gaussian, Markov Processes

In the thermodynamic limit the stochastic descriptions of both equilibrium and asymptotically stable steady state are stationary, Gaussian, and Markovian. This result is predicated on the underlying statistical assumption in Chapter 4 that the transition probabilities depend only on the instantaneous values of the extensive variables. If this is true on the time scale of interest, then the extensive variables form a statistically complete set on that time scale. If a set of extensive variables is not complete, then we can imagine including additional extensive variables to complete the description. Indeed, this idea lies at the basis of the hierarchical picture of stochastic processes that we have been using. Thus at some level of description, the use of stationary, Gaussian, Markov ensembles for both equilibrium and steady states should be justified.

In this section we raise the inverse problem, namely what happens when variables are ignored in a stationary, Gaussian, Markov process? In other words, what type of stochastic picture results from the contraction of a stationary, Gaussian, Markov process? This is an important question since it is not always possible to keep track of a complete set of extensive variables. Indeed, for certain systems one does not even know what the relevant extensive variables are, let alone how to keep track of them. If these unknown variables do not change on a time scale which is rapid with respect to the variables that are being followed, the resulting stochastic description will not be Markovian.

To investigate this phenomenon, let us assume that the variable $\mathbf{a}$ is a stationary, Gaussian, Markov process. As such, it might represent a vector of deviations of extensive variables from their equilibrium or steady-state average values. Thus we have

$$d\mathbf{a}/dt = H\mathbf{a} + \tilde{\mathbf{f}}, \tag{9.3.1}$$

with $\langle \tilde{\mathbf{f}}(t) \rangle = \mathbf{0}$ and

$$\langle \tilde{\mathbf{f}}(t)\tilde{\mathbf{f}}^T(t') \rangle = \gamma\delta(t - t'). \tag{9.3.2}$$

The single-time covariance, σ, satisfies the fluctuation–dissipation theorem

$$H\sigma + \sigma H^T = -\gamma. \tag{9.3.3}$$

We will assume that our contracted description consists of a subset of some invertible, linear combination of these variables $\mathbf{a}' = G\mathbf{a}$. As we saw in Section 8.2, such a linear combination is still a stationary, Gaussian, Markov process and satisfies Eq. (9.3.3) with the transformed matrices $H' = GHG^{-1}$ and $\gamma' = G\gamma G^T$. For simplicity of notation, we will assume that the $\mathbf{a}$ variables have already undergone this linear transformation, and that our contracted description consists of the final k variables, $a_{l+1}, \ldots, a_{l+k}$, while the first l of the variables are to be ignored. Writing the contracted variables as the column vector $\mathbf{a}_2$ and the ignored variables as $\mathbf{a}_1$, Eq. (9.3.1) takes the form

$$\begin{aligned} d\mathbf{a}_1/dt &= H_{11}\mathbf{a}_1 + H_{12}\mathbf{a}_2 + \tilde{\mathbf{f}}_1 \\ d\mathbf{a}_2/dt &= H_{21}\mathbf{a}_1 + H_{22}\mathbf{a}_2 + \tilde{\mathbf{f}}_2. \end{aligned} \tag{9.3.4}$$

We can eliminate $\mathbf{a}_1$ from Eq. (9.3.4) by solving the first of these equations for $\mathbf{a}_1$, treating $\mathbf{a}_2$ as a known function of time. This gives

$$\mathbf{a}_1(t) = \exp(H_{11}t)\mathbf{a}_1(0) + \int_0^t \exp[H_{11}(t-s)][H_{12}\mathbf{a}_2(s) + \tilde{\mathbf{f}}_1(s)]\,ds. \tag{9.3.5}$$

Substituting this expression into the second of Eqs. (9.3.4) then gives

$$\begin{aligned} d\mathbf{a}_2/dt &= H_{21}\exp(H_{11}t)\mathbf{a}_1(0) \\ &\quad + \int_0^t H_{21}\exp[H_{11}(t-s)][H_{12}\mathbf{a}_2(s) + \tilde{\mathbf{f}}_1(s)]\,ds + H_{22}\mathbf{a}_2(t) + \tilde{\mathbf{f}}_2. \end{aligned} \tag{9.3.6}$$

By eliminating $\mathbf{a}_1$, we have introduced memory effects in the form of the integral operator in Eq. (9.3.6). Notice that the value of $\mathbf{a}_1$ at $t = 0$, which is a Gaussian random variable, also appears in Eq. (9.3.6). Both the integral operator and the initial value of $\mathbf{a}_1$ are essential features of the contracted stochastic process. The basic structure of the contracted stochastic process becomes clearer if we write Eq. (9.3.6) in the form

$$d\mathbf{a}_2/dt = \int_0^t H(t-s)\mathbf{a}_2(s)\,ds + \tilde{\mathbf{F}}(t), \tag{9.3.7}$$

where

$$H(t-s) \equiv 2H_{22}\delta(t-s) + H_{21}\exp[H_{11}(t-s)]H_{12} \tag{9.3.8}$$

and

$$\tilde{\mathbf{F}}(t) \equiv \int_0^t H_{21}\exp[H_{11}(t-s)]\tilde{\mathbf{f}}_1(s)\,ds + H_{21}\exp(H_{11}t)\mathbf{a}_1(0) + \tilde{\mathbf{f}}_2(t). \tag{9.3.9}$$

The contracted description of $\mathbf{a}_2$ in Eqs. (9.3.7)–(9.3.9) is stationary since the underlying ensemble for $\mathbf{a}$ is stationary. Thus the single-time average of $\mathbf{a}_2$ is time independent and its two-time correlation function is given by the usual exponential formula in Eq. (1.8.15). The process is also Gaussian since the random term in Eq. (9.3.7) is an integral over Gaussian variables. Indeed, using the Laplace transform,

$$\hat{\mathbf{a}}_2(p) \equiv \int_0^\infty \exp(-pt)\mathbf{a}_2(t)\,dt, \tag{9.3.10}$$

Eq. (9.3.7) can be rewritten as

$$p\hat{\mathbf{a}}_2(p) - \mathbf{a}_2(0) = \hat{H}(p)\hat{\mathbf{a}}_2(p) + \hat{\mathbf{F}}(p), \tag{9.3.11}$$

where the time derivative in Eq. (9.3.7) was integrated by parts and the convolution theorem was used to rewrite the integral on the right-hand side. Solving Eq. (9.3.11) gives

$$\hat{\mathbf{a}}_2(p) = [p - \hat{H}(p)]^{-1}[\mathbf{a}_2(0) + \hat{\mathbf{F}}(p)]. \tag{9.3.12}$$

Thus the Laplace transform of $\mathbf{a}_2$ is a linear combination of the Gaussian processes $\mathbf{a}_2(0)$ and $\hat{\mathbf{F}}(p)$, and so $\mathbf{a}_2$ is also a Gaussian process. Since the contracted process $\mathbf{a}_2$ is stationary and Gaussian, it cannot also be a Markov process unless it solves a linear equation similar to Eq. (9.3.1). Reexamining Eqs. (9.3.7)–(9.3.9) we see that this is the case if and only if H_{21} vanishes. Thus as long as the variables $\mathbf{a}_1$ and $\mathbf{a}_2$ are coupled dynamically, the contracted description will be stationary and Gaussian, but not Markovian.

The properties of the random component of the time derivative, $\tilde{\mathbf{F}}(t)$, in Eq. (9.3.9) are not difficult to obtain. Since $\tilde{\mathbf{f}}_j(t)$ and $\mathbf{a}_1(0)$ are Gaussian, it follows that $\tilde{\mathbf{F}}$ is a Gaussian random variable. Furthermore, since these random variables have zero mean, it follows from Eq. (9.3.9) that

$$\langle \tilde{\mathbf{F}}(t) \rangle = \mathbf{0}. \tag{9.3.13}$$

The covariance of $\tilde{\mathbf{F}}$ can be obtained from the underlying equations (9.3.1)–(9.3.3) which govern the uncontracted process. Using Eq. (9.3.9) we find that

$$\begin{aligned}\langle \tilde{\mathbf{F}}(t)\tilde{\mathbf{F}}^T(t') \rangle &= \int_0^t ds \int_0^{t'} ds'\, H_{21} \exp[H_{11}(t-s)] \langle \tilde{\mathbf{f}}_1(s)\tilde{\mathbf{f}}_1^T(s') \rangle \\ &\quad \times \exp[H_{11}^T(t'-s')]H_{21}^T + \int_0^t ds\, H_{21} \exp[H_{11}(t-s)] \\ &\quad \times \langle \tilde{\mathbf{f}}_1(s)\tilde{\mathbf{f}}_2^T(t') \rangle + \int_0^{t'} ds' \langle \tilde{\mathbf{f}}_2(t)\tilde{\mathbf{f}}_1^T(s') \rangle \exp[H_{11}(t'-s')]H_{21}^T \\ &\quad + \langle \tilde{\mathbf{f}}_2(t)\tilde{\mathbf{f}}_2^T(t') \rangle + H_{21}\exp[H_{11}t] \\ &\quad \times \langle \mathbf{a}_1(0)\mathbf{a}_1^T(0) \rangle \exp[H_{11}^T t']H_{21}^T.\end{aligned} \tag{9.3.14}$$

Notice that the cross terms between $\mathbf{a}_1(0)$ and $\tilde{\mathbf{f}}_j(t)$ do not appear in Eq. (9.3.14) because of the fact that $\langle \mathbf{a}(0)\mathbf{f}^T(t) \rangle$ vanishes for $t > 0$. If we define the four matrices $\gamma_{11}, \gamma_{12}, \gamma_{21}, \gamma_{22}$ by

$$\langle \tilde{\mathbf{f}}_i(t)\tilde{\mathbf{f}}_j(t') \rangle \equiv \gamma_{ij}\delta(t-t'), \tag{9.3.15}$$

then Eq. (9.3.14) can be integrated directly. For $t' > t$, the first term becomes

$$\begin{aligned}A &= H_{21} \int_0^t ds \exp[H_{11}(t-s)]\gamma_{11} \exp[H_{11}^T(t'-s)]H_{21}^T \\ &= H_{21} \left\{ \int_0^t dx \exp[H_{11}x]\gamma_{11} \exp[H_{11}^T x] \right\} \exp[H_{11}^T(t'-t)]H_{21}^T \\ &= H_{21}\{\sigma^* - \exp[H_{11}t]\sigma^* \exp[H_{11}^T t]\} \exp[H_{11}^T(t'-t)]H_{21}^T \\ &= H_{21}\sigma^* \exp[H_{11}^T(t'-t)]H_{21}^T - H_{21}\exp[H_{11}t]\sigma^* \exp[H_{11}^T t']H_{21}^T,\end{aligned} \tag{9.3.16}$$

where to get from the first to the second line we made the change of variables $x = t - s$, and to get from the second to the third line we have used the identities in Eqs. (1.8.10), (1.8.12), and (1.8.14) and have defined σ^* by

$$H_{11}\sigma^* + \sigma^* H_{11}^T = -\gamma_{11}. \tag{9.3.17}$$

The other integrals in Eq. (9.3.14) are straightforward. The second one vanishes when $t' > t$ and the third one is

$$\begin{aligned} B &= \int_0^{t'} ds' \gamma_{21}\delta(t - s')\exp[H_{11}(t' - s')]H_{21}^T \\ &= \gamma_{21}\exp[H_{11}(t' - t)]H_{21}^T \end{aligned} \tag{9.3.18}$$

for $t' > t$. Combining all these expressions gives for $t' > t$

$$\begin{aligned} \langle \tilde{\mathbf{F}}(t)\tilde{\mathbf{F}}^T(t')\rangle = \{&H_{21}\sigma^* \exp[H_{11}^T(t' - t)]H_{21}^T \\ &- H_{21}\exp[H_{11}t]\sigma^*\exp[H_{11}^T t']H_{21}^T \\ &+ \gamma_{21}\exp[H_{11}^T(t' - t)]H_{21}^T \\ &+ H_{21}\exp[H_{11}t]\sigma_{11}\exp[H_{11}^T t']H_{21}^T + \gamma_{22}\}\delta(t' - t), \end{aligned} \tag{9.3.19}$$

where $\sigma_{11} \equiv \langle \mathbf{a}_1(0)\mathbf{a}_1^T(0)\rangle$ is the single-time covariance of the ignored variables. The comparable expression for $t' < t$ can be obtained by taking the transpose of Eq. (9.3.19) and relabeling $t' \leftrightarrow t$. For $t = t'$ the same methods show that the third term in Eq. (9.3.19) is replaced by $(\gamma_{21}H_{21}^T + H_{21}\gamma_{12})/2$.

An interesting special case of the covariance formula occurs when the contracted and ignored variables are uncorrelated, i.e., when

$$\sigma_{12} \equiv \langle \mathbf{a}_1(0)\mathbf{a}_2^T(0)\rangle \quad \text{and} \quad \sigma_{21} \equiv \langle \mathbf{a}_2(0)\mathbf{a}_1^T(0)\rangle \tag{9.3.20}$$

vanish. To see what Eq. (9.3.19) becomes in this case, we write the fluctuation–dissipation theorem in Eq. (9.3.3) in its component form as

$$\begin{aligned} H_{11}\sigma_{11} + H_{12}\sigma_{21} + \sigma_{11}H_{11}^T + \sigma_{12}H_{12}^T &= -\gamma_{11} \\ H_{11}\sigma_{12} + H_{12}\sigma_{22} + \sigma_{11}H_{21}^T + \sigma_{12}H_{22}^T &= -\gamma_{12} \\ H_{21}\sigma_{11} + H_{22}\sigma_{21} + \sigma_{21}H_{11}^T + \sigma_{22}H_{12}^T &= -\gamma_{21} \\ H_{21}\sigma_{12} + H_{22}\sigma_{22} + \sigma_{21}H_{21}^T + \sigma_{22}H_{22}^T &= -\gamma_{22}. \end{aligned} \tag{9.3.21}$$

When σ_{12} and σ_{21} vanish, Eq. (9.3.21) reduces to

$$\begin{aligned} H_{11}\sigma_{11} + \sigma_{11}H_{11}^T &= -\gamma_{11} \\ H_{12}\sigma_{22} + \sigma_{11}H_{21}^T &= -\gamma_{12} \\ H_{21}\sigma_{11} + \sigma_{22}H_{12}^T &= -\gamma_{21} \\ H_{22}\sigma_{22} + \sigma_{22}H_{22}^T &= -\gamma_{22}. \end{aligned} \tag{9.3.22}$$

We recognize that the first of these equations is identical to Eq. (9.3.17), which is the equation satisfied by σ^*. Thus under these conditions $\sigma^* = \sigma_{11}$. Consequently, the second and fourth terms in Eq. (9.3.19) cancel one another. If we eliminate γ_{21} from the third term in Eq. (9.3.19) using Eq. (9.3.22), we find that

$$\langle \tilde{\mathbf{F}}(t)\tilde{\mathbf{F}}^T(t')\rangle = \gamma_{22}\delta(t'-t) - \sigma_{22}H_{12}^T \exp[H_{11}^T(t'-t)]H_{21}^T. \quad (9.3.23)$$

Recalling the form of the integral kernel in Eq. (9.3.8), we see that Eq. (9.3.23) can also be written in the form

$$\langle \tilde{\mathbf{F}}(t)\tilde{\mathbf{F}}^T(t')\rangle = -(H_{22}\sigma_{22} - \sigma_{22}H_{22}^T)\delta(t'-t) - \sigma_{22}H^T(t'-t), \quad (9.3.24)$$

where we have used the fact in Eq. (9.3.22) that

$$H_{22}\sigma_{22} + \sigma_{22}H_{22}^T = -\gamma_{22}. \quad (9.3.25)$$

Equations (9.3.24) and (9.3.25) show that the covariance of the random force is determined by a knowledge of γ_{22}, H_{22}, and the integral kernel, $H(t'-t)$. Indeed, Eq. (9.3.24) can be written more symmetrically as

$$\operatorname{sym}\langle \tilde{\mathbf{F}}(t)\tilde{\mathbf{F}}^T(t')\rangle = -\operatorname{sym} H(t'-t)\sigma_{22}, \quad (9.3.26)$$

where sym represents the symmetric part of the matrix. If the contracted and ignored variables are correlated at equal times, these simple formulas are no longer valid, and the complete Eq. (9.3.19) must be used.

If the ignored variables relax rapidly, then we have a situation which is equivalent to the quasi-steady state treated in Section 9.2. In this case H_{11} scales like $\bar{H}_{11}/\tau_1$, where τ_1 is a rapid time scale characteristic of the ignored variables. Using the identity in Eq. (9.2.32) in Eqs. (9.3.7) and (9.3.8) then gives

$$d\mathbf{a}_2/dt = (H_{22} - H_{21}H_{11}^{-1}H_{12})\mathbf{a}_2 + \tilde{\mathbf{F}}. \quad (9.3.27)$$

Using Eq. (9.3.14) one can obtain the expression for the covariance of $\tilde{\mathbf{F}}$ in the limit that $\tau_1 \to 0$. Being careful to use the discontinuities in the second and third terms in Eqs. (9.3.14) at $t = t'$, one verifies that

$$\lim_{\tau\to 0} \langle \tilde{\mathbf{F}}(t)\tilde{\mathbf{F}}^T(t')\rangle = (\gamma_{22} - \gamma_{21}H_{11}^{T-1}H_{21}^T - H_{21}H_{11}^{-1}\gamma_{12}$$
$$+ H_{21}H_{11}^{-1}\gamma_{11}H_{11}^{T-1}H_{21}^T)\delta(t'-t). \quad (9.3.28)$$

Equations (9.3.27) and (9.3.28) are identical in form to the expressions for the quasi-steady state obtained in Eqs. (9.2.35) and (9.2.37). Again we notice that unless $H_{21}H_{11}^{-1}$ is of order τ_1, the contracted stochastic process does not have the canonical form, even in the limit that $\tau_1 \to 0$. Similarly, unless σ_{12} or H_{12} can be neglected, the equation satisfied by σ_{22} in Eq. (9.3.21) does not have the form of the fluctuation–dissipation theorem in Eq. (9.3.22). Nonetheless, it will satisfy the fluctuation–dissipation theorem

$$(H_{22}^{-1} - H_{21}H_{11}^{-1}H_{12})\sigma_{22} + \sigma_{22}(H_{22} - H_{21}H_{11}^{-1}H_{12})^T = -\gamma \quad (9.3.29)$$

with γ the matrix on the right-hand side of Eq. (9.3.28). Thus the elimination of fast variables in a stationary, Gaussian, Markov process leads to a contracted process of the slow variables that is also a stationary, Gaussian, Markov process.

The contraction procedure must be handled somewhat differently if the slow variables are conserved by the complete equations. In this case there will be no characteristic relaxation, τ_2, associated with the slow variables. Thus the mechanics of the contraction are somewhat different, and in honor of its originator we will call it the Fox contraction. The details are easier to carry out if we use the simplest possible general representation of a stationary Gaussian, Markov process. Recall that for independent variables the single-time covariance matrix, σ, is positive definite. This means that we can linearly transform variables using Eqs. (8.2.34)–(8.2.40) to obtain a new set such that $\sigma''_{ij} = k_B \sigma_i \delta_{ij}$, where σ_i is positive. Applying the second linear transformation, $G_{ij} = \sigma_i^{-1/2} \delta_{ij}$, the single-time covariance matrix is

$$\sigma'_{ij} = k_B \delta_{ij}. \tag{9.3.30}$$

The resulting transformed stochastic process $\mathbf{a}'$ is still stationary, Gaussian, and Markovian and satisfies an equation

$$d\mathbf{a}'/dt = (A' + S)\mathbf{a}' + \tilde{\mathbf{f}}', \tag{9.3.31}$$

where A' is an antisymmetric matrix and S is symmetric. Because the process is stationary, the matrix S is negative semi-definite. Thus S can be diagonalized by an orthogonal transformation G', i.e.,

$$-\Lambda_{ij} \equiv (G'SG'^{-1})_{ij} = -\lambda_i \delta_{ij}, \tag{9.3.32}$$

where $\lambda_i \geq 0$. According to Eqs. (8.2.36) and (8.2.37) the transformed variables $\mathbf{a} \equiv G'\mathbf{a}'$ satisfy the equation

$$d\mathbf{a}/dt = A\mathbf{a} - \Lambda\mathbf{a} + \tilde{\mathbf{f}}, \tag{9.3.33}$$

where $A \equiv G'A'G'^{-1}$ is still antisymmetric since $G'^T = G'^{-1}$ for an orthogonal transformation. For the same reason, the transformation equation for σ derived in (8.2.40) shows that σ still satisfies Eq. (9.3.30). To obtain the covariance of $\tilde{\mathbf{f}}$ we can use the fluctuation–dissipation theorem which, since σ is a multiple of the identity matrix, gives

$$-\gamma = (A - \Lambda)\sigma + \sigma(A^T - \Lambda^T) = k_B[(A + A^T) - (\Lambda + \Lambda^T)] = -2k_B\Lambda. \tag{9.3.34}$$

Hence

$$\gamma_{ij} = 2k_B \lambda_i \delta_{ij} \tag{9.3.35a}$$

and

$$\sigma_{ij} = k_B \delta_{ij}. \tag{9.3.35b}$$

Equations (9.3.32), (9.3.33), and (9.3.35) provide a diagonal representation for any stationary, Gaussian, Markov process.

If a subset of the variables $\mathbf{a}$ are conserved, then the relaxation matrix Λ has zeros along the diagonal for these variables. Thus splitting Eq. (9.3.33) into the conserved variables, $\mathbf{a}_2$, and the nonconserved variables, $\mathbf{a}_1$, in the usual way, we have

$$\begin{aligned} d\mathbf{a}_1/dt &= A_{11}\mathbf{a}_1 + A_{12}\mathbf{a}_2 - \Lambda_{11}\mathbf{a}_1 + \tilde{\mathbf{f}}_1 \\ d\mathbf{a}_2/dt &= A_{21}\mathbf{a}_1 + A_{22}\mathbf{a}_2. \end{aligned} \tag{9.3.36}$$

Notice that the equation for the conserved variables has no random term since according to Eq. (9.3.35a) γ_{22} and γ_{21} vanish. Indeed, in this representation the dissipation function $\hat{\Phi}$ can be seen from Eqs. (8.3.16), (8.3.18) and (8.3.22) to be

$$\hat{\Phi} = \chi^T L \chi = -\mathbf{a}_1^T \Lambda \mathbf{a}_1 = -\mathbf{a}_1^T \Lambda_{11} \mathbf{a}_1 \geq 0. \tag{9.3.37}$$

Thus the antisymmetric matrices do not lead to dissipation and, so, the equation for $\mathbf{a}_2$ involves no transport coefficients and contains no characteristic relaxation time.

Let us adopt the point of view that the characteristic times $\tau_i \equiv \lambda_i^{-1}$ for the nonconserved variables are all rapid and then solve the first of Eqs. (9.3.36) so as to eliminate the rapid variables from the second equation. Multiplying by $\exp(\Lambda_{11} t)$ and integrating, we see that the first equation is equivalent to the integral equation

$$\begin{aligned} \mathbf{a}_1(t) &= \exp(-\Lambda_{11} t)\mathbf{a}_1(0) \\ &\quad + \int_0^t \exp[-\Lambda_{11}(t-s)]\{A_{11}\mathbf{a}_1(s) + A_{12}\mathbf{a}_2(s) + \tilde{\mathbf{f}}_1(s)\}\, ds. \end{aligned} \tag{9.3.38}$$

To obtain a systematic expansion of the solution to Eq. (9.3.38) in terms of the relaxation time matrix, we replace $\mathbf{a}_1(s)$ in the integrand using the entire expression for $\mathbf{a}_1$ on the right-hand side of the equation. This gives

$$\begin{aligned} \mathbf{a}_1(t) &= \exp(-\Lambda_{11} t)\mathbf{a}_1(0) + \int_0^t \exp[-\Lambda_{11}(t-s)]\Big\{A_{11}\exp(-\Lambda_{11} s)\mathbf{a}_1(0) \\ &\quad + A_{11}\int_0^s \exp[-\Lambda_{11}(s-s')][A_{11}\mathbf{a}_1(s') + A_{12}\mathbf{a}_2(s') \\ &\quad + \tilde{\mathbf{f}}_1(s')]\, ds' + A_{12}\mathbf{a}_2(s) + \tilde{\mathbf{f}}_1(s)\Big\}\, ds. \end{aligned} \tag{9.3.39}$$

Repeating this substitution using Eq. (9.3.39) and its subsequent iterates leads to the series expansion

$$\mathbf{a}_1(t) = \exp(-\Lambda_{11}t)\mathbf{a}_1(0) + \int_0^t \exp[-\Lambda_{11}(t-s)]\{A_{11}\exp(-\Lambda_{11}s)\mathbf{a}_1(0)$$
$$+ A_{12}\mathbf{a}_2(s) + \tilde{\mathbf{f}}_1(s)\}\, ds + \int_0^t \exp[-\Lambda_{11}(t-s)]A_{11}$$
$$\times \int_0^s \exp[-\Lambda_{11}(s-s')]\{A_{11}\exp(-\Lambda_{11}s')\mathbf{a}_1(0)$$
$$+ A_{12}\mathbf{a}_2(s') + \tilde{\mathbf{f}}_1(s')\}\, ds\, ds' + \cdots. \tag{9.3.40}$$

This series is a systematic expansion in the powers of the characteristic times, τ_i, of the fast variables. Indeed, recalling the identity in Eq. (9.2.32), we can write

$$\lim_{\tau_i \to 0} \exp[-\Lambda_{11}(t-s)] = 2\Lambda_{11}^{-1}\delta(t-s). \tag{9.3.41}$$

Since Λ_{11}^{-1} is a diagonal matrix with the τ_i along the diagonal, it follows that the nested integrals in Eq. (9.3.40) involve increasing powers of the τ_i. In the limit that $\tau_i \to 0$, the terms in Eq. (9.3.40) that involve $\mathbf{a}_1(0)$ all vanish except at $t = 0$. Thus, to lowest order in the τ_i, Eq. (9.3.41) shows that for $t > 0$

$$\mathbf{a}_1(t) = \Lambda_{11}^{-1}[A_{12}\mathbf{a}_2(s) + \tilde{\mathbf{f}}_1(s)]. \tag{9.3.42}$$

Equation (9.3.42) is the lowest-order term in the Fox contraction procedure. The second- and higher-order terms can be obtained systematically using Eq. (9.3.40).

The contracted description which results from the Fox procedure is obtained by substituting Eq. (9.3.42) into the equation for $d\mathbf{a}_2/dt$ in Eq. (9.3.36). This gives

$$d\mathbf{a}_2/dt = (A_{21}\Lambda_{11}^{-1}A_{12} + A_{22})\mathbf{a}_2 + A_{21}\Lambda_{11}^{-1}\tilde{\mathbf{f}}_1. \tag{9.3.43}$$

Notice that this equation for the conserved variables, which in the complete description involved no dissipation, has inherited both a relaxation matrix and a random term from the fast variables. Indeed, since the antisymmetry of A implies that $A_{12}^T = -A_{21}$, it is easy to see that $A_{21}\Lambda_{11}^{-1}A_{12}$ is a symmetric matrix. Although it is of order τ_i, it cannot be neglected because the conserved variables do not relax by direct coupling among themselves. Indeed, this symmetric matrix is the matrix of transport coefficients that are induced by the rapid relaxation of the fast variables. Notice that Eq. (9.3.43) is the stochastic differential equation of a stationary, Gaussian, Markov process. Thus this feature of the original equations is again preserved. Using Eq. (9.3.35) it is easy to see that the covariance of the random term $\tilde{\mathbf{f}} \equiv A_{21}\Lambda_{11}^{-1}\tilde{\mathbf{f}}_1$ in Eq. (9.3.43) is

$$\langle\tilde{\mathbf{f}}(t)\tilde{\mathbf{f}}^T(t')\rangle = A_{21}\Lambda_{11}^{-1}\langle\tilde{\mathbf{f}}_1(t)\tilde{\mathbf{f}}_1^T(t')\rangle\Lambda_{11}^{-1}A_{21}^T$$
$$= -2k_B A_{21}\Lambda_{11}^{-1}A_{12}\delta(t-t'). \tag{9.3.44}$$

Since Eq. (9.3.35b) shows that the single-time covariance, σ_{11}, is the identity times Boltzmann's constant, we can easily verify using Eq. (9.3.44) that the fluctuation–dissipation theorem holds for the contracted process. In the next sections we use the Fox contraction procedure to derive fluctuating hydrodynamics and explicit expressions for the transport coefficients from the Boltzmann equation.

9.4. Derivation of the Hydrodynamic Level of Description from the Boltzmann Level

It is important to understand how the various levels of description in statistical nonequilibrium thermodynamics are connected to one another. An elegant way of relating the Boltzmann level of description to the hydrodynamic level was discovered by Fox a number of years ago. It is based on the contraction procedure described in the previous section, and when applied in an equilibrium ensemble to the Boltzmann level of description it gives rise to the hydrodynamic level of description at equilibrium. Since it is equivalent to the Chapman–Enskog procedure for deriving transport coefficients from the Boltzmann equation, we will refer to it as the Fox–Chapman–Enskog contraction.

The mechanistic statistical picture of nonequilibrium thermodynamics at the Boltzmann level of description was derived in Sections 3.2 and 3.3. The underlying stochastic differential equations that are satisfied by the phase space number density, $\rho(\mathbf{r}, \mathbf{v}, t)$, are given in Eqs. (3.3.28), (3.3.30), and (3.3.37). In a stationary equilibrium ensemble they reduce to the linear theory of Onsager, as outlined in Section 2.11. The stationary, Gaussian, Markov process in that case is $\Delta\rho(\mathbf{r}, \mathbf{v}, t) = \rho(\mathbf{r}, \mathbf{v}, t) - \rho^e(v)$ where

$$\begin{aligned}\rho^e(v) &= \rho^e(2\pi k_B T^e/m)^{-3/2} \exp(-mv^2/2k_B T^e) \\ &\equiv \rho^e P^e(v)\end{aligned} \tag{9.4.1}$$

is the equilibrium Maxwell distribution with T^e the equilibrium temperature. It is convenient to introduce the thermodynamic force, X, defined in Eq. (2.11.14), in place of $\Delta\rho$. Specifically, we choose

$$h(\mathbf{v}, t) \equiv \Delta\rho(\mathbf{v}, t)/\rho^e(v) = -X(\mathbf{v}, t)/k_B. \tag{9.4.2}$$

In terms of this variable the relaxation equation (2.11.25) for the density can be written

$$\partial h/\partial t = -\mathbf{v}\cdot\nabla_{\mathbf{r}} h - \int [\kappa(\mathbf{v}, \mathbf{v}_1) h_1(\mathbf{v}_1)] \rho^e(v_1)\, d\mathbf{v}_1 + \tilde{\mathbf{f}}, \tag{9.4.3}$$

where the form of the kernel $\kappa(\mathbf{v}, \mathbf{v}_1)$ can be determined from Eq. (2.11.8) to be

$$\kappa(\mathbf{v}, \mathbf{v}_1) = -\hat{\sigma}_T |\mathbf{v} - \mathbf{v}_1| [T_1' + T' - T - T_1], \tag{9.4.4}$$

with the translation operators defined in Eqs. (2.11.6) and (2.11.7). The covariance of the Gaussian random term, $\tilde{f}$, in Eq. (9.4.3) can be obtained from Eqs. (2.11.15) and (2.11.26) and is

$$\langle \tilde{f}(\mathbf{r}, \mathbf{v}, t) \tilde{f}(\mathbf{r}', \mathbf{v}_1, t') \rangle = 2k_B \kappa(\mathbf{v}, \mathbf{v}_1) \delta(\mathbf{r} - \mathbf{r}') \delta(t - t'). \tag{9.4.5}$$

To put Eqs. (9.4.3)–(9.4.5) in a form suitable for applying the Fox contraction procedure, it is useful to introduce the linear operator

$$K[h] \equiv \rho^e \int \{\kappa(\mathbf{v}, \mathbf{v}_1) h(\mathbf{v}_1)\} P^e(v_1) \, d\mathbf{v}_1. \tag{9.4.6}$$

This operator is a weighted integral operator with the normalized weight function, $P^e(v)$. Alternatively, if we introduce the differential $d\mu = P^e(v) \, d\mathbf{v}$, we can think of the weighted integral as a Stieltjes integral. In either case, it is natural to define the inner product

$$(g, h) = \int g^*(\mathbf{v}) h(\mathbf{v}) P^e(v) \, d\mathbf{v}, \tag{9.4.7}$$

where the asterisk implies the complex conjugate. With this inner product we can use Eq. (9.4.6) to write

$$(g, K[h]) = \int \int g^*(\mathbf{v}) P^e(v) \rho^e(v_1) \kappa(\mathbf{v}, \mathbf{v}_1) h(\mathbf{v}_1) \, d\mathbf{v} \, d\mathbf{v}_1. \tag{9.4.8}$$

Referring back to Eqs. (2.11.8) and (2.11.15), we see that

$$\rho^e(v) \rho^e(v_1) \kappa(\mathbf{v}, \mathbf{v}_1) = k_B L^s(\mathbf{v}, \mathbf{v}_1), \tag{9.4.9}$$

where L^s is the symmetric part of the Onsager functional for the Boltzmann equation. Thus we can write

$$(g, K[h]) = k_B \rho^{e-1} \int \int g^*(\mathbf{v}) L^s(\mathbf{v}, \mathbf{v}_1) h(\mathbf{v}_1) \, d\mathbf{v} \, d\mathbf{v}_1. \tag{9.4.10}$$

Separating $g(\mathbf{v})$ into its real and imaginary parts and using the fact derived in Eq. (2.11.20) that L^s is symmetric for real functions gives

$$\int g^*(\mathbf{v}) L^s(\mathbf{v}, \mathbf{v}_1) h(\mathbf{v}_1) \, d\mathbf{v} \, d\mathbf{v}_1 = \left(\int h^*(\mathbf{v}) L^s(\mathbf{v}, \mathbf{v}_1) g(\mathbf{v}_1) \, d\mathbf{v} \, d\mathbf{v}_1 \right)^* . \tag{9.4.11}$$

Combining this with the previous equation, we conclude that

$$(g, K[h]) = (h, K[g])^*, \tag{9.4.12}$$

which means that the operator K is Hermitian and, thus, that its eigenvalues are real. Indeed, its eigenvalues are greater than or equal to zero, since we can use the identity in Eq. (2.11.23) to write

$$(h, K[h]) = \frac{\rho^e}{4} \int\int \hat{\sigma}_T |h(\mathbf{v}) + h(\mathbf{v}_1) - h(\mathbf{v}') - h(\mathbf{v}_1')|^2$$
$$\times |\mathbf{v} - \mathbf{v}_1| P^e(v) P^e(v_1)\, d\mathbf{v}\, d\mathbf{v}_1 \geq 0. \tag{9.4.13}$$

Because K is Hermitian, it will have a complete set of eigenfunctions. Indeed, it is known for interaction potentials of the form

$$u(r) = r^{-n}, \tag{9.4.14}$$

with $n \geq 2$, that the eigenvalues are discrete and have no accumulation points. Thus if ζ_i, $i = 1, 2, \ldots$ are a complete orthonormal set of eigenfunctions of K, i.e.,

$$K[\zeta_i] = \lambda_i \zeta_i \tag{9.4.15}$$

with

$$(\zeta_i, \zeta_j) = \delta_{ij}, \tag{9.4.16}$$

it follows that except for the zero eigenvalues

$$\lambda_i > 0. \tag{9.4.17}$$

From Eqs. (9.4.13)–(9.4.16) we see in fact that

$$\lambda_i = \frac{\rho^e}{4} \int\int \hat{\sigma}_T |\zeta_i(\mathbf{v}) + \zeta_i(\mathbf{v}_1) - \zeta_i(\mathbf{v}') - \zeta_i(\mathbf{v}_1')|^2 |\mathbf{v} - \mathbf{v}_1| P^e(v) P^e(v_1)\, d\mathbf{v}\, d\mathbf{v}_1. \tag{9.4.18}$$

Since the integrand in Eq. (9.4.18) vanishes if and only if

$$\zeta_i(\mathbf{v}) + \zeta_i(\mathbf{v}_i) = \zeta_i(\mathbf{v}') + \zeta_i(\mathbf{v}_1'), \tag{9.4.19}$$

we further see that the zero eigenvectors are linear combinations of the five collisional invariants, namely, mass, the linear momenta, and the energy, as found in Section 2.9. Using the inner product in Eq. (9.4.7) it is easy to check that the orthonormal set based on these five functions is

$$\begin{aligned} \zeta_1 &= 1 \\ \zeta_\alpha &= (m/k_B T^e)^{1/2} v_\alpha \\ \zeta_5 &= (2/3)^{1/2}(mv^2/2k_B T^e - 3/2), \end{aligned} \tag{9.4.20}$$

where $\alpha = x, y, z$. Checking that these functions are orthonormal is straightforward and requires only the integrals

$$\begin{aligned} \int P^e(v) v_\alpha v_\beta\, d\mathbf{v} &= (k_B T^e/m)\delta_{\alpha\beta} \\ \int P^e(v)(v^2/2) v_\alpha v_\beta\, d\mathbf{v} &= \frac{5}{2}(k_B T^e/m)^2 \delta_{\alpha\beta}. \end{aligned} \tag{9.4.21}$$

As is well known, any bounded, linear operator can be expressed in terms of its eigenfunctions and eigenvalues. Indeed, the kernel of the integral operator K has the representation

$$\rho^e \kappa(\mathbf{v}, \mathbf{v}_1) = \sum_j \lambda_j \zeta_j(\mathbf{v}) \zeta_j(\mathbf{v}_1). \tag{9.4.22}$$

This follows using the orthonormality relationship in Eq. (9.4.16), and it is easy to verify that Eq. (9.4.22) leads to the correct eigenfunctions and eigenvalues for K. Because the eigenfunctions form a complete set, we can represent the solution, h, to the stochastic differential equation as the series

$$h(\mathbf{r}, \mathbf{v}, t) = \sum_i a_i(\mathbf{r}, t) \zeta_i(\mathbf{v}). \tag{9.4.23}$$

Similarly, we can expand the random term, $\tilde{f}$, in the form

$$\tilde{f}(\mathbf{r}, \mathbf{v}, t) = \sum_i \tilde{F}_i(\mathbf{r}, t) \zeta_i(\mathbf{v}). \tag{9.4.24}$$

The covariance of the $\tilde{F}_i$ can be obtained by comparing the covariance of $\tilde{f}$ in Eq. (9.4.5) using the representation for $\kappa(\mathbf{v}, \mathbf{v}_1)$ in Eq. (9.4.22) with that of $\tilde{f}$ on the left-hand side of Eq. (9.4.24). This gives

$$\langle \tilde{f}(\mathbf{r}, \mathbf{v}, t) \tilde{f}(\mathbf{r}', \mathbf{v}_1, t') \rangle = (2k_B/\rho^e) \sum_j \lambda_j \zeta_j(\mathbf{v}) \zeta_j(\mathbf{v}_1) \delta(\mathbf{r} - \mathbf{r}') \delta(t - t') \tag{9.4.25a}$$

and

$$\langle \tilde{f}(\mathbf{r}, \mathbf{v}, t) \tilde{f}(\mathbf{r}', \mathbf{v}_1, t') \rangle = \sum_{i,j} \langle \tilde{F}_i(\mathbf{r}, t) \tilde{F}_j(\mathbf{r}', t') \rangle \zeta_i(\mathbf{v}) \zeta_j(\mathbf{v}_1), \tag{9.4.25b}$$

from which we conclude that

$$\langle \tilde{F}_i(\mathbf{r}, t) \tilde{F}_j(\mathbf{r}', t) \rangle = (2k_B/\rho^e) \lambda_i \delta_{ij} \delta(\mathbf{r} - \mathbf{r}') \delta(t - t'). \tag{9.4.26}$$

We can put the fluctuating Boltzmann equation into the diagonal form described in Section 9.3 using these relationships. Thus substituting the eigenfunction expansions of h and $\tilde{f}$ into the fluctuating Boltzmann equation (9.4.3), we obtain

$$\sum_i (\partial a_i/\partial t) \zeta_i(\mathbf{v}) = -\sum_i \mathbf{v} \cdot (\nabla_{\mathbf{r}} a_i) \zeta_i(\mathbf{v}) - \sum_i \lambda_i a_i \zeta_i(\mathbf{v}) + \sum_i \tilde{F}_i \zeta_i(\mathbf{v}). \tag{9.4.27}$$

Forming the inner product of both sides with ζ_j and using the orthonormality condition, we conclude that

$$\partial a_j/\partial t = \sum_i -(\zeta_j, \mathbf{v}\zeta_i) \cdot \nabla_{\mathbf{r}} a_i - \lambda_j a_j + \tilde{F}_j. \tag{9.4.28}$$

Recalling that $\zeta_\alpha = (m/k_B T^e)^{1/2} v_\alpha$, we can write the constants in the streaming terms in Eq. (9.4.28) as

$$A_{ji}^\alpha \equiv -(\zeta_j, \zeta_\alpha \zeta_i)(k_B T^e/m)^{1/2}. \tag{9.4.29}$$

Consequently, if we define

$$A_{ji} \equiv A_{ji}^{\alpha} \nabla_{\alpha}, \tag{9.4.30}$$

where $\nabla_\alpha \equiv \partial/\partial x, \partial/\partial y, \partial/\partial z$ and the repeated index α is to be summed over, Eq. (9.4.28) takes the form

$$\partial a_j/\partial t = \sum_i A_{ji} a_i - \lambda_j a_j + \tilde{F}_j, \tag{9.4.31}$$

with

$$\langle \tilde{F}_i(\mathbf{r}, t) \tilde{F}_j(\mathbf{r}', t') \rangle = (2k_B/\rho^e) \lambda_i \delta_{ij} \delta(\mathbf{r} - \mathbf{r}') \delta(t - t'). \tag{9.4.32}$$

Because the eigenfunctions, ζ_j, are real, it is obvious from Eq. (9.4.29) that the matrix A_{ji}^{α} is symmetric. The differential operators, ∇_α, however, are antisymmetric, as we discovered in Section 2.4 [cf. Eq. (2.4.44)]. Thus it follows from Eq. (9.4.30) that A_{ij} is an antisymmetric matrix, which puts the fluctuating Boltzmann equation into the diagonal form described in Eq. (9.3.33) of the previous section.

Since the first five eigenvalues λ_1, λ_β, λ_5 vanish, Eq. (9.4.31) can be written out more explicitly as

$$\begin{aligned} \partial a_1/\partial t &= A_{1i}^{\alpha} \nabla_\alpha a_i \\ \partial a_\beta/\partial t &= A_{\beta i}^{\alpha} \nabla_\alpha a_i \\ \partial a_5/\partial t &= A_{5i}^{\alpha} \nabla_\alpha a_i \\ \partial a_j/\partial t &= A_{ji}^{\alpha} \nabla_\alpha a_i - \lambda_j a_j + \tilde{F}_j \end{aligned} \tag{9.4.33}$$

with $j > 5$ and the summation convention now being used on all repeated indices. Equations (9.4.32) and (9.4.33) are in the correct form to apply the Fox contraction procedure since the expansion coefficients of the conserved variables a_1, a_β, and a_5 relax only through an antisymmetric coupling to the other coefficients. In fact, it is known that $\tau_j = 1/\lambda_j$ for $j > 5$ is proportional to the mean free time, which in typical gases at room temperature and pressure is of the order of 10^{-10} s. Thus there is a clear separation of time scales with a_j for $j > 5$ being the fast variables.

To place a physical interpretation on the contraction procedure, we need to clarify the meaning of the expansion coefficients. Taking the scalar product of ζ_j with h using the right-hand side of Eq. (9.4.23), we see that

$$a_j(\mathbf{r}, t) = \int h(\mathbf{r}, \mathbf{v}, t) \zeta_j(\mathbf{v}) P^e(v)\, d\mathbf{v}. \tag{9.4.34}$$

Recalling the definition of h in Eq. (9.4.2), we see further that

$$a_j(\mathbf{r}, t) = \int \zeta_j(\mathbf{v}) [\rho(\mathbf{r}, \mathbf{v}, t) - \rho^e(v)]\, d\mathbf{v}/\rho^e. \tag{9.4.35}$$

For $j = 1$, β, 5 the eigenfunctions ζ_j are the five collisional invariants. Substituting the explicit expressions for them given in Eq. (9.4.20) into Eq. (9.4.35) and recalling the definitions of the mass density, $\langle \rho(\mathbf{r}, t) \rangle$, the velocity, $\langle \mathbf{v}(\mathbf{r}, t) \rangle$, and the internal energy density, $\langle e(\mathbf{r}, t) \rangle$, given in section 2.10, we find that

$$
\begin{aligned}
a_1(\mathbf{r}, t) &= (\langle \rho(\mathbf{r}, t) \rangle - \langle \rho \rangle^e)/m\rho^e \\
a_\beta(\mathbf{r}, t) &= (mk_B T^e)^{-1/2} \langle \mathbf{v}(\mathbf{r}, t) \rangle \langle \rho(\mathbf{r}, t) \rangle / \rho^e \\
a_5(\mathbf{r}, t) &= \left(\frac{2}{3}\right)^{1/2} (k_B T^e \rho^e)^{-1} \left[\langle e(\mathbf{r}, t) \rangle - \langle e \rangle^e - \frac{3k_B T^e}{2m} (\langle \rho(\mathbf{r}, t) \rangle - \langle \rho^e \rangle) \right],
\end{aligned}
\tag{9.4.36}
$$

where $\langle \rho \rangle^e = m\rho^e$ is the equilibrium mass density and $\langle e \rangle^e = 3k_B T \langle \rho^e \rangle / 2m$ is the equilibrium internal energy density. Thus a_1–a_5 are combinations of the conserved densities for the Boltzmann equation. If we revert to the notation for these quantities used at the hydrodynamic level of description, then we can rewrite Eqs. (9.4.36) as

$$
\begin{aligned}
\Delta \rho(\mathbf{r}, t) &= m\rho^e a_1(\mathbf{r}, t) \\
\Delta p_\beta(\mathbf{r}, t) &= \rho^e (mk_B T^e)^{1/2} a_\beta(\mathbf{r}, t) \\
\Delta e(\mathbf{r}, t) &= \frac{3k_B T^e \rho^e}{2} \left[a_1 + \left(\frac{2}{3}\right)^{1/2} a_5 \right].
\end{aligned}
\tag{9.4.37}
$$

According to these identities, the slow variables in the Fox contraction are linear combinations of the densities of the extensive variables at the hydrodynamic level of description. Thus the contraction procedure will lead directly to the hydrodynamic level.

Writing out Eq. (9.4.33) in somewhat more detail, the equations satisfied by the hydrodynamic variables a_1, a_β, a_5 are

$$
\begin{aligned}
\partial a_1/\partial t &= A^\alpha_{11} \nabla_\alpha a_1 + A^\alpha_{1\kappa} \nabla_\alpha a_\kappa + A^\alpha_{15} \nabla_\alpha a_5 + \sum_{i>5} A^\alpha_{1i} \nabla_\alpha a_i \\
\partial a_\beta/\partial t &= A^\alpha_{\beta 1} \nabla_\alpha a_1 + A^\alpha_{\beta\kappa} \nabla_\alpha a_\kappa + A^\alpha_{\beta 5} \nabla_\alpha a_5 + \sum_{i>5} A^\alpha_{\beta i} \nabla_\alpha a_i \\
\partial a_5/\partial t &= A^\alpha_{51} \nabla_\alpha a_1 + A^\alpha_{5\kappa} \nabla_\alpha a_\kappa + A^\alpha_{55} \nabla_\alpha a_5 + \sum_{i>5} A^\alpha_{5i} \nabla_\alpha a_i
\end{aligned}
\tag{9.4.38}
$$

The first three terms on the right-hand side of each of these equations are the conservative, or Euler, terms. To see this, we need to evaluate the A^α_{ij}. Using the definition in Eq. (9.4.29), we see that A^α_{ij} is symmetric in i and j. Thus we need only consider A^α_{ij} for $i \leq j$. Recalling that $\zeta_1 = 1$, Eq. (9.4.29) implies that

$$
A^\alpha_{1j} = -(k_B T^e/m)^{1/2} (\zeta_\alpha, \zeta_j)^{1/2} = -(k_B T^e/m)^{1/2} \delta_{\alpha j} \tag{9.4.39}
$$

since the eigenfunctions are orthonormal. To evaluate $A^\alpha_{\beta j}$ we combine Eqs. (9.4.20) and (9.4.29) to obtain

$$
A^\alpha_{\beta j} = -(v_\beta, \zeta_\alpha \zeta_j) = -(m/k_B T^e)^{1/2} \int v_\beta v_\alpha \zeta_j(\mathbf{v}) P^e(v) \, d\mathbf{v}. \tag{9.4.40}
$$

For $j = \kappa$ we have, therefore, that

$$A^{\alpha}_{\beta\kappa} = -(m/k_B T^e) \int v_\beta v_\alpha v_\kappa P^e(v)\, d\mathbf{v} = 0 \tag{9.4.41}$$

since $v_\beta v_\alpha v_\kappa$ is an odd function and $P^e(v)$ is an even function of $\mathbf{v}$. Similarly,

$$A^{\alpha}_{\beta 5} = -(2m/3k_B T^e)^{1/2}\left[\int (mv^2 v_\alpha v_\beta/2k_B T^e)P^e(v)\, d\mathbf{v} - \int (3v_\alpha v_\beta/2)P^e(v)\, d\mathbf{v}\right]. \tag{9.4.42}$$

The values of the integrals in Eq. (9.4.42) are given in Eq. (9.4.21) and lead to the expression

$$A^{\alpha}_{\beta 5} = -(2k_B T^e/3m)^{1/2}\, \delta_{\alpha\beta}. \tag{9.4.43}$$

Finally, we find that

$$A^{\alpha}_{55} = -\int v_\alpha \zeta_5^2(\mathbf{v}) P^e(v)\, d\mathbf{v} = 0, \tag{9.4.44}$$

which vanishes because the integrand is obviously an odd function. Using these expressions, Eqs. (9.4.38) can be reduced to

$$\begin{aligned}
\partial a_1/\partial t &= -(k_B T/m)^{1/2}\nabla_\alpha a_\alpha \\
\partial a_\beta/\partial t &= -(k_B T/m)^{1/2}\nabla_\beta a_1 - (2k_B T/3m)^{1/2}\nabla_\beta a_5 + \sum_{i>5} A^{\alpha}_{\beta i}\nabla_\alpha a_i \\
\partial a_5/\partial t &= -(2k_B T/3m)^{1/2}\nabla_\alpha a_\alpha + \sum_{i>5} A^{\alpha}_{5i}\nabla_\alpha a_i,
\end{aligned} \tag{9.4.45}$$

where we continue to use the summation convention for repeated Greek indices.

The five equations in (9.4.45) describe the conserved quantities. They are coupled to the nonconserved quantities through the remaining equations in (9.4.33). To eliminate these variables from Eqs. (9.4.45), we apply the Fox contraction procedure described in Section 9.3. For this problem $\mathbf{a}_2$ represents the five conserved components, namely, a_1, a_β, a_5, while there are an infinite number of fast components in $\mathbf{a}_1$. According to the Fox procedure, the conserved variables satisfy Eq. (9.3.43). Using Eqs. (9.4.45), we find that for this problem the Fox contraction is

$$\begin{aligned}
\partial a_1/\partial t &= -(k_B T/m)^{1/2}\nabla_\alpha a_\alpha \\
\partial a_\beta/\partial t &= -(k_B T/m)^{1/2}\nabla_\beta a_1 - (2k_B T/3m)^{1/2}\nabla_\beta a_5 \\
&\quad + S^{\alpha\kappa}_{\beta\gamma}\nabla_\alpha\nabla_\kappa a_\gamma + S^{\alpha\kappa}_{\beta 5}\nabla_\alpha\nabla_\kappa a_5 + \tilde{f}_\beta. \\
\partial a_5/\partial t &= -(2k_B T/3m)^{1/2}\nabla_\alpha a_\alpha + S^{\alpha\kappa}_{5\gamma}\nabla_\alpha\nabla_\kappa a_\gamma + S^{\alpha\kappa}_{55}\nabla_\alpha\nabla_\kappa a_5 + \tilde{f}_5,
\end{aligned} \tag{9.4.46}$$

where the tensor coefficients $S^{\alpha\kappa}_{ij}$ and random terms are given by

$$S^{\alpha\kappa}_{ij} \equiv \sum_{k>5} (A^{\alpha}_{ik} A^{\kappa}_{kj}/\lambda_k) \tag{9.4.47}$$

$$\tilde{f}_j \equiv \sum_{k>5} (A^{\alpha}_{jk}/\lambda_k)\nabla_\alpha \tilde{f}_k. \tag{9.4.48}$$

For spherically symmetric intermolecular potentials, which is the only case we consider, it is easy to see from the expression for $S^{\alpha\kappa}_{ij}$ that the terms $S^{\alpha\kappa}_{\beta 5}$ and $S^{\alpha\kappa}_{5\beta}$ are identically zero. We need to rely on the fact that spherical symmetry implies that the eigenfunctions, ζ_j, are either odd or even functions of $\mathbf{v}$. Consider, for example,

$$S^{\alpha\kappa}_{\beta 5} = \sum_k (A^{\alpha}_{\beta k} A^{\kappa}_{k5}/\lambda_k). \tag{9.4.49}$$

According to Eq. (9.4.40),

$$A^{\alpha}_{\beta k} = -(m/k_B T^e)^{1/2} \int v_\beta v_\alpha \zeta_k(\mathbf{v}) P^e(v)\, d\mathbf{v} \tag{9.4.50}$$

which is nonzero only if ζ_k is even. The function ζ_5, however, is an even function. Thus

$$A^{\kappa}_{k5} \equiv -\int v_\kappa \zeta_k(\mathbf{v}) \zeta_5(\mathbf{v}) P^e(v)\, d\mathbf{v} \tag{9.4.51}$$

is nonzero only when ζ_k is an odd function. As a consequence, Eq. (9.4.49) shows that $S^{\alpha\kappa}_{\beta 5}$ vanishes identically.

These contracted equations can be cast in a more familiar form by changing to the hydrodynamic variables. The linear transformations that accomplish this are given in Eqs. (9.4.36) and (9.4.37). After a bit of algebra they allow us to write Eqs. (9.4.46) as

$$\begin{aligned} \partial \Delta\rho/\partial t &= -\nabla \cdot \Delta\mathbf{p} \\ \partial \Delta\mathbf{p}/\partial t &= -\nabla p^e + \nabla \cdot \sigma' + \tilde{\mathbf{f}}_m \\ \partial \Delta e/\partial t &= -h^e \nabla \cdot \mathbf{v} - \nabla \cdot \mathbf{j}_q + f_e, \end{aligned} \tag{9.4.52}$$

where $\mathbf{v} \equiv \Delta\mathbf{p}/m\rho^e$ is the velocity field, $p^e = 2e/3$ is the local equilibrium pressure as given by the virial theorem, and h^e is the equilibrium value of the enthalpy density. The nonequilibrium portion of the stress tensor, σ', and the heat flux vector, $\mathbf{j}_q$, are given by

$$\sigma'_{\beta\alpha} \equiv m\rho^e S^{\alpha\kappa}_{\beta\gamma} \nabla_\kappa v_\gamma \tag{9.4.53}$$

$$j_{q\alpha} \equiv -(3k_B\rho^e/2) S^{\alpha\kappa}_{55} \nabla_\kappa T. \tag{9.4.54}$$

In Eq. (9.4.54) we have used the fact that the internal energy is all kinetic to write $e = (3k_B T/2)\rho^e$ along with the definition of a_5 in Eq. (9.4.36), i.e.,

$$\left(\frac{3}{2}\right)^{1/2} k_B T^e \rho^e a_5 = \left(\Delta e - \frac{3k_B T^e}{2m}\Delta\rho\right) = (3k_B\rho^e/2)\Delta T, \tag{9.4.55}$$

where ΔT is the change in the temperature from its equilibrium value, T^e.

We recognize in Eq. (9.4.52) the conservation equations for the hydrodynamic variables, linearized around equilibrium, with random terms characteristic of the linear Onsager theory. The dissipative parts of these equations are contained in the constitutive relations in Eqs. (9.4.53) and (9.4.54). If we identify the coefficients in these equations as the fourth-order viscosity tensor and the second-order thermal conductivity tensor, we can write them in the notation of Section 6.4 as

$$\begin{aligned}\sigma' &= 2\hat{\eta}\nabla\mathbf{v} \\ \mathbf{j}_q &= -\hat{\kappa}\nabla T\end{aligned} \tag{9.4.56}$$

with

$$2\hat{\eta}_{\alpha\beta,\kappa\gamma} \equiv m\rho^e S_{\beta\gamma}^{\alpha\kappa} \quad \text{and} \quad \hat{\kappa}_{\alpha\gamma} \equiv (3k_B\rho^e/2)S_{55}^{\alpha\gamma}. \tag{9.4.57}$$

Since we have assumed that the intermolecular potential is spherically symmetric, the viscosity and heat flux vector will have the symmetry appropriate for an isotropic fluid. According to Eqs. (6.3.14) and (6.4.18) we anticipate, therefore, that

$$2\hat{\eta}_{\beta\gamma,\kappa\alpha} = \eta^e\left(\delta_{\beta\gamma}\delta_{\kappa\alpha} + \delta_{\beta\kappa}\delta_{\gamma\alpha} - \frac{2}{3}\delta_{\beta\alpha}\delta_{\gamma\kappa}\right) \tag{9.4.58}$$

$$\hat{\kappa}_{\alpha\gamma} = \kappa^e\delta_{\alpha\gamma}, \tag{9.4.59}$$

where η^e and κ^e are the equilibrium values of the viscosity coefficient and the thermal conductivity. Because of the absence of internal states, e.g., rotational or vibrational degrees of freedom, the bulk viscosity coefficient vanishes, as was discussed at the end of Section 6.4.

Since the Fox–Chapman–Enskog contraction is based on the Boltzmann equation, it is not necessary to assume symmetry properties for the transport coefficient. Indeed, tracing back the definitions of $\hat{\eta}$ and $\hat{\kappa}$ to Eqs. (9.4.29) and (9.4.47), we find explicit expressions for these tensors in terms of the eigenvalues and eigenvectors of the linearized collision operator. In the next section we continue this development by explicitly demonstrating isotropic formulas for $\hat{\eta}$ and $\hat{\kappa}$. For so-called Maxwell molecules, which interact through a $1/r^4$ potential, we exhibit simple analytical expressions for both the viscosity coefficient and the thermal conductivity.

We close this section by showing that the contracted stochastic process in Eq. (9.4.52) is identical with that given by the equilibrium Onsager theory. To do so, we recall that in Chapter 4 it was shown that the mechanistic statistical theory reduces to the linear Onsager theory at equilibrium. As applied to the hydrodynamic variables, the basic equations of the mechanistic theory are given in Section 6.5. To compare with Eq. (9.4.52) we need to linearize the average equations (6.5.1)–(6.5.3) around equilibrium and add them to Eqs. (6.5.4)–(6.5.6), which describe the fluctuations. Moreover, to come into correspondence with Eq. (9.4.52) we need to replace the total energy density, ε, in Eqs. (6.5.2) and (6.5.5) with the internal energy density, e. This can be done using the change of variable in Eqs. (6.5.32) and (6.5.33). Near equilibrium these equations show that in the absence of an external potential

$$\bar{\varepsilon} - \varepsilon^e = e - e^e \equiv \Delta e, \quad \delta\varepsilon = \delta e. \tag{9.4.60}$$

With this change of variable, Eqs. (6.5.1)–(6.5.6) take the form

$$\begin{aligned} \partial \Delta\rho/\partial t &= -\nabla \cdot \Delta\mathbf{p} \\ \partial \Delta e/\partial t &= -h^e \nabla \cdot \mathbf{v} - \nabla \cdot K^e \nabla T^{-1} + \tilde{f}_e \\ \partial \Delta\mathbf{p}/\partial t &= -\nabla p^e + 2\nabla \cdot \eta^e \mathring{e} + \nabla \cdot \zeta^e \nabla\mathbf{v} + \tilde{\mathbf{f}}_m. \end{aligned} \tag{9.4.61}$$

It is easy to identify corresponding terms in Eqs. (9.4.52) and (9.4.61). The density equations are obviously the same as are the conservative terms in the energy and momentum equations. If we write $K^e/T^{e2} \equiv \kappa^e$, then the heat flux in the second equation in Eq. (9.4.61) is seen to be the same as that in Eq. (9.4.52). Next, if we set $\zeta^e \equiv 0$ and recall from Eq. (6.5.14) that the traceless, symmetric part of the rate of strain tensor is

$$\mathring{e}_{\beta\alpha} = \frac{1}{2}\left[\delta_{\beta\gamma}\delta_{\kappa\alpha} + \delta_{\beta\kappa}\delta_{\gamma\alpha} - \frac{2}{3}\delta_{\beta\alpha}\delta_{\gamma\kappa}\right](\nabla\mathbf{v})_{\gamma\kappa}, \tag{9.4.62}$$

we see that the expression for σ' in Eqs. (9.4.56) and (9.4.52) can be written

$$\sigma' = 2\eta^e \mathring{e}. \tag{9.4.63}$$

Thus the viscous stress terms are identical. Finally, we can identify the random terms in the two equations since both stochastic processes are stationary, Gaussian, Markov processes. Since the nonrandom terms are identical, the equilibrium fluctuation–dissipation theorem implies that the covariance of the random terms will be the same. Thus the contracted hydrodynamic description is identical to the usual hydrodynamic-level description.

The fact that the hydrodynamic level of description is imbedded in the Boltzmann level of description is an important result that illustrates how deeply the ideas of Onsager and Boltzmann are interwoven in nonequilibrium statistical thermodynamics. Other results of this kind are known. For example, the Langevin description of Brownian motion can be deduced by evaluating the random force induced by a fluctuating fluid through the random stress tensor which acts on the Brownian particle. Results like this are special cases of the general contraction procedures developed in Sections 9.2 and 9.3 and provide a bridge between the various levels of description of nonequilibrium statistical thermodynamics.

9.5. Evaluation of Transport Coefficients

The Fox–Chapman–Enskog contraction, which was described in the previous section, provides a derivation of fluctuating hydrodynamics at equilibrium. It also provides explicit expressions for the transport coefficients for momentum and internal energy. The use of kinetic theory to evaluate transport coefficients constitutes an important chapter in statistical mechanics, which was first developed by Maxwell and Boltzmann. The systematic use of the Boltzmann collision operator for obtaining the viscosity and heat conductivity was begun by Hilbert and carried to completion by Chapman and Enskog. The Chapman–Enskog expressions for the transport coefficients are based on a sequence of approximate solutions to the Boltzmann equation that result in a set of inhomogeneous linear equations. When expressed in terms of the hydrodynamic variables, the lowest-order approximation gives the Euler equations, while the next approximation leads to the Navier–Stokes equations. The explicit expressions for the viscosity and thermal conductivity obtained by the Chapman–Enskog method are identical to those given by the Fox contraction of the fluctuating Boltzmann equation. In this section we fulfill the promise made in Chapter 2 of obtaining explicit expressions for the transport coefficient in terms of the scattering cross section.

According to Eqs. (9.4.57) and (9.4.47) the viscosity tensor, $\hat{\eta}$, and the heat conductivity tensor, $\hat{\kappa}$, can be written in the form

$$\begin{aligned} 2\hat{\eta}_{\alpha\beta,\kappa\gamma} &= m\rho^e \sum_{k>5} A^{\alpha}_{\beta k} A^{\kappa}_{k\gamma}/\lambda_k \\ \hat{\kappa}_{\alpha\gamma} &= (3k_B\rho^e/2) \sum_{k>5} A^{\alpha}_{5k} A^{\gamma}_{k5}/\lambda_k. \end{aligned} \tag{9.5.1}$$

Thus to evaluate these quantities we must have a knowledge of the matrix elements

$$A^{\alpha}_{jk} = -(\zeta_j, v_\alpha \zeta_k) \tag{9.5.2}$$

for $j = \beta$ or 5 and $k > 5$. This requires some knowledge of the eigenfunctions corresponding to the nonconserved quantities. Recall that the eigenfunctions are determined by

$$\begin{aligned} K[\zeta_k] &\equiv -\int \hat{\sigma}_T |\mathbf{v} - \mathbf{v}'| [\zeta_k(\mathbf{v}') + \zeta_k(\mathbf{v}'_1) - \zeta_k(\mathbf{v}) - \zeta_k(\mathbf{v}_1)] \rho^e(v_1)\, d\mathbf{v}_1 \\ &= \lambda_k \zeta_k. \end{aligned} \tag{9.5.3}$$

If we operate on both sides of this equation with a rotation operator, R, we can easily check that

$$RKR^{-1}(R\zeta_k) = \lambda_k R\zeta_k. \tag{9.5.4}$$

Because we have assumed that the potential is spherically symmetric, the rotated collision operator RKR^{-1} is the same as the unrotated collision operator, K. Thus $RKR^{-1} = K$ and, accordingly, Eq. (9.5.3) implies that

$$K(R\zeta_R) = \lambda_k(R\zeta_k). \tag{9.5.5}$$

Equation (9.5.5) shows that $R\zeta_k$ is also an eigenvector with eigenvalue λ_k.

This means that the eigenvectors form a representation of the rotational group in three dimensions and that we can classify them according to the irreducible representations of the rotation group. Thus, as in the quantum theory calculation of the energy eigenfunctions in a central force field, we can write

$$\zeta_k(\mathbf{v}) = R_{rl}(v)\, Y^l_m(\theta, \phi), \tag{9.5.6}$$

where Y^l_m are the usual spherical harmonics and R_{rl} depends only on the magnitude of v. The numbers r and l, which index the eigenfunctions, depend on k, and for given values of r and l there are $2l + 1$ degenerate eigenvalues corresponding to $-l \leq m \leq +l$.

Instead of the spherical harmonics it is convenient to use Cartesian tensors as a basis set for the angular coordinates. For a given l these are the $2l + 1$ homogeneous, symmetric, and traceless tensors of order l formed from the components of the three-vector, v_α. They can be obtained by differentiation from the formula

$$\langle v_{\alpha_1} v_{\alpha_2} \dots v_{\alpha_l} \rangle = \frac{(-1)^l v^{2l+1}}{1 \cdot 3 \cdot 5 \dots (2l-1)} \frac{\partial^l v^{-1}}{\partial v_{\alpha_1} \dots \partial v_{\alpha_l}}. \tag{9.5.7}$$

Using Eq. (9.5.7) it is easy to see that these tensors are symmetric in all indices and traceless when any trace of the form $\langle v_{\alpha 1} v_{\alpha 1} v_{\alpha 3} \dots v_{\alpha l} \rangle$ is taken. Considered as functions of the Cartesian coordinates of $\mathbf{v}$, Eq. (9.5.7) also makes it clear that these tensors are homogeneous polynomials of degree l, i.e., $\langle (\lambda v_{\alpha 1})(\lambda v_{\alpha 2}) \dots (\lambda v_{\alpha l}) \rangle = \lambda^l \langle v_{\alpha 1} v_{\alpha 2} \dots v_{\alpha l} \rangle$. When considered as functions of the angular part of the spherical coordinates, θ and ϕ, they can be shown to be orthogonal, linear combinations of Y^l_m.

Since we are interested in the form of the viscosity and heat flux tensors referred to Cartesian axes, we will use the Cartesian tensors in Eq. (9.5.7) as the basis for the angular part of the eigenfunctions ζ_k. Because they are homogeneous polynomials of order l, we write ζ_k as

$$\zeta_k(\mathbf{v}) = R_{rl}(v) v^{-l} \langle v_{\alpha_1} v_{\alpha_2} \dots v_{\alpha_l} \rangle. \tag{9.5.8}$$

Using Eq. (9.5.7) it is easy to see that the first few symmetric, traceless Cartesian tensors are

$$\begin{aligned} \langle v_{\alpha_1} \rangle &= v_{\alpha_1} \\ \langle v_{\alpha_1} v_{\alpha_2} \rangle &= v_{\alpha_1} v_{\alpha_2} - v^2 \delta_{\alpha_1 \alpha_2}/3 \\ \langle v_{\alpha_1} v_{\alpha_2} v_{\alpha_3} \rangle &= v_{\alpha_1} v_{\alpha_2} v_{\alpha_3} - \frac{v^2}{5} [v_{\alpha_1} \delta_{\alpha_2 \alpha_3} + v_{\alpha_2} \delta_{\alpha_3 \alpha_1} + v_{\alpha_3} \delta_{\alpha_1 \alpha_2}]. \end{aligned} \tag{9.5.9}$$

We need these expressions to calculate the matrix elements in Eq. (9.5.2). First, we express the zero-eigenvalue eigenfunctions in Eq. (9.4.20) in this form. Using Eqs. (9.4.20), (9.5.8), and (9.5.9) it is easy to set that

$$\begin{aligned}
\zeta_1(\mathbf{v}) &\equiv R_{00}(v) = 1 \\
\zeta_\alpha(\mathbf{v}) &\equiv R_{01}(v)v^{-1}\langle v_\alpha\rangle; \quad R_{01} \equiv v(m/k_B T^e)^{1/2} \\
\zeta_5(\mathbf{v}) &\equiv R_{10}(v) = (2/3)^{1/2}(mv^2/2k_B T^e - 3/2).
\end{aligned} \tag{9.5.10}$$

We can take advantage of the orthogonality of the symmetric, traceless Cartesian tensors in simplifying the expressions for A^α_{jk}. Note first that Eqs. (9.5.2) and (9.5.10) imply the equality

$$\begin{aligned}
A^\alpha_{\beta k} &= -(m/k_B T^e)^{1/2} \int \zeta_k(\mathbf{v}) v_\alpha v_\beta P^e(v)\, d\mathbf{v} \\
&= -(m/k_B T^e)^{1/2} \int \zeta_k(\mathbf{v})(\langle v_\alpha v_\beta\rangle + v^2\delta_{\alpha\beta}/3)P^e(v)\, d\mathbf{v},
\end{aligned} \tag{9.5.11}$$

where the second equality follows from Eq. (9.5.9). Now Eq. (9.5.10) shows that v^2 is a linear combination of the eigenfunctions ζ_1 and ζ_5, which are orthogonal to ζ_k for $k > 5$. Thus v^2 makes no contribution to the integral in Eq. (9.5.11). Furthermore, writing $\zeta_k(\mathbf{v})$ in the form of Eq. (9.5.8), the orthogonality of the Cartesian tensors implies that the only nonzero terms for $k > 5$ are

$$A^\alpha_{\beta k} = -(m/k_B T^e)^{1/2} \int \langle v_\alpha v_\beta\rangle \langle v_{\alpha_1} v_{\alpha_2}\rangle v^{-2} R_{r2}(v) P^e(v)\, d\mathbf{v}, \tag{9.5.12}$$

where the label k represents the indices r and α_1, α_2. Similarly, the expression for A^α_{5k} can be written using Eqs. (9.5.2) and (9.5.10) as

$$A^\alpha_{5k} = -\int \zeta_k(\mathbf{v})[(2/3)^{1/2}(mv^2/2k_B T^e - 3/2)]\langle v_\alpha\rangle P^e(v)\, d\mathbf{v} \tag{9.5.13}$$

since $v_\alpha = \langle v_\alpha\rangle$. However, Eq. (9.5.10) shows that ζ_α is proportional to $\langle v_\alpha\rangle$, which implies that the second term in the square brackets of Eq. (9.5.13) integrates to zero for $k > 5$. Integrating over the angular coordinates of the first term we see that only when $\zeta_k = R_{r1}v^{-1}\langle v_{\alpha_1}\rangle$ will there be a nonzero contribution. Hence

$$A^\alpha_{5\alpha} = -(2/3)^{1/2}(m/2k_B T^e) \int \langle v_\alpha\rangle \langle v_{\alpha_1}\rangle R_{r1}(v) v P^e(v)\, d\mathbf{v}. \tag{9.5.14}$$

The integral expressions in Eqs. (9.5.12) and (9.5.14) can be simplified using the angular integrals

$$\int \langle v_\alpha\rangle \langle v_{\alpha_1}\rangle\, d\Omega = \frac{4\pi v^2}{3}\delta_{\alpha\alpha_1} \tag{9.5.15}$$

and

$$\int \langle v_\alpha v_\beta\rangle \langle v_{\alpha_1} v_{\alpha_2}\rangle\, d\Omega = \frac{4\pi v^4}{15}\left[\delta_{\alpha\alpha_1}\delta_{\beta\alpha_2} + \delta_{\alpha\alpha_2}\delta_{\beta\alpha_1} - \frac{2}{3}\delta_{\alpha\beta}\delta_{\alpha_1\alpha_2}\right], \tag{9.5.16}$$

where $d\Omega$ is the element of solid angle and the integrals extend over the sphere of radius v. These expressions are straightforward to verify using spherical polar coordinates and the definitions of $\langle v_\alpha \rangle$ and $\langle v_\alpha v_\beta \rangle$ in Eq. (9.5.9). With these expressions it is possible to carry out the angular integrations in Eqs. (9.5.12) and (9.5.14) to obtain

$$A^\alpha_{\beta k} = -\frac{1}{15}(m/k_B T^e)^{1/2} \int v^2 R_{r2}(v) P^e(v)\, d\mathbf{v} \times \left(\delta_{\alpha\alpha_1}\delta_{\beta\alpha_2} + \delta_{\alpha\alpha_2}\delta_{\beta\alpha_1} - \frac{2}{3}\delta_{\alpha\beta}\delta_{\alpha_1\alpha_2}\right) \tag{9.5.17}$$

$$A^\alpha_{5k} = -(2/3)^{1/2}(m/6k_B T^e) \int v^3 R_{r1}(v) P^e(v)\, d\mathbf{v}\, \delta_{\alpha\alpha_1}. \tag{9.5.18}$$

Finally, if we define the integrals

$$\begin{aligned} I_{r2} &\equiv -\frac{1}{15}(m/k_B T^e)^{1/2} \int v^2 R_{r2}(v) P^e(v)\, d\mathbf{v} \\ I_{r1} &\equiv -(2/3)^{1/2}(m/6k_B T^e) \int v^3 R_{r1}(v) P^e(v)\, d\mathbf{v}, \end{aligned} \tag{9.5.19}$$

we can use Eq. (9.5.1) to write the viscosity and heat conductivity tensors as

$$\begin{aligned} 2\eta_{\alpha\beta,\kappa\gamma} &= m\rho^e\left(\sum_{r=0}^{\infty} I_{r2}^2/\lambda_{r2}\right) G_{\alpha\beta,\kappa\gamma} \\ \kappa_{\alpha\gamma} &= (3k_B\rho^e/2)\left(\sum_{r=0}^{\infty} I_{r1}^2/\lambda_{r1}\right)\delta_{\alpha\alpha_1}\delta_{\alpha_1\gamma}, \end{aligned} \tag{9.5.20}$$

where

$$G_{\alpha\beta,\kappa\gamma} \equiv \sum_{\alpha_1,\alpha_2}\left(\delta_{\alpha\alpha_1}\delta_{\beta\alpha_2} + \delta_{\alpha\alpha_2}\delta_{\beta\alpha_1} - \frac{2}{3}\delta_{\alpha\beta}\delta_{\alpha_1\alpha_2}\right) \times \left(\delta_{\kappa\alpha_1}\delta_{\gamma\alpha_2} + \delta_{\kappa\alpha_2}\delta_{\gamma\alpha_1} - \frac{2}{3}\delta_{\kappa\gamma}\delta_{\alpha_1\alpha_2}\right). \tag{9.5.21}$$

The sum in Eq. (9.5.21) runs over the $2 \times 2 + 1 = 5$ independent symmetric, traceless Cartesian tensors $\langle v_{\alpha_1} v_{\alpha_2} \rangle$. To simplify the tensor G, we use the fact that when the tensor

$$\Delta_{\alpha\alpha_1,\beta\alpha_2} = \delta_{\alpha\alpha_1}\delta_{\beta\alpha_2} + \delta_{\alpha\alpha_2}\delta_{\beta\alpha_1} - \frac{2}{3}\delta_{\alpha\beta}\delta_{\alpha_1\alpha_2}, \tag{9.5.22}$$

operates on a rank-two tensor, H, it yields twice its traceless, symmetric part, $\mathring{H}$. Indeed, one easily verifies that

$$\Delta_{\alpha\alpha_1,\beta\alpha_2} H_{\alpha_2\alpha_1} = (H_{\alpha\beta} + H_{\beta\alpha}) - \frac{2}{3}H_{\gamma\gamma}\delta_{\alpha\beta} = 2\mathring{H}_{\alpha\beta}. \tag{9.5.23}$$

Therefore, when restricted to the five-dimensional space of symmetric, traceless tensors, Δ is twice the identity. Thus, symbolically, $\Delta^2 = 4I = 2\Delta$, which in conjunction with Eq. (9.5.21) implies that

$$G_{\alpha\beta,\kappa\gamma} = 2\left(\delta_{\alpha\kappa}\delta_{\beta\gamma} + \delta_{\alpha\kappa}\delta_{\beta\gamma} - \frac{2}{3}\delta_{\alpha\beta}\delta_{\kappa\gamma}\right). \tag{9.5.24}$$

Taking this into account, the viscosity and heat conductivity tensors take the form

$$\begin{aligned} 2\eta_{\alpha\beta,\kappa\gamma} &= 2m\rho^e\left(\sum_{r=0}^{\infty} I_{r2}^2/\lambda_{r2}\right)\Delta_{\alpha\kappa,\beta\gamma} \\ \kappa_{\alpha\gamma} &= (3k_B\rho^e/2)\left(\sum_{r=0}^{\infty} I_{r1}^2/\lambda_{r1}\right)\delta_{\alpha\gamma}. \end{aligned} \tag{9.5.25}$$

This verifies the isotropy of the tensorial coefficients, which was assumed in the previous section, and gives the expressions

$$\begin{aligned} \eta^e &= 2m\rho^e\left(\sum_{r=0}^{\infty} I_{r2}^2/\lambda_{r2}\right) \\ \kappa^e &= (3k_B\rho^e/2)\left(\sum_{r=0}^{\infty} I_{r1}^2/\lambda_{r1}\right) \end{aligned} \tag{9.5.26}$$

for the viscosity and thermal conductivity of a low-pressure gas.

Explicit evaluation of the transport coefficients using Eq. (9.5.26) requires a knowledge of the eigenfunctions and eigenvalues of the collision operator for $k > 5$. In general, one has only qualitative information about these quantities, except for certain special cases. The most important of these is for so-called Maxwell molecules, which repel each other with a potential

$$u(r) = \varepsilon/r^4. \tag{9.5.27}$$

As Maxwell discovered, this potential has the useful property that the cross section is proportional to $|\mathbf{v} - \mathbf{v}'|^{-1}$. Thus the kernel of the linearized Boltzmann equation, which involves the operator $\hat{\sigma}_T|\mathbf{v} - \mathbf{v}'|$, is independent of the velocity. To prove this, one needs to use the solution to Newton's equation for the trajectory of the two-particle scattering process described in Section 2.7.

In Fig. 9.1 a trajectory that corresponds to a collision of identical particles of mass m, impact parameter b, and a magnitude of relative velocity g is depicted. To represent the trajectory it is convenient to use the coordinates ρ, ϕ in the plane of the relative motion, which we recall is perpendicular to the direction of the angular momentum vector, $\mathbf{l}$. The coordinates of the trajectory at a precollision point, $\boldsymbol{\rho}$, at the distance of closest approach, $\boldsymbol{\rho}_{\min}$, and at infinite time, $\boldsymbol{\rho}_{\infty}$, are illustrated in the figure. Using conservation of angular momentum, the differential equation of the trajectory of the relative particle is found to be given by

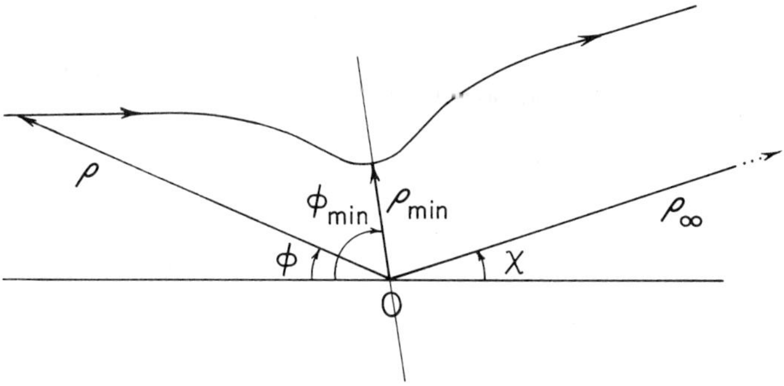

FIGURE 9.1. A sketch of the trajectory of a two body collision in the reference frame attached to a particle at the origin, 0. The relative separation vector is $\boldsymbol{\rho}$ and the angular coordinate is ϕ. The values of these variables at the distance of closest approach are $\boldsymbol{\rho}_{\min}$ and $\phi_{\min}$, and the scattering angle is χ.

$$d\phi/d\rho = +b/\rho^2\left[1 - \frac{4u(\rho)}{mg^2} - \frac{b^2}{\rho^2}\right]^{1/2}. \tag{9.5.28}$$

The plus sign corresponds to the relative particle moving from left to right in the figure. Because of spherical symmetry of the potential, $u(\rho)$, the geometry of the collision is symmetric around $\boldsymbol{\rho}_{\min}$. Thus the polar angle satisfies $\phi_\infty = 2\phi_{\min}$. The geometry in Fig. 9.1, therefore, shows that the scattering angle is

$$\chi = |\pi - 2\phi_{\min}|. \tag{9.5.29}$$

According to the kinematic formula in Eq. (9.5.29), to calculate the scattering angle we need to integrate Eq. (9.5.28) between $\rho_{\min}$ and ρ_∞. For a reciprocal power law potential, i.e., $u(\rho) = \varepsilon/\rho^n$, this gives

$$\phi_{\min} = b\int_{\rho_{\min}}^{\infty} \frac{d\rho}{\rho^2\left[1 - \dfrac{4\varepsilon}{mg^2\rho^n} - \dfrac{b^2}{\rho^2}\right]^{1/2}}. \tag{9.5.30}$$

The value of $\rho_{\min}$ is a turning point of the motion and is the largest value of ρ for which the denominator in the integral vanishes, i.e.,

$$\left[1 - \frac{4\varepsilon}{mg^2\rho_{\min}^n} - \frac{b^2}{\rho_{\min}^2}\right] = 0. \tag{9.5.31}$$

To obtain the cross section from Eq. (9.5.30), we recall that the cross section is expressed in Eq. (2.7.9) as

$$\sigma(\Omega, g) = [b(\chi)/\sin\chi]\,|db(\chi)/d\chi|. \tag{9.5.32}$$

Equation (9.5.30) gives $\phi_{\min}$ as a function of g and b and Eq. (9.5.29) gives χ as a function of $\phi_{\min}$. Thus, inverting these expressions to obtain $b(\chi, g)$, we can, in principle, obtain the cross section using Eq. (9.5.32). In practice, this cannot be done analytically. However, by an appropriate change of variables we can discover the g-dependence of the cross section. To this end we use in place of ρ and b the quantities

$$x = b/\rho, \quad y = (m/4\varepsilon n)^{1/n} b g^{2/n}, \tag{9.5.33}$$

which, when substituted in Eqs. (9.5.30) and (9.5.31), give

$$\phi_{\min}(y) = \int_0^{x_{\min}} dx \left[1 - x^2 - \frac{1}{n}\left(\frac{x}{y}\right)^n\right]^{-1/2} \tag{9.5.34}$$

and

$$1 - x_{\min}^2 - \frac{1}{n}\left(\frac{x_{\min}}{y}\right)^n = 0. \tag{9.5.35}$$

In terms of these variables, the cross section is given by

$$\sigma(\Omega, g) = (4n\varepsilon/m)^{2/n} g^{-4/n} y(\chi) |dy(\chi)/d\chi| / \sin\chi. \tag{9.5.36}$$

Since neither of Eqs. (9.5.34) and (9.5.35) depends on g, nor does the relationship between χ and $\phi_{\min}$, it follows from Eq. (9.5.36) that the cross section for an inverse power law potential has the form

$$\sigma(\Omega, g) = \sigma(\chi) g^{-4/n}. \tag{9.5.37}$$

Using Eq. (9.5.37) we see that the cross section of Maxwell molecules, for which $n = 4$, is inversely proportional to $g = |\mathbf{v} - \mathbf{v}'|$. Consequently, $\hat{\sigma}_T |\mathbf{v} - \mathbf{v}'|$ is independent of the velocity. This greatly simplifies the Boltzmann collision operator and, in fact, permits a solution of the eigenvalue problem in terms of polynomials. Indeed, for Maxwell molecules the linearized collision operator takes the form

$$K[\zeta_k] = \int\int d\Omega\, \sigma(\chi) [\zeta_k(\mathbf{v}') + \zeta_k(\mathbf{v}_1') - \zeta_k(\mathbf{v}) - \zeta_k(\mathbf{v}_1)] \rho^e(v_1)\, d\mathbf{v}_1, \tag{9.5.38}$$

where we have written out explicitly the integral over the scattering angle. Using the form of the eigenfunctions in Eq. (9.5.8), Eq. (9.5.38) can be reduced to a linear, integral equation solved by $R_{rl}(v)$, the angle-independent factor of the eigenfunction, ζ_k. In a manner similar to the solution of the radial part of the hydrogen atom Schrödinger equation in terms of associated Laguerre polynomials, one can solve this linear integral equation in terms of Sonine polynomials. One finds that

$$R_{rl}(v) = N_{rl} (mv^2/2k_B T^e)^{l/2} S_{l+1/2}^{(r)}(mv^2/2k_B T^e) \tag{9.5.39}$$

$$S_l^{(r)}(x) = \frac{(-1)^j \Gamma(l + r + 1) x^j}{\Gamma(l + j + 1)(r - j)!}, \tag{9.5.40}$$

with N_{rl} a normalization constant and Γ the gamma function. The first few normalized, nonangular eigenfunctions for Maxwell molecules are

$$\begin{aligned} R_{00}(v) &= 1 \\ R_{01}(v) &= (m/k_B T^e)^{1/2} v \\ R_{02}(v) &= 2^{1/2}(m/2k_B T)v^2 \\ R_{10}(v) &= (2/3)^{1/2}(mv^2/2k_B T^e - 3/2) \\ R_{11}(v) &= (4/5)^{1/2}(m/k_B T^e)^{3/2} v(v^2 - 5k_B T^e/m). \end{aligned} \tag{9.5.41}$$

Notice that the expressions for R_{00}, R_{01}, and R_{10} agree with those obtained for the zero-eigenvalue eigenfunctions in Eq. (9.5.10). Using these functions and Eq. (9.4.18) it can be shown that their eigenvalues are

$$\begin{aligned} \lambda_{00} &= \lambda_{01} = \lambda_{10} = 0 \\ \lambda_{11} &= \lambda_{02} = \rho^e \pi \int_0^\pi \sigma(\chi) \sin^3 \chi \, d\chi. \end{aligned} \tag{9.5.42}$$

We have gone to the trouble of describing eigenfunctions and eigenvalues of the collision operator for Maxwell molecules so that we can calculate explicitly the viscosity and thermal conductivity. The expressions that we need to evaluate are given in Eqs. (9.5.19) and (9.5.26). The key to evaluating the integrals I_{r2} and I_{r1} is that the eigenfunctions R_{rl} are orthogonal for different values of r and the same value of l. However, according to Eq. (9.5.41), $R_{02}(v)$ is proportional to v^2. Thus since

$$I_{r2} = -\frac{1}{15}(m/k_B T^e)^{1/2} \int v^2 R_{r2}(v) P^e(v) \, d\mathbf{v}, \tag{9.5.43}$$

it follows that

$$\begin{aligned} I_{r2} &= -\frac{1}{15}(m/k_B T^e)^{1/2} \int v^2 2^{1/2}(m/2k_B T^e) v^2 P^e(v) \, d\mathbf{v} \delta_{r0} \\ &= -(k_B T^e/2m)^{1/2} \delta_{r0}, \end{aligned} \tag{9.5.44}$$

where we have used the standard integral

$$\int v^{2n} P^e(v) \, d\mathbf{v} = 1 \cdot 3 \cdot 5 \ldots (2n + 1)(k_B T^e/m)^n, \tag{9.5.45}$$

for $n = 2$. Similarly, using the facts that v^3 is a linear combination of R_{01} and R_{11} [cf. Eq. (9.5.41)] and that

$$I_{r1} = -(2/3)^{1/2}(m/6k_B T^e) \int v^3 R_{r1}(v) P^o(v) \, d\mathbf{v}, \tag{9.5.46}$$

we see that I_{r1} is nonzero only for $r = 1$. Indeed,

$$I_{r1} = -(2/3)^{1/2}(m/6k_BT^e)\int v(v^2 - 5k_BT^e/m)R_{11}(v)P^e(v)\,d\mathbf{v}\,\delta_{r1}. \quad (9.5.47)$$

Substituting R_{11} from Eq. (9.5.41) into Eq. (9.5.46) and integrating using Eq. (9.5.45) one obtains

$$I_{r1} = -(5k_BT^e/3m)^{1/2}\delta_{r1}. \quad (9.5.48)$$

The simplicity of Eqs. (9.5.44) and (9.5.48) for Maxwell molecules makes it easy to obtain explicit expressions for the viscosity coefficient and the thermal conductivity. Substituting the expressions for I_{r1} and I_{r2} into Eq. (9.5.26), one finds that

$$\begin{aligned}\eta^e &= k_BT^e \Big/ \pi\int_0^\pi \sigma(\chi)\sin^3\chi\,d\chi\\ \kappa^e &= 5k_B^2T^e \Big/ 2\pi\int_0^\pi \sigma(\chi)\sin^3\chi\,d\chi.\end{aligned} \quad (9.5.49)$$

Thus to obtain numerical values of the transport coefficients, one need only perform the integral over $\sigma(\chi)$ in Eq. (9.5.49). Comparing Eqs. (9.5.36) and (9.5.37), we see for Maxwell molecules that $\sigma(\chi)$ has the form

$$\sigma(\chi) = (\varepsilon/m)^{1/2}G(\chi), \quad (9.5.50)$$

where G is independent of the strength of the potential, ε, and the mass. The value of the integral over $\sigma(\chi)$ was first calculated numerically by Maxwell, who found that

$$\int_0^\pi \sigma(\chi)\sin^3\chi\,d\chi = 1.85(\varepsilon/m)^{1/2}. \quad (9.5.51)$$

Thus for Maxwell molecules

$$\begin{aligned}\eta^e &= 0.172k_BT(m/\varepsilon)^{1/2}\\ \kappa^e &= 0.430k_B^2T(m/\varepsilon)^{1/2}.\end{aligned} \quad (9.5.52)$$

Notice that Eq. (9.5.52) predicts that both η^e and κ^e are independent of the density of the gas. Maxwell was the first to notice this fact, which he later verified experimentally for the viscosity coefficient.

The expressions we have obtained in this section for the viscosity and thermal conductivity are identical to those given by the classical Chapman–Enskog procedure. The work in the previous sections of this chapter makes it clear that this sort of contraction is a special case of the reduction of variables associated with any fast and slow variables that are coupled dynamically. The existence of two well-separated time scales is common in real systems, and these contraction procedures can be used to obtain expressions for the transport coefficients for other slow variables in terms of the dissipative parameters for the fast variables.

Comparisons of the Chapman–Enskog expressions for the transport coefficients with experimental values have been quite successful. Although the real intermolecular potentials between molecules are often not known accurately, by adjusting parameters in plausible models of potential it has been possible to fit the temperature dependence of the viscosity and thermal conductivity for the inert gases over a broad range of temperatures. For these gases one also finds that the *Eucken formula*, $\kappa^e/\eta^e = 5k_B/2$, which is predicted by Eq. (9.5.49) to hold exactly for Maxwell molecules, holds rather well. The success of the Boltzmann equation in describing hydrodynamic relaxation processes for dilute gases is a prime example of the utility of the hierarchical approach to nonequilibrium thermodynamics.

9.6. Rate Constants for Rapid Bimolecular Chemical Reactions

There is another way in which the idea of a contraction can be used to calculate transport coefficients. One of the best examples of this is provided by rapid bimolecular reactions in condensed phases. The rate of a bimolecular reaction is governed in part by the rate at which two chemical species come together. In the gas phase this is determined by the number of collisions per unit time at a given relative speed, g, and the total reactive cross section, $\sigma_r(g)$. Indeed, for a gas in which the velocity distribution is Maxwellian, it is not difficult to derive an expression for the rate constant for the bimolecular reaction

$$\mathrm{A} + \mathrm{B} \rightleftarrows \mathrm{C}. \tag{9.6.1}$$

In terms of the cross section the forward rate constant is

$$k^+ = 4\pi(\mu/2\pi k_B T)^{3/2} \int_0^\infty \sigma_r(g) g^3 \exp(-\mu g^2/2k_B T)\, dg, \tag{9.6.2}$$

where $\mu \equiv m_A m_B/(m_A + m_B)$ is the reduced mass. Equation (9.6.2) is the analogue of the Chapman–Enskog expressions for the viscosity and heat conductivity derived in the previous section. We can estimate the size of the rate constant if we assume that reaction occurs upon each collision. A good model cross section for this case is the one obtained in Section 2.7 for hard spheres, namely,

$$\sigma_r = 4\pi R^2, \tag{9.6.3}$$

where $R = R_A + R_B$ is the sum of the hard-sphere radii. For this cross section the integral in Eq. (9.6.2) can be evaluated and one finds that

$$k^+ = 4\pi R^2 (8k_B T/\pi\mu)^{1/2}. \tag{9.6.4}$$

The cgs units for the bimolecular rate constant are cm^3 molecule^{-1} s^{-1}, and to express it in the more common units of molar^{-1} s^{-1} one must multiply Eq. (9.6.4) by 6.02×10^{20}. Thus in these units one has

$$k^+ = 24.1\pi R^2(8k_B T/\pi\mu)^{1/2} 10^{20}\ M^{-1}\ \mathrm{s}^{-1}. \tag{9.6.5}$$

Using $T = 300$ K, $R = 1 \times 10^{-8}$ cm, and $\mu = 6 \times 10^{-23}$ g gives an estimated gas-phase rate constant of $3 \times 10^{10}\ M^{-1}\ \mathrm{s}^{-1}$. This is typical of measured rate constants for rapid gas-phase chemical reactions. For example, the measured rate constant for the gas-phase recombination of ethyl radicals,

$$2\mathrm{CH_3CH_2^{\cdot}} \rightarrow \mathrm{CH_3(CH_2)_2CH_3}, \tag{9.6.6}$$

is $2 \times 10^{10}\ M^{-1}\ \mathrm{s}^{-1}$ at room temperature.

In condensed phases such as solutions, solids, or membranes the situation is quite different. Even though one expects the velocity distribution to be Maxwellian under normal circumstances, bimolecular reaction rate constants seldom exceed $4 \times 10^9\ M^{-1}\ \mathrm{s}^{-1}$. Indeed, the rate constant for ethyl radical recombination is reduced by a factor of 20 in aqueous solution. The explanation for this goes back to Smoluchowski, who noticed that in condensed phases the rate at which reactants diffuse together might limit the reaction rate for rapid reactions. This effect decreases the average spatial distribution of reactants around one other. For a reaction that occurs instantaneously, Smoluchowski used the diffusion equation to calculate the steady-state rate constant

$$k^+ = 4\pi(D_\mathrm{A} + D_\mathrm{B})R, \tag{9.6.7}$$

where D_A and D_B are the diffusion constants for A and B. R is usually called the encounter radius and is roughly the sum of the radii of A and B. If we multiply Eq. (9.6.7) by 6.02×10^{20}, we can estimate the size of the bimolecular rate constant in solution in molar units. Using $D_\mathrm{A} + D_\mathrm{B} = 2 \times 10^{-5}\ \mathrm{cm^2\ s^{-1}}$, which is typical for aqueous solution, and $R = 4 \times 10^{-8}$ cm gives $k^+ = 6 \times 10^9\ M^{-1}\ \mathrm{s}^{-1}$. This is in the range of experimental values for rapid reactions in solution.

From the point of view of nonequilibrium thermodynamics, the interesting thing about Smoluchowski's expression for the rate constant is that it depends on the diffusion constants. The expression is distinctly different from the Chapman–Enskog expression in Eq. (9.6.2), which depends on the collisional cross section, and, as we have seen, gives rate constants that are smaller than predicted by the Boltzmann equation. The reason for this is that the Boltzmann equation is valid only for dilute gases and does not take into account dynamical effects due to the solvent.

Nonetheless, the effect of diffusion on rapid reactions can also be understood by means of the contraction of a stochastic process. The appropriate contraction is from the hydrodynamic level of description to the thermodynamic level. To see this, we recall that at the thermodynamic level of description the rate of the elementary bimolecular reaction in Eq. (9.6.1) can be written, on the average, as

$$\bar{V}^+ = k^+ \bar{\rho}_A \bar{\rho}_B. \tag{9.6.8}$$

At the thermodynamic level k^+ is treated simply as a constant whose value must be obtained from experiment. We can, however, take into account spatial variation of the reaction rate by going to the hydrodynamic level of description. In a solution molecules generally can react over a range of separations. Thus it makes sense to introduce an intrinsic reactivity, k^0, such that $k^0(\mathbf{r}, \mathbf{r}')\,d\mathbf{r}$ is the reaction rate constant when an A molecule is at $\mathbf{r}$ and a B molecule is at $\mathbf{r}'$. In terms of the intrinsic reactivity, we can write the average reaction rate at $\mathbf{r}$ as

$$\bar{V}^+ = \int k^0(\mathbf{r}, \mathbf{r}')\rho_{AB}(\mathbf{r}, \mathbf{r}', t)\,d\mathbf{r}', \tag{9.6.9}$$

where $\rho_{AB}(\mathbf{r}, \mathbf{r}', t)$ is the pair distribution function. The pair distribution function is related to the joint probability density, $P_{AB}^{(2)}$, of finding an A molecule at $\mathbf{r}$ and a B molecule at $\mathbf{r}'$ by the formula

$$P_{AB}^{(2)}(\mathbf{r}, \mathbf{r}', t) = \rho_{AB}(\mathbf{r}, \mathbf{r}', t)/N_A(t)N_B(t), \tag{9.6.10}$$

where $N_A(t)$ and $N_B(t)$ are the total numbers of A and B molecules in the solution.

To relate the hydrodynamic level of description to the thermodynamic level, we need to introduce the nonequilibrium radial distribution function. In analogy to the equilibrium radial distribution function discussed in Section 6.9, the nonequilibrium radial distribution function, $g_{AB}^{(2)}(\mathbf{r}, \mathbf{r}, t)$, is defined by

$$\rho_{AB}(\mathbf{r}, \mathbf{r}', t) = \bar{\rho}_A(\mathbf{r}, t)\bar{\rho}_B(\mathbf{r}', t)g_{AB}^{(2)}(\mathbf{r}, \mathbf{r}', t). \tag{9.6.11}$$

It is easy to find a physical interpretation of the radial distribution function. Indeed, ρ_{AB} is related by Eq. (9.6.10) to the joint probability of finding an A and a B molecule at $\mathbf{r}$ and $\mathbf{r}'$ and $\bar{\rho}_A(\mathbf{r}, t)$ is proportional to the probability density of finding an A molecule at $\mathbf{r}$. Thus their ratio is the density of B molecules at $\mathbf{r}'$ conditioned on the presence of a molecule of A at $\mathbf{r}$, i.e.,

$$\rho_B(\mathbf{r}|\mathbf{r}', t) = \bar{\rho}_B(\mathbf{r}', t)g_{AB}^{(2)}(\mathbf{r}, \mathbf{r}', t). \tag{9.6.12}$$

In other words, the radial distribution function is the proportionality factor between the bulk density of B molecules at $\mathbf{r}'$ and the density of B molecules at $\mathbf{r}'$ conditioned on the presence of an A molecule at $\mathbf{r}$.

In an isotropic, uniform solution both $\bar{\rho}_A$ and $\bar{\rho}_B$ are independent of position, and their radial distribution function will depend only on $|\mathbf{r} - \mathbf{r}'| = r$. In this case, using Eqs. (9.6.10) and (9.6.9) we can write

$$\bar{V}^+ = \int k^0(r)g_{AB}^{(2)}(r, t)\,d\mathbf{r}\,\bar{\rho}_A\bar{\rho}_B, \tag{9.6.13}$$

where we have assumed that k^0 depends only on r. Comparing the hydrodynamic-level expression for the rate in Eq. (9.6.13) with Eq. (9.6.8), we see that

$$k^+ = \int k^0(r) g^{(2)}_{AB}(r,t)\, d\mathbf{r}. \tag{9.6.14}$$

Equation (9.6.14) shows that the rate constant is independent of time only if the radial distribution function is. This will be true either in an equilibrium or steady-state ensemble, and in those cases one can write

$$k^+ = 4\pi \int_0^\infty k^0(r) g^{(2)}_{AB}(r) r^2\, dr. \tag{9.6.15}$$

Equations (9.6.14) and (9.6.15) provide expressions for the bimolecular rate constant in terms of the intrinsic reactivity function, $k^0(r)$. This function depends on the molecular nature of the reaction event and can, in principle, be calculated from quantum or classical mechanics. A model reactivity, which is equivalent to Smoluchowski's notion of an encounter radius, is

$$k^0(r) = k^0 \delta(r - R)/4\pi r^2. \tag{9.6.16}$$

This reactivity restricts the reaction to occur at the encounter radius, R, and will be called the Smoluchowski reactivity. When used in Eq. (9.6.15), it leads to the simple expression

$$k^+ = k^0 g^{(2)}_{AB}(R). \tag{9.6.17}$$

Other more realistic reactivity functions have been used to describe chemical reactions, including the Föster reactivity for dipolar energy transfer

$$k^0(r) = (1/\tau)(r_0/r)^6, \tag{9.6.18}$$

and a reactivity function based on electron overlap, which has the form

$$k^0(r) = k^0(T)\exp(-r/r_0). \tag{9.6.19}$$

For simplicity we restrict ourselves here to the Smoluchowski reactivity.

To evaluate the expressions for the rate constant, we use the hydrodynamic level of description to calculate the radial distribution function. To do this we need to generalize the relationship between the density–density correlation function and the radial distribution function given in Eq. (6.9.32). Since hydrodynamic fluctuation theory is classical, the phase function that gives the number density of molecule A at point $\mathbf{r}$ is simply

$$\rho_A(\mathbf{r}) = \sum_{i \in A} \delta(\mathbf{r} - \mathbf{r}_i), \tag{9.6.20}$$

where the sum is over all the $N_A(t)$ A molecules. Similarly, the density of B molecules is

$$\rho_B(\mathbf{r}) = \sum_{j \in B} \delta(\mathbf{r} - \mathbf{r}_j). \tag{9.6.21}$$

The average density is obtained by averaging these functions over phase space, e.g.,

$$\bar{\rho}_{\mathrm{A}}(\mathbf{r},t) = \int \cdots \int P_{\mathrm{AB}}(\mathbf{r}_1,\ldots,\mathbf{r}_{N_{\mathrm{A}}};\mathbf{r}'_1,\ldots,\mathbf{r}'_{N_{\mathrm{B}}},t)\rho_{\mathrm{A}}(\mathbf{r})\,d\mathbf{r}_1\ldots d\mathbf{r}'_{N_{\mathrm{B}}}, \quad (9.6.22)$$

where P_{AB} is the probability density in configuration space. Substituting Eq. (9.6.20) into Eq. (9.6.22) and using the fact that the A molecules are identical, we see that

$$\bar{\rho}_{\mathrm{A}}(\mathbf{r},t) = N_{\mathrm{A}}(t)P_{\mathrm{A}}(\mathbf{r},t), \quad (9.6.23)$$

where

$$P_{\mathrm{A}}(\mathbf{r},t) \equiv \int \cdots \int P_{\mathrm{AB}}(\mathbf{r},\ldots,\mathbf{r}_{N_{\mathrm{A}}};\mathbf{r}'_1,\ldots,\mathbf{r}'_{N_{\mathrm{B}}},t)\,d\mathbf{r}_2\ldots d\mathbf{r}'_{N_{\mathrm{B}}} \quad (9.6.24)$$

is the probability density of finding any A molecule at $\mathbf{r}$. We can also calculate the density–density correlation function, namely,

$$\begin{aligned}\langle \delta\rho_{\mathrm{A}}(\mathbf{r},t)\delta\rho_{\mathrm{B}}(\mathbf{r}',t)\rangle &= \langle [\rho_{\mathrm{A}}(\mathbf{r}) - \bar{\rho}_{\mathrm{A}}(\mathbf{r})][\rho_{\mathrm{B}}(\mathbf{r}') - \bar{\rho}_{\mathrm{B}}(\mathbf{r}')]\rangle \\ &= \langle \rho_{\mathrm{A}}(\mathbf{r})\rho_{\mathrm{B}}(\mathbf{r}')\rangle - \bar{\rho}_{\mathrm{A}}(\mathbf{r})\bar{\rho}_{\mathrm{B}}(\mathbf{r}'), \end{aligned} \quad (9.6.25)$$

using the configuration space average. Indeed, Eqs. (9.6.20), (9.6.21), and (9.6.25) show that

$$\langle \delta\rho_{\mathrm{A}}(\mathbf{r},t)\delta\rho_{\mathrm{B}}(\mathbf{r}',t)\rangle = N_{\mathrm{A}}(t)N_{\mathrm{B}}(t)P^{(2)}_{\mathrm{AB}}(\mathbf{r},\mathbf{r}',t) - \bar{\rho}_{\mathrm{A}}(\mathbf{r},t)\bar{\rho}_{\mathrm{B}}(\mathbf{r}',t), \quad (9.6.26)$$

where

$$P^{(2)}_{\mathrm{AB}}(\mathbf{r},\mathbf{r}',t) \equiv \int \cdots \int P_{\mathrm{AB}}(\mathbf{r},\ldots,\mathbf{r}_{N_{\mathrm{A}}};\mathbf{r}',\ldots,\mathbf{r}'_{N_{\mathrm{B}}},t)\,d\mathbf{r}_2\ldots d\mathbf{r}_{N_{\mathrm{A}}}\,d\mathbf{r}'_2\ldots d\mathbf{r}'_{N_{\mathrm{B}}} \quad (9.6.27)$$

is clearly the joint probability density for an A molecule at $\mathbf{r}$ and a B molecule at $\mathbf{r}'$. Recalling that Eqs. (9.6.10) and (9.6.11) give a relationship between $P^{(2)}_{\mathrm{AB}}$ and $g^{(2)}_{\mathrm{AB}}$, Eq. (9.6.26) can be rewritten

$$\langle \delta\rho_{\mathrm{A}}(\mathbf{r},t)\delta\rho_{\mathrm{B}}(\mathbf{r}',t)\rangle = \bar{\rho}_{\mathrm{A}}(\mathbf{r},t)\bar{\rho}_{\mathrm{B}}(\mathbf{r},t)[g^{(2)}_{\mathrm{AB}}(\mathbf{r},\mathbf{r}',t) - 1]. \quad (9.6.28)$$

Comparable manipulations for A alone give

$$\langle \delta\rho_{\mathrm{A}}(\mathbf{r},t)\delta\rho_{\mathrm{A}}(\mathbf{r}',t)\rangle = \bar{\rho}_{\mathrm{A}}(\mathbf{r},t)\delta(\mathbf{r}-\mathbf{r}') + \bar{\rho}_{\mathrm{A}}(\mathbf{r},t)\bar{\rho}_{\mathrm{A}}(\mathbf{r}',t)[g^{(2)}_{\mathrm{AA}}(\mathbf{r},\mathbf{r}',t) - 1]. \quad (9.6.29)$$

Equations (9.6.28) and (9.6.29) provide the necessary relationships among the radial distribution function, the average density, and the density–density correlation function.

The simplest chemical reaction for which we can use this contraction procedure is the dimerization reaction

$$\mathrm{A} + \mathrm{A} \rightleftarrows \mathrm{A}_2. \quad (9.6.30)$$

A good example of this is the recombination reaction of iodine atoms. Atomic iodine can be produced in organic solvents by photolysis of I_2. The overall processes are, therefore,

$$h\nu + I_2 \rightarrow I + I$$
$$I + I \rightleftarrows I_2. \tag{9.6.31}$$

Steady illumination of this system produces a steady-state concentration of I. Experimentally, the bimolecular reaction rate is determined from the fact that the production of I by photolysis balances its loss in the bimolecular reaction, since back reaction from I_2 is negligible. The rate constant for iodine recombination has been measured in this way to be about $10^{10}\ M^{-1}\ \mathrm{s}^{-1}$, and diffusion is known to have an effect on its value. For iodine, the recombination reaction is caused by electron overlap. Thus reaction probably occurs only when the atoms are within a molecular diameter of each other. The Smoluchowski reactivity in Eq. (9.6.16) provides a reasonable model of this, and thus we can use Eq. (9.6.17) to express the rate constant as

$$k = k^0 g_{\mathrm{AB}}^{(2)}(R), \tag{9.6.32}$$

where R is the encounter radius. For iodine recombination the intrinsic rate constant, k^0, can be estimated using scattering theory. For identical molecules one must divide the expression in Eq. (9.6.4) by 2, i.e.,

$$k^0 = 8\pi R^2 (k_{\mathrm{B}} T/\pi m)^{1/2}, \tag{9.6.33}$$

where m is the mass of the atom. At 298 K a collision radius of $R = 6$ Å gives the value $k^0 = 7.2 \times 10^{-10}\ \mathrm{cm}^3\ \mathrm{molecule}^{-1}\ \mathrm{s}^{-1} = 4.3 \times 10^{11}\ M^{-1}\ \mathrm{s}^{-1}$.

To obtain the observed rate constant using Eq. (9.6.32), we need to know the radial distribution function. In Section 6.6 the Fourier transform of the steady-state density–density correlation function was calculated for the reaction scheme in Eq. (9.6.31). The result is given in Eq. (6.6.44). For the iodine recombination reaction, the back reaction and the term due to the illumination intensity in Eq. (6.6.44) can be neglected. Under these conditions the density–density correlation function reduces to the simpler result given in Eq. (6.6.26), namely,

$$\langle \delta\rho_{\mathrm{A}}(\mathbf{r})\delta\rho_{\mathrm{A}}(\mathbf{r}')\rangle^{ss} = \rho_{\mathrm{A}}^{ss}\delta(\mathbf{r} - \mathbf{r}') - \frac{k^+ \rho_{\mathrm{A}}^{ss2} \exp(-\xi|\mathbf{r} - \mathbf{r}'|)}{2\pi D_{\mathrm{A}}|\mathbf{r} - \mathbf{r}'|} \tag{9.6.34}$$

with $\xi^{-1} = (D_{\mathrm{A}}/4k^+\rho_{\mathrm{A}}^{ss})^{1/2}$. Comparing this with Eq. (9.6.29), the radial distribution function for the iodine atoms is seen to be

$$g_{\mathrm{AA}}^{(2)}(r) = 1 - \frac{k^+ \exp(-\xi r)}{2\pi D_{\mathrm{A}} r}. \tag{9.6.35}$$

Finally, using Eq. (9.6.32) we obtain the rate constant as

$$k^+ = k^0\left(1 - \frac{k^+ \exp(-\xi R)}{2\pi D_{\mathrm{A}} R}\right). \tag{9.6.36}$$

Notice that this is not simply an expression for k^+, but an equation to be solved for k^+ since the correlation length, ξ^{-1}, depends explicitly on k^+. Equation (9.6.36) can be partially solved for k^+ by writing it in the form

$$k^+ = \frac{2\pi D_A R \exp(\xi R) k^0}{2\pi D_A R \exp(\xi R) + k^0}. \tag{9.6.37}$$

This expression turns out to be especially convenient for obtaining numerical solutions.

An important special case of Eq. (9.6.37) occurs in dilute solution where the correlation length, $\xi^{-1} = (D_A/4k^+ \rho_A^{ss})^{1/2}$, approaches infinity. This means that the factor $\exp(\xi R)$ can be replaced by unity. Thus the right-hand side of Eq. (9.6.37) becomes independent of k^+, and the explicit solution for k^+ is simply

$$k^+ = 2\pi D_A R k^0/(2\pi D_A R + k^0). \tag{9.6.38}$$

The diffusion constant for I atoms in carbon tetrachloride has been measured to be 4.2×10^{-5} cm^2 s^{-1} at 298 K. Using this number, an encounter radius of $R = 6$ Å, and the value of $k^0 = 7.2 \times 10^{-10}$ cm^3 molecule^{-1} s^{-1} based on Eq. (9.6.33), Eq. (9.6.38) gives a value of $k^+ = 1.5 \times 10^{-11}$ cm^3 molecule^{-1} s^{-1} $= 9.3 \times 10^9$ M^{-1} s^{-1}. This is in reasonable agreement with the experimental value of 8.2×10^9 M^{-1} s^{-1} measured by Noyes.

The dimerization reaction is said to be a *diffusion-controlled* reaction if $k^0 \gg 2\pi D_A R$. In this case Eq. (9.6.38) simplifies further to

$$k^+ = 2\pi D_A R. \tag{9.6.39}$$

Thus the rate constant for a diffusion-controlled reaction depends only on the rate at which reactants diffuse together and is independent of the chemical reaction mechanism. Equation (9.6.39) is the analogue for identical reactants of the Smoluchowski expression in Eq. (9.6.7). If, on the other hand, the intrinsic reaction rate constant is small, i.e., $k^0 \ll 2\pi D_A R$, then Eq. (9.6.38) reduces to

$$k^+ = k^0. \tag{9.6.40}$$

In this case the chemical mechanism determines the rate constant and the chemical process is said to be *reaction controlled.*

The contraction procedure outlined above can be used to calculate bimolecular rate constants for more complicated reaction schemes. A rather general scheme, which includes as limiting cases the bimolecular reaction $A + B \rightarrow$ products and the quenching of excited state fluorescence, is the following:

$$\begin{aligned} h\nu + \bar{A} &\longrightarrow A \\ A &\xrightarrow{k_u} \bar{A} + h\nu' \\ A + B &\xrightarrow{k^+} \bar{A} + B'. \end{aligned} \tag{9.6.41}$$

As long as $\bar{\mathrm{A}}$ is in great excess, its average concentration can be treated as constant. Only the final step in Eq. (9.6.41) is bimolecular, and using the steady-state radial distribution function $g_{\mathrm{AB}}^{(2)}(r)$ for this reaction scheme, we can calculate a value of k^+. To simplify the calculation we will ignore fluctuations in $\bar{\mathrm{A}}$. This leaves only fluctuations of A and B to be considered, since the mechanism does not couple their fluctuations to those of B$'$. Adding a uniform source of A molecules, K, and B molecules, K', the average hydrodynamic equations corresponding to the above mechanism are

$$\begin{aligned}\partial\bar{\rho}_{\mathrm{A}}/\partial t &= D_{\mathrm{A}}\nabla^2\bar{\rho}_{\mathrm{A}} - k^+\bar{\rho}_{\mathrm{A}}\bar{\rho}_{\mathrm{B}} - k_u\bar{\rho}_{\mathrm{A}} + K\\ \partial\bar{\rho}_{\mathrm{B}}/\partial t &= D_{\mathrm{A}}\nabla^2\bar{\rho}_{\mathrm{B}} - k^+\bar{\rho}_{\mathrm{A}}\bar{\rho}_{\mathrm{B}} + K',\end{aligned} \tag{9.6.42}$$

where the ρ's are number densities. Equations (9.6.42) admit a uniform steady state with

$$\begin{aligned}\rho_{\mathrm{A}}^{ss} &= K/(k^+\rho_{\mathrm{B}}^{ss} + k_u)\\ \rho_{\mathrm{B}}^{ss} &= k_u K'/k^+(K - K').\end{aligned} \tag{9.6.43}$$

According to the hydrodynamic fluctuation theory developed in Chapter 6, fluctuations in the density around the steady-state values satisfy

$$\begin{aligned}\partial\delta\rho_{\mathrm{A}}/\partial t &= D_{\mathrm{A}}\nabla^2\delta\rho_{\mathrm{A}} - k^+\rho_{\mathrm{A}}^{ss}\delta\rho_{\mathrm{B}} - k^+\rho_{\mathrm{B}}^{ss}\delta\rho_{\mathrm{A}} - k_u\delta\rho_{\mathrm{A}} + \tilde{f}_{\mathrm{A}}\\ \partial\delta\rho_{\mathrm{B}}/\partial t &= D_{\mathrm{B}}\nabla^2\delta\rho_{\mathrm{B}} - k^+\rho_{\mathrm{A}}^{ss}\delta\rho_{\mathrm{B}} - k^+\rho_{\mathrm{B}}^{ss}\delta\rho_{\mathrm{A}} + \tilde{f}_{\mathrm{B}}.\end{aligned} \tag{9.6.44}$$

As we discovered in Section 6.6, it is easy to solve for the steady-state density–density correlation function using the spatial Fourier transform. Because the steady state is uniform, Fourier transformation of Eqs. (9.6.44) gives

$$\begin{aligned}\partial\delta\hat{\rho}_{\mathrm{A}}/\partial t &= -(D_{\mathrm{A}}k^2 + k^+\rho_{\mathrm{B}}^{ss} + k_u)\delta\hat{\rho}_{\mathrm{A}} - k^+\rho_{\mathrm{A}}^{ss}\delta\hat{\rho}_{\mathrm{B}} + \tilde{f}_{\mathrm{A}}\\ \partial\delta\hat{\rho}_{\mathrm{B}}/\partial t &= -k^+\rho_{\mathrm{B}}^{ss}\delta\rho_{\mathrm{A}} - (D_{\mathrm{B}}k^2 + k^+\rho_{\mathrm{A}}^{ss})\delta\hat{\rho}_{\mathrm{B}} + \tilde{f}_{\mathrm{B}}.\end{aligned} \tag{9.6.45}$$

This 2×2 relaxation matrix $\hat{H}(k)$ can be read off Eq. (9.6.45), while the covariance of the random terms can be obtained in the usual way from the canonical formula. One finds (cf. Section 6.6) that

$$\langle \tilde{f}_i(\mathbf{k}, t)\tilde{f}_j(\mathbf{k}', t')\rangle = \hat{\gamma}_{ij}(k)\delta(\mathbf{k} + \mathbf{k}')\delta(t - t')/(2\pi)^3, \tag{9.6.46}$$

where

$$\begin{aligned}\gamma_{\mathrm{AA}} &= k^+\rho_{\mathrm{A}}^{ss}\rho_{\mathrm{B}}^{ss} + k_u\rho_{\mathrm{A}}^{ss} + 2D_{\mathrm{A}}\rho_{\mathrm{A}}^{ss}k^2\\ \gamma_{\mathrm{AB}} &= \gamma_{\mathrm{BA}} = k^+\rho_{\mathrm{A}}^{ss}\rho_{\mathrm{B}}^{ss}\\ \gamma_{\mathrm{BB}} &= k^+\rho_{\mathrm{A}}^{ss}\rho_{\mathrm{B}}^{ss} + 2D_{\mathrm{B}}\rho_{\mathrm{B}}^{ss}k^2.\end{aligned} \tag{9.6.47}$$

Just as in Section 6.6, the steady-state density–density correlation function can be written as

$$\langle \delta\rho_i(\mathbf{k})\delta\rho_i(\mathbf{k}')\rangle^{ss} = \hat{\sigma}_{ij}(k)\delta(\mathbf{k} + \mathbf{k}')/(2\pi)^3, \tag{9.6.48}$$

where $\hat{\sigma}(k)$ satisfies the fluctuation–dissipation theorem

$$\hat{H}(k)\hat{\sigma}(k) + \hat{\sigma}(k)\hat{H}(k)^T = -\hat{\gamma}(k). \tag{9.6.49}$$

Equation (9.6.49) is equivalent in form to the three linear equations for σ_{AA}, σ_{AB}, and σ_{BB}, which have been written out explicitly in Eq. (7.3.42). The solution of these equations is tedious, but straightforward, and after a good deal of algebra one obtains

$$\hat{\sigma}_{AB}(k) = \frac{k^+ \rho_A^{ss} \rho_B^{ss}}{D'(\beta + k^2)} + \frac{k^+ \rho_A^{ss} \rho_B^{ss} \alpha(k^2 + \gamma)}{2D'(\beta + k^2)(\lambda_1 + k^2)(\lambda_2 + k^2)}, \tag{9.6.50}$$

where $D' = D_A + D_B$, $\alpha = (K_1 D_B + k^+ \rho_A^{ss} D_A)/D_A D_B$, $K_1 = k_u + k^+ \rho_B^{ss}$, $\beta = (K_1 + k^+ \rho_A^{ss})/D'$, $\gamma = 2k_u k^+ \rho_A^{ss}/\alpha D_A D_B$, $\lambda_1 = (\alpha/2)(1 + [1 - 2\gamma/\alpha]^{1/2})$, $\lambda_2 = (\alpha/2)(1 - [1 - 2\gamma/\alpha]^{1/2})$.

To obtain the radial distribution function we need first to invert the Fourier transform in Eq. (9.6.50). Using tabulated integrals the inverse transform of Eq. (9.6.50) gives

$$\begin{aligned}\langle \delta\rho_A(\mathbf{r})\delta\rho_B(\mathbf{r}')\rangle^{ss} = \frac{-k^+ \rho_A^{ss} \rho_B^{ss}}{4\pi D'|\mathbf{r} - \mathbf{r}'|} \Bigg\{ &\left[\frac{\alpha(\beta - \gamma)}{2(\lambda_1 - \beta)(\lambda_2 - \beta)} + 1\right] \exp(-|\mathbf{r} - \mathbf{r}'|\beta^{1/2}) \\ &+ \left[\frac{\alpha(\lambda_1 - \gamma)}{2(\lambda_1 - \beta)(\lambda_1 - \lambda_2)}\right] \exp(-|\mathbf{r} - \mathbf{r}'|\lambda_1^{1/2}) \\ &+ \left[\frac{\alpha(\lambda_2 - \gamma)}{2(\lambda_2 - \beta)(\lambda_2 - \lambda_1)}\right] \exp(-|\mathbf{r} - \mathbf{r}'|\lambda_2^{1/2}) \Bigg\}. \end{aligned} \tag{9.6.51}$$

There are three correlation lengths, i.e., $\beta^{-1/2}$, $\lambda_1^{-1/2}$, and $\lambda_2^{-1/2}$, in this formula, one for each of the independent covariances in Eq. (9.6.49). To obtain the radial distribution function we substitute Eq. (9.6.51) into Eq. (9.6.28) to find that

$$g_{AB}^{(2)}(r) = 1 - \frac{k^+}{4\pi D' r C(r)}, \tag{9.6.52}$$

with

$$\begin{aligned} C(r) = \Bigg\{ &\left[\frac{\alpha(\beta - \gamma)}{2(\lambda_1 - \beta)(\lambda_2 - \beta)} + 1\right] \exp(-r\beta^{1/2}) \\ &+ \left[\frac{\alpha(\lambda_1 - \gamma)}{2(\lambda_1 - \beta)(\lambda_1 - \lambda_2)}\right] \exp(-r\lambda_1^{1/2}) \\ &+ \left[\frac{\alpha(\lambda_2 - \gamma)}{2(\lambda_2 - \beta)(\lambda_2 - \lambda_1)}\right] \exp(-r\lambda_2^{1/2}) \Bigg\}^{-1}. \end{aligned} \tag{9.6.53}$$

Finally, using Eq. (9.6.32) to determine the rate constant, we obtain the expression

$$k^+ = k^0\left(1 - \frac{k^+}{4\pi D'RC(R)}\right), \tag{9.6.54}$$

or, rearranging,

$$k^+ = \frac{4\pi D'RC(R)k^0}{4\pi D'RC(R) + k^0}. \tag{9.6.55}$$

To apply Eq. (9.6.55) to the generic bimolecular reaction A + B → products, we need to eliminate the coupling to the unimolecular process in Eq. (9.6.41). This can be done by setting $k_u = 0$. Referring to the expressions below Eq. (9.6.50) we see that this gives $\gamma = 0$, $\lambda_2 = 0$, $\lambda_1 = \alpha$, and

$$\alpha = k^+(\rho_B^{ss}D_B + \rho_A^{ss}D_A)/D_A D_B, \quad \beta = k^+(\rho_B^{ss} + \rho_A^{ss})/D'. \tag{9.6.56}$$

Making these substitutions in Eq. (9.6.55) yields

$$C(R) = \left[\frac{\alpha - 2\beta}{2(\alpha - \beta)}\exp(-R\beta^{1/2}) + \frac{\alpha}{2(\alpha - \beta)}\exp(-R\alpha^{1/2})\right]^{-1}. \tag{9.6.57}$$

In dilute solution, Eq. (9.6.56) shows that both α and β approach zero. Thus the exponential factors approach unity and $C(R) = 1$. In this case the rate constant can be written

$$k^+ = 4\pi D'Rk^0/(4\pi D'R + k^0). \tag{9.6.58}$$

This expression was first derived by Collins and Kimball using a generalization of Smoluchowski's method. When the intrinsic reaction rate constant is large, i.e., $k^0 \gg 4\pi D'R$, the Collins–Kimball expression reduces to

$$k^+ = 4\pi D'R, \tag{9.6.59}$$

which is the formula originally obtained by Smoluchowski.

In more concentrated solutions one must take into account the correction term, $C(R)$, in Eq. (9.6.57), and the expression for the diffusion-controlled rate constant becomes

$$k^+ = 4\pi D'RC(R). \tag{9.6.60}$$

Notice that the correlation lengths $\alpha^{-1/2}$ and $\beta^{-1/2}$ depend on the densities of A and B. According to Eq. (9.6.56) the correlation lengths are related to the root mean square distance that an A or B molecule diffuses in its bimolecular lifetime. Indeed, if A is in low concentration, then

$$\alpha^{-1/2} = (D_B/k^+\rho_A^{ss})^{1/2}, \quad \beta^{-1/2} = (D'/k^+\rho_B^{ss})^{1/2}. \tag{9.6.61}$$

The average lifetime of an A molecule before undergoing reaction with a B molecule is $\tau_B = 1/k^+\rho_B^{ss}$. Thus remembering formula (1.4.4) for the mean square distance that a molecule diffuses in time τ_B, we see that the correlation lengths are the square root of these distances for the B molecule and for the relative motion. The effect of these correlations is to increase the size of the rate constant with respect to the dilute solution value in Eq. (9.6.59). Indeed, using Eq. (9.6.57) it is not difficult, checking the special cases $\beta \geq \alpha$, $\beta < \alpha \leq 2\beta$, and $2\beta < \alpha$, to see that

$$1/C(R) \leq \exp(-\xi R) \leq 1, \tag{9.6.62}$$

where ξ is the smaller of $\alpha^{1/2}$ and $\beta^{1/2}$. Eq. (9.6.62) shows that the correction factor to Smoluchowski's result in Eq. (9.6.60) is always greater than or equal to one. Thus the diffusion-controlled rate constant is increased from its dilute solution value by a reduction in the spatial range over which correlations affect the radial distribution function.

There is a comparable effect due to the unimolecular lifetime that occurs even in dilute solution. Taking the limits $\rho_{\mathrm{B}}^{ss} \to 0$ and $\rho_{\mathrm{A}}^{ss} \to 0$ in Eq. (9.6.53) using the expressions below Eq. (9.6.50), one again obtains Eq. (9.6.57) with

$$\alpha^{-1/2} = (D_{\mathrm{A}}/k_u)^{1/2}, \quad \beta^{-1/2} = (D'/k_u)^{1/2}. \tag{9.6.63}$$

Since the unimolecular lifetime is $\tau_u = 1/k_u$, we see that the correlation lengths are the root mean square distances associated with the A molecule and the relative motion during the unimolecular lifetime of A. This, again, increases the size of the diffusion-controlled rate constant with respect to its limiting value in Eq. (9.6.59).

Both the unimolecular lifetime effect and the bimolecular lifetime effect are important in fluorescence quenching. According to the reaction scheme (9.6.41), A is the excited state of a fluorophore $\bar{\mathrm{A}}$ which undergoes fluorescence in the unimolecular reaction. The bimolecular reaction quenches the excited state and in such experiments B is referred to as a quencher molecule. The intensity of fluorescence at steady state is proportional to the steady-state density of the excited A molecules. According to Eq. (9.6.43) the ratio of ρ_{A}^{ss} in the absence of quencher to $\rho_{\mathrm{A}0}^{ss}$ in the presence of quencher is

$$\rho_{\mathrm{A}0}^{ss}/\rho_{\mathrm{A}}^{ss} = (k^+ \rho_{\mathrm{B}}^{ss} + k_u)/k_u. \tag{9.6.64}$$

Thus the ratio of the intensity of fluorescence, I_0, in the absence of quencher to that in the presence of quencher, I, is

$$I_0/I = 1 + (k^+/k_u)\rho_{\mathrm{B}}^{ss}. \tag{9.6.65}$$

Eq. (9.6.65) is called the *Stern–Volmer* equation and shows that a graph of I_0/I versus ρ_{B}^{ss} will have a slope equal to k^+/k_u. Since the fluorescence lifetime, $\tau_u = 1/k_u$, can be measured independently, the Stern–Volmer equation provides a way to determine k^+. If k^+ is a constant, then the graph will be a straight line. For a diffusion-controlled reaction we anticipate a positive curvature in these plots at higher concentrations due to the bimolecular lifetime effect. If the fluorescence lifetime is short, we also anticipate deviations from the Smoluchowski formula for k^+ due to the unimolecular lifetime effect.

To compare the expression for k^+ in Eq. (9.6.60) with experimental Stern–Volmer plots, we can simplify the analysis using the fact that the excited state is in low concentration. Setting ρ_A^{ss} equal to zero, we again recover Eq. (9.6.57) with

$$\alpha = (k_u + k^+ \rho_B^{ss})/D_A, \quad \beta = (k_u + k^+ \rho_B^{ss})/D'. \tag{9.6.66}$$

Substituting these expressions for α and β into Eq. (9.6.57) and the resulting expression for $C(R)$ into Eq. (9.6.60) gives the rate constant

$$k^+ = 4\pi D'R\{2D_B/[(D_B - D_A)\exp(-\beta^{1/2}R) + D'\exp(-\alpha^{1/2}R)]\}. \tag{9.6.67}$$

Because α and β depend on k^+, Eq. (9.6.67) must be solved for k^+. Although it is a transcendental equation, it can be solved rather easily by iteration. Given values of D_A, D_B, k_u, and ρ_B^{ss}, one uses as a first approximation to k^+ the value $k^{(1)} = 4\pi D'R$. Equations (9.6.66) then give values for $\alpha^{(1)}$ and $\beta^{(1)}$, which then substituted in Eq. (9.6.67) give the second approximation, $k^{(2)}$. Repeating this iteratively until the ratio $|k^{(n)} - k^{(n-1)}|/k^{(n-1)}$ is sufficiently small yields a good approximation to the solution

$$k^+ = \lim_{n\to\infty} k^{(n)}. \tag{9.6.68}$$

This iterative solution has been carried out for the quenching of the fluorescence of the organic molecule 9-vinylanthracene by molecular oxygen in the solvent dodecane. The results are plotted in Fig. 9.2 according to the Stern–Volmer equation (9.6.65) along with experimental values for the same

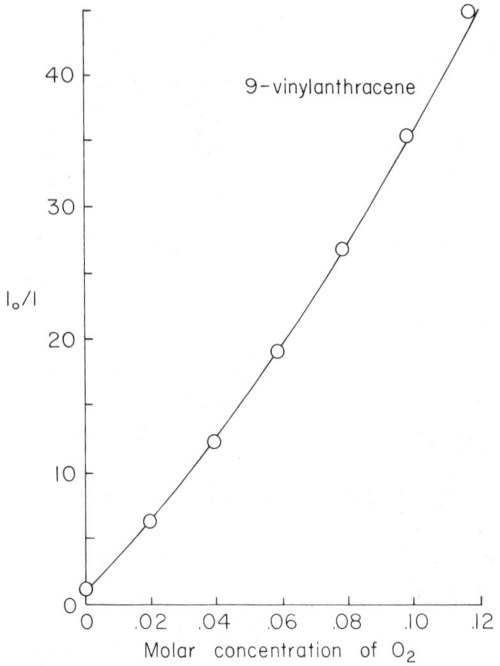

FIGURE 9.2. Stern-Volmer plot of fluorescence intensity for 9-vinylanthracene in the absence of the quencher oxygen (I_0) to the intensity at the indicated molar concentration (I). The line is based on Eq. (9.6.66) and the data are taken from J.R. Lakowicz and G. Weber, *Biochemistry*, **12**, 4161 (1973).

system. The diffusion constants for 9-vinylanthracene and O_2 have been estimated to be 9×10^{-6} cm^2 s^{-1} and 4×10^{-5} cm^2 s^{-1} in this solvent and the fluorescence lifetime $\tau_u = 1/k_u$ has been measured to be 11 ns. The full line in the figure was calculated by iteration using these values and a value for the encounter radius of $R = 3.6$ Å. The positive curvature in the Stern–Volmer plot is in good agreement with the calculated bimolecular lifetime effect. In dilute solution the correction factor to the Smoluchowski expression for the rate constant, $4\pi D'R$, is found to be 1.08. This reflects the short fluorescence lifetime of 9-vinylanthracene and is an example of the unimolecular lifetime effect discussed above.

References

Contractions of the Description

R.F. Fox, *Contributions to the Theory of Nonequilibrium Thermodynamics*, Doctoral dissertation, Rockefeller University, 1969.

M. Bixon and R. Zwanzig, Boltzmann–Langevin equation and hydrodynamic fluctuations, *Phys. Rev.* **187**, 267–272 (1969).

N.G. van Kampen, Elimination of fast variables, *Phys. Reports* **124**, 69–160 (1985).

J. Keizer, On the relationship between fluctuating irreversible thermodynamics and "extended" irreversible thermodynamics, *J. Stat. Phys.* **31**, 485–497 (1983).

Contracted Description of Brownian Motion

R.F. Fox and G.E. Uhlenbeck, Contributions to non-equilibrium thermodynamics. I. Theory of hydrodynamic fluctuations, *Phys. Fluids* **13**, 1893–1902 (1970).

E.H. Hauge and A. Martin-Löf, Fluctuating hydrodynamics and Brownian motion, *J. Stat. Phys.* **7**, 259–281 (1973).

D.H. Berman, The fluctuation–dissipation theorem for contracted descriptions of Markov processes, *J. Stat. Phys.* **20**, 57–81 (1979).

M. Medina-Noyola and A. Vizcarra-Rendon, Electrolyte friction and the Langevin equation for charged Brownian particles, *Phys. Rev. A* **32**, 3596–3605 (1985).

Contractions of the Boltzmann Equation

R.F. Fox and G.E. Uhlenbeck, Contributions to non-equilibrium thermodynamics. II. Fluctuation theory for the Boltzmann equation, *Phys. Fluids*, **13**, 2881–2890 (1970).

K.T. Mashiyama and H. Mori, Origin of the Landau-Lifshitz hydrodynamic fluctuations in nonequilibrium systems and a new method for reducing the Boltzmann equation, *J. Stat. Phys.* **18**, 385–407 (1978).

The Linearized Boltzmann Equation

S. Chapman and T.G. Cowling, *The Mathematical Theory of Non-Uniform Gases* (Cambridge University, Cambridge, 1970).

P. Resibois and M. de Leener, *Classical Kinetic Theory of Fluids* (Wiley-Interscience, New York, 1977).

G.E. Uhlenbeck and G.W. Ford, *Lectures in Statistical Mechanics* (American Mathematical Society, Providence, 1963), Chapters IV–VI.

L. Waldman, in *Handbuch der Physik*, Vol. 12, S. Flugge, ed. (Springer-Verlag, Berlin, 1958), pp. 366–493.

C.S. Wang Chang and G.E. Uhlenbeck, The kinetic theory of gases, in *Studies in Statistical Mechanics*, Vol. 5, J. de Boer and G.E. Uhlenbeck, eds. (North-Holland, Amsterdam, 1970).

Diffusion Effects on Chemical Reactions

M. von Smoluchowski, Versuch einer mathematischen theorie der koagulationskinetik kolloider lösungen, *Z. Phys. Chem.* **92**, 129–168 (1917).

R.M. Noyes, Effects of diffusion rates on chemical kinetics, *Prog. React. Kin.* **1**, 128–160 (1961).

J. Keizer, Nonequilibrium statistical thermodynamics and the effect of diffusion on chemical reaction rates, *J. Phys. Chem.* **86**, 5052–5067 (1982).

J. Keizer, Diffusion effects on rapid bimolecular chemical reactions, *Chem. Rev.* **87**, 167–180 (1987).

CHAPTER 10

Nonstationary Processes: Transients, Limit Cycles, and Chaotic Trajectories

10.1. Introduction

In the preceding chapters we have focused attention on the statistical thermodynamics of systems which are near equilibrium or near stable, nonequilibrium steady states. There are, however, many important examples of nonequilibrium systems that are nonstationary and exhibit nonlinear transients, periodic orbits, and bounded, aperiodic motion. The time course of variables in nonstationary systems is usually studied with the aid of differential equations. For macroscopic systems these equations correspond to the conditional average, and the three types of nonstationary behavior just mentioned represent the nonstationary average trajectories in physical ensembles. In Section 5.4 we have already studied an example of this sort, namely, the nonlinear isomerization reaction with the mechanism

$$\mathrm{A} + \mathrm{B} \rightleftarrows 2\mathrm{B}. \tag{10.1.1}$$

There we found that the average density of B molecules, $\bar{\rho}_{\mathrm{B}}(\rho_{\mathrm{B}}^{0}, t)$, conditioned on the initial density, ρ_{B}^{0}, solved the canonical differential equation, which has the solution

$$\bar{\rho}_{\mathrm{B}}(\rho_{\mathrm{B}}^{0}, t) = \rho_{\mathrm{B}}^{e}[1 - (1 - \rho_{\mathrm{B}}^{e}/\rho_{\mathrm{B}}^{0})\exp(-\lambda t)]^{-1}. \tag{10.1.2}$$

If ρ_{B}^{0} differs significantly from the asymptotic equilibrium value, ρ_{B}^{e}, then the average trajectory is strongly affected by the nonlinearities in the canonical equation.

The fact that nonstationary trajectories depend on the initial condition and time, often in a complex fashion, makes the calculation of fluctuations around the conditional average difficult. If the average equations are linear, then, as we saw in Section 5.3, the covariance matrix can be expressed analytically in terms of the average trajectories and the relaxation times. We were even able to do this for the nonlinear isomerization reaction in Eq. (10.1.1), although the analytical expression for the conditional variance, $\sigma(\rho_{\mathrm{B}}^{0}, t)$, in Eq. (5.4.16) is a bit complicated. Nonetheless, for initial nonequilibrium conditions, i.e., $\rho_{\mathrm{B}}^{0} \neq \rho_{\mathrm{B}}^{e}$, the variance of the density fluctuations, as shown in Fig. 5.4, differs significantly from the linear, equilibrium behavior. On the basis of this calculation, we can anticipate a rich behavior for the statistical properties of nonstationary systems.

The systematic investigation of the average behavior of periodic and aperiodic systems is relatively recent. The average behavior of such systems is sufficiently fascinating that little experimental work has been reported on fluctuations in these systems. Moreover, even though the theoretical framework of statistical nonequilibrium thermodynamics developed in Chapter 4 is applicable to nonstationary phenomena, calculations with the theory are difficult. Thus there are few real physical systems for which the statistical behavior is understood even at the thermodynamic level of description. Because of this, our description of nonstationary systems in this chapter is necessarily incomplete. Indeed, in this chapter we simply introduce some general methods that can be applied to nonstationary systems and describe some of the interesting nonstationary behavior that is already known.

From the mechanistic point of view, the first step in understanding the behavior of any system is to uncover the molecular processes which dominate the behavior of the system. Because of the complex processes which occur in periodic and aperiodic systems, this is often a difficult task. In the following sections we will discuss three systems that have yielded to this approach. One is the homogeneous Belousov–Zhabotinsky chemical reaction mixture in which nonlinear transients, oscillations, and chaotic behavior have been observed. The second is a heterogeneous electrochemical oscillator called the beating mercury heart, and the third is the pancreatic β-cell, a living oscillatory system, which is found in the islets of Langerhans.

10.2. Nonstationary Systems and Nonlinear Transients

Systems that can support a stationary ensemble may exhibit nonstationary behavior during the approach to steady state. This is the case even for a system near equilibrium if the single-time probability density, W_1, differs from its equilibrium form. In a nonstationary ensemble, the single-time probability density depends explicitly on time, while the two-time conditional probability density, P_2, depends explicitly on two times. The time evolution of W_1 is determined by P_2 and if $\mathbf{n}$ represents a vector of extensive variables or their densities, then Eq. (1.3.20) gives

$$W_1(\mathbf{n}, t) = \int W_1(\mathbf{n}^0, t^0) P_2(\mathbf{n}^0, t^0 | \mathbf{r}, t)\, d\mathbf{n}^0. \tag{10.2.1}$$

In Section 7.3 we used this expression to discuss the evolution of a nonstationary initial probability density in an ensemble with multiple steady states. Because the conditional average trajectory $\bar{\mathbf{n}}(\mathbf{n}^0, t)$ satisfies the macroscopic equations of motion, the conditional probability remains concentrated on the domain of attraction occupied by $\mathbf{n}^0$. As time proceeds, the single-time probability density breaks up into a sum of Gaussians centered at the stable steady states and weighted by the initial fraction of the probability density in each domain of attraction. When only one domain of attraction exists, a unique stationary probability density exists in the stationary ensemble. This analysis depends on the existence of domains of attraction that are large with respect to molecular fluctuations. If a domain of attraction is too small, which, as we noted in Section 7.5, is the case close to a critical point, then this picture breaks down.

A system of nonlinear ordinary or partial differential equations can exhibit behavior that is significantly more complicated than multiple steady states with large domains of attraction. Orbits called limit cycles, which are closed one-dimensional trajectories corresponding to sustained oscillations, can exist as well as orbits called chaotic attractors, which are continuous, locally dense, and nonperiodic. Under appropriate conditions, this rich behavior of nonlinear differential equations may represent the average trajectories of real physical, chemical, or biological systems.

Since molecular fluctuations are coupled to the average, it is important to understand the way in which nonstationary dynamical processes affect the fluctuations. To begin this, we recall that the two-time joint probability density is given by

$$W_2(\mathbf{n}, t; \mathbf{n}', t') = W_1(\mathbf{n}, t) P_2(\mathbf{n}, t | \mathbf{n}', t'). \tag{10.2.2}$$

From Eqs. (10.2.1) and (10.2.2) it follows that to understand two-time measurements we need only know the probability density, $W_1(\mathbf{n}^0, t^0)$, at some initial time, t_0, and the conditional probability density. In the limit of a large system, the behavior of P_2 for an arbitrary ensemble is given in Section 4.4. The conditional probability density is Gaussian with an average satisfying an equation of the form

$$d\bar{\mathbf{n}}/dt = \sum_{\kappa} \boldsymbol{\omega}_\kappa (\bar{V}_\kappa^+ - \bar{V}_\kappa^-) + \bar{S} \equiv \mathbf{R}(\bar{\mathbf{n}}, t), \tag{10.2.3}$$

where $\bar{S}$ represents the nondissipative contributions to the time derivative, which may depend explicitly on the time. Equation (10.2.3) is to be solved with the initial condition $\bar{\mathbf{n}}(\mathbf{n}^0, t^0) \equiv \mathbf{n}^0$, corresponding to the fact that $\bar{\mathbf{n}}$ represents the conditional average. The covariance matrix of the Gaussian conditional density also satisfies a differential equation, namely,

$$d\sigma/dt = H\sigma + \sigma H^T + \gamma, \tag{10.2.4}$$

where γ is the non-negative definite, symmetric matrix

$$\gamma_{ij}(\bar{\mathbf{n}}(\mathbf{n}^0, t)) \equiv \sum_{\kappa} \omega_{\kappa i} (\bar{V}_\kappa^+ + \bar{V}_\kappa^+) \omega_{\kappa j} \tag{10.2.5}$$

and the matrix H is the Jacobian of $\mathbf{R}$, i.e.,

$$H_{ij}(\bar{\mathbf{n}}(\mathbf{n}^0, t)) = \partial R_i(\bar{\mathbf{n}}, t)/\partial \bar{n}_j. \tag{10.2.6}$$

Because the conditional probability density at $t = t_0$ is a delta function, i.e., $P_2(\mathbf{n}^0, t^0|\mathbf{n}, t^0) = \delta(\mathbf{n} - \mathbf{n}^0)$, the initial conditions for solving Eq. (10.2.4) are $\sigma_{ij}(\mathbf{n}^0, t^0) = 0$.

While Eqs. (10.2.3)–(10.2.6) provide a means of describing the time dependence of the fluctuations, in practice the equations are impossible to solve in terms of known functions when the rate expression $\mathbf{R}(\bar{\mathbf{n}}, t)$ is nonlinear. Thus one is reduced to solving the equations numerically. There are a number of computational methods which generate reliable numerical solutions to nonlinear ordinary differential equations. One of the most versatile was developed by Gear. It has the feature that it can be used with equations that involve widely differing time scales. As a consequence, it can be used to treat a variety of nonstationary problems.

To obtain the conditional probability density one needs to solve both Eqs. (10.2.3) and (10.2.4). These are coupled equations since the matrix H and the matrix γ in Eq. (10.2.4) depend on the solution to Eq. (10.2.3). The coupling between the equations, however, is only in one direction and the average equation (10.2.4) does not depend on the solution to the equation for the covariance. As a consequence, the average equation can be solved first and then used as input in the solution of the covariance equation.

There are two ways in which the covariance equation can be simplified before it is integrated. The first is to eliminate the redundancy $\sigma_{ij} = \sigma_{ji}$ caused by the symmetry of the covariance matrix. The diagonal and upper off-diagonal elements suffice to specify a symmetric matrix, which means that Eq. (10.2.4) contains only $k(k + 1)/2$ independent equations. To eliminate the extra equations we use the column vector $\boldsymbol{\sigma}$ with the $k(k + 1)/2$ components

$$\sigma_{11}, \sigma_{12}, \ldots, \sigma_{1k}, \sigma_{22}, \sigma_{23}, \ldots, \sigma_{2k}, \ldots, \sigma_{kk}. \tag{10.2.7}$$

In terms of this vector Eq. (10.2.4) takes the form

$$d\boldsymbol{\sigma}/dt = h\boldsymbol{\sigma} + \mathbf{g}, \tag{10.2.8}$$

where h is a $k(k + 1)/2 \times k(k + 1)/2$ matrix with components that are linear combinations of the elements of H, and the $k(k + 1)/2$ column vector $\mathbf{g}$ has components

$$\gamma_{11}, \gamma_{12}, \ldots, \gamma_{1k}, \gamma_{22}, \gamma_{23}, \ldots, \gamma_{2k}, \ldots, \gamma_{kk}. \tag{10.2.9}$$

A further simplification comes from the fact, discussed in Sections 4.7 and 4.8, that fluctuations in the extensive variables scale with the square root of the size of a system, Ω. As we have seen, this implies that the matrix γ scales in proportion to the size of a system. Thus if the vector $\mathbf{n}$ represents extensive variables, then the vector $\mathbf{g}$ is proportional to Ω, whereas if $\mathbf{n}$ represents the densities of extensive variables, $\mathbf{g}$ is proportional to Ω^{-1}. We can eliminate this scale factor by defining $\bar{\boldsymbol{\sigma}} = \boldsymbol{\sigma}\Omega^{-1}$, if $\mathbf{n}$ represents extensive variables, or $\bar{\boldsymbol{\sigma}} = \boldsymbol{\sigma}\Omega$, if $\mathbf{n}$ represents densities. In either case Eq. (10.2.8) implies that

$$d\bar{\boldsymbol{\sigma}}/dt = h\bar{\boldsymbol{\sigma}} + \bar{\mathbf{g}}, \tag{10.2.10}$$

where $\bar{\mathbf{g}} \equiv \Omega^{-1}\mathbf{g}$ or $\bar{\mathbf{g}} = \Omega\mathbf{g}$, depending on whether $\mathbf{n}$ represents extensive variables or densities of extensive variables. None of the quantities in Eq. (10.2.10) depend on the system size, and the solution $\bar{\boldsymbol{\sigma}}$ will provide the properly scaled covariances in the limit that $\Omega \to \infty$.

As an example of this procedure, consider the *o*-cresolphthalein (OCP) ionization reaction discussed in Section 7.4,

$$\text{OCP} \rightleftarrows \text{OCP}^- + \text{H}^+. \tag{10.2.11}$$

When the basic form is stimulated to absorb light, the average values of the number of ionized *o*-cresolphthalein molecules, N, and the internal energy of the system, E, satisfy Eqs. (7.4.25) and (7.4.26), i.e.,

$$\begin{aligned} d\bar{N}/dt &= Vk^-[K(\rho_0 - \bar{\rho}) - \bar{\rho}_H\bar{\rho}] \\ d\bar{E}/dt &= A(\bar{\rho})I_0 - \beta(\bar{T} - T_R), \end{aligned} \tag{10.2.12}$$

where I_0 is the intensity of the radiation, T is the temperature, ρ is the density of OCP, and ρ_H is the density of hydrogen ions. Under appropriate conditions this system exhibits multiple steady states, as shown in Fig. 7.6. Kramer and Ross have used a method which allowed them to examine the dynamics of systems that are initially close to the central unstable state. This produces nonlinear transients as the system relaxes towards one or the other of the two stable states. Several examples of these transients are shown in Fig. 7.7. This figure shows the measured increase in absorbance as the high-absorbance steady state is approached from three points near the unstable steady state. The variance in absorbance as a function of time from these three points is shown in Fig. 10.1. Notice that there is a large initial variance due to the experimental uncertainty in establishing the initial state. This corresponds to a broad initial distribution, $W_1(\mathbf{n}, 0)$. This ambiguity increases transiently and then decreases to the limit of the experimental apparatus as the stable steady state is approached.

A simplified analysis of this transient phenomenon was given in Section 7.4. Using the full nonlinear equations and the numerical methods described above, a more complete analysis of the transients can be given. If we choose as extensive variables the number and energy densities, ρ and e, then the average equations (10.2.12) are

$$\begin{aligned} d\bar{\rho}/dt &- k^-[K(\bar{e})(\rho_0 - \bar{\rho}) - \bar{\rho}_H\bar{\rho}] \\ d\bar{e}/dt &= A(\bar{\rho})i_0 - \bar{\beta}(\bar{e} - e_R), \end{aligned} \tag{10.2.13}$$

where for simplicity we have written $E = C_V T$, with C_V the constant volume heat capacity. Thus $\bar{\beta} = \beta/C_V$, $i_0 = I_0/V$, $e_R = C_V T_R/V$, with V the volume of the system. The equilibrium constant, $K(\bar{e})$, and the absorbance, $A(\bar{\rho})$, have the form

$$\begin{aligned} K(\bar{e}) &= D\exp(-B/\bar{e}) \\ A(\bar{\rho}) &= 1 - \exp(-F\bar{\rho}), \end{aligned} \tag{10.2.14}$$

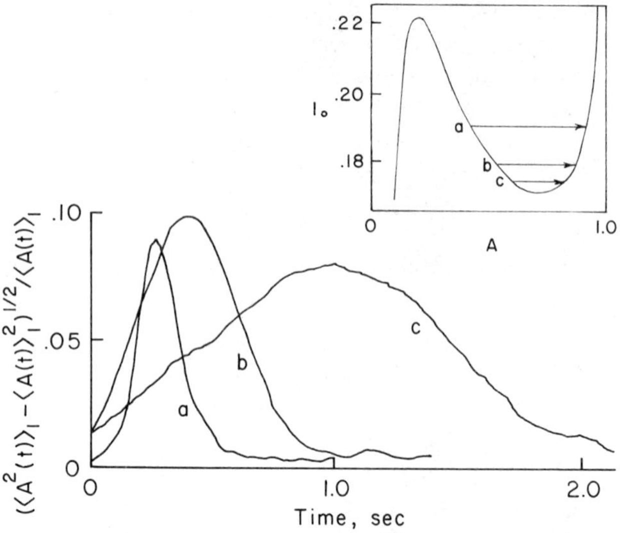

FIGURE 10.1. The standard deviation of the absorbance in the *o*-cresolphthalein system as a function of time. The initial conditions for the curves *a*, *b*, and *c* are close to the unstable steady states, as shown in the inset. The large size of the variance can be attributed to uncertainties in the initial conditions. Data taken from J. Kramer and J. Ross, *J. Chem. Phys.* **83**, 6234 (1985).

with B, D, and F constants. As usual, the fluctuations satisfy linearized equations based on Eq. (10.2.13), which are

$$\begin{aligned} d\delta\rho/dt &= -k^-[K(\bar{e}) + \bar{\rho}_{\mathrm{H}}]\delta\rho - \frac{k^- BK(\bar{e})}{\bar{e}^2}\delta e + \tilde{f}_1 \\ d\delta e/dt &= F\exp(-F\bar{\rho})i_0\delta\rho - \bar{\beta}\delta e + \tilde{f}_2, \end{aligned} \tag{10.2.15}$$

and the covariance of the random terms has the form $\gamma\delta(t - t)$, with

$$\gamma = \begin{pmatrix} k^-[K(\bar{e})(\bar{\rho}_0 - \bar{\rho}) + \bar{\rho}_{\mathrm{H}}\bar{\rho}]V^{-1} & 0 \\ 0 & 2\bar{\beta}k_B e_R^2/C_V \end{pmatrix}. \tag{10.2.16}$$

The entries in the matrix in Eq. (10.2.16) have the canonical form, with the diagonal energy term taken from Eq. (7.4.32) and rewritten in the present notation. The off-diagonal entries vanish because the chemical reaction conserves energy and heat transport conserves molecule numbers.

To apply the numerical procedure described above to this problem, we need to integrate Eqs. (10.2.13) and use that solution to obtain the terms in Eq. (10.2.4) for the covariance. The missing ingredient in that equation is the matrix H, which can be read off Eq. (10.2.15), i.e.,

$$H = \begin{pmatrix} -k^-[K(\bar{e}) + \bar{\rho}_{\mathrm{H}}] & -k^- BK(\bar{e})/\bar{e}^2 \\ F\exp(-F\bar{\rho})i_0 & -\bar{\beta} \end{pmatrix}. \tag{10.2.17}$$

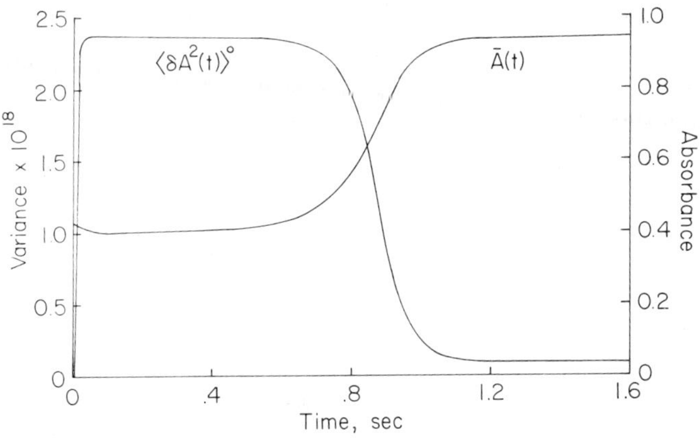

FIGURE 10.2. The conditional variance and conditional average absorbance as calculated from Eqs. (10.2.13)–(10.2.16) for the *o*-cresolphthalein system of Kramer and Ross. The initial conditions are similar to those for curve *a* in Fig. 10.1. The small size of the covariance shows that conditional fluctuations are not responsible for the experimental observations in Fig. 10.1.

Explicitly writing out Eq. (10.2.4) for this problem we find that the scaled covariances $\bar{\sigma}_1 \equiv \sigma_{11}V$, $\bar{\sigma}_2 \equiv \sigma_{12}V$, $\bar{\sigma}_3 \equiv \sigma_{22}V$ satisfy the three coupled equations

$$\begin{aligned} d\bar{\sigma}_1/dt &= 2H_{11}\bar{\sigma}_1 + 2H_{12}\bar{\sigma}_2 + \bar{g}_1 \\ d\bar{\sigma}_2/dt &= H_{21}\bar{\sigma}_1 + (H_{11} + H_{22})\bar{\sigma}_2 + H_{12}\bar{\sigma}_3 \\ d\bar{\sigma}_3/dt &= 2H_{21}\bar{\sigma}_2 + 2H_{22}\bar{\sigma}_3 + \bar{g}_3, \end{aligned} \tag{10.2.18}$$

where $\bar{g}_1 \equiv \gamma_{11}V$, $\bar{g}_3 \equiv \gamma_{22}V$.

It is easy to generate solutions to Eqs. (10.2.15) and (10.2.18) using the Hindmarsh version of the Gear algorithm. Results of such a calculations are given in Fig. 10.2, which illustrates both average values of the absorbance, $A(\bar{\rho})$, and its conditional fluctuations. The fluctuations in A are related to $\bar{\sigma}_1$ by the relationship

$$\begin{aligned} \langle \delta A^2(t) \rangle^0 &= (dA/d\bar{\rho})^2 \langle \delta\rho^2(t) \rangle^0 \\ &= (dA/d\bar{\rho})^2 V^{-1} \bar{\sigma}_1(t), \end{aligned} \tag{10.2.19}$$

where the superscript zero reminds us that these are conditional fluctuations with $\bar{\sigma}_1(0) = \bar{\sigma}_2(0) = \bar{\sigma}_3(0) \equiv 0$. Most of the parameters in this calculation are taken from the experiments of Kramer and Ross, while the rate constant was set equal to its diffusion-controlled value of 5×10^{13} cm^3 mol^{-1} s^{-1}. The initial conditions for the averages have been set within a few percent of the values of the unstable steady state. Notice that the variance in the absorbance is extremely small, even though it exhibits a strongly nonexponential relaxation to the steady state. It is obvious from this figure that the variances observed by Kramer and Ross in Fig. 10.1 for relaxation to the steady state cannot be explained by an anomalous growth of fluctuations associated with the conditional density, P_2. However, they can be explained by the propagation of fluctuations in the initial probability density.

In order to study this effect, we need to use the evolution equation

$$W_1(\mathbf{n}, t) = \int W_1(\mathbf{n}^0, t^0) P_2(\mathbf{n}^0, t^0 | \mathbf{n}, t)\, d\mathbf{n}^0. \tag{10.2.20}$$

Since for the o-cresolphthalein system we have generated a numerical solution for the Gaussian conditional density, Eq. (10.2.20) can be integrated directly to obtain the single-time probability density. The result, of course, depends on $W_1(\mathbf{n}^0, t^0)$. To agree with the experimental situation we have assumed that $W_1(\mathbf{n}^0, t^0)$ is uniform on a region with a width in density of 0.76% of the initial value. The average values are taken in the domain of attraction of the high-absorbance state and very close to the unstable state, as described in the previous paragraph. The resulting variance of the absorbance is shown as a function of time in Fig. 10.3. Notice that the variance increases to a maximum value and then decreases as the steady state is approached, in agreement with the experimental results in Fig. 10.1.

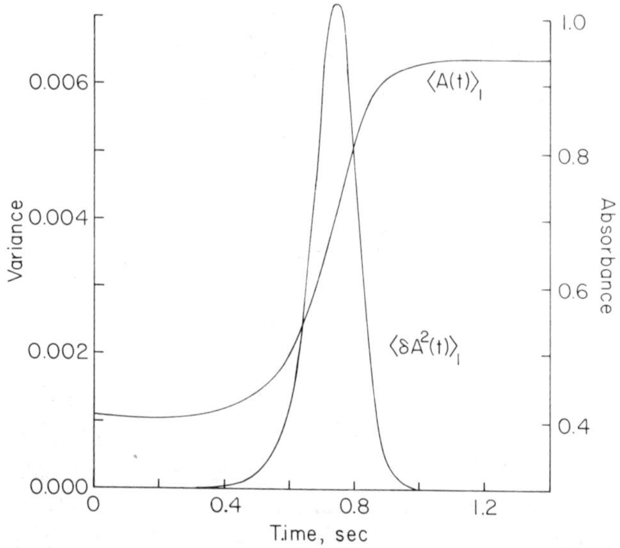

FIGURE 10.3. The single-time average and variance of the absorbance for the o-cresolphthalein system of Kramer and Ross based on Eqs. (10.2.21) and (10.2.26). The initial probability density is uniform on a region close to the initial point a in the inset of Fig. 10.1.

To understand this behavior, we need to analyze how the mean and covariance propagate as a function of time. Multiplying both sides of Eq. (10.2.20) by $\mathbf{n}$ and integrating over $\mathbf{n}$, we see that the average is given by

$$\langle \mathbf{n}(t) \rangle_1 = \int W_1(\mathbf{n}^0, t^0)\bar{\mathbf{n}}(\mathbf{n}^0, t)\, d\mathbf{n}^0, \tag{10.2.21}$$

where the subscript emphasizes that this is the single-time average. Thus the single-time average equals the average over the initial condition of the conditional average. The covariance of the fluctuations for the single-time probability density is also defined in the usual way and is

$$\begin{aligned}\langle \delta\mathbf{n}(t)\delta\mathbf{n}^T(t) \rangle_1 &\equiv \langle [\mathbf{n}(t) - \langle \mathbf{n}(t) \rangle_1][\mathbf{n}(t) - \langle \mathbf{n}(t) \rangle_1]^T \rangle_1 \\ &= \langle \mathbf{n}(t)\mathbf{n}^T(t) \rangle_1 - \langle \mathbf{n}(t) \rangle_1 \langle \mathbf{n}^T(t) \rangle_1. \end{aligned} \tag{10.2.22}$$

Like the average, the first term on the right-hand side of Eq. (10.2.22) can be written in terms of conditional averages by multiplying Eq. (10.2.20) by $\mathbf{n}\mathbf{n}^T$ and integrating. This gives

$$\langle \mathbf{n}(t)\mathbf{n}^T(t) \rangle_1 = \int W_1(\mathbf{n}^0, t^0)\overline{\mathbf{n}(t)\mathbf{n}^T(t)}^0\, d\mathbf{n}^0, \tag{10.2.23}$$

where we have introduced the notation

$$\overline{\mathbf{n}(t)\mathbf{n}^T(t)}^0 \equiv \int P_2(\mathbf{n}^0, t^0|\mathbf{n}, t)\mathbf{n}\mathbf{n}^T\, d\mathbf{n} \tag{10.2.24}$$

for the conditional covariance of $\mathbf{n}$. This covariance can, in turn, be written in terms of the covariance of the conditional fluctuations. If we substitute $\mathbf{n}(t) = \bar{\mathbf{n}}^0(t) + \delta\mathbf{n}(t)$, where $\bar{\mathbf{n}}^0(t) \equiv \bar{\mathbf{n}}(\mathbf{n}^0, t)$, into Eq. (10.2.24), we find that

$$\overline{\mathbf{n}(t)\mathbf{n}^T(t)}^0 = \bar{\mathbf{n}}^0(t)\bar{\mathbf{n}}^{0T}(t) + \sigma^0(t^0, t), \tag{10.2.25}$$

where $\sigma^0(t^0, t) = \langle \delta\mathbf{n}(t)\delta\mathbf{n}(t) \rangle^0$ is the conditional covariance. Substituting Eqs. (10.2.25), (10.2.23) and (10.2.21) into Eq. (10.2.22) then gives

$$\begin{aligned}\langle \delta\mathbf{n}(t)\delta\mathbf{n}^T(t) \rangle_1 = &\int W_1(\mathbf{n}^0, t^0)\sigma^0(t^0, t)\, d\mathbf{n}^0 + \int W_1(\mathbf{n}^0, t^0)\bar{\mathbf{n}}^0(t)\bar{\mathbf{n}}^{0T}(t)\, d\mathbf{n}^0 \\ &- \int W_1(\mathbf{n}^0, t^0)\bar{\mathbf{n}}^0(t)\, d\mathbf{n}^0 \int W_1(\mathbf{n}^0, t^0)\bar{\mathbf{n}}^{0T}(t)\, d\mathbf{n}^0. \end{aligned} \tag{10.2.26}$$

Equation (10.2.26) separates the covariance of the single-time fluctuations into two distinct contributions. The first term represents the covariance of the fluctuations in a conditional ensemble averaged over the initial distribution of fluctuations. The remaining two terms represent the way in which the conditional average trajectories propagate the initial covariance. This is obvious at $t = t^0$, since then the conditional covariance, σ^0, vanishes and only the second and third terms remain. As we saw in the o-cresolphthalein system, the conditional covariance is often negligible. In this case, the first term in Eq. (10.2.26) can be set equal to zero and one has

$$\langle \delta\mathbf{n}(t)\delta\mathbf{n}^T(t)\rangle_1 = \int W_1(\mathbf{n}^0, t^0)\bar{\mathbf{n}}^0(t)\bar{\mathbf{n}}^{0T}(t)\,d\mathbf{n}^0 - \int W_1(\mathbf{n}^0, t^0)\bar{\mathbf{n}}^0(t)\,d\mathbf{n}^0 \int W_1(\mathbf{n}^0, t^0)\bar{\mathbf{n}}^{0T}(t)\,d\mathbf{n}^0. \quad (10.2.27)$$

Under this condition the change in the variance is determined exclusively by the conditional average and the initial distribution of probability.

As long as the conditional covariance is small with respect to the widths of the single-time probability density, it is possible to obtain an accurate approximation to $W_1(\mathbf{n}, t)$ in terms of $W_1(\mathbf{n}^0, t^0)$ and the conditional average. Since the conditional density is Gaussian, it can be approximated by a delta function centered at $\bar{\mathbf{n}}(\mathbf{n}^0, t)$ whenever the covariance is sufficiently small. Thus the evolution equation (10.2.20) can be written

$$W_1(\mathbf{n}, t) = \int W_1(\mathbf{n}^0, t^0)\delta(\mathbf{n} - \bar{\mathbf{n}}(\mathbf{n}^0, t))d\mathbf{n}^0. \quad (10.2.28)$$

To perform this integral, we assume that the functional relationship between $\bar{\mathbf{n}}$ and $\mathbf{n}^0$ is invertible. Thus we can write

$$\mathbf{n}^0(\bar{\mathbf{n}}, t) = \mathbf{n}^0 \quad \text{and} \quad d\mathbf{n}^0 = |\partial\mathbf{n}^0/\partial\bar{\mathbf{n}}|\,d\bar{\mathbf{n}}, \quad (10.2.29)$$

where $|\partial\mathbf{n}^0/\partial\bar{\mathbf{n}}|$ is the absolute value of the determinant of the Jacobian matrix. With the change of variable from $\mathbf{n}^0$ to $\bar{\mathbf{n}}$ the integral over the delta function is straightfoward and gives

$$W_1(\mathbf{n}, t) = W_1(\mathbf{n}^0(\mathbf{n}, t), t^0)|\partial\mathbf{n}^0/\partial\mathbf{n}|. \quad (10.2.30)$$

According to Eq. (10.2.30), the probability density at $\mathbf{n}$ at time t is the same as its pre-image point $\mathbf{n}^0$ at time t^0 multiplied by a scale factor which takes into account the change in volume of state space in the interval t^0 to t.

As an example of this, consider the bimolecular dimerization reaction discussed in the preceding section. If we take the number density of B to be the variable of interest, then $n = \rho_B$ and the solution to the conditional average equation is given in Eq. (10.1.2). Inverting this equation gives

$$n^0 = n^e[1 - (1 - n^e/n)\exp(\lambda t)]^{-1}, \quad (10.2.31)$$

where n^e is the equilibrium density and $t^0 = 0$. The Jacobian is

$$|dn^0/dn| = \pm(n^e/n)^2\exp(\lambda t)[1 - (1 - n^e/n)\exp(\lambda t)]^{-2} \quad (10.2.32)$$

with the + or − sign depending on the denominator. Notice that the denominator vanishes in Eqs. (10.2.31) and (10.2.32) whenever

$$n = n^*(t) = n_e/[1 - \exp(-\lambda t)], \tag{10.2.33}$$

which is always greater than or equal to n^e. Thus the + sign is appropriate for values of $n < n^*(t)$ and the − sign for $n > n^*(t)$. According to Eq. (10.2.30), the single-time probability density at time t can be written

$$W_1(n,t) = W_1(n^e|1 - (1 - n^e/n)e^{\lambda t}|^{-1}, 0)(n^e/n)^2 e^{\lambda t}|1 - (1 - n^e/n)e^{\lambda t}|^{-2}, \tag{10.2.34}$$

except at $n = n^*(t)$ where the denominators vanish.

Graphs of the single-time probability density in Eq. (10.2.34) for $t = 2/\lambda$, $5/\lambda$, and $10/\lambda$ based on an initial distribution that is uniform on $0.1 \leq n/n^e \leq 0.25$ are given in Fig. 10.4. At $t = 2/\lambda$ the probability distribution has broadened considerably since the members of the ensemble with initial values of n near 0.1 lag behind those systems with n^0 near 0.25. At $t = 5/\lambda$ all the systems are close to equilibrium and by $t = 10/\lambda$, the distribution is so sharp it cannot be distinguished from a vertical line on the scale of this graph. Nonetheless, even at $t = 10/\lambda$ the distribution is not so sharp that molecular fluctuations are yet important. As time proceeds, however, the single-time variance becomes comparable to the conditional variance and at that point Eq. (10.2.20) must be used to calculate the probability density without introducing approximations.

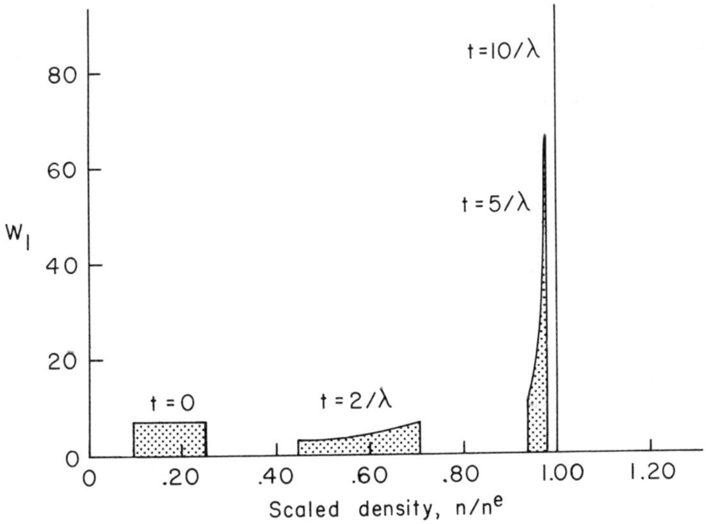

FIGURE 10.4. The time course of the single-time probability density for the reaction A + B ⇄ 2B based on the approximation in Eq. (10.2.34). The probability density first spreads and then contracts as the final equilibrium density is approached. Time is measured in multiples of the characteristic time $1/\lambda$.

10.3. Limit Cycle Oscillations

Many systems exhibit persistent nonstationary behavior. Of the variety of nonstationary behaviors that have been observed, perhaps the most striking are undamped oscillations in dissipative systems. While oscillations in simple mechanical systems, such as a pendulum, are easy to understand, oscillations in chemical and biological systems may seem mysterious since the underlying equations of motion are purely dissipative. Of course all observable systems are dissipative to some extent, and even a pendulum clock uses energy stored in a spring, transmitted through the escapement, to keep regular time.

The first oscillating chemical reaction to be understood mechanistically involved the oxidation and bromination of malonic acid. The reaction mixture is made up of a dilute aqueous solution of BrO_3^- (bromate), Br^- (bromide), cerium ions, and $CH_2(COOH)_2$ (malonic acid) that is about 1 *M* in sulfuric acid. The oscillations are easily visualized by adding a few drops of the oxidation–reduction indicator ferroin. This gives the solution an orange color when most of the cerium is Ce^{4+} and a blue color when it is mostly Ce^{3+}. This interesting reaction mixture was discovered accidentally by a Russian biochemist, B.P. Belousov. After being suppressed from the scientific literature for over a decade, it was finally brought to the attention of a wide audience by A.M. Zhabotinsky. The reaction is now usually referred to as the Belousov or the Belousov–Zhabotinsky reaction.

When the Belousov reaction is carried out in a flask, oscillations can be observed in the concentration of Br^- using a bromide-sensitive electrode and in the Ce^{3+}/Ce^{4+} ratio using a platinum electrode. The oscillations appear after a brief induction period and have an amplitude and frequency that depends on the initial concentrations in the solution. After a period of time that can be as long as an hour, the oscillations begin to decrease in amplitude and finally disappear. It is possible to maintain the oscillations *in perpetuity* by feeding the reactants in separate input lines into a continuously stirred tank reactor (CSTR), as described in Section 5.5. An example of the sort of oscillations that are observed in a CSTR is given in Fig. 10.5. That figure shows oscillations, not in the Belousov reaction mixture, but in the so-called chlorite–thiosulfate oscillator. This system consists of separate inflows of ClO_2^- (chlorite) and $S_2O_3^{2-}$ (thiosulfate) in aqueous solution at pH = 4. These oscillations will persist as long as the reaction tank is continuously fed and stirred. Eight different types of oscillations are shown, each corresponding to a different residence time, or pumping rate, for the reactor.

Another type of system that undergoes oscillations is the so-called beating mercury heart. This is an electrochemical–mechanical oscillator that has been known in various forms for over 100 years. In its simplest form it consists of a drop of mercury in a watch glass that has been covered by a 2–3 *M* aqueous solution of sulfuric acid containing dilute dichromate ions. When a rigidly

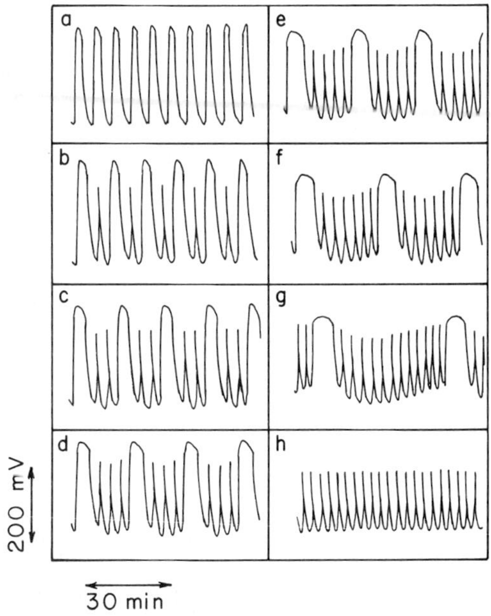

FIGURE 10.5. Oscillations in the chlorite–thiosulfate reaction as detected by electrode potential measurements in a CSTR. Panels a–h show the complex oscillations that are observed when the residence time is increased in steps from 360 s to 2960 s. From I. Epstein, *J. Phys. Chem.* **88**, 187 (1984).

fixed iron nail is adjusted to touch the side of the mercury, the mercury is set into an oscillatory motion that is accompanied by oscillations in the electrical potential between the iron nail and the mercury. The motion is caused by the changing shape of the mercury surface due to a change in surface tension.

A variant of this system, which is more easily studied in the laboratory, is shown in Fig. 10.6. In this geometry the change of surface shape is restricted to a rising and flattening of the mercury meniscus, and the iron nail is replaced by a tungsten tip. Shape oscillations are observed if the aqueous solution contains dilute sodium hydroxide and the tungsten is connected to a sufficiently negative voltage source, for example, a piece of corroding aluminum. A typical oscilloscope tracing, which gives the electrical potential between the mercury and the tungsten as a function of time, is shown in Fig. 10.6. These oscillations are reproducible and have a period of about 75 ms. The height of the menicus as a function of time was measured frame by frame from pictures taken by a high-speed movie camera and is also shown in the figure.

There are numerous other examples of dissipative systems that undergo sustained oscillations. The mathematical term for such oscillations is *limit cycle*. Limit cycles are isolated, autonomous trajectories which are finite in length and which return to the same point in a finite time. *Autonomous* implies that the trajectories develop from differential equations that do not depend explicitly on time, while *isolated* implies that all points on the trajectory are separated from other recurrent trajectories. Thus they have the property that

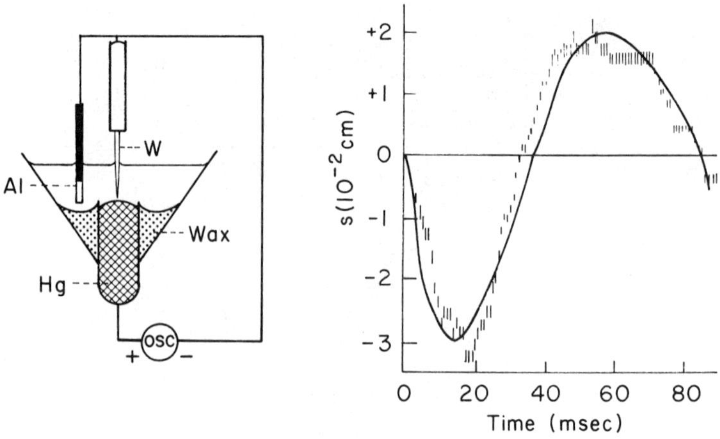

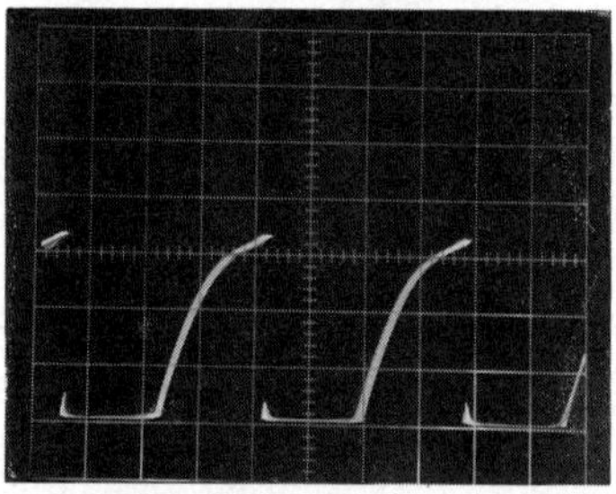

FIGURE 10.6. Oscillations in the beating mercury heart system. The diagram of the experimental apparatus illustrates the inert tungsten tip (W) attached to a piece of aluminum corroding in a NaOH solution. When the tip is correctly positioned in contact with the column of mercury, the surface of the mercury rises and falls periodically as shown in the accompanying graph. The curve represents the value of the separation between the mercury and tungsten calculated from the model of Keizer, Lin, and Rock, and the vertical lines are the results of measurements using rapid motion picture photography. The electrical potential recorded by the oscilloscope (osc) is shown in the bottom photograph. Taken from J. Keizer, P.A. Rock, and S.-W. Lin, *J. Am. Chem. Soc.* **101**, 5637 (1979).

for some τ

$$\mathbf{n}(\mathbf{n}^0, \tau) = \mathbf{n}^0 \tag{10.3.1}$$

if $\mathbf{n}^0$ is on the limit cycle. Because a limit cycle is autonomous, each point on the limit cycle recurs with the same characteristic time, τ. Indeed, for an autonomous system of equations, each point $\mathbf{n}^0$ determines a unique trajectory, which satisfies

$$\mathbf{n}(\mathbf{n}^0, t + \tau) = \mathbf{n}(\mathbf{n}(\mathbf{n}^0, t), \tau) = \mathbf{n}(\mathbf{n}(\mathbf{n}^0, \tau), t). \tag{10.3.2}$$

If n^* is another point on the limit cycle, then $\mathbf{n}^* = \mathbf{n}(\mathbf{n}^0, t)$ for some finite t. Using this and Eq. (10.3.1), the final equality in Eq. (10.3.2) can be rewritten

$$\mathbf{n}(\mathbf{n}^*, \tau) = \mathbf{n}(\mathbf{n}^0, t) = \mathbf{n}^*, \tag{10.3.3}$$

which shows that the $\mathbf{n}^*$ also recurs in time τ. The recurrence time is obviously not unique since Eq. (10.3.2) also shows that any integral multiple of τ is also a recurrence time. The smallest such time, which is guaranteed to be positive by the finiteness of the trajectory, is called the *period* of the limit cycle.

Steady states, which are a degenerate case of a limit cycle, have the property that $\mathbf{n}^{ss}(\mathbf{n}^{ss}, t) = \mathbf{n}^{ss}$ for all t. With a change in a parameter of the differential equation, a trajectory that is a steady state can abruptly change to a limit cycle. If this is associated with the critical point of a pair of complex eigenvalues of the relaxation matrix, then this is called a *Hopf bifurcation* of the steady state to a limit cycle.

Limit cycles can be stable or unstable, attractive or repelling, just like steady states. Limit cycles are invariant sets, in the sense that for any $\mathbf{n}^0$ on the limit cycle, $\mathbf{n}(\mathbf{n}^0, t)$ is also on the limit cycle. In general, an invariant set, M, is *stable* if the shortest distance between $\mathbf{n}(\mathbf{n}', t)$ and M is less than some increasing function of the distance between the initial point $\mathbf{n}'$ and M (which vanishes at zero). For a limit cycle this implies that the distance between the trajectory which begins at $\mathbf{n}'$ and the limit cycle itself is bounded. To be *attractive*, an invariant set must have the property that for any initial point $\mathbf{n}'$ close to M the shortest distance between $\mathbf{n}(\mathbf{n}', t)$ and M is smaller than some function of time which decreases to zero. Thus points near an attractive limit cycle ultimately approach the limit cycle. A limit cycle which is both stable and attractive is called *orbitally asymptotically stable*. Such a limit cycle is not asymptotically stable in the Liapunov sense described in Section 7.2, since distinct points on or near the limit cycle never approach one another asymptotically.

A simple mathematical example which exhibits limit cycles in the plane is given by the equations

$$\begin{aligned} \dot{x} &= x(x^2 + y^2 - r_1^2)(x^2 + y^2 - r_2^2) - y/\tau \\ \dot{y} &= y(x^2 + y^2 - r_1^2)(x^2 + y^2 - r_2^2) + x/\tau, \end{aligned} \tag{10.3.4}$$

with $r_1 < r_2$ and τ positive. The forms of the solutions to this equation are easier to visualize after changing variables to polar coordinates $x = r\cos\theta$ and $y = r\sin\theta$. This gives

$$\begin{aligned} d\theta/dt &= 1/\tau \\ dr/dt &= r(r^2 - r_1^2)(r^2 - r_2^2). \end{aligned} \tag{10.3.5}$$

The first of these equations has the immediate solution

$$\theta(\theta^\circ, t) = t/\tau + \theta^\circ, \tag{10.3.6}$$

while the second can be integrated using standard integrals to obtain an equation for t as a function of r.

The right-hand side of the radial equation has roots at $r = 0, r_1$, and r_2. The first of these corresponds to an unstable steady state, since for r slightly larger than zero

$$dr/dt = r_1^2 r_2^2 r, \tag{10.3.7}$$

so that r moves away from zero. Since the time derivative of r vanishes on the circles $r = r_1$ and $r = r_2$ and Eq. (10.3.6) shows that θ increases by 2π whenever t increases by $2\pi\tau$, it follows that both these circles are limit cycles with period τ. The larger circle, however, is an unstable limit cycle. This can be seen by linearizing the radial equation around r_2, which gives

$$d\Delta r/dt = 2r_2^2(r_2^2 - r_1^2)\Delta r. \tag{10.3.8}$$

Since r_2 is larger than r_1, it follows that small deviations from $r = r_2$ increase exponentially. For $r < r_1$, the motion is bounded between 0 and r_1 and trajectories spiral out to the circle at r_1. Indeed, the limit cycle at r_1 is stable and attracting as indicated by the linear relaxation equation

$$d\Delta r/dt = 2r_1^2(r_1^2 - r_2^2)\Delta r. \tag{10.3.9}$$

The phase plane for this example with a schematic representation of trajectories is given in Fig. 10.7.

Although sustained oscillations have been observed in many dissipative systems, the molecular mechanisms for most oscillations are not well understood. Of those for which a molecular mechanism has been proposed, the one for the beating mercury heart in Fig. 10.6 is probably the simplest. In alkaline solution that mechanism appears to involve four elementary processes: the conduction of electrons from the tungsten tip to the mercury; the reduction

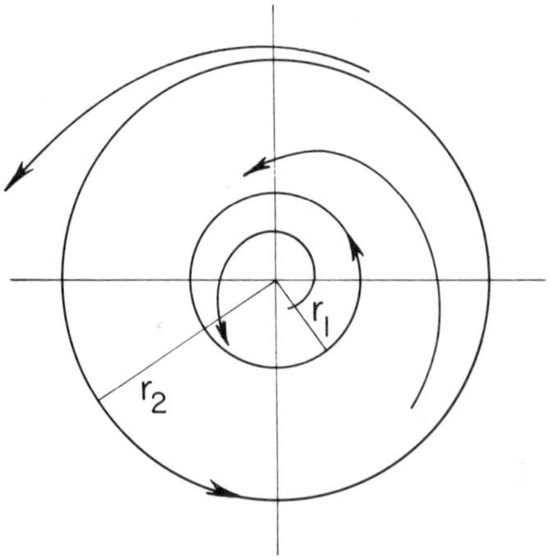

FIGURE 10.7. Schematic phase plane diagram of trajectories for the limit cycle oscillator in Eq. (10.3.5). The origin is an unstable steady state, while the circles at r_1 and r_2 are stable and unstable limit cycles, respectively.

of molecular oxygen on the mercury surface; the desorption of hydroxide ions from the mercury surface; and the movement of mercury in response to its changing surface tension.

As we saw in Section 5.7, the electrical potential of a metal surface is determined by the electrochemical processes occurring on it. In an aqueous solution of sodium hydroxide, the most important process is the reduction of dissolved molecular oxygen to hydroxide ion. The net reaction is a complex set of elementary processes with the stoichiometry

$$O_2(aq) + 2H_2O + 4e^- = 4OH^-(aq). \tag{10.3.10}$$

The rate-limiting step is thought to be the formation of a hydrogen peroxide intermediate. This reaction has a large overpotential and proceeds extremely slowly on a free mercury surface, whose potential is about -0.14 V with respect to a mercury–mercury oxide reference electrode. When the mercury makes contact with the tungsten tip, as indicated in Fig. 10.6, its potential is lowered by -1.27 V to that of the corroding aluminum. This increases the rate of the reduction reaction dramatically. The decrease in voltage has the further effect of changing the surface tension of the mercury, causing the surface to change its shape. This is called the electrocapillary effect. This change is reversed by the appearance of OH^- ions on the surface as a result of the reaction, which provides the switch for reversing the motion of the surface.

This molecular mechanism easily translates into a set of first-order differential equations for the voltage difference between the mercury and the tungsten tip, v; the position of the mercury surface as determined by the separation, s, between the mercury and the tungsten tip; and the concentration of hydroxide ions absorbed on the mercury, c. The equations are

$$ds/dt = L[\gamma(c, v) - \gamma(s)] \tag{10.3.11}$$

$$dc/dt = i_0 \exp(-v/v_0) - D(c - c_0) \tag{10.3.12}$$

$$C dv/dt = i_0 \exp(-v/v_0) - \sigma(s)v. \tag{10.3.13}$$

In the first equation, γ represents the surface tension, with $\gamma(c, v)$ the local equilibrium value determined by the electrocapillary function, and $\gamma(s)$ the value that corresponds to equilibrium at the separation s. Equation (10.3.11) is obtained from the linear Onsager equation connecting the flux (in this case the time rate of change of area) with the force (the deviation of the surface tension from its equilibrium value). For small-amplitude motions, the mercury meniscus has the shape of a spherical cap whose height is a monotone function of the surface area. The second equation includes the Tafel expression, derived in Section 5.7, for the increase in surface concentration of hydroxide ions due to the reduction process along with a term which accounts for the desorption of the ions. Finally, Eq. (10.3.13) describes the voltage and involves the capacitive current Cdv/dt, the electrochemical current, and Onsager's form of

Ohm's law in which $\sigma(s)$ is the conductivity between the tungsten and mercury as a function of distance. The conductivity is appreciable, of course, only when the tungsten and mercury touch, i.e., only when $s < 0$. The simplest model for this is a Heaviside function

$$\sigma(s) = H(s)/R, \tag{10.3.14}$$

where R is the contact resistance. Because CR is the order of a few tenths of a microsecond, it follows from Eq. (10.3.13) that when the tip contacts the surface, a short circuit occurs within a microsecond and the voltage difference, v, will vanish.

The rapid time scale on which the short circuit occurs makes it possible to obtain explicit solutions for the voltage and concentration. Indeed, when the tungsten and mercury are not touching we set $\sigma(s) = 0$ and notice that Eq. (10.3.13) can be integrated directly to give

$$v(t) = v_0 \ln\{(i_0 t/cv_0) + \exp[v(0)/v_0]\}. \tag{10.3.15}$$

Substituting this into Eq. (10.3.12), we obtain an inhomogeneous linear equation whose solution is easily seen to be

$$c(t) = \exp(-Dt)\left(c(0) + i_0 \int_0^t \exp(D\tau)\{\exp[v(0)/v_0] + i_0\tau/cv_0\}^{-1}\, d\tau\right). \tag{10.3.16}$$

On the other hand, when the tip and mercury are touching ($s \leq 0$), one has short circuit and within a microsecond the solution to the voltage equation becomes

$$v(0) = 0. \tag{10.3.17}$$

Substituting this into Eq. (10.3.12) one again obtains a homogeneous linear equation, whose solution with $c_0 = 0$ is

$$c(t) = \exp(-Dt)c(0) + (i_0/D)[1 - \exp(-Dt)]. \tag{10.3.18}$$

To find the separation as a function of time, one needs to integrate Eq. (10.3.11) using the explicit solutions given above. Since changes in separation are a few tenths of a millimeter, one can use the Taylor series expansion $\gamma(s) = \gamma_0 + Ys$, where γ_0 is the surface tension when $s = 0$. The electrocapillary function, $\gamma(c, v)$, can be estimated from equilibrium measurements. For hydroxide ion on mercury it has the parabolic form

$$\gamma(c, v) = \gamma_m(c) - C[v - v_m(c)]^2/2, \tag{10.3.19}$$

where $\gamma_m(c)$ is the surface tension at the maximum in the electrocapillary curve, v_m is the voltage at the maximum, and C is the capacitance. According to experiments, γ_m is a decreasing function of the concentration of absorbed OH^- ions as is the voltage, v_m. Using these assumptions, the separation equation can be integrated to give

$$s(t) = \exp(-LYt)\left(s(0) + \int_0^t \exp(LY\tau)\{\gamma[c(\tau), v(\tau)] - \gamma_o\}\, d\tau \right). \quad (10.3.20)$$

Equations (10.3.15)–(10.3.20) reduce the solution of the differential equations to quadrature on a time scale slower than a microsecond. To determine the solution starting at a point $s(0)$, $c(0)$, $v(0)$, one uses Eqs. (10.3.15), (10.3.16), and (10.3.20) if $s(0) > 0$ and Eqs. (10.3.17), (10.3.18), and (10.3.20) if $s(0) \leq 0$. The solution continues on the appropriate branch until $s(t)$ equals zero, at which time the other solution is used with the new initial conditions $s(0) = 0$, $c(0) = c(t)$, and $v(0) = v(t)$.

The nature of the solutions to these equations depends critically on the positioning of the tungsten tip. This dependence is contained in the parameter γ_0, which is the surface tension that allows the mercury meniscus to just touch the tungsten. When this parameter is large, the mercury cannot reach the tip and the voltage difference $v^{ss} = 1.27$ V of the free merucry surface develops at steady state. This can be seen from the differential equations (10.3.11)–(10.3.13) if the reverse anodic reaction is included. On the other hand, when the parameter γ_0 is small, the only possible steady state is one with the tip and mercury touching and $v^{ss} = 0$. For an intermediate range of placements of the tip, three steady states exist, one with $s^{ss} > 0$, which is stable, and two unstable states, one with $s^{ss} \simeq 0$ and one with $s^{ss} \leq 0$. These are illustrated in Fig. 10.8, where the phase space has been projected onto the plane $s = 0$. The dashed line indicates the *null cline* for the separation, i.e., the curve on which $ds/dt = 0$, while the full line represents the null cline for the concentration. If the solutions in Eqs. (10.3.15)–(10.3.20) are pieced together for a value γ_0 in the intermediate range, one discovers an orbitally asymptotically stable limit cycle which, in projection, encircles the two unstable states. The oscillations of v, c, and s calculated in this way are also shown in Fig. 10.8. As long as the amplitude of the oscillations is not too large, the detailed comparison of such calculations with experiment is quite good.

The Belousov–Zhabotinsky reaction is another oscillating system for which the molecular mechanism has been investigated extensively. The overall reaction is

$$\begin{aligned} 3BrO_3^- + 5CH_2(COOH)_2 + 3H^+ &= 3BrCH(COOH)_2 \\ &+ 2HCOOH + 4CO_2 + 5H_2O. \end{aligned} \quad (10.3.21)$$

Equation (10.3.21) represents the net effect of more than a dozen elementary chemical reactions. The complexity of this system is a result of the manifold oxidation states of bromine, which range from $+5$ in bromate (BrO_3^-) to -1 in bromide (Br^-). According to Field, Körös, and Noyes, when sufficient bromide is present the following reactions, collectively referred to as *process A*, occur:

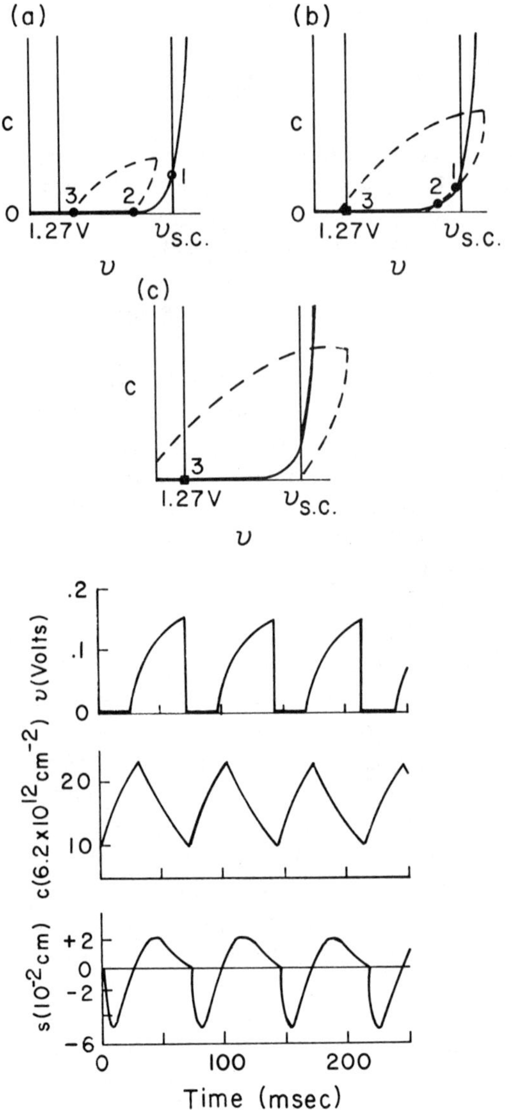

FIGURE 10.8. The schematic diagrams (a)–(c) show the null clines for the separation (dashed line) and the concentration (full line) for the beating mercury heart model in Eqs. (10.3.11)–(10.3.13) projected onto the plane $s = 0$. The circles represent the steady states for three different positionings of the tungsten tip. In (a) the tip is far into the mercury and only steady state 1 at the short-circuited voltage, $v_{s.c.}$, is stable. In (b) the tip is positioned correctly for oscillations, and steady states 1 and 2 are unstable and 3 is stable. In (c) the tip is far removed from the mercury, and steady state 3 is stable at the voltage of the free tungsten. The oscillations in concentration (c), voltage (v), and separation (s) calculated from the differential equations with the tungsten positioned as in (b) are shown in the lower panel. Taken from J. Keizer, P.A. Rock, and S.-W. Lin, *J. Am. Chem. Soc.* **101**, 5637 (1979).

$$BrO_3^- + Br^- + 2H^+ \rightarrow HBrO_2 + HOBr$$
$$HBrO_2 + Br^- + H^+ \rightarrow 2HOBr$$
$$HOBr + Br^- + H^+ \rightarrow Br_2 + H_2O$$
$$Br_2 + CH_2(COOH)_2 \rightarrow BrCH(COOH)_2 + Br^- + H^+. \qquad (10.3.22)$$

These are elementary complex reactions whose net effect is the bromination of malonic acid, $CH_2(COOH)_2$, by molecular bromine. The transient intermediates in this process are bromous acid ($HBrO_2$) and hypobromous acid (HOBr).

Process A removes Br^- from solution. If the reaction mixture does not contain enough Br^- and the cerium is predominantly in the reduced state, Ce^{3+}, then the predominant reactions are *process B*:

$$BrO_3^- + HBrO_2 + H^+ \rightleftarrows 2BrO_2^{\cdot} + H_2O$$

$$BrO_2^{\cdot} + Ce^{3+} + H^+ \rightleftarrows HBrO_2 + Ce^{4+}$$

$$2HBrO_2 \rightarrow BrO_3^- + HOBr + H^+$$

$$HOBr + CH_2(COOH)_2 \rightarrow BrCH(COOH)_2 + H_2O. \qquad (10.3.23)$$

Process B also leads to the bromination of malonic acid but with hypobromous acid as the bromination reagent. Notice that the net effect of the first two reactions in process B is to produce bromous acid autocatalytically, while the third one destroys it. Using rate constants for these reactions and the steady-state contraction procedure to eliminate the bromous acid, one finds that the concentration of $HBrO_2$ is about 10^{-6} *M*. This is about 10^4 times larger than the concentration supported by process A.

According to the Field, Körös, Noyes mechanism, during oscillations the reaction mixture switches back and forth between process B and process A. In their mechanism the first step in process B is rate determining. Its rate is proportional to the bromous ion concentration, i.e.,

$$V_B^+ = k[BrO_3^-][HBrO_2][H^+], \qquad (10.3.24)$$

where the square brackets represent molar concentrations. Thus as long as the Br^- and Ce^{4+} ion concentrations are small, process B dominates because of the high relative concentration of $HBrO_2$. This continues until enough of the product Ce^{4+} builds up that it begins to oxidize the malonic acid and bromomalonic acid at a significant rate. The net reactions for this are

$$6Ce^{4+} + CH_2(COOH)_2 + 2H_2O \rightarrow 6Ce^{3+} + HCOOH + 2CO_2 + 6H^+$$

$$4Ce^{4+} + BrCH(COOH)_2 + 2H_2O$$

$$\rightarrow 4Ce^{3+} + HCOOH + 2CO_2 + 5H^+ + Br^-. \qquad (10.3.25)$$

The second of these reactions is the switching reaction that turns off process B indirectly by turning on process A. Indeed, Eq. (10.3.22) shows that Br^- stimulates the destruction of $HBrO_2$, which untimately reduces the rate of process B by nearly four orders of magnitude. When the first three reactions in (10.3.22) have removed enough Br^-, process A stops and process B takes over.

Field and Noyes have translated this mechanism into a set of elementary processes for the species $X = HBrO_2$, $Y = Br^-$, $Z = Ce^{4+}$, $P = HOBr$, and $A = BrO_3^-$. Treating the hypobromous acid as inert and the bromate concentration as constant they write

$$\begin{aligned} A + Y &\xrightarrow{k_1} X + P \\ X + Y &\xrightarrow{k_2} 2P \\ A + X &\xrightarrow{k_3} 2X + Z \\ 2X &\xrightarrow{k_4} A + P \\ Z &\xrightarrow{k_5} fY. \end{aligned} \tag{10.3.26}$$

The reactions are only a caricature of the molecular process represented in Eqs. (10.3.22)–(10.3.25). The final reaction, for example, is an attempt to incorporate the switching reactions in Eq. (10.3.25) and introduces a stoichiometric coefficient, f, whose value is usually taken as unity. Field and Noyes called this model the *Oregonator*. The corresponding differential equations have been solved numerically and, in appropriate ranges of parameter space, lead to limit cycle oscillations that closely resemble the observed behavior in this system. We return to this interesting chemical system in Section 10.6 where chaotic behavior is discussed.

10.4. Fluctuations on Limit Cycles

Limit cycle oscillations involve periodic trajectories. This persistent time dependence usually defies detailed analytic treatment. Thus it is necessary to apply the sort of numerical analysis that is described for transient, nonlinear trajectories in Section 10.2. As we saw there, it is important to differentiate between fluctuations in initial conditions, which may be quite large, and fluctuations due to molecular processes, which may be quite small. The combined effect of both types of statistical uncertainty is contained in Eq. (10.2.26). The second term is caused by uncertainty in the initial conditions and is the simplest one to analyze.

For simplicity let us assume that the ensemble is prepared so that at time t^0 all systems are in the domain of attraction of an orbitally asymptotically stable limit cycle. This implies that the single-time probability, $W_1(\mathbf{n}^0, t^0)$, is concentrated on the domain of attraction and that all the conditional average trajectories, $\bar{\mathbf{n}}(\mathbf{n}^0, t)$, are attracted to the limit cycle after some initial period of time. The contribution of these trajectories to the single-time covariance is given by Eq. (10.2.27), i.e.,

$$\begin{aligned} \langle \delta\mathbf{n}(t)\delta\mathbf{n}^T(t)\rangle_1 = \int W_1(\mathbf{n}^0, t^0)\bar{\mathbf{n}}^0(t)\bar{\mathbf{n}}^{0T}(t)\,d\mathbf{n}^0 - \int W(\mathbf{n}^0, t)\bar{\mathbf{n}}^0(t)\,d\mathbf{n}^0 \\ \times \int W(\mathbf{n}^0, t^0)\bar{\mathbf{n}}^{0T}(t)\,d\mathbf{n}^0. \end{aligned} \tag{10.4.1}$$

Under the conditions above, $\bar{\mathbf{n}}^0(t) \equiv \bar{\mathbf{n}}(\mathbf{n}^0, t)$ will approach the limit cycle for large t. The limit cycle, however, is periodic with period τ. Thus, asymptotically, the contribution to the covariance in Eq. (10.4.1) will also be periodic with an identical period. The nature of the resulting function depends crucially on how the ensemble is spread over the domain of attraction at time t^0.

A simple example illustrates the complex nature of this dependence. Assume that at time t^0 only two initial states are represented in the ensemble. One, $\mathbf{n}_1^0$, is on the limit cycle and another, $\mathbf{n}_2^0$, is in the domain of attraction, but not on the limit cycle. The initial probability density is, therefore,

$$W_1(\mathbf{n}^0, t^0) = \frac{1}{2}[\delta(\mathbf{n}^0 - \mathbf{n}_1^0) + \delta(\mathbf{n}^0 - \mathbf{n}_2^0)]. \tag{10.4.2}$$

Substituting this into Eq. (10.4.1) shows that the covariance is

$$\langle \delta\mathbf{n}(t)\delta\mathbf{n}^T(t)\rangle_1 = \frac{1}{4}[\bar{\mathbf{n}}(\mathbf{n}_1^0, t) - \bar{\mathbf{n}}(\mathbf{n}_2^0, t)][\bar{\mathbf{n}}(\mathbf{n}_1^0, t) - \bar{\mathbf{n}}(\mathbf{n}_2^0, t)]. \tag{10.4.3}$$

This will be a large number at t^0 if $\mathbf{n}_2^0$ is appreciably distant from the limit cycle. As time proceeds, $\bar{\mathbf{n}}(\mathbf{n}_1^0, t)$ stays on the limit cycle and returns to $\mathbf{n}_1^0$ after each period of time τ. The trajectory beginning at $\mathbf{n}_2^0$ will approach the limit cycle and after an interval of time, $t(\mathbf{n}_2^0)$, will be indistinguishable from the limit cycle. Although $\mathbf{n}(\mathbf{n}_2^0, t)$ can never equal $\mathbf{n}(\mathbf{n}_1^0, t)$, if the two points are chosen correctly they can become identical in the limit that $t \to \infty$. In this case the covariance in Eq. (10.4.3) vanishes as $t \to \infty$. On the other hand, if the two points can be chosen so that asymptotically $\bar{\mathbf{n}}(\mathbf{n}_2^0, t)$ approaches a point on the limit that is distant from $\bar{\mathbf{n}}(\mathbf{n}_1^0, t)$, then the variance will stay large and change periodically. Although this example is artifical, it typifies the complex and persistent effects that initial conditions can have on the statistical properties of systems with limit cycles. Indeed, if $\mathbf{n}_1^0$ and $\mathbf{n}_2^0$ are both off of the limit cycle, then the contribution of these trajectories to the covariance will depend on the relative phases of the two points as they asymptotically approach the limit cycle.

For initial ensembles that consist of systems close to the limit cycle with a narrow range of initial values, the sort of behavior described above will not occur. There may be a slight increase or decrease of the covariance associated with the approach to the limit cycle, but nothing appreciable. If the range of initial values is small and the ensemble is far from the limit cycle, ambiguity in final positions will develop if a separation in phase angle occurs during the approach to the limit cycle. This will not always be the case, as the limit cycle described by Eqs. (10.3.5) and illustrated in Fig. 10.7 shows. In that example, the rate of change of phase is independent of the distance from the limit cycle. Thus a set of points with a total range of initial phases of $\Delta\theta_0$ will maintain the same range of phase values as time proceeds. It should be kept in mind that this sort of contribution to the statistical properties of oscillating systems is relevant only for repetitions of the same experiment. If a single experiment is carried out, or repeated experiments in which the initial conditions are different but well characterized experimentally, these effects due to the initial distribution are irrelevant.

It is more difficult to obtain a general qualitative understanding of the effect of conditional fluctuations on limit cycle oscillations. Their contribution to the single-time covariance is given by the first term in Eq. (10.2.26), i.e.,

$$\langle \delta\mathbf{n}(t)\delta\mathbf{n}^T(t)\rangle_1 = \int W(\mathbf{n}^0, t^0)\sigma^0(t^0, t)\, d\mathbf{n}^0, \tag{10.4.4}$$

where $\sigma^0(t^0, t)$ is the covariance conditioned on the precise initial state $\mathbf{n}^0$ at t^0. As we saw in Section 10.2, numerical integration is usually required to obtain the time dependence of the conditional covariance. On the other hand, the equation which describes the covariance, Eq. (10.2.4), depends explicitly on the solution to the conditional average equation. Thus, asymptotically, points which start in the domain of attraction of the limit cycle will develop a periodic component to the relaxation matrix $H(\mathbf{n}^0, t)$ and the random term $\gamma(\mathbf{n}^0, t)$. While the period of this oscillatory component will be independent of $\mathbf{n}^0$, its phase on the limit cycle may depend strongly on $\mathbf{n}^0$. If the probability distribution is narrow and the initial points are close to the limit cycle, then $\sigma^0(t^0, t)$ will be relatively independent of $\mathbf{n}^0$. Thus one can estimate the behavior of that term using a representative point $\mathbf{n}^0$ since

$$\langle \delta\mathbf{n}(t)\delta\mathbf{n}^T(t)\rangle_1 \simeq \int W(\mathbf{n}^0, t^0)\, d\mathbf{n}^0 \sigma^0(t^0, t) = \sigma^0(t^0, t). \tag{10.4.5}$$

The detailed behavior of the conditional covariance for a point initially on a limit cycle depends on the underlying differential equations and the molecular processes which give rise to those equations.

Nonetheless, a qualitative picture of how the fluctuations change in time can be obtained using the mathematical model in Section 10.3. In polar coordinates the two variables, θ and r, are uncoupled by the differential equations in Eq. (10.3.5). Let us assume that these equations describe the conditional average of the densities of some appropriately transformed extensive variables x and y. Then fluctuations would be described by the equations

$$\begin{aligned} d\delta\theta/dt &= \tilde{f}_\theta \\ d\delta r/dt &= -r_1(r_2^2 - r_1^2)\delta r + \tilde{f}_r, \end{aligned} \tag{10.4.6}$$

where we have taken the initial point θ^0, r^0 on the stable limit cycle at r_1. The solutions for the two variances as a function of time are obviously

$$\langle \delta\theta^2(t)\rangle = \int_0^t \gamma_{\theta\theta}(s)\, ds \tag{10.4.7}$$

and

$$\langle \delta r^2(t)\rangle = \int_0^t \exp[-\lambda(t - s)]\gamma_{rr}(s)\, ds, \tag{10.4.8}$$

where $\lambda \equiv r_1(r_2^2 - r_1^2)$ is positive since $r_1 < r_2$. The strengths of the random terms are defined by

$$\langle \tilde{f}_i(t)\tilde{f}_i(t')\rangle = \gamma_{ii}(t)\delta(t - t') \tag{10.4.9}$$

and depends explicitly on the conditional averages $\bar{\theta}^0(t) = \theta^0 + t/\tau$ and $\bar{r}^0(t) = r_1$.

The functional form of the random terms $\gamma_{\theta\theta}$ and γ_{rr} must be obtained from their form for the Cartesian variables x and y. If we assume that x and y represent densities of extensive variables, which must be positive, we can write them hueristically as $x = x^* + r_1 \cos\theta$, $y = y^* + r_1 \sin\theta$, where x^* and y^* are large enough to keep x and y positive. To understand the scaling of the covariance of the random terms, we need to transform the random terms γ_{xx}, γ_{xy}, and γ_{yy} into the random terms for θ and r. Ths is accomplished by a similarity transformation, as described in Eq. (8.2.39). Thus the term $\gamma_{\theta\theta}$ and γ_{rr} will be a lincar combination of the terms for x and y and will scale inversely with the volume. Hence we can write

$$\begin{aligned}\gamma_{\theta\theta} &= \bar{\gamma}_{\theta\theta} V^{-1}\\ \gamma_{rr} &= \bar{\gamma}_{rr} V^{-1},\end{aligned} \tag{10.4.10}$$

where the overbars represent terms independent of the system size. Since we have taken $\bar{x}^0 = x^* + r_1 \cos\bar{\theta}^0$ and $\bar{y}^0 = y^* + r_1 \sin\theta^0$, the time dependence of $\bar{\gamma}_{\theta\theta}$ and $\bar{\gamma}_{rr}$ will involve constant terms plus powers of $\sin(\theta^0 + t/\tau)$ and $\cos(\theta^0 + t/\tau)$.

This is enough qualitative information to discuss the time dependence of the conditional covariance. Because λ is positive, Eq. (10.4.8) has a limit as $t \to \infty$, even if the periodic sine and cosine terms have a significant amplitude. This corresponds to the stable transverse motion of the average trajectories close by the limit cycle. The phase fluctuations, on the other hand, are tangential to the limit cycle. Since they are marginally stable, they do not relax. Thus Eq. (10.4.7) has an initial behavior

$$\langle \delta\theta^2(t)\rangle = \gamma_{\theta\theta} t, \tag{10.4.11}$$

which persists for the constant terms in the integrand. The length of time required for this linear divergence to become important depends on the value of the constant term in $\gamma_{\theta\theta}$. The effect of this divergence would be a loss of phase coherence in an initially coherent ensemble with a standard deviation proportional to $t^{1/2}$. Since $\gamma_{\theta\theta}$ scales inversely with system size, the time scale for the divergence is extremely long. As a consequence, such a dephasing would be almost impossible to observe. Indeed, no such dephasing has been reported experimentally.

There is, however, experimental evidence that fluctuations in the oscillatory regime of the Gunn diode are anomalously large. In Section 7.6 we examined the low-frequency voltage fluctuations below the oscillation threshold in

GaAs and found that they increased in an unbounded way as the critical voltage was approached, in agreement with experiment. Above the critical point microwave oscillations are observed, and the time average of the variance of the voltage fluctuations remain about six orders of magnitude greater than at equilibrium. This phenomenon is not presently understood.

10.5. Chaotic Trajectories

A limit cycle is an isolated, periodic trajectory that is closed and bounded. According to a theorem of Bendixson, limit cycles are the only closed, bounded attractors which can result from first-order ordinary differential equations in two variables. For three or more variables the situation is much richer and the classification of the possible types of dynamical behavior is still incomplete. For systems of three and more variables, closed, bounded attractors are known that can be either quasi-periodic or aperiodic. Limit sets of this sort have been found for both dissipative and nondissipative systems. The geometrical structure of these limit sets is complicated, in some cases too complicated to be described by analytic geometry. The so-called *strange attractors* are composed of an infinite number of points but cannot be represented by piecewise differentiable functions. Another class of attractors, the *chaotic attractors*, have the property that trajectories on the attractor are extremely sensitive to initial conditions. Indeed, points that are initially close together tend to diverge from one another as time goes on. While it appears that all chaotic attractors are also strange attractors, the converse is known to be false.

Because of the complicated geometric structure and dynamical properties of chaotic attractors, our knowledge of their properties has come primarily from the numerical solution of ordinary differential equations. One of the simplest sets of differential equations to exhibit a chaotic attractor is the Rössler equations

$$\begin{aligned} dx/dt &= -y - z \\ dy/dt &= x + ey \\ dz/dt &= f + xz - gz, \end{aligned} \tag{10.5.1}$$

where e, f, and g are positive constants. While these equations have a structure that is similar to that of mass action equations for the number density, they do not have the canonical form and are *ad hoc*. They are interesting, nonetheless, because they provide a simple example of a chaotic attractor. For the parameter values $e = f = 0.2$ and g in the range $2.6 \leq g \leq 4.2$, the Rössler equations possess orbitally stable limit cycles. In this parameter range, the shape and period of the limit cycle undergoes a sequence of bifurcations known as *period doubling*. This is shown by the projection of the limit cycle

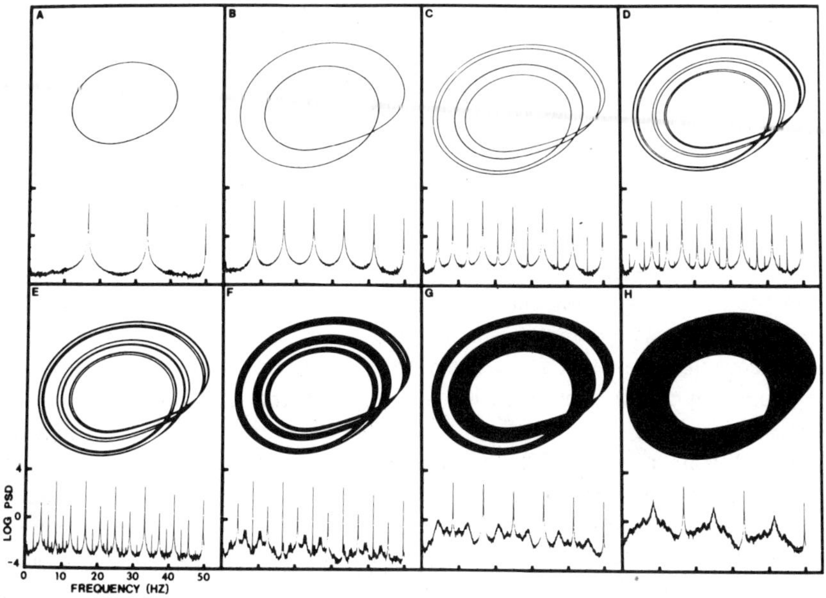

FIGURE 10.9. Plots of solutions of the Rössler equations (10.5.1) projected on the x–y plane after transients have died out. Directly below each trajectory is the corresponding power spectral density (*PSD*) as a function of frequency. Taken from J. Crutchfield, D. Framer, N. Packard, R. Shaw, G. Jones, and R.J. Donnelly, *Phys. Letters* **76A**, 1 (1980).

in the x–y plane in Fig. 10.9. The period abruptly doubles three times from about 0.06 s to about 0.24 s in the first four frames of the figure.

Below each trajectory in Fig. 10.9 is the numerically calculated power spectral density of the variable $z(t)$. Here the power spectral density, $P(\omega)$, is defined as the limit

$$\lim_{T\to\infty} \frac{1}{T}\int_0^T z(t)\,dt = \int_{-\infty}^{+\infty} P(\omega)\,d\omega, \tag{10.5.2}$$

if it exists. It can also be defined by the Fourier transform of $z(t)$, if it exists. For a stationary process the Wiener–Khintchine theorem shows that $P(\omega)$ is related to the power spectrum defined in Section 1.8. While this is not true for a nonstationary process, the power spectral density in Eq. (10.5.2) may still be a useful quantity since it provides a resolution of the long-time average of the trajectory into its component frequencies. In the first frame of Fig. 10.9, peaks are visible at the primary oscillation frequency $\omega_1 = 16\ \text{s}^{-1}$ accompanied by the harmonic frequencies $2\omega_1$ and $3\omega_2$. The amplitudes of the background frequencies are associated with errors in the analog integration procedure and are small. During the period doubling sequence the frequency $\omega_{1/2} = 8\ \text{s}^{-1}$

first occurs along with its higher harmonics, followed by $\omega_{1/4}$ and then $\omega_{1/8}$ as the parameter g is increased towards 4.2.

Just above $g = 4.2$ an entirely new behavior develops which is characteristic of a chaotic attractor. The thin, well-defined trajectories of the limit cycles merge into a near continuum, which is so dense that it appears as black bands in projection. In this parameter range the background frequencies become amplified significantly, indicating that the motion has a noisy character usually associated with stochastic processes. The noise, however, is many orders of magnitude greater than one usually attributes to molecular fluctuations. Since the noise arises from the solution of the deterministic equation (10.5.1), it is referred to as *deterministic chaos.*

The complex geometric structure of the Rössler attractor can be visualized by examining its intersection with a plane surface that cuts through a branch of the trajectories. A cut like this through an attractor is called a *Poincaré section.* A Poincaré section transverse to a limit cycle gives at most a finite number of distinct points. A similar Poincaré section through a chaotic attractor is significantly more complex. A schematic diagram of a cut through the Rössler attractor is shown in Fig. 10.10. Notice the highly folded geometrical arrangement of the sheets of the attractor. If we take a second Poincaré section perpendicular to the original section, the detailed arrangement of the sheets is as illustrated by the dots in Fig. 10.10.

In the more detailed renderings of the Rössler attractor by Shaw, given in the book by Abraham and Shaw in the list of references at the end of this chapter, the pattern of dots in Fig. 10.10 is reminiscent of the Cantor set. Recall that the Cantor set is the set of points that results from removing the middle third from the unit interval and successively the middle third from the remaining intervals, *ad infinitum.* Indeed, increasing the resolution of the calculation of trajectories for the Rössler attractor, Shaw shows that the fine-grained structure of the regions shown schematically in Fig. 10.10 has an appearance similar to the coarse-grained structure in the figure. The emergence of similar geometrical structure at finer and finer levels of resolution is termed *self-similarity*, and objects which are self-similar have been dubbed *fractals* by Mandelbrot.

Because chaotic attractors are fractals, one needs to introduce a new measure of their size. Since a limit cycle is an isolated, finite-length trajectory it is obviously one-dimensional. It is not so obvious, however, how to assign a dimension to a fractal such as the Cantor set. Kolmogorov introduced the notion of capacity, which is now commonly referred to as the *fractal dimension*, to solve this difficulty. The fractal dimension is defined as

$$d_F = \lim_{\varepsilon \to 0} \ln N(\varepsilon)/\ln(1/\varepsilon), \tag{10.5.3}$$

where $N(\varepsilon)$ is the minimum number of cubes of side ε which are required to cover the attractor. For an attractor involving k variables, the cubes are

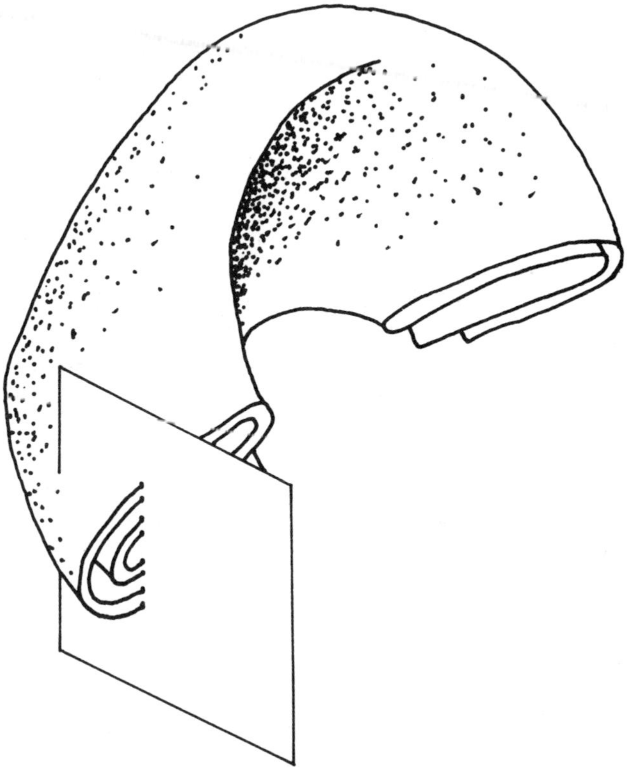

FIGURE 10.10. A schematic diagram of the Rössler attractor, illustrating the folding of the surface and Poincaré sections on the left and right. The plane perpendicular to the Poincaré section (a Lorentz section) helps to illustrate the fractal nature of the folding. After R.H. Abraham and C.D. Shaw, *Dynamics—The Geometry of Behavior. Part II. Chaotic Behavior* (Aerial Press, Santa Cruz, CA, 1983).

k-dimensional. For a steady state, which is a point, it is obvious that $N(\varepsilon) = 1$. For a limit cycle, which is a line, the minimum number of cubes required to cover the cycle multiplied by the side of a cube is the order of the length of the cycle. Thus $N(\varepsilon) \sim \varepsilon^{-1}$ for a limit cycle. Applying Eq. (10.5.3), we see that the fractal dimension of a steady state is zero, while a limit cycle has fractal dimension one, just as we expect. The Cantor set, however, has fractal dimension $d_F = \ln 2/\ln 3 \simeq 0.630$. This can be seen by choosing cubes of side $\varepsilon = (1/3)^m$ and letting $m \to \infty$. Drawing a picture of the successive steps in the construction of the Cantor set, it is easy to see that $N(\varepsilon) = 2^m$ so that

$$d_F = \lim_{m\to\infty} \ln 2^m/\ln 3^m = \ln 2/\ln 3. \qquad (10.5.4)$$

Thus the fractal dimension of the Cantor set is intermediate between that of a point and that of a line, as we might except on the basis of its self-similar structure. The fractal dimension of the Rössler attractor is between 2 and 3.

The dynamical origin of the complex geometrical structure of chaotic attractors is the exquisite sensitivity of trajectories on the attractor to initial conditions. We have seen this sort of sensitivity for trajectories beginning near unstable states in Section 10.2. For the trajectories calculated in Fig. 10.3, this sensitivity is transient and after a while all trajectories converge to the stable steady state. This is not true for a chaotic attractor, on which trajectories that are initially close tend to separate exponentially.

An indicator of the sensitivity to initial conditions is provided by Liapunov numbers. Liapunov numbers give a measure of the average rate of separation of nearby points along a trajectory. They are defined most easily for so-called *p-dimensional maps.* A p-dimensional map is a map $\mathbf{F}$

$$\mathbf{n}_{k+1} = \mathbf{F}(\mathbf{n}_k) \tag{10.5.5}$$

of a p-dimensional vector space into itself. Such maps are useful in many contexts and provide a natural representation for the way in which digital computers solve ordinary differential equations. For example, the autonomous equation

$$d\mathbf{n}/dt = \mathbf{R}(\mathbf{n}) \tag{10.5.6}$$

can be solved numerically by using small time steps Δt and writing

$$\mathbf{n}_{k+1} = \mathbf{n}_k + \mathbf{R}(\mathbf{n}_k)\Delta t, \tag{10.5.7}$$

where $\mathbf{n}_{k+1}$ is the value after $k + 1$ time steps. For Δt small enough, the iterated solution to the p-dimensional map in Eq. (10.5.5) with

$$\mathbf{F}(\mathbf{n}) = \mathbf{n} + \mathbf{R}(\mathbf{n})\Delta t \tag{10.5.8}$$

provides a good approximation to the solution of Eq. (10.5.6).

The Liapunov numbers are constructed for a p-dimensional map in the following way. First one forms the Jacobian matrix of the map, $J_{ij} = \partial F_i/\partial n_j$, and the product Jacobian matrix

$$J^{(k)}(\mathbf{n}_0) \equiv J(\mathbf{n}_k)J(\mathbf{n}_{k-1})\dots J(\mathbf{n}_0). \tag{10.5.9}$$

The eigenvalues of this matrix, which by the chain rule of differentiation is the Jacobian of the map $\mathbf{F}$ repeated k times, are then calculated. Ordering the *magnitudes* of the eigenvalues

$$\lambda_1(k) \geq \lambda_2(k) \geq \cdots \geq \lambda_p(k) \geq 0, \tag{10.5.10}$$

the *Liapunov numbers* are defined by

$$\lambda_i(\mathbf{n}^0) \equiv \lim_{k\to\infty} \left(\prod_{n=0}^{k} \lambda_i(n) \right)^{1/k} \tag{10.5.11}$$

Thus the Liapunov numbers are the geometric mean of the magnitude of the eigenvalues of the Jacobian after an infinite number of repetitions of the map **F**. This provides a measure of the average way in which points move during the iterations. Indeed, if we consider the map in Eq. (10.5.8) that is associated with the autonomous differential equation (10.5.6), then the Jacobian is

$$J = I + H\Delta t, \tag{10.5.12}$$

with $H_{ij} = \partial R_i/\partial n_j$ the linearized rate matrix. The eigenvalues of J are obviously

$$\lambda_i = 1 + \bar{\lambda}_i \Delta t, \tag{10.5.13}$$

where $\bar{\lambda}_i$ is an eigenvalue of H. We know that if $\bar{\lambda}_i$ is positive, then for short times points move apart like $\exp(\bar{\lambda}_i t)$, while if $\bar{\lambda}_i$ is negative, then points approach each other exponentially. Thus if any of the eigenvalues λ_i are greater than one, there will exist nearby trajectories which diverge for a short period of time. The Liapunov numbers in Eq. (10.5.11) are the geometric mean of these quantities along the entire trajectory. Thus if any of the Liapunov numbers are greater than unity, the typical local behavior of the entire trajectory is for points to diverge from one another. Notice that according to Eq. (10.5.11), the Liapunov numbers are dependent on the starting point, $\mathbf{n}_0$.

On a chaotic attractor the largest Liapunov numbers are greater than one. Although this implies a local divergence of trajectories, the complete trajectories are confined to an attractor of finite extent. Thus globally the trajectories never get too far away from one another. It is the strong tendency of points which are initially close together to separate over time that is the origin of the name chaotic attractor and the origin of the noisy appearance of their power spectra. In the next section we discuss two chaotic attractors that have been discovered in the laboratory, one in a chemical system and the other in a living cell.

10.6. Chaos in Complex Systems

Chaotic attractors have been observed in a number of physical, chemical, and biological systems. In this section we focus attention on the Belousov reaction, where period doubling bifurcations and chaos are well documented experimentally, and on glucose-stimulated electrical activity in the pancreatic β-cell, for which model calculations have uncovered a chaotic attractor. Chaotic behavior also has been observed both experimentally and in numerical calculations using simple models of hydrodynamic flows, including couette flow between concentric rotating cylinders and convection induced by thermal gradients, as well as in certain mechanical devices. Chaos is often conceived as a paradigm for turbulent motion in fluids, and a variety of bifurcation sequences leading to chaotic attractors have been proposed as routes for

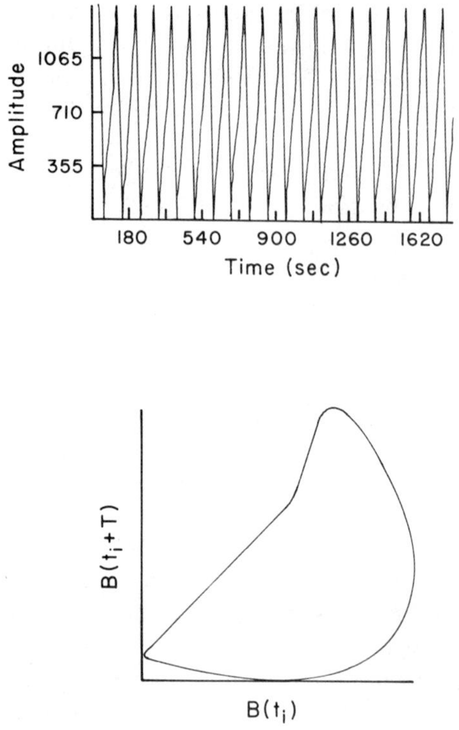

FIGURE 10.11. Oscillations in the potential of a bromide-sensitive electrode used to monitor the Belousov reaction in a CSTR with a residence time of 0.49 h. The lower panel shows a two-variable phase portrait of the limit cycle using the electrode potential, $B(t)$, and its time translation, $B(t + 8.8\text{ s})$, as variables. Taken from J.-C. Roux, J.S. Turner, W.D. McCormick, and H.L. Swinney, in *Nonlinear Problems: Present and Future*, A. Bishop, D. Campbell, and B. Nicolaenko, eds. (North-Holland, Amsterdam, 1982), p. 409.

the development of turbulence. This is presently an active area of research, and the detailed connection between chaos and turbulence remains to be elucidated.

The most covincing documentation of a chaotic attractor in an experimental system is for the Belousov reaction. The mechanism of this reaction, which involves the oxidation and bromination of malonic acid, is summarized in Section 10.3. This reaction is studied most easily in a CSTR. This permits the reaction mixture to be maintained as an open system with constant inputs. Experimentally, the constancy of the input flows is readily maintained with a peristaltic pump and uniform mixing can be achieved with magnetic stirrers or other standard devices. Figures 10.11 and 10.12 show the results of two experimental measurements by Swinney and colleagues of the potential of a bromide-selective electrode for a CSTR with identical input concentrations (0.25 *M* malonic acid, 0.14 *M* potassium bromide, 8.3×10^{-4} *M* Ce^{3+}, and 0.2 *M* sulfuric acid) and residence times of $\tau = 0.49$ h and $\tau = 0.90$ h. With a short residence time (Fig. 10.11), the reaction mixture exhibits a stable periodic motion with a period of about 83 s. This is reflected in the power spectrum, which has a peak at the principal frequency $\omega_0 = 0.012\ s^{-1}$ and peaks at the higher harmonics $2\omega_0$, $3\omega_0$, and $4\omega_0$. The background noise in the spectrum is very small and is due to instrumental uncertainties.

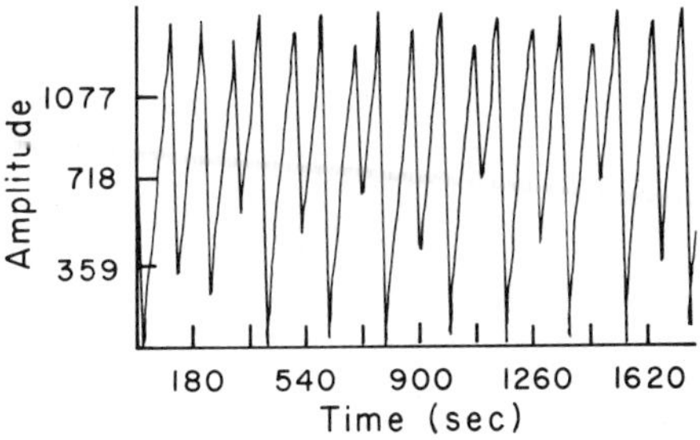

FIGURE 10.12. Chaos in the Belousov reaction measured using a bromide-sensitive electrode under the same conditions as in Fig. 10.11 with a residence time of 0.90 h. The dashed line shows the location of the Poincaré section used to calculate the return map in Fig. 10.13. See the legend of Fig. 10.11 for other details.

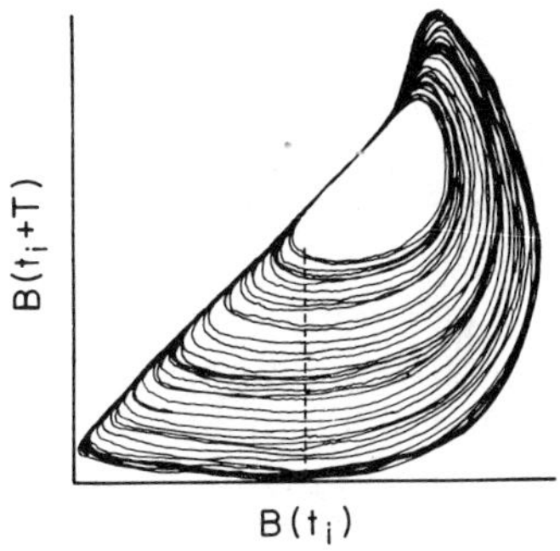

The time record of the potential at the longer residence time in Fig. 10.12 is clearly not periodic. This is also reflected in the power spectrum. The spectrum is suggestive of white noise and is similar to what is found for the Rössler attractor in the chaotic regime in Fig. 10.9. The behavior of the system for residence times intermediate between and beyond these two extremes has also been examined. One finds alternating regimes of residence times of width about 0.12 h in which periodic and chaotic behavior is observed. As the residence time is increased, the chaotic trajectories pick up additional small-amplitude oscillations like those seen in the upper panel of Fig. 10.12. This bounded, noisy, aperiodic motion is characteristic of chaotic attractors, and these experiments strongly support the contention that the Belousov reaction in CSTR posseses a chaotic attractor.

Although only a single variable, the bromide potential $B(t)$, was measured in these experiments, its complete time record can be used to construct a more complete representation of all the independent variables that are changing in the system. One way to do this is to calculate the derivatives dB/dt, d^2B/dt^2, etc. If we recall that the voltage reflects primarily the concentration of bromide ions, then the derivatives will represent some nonlinear functions of the variables which cause the bromide ion concentration to change. For example, referring back to the Field–Noyes model of the Belousov reaction in

Eq. (10.3.26), the time derivative of the bromide concentration Y is given by

$$dY/dt = -k_1AY - k_2XY + k_5Z, \tag{10.6.1}$$

where X is the concentration of bromous acid and Z is the concentration of ceric ion. Using Eq. (10.6.1) we see that the second derivative, d^2Y/dt^2, is a different nonlinear combination of Y, X, and Z. Thus by using enough time derivatives as the primary variables we should be able to obtain a transformed representation of the complete motion in a state space.

This technique is often limited by inherent inaccuracies in numerical differentiation of the data. Thus another technique is to introduce as a new variable of t the measured variable itself translated in time by a fixed amount, T. When this was done using $T = 8.8$ s for the two experiments in Figs. 10.11 and 10.12, the lower panels in the figures were obtained. As long as T is much shorter than any of the characteristic periodicities in the data, the shape of the resulting two-variable trajectory is qualitatively the same. By introducing the three variables $B(t)$, $B(t + 8.8\text{ s})$, and $B(t + 17.6\text{ s})$, a representation of a three-variable trajectory was obtained. The success of the Field–Noyes model in predicting the periodic behavior in the Belousov reaction suggests that only three independent variables are required to describe this system. Thus a three-variable trajectory probably gives a good representation of the geometric shape of the attractor.

Using the three-variable trajectory constructed in this way, one can obtain Poincaré sections as previously described in Section 10.5 for the Rössler attractor. For the Belousov reaction these sections are extremely narrow and have the appearance of a thin line. They are not, however, continuous lines and appear to have a fractal character like that of the Cantor set. Thus one sees in the Poincaré sections for the Belousov reaction gaps where the trajectory is never found and regions of high density in which the trajectory frequently appears.

From these Poincaré sections one can construct a *return map*. The return map plots the next value of one of the variables when it crosses through the Poincaré section, expressed as a function of its previous value. Thus if t_1, t_2, $t_3, \ldots, t_k, \ldots$ are the times of intersection of the trajectory with the Poincaré section, one plots $B_{k+1} \equiv B(t_{k+1})$ versus $B_k \equiv B(t_k)$. The return map for the Poincaré section indicated by the dotted line intersecting the two-dimensional projection of the chaotic trajectory in Fig. 10.12 is shown in Fig. 10.13.

The return map provides a useful way of thinking about chaotic trajectories. In cases like the Belousov reaction, where the map is approximately single-valued, it can be thought of as a one-dimensional map of the real line into itself. We have already encountered such maps in p dimensions as a representation of the way in which digital computers solve ordinary differential equations. In the present context, such a map reduces the complex dynamics of a multivariable system to a single-valued function of a real variable. The resulting function $F(n)$ produces a discrete set of points, n_k, starting from an

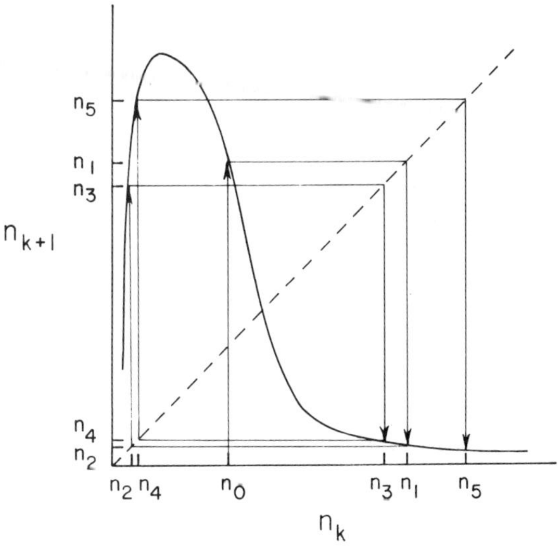

FIGURE 10.13. The return map for the Belousov reaction in a CSTR under the conditions described in Fig. 10.12. The reconstruction of the return points for the trajectory using the dashed line $n_{k+1} = n_k$ is described in the text. Data taken from Roux, Turner, McCormick, and Swinney (see Fig. 10.11).

initial point, n_0, according to the mapping

$$n_{k+1} = F(n_k). \tag{10.6.2}$$

The intersection of the 45° line $G(n) = n$ with $F(n)$ locates the *fixed points* of the map, that is, the points where $n_f = F(n_f)$. Such points may be stable or unstable attractors of the map depending on whether or not points near n_f converge towards n_f during the course of iteration of the map. The successive iterations of the map, i.e., n_0, $n_1 = F(n_0)$, $n_2 = F(F(n_0))$, etc., correspond to snapshots of the variable n on the underlying multivariable trajectory taken whenever the Poincaré section is crossed.

The iterated maps, such as $F_2(n) \equiv F(F(n))$, are also one-dimensional maps, and they may also have fixed points. A little thought shows that fixed points of F correspond to limit cycles in the underlying dynamics whose period is one iteration, while the fixed points of F_2 correspond to limit cycles whose period is two iterations. The phenomenon of period doubling, discussed for the Rössler attractor in the previous section, is well known for one-dimensional maps. The first such bifurcation occurs when a small change in a parameter of the map destroys the stability of a fixed point of F while simultaneously a fixed point of the map F_2 becomes stable.

A simple one-dimensional map which exhibit a period-doubling bifurcation sequence is the *logistic map*

$$n_{k+1} = rn_k(1 - n_k) \equiv F(n_k). \tag{10.6.3}$$

If the parameter r is restricted to $0 < r \leq 4$, this is a map of the interval [0, 1] into itself. It is easy to see that this map has a fixed point at $n = 0$ and for

$r > 1$ a second fixed point at $n = 1 - 1/r$. If $r < 1$, then for n_k in $[0, 1]$ one has the inequality

$$n_{k+1} = F(n_k) \leq rn_k < n_k. \tag{10.6.4}$$

Under these conditions the map is a contraction mapping and the iterations for any value of n_0 converge to $n = 0$. Thus the fixed point at $n = 0$ is stable. To analyze the behavior of the logistic map for $1 < r < 4$, one can write a short program on a hand-held calculator to iterate the map. For $1 < r \leq 3$ one finds that the fixed point $n = 1 - 1/r$ is stable for all choices of n on the interval (0, 1). For r slightly greater than 3 the map has a stable period-two attractor that depends on the value of r. Thus $r = 3$ is the first bifurcation point for period doubling. A period-four attractor appears for $r = 3.45$, which locates the second bifurcation point. In the range $3.45 \leq r \leq 3.57$, attractors of period 2^k are successively replaced by attractors of period 2^{k+1}. Finally, for r greater than 3.57 one finds an aperiodic attractor which is analogous to the chaotic attractor found for the Belousov reaction in the CSTR. One-dimensional maps with a single maximum of the sort given by the logistic map have been investigated in great detail.

The experimentally determined one-dimensional map for the Belousov reaction in Fig. 10.13 also exhibits chaos. This can be seen graphically using the 45° line to help follow the iterations. Picking a value of n_0, one finds $n_1 = F(n_0)$ using the map, which corresponds to the first vertical arrow in the figure. This value is then used as the initial point for the next iteration by translating it along the horizontal line to the 45° line. That makes the abscissa of this point equal to n_1, and thus the value of $n_2 = F(F(n_0))$ is given by the second arrow. Continuing this process produces a bounded, aperiodic sequence which is the trajectory of the iterated map. The Liapunov number of this map can be calculated using the definition in Eq. (10.5.11). For a one-dimensional map the Jacobian is a scalar, namely, dF/dn, which is the slope. Thus the chain rule the Jacobian of the kth iterate is

$$dF_k/dn_0 = (dF/dn_k)(dF/dn_{k-1})\dots(dF/dn_0). \tag{10.6.5}$$

Since this scalar is also the eigenvalue of the Jacobian, the single Liapunov number for the map is

$$\lambda(n_0) = \lim_{k\to\infty} \left(\prod_{l=0}^{k} |dF/dn_l| \right)^{1/k}. \tag{10.6.6}$$

By estimating the slope of the map in Fig. 10.13 at each of the points n_0, n_1, $\dots$, n_k a Liapunov number of about $\frac{1}{2}$ has been calculated. Since a positive Liapunov number indicates chaos, the experimental one-dimensional map is chaotic. This provides a further indication that the Belousov reaction in the CSTR possesses a chaotic attractor.

Another system in which chaotic attractors appear to play a role is the pancreatic β-cell. The β-cell is an endocrine cell in the islets of Langerhans of

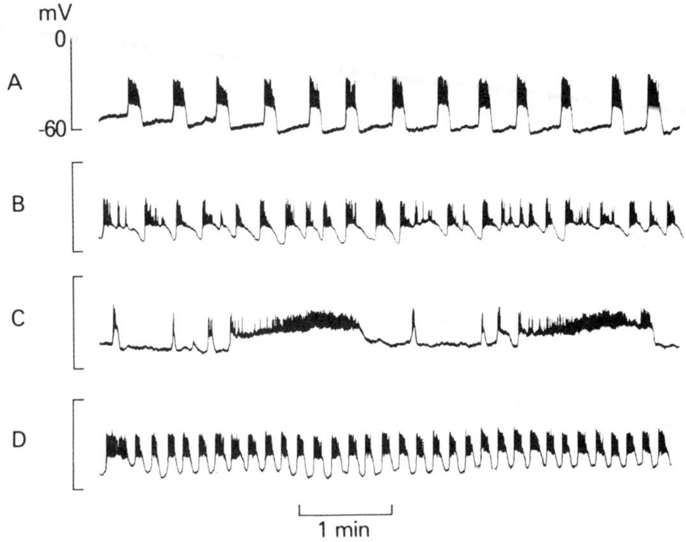

FIGURE 10.14. Burst oscillations in the membrane potential of the pancreatic β-cell after 30-min exposure to 11.1 mM glucose. Panel A is for islets taken from a mouse fed the NIH diet, while the irregular bursting in panels B and C is from mice on the Charles River (CR) diet. Panel D shows that regular bursting returns to CR mice after 30 days on the NIH diet. Taken from P. Lebrun and I. Atwater, *Biophys. J.* **48**, 529 (1985).

the mammalian pancreas. It is the cell that synthesizes and releases insulin. Using microelectrodes, electrophysiologists have been able to investigate the electrical potential across the plasma membrane of the β-cell in intact islets. In the absence of glucose the membrane potential has a resting value of about -70 mV. When glucose in concentrations of the order of 1.6×10^{-2} M is added to the solution that bathes an islet, a complicated pattern of electrical activity called bursting is observed. This is shown in Fig. 10.14. The bursts are a periodic change in electrical potential characterized by a silent phase and a spiking phase. In normal rat or mouse β-cells the period is 15–20 s. Raising the glucose concentration above 2×10^{-2} M eliminates the silent phase and only the spiking portion of the periodic signal is observed. Under conditions that lead to bursting, the duration of the spiking phase has been correlated with the release of the hormone insulin from the β-cell. The mechanism for this appears to involve an increase in intracellular calcium associated with bursting, although the precise details of this process are as yet unknown. Insulin is an important regulator of sugar metabolism and the dynamical characteristics of the β-cell seem to provide the control mechanism for this process.

The electrical behavior of β-cells under certain conditions differs significantly from the regular bursting shown in the top record of Fig. 10.14. The membrane potential records B and C in that figure were recorded from β-cells in mice bred at the Charles River Station (CR mice), while record A comes from the same strain of mice bred at the National Institutes of Health (NIH mice). The only differences between the two sets of mice are small differences in diet. These differences, however, lead to marked differences in the bursting patterns, which are distinctly irregular for the CR mice. That diet is the cause of these differences is shown by the control experiment in record D, in which the bursting of a CR mouse β-cell has become regular after one month on the NIH diet. These empirical observations are reminiscent of the qualitative differences between a limit cycle attractor and a chaotic attractor. Presuming that a change in diet can lead to changes in the physical parameters involved in the electrical activity of the β-cell, it is intriguing to imagine that records B and C represent a chaotic attractor in the dynamics of the β-cell.

Using the elementary processes that electrophysiologists have uncovered in the β-cell, it has been possible to develop a simple model of the bursting process. It has been established that current is carried through the plasma membrane of the β-cell by a variety of ion channels. The most important ones for bursting seem to be voltage-dependent calcium ion and potassiom ion channels and a postassium channel, whose conductance is activated by both calcium and an increase in voltage. A third type of potassium channel, which is inhibited by 10^{-2} M glucose, appears to determine the -70 mV resting potential of the plasma membrane.

Under physiological conditions the concentration of free Ca^{2+} within the β-cell is about 10^{-7} M, while outside the cell the concentration is about 10^{-3} M. The potassium ion concentration, on the other hand, is high inside (about 10^{-1} M) and low outside (about 5×10^{-3} M). Since the ions are transported down their electrochemical gradients, potassium tends to produce an outward current and calcium an inward current. The voltage dependence of these channels is similar to that of the K^+ and Na^+ channels in nerve axons, and the rapid spikes that occur during bursting in the β-cell are analogous to the action potentials of a nerve cell. In the β-cell, however, the inward current of Ca^{2+} replaces the inward current of Na^+ in the axon.

The trigger for the bursting activity, which is absent in nerve axons, is the Ca^{2+}-activated potassium conductance. When calcium is present in high enough concentrations, these channels are substantially activated and permit an outward current of potassium that tends to lower the membrane potential. Calcium-activated potassium channels are found in many different tissues and the characteristics of these channels in rat muscle cells were discussed in Section 5.9. Recently, these channels have been isolated from pancreatic β-cells and have also been studied using the patch clamp technique.

A simplified description of the β-cell membrane potential, which focuses on the bursting process, can be constructed using the channels described above. To account for the glucose sensitivity of the bursting, it is necessary to introduce a glucose-dependent removal of Ca^{2+} from the cell. Experiments suggest that this is due to uptake of Ca^{2+} by internal organelles, probably the mitochondria. Changes in the potassium concentration inside the cell, on the other hand, can be neglected since the internal concentration of K^+ is so large that it is effectively constant during bursting. If C is used to represent the capacitance of the plasma membrane and ϕ is the difference in potential, $\phi_{\text{in}} - \phi_{\text{out}}$, then the condition of charge conservation in Eq. (5.7.3) provides the differential equation

$$C d\phi/dt + i_{\text{Ca}} + i_{\text{K}} + i_L = 0, \tag{10.6.7}$$

where i_{Ca} and i_{K} are currents through the calcium and potassium channels and i_L is a leak current due to other ion channels. Using the elementary processes for ion currents developed in Section 5.8, it is possible to express these currents in terms of conductances, G_j:

$$\begin{aligned} i_{\text{Ca}} &= G_{\text{Ca}}(\phi - \phi_{\text{Ca}}) \\ i_{\text{K}} &= G_{\text{K}}(\phi - \phi_{\text{K}}) \\ i_L &= G_L(\phi - \phi_L), \end{aligned} \tag{10.6.8}$$

where from Eq. (5.8.14) the Nernst potentials are

$$\phi_j = (k_B T/ez_j)\ln[a_j^{\text{out}}/a_j^{\text{in}}], \tag{10.6.9}$$

with a_j^{out} and a_j^{in} the activities of the ion inside and outside the cell and ez_j the charge of the ion.

In the so-called minimal model of the β-cell, the conductance of the two kinds of potassium channels is written

$$G_{\text{K}} = \bar{G}_{\text{K,HH}} n^4 + \bar{G}_{\text{K,Ca}} \rho_{\text{Ca}}/(\rho_{\text{Ca}} + K). \tag{10.6.10}$$

The first term corresponds to the Hodgkin–Huxley conduction mechanism described in Section 5.8 for a voltage-activated potassium channel, while the second term represents the potassium channels whose conductance is activated by an increase in the number density, ρ_{Ca}, of Ca^{2+} ions inside the cell. K is the binding constant of Ca^{2+} to the channel. The Ca^{2+} conductance is also modeled after the Na^+ conductance of the Hodgkin–Huxley model, and in the minimal model

$$G_{\text{Ca}} = \bar{G}_{\text{Ca,HH}} m^3 h. \tag{10.6.11}$$

As described in Section 5.8, the quantities n, m, and h represent the fraction of the channels in various states. They solve the linear equations (5.8.46)–(5.8.48), which we write in the form

$$\begin{aligned} dn/dt &= [n_\infty - n]/\tau_n \\ dm/dt &= [m_\infty - m]/\tau_m \\ dh/dt &= [h_\infty - h]/\tau_h, \end{aligned} \tag{10.6.12}$$

where the parameters n_∞, τ_n, etc., are explicitly voltage dependent.

To complete the minimal model one needs an equation to describe the change in density of free calcium ions inside the cell. Since the calcium ions are divalent, they rapidly bind to negative charges within the cell and only a fraction, f, of the Ca^{2+} is actually free in the cytoplasm. Using the chain rule we can write

$$d\rho_{Ca}/dt = (d\rho_{Ca}/d\rho_{Ca}^T)(d\rho_{Ca}^T/dt) = f(d\rho_{Ca}^T/dt), \tag{10.6.13}$$

where ρ_{Ca}^T is the total density of bound and free Ca^{2+}. The time rate of change in the total density of Ca^{2+} is given by

$$d\rho_{Ca}^T/dt = i_{Ca}/2e - k_{Ca}\rho_{Ca}, \tag{10.6.14}$$

where the factor $1/2e$ takes into account the divalent charge of calcium and the second term represents the glucose-dependent loss of Ca^{2+} from the cytoplasm. Combining Eqs. (10.6.13) and (10.6.14), the equation that governs the density of free Ca^{2+} in the minimal model is

$$d\rho_{Ca}/dt = f(i_{Ca}/2e - k_{Ca}\rho_{Ca}). \tag{10.6.15}$$

Equations (10.6.7)–(10.6.15) provide a minimal mathematical characterization of bursting in the β-cell.

Parameter values in the minimal model that correspond to physiological conditions give rise to burst oscillations that closely resemble the experimental results in Fig. 10.15A. Typical burst oscillations obtained by numerical integration using the Gear algorithm are shown in Fig. 10.15 along with oscillations in the Ca^{2+} concentrations. If the parameters used to obtain that figure are modified by changing the temperature to 17°C, a different type of

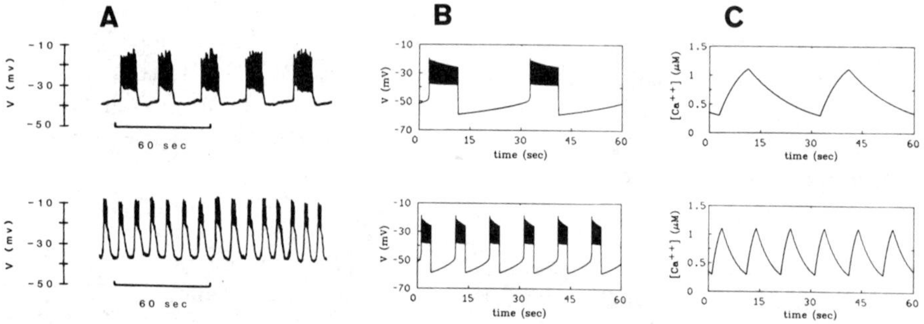

FIGURE 10.15. Experimental (A) and theoretical (B and C) oscillations in the pancreatic β-cell based on the minimal model in Eqs. (10.6.7)–(10.6.15). Panel A shows membrane potential oscillations in two different β-cells, both perfused with 11.1 mM glucose. The calculations in panels B and C show the membrane potential and calcium ion oscillations using physiological values of the parameters in the minimal model. In the upper curves the parameter $f = 0.016$, while in the lower curves $f = 0.048$. Taken from I. Atwater and J. Rinzel, in *Ionic Channels in Cells and Model Systems*, R. Latorre, ed. (Plenum Publishing, New York, 1986), p. 353.

periodic bursting is observed that changes into chaotic bursting, chaotic beating, and periodic beating as the parameter k_{Ca} is increased from 0.038 to 0.045. A typical time record of the membrane potential during chaotic bursting at $k_{Ca} = 0.040$ is shown in Fig. 10.16 along with an approximate one-dimensional map for the Ca^{2+} density under those conditions.

The one-dimensional map, shown as dots in the figure, was constructed by calculating the calcium densities C_{k+1} at which the membrane potential crosses -45 mV in the positive direction on trajectories that begin at C_k, a voltage of -45 mV, and fixed values of m, n, and h. This map has a single maximum similar to that of the experimental one-dimensional return map for the Belousov reaction in Fig. 10.13. The open circles in Fig. 10.16 show the calculated sequence of calcium densities at the -45-mV upcrossings taken from chaotic bursting in the upper panel of the figure. This is the return map for the calcium density, and it is seen to correspond closely to the approximate one-dimensional map. The geometrical pattern of the return sequence shown in Fig. 10.16 has the expected fractal appearance, suggesting the existence of a fractal attractor in the dynamics of the minimal model.

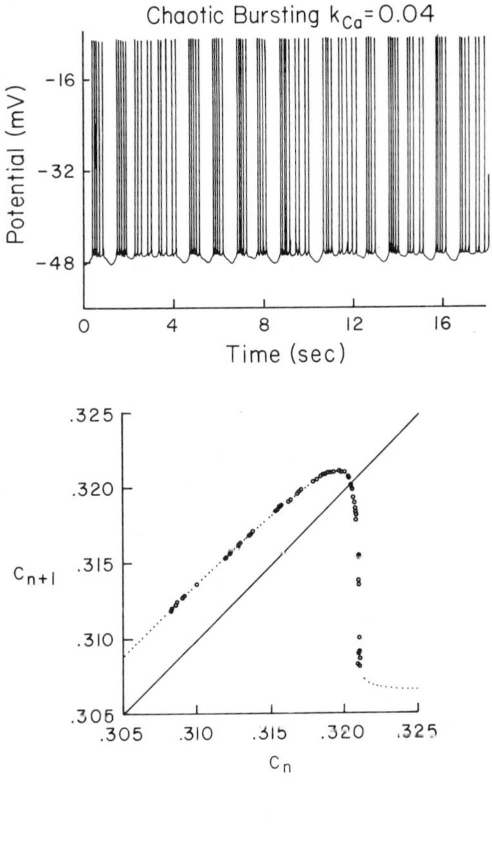

FIGURE 10.16. Chaotic bursting in the minimal model of the pancreatic β-cell with the calcium uptake rate constant $k_{Ca} = 0.04$. The upper panel shows the time record of the membrane potential, while the lower panel shows the return map under these conditions, constructed as described in the text. Taken from T. Chay and J. Rinzel, *Biophys. J.* **47**, 357 (1985).

If the parameter k_{Ca} is decreased from 0.04500 down to 0.04187, the calculated trajectories exhibit a sequence of period-doubling bifurcations. At $k_{Ca} = 0.045$ one has a stable limit cycle in which a single spike, or beat, of electrical activity repreats itself every 0.18 s. At $k_{Ca} = 0.043$ the oscillation has two beats and a period of 0.36 s, while at $k_{Ca} = 0.04196$ and 0.04187 there are four beats and eight beats, respectively, which occur with periods of 0.73 s and 1.45 s. Finally when $k_{Ca} = 0.04150$ the trajectories are no longer periodic and the beats appear to occur at random.

The correspondence between the minimal model for bursting in Eqs. (10.6.7)–(10.6.15) and the electrophysiology of the β-cell is incomplete. Nonetheless, the existence of irregular bursting in the β-cell and the parallel evidence for a chaotic attractor in the dynamics of the minimal model is striking. Indeed, irregular behavior, like that of the β-cell, is well known for many complex physiological processes, and an understanding of chaotic attractors may be important ultimately in the control of diabetes, heart arrhythmias, and other medical disorders.

10.7. Molecular Fluctuations versus Deterministic Chaos

Chaotic attractors are inherently deterministic; that is, trajectories on the attractor can be determined by integrating differential equations starting from precise initial conditions. Nonetheless, because a chaotic attractor has at least one Liapunov number greater than unity, trajectories tend to diverge from one another locally and slight deviations in initial conditions ultimately lead to large variations in calculated values. As we have seen, this is reflected in the broad-band noise seen in the power spectrum of chaotic trajectories, as illustrated for the Rössler attractor in Fig. 10.9. This noise arises because of the structure of the phase space in which the dynamics takes place and is a consequence of the average effect of molecular processes.

This sort of deterministic noise differs from molecular noise in that it appears only in the average description of a conditional ensemble. Indeed, if the initial condition, $\mathbf{n}^0$, is precisely given, then deterministic noise represents only the time record of the conditional average, $\bar{\mathbf{n}}(\mathbf{n}^0, t)$. The fact that the power spectrum of such a trajectory appears noisy reflects the continuous distribution of frequencies inherent in chaotic motion. This, of course, is the same reason that the power spectra of the random terms, $\tilde{f}_i$, in the conditional average equation appear to be noisy. The power spectrum of a chaotic trajectory, however, is many orders of magnitude larger than the power spectra of the ensembles encountered thus far, except very close to critical points. This reflects the fact that on a chaotic attractor, deviations from the time-averaged trajectory are large, while molecular fluctuations at asymptotically stable steady states are small.

There are other differences between deterministic chaos and molecular fluctuations. The range of frequencies contained in the time record of low-dimensional chaotic attractors is much smaller than that for molecular fluctuations. As we discovered in Section 5.1, the characteristic frequency below which Johnson noise appears to be white is about 10^{10} s^{-1}. Compare this with the maximum frequency of about 20 s^{-1} in the calculated power spectra of the Rössler attractor in Fig. 10.9. This difference is due to the different origins of deterministic and molecular noise. Deterministic noise arises directly from the conditional average equations and, therefore, reflects the size of rate constants and transport coefficients in those equations. Molecular noise, on the other hand, comes from uncertainties inherent in molecular events that occur on a time scale many orders of magnitude faster than the average. Indeed, in Chapter 9 we worked out in great detail the manner in which collisional time scale events at the Boltzmann level of description lead both to the Newtonian hydrodynamic equations on the conditional average and to small, rapid fluctuations about the average. Although the details of this sort of contraction are understood presently only for dilute gases, it seems clear that even in condensed phases the separation between the average transport equations and molecular fluctuations arises in a similar fashion.

Despite the fact that molecular fluctuations and deterministic chaos arise in different ways from molecular processes, they are, nonetheless, coupled dynamically. Remember that in a single dynamical measurement the observable is neither the conditional average nor the fluctuations but the time record of an observable, $\mathbf{n}(t)$. This record represents one of many possible trajectories from the initial ensemble. For example, in a conditional ensemble the measured variable is $\mathbf{n}(t) = \bar{\mathbf{n}}(\mathbf{n}^0, t) + \delta\mathbf{n}(t)$, where $\delta\mathbf{n}(t)$ is the conditional fluctuation. Thus even in a conditional ensemble the stochastic behavior of $\mathbf{n}(t)$ depends both on the behavior of the deterministic trajectory and the fluctuations around it. Moreover, as we saw in Section 10.1, the statistical distribution of the conditional fluctuations depends explicitly on the deterministic trajectory. The calculation for the *o*-cresophthalein reaction illustrated in Fig. 10.2, however, shows that for a transient relaxation process this dynamical effect is short-lived.

This sort of coupling is not short-lived on a chaotic attractor. Because the conditional average is continually forcing the fluctuations, the conditional average has the potential to increase the size of the conditional fluctuations. To see this, consider the differential equations for the Rössler attraction given in Section 10.5, i.e.,

$$\begin{aligned} dx/dt &= -y - z \\ dy/dt &= x + ey \qquad (10.7.1) \\ dz/dt &= f + xz - gz. \end{aligned}$$

Although these equations do not correspond to any physical system, for the sake of illustration we will treat them as if they describe the conditional average of the three stochastic variables x, y, and z. The conditional fluctuations then would satisfy the linearized equations

$$\begin{aligned} d\delta x/dt &= -\delta y - \delta z + \tilde{f}_x \\ d\delta y/dt &= \delta x + e\delta y + \tilde{f}_y \\ d\delta z/dt &= z\delta x + (x-g)\delta z + \tilde{f}_z. \end{aligned} \tag{10.7.2}$$

The coefficients of δx and δz in the third equation depend on time, which makes the matrix H an explicit function of the chaotic variables. This is significant because the Liapunov numbers for a set of ordinary differential equations are determined by the eigenvalues of H. Indeed, in Section 10.5 we saw that a Liapunov number greater than unity corresponds to the largest time-averaged eigenvalue of H being positive. In other words, for the majority of the time on the Rössler attractor the linearized matrix

$$H(t) = \begin{pmatrix} 0 & -1 & -1 \\ 1 & e & 0 \\ z(t) & 0 & x(t)-g \end{pmatrix} \tag{10.7.3}$$

has a largest eigenvalue which is positive.

At a given instant of time, t^0, the existence of a positive eigenvalue implies that the covariance of the fluctuations will have a component which is increasing. This can be seen by solving Eq. (10.2.4) for the covariance matrix,

$$d\sigma/dt = H(t)\sigma + \sigma H^T(t) + \gamma(t). \tag{10.7.4}$$

If $\sigma(t^0)$ is the conditional covariance at time t^0, then the solution at time t is

$$\sigma(t) = \phi(t,t^0)\sigma(t^0)\phi^T(t,t^0) + \int_{t^0}^{t} \phi(\tau,t^0)\gamma(\tau)\phi^T(\tau,t^0)\,d\tau, \tag{10.7.5}$$

where ϕ is the time-ordered exponential defined in Eq. (4.4.16), namely,

$$\phi(t,t^0) \equiv \mathrm{P}\exp\left[\int_{t^0}^{t} H(\tau)\,d\tau\right]. \tag{10.7.6}$$

For short intervals of time we can approximate the time-ordered exponential by

$$\phi(t,t^0) \simeq \exp[H(t^0)(t-t_0)]. \tag{10.7.7}$$

If one of the eigenvalues of H is positive, then for short times both terms in Eq. (10.7.5) have components that grow exponentially.

Although this exponential growth of the conditional covariance occurs most of the time, it is not always the same eigenmode that is growing. Since $H(t)$ is changing, at each instant the eigenvectors of $H(t)$ will correspond to different linear combinations of δx, δy, and δz. Indeed, because a chaotic attractor is bounded, it is not possible for the same mode to continue to grow exponentially. Because the largest eigenvalue of $H(t)$ is positive for long intervals of time, it is possible that in the long run conditional fluctuations will become comparable to the average values. Because of the complexity of the dynamics of chaotic attractors, detailed numerical calculations, like those described in Section 10.2, are necessary to determine if this is so and on what time scale, if any, this occurs.

Due to dispersion in initial conditions, fluctuations in a single-time, non-conditional ensemble are more complicated to describe than conditional fluctuations. In Section 10.2 the effect of dispersion in initial conditions was explored for two transient systems relaxing from near an unstable state to a stable state. In those systems the single-time variance was originally large and transiently increased in size. As the stable steady states were approached, the variances ultimately decreased to their infinitesimally small steady-state values. Even if the conditional covariance were to remain small, the single-time covariance would not remain small on a chaotic attractor due to the effect of dispersion in initial conditions, which is described by Eq. (10.2.27). Because of the great sensitivity of conditional trajectories to initial conditions, the single-time covariance will quickly become large on a chaotic attractor. Thus, *a'priori*, one expects that statistical fluctuations on chaotic attractors will always be large.

References

Nonlinear Transients

J. Kramer and J. Ross, Stabilization of unstable states, relaxation, and critical slowing down in a bistable system, *J. Chem. Phys.* **83**, 6234–6241 (1985).

J. Keizer, A theory of spontaneous fluctuations in macroscopic systems, *J. Chem. Phys.* **63**, 398–403 (1975).

A.C. Hindmarsh, Ordinary differential equation solver, Lawrence Livermore Laboratory, Report UCID-30001 (1974).

Limit Cycles: Oscillations and Fluctuations

W. Hahn, *The Stability of Motion* (Springer-Verlag, Berlin, 1967).

L.S. Pontryagin, *Ordinary Differential Equations* (Addison-Wesley, Reading, MA, 1962), Chapter 5.

R.H. Abraham and C.D. Shaw, *Dynamics—The Geometry of Behavior. Part 1: Periodic Behavior* (Aerial Press, Santa Cruz, CA, 1981).

Chemical and Electrochemical Oscillations

R.J. Field, E. Körös, and R.M. Noyes, Oscillations in chemical systems. II. Thorough analysis of temporal oscillations in the bromate–cerium–malonic acid system, *J. Am. Chem. Soc.* **94**, 8649–8664 (1972).

I.R. Epstein, Oscillations and chaos in chemical systems, *Physica* **7D**, 47–58 (1983).
R.J. Field and M. Burger, *Oscillations and Travelling Waves in Chemical Systems* (Wiley, New York, 1985).
J. Wajtowicz, Oscillatory behavior in electrochemical systems, in *Modern Aspects of Electrochemistry*, J. Bockris and B. Conway, eds. (Plenum Press, New York, 1973), pp. 47–120.
S.-W. Lin, J. Keizer, P.A. Rock, and H. Stenschke, On the mechanism of oscillations in the "beating mercury heart," *Proc. Nat. Acad. Sci. U.S.A.* **71**, 4477–4481 (1974).
J. Keizer, P.A. Rock, and S.W. Lin, Analysis of the oscillations in "beating mercury heart" systems, *J. Am. Chem. Soc.* **101**, 5637–5649 (1979).

Chaotic Attractors

R.H. Abraham and C.D. Shaw, *Dynamics—The Geometry of Behavior. Part II. Chaotic Behavior* (Aerial Press, Santa Cruz, CA, 1983).
J.D. Farmer, E. Ott, and J. Yorke, The dimension of chaotic attractors, *Physica* **7D**, 153–180 (1983).
O.E. Rössler, An equation for continuous chaos, *Physics Letters* **57A**, 397–398 (1976).
J.L. Hudson and J.C. Mankin, Chaos in the Belousov–Zhabotinski reaction, *J. Chem. Phys.* **74**, 6171–6177 (1981).
H.L. Swinney, Observations of order and chaos in nonlinear systems, *Physica* **7D**, 3–15 (1983).
J.-C. Roux, J.S. Turner, W.D. McCormick, and H.L. Swinney, Experimental observations of complex dynamics in chemical reaction, in *Nonlinear Problems: Present and Future*, A. Bishop, D. Campbell, and B. Nicolaenko, eds. (North-Holland, Amsterdam, 1982), pp. 409–422.
A.J. Lichtenberg and M.A. Lieberman, *Regular and Stochastic Motion* (Springer-Verlag, New York, 1982).
B. Mandelbrot, *Fractals* (W.H. Freeman, San Franciso, 1982).
H.L. Swinney and J. P. Gollub, *Hydrodynamic Instabilities and the Transition to Turbulence*, 2nd ed. (Springer-Verlag, Berlin, 1985).

Pancreatic β-cell

I. Atwater, C.M. Dawson, A. Scott, G. Eddlestone, and E. Rojas, The nature of the oscillatory behavior in electrical activity for the pancreatic β-cell, in *Biochemistry and Biophysics of the Pancreatic β-Cell* (Georg Thieme Verlag, New York, 1980), pp. 100–107.
T. Chay and J. Keizer, Minimal model for membrane oscillations in the pancreatic β-cell, *Biophys. J.* **42**, 181–190 (1983).
T. Chay and J. Rinzel, Bursting, beating, and chaos in an excitable membrane model, *Biophys. J.* **47**, 357–366 (1985).
I. Atwater and J. Rinzel, The β-cell bursting pattern and intracellular calcium, in *Ionic Channels in Cells and Model Systems*, R. Latorre, ed. (Plenum Publishing, New York, 1986), pp. 353–362.

Index